Springer Series in
MATERIALS SCIENCE 47

Springer
Berlin
Heidelberg
New York
Barcelona
Hong Kong
London
Milan
Paris
Singapore
Tokyo

Springer Series in

MATERIALS SCIENCE

The Springer Series in Materials Science covers the complete spectrum of materials physics, including fundamental principles, physical properties, materials theory and design. Recognizing the increasing importance of materials science in future device technologies, the book titles in this series reflect the state-of-the-art in understanding and controlling the structure and properties of all important classes of materials.

29 **Elements of Rapid Solidification**
Fundamentals and Applications
Editor: M. A. Otooni

30 **Process Technology for Semiconductor Lasers**
Crystal Growth and Microprocesses
By K. Iga and S. Kinoshita

31 **Nanostructures and Quantum Effects**
By H. Sakaki and H. Noge

32 **Nitride Semiconductors and Devices**
By H. Morkoç

33 **Supercarbon**
Synthesis, Properties and Applications
Editors: S. Yoshimura and R. P. H. Chang

34 **Computational Materials Design**
Editor: T. Saito

35 **Macromolecular Science and Engineering**
New Aspects
Editor: Y. Tanabe

36 **Ceramics**
Mechanical Properties, Failure Behaviour, Materials Selection
By D. Munz and T. Fett

37 **Technology and Applications of Amorphous Silicon**
Editor: R. A. Street

38 **Fullerene Polymers and Fullerene Polymer Composites**
Editors: P. C. Eklund and A. M. Rao

39 **Semiconducting Silicides**
Editor: V.E. Borisenko

40 **Reference Materials in Analytical Chemistry**
A Guide for Selection and Use
Editor: A. Zschunke

41 **Organic Electronic Materials**
Conjugated Polymers and Low Molecular Weight Organic Solids
Editors: R. Farchioni and G. Grosso

42 **Raman Scattering in Materials Science**
Editors: W. H. Weber and R. Merlin

43 **The Atomistic Nature of Crystal Growth**
By B. Mutaftschiev

44 **Thermodynamic Basis of Crystal Growth**
P–T–X Phase Equilibrium and Nonstoichiometry
By J. Greenberg

45 **Thermoelectrics**
Basic Principles and New Materials Developments
By G.S. Nolas, J. Sharp, and H. J. Goldsmid

46 **Fundamental Aspects of Silicon Oxidation**
Editor: Y. J. Chabal

47 **Disorder and Order in Strongly Nonstoichiometric Compounds**
Transition Metal Carbides, Nitrides and Oxides
By A.I. Gusev, A.A. Rempel, and A.J. Magerl

48 **The Glass Transition**
Relaxation Dynamics in Liquids and Disordered Materials
By E. Donth

49 **Alkali Halides**
A Handbook of Physical Properties
By D. B. Sirdeshmukh, L. Sirdeshmukh, and K. G. Subhadra

Series homepage – http://www.springer.de/phys/books/ssms/

Volumes 1–28 are listed at the end of the book.

A.I. Gusev A.A. Rempel A.J. Magerl

Disorder and Order in Strongly Nonstoichiometric Compounds

Transition Metal Carbides, Nitrides and Oxides

With 205 Figures and 107 Tables

Springer

Professor Dr. Sc. Alexandr I. Gusev
Professor Dr. Sc. Andrej A. Rempel
Ural Division of the Russian Academy of Sciences, Institute of Solid State Chemistry
Pervomaiskaya 91, GSP-145, 620219 Ekaterinburg, Russia
e-mail: gusev@ihim.uran.ru; rempel@itap.physik.uni-stuttgart.de

Professor Dr. Andreas J. Magerl
Lehrstuhl für Kristallographie und Strukturphysik, Universität Erlangen-Nürnberg
Bismarckstraße 10, 91054 Erlangen, Germany
e-mail: andreas.magerl@krist.uni-erlangen.de

Series Editors:

Prof. Robert Hull
University of Virginia
Dept. of Materials Science and Engineering
Thornton Hall
Charlottesville, VA 22903-2442, USA

Prof. H. Sakaki
Institute of Industrial Science
University of Tokyo
7-22-1 Roppongi, Minato-ku
Tokyo 106, Japan

Prof. R. M. Osgood, Jr.
Microelectronics Science Laboratory
Department of Electrical Engineering
Columbia University
Seeley W. Mudd Building
New York, NY 10027, USA

Prof. Alex Zunger
NREL
National Renewable Energy Laboratory
1617 Cole Boulevard
Golden Colorado 80401-3393, USA

ISSN 0933-033x

ISBN 3-540-41817-2 Springer-Verlag Berlin Heidelberg New York

Library of Congress Cataloging-in-Publication Data applied for

Die Deutsche Bibliothek – CIP-Einheitsaufnahme
Gusev, Aleksandr I.:
Disorder and order in strongly nonstoichiometric compounds : transition metal carbids, nitrides and oxides / A. I. Gusev ; A. A. Rempel ; A. J. Magerl. - Berlin ; Heidelberg ; New York ; Barcelona ; Hong Kong ; London ; Milan ; Paris ; Singapore ; Tokyo : Springer, 2001
(Springer series in materials science ; 47)
(Physics and astronomy online library)
ISBN 3-540-41817-2

Springer-Verlag Berlin Heidelberg New York
a member of BertelsmannSpringer Science+Business Media GmbH
http://www.springer.de

Printed in Germany

Typesetting: Camera-ready copy from the authors
Cover concept: eStudio Calamar Steinen
Cover production: *design & production* GmbH, Heidelberg

Printed on acid-free paper SPIN: 10779025 57/3020/cu 5 4 3 2 1 0

"L'ordre est le plaisir de la raison,
mais le désordre est le délice de l'imagination"
d'après Paul Clodel

"Nur wer denkt, irrt auch"
Horst Friedrich

"Something there is that doesn't love a wall..."
After Robert Frost

Preface

In the last decades of the 20th century the physics of condensed matter significantly extended its sphere of interest to include new groups of materials in addition to those traditionally studied. One of these groups comprises a variety of nonstoichiometric compounds. Physicists paid interest to the phenomenon of nonstoichiometry only after the discovery of high-temperature superconductivity, although solid-state chemistry has been dealing with nonstoichiometry for at least forty years.

Nonstoichiometry occurs widely in solid-state compounds and is most pronounced in strongly nonstoichiometric compounds, such as cubic carbides, nitrides and oxides of transition metals and related materials. Carbides, being those strongly nonstoichiometric compounds that have received most study, have been known for nearly a hundred years, but their nature is still not clearly understood. This is especially true of nonstoichiometry and the phenomena of disorder and order directly related to nonstoichiometry.

This book summarizes the 20-year experience of the authors working on the problems of disorder and order in strongly nonstoichiometric compounds. Our studies were concerned mainly with crystallographic and symmetry aspects of ordering, thermodynamics of disorder–order phase transformations, and the effect of ordering on properties of strongly nonstoichiometric compounds. It did not take long to understand that the phenomena associated with the redistribution of interstitial atoms and structural vacancies in nonstoichiometric compounds represent an excellent model of analogous phenomena arising in solid-state systems with mutual substitution of components. Therefore methods of analysis and description proposed and developed for nonstoichiometric compounds are applicable to a number of systems with substitutional disorder.

In 1980, when we embarked on this work, ordering in nonstoichiometric compounds was assumed to be a rare and accidental effect, which hardly deserved close attention. Now many people understand that ordering represents a general regular phenomenon in all nonstoichiometric compounds. Ordering effects, which were not even mentioned, or were referred to as something insignificant and negligible just 20 years ago, proved to be numerous and comparable in magnitude with the change of properties in the whole homogeneity interval of nonstoichiometric compounds. Phase diagrams of binary systems of transition metals with carbon and nitrogen, which remained unchanged from the time they were plotted in 1950–60s and containing one or two nonstoichiometric compounds, were enriched with many ordered phases.

Unfortunately, the accumulated experimental and theoretical knowledge has not so far been disseminated widely.

Engineers concerned with commercial production and use of carbides and nitrides, and even many scientists, know very little or nothing about ordering and its relation to nonstoichiometry. English, German and French scientific literature has had no monographs dedicated at least partly to ordering of nonstoichiometric compounds. However, hundreds of original papers on order and disorder in those compounds have been published in scientific journals. But even the latest books on carbides do not contain data on ordering. For example, one book[1], which was published in 1996, discusses carbides of transition metals, but the overwhelming majority of the reported experimental results were obtained between 1950 and 1975. Out of 2090 works cited in Upadhyaya's book, only 21 refer to studies performed in the period from 1981 to 1985. No references are made to studies performed after 1985. Any issues of short- or long-range order and their relation to nonstoichiometry are not even mentioned. A handbook[2] issued in 1996 contains only two sentences dedicated to ordering in carbides: "When the concentration of carbon-atom vacancies is high, long-range ordering in their arrangement within the metal lattice is usually observed. The effect of this ordering on the structure and physical properties of the carbide may be considerable in some cases but the mechanisms of formation, the structural and bonding arrangement are still not well defined."

---

[1] G. S. Upadhyaya: *Nature and Properties of Refractory Carbides* (Nova Science Publishers, New York 1996) 545 pp.

[2] H. O. Pierson: *Handbook of Refractory Carbides and Nitrides: Properties, Characteristics, Processing and Applications* (Noyes Publications: Westwood, 1996) 318 pp.

The only books[3] in the scientific literature describing disorder–order transitions in strongly nonstoichiometric compounds were published in Russian and are inaccessible to readers outside the former Soviet Union.

The present monograph is intended to bridge this gap. Reading this book will require a certain effort, the knowledge of basic postulates of thermodynamics and solid-state theory, and a command of a body of mathematics. The point is that ordering theory is not so simple that it can be mastered without a profound scientific training. We have done our best to present today's understanding of the mutual relationship between disorder, order and nonstoichiometry in a solid, taking nonstoichiometric compounds as an example. Hopefully this book will direct the reader's attention to interesting and by far incomplete studies of a real solid. We think it will be interesting and useful for your professional research. Of course, this book does not cover all issues of nonstoichiometry, disorder and order in solids. This means that we shall continue our work and, with time, interested readers will be able to read a new book about defects, nonstoichiometry and disorder–order transformations in crystals.

Ekaterinburg–Erlangen, May 2001 *A. Gusev, A. Rempel, A Magerl*

---

[3] A. I. Gusev, A. A. Rempel: *Structural Phase Transitions in Nonstoichiometric Compounds* (Nauka, Moscow 1988) 308 pp. (in Russian)
A. I. Gusev: *Physical Chemistry of Nonstoichiometric Refractory Compounds* (Nauka, Moscow 1991) 286 pp. (in Russian)
A. A. Rempel: *Effects of Ordering in Nonstoichiometric Interstitial Compounds* (Nauka, Ekaterinburg 1992) 232 pp. (in Russian)
V. N. Lipatnikov, A. I. Gusev: *Ordering of Titanium and Vanadium Carbides* (Ural Division of the Russian Academy of Sciences, Ekaterinburg 2000) 266 pp. (in Russian)

Contents

List of Main Notation

$\boldsymbol{a}_1, \boldsymbol{a}_2, \boldsymbol{a}_3; \boldsymbol{a}, \boldsymbol{b}, \boldsymbol{c}$	basis vectors of the direct lattice (fundamental translation vectors of crystal lattice)
a, b, c	lengths of basis vectors
a_{B1}	lattice constant (period) of cubic unit cell with $B1$ structure
$\boldsymbol{b}_1, \boldsymbol{b}_2, \boldsymbol{b}_3$	basis vectors of reciprocal lattice
$\boldsymbol{B}_\omega(\varphi)$	rotation matrix
B_ℓ, B_s	interchange energies in the liquid and solid phases
c	concentration
C_p, C_V	heat capacity at constant pressure and at constant volume
d	density
d_e	electron density
d_g	grain size
E	lattice configurational energy
E^{at}	atomization energy
E_F	Fermi energy
F	free energy
$g(\omega)$	frequency distribution function
G	Gibbs free energy
G_s^e	excess free energy of mixing
$h, \hbar = h/2\pi$	Planck's constant
h_k, h_i	symmetry operation (rotation symmetry element) of a cubic point group $m\bar{3}m$
H_k, H_i	symmetry operation (rotation symmetry element) of a hexagonal point group $6/mmm$
H_V	microhardness
ΔH_{trans}	heat of phase transformation
k_B	Boltzmann constant
$\boldsymbol{k}$	wave vector
$\{\boldsymbol{k}_s\}$	wave-vector star
$\boldsymbol{k}_s^{(j)}$	ray of the star $\{\boldsymbol{k}_s\}$ (superstructural vector that belongs to the star $\{\boldsymbol{k}_s\}$)
m_0	free electron mass
m^*	effective electron mass
$n(\boldsymbol{r}), n(x_I, y_I, z_I)$	distribution function of atoms on the crystal lattice sites
N_A	Avogadro's number
$N(E_F)$	density of electronic states on the Fermi level

p	pressure
P, $P_i^{(s)}$	probability
q	charge
Q	activation energy
$\boldsymbol{r}$	radius-vector that determines the position of the crystal lattice site
R	molar gas constant
$R^{(s)}$	the number of the ordering lattice sites in an s-type figure
$R_{\square}^{\min}$	minimum admissible radius of vacancy
$R(\theta)$	angle resolution function
s	type of figure
S_c	configuration entropy
ΔS_{trans}	entropy of phase transformation
T	temperature
T_c	superconducting critical temperature
T_m	melting temperature
T_{trans}	phase transformation (transition) temperature
$V_{\square}^{\min}$	minimum admissible volume of vacancy
x_I , y_I , z_I	coordinates of the sites $\boldsymbol{r}$ of Ising lattice
y	relative content of interstitial atoms X in nonstoichiometric compounds MX_y
$y^{(s)}$	overestimate coefficient
z_j	coordination number of the j-th coordination sphere
z_c	critical structural vacancy concentration
Z	statistical sum
α, β, γ	interaxial (lattice) angles
α	linear thermal expansion coefficient
α_j	short-range order parameter for j-th coordination sphere
β_i	thermal coefficient of the shear modulus
γ	electronic heat capacity coefficient
$\delta(x)$	Dirac delta-function
δ_{ik}	Kronecker symbol
$\varepsilon_i^{(s)}$	energy of the i-configuration of the s-type figure
η	long-range order parameter
θ_D, θ_E	characteristic Debye and Einstein temperatures
κ_T	isothermal compressibility
λ	wavelength
λ^*	normalized positron annihilation rate
λ_c	electron-phonon coupling constant
λ_L	Lagrangian multiplier
$\lambda_i^{(s)}$	multiplicity of i-configuration of the s-type figure
μ	chemical potential
μ^*	Coulomb pseudopotential
ρ	electrical resistivity

σ_d	displacement cross-section
τ	positron lifetime
$\varphi_s^{(j)}$	phase shift of the concentration wave
Φ	thermodynamic potential
Φ_{irr}	irradiation dose
Φ_L	Lagrange function
Φ_s	the number of lattice sites constituting a unit cell of the ordered crystal
χ	magnetic susceptibility
χ_d^-	Landau diamagnetism of conduction electrons
χ_p^-	Pauli paramagnetism of conduction electrons
χ_{orb}	Van Vleck orbital paramagnetism
$\psi(\boldsymbol{r})$	wave function
ω	angular frequency
$\inf(x)$	inferior boundary
$\sup(x)$	superior boundary
$\subset$	subset ($G_D \subset G$ means G_D is subset of the set G)
$\in$	element ($i \in s$ means i is an element of the set s)
□	structural vacancy

"Deux dangers ne cessent de menacer le monde:
l'ordre et le désordre"
d'après Paul Valery

1. Introduction

Chemical compounds, which amount to hundreds of thousands in number, include a group of compounds that are largely nonstoichiometric under normal conditions. This group typically includes carbides, nitrides, oxides, borides, hydrides, metal sulfides, and a variety of ternary compounds. A long listing of the compounds itself suggests the wide occurrence of nonstoichiometry. Nonstoichiometry had long been considered more a curiosity than a real problem of science until N. S. Kurnakov, at the beginning of 20th century, recognized the practical importance of the ideas developed by C. L. Bertholett at the beginning of 19th century. Today nonstoichiometry plays an important role in solid-state research. Let us consider briefly the main concepts associated with nonstoichiometry.

The concept of a crystal as a three-dimensional periodic and fully ordered assembly of particles represents the physical model of an ideal solid. Real crystals often contain various types of defects.

Pioneering X-ray diffraction investigations showed that the angular distribution of the radiation intensity does not quite correlate with a strictly periodic three-dimensional distribution of atoms in the lattice. Therefore real crystals were assumed to have a mosaic structure and to comprise blocks typically 1 μm in size and which were slightly misoriented with respect to each other. Clearly, the regular periodic arrangement of atoms is disturbed along the block boundaries. Plastic deformation of crystals also results in a widening of diffraction profiles as a result of a disturbed arrangement of atomic planes (the appearance of stacking faults), a microdeformation of the lattice or a decrease of the sizes of crystallites.

Numerous investigations since N. S. Kurnakov [1–5] and G. Tammann [6, 7] have revealed the presence of two types of phases in intermetallics. Properties of the same phase change monotonically within the homogeneity interval and the composition corresponding to a certain compound cannot be identified. The "property–composition" curves of other phases exhibit a singularity within the homogeneity interval. The phase composition, which corresponds to the singular point, is stoichiometric, i.e. can be represented by simple integer coefficients.

Another group of investigations, which dealt mainly with diffusion and electrical conductivity of ionic crystals, suggests that the ideal crystal structure is disturbed, because ions move to interstices or because unoccupied sites (vacancies) are formed. Chlorides of alkali metals, e.g., NaCl, are often taken to exemplify crystal compounds with an ideal stoichiometric composition. However, when heated in sodium or potassium vapors, NaCl crystals alter their composition and properties (turn yellow; become a semiconductor), while the crystal structure remains unchanged. Heating causes the Na/Cl ratio to exceed 1. The surplus of sodium is 10^{16}–10^{17} Na atoms per 1 cm^3 with respect to the stoichiometric composition [8].

Thus, examination of solids, performed on different substances and with different methods, reveals the presence of defects and deviations from stoichiometry in real crystals. The geometrical classification of defects takes into account the dimensionality of the lattice distortions caused by these defects. Zero-dimensional (0D) defects are represented by point defects, for example, by vacancies or impurity atoms. The aggregate of two, three or more vacancies is also a point defect. However, quantitative changes eventually turn to qualitative ones as a result of the vacancy aggregation. 0D defects are responsible for atomic displacements in the crystal lattice. The atomic displacement field vanishes with distance from a point defect but vanishes very slowly and remains observable at a distance ten or more interatomic spacings.

If vacancies are located in the same atomic plane and form a flat aggregate of them, this actually corresponds to a break of the atomic plane along some line, which is the aggregate perimeter. This defect has microscopic dimensions along its perimeter and therefore is one-dimensional (1D) or linear. Such a linear defect is called an edge dislocation. Another frequent type of linear defect is a screw dislocation. Dislocations are present in almost all crystals. Their concentration depends on the method used for production of materials and varies within broad limits. For example, in metals that underwent severe plastic deformation the dislocation concentration consists of 10^{15}–10^{16} m^{-2} [9–11]. Only crystals prepared for semiconductor applications are free from dislocations.

Two-dimensional (2D) or surface defects are represented by stacking faults, grain boundaries (interfaces), boundaries of twins and domains, antiphase boundaries in ordered alloys, and also the external surfaces of crystals. Three-dimensional defects (3D) are any distortions of the crystal lattice, which are macroscopic in three dimensions. They include precipitates of crystalline or amorphous inclusions, and discontinuities (cracks, voids) in the crystal.

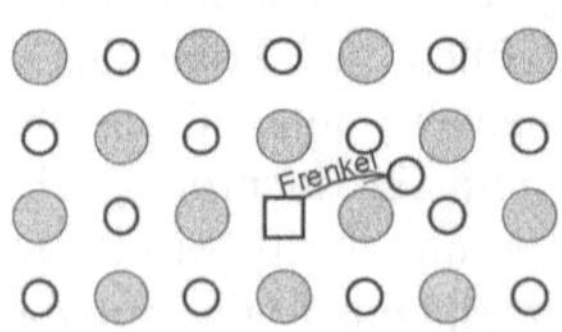

Fig. 1.1. Frenkel defect in a binary compound. In a Frenkel defect an atom is removed from its lattice site to an interstitial position

Nonstoichiometry in solids can be achieved by a point defect called a structural or constitutional vacancy. The formation of a structural vacancy leads to a change in the number of atoms in the crystal. A structural vacancy is different from a Frenkel, Schottky and anti-Schottky defect, which are also point defects in a crystal but do not change the stoichiometry of the crystal. The Frenkel defect represents a "vacancy–interstitial" pair. It is formed when an atom moves from its lattice site to an interstice (Fig. 1.1). This can be achieved, for example, by high-energy particle irradiation.

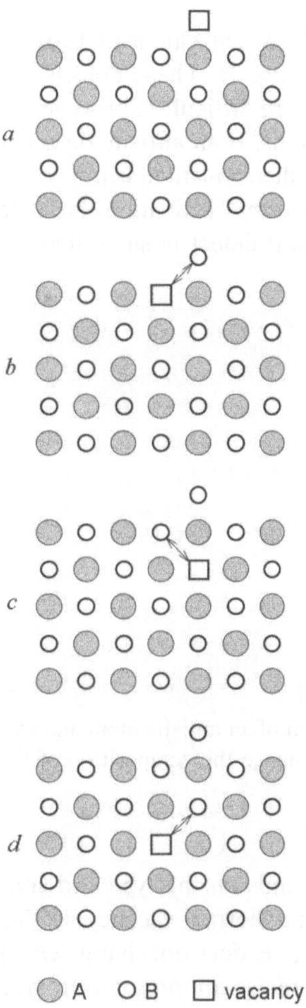

Fig. 1.2. Schottky defect in a binary compound AB. The arrows indicate the exchange of the atom B and vacancy □. The formation of a Schottky defect is a result of vacancy diffusion from the surface into the crystal. Parts (*a*), (*b*), (*c*), and (*d*) show the diffusional process step by step. A Schottky defect is a pair comprising an atom on the surface and a vacancy in the crystal

Transfer of a vacancy from the crystal surface to the lattice site causes the formation of a "vacancy–surface atom" pair, i.e. a Schottky defect (Fig. 1.2) or a thermal vacancy. Anti-Schottky defects appear when an atom moves from the crystal surface to an interstice.

In compounds with atoms of two species A and B, Frenkel defects may appear either in one of the sublattices (A or B) or in both sublattices. If only Schottky defects are formed in binary compounds then they are formed in both sublattices and with equal concentration. Consequently, these binary compounds preserve stoichiometry and homogeneity.

Another type of point defect in binary compounds which can lead to non-stoichiometry is represented by an atom A in a sublattice B. These defects are referred to as antistructural or antisite. Antisite defects are difficult to observe but they are common in alloys or substitutional solid solutions. If an antisite defect is accompanied by two Schottky defects in one sublattice then so-called triple defect is formed (Fig. 1.3). Triple defect can be formed by thermal activation, because the stoichiometry is preserved in this case. Such a point defect is suggested for intermetallics [12].

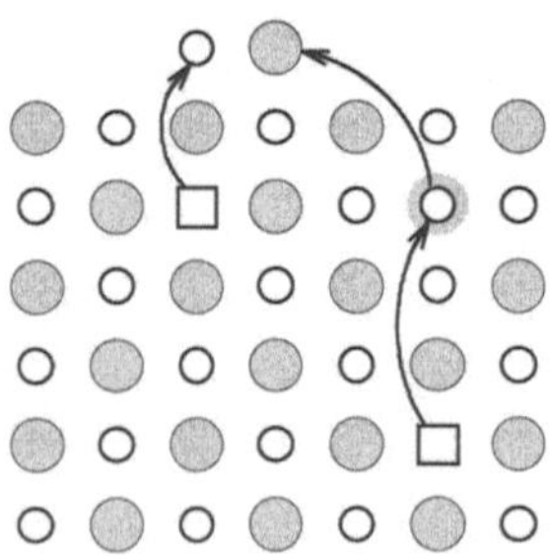

Fig. 1.3. Triple defect in a binary compound as a combination of an antisite atom and two Schottky defects. The formation of triple defects does not change the composition of the binary compound

The effect of point defects on crystal properties depends on the type and concentration of the defects. As an example, let us consider a thermal vacancy. In this case the total number of atoms N in a crystal is constant, i.e. does not change when vacancies appear. In accordance with [13,14], if n vacancies are present, the free energy of the crystal is changed by

$$\Delta F = nE - T\Delta S\,, \tag{1.1}$$

where E is the vacancy formation energy and ΔS is the change of the entropy caused by vacancy formation. Neglecting the vibrational part of the entropy, the ΔS value coincides with the configuration entropy and equals $\Delta S = k_B \ln W$, where

$W = (N + n)!/(n!N!)$ denotes the number of different location options of n vacancies and N atoms at $(N + n)$ sites. Using the Stirling formula $\ln(N!) \approx N\ln N - N$ (when N is large), we have $\Delta F = nE + k_B T[N\ln(N/(N + n)) + n\ln(n/(N + n))]$. In equilibrium conditions the change of the free energy versus the number of vacancies becomes a minimum. Therefore $\partial\Delta F/\partial n = 0$ and the equilibrium concentration of vacancies $c = n/(N + n)$ is

$$c = \exp(-E/k_B T)\,. \tag{1.2}$$

From (1.2) it follows that the equilibrium concentration of vacancies is an exponential function of temperature. Hence, absolutely defect free crystals do not exist at temperatures $T > 0$ K. In the case of metals (Cu, Au, Ag) the vacancy formation energy E is about 1 eV [15]. At a temperature of $T = 300$ K the vacancy concentration $c \approx 10^{-17}$ corresponds to this value E. Assuming that vacancies are distributed uniformly in the crystal, the vacancy-to-vacancy distance is the inverse of the number of vacancies per unit length, i.e. $r \sim (N_A c\rho/A)^{-1/3}$ (A being the atomic mass, ρ the density, and N_A the Avogadro number). For a vacancy concentration $c \approx 10^{-17}$, the intervacancy distance is estimated to exceed 0.1 mm. The interaction between vacancies in this case and their effect on crystal properties is very small.

The aggregation of vacancies in metals reduces the crystal energy. For example, 24 bonds are broken if two isolated vacancies are formed in an fcc lattice, while only 23 bonds are broken if one bivacancy appears. Indeed, calculations show that the bivacancy formation energy E_{2v} is smaller than the energy E_{1v} required to form an isolated vacancy. In metals the bivacancy binding energy $E_{v\text{-}v} = 2E_{1v} - E_{2v}$ usually corresponds to some tenths of an electron-volt. Therefore at temperatures close to the melting point nearly one percent of all vacancies are paired and in addition the diffusion coefficient increases. In nonstoichiometric compounds the structural vacancies binding energy is smaller than the binding energy between a structural vacancy and a neighbor atom and therefore structural vacancies repel each other.

Unlike vacancies, interstitial atoms can have different equilibrium configurations. Fcc, hcp and bcc lattices have octahedral and tetrahedral interstices, which can host interstitials like H, C, N, O, etc. Self interstitials in metals are located pairwise in the form of dumbbells. Interstitials may be neutral or charged in ionic crystals. Charge effects arising in ionic crystals are responsible for the great variety of atomic configurations of defects.

Since absolutely defect free crystals do not exist at temperatures $T > 0$ K, the presence of defects by itself is not an indicator of nonstoichiometry. Significant evidence of nonstoichiometry is the disagreement between the chemical composition of a compound and the concentration of lattice sites occupied by the compound components. Nonstoichiometry is possible only in two- and multicomponent substances.

Perfect stoichiometric composition of crystalline compounds is rather the exception than the rule, because defects are always present in real crystals. At the same time, in most simple substances, binary and more complex compounds, the concentration of structural vacancies is relatively small and does not exceed 0.01 at.% at 300 K, while the distance between nearest defects equals tens of micrometers. In such cases deviation from stoichiometry may be neglected.

However, in some compounds the concentration of vacancies is high under standard conditions. Probably the most familiar among these compounds is wustite FeO which is not available in stoichiometric form. It always contains excess oxygen, which is due to the presence of vacancies in the iron sublattice [16]. For example, at 1300 K the wustite composition is $Fe_{0.88}O$. At high temperatures wustite has a homogeneity interval from $Fe_{0.84}O$ to $Fe_{0.96}O$ and it has no stoichiometric composition. The mismatch between the homogeneity interval of compounds and their stoichiometric composition is a commonly occurring phenomenon. Considerable deviation from stoichiometry, which is accompanied by the formation of vacancies in the metal sublattice, is observed in iron $Fe_{0.85}S$ and copper $Cu_{1.73}S$ sulfides with the *B*8 (NiAs) type structure [17]. Large deviations from stoichiometry with vacancies in the oxygen sublattice are characteristic of higher oxides of transition metals: TiO_2 , V_2O_5 , CeO_2 , UO_2 , MoO_3 , WO_3 , etc.

Structural vacancies can be formed not only in binary, but also in ternary and more complex compounds. From the crystallographical point of view the presence of structural vacancies is due to the mismatch between the chemical composition, i.e. the relative number of atoms of different species, and the relative number of sites in different crystal sublattices occupied by these atoms (Fig. 1.4). From the physical or chemical point of view the reason for structural vacancy formation is the following. If the chemical composition does not correspond to the crystal structure and atoms from the sublattice of the one species cannot occupy the sites of the sublattice of the other species, i.e. the formation of antisite defects is not energetically possible, then structural vacancies are formed in a crystal.

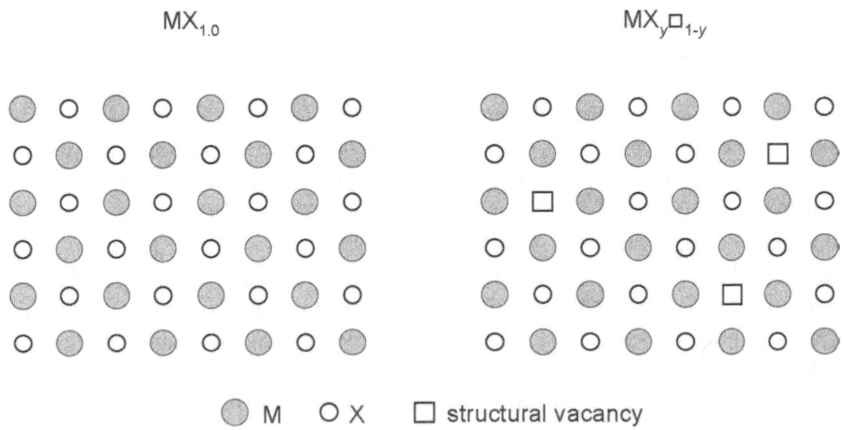

Fig. 1.4. Structural vacancy in a nonstoichiometric compound: the compound $MX_{1.0}$ without structural vacancies and the compound $MX_y\square_{1-y}$ with structural vacancies

The term "structural vacancy" was probably used for the first time by researchers [18] who described self-diffusion in carbides. Later it was used in the description of the defect structure of nonstoichiometric carbides, nitrides and oxides [13, 14, 19–38]. Currently this term is popular in the discussion of nonstoichiometric compounds. A specific feature of structural vacancies is that their concentration is directly related to the chemical composition of nonstoichiometric compounds and it can reach up to the range of a dozen atomic percent. In nonstoichiometric compounds structural vacancies represent analogs of atoms, i.e. are quasiparticles. In their sublattice they perform the same function as atoms of the same sublattice. Most nonstoichiometric compounds contain structural vacancies only in the nonmetallic sublattice. There are also nonstoichiometric compounds that contain structural vacancies in two sublattices. Depending on the compound composition, the concentration of structural vacancies in the sublattices may be equal or different. The presence of structural vacancies both in the metal and oxygen sublattices is characteristic of cubic titanium and vanadium monoxides [39, 40], niobium nitride, and some other nitrides. The concurrent presence of structural vacancies in niobium and oxygen sublattices is also characteristic of cubic niobium monoxide, if its structure is considered as a derivative of the $B1$-type structure [41]. However, there is the opinion that niobium monoxide has no structural vacancies because it has no homogeneity interval and a vacancy disorder in it is not observed, i.e. the free sites are never occupied.

An intensive study of oxides, sulfides and selenides of transition metals, which was performed in the 1930–40s revealed the existence of nonstoichiometric phases with wide homogeneity intervals. For example, cubic titanium monoxide TiO is homogeneous within the interval from $TiO_{0.65\text{-}0.80}$ to $TiO_{1.25\text{-}1.33}$. A very high content of vacancies in these phases raised the question as to the distribution of vacancies in the crystal lattice. Later it was shown that cubic carbides of IV- and V-group transition metals also have extremely wide intervals of homogeneity.

Formally, vacant sites of the crystal lattice behave themselves as atoms occupying sites in the same lattice. For this reason structural vacancies are viewed as an analog of atoms rather than "holes" in the crystal lattice. Deviation from stoichiometry and the resulting interval of homogeneity can be considered as a substitutional solution with atoms and vacancies being its components. Thus, vacancies and atoms form a substitutional solution in the structure of nonstoichiometric compounds. This substitutional solution may be disordered or ordered. The concept of a vacant site being a structural element of crystals similar to an occupied site [42–44] was introduced in the 1960–70s, when issues of nonstoichiometry and ordering in oxides, sulfides and chalcogenides were vigorously discussed in the literature [43–45].

In substitutional solutions the crystal lattice preserves its symmetry even in the presence of disorder. Indeed, one cannot tell which atomic species occupies the given site $\boldsymbol{r}$ in substitutional solutions. However, the probability that the given site $\boldsymbol{r}$ is occupied by an atom of a specific species is known. Put another way, a certain probability array, which includes all elements of the crystal symmetry, is available for substitutional solutions.

The analysis of the structure of nonstoichiometric compounds allows one to distinguish between two opposite tendencies: ordering and disordering. The ordered distribution of vacancies is more probable at low temperatures, while the disordered distribution takes place at high temperatures, when entropy makes a relatively large contribution to the free energy of nonstoichiometric compounds. Fully ordered and fully disordered distributions represent the limiting states of nonstoichiometric compounds. One or several ordered phases, which can also have a homogeneity interval, appear in the homogeneity range of nonstoichiometric compounds as a result of ordering. The maximum degree of long-range order, which is equal to unity, is achieved in ordered phases with the stoichiometric composition. Intermediate states, including short-range order, exist between the disordered distribution of structural vacancies and their long-range order.

Thus, the phenomenon of nonstoichiometry is intimately connected with ordering and disordering processes. Moreover, it is nonstoichiometry that predetermines the disordered or ordered distribution of atoms and vacancies in the structure of nonstoichiometric compounds. This book is dedicated to various aspects of nonstoichiometry, disorder and order, taking strongly nonstoichiometric compounds like carbides, nitrides and oxides of IV- and V-group transition metals and also alloys and substitutional solutions as an example.

References

1 N. S. Kurnakow: Über die gegenseitigen Verbindungen der Metalle, Z. Anorg. Chemie **23**, 439–462 (1900)

2 N. S. Kurnakow: Verbindung und chemisches Individuum, Z. Anorg. Chemie **88**, 109–127 (1914)

3 N. S. Kurnakov: Compound and chemical individual, Izv. Imperator. Akad. Nauk **8**, series 6, No 5, 321–338 (1914) (in Russian)

4 N. S. Kurnakov: Continuity of chemical transformations of substance, Uspekhi Fiz. Nauk **4**, 339–356 (1924) (in Russian)

5 N. S. Kurnakow: Singuläre Punkte chemischer Diagramme, Z. Anorg. Chemie **146**, 69–102 (1925)

6 G. Tammann: Über die Ermittelung der Zusammensetzung chemischer Verbindungen ohne Hilfe der Analyse, Z. Anorg. Chemie **37**, 303–313 (1903)

7 G. Tammann: *Die chemischen und galvanischen Eigenschaften von Mischkristallreihen und ihre Atomverteilung: ein Beitrag zur Kenntnis der Legierungen. Zum Gedächtnis der Entdeckung des Isomorphismus vor 100 Jahren* (Voss, Leipzig 1919) 239 SS

8 R. Collongues: *La Non Stoechiométrie* (Masson et Cie, Paris 1971) 230 pp.

9 R. Z. Valiev, A. V. Korznikov, R. R. Mulyukov: Structure and properties of metal materials having a submicrocrystalline structure, Fiz. Metall. Metalloved. **73**, 70–86 (1992) (in Russian)

10 A. I. Gusev: Effects of the nanocrystalline state in solids, Uspekhi Fiz. Nauk **168**, 55–83 (1998) (in Russian). (Engl. transl.: Physics – Uspekhi **41**, 49-76 (1998))
11 A. I. Gusev: *Nanocrystalline Materials: Production and Properties* (Ural Division of the RAS, Ekaterinburg 1998) 200 pp. (in Russian)
12 R. J. Wasilewski: Structure defects in CsCl intermetallic compounds – I. Theory, J. Phys. Chem. Solids **29**, 39–49 (1968)
13 A. I. Gusev, A. A. Rempel: *Thermodynamics of Structural Vacancies in Nonstoichiometric Interstitial Phases* (Ural Scientific Center, Sverdlovsk 1987) 114 pp. (in Russian)
14 A. I. Gusev, S. I. Alyamovskii, Yu. G. Zainulin, G. P. Shveikin: Structural vacancies in compounds of variable composition, Uspekhi Khimii **55**, 2067–2085 (1986) (in Russian). (Engl. transl.: Russian Chemical Review **55**, 1175–1185 (1986))
15 R. W. Balluffi: Vacancy defect mobilities and binding energies obtained from annealing studies, J. Nucl. Mater. **69–70**, 240–263 (1978)
16 L. S. Darken, R. W. Gurry: The system iron–oxygen. I. The wustite field and related equilibria, J. Amer. Chem. Soc. **67**, 1398–1412 (1945)
17 H. Krebs: *Grundzüge der Anorganischen Kristallchemie* (Ferdinand Enke Verlag, Stuttgart 1968) 380 pp.
18 R. A. Andrievski, K. P. Gurov: Self-diffusion in interstitial phases, Fiz. Metall. Metalloved. **26**, 818–822 (1968) (in Russian)
19 R. A. Andrievski, K. P. Gurov: Self-diffusion of nonmetal atoms in interstitial phases, Fiz. Metall. Metalloved. **39**, 57–61 (1975) (in Russian)
20 A. V. Korobeinikova, V. I. Fadeyeva, L. A. Reznitskii: Analysis of distribution of structural vacancies in iron γ-oxide, Zh. Strukturnoi Khimii **17**, 860–865 (1976) (in Russian)
21 A. I. Gusev: Statistical description of structural vacancy formation in compounds of variable composition such as zirconium nitride and carbide, Zh. Fiz. Khimii **53**, 1379–1383 (1979) (in Russian). (Engl. transl.: Russ. J. Phys. Chem. **53**, 780–782 (1979))
22 A. I. Gusev: Calculations of some thermodynamic characteristics of structural vacancies in high-melting compounds of the type zirconium and niobium carbides, Teplofizika Vysokikh Temp. **17**, 1232–1235 (1979) (in Russian). (Engl. transl.: High Temp. **17**, 1020–1023 (1979))
23 B. Gillot, F. Bouton: Correlation between IR spectra, X-ray diffraction, and distribution of structural vacancies in Fe_3O_4-type spinels, J. Solid State Chem. **32**, 303–310 (1980)
24 A. I. Gusev: Energy characteristics of the structural vacancies in titanium and zirconium nitrides, Zh. Neorgan. Khimii **25**, 629–633 (1980) (in Russian). (Engl. transl.: Russ. J. Inorg. Chem. **25**, 345–347 (1980))
25 A. I. Gusev: The energy of formation of structural vacancies in compounds with doubly defective structures in relation to vanadium and titanium monoxides, Zh. Fiz. Khimii **54**, 773–775 (1980) (in Russian). (Engl. transl.: Russ. J. Phys. Chem. **54**, 445–446 (1980))
26 A. I. Gusev: Structural vacancies in compounds of variable composition at high pressure, Zh. Fiz. Khimii **54**, 2211–2216 (1980) (in Russian). (Engl. transl.: Russ. J. Phys. Chem. **54**, 1263–1266 (1980))
27 A. I. Gusev: Characteristics of the formation of structural vacancies in vanadium carbide and thermodynamic properties of defect-free vanadium carbide, Zh. Fiz.

Khimii **57**, 1382–1386 (1983) (in Rusian). (Engl. transl.: Russ. J. Phys. Chem. **57**, 837–839 (1983))
28 A. V. Gubanov, A. L. Ivanovsky, G. P. Shveikin, D. E. Ellis: Vacancies and the energy spectrum of refractory metal compounds: TiC and TiO, J. Phys. Chem. Solids **45**, 719–730 (1984)
29 A. I. Gusev: Structural vacancies in nonstoichiometric compounds at high pressure. Thermodynamic model, Phys. Stat. Sol. (a) **85**, 159–166 (1984)
30 D. L. Novikov, A. L. Ivanovsky, V. A. Gubanov: The influence of structural vacancies and impurities on the electronic structure of TiC(100), Philosoph. Mag. B **63**, 1409–1418 (1991)
31 A. I. Gusev, A. A. Rempel: *Structural Phase Transformations in Nonstoichiometric Compounds* (Nauka, Moscow 1988) 308 pp. (in Russian)
32 V. I. Ivanchenko, E. P. Trofimova, A. A. Lisenko, E. A. Zhurakovskii: Electronic structure of zirconium nitrides with nitrogen and zirconium structural vacancies, Metallofizika **12**, 14–21 (1990) (in Russian)
33 S. Tsurekawa, H. Yoshinaga: Identification of long range ordered structure in $TiC_{0.59}$ by transmission electron microscopy, J. Japan Inst. Metals **56**, 133–141 (1992)
34 A. A. Rempel: *Ordering Effects in Nonstoichiometric Interstitial Compounds* (Nauka, Ekaterinburg 1992) 232 pp. (in Russian)
35 M. J. Puska, M. Šob, G. Brauer, T. Kornonen: First-principles calculation of positron lifetimes and affinities in perfect and imperfect transition metal carbides and nitrides, Phys. Rev. B **49**, 10947–10958 (1994)
36 G. Brauer, W. Anwand, E.-M. Nicht, P. G. Coleman, A. P. Knights, H. Schut, G. Kögel, N. Wagner: Positron studies of polycrystalline TiC, J. Phys.: Condens. Matter **7**, 9091–9099 (1995)
37 V. V. Il'yasov, I. Ya. Nikiforov: Effect of the degree of ordering of structural vacancies on the fine structure of the valance band maximum in cubic boron nitride, Fiz. Tverd. Tela **39**, 1064–1065 (1997) (in Russian)
38 T. Ogawa, F. Kobayashi, T. Sato, R. G. Haire: Actinide nitrides and nitride-halides in high-temperature systems, J. Alloys Comp. **271–273**, 347–354 (1998)
39 S. Andersson, B. Collen, U. Kuylenstierna, A. Magneli: Phase analysis studies on the titanium - oxygen system, Acta Chem. Scand. **11**, 1641–1647 (1957)
40 S. Westman, C. Nordmark: Phase analysis studies on the vanadium–oxygen system within $VO_{0.25}$–$VO_{1.5}$ region at 800° C , Acta Chem. Scand. **14**, 465–470 (1960)
41 G. Brauer: Die Oxide Niobs, Z. Anorg. Allgem. Chem. **248**, 1–31 (1941)
42 J. S. Anderson: The thermodynamics and theory of nonstoichiometric compounds, in *Problems of Nonstoichiometry*, A. Rabenau (Ed.) (North-Holland Publ., Amsterdam-London 1970) pp. 1–76
43 J. S. Anderson: Nonstoichiometric and ordered phases: thermodynamic considerations, in *The Chemistry of Extended Defects in Non-Metallic Solids*, L. Eyring, M. O'Keeffe (Eds.) (North-Holland Publ., Amsterdam-London 1970) pp. 1–20
44 J. S. Anderson: Defect chemistry and non-stoichiometric compounds, in *Modern Aspects of Solid State Chemistry*, C. N. R. Rao (Ed.) (Plenum Press, New York 1970) pp. 29–105
45 A. Rabenau (Ed.): *Problems of Nonstoichiometry* (North-Holland Publ., Amsterdam-London 1970) 292 pp.

2. Crystal Symmetry

This chapter briefly describes basic notions of the theory of crystal symmetry and space groups. They will be repeatedly referred to in subsequent discussion of the structure of nonstoichiometric compounds, alteration of symmetry and properties of these compounds during ordering.

The first section deals with point and space groups of the crystal symmetry. The second section is dedicated to such important notions of solid-state theory as the reciprocal lattice, the Brillouin zone, and the wave-vector star. The main relations of the diffraction theory necessary for understanding specific features of superstructural diffraction patterns are presented in the third section.

A crystal structure is a three-dimensional periodic ensemble of particles. The regular and periodic arrangement of particles in crystals determines the symmetry of their physical properties. Symmetry of a physical system means invariance of its properties under certain transformations. The structure and symmetry of crystals are described using space groups. All 230 space groups were worked out simultaneously and independently by E. S. Fedorov [1–9] and A. Schoenflies [10–13] in the period from 1885 to 1894. Space groups are described comprehensively in, e.g. [14–17]. The basic notions of crystal symmetry theory and various aspects of usage of symmetry groups in solid-state physics are discussed in detail in the treatises [18–23].

2.1 Point and Space Groups

Geometrical symmetry is the property of a crystal or a crystal structure of being invariant under some isometric transformation of the space in which the crystal is embedded. These transformations are called symmetry operations. Subspaces which are mapped onto themselves under the transformation are used to characterize the transformation (symmetry elements). Symmetry transformations can be classified with respect to invariant sets: simple (fixed-point) and infinite (fixed-point free) transformations. Simple or point (finite) symmetry transformations are those which map at least one point of the space to itself, i.e. a fixed point. Usually one of those points is taken as the origin of coordinates. Neither point of a figure remains in place during infinite (space) transformations. Finite symmetry transformations correspond to the symmetry of ideal crystalline polyhedra, while infinite transformations correspond to the symmetry of crystalline structures.

Simple Symmetry Elements. Simple symmetry operations include the identity transformation $\boldsymbol{E}$, rotations described by n-fold ($n = 1, 2, 3, 4$, and 6) symmetry axes, symmetry planes m, the symmetry center (inversion) $\boldsymbol{i}$ or $\bar{1}$, and inversion symmetry axes (rotoinversions). These axes and planes combined with the information on the type of the symmetry operation (e.g. four-fold rotation) are called the symmetry elements corresponding to the symmetry transformation.

The position of any arbitrary point (site) of a lattice is determined by the radius-vector $\boldsymbol{r}$. Its coordinates with respect to a chosen basis are written as a column vector of $\mathbb{R}^3$: $\boldsymbol{r} = (x, y, z)^{\mathrm{T}}$. For convenience the T (for transposed) is omitted in text lines. The identity transformation $\boldsymbol{E}$ transforms each point $\boldsymbol{r}$ to itself:

$$\boldsymbol{E}\boldsymbol{r} = \boldsymbol{r} \tag{2.1}$$

or

$$\boldsymbol{E}\begin{pmatrix} x \\ y \\ z \end{pmatrix} = \begin{pmatrix} x \\ y \\ z \end{pmatrix}, \tag{2.2}$$

i.e. the matrix $\boldsymbol{E}$ is the unit matrix

$$\boldsymbol{E} = \begin{pmatrix} 1 & 0 & 0 \\ 0 & 1 & 0 \\ 0 & 0 & 1 \end{pmatrix}. \tag{2.3}$$

The symmetry axis is a straight line, about which the figure is mapped onto itself when turned through some angle. The symmetry order n of an axis determines how many times the figure is mapped onto itself while making a complete revolution about the axis. Rotations through 60°, 90°, 120°, 360° and multiples thereof are possible in crystals about axes of six-, four-, three-, two-, and one-fold symmetry respectively. Axes of five-fold symmetry are impossible in crystals comprising regularly repeated unit cells. However, five-fold symmetry was found in quasicrystals formed during rapid quenching of alloys of the Al–Mn system and other similar systems [24, 25]. Symmetry axes of any order (from first to infinite order) are possible in living objects and geometrical figures. Any figure has a one-fold symmetry axis, while symmetry axes of infinite order are inherent in figures (bodies) of revolution. Ideal amorphous bodies and liquids also possess infinite-fold symmetry axes.

Rotation through an angle φ about an axis, which is determined by the unit vector $\boldsymbol{\omega}$, moves the point $\boldsymbol{r}$ to the point $\boldsymbol{r}'$. The coordinates of the point $\boldsymbol{r}'$ can be found from the expression

$$\boldsymbol{r}' = \boldsymbol{B}_{\omega}(\varphi)\boldsymbol{r}, \tag{2.4}$$

where $\boldsymbol{B}_{\omega}(\varphi)$ is the rotation matrix.

The symmetry plane m divides the figure into two equal mirror parts. In the general case any arbitrary mirror-reflection plane transforms space points by the optical laws of constructing virtual images in a flat mirror, i.e.

$$\boldsymbol{r} = \boldsymbol{\sigma r}, \tag{2.5}$$

where $\boldsymbol{\sigma}$ is the reflection matrix. For example, reflection in coordinate planes is given by the matrices $\sigma_x \equiv m_x \equiv m[100]$, $\sigma_y \equiv m_y \equiv m[010]$, and $\sigma_z \equiv m_z \equiv m[001]$. The subscript shows which axis the reflection plane is perpendicular to. These matrices are of the form

$$\sigma_x = \begin{pmatrix} -1 & 0 & 0 \\ 0 & 1 & 0 \\ 0 & 0 & 1 \end{pmatrix}, \quad \sigma_y = \begin{pmatrix} 1 & 0 & 0 \\ 0 & -1 & 0 \\ 0 & 0 & 1 \end{pmatrix}, \quad \sigma_z = \begin{pmatrix} 1 & 0 & 0 \\ 0 & 1 & 0 \\ 0 & 0 & -1 \end{pmatrix}. \tag{2.6}$$

The identity transformation, axes and planes of symmetry represent point elements of one-fold symmetry. The center of symmetry and inversion axes are elements of two-fold symmetry.

The center of symmetry (inversion or center of inversion) $\boldsymbol{i} \equiv \bar{1}$ is called a singular point inside a figure, which maps any point to its mirror-symmetrical point with respect to the origin of coordinates,

$$\boldsymbol{ir} = -\boldsymbol{r}, \tag{2.7}$$

where $\boldsymbol{i} = \begin{pmatrix} -1 & 0 & 0 \\ 0 & -1 & 0 \\ 0 & 0 & -1 \end{pmatrix}$. The inversion $\boldsymbol{i} \equiv \bar{1}$ is the simplest second-order transformation. The center of symmetry of a figure is the point where the even-order symmetry axis crosses the symmetry plane perpendicular to this axis. A combination of the n-order symmetry axis and a symmetry plane perpendicular thereto is designated n/m, for example, $2/m$, $3/m$, $4/m$, and $6/m$.

The other second-order transformations $\bar{2}$, $\bar{3}$, $\bar{4}$, and $\bar{6}$ represent a combination of rotation and inversion. They are called inversion axes. The inversion revolution $\bar{2}$ is equivalent to the operation of a symmetry plane, i.e. $\bar{2} \equiv m$. Inversion rotations represent revolution about the corresponding symmetry axis and simultaneous reflection through the center of symmetry.

Point Groups and Generators. The symmetry of any crystalline polyhedron is described by a set of symmetry elements 1, 2, 3, 4, 6, and $\bar{1}$; $m \equiv \bar{2}$, $\bar{3}$, $\bar{4}$, $\bar{6}$. The set of symmetry operations of an ideal crystalline polyhedron, i.e. transformations, by which this polyhedron is mapped onto itself, form a point group of the crystal symmetry. The term "point group" originates from the fact that given any symmetry transformations of a polyhedron, at least one point, where all symmetry elements comprising a group intersect, remains in place. The number of different symmetry operations, which make up a group, is referred to as the group order.

Point groups include rotations and rotoinversions. Various combinations of these operations form 32 independent three-dimensional crystallographic point groups. All symmetry elements of any group can be determined from the group generators. The group generators are no more than three in number. The Schoenflies notation is used in theoretical physics and spectroscopy for symmetry elements and point groups: C_n is the figure rotation axis of order n; σ_v the vertical symmetry plane, which contains the figure axis; σ_d the diagonal symmetry plane, which contains the figure axis and bisects the angle between two axes of two-fold symmetry perpendicular to the figure axis (in some cases the σ_d plane may contain other symmetry elements); σ_h the horizontal symmetry plane perpendicular to the figure plane; $S_n = \sigma_h C_n$ the noneigenaxis rotation (in terms of the international system it is referred to as the rotoinversion or the inversion axis); $i \equiv S_2$ the inversion point (center); and C_1 the identity transformation (unity).

In total, there are 15 types of three-dimensional point groups: 12 crystallographic and three noncrystallographic groups. The generators of the 12 types of crystallographic point groups in the Schoenflies notation are given in Table 2.1.

Table 2.1. Generators of 12 crystallographic point groups

Group	Generators	Comments
C_n	C_n	C_1 – group of order 1 for the systems without symmetry
S_n	S_n	Group S_2 is equivalent to the group C_i
C_{nv}	C_n , σ_v	
C_{nh}	C_n , σ_h	Group C_s is equivalent to the group C_h (C_{1h})
D_n	C_n , C_2	
D_{nd}	C_n , C_2 , σ_d	
D_{nh}	C_n , C_2 , σ_h	
T	C_2^x , C_3^{xyz}	Tetrahedron axes group (order of group is equal to 12) is the main subgroup of all cubic groups
T_d	S_4^z , C_3^{xyz}	Group of full symmetry of the tetrahedron (order of group is equal to 24)
T_h	C_2^z , C_3^{xyz} , i	
O	C_4^z , C_3^{xyz}	Octahedron group (order of group is equal to 24)
O_h	C_4^z , C_3^{xyz} , i	Group of full symmetry of the cube (order of group is equal to 48)

Crystallographic and noncrystallographic point groups are classes of general point groups, which are described elsewhere [14]. Of greatest interest among noncrystallographic point groups is the rotation group of the icosahedron I and the group of the full icosahedron symmetry I_h. The groups I and I_h have orders of 60 and 120 respectively. The group I contains simple rotations only, while the group I_h includes simple and inversion rotations. Generators of the group I are two-, three-, and five-fold symmetry axes, while generators of the group I_h include inversion, in addition. Symmetry elements of each icosahedral group contain six rotation axes of five-fold symmetry located at an angle of $\arccos[1/(2\tau-1)] \approx 63.43°$ to one another ($\tau^2 = \tau + 1$, $\tau > 0$). Special forms of icosahedral groups are noncrystallographic regular polyhedra: an icosahedron and a pentagon-dodecahedron. The icosahedral symmetry groups have only be used in solid-state physics since five-fold symmetry was detected experimentally in 1984 in rapidly quenched Al–Mn alloys [24, 25]. Now they are used on a wide scale in studies of the structure and symmetry of quasicrystals.

The last fifteenth type of point groups are represented by continuous (finite) groups, which include infinite symmetry axes. Point groups of cylindrical and spherical systems comprise continuous groups.

Crystallographic groups can be classified with respect to various criteria. The common classification, resulting in 32 classes, uses the idea of congruent arrangements of symmetry elements. Formally this means that two groups of matrices $\{\boldsymbol{M}_j\}$ and $\{\boldsymbol{N}_j\}$ are *geometrically equivalent* if there exists a nonsimilar matrix $\boldsymbol{A}$ such that $\boldsymbol{N}_j = \boldsymbol{A}^{-1}\boldsymbol{M}_j\boldsymbol{A}$ for all j. All 32 crystallographic point groups represent subgroups of continuous (finite) symmetry groups.

The international system of notation is adopted in crystallography, unlike in solid-state physics. Alternatively it is referred to as the Hermann–Mauguin system. In terms of this system a symmetry group is labeled by a set of generators, which are fixed by reference to the symmetry classes of lattice symmetry directions. According to the seven groups of lattices (holohedries) there are seven types of these classes. For each holohedry a conventional basis of lattice vectors is defined by standardization rules and thus the description of the point groups by matrices is fixed as far as can be done by symmetry. Further standardization rules use the shortest lattice vectors and their scalar products.

For example, generators of the point group $2/m$ are the two-fold symmetry axis and a symmetry plane perpendicular to this axis. As there is only one class of lattice symmetry directions in a monoclinic (see below) group which conventionally [14] is chosen as the set {±[010]} the generating matrices are fixed as 2[010] and m[010]. In the text that follows, according to Kovalev [16], the direction [001] is preferred for monoclinic groups.

As any crystallographic group refers to a lattice with the symmetry of one of the holohedries there are seven descriptive systems for the point groups, which are called crystal systems. They are named like the holohedries: a = triclinic, m = monoclinic, o = orthorhombic, t = tetragonal, r = rhombohedral, h = hexagonal and c = cubic. As all rhombohedral symmetries can also be described as integral matrices with respect to a hexagonal basis, both systems are combined to form the hexagonal-rhombohedral crystal family and only for comparison with cubic groups is a rhombohedral description chosen. Table 2.2 gives all the point groups and lists all the elements of symmetry and generators in accordance with the notation adopted in [16, 17]: 48 elements of the full cubic symmetry group $m\bar{3}m$ (O_h) are sequentially labeled h_1 to h_{48}, while 24 elements of the hexagonal group $6/mmm$ (D_{6h}) are designated from H_1 to H_{24}.

Table 2.2. Point groups of crystals

Crystal system	Group symbol		Symmetry elements of the point group	Generators of group	Number of	
	International	Schoenflies			classes	space groups
1	2	3	4	5	6	7
Triclinic	1	C_1	h_1	h_1	1	1
	$\bar{1}$	C_i (S_2)	h_1, h_{25}	h_{25}	2	1
Monoclinic	2	C_2	h_1, h_4	h_4	2	3
	m	C_s (C_h)	h_1, h_{28}	h_{28}	2	4
	$2/m$	C_{2h}	h_1, h_4, h_{25}, h_{28}	h_4, h_{25}	4	6
Orthorhombic	222	D_2	h_1–h_4	h_2, h_4	4	9
	$mm2$	C_{2v}	h_1, h_4, h_{26}, h_{27}	h_4, h_{27}	4	22
	mmm	D_{2h}	h_1–h_4, h_{25}–h_{28}	h_2, h_3, h_{25}	8	28
Trigonal (rhombohedral)	3	C_3	H_1, H_3, H_5	H_3	3	4
	$\bar{3}$	C_{3i} (S_6)	$H_1, H_3, H_5, H_{13}, H_{15}, H_{17}$	H_3, H_{13}	6	2
	322	D_3	$H_1, H_3, H_5, H_8, H_{10}, H_{12}$	H_3, H_8	3	7
	$3mm$	C_{3v}	$H_1, H_3, H_5, H_{20}, H_{22}, H_{24}$	H_3, H_{22}	3	6
	$\bar{3}m$	D_{3d}	$H_1, H_3, H_5, H_8, H_{10}, H_{12}, H_{13}, H_{15}, H_{17}, H_{20}, H_{22}, H_{24}$	H_3, H_8, H_{13}	6	6
Tetragonal	4	C_4	h_1, h_4, h_{14}, h_{15}	h_{14}	4	6
	$\bar{4}$	S_4	h_1, h_4, h_{38}, h_{39}	h_{38}	4	2
	$4/m$	C_{4h}	$h_1, h_4, h_{14}, h_{15}, h_{25}, h_{28}, h_{38}, h_{39}$	h_{14}, h_{28}	8	6
	422	D_4	h_1–h_4, h_{13}–h_{16}	h_{14}, h_{16}	5	10
	$4mm$	C_{4v}	$h_1, h_4, h_{14}, h_{15}, h_{26}, h_{27}, h_{37}, h_{40}$	h_{14}, h_{27}	5	12
	$\bar{4}2m$	D_{2d}	h_1–h_4, h_{37}–h_{40}	h_{38}, h_{40}	5	12
	$4/mmm$	D_{4h}	h_1–h_4, h_{13}–h_{16}, h_{25}–h_{28}, h_{37}–h_{40}	h_{14}, h_{27}, h_{28}	10	20
Cubic	23	T	h_1–h_{12}	h_4, h_5	4	5
	$m3$	T_h	h_1–h_{12}, h_{25}–h_{36}	h_4, h_5, h_{25}	8	7
	$\bar{4}3m$	T_d	h_1–h_{12}, h_{37}–h_{48}	h_9, h_{38}	5	6
	432	O	h_1–h_{24}	h_9, h_{14}	5	8
	$m3m$	O_h	h_1–h_{48}	h_9, h_{14}, h_{25}	10	10

(*Continued*)

Table 2.2 (continued)

1	2	3	4	5	6	7
Hexagonal	6	C_6	H_1–H_6	H_2	6	6
	$\bar{6}$	C_{3h} (S_3)	H_1, H_3, H_5, H_{14}, H_{16}, H_{18}	H_{18}	6	1
	$6/m$	C_{6h}	H_1–H_6, H_{13}–H_{18}	H_2, H_{16}	12	2
	622	D_6	H_1–H_{12}	H_2, H_{11}	6	6
	$6mm$	C_{6v}	H_1–H_6, H_{19}–H_{24}	H_2, H_{20}	6	4
	$\bar{6}2m$	D_{3h}	H_1, H_3, H_5, H_8, H_{10}, H_{12}, H_{14}, H_{16}, H_{18}, H_{19}, H_{21}, H_{23}	H_{16}, H_{18}	6	4
	$6/mmm$	D_{6h}	H_1–H_{24}	H_2, H_{16}, H_{20}	12	4

Any crystallographic point group represents a subgroup of both or either of these groups. In the cubic group $m3m$ (O_h) the first 24 symmetry elements h_1–h_{24} correspond to rotations, the symmetry element h_{25} corresponds to inversion, and the other elements are rotoinversions, i.e. $h_{24+k} = h_k h_{25}$ (k = 1–24, integer). Analogously, in the hexagonal group $6/mmm$ (D_{6h}), where inversion is the element H_{13}, $H_{12+k} = H_k H_{13}$ (k = 1–12, integer).

Point groups C_1, $C_i \equiv S_2$, C_n, S_n, and $C_s \equiv C_h$ are cyclic groups and thus formed by one generator. The other point groups are formed by two or more generators. Elements of higher-symmetry groups can be determined if one finds an adjacent class, which is formed by a subgroup corresponding to the figure axis with respect to the additional generator. If two additional generators are assigned, the adjacent class of the cyclic subgroup with respect to one of the generators is formed first. As a result, an extended subgroup of the point group is constructed. The adjacent class of the last subgroup with respect to the other additional generator is formed second. The final result is the full group.

Let us take the $\bar{3}m$ (D_{3d}) group as an example and consider how generators of this group are used to construct all its symmetry elements. Generators of the group D_{3d} are the axes $C_3 = 3_{xxx} = 3[111]$ and $C_2 = 2_{\bar{x}y} = 2[-110]$ and the diagonal symmetry plane $\sigma_d = m[-110]$ (see Table 2.1). The first generator C_3 forms an analogous cyclic subgroup with elements $\{E, C_3, C_3^2\}$ or, using the notation [16, 17], $\{h_1, h_5, h_9\}$ or $\{H_1, H_3, H_5\}$. The adjacent class with respect to the second generator C_2 is the set of three axes C_2, which are rotated relative to one another through 120°. In other words, we have the subgroup $\{C_2, C_2', C_2''\} \equiv \{h_{13}, h_{17}, h_{45}\} \equiv \{H_8, H_{10}, H_{24}\}$. These two subgroups comprise the extended subgroup $D_3 = \{E, 2C_3, 3C_2\} \equiv \{h_1, h_5, h_9, h_{13}, h_{17}, h_{45}\} \equiv \{H_1, H_3, H_5, H_8, H_{10}, H_{24}\}$, which is the adjacent class to the inversion i (the inversion results from the product $C_2\sigma_d$). Allowing for the inversion, we obtain the full group $D_{3d} = \{E, 2C_3, 3C_2, i, 2S_3, 3\sigma_d\} \equiv \{h_1, h_5, h_9, h_{13}, h_{17}, h_{21}, h_{25}, h_{29}, h_{33}, h_{37}, h_{41}, h_{45}\} \equiv \{H_1, H_3, H_5, H_8, H_{10}, H_{12}, H_{13}, H_{15}, H_{17}, H_{20}, H_{22}, H_{24}\}$. Indeed, multiplication of rotation matrices of the generators h_5, h_{13} and h_{25} or H_3, H_8 and H_{13} of the group $\bar{3}m$ (D_{3d}) gives rotation matrices of all other elements of this group.

Geometrical images of all the symmetry elements are described elsewhere [16]. Table 2.3 gives international designations for each symmetry element h or H, the corresponding rotation matrices, and coordinates of the point $\boldsymbol{r}'$, which result when the point $\boldsymbol{r} = (x, y, z)$ undergoes a space transformation by the corresponding group element. Using rotation matrices and (2.4), an arbitrary point $\boldsymbol{r} = (x, y, z)$ can be space-transformed to the point $\boldsymbol{r}' = (x', y', z')$, as with group elements h or H.

Table 2.3. Rotation matrices $\boldsymbol{B}_{\omega}(\varphi)$ and rotation symmetry elements h_i and H_i of the cubic and hexagonal groups

h_i	$h_1 - 1$	$h_2 - 2_x$	$h_3 - 2_y$	$h_4 - 2_z$	$h_5 - 3^2_{111}$	$h_6 - 3_{11\bar{1}}$
	(x, y, z)	$(x, \bar{y}, \bar{z})$	$(\bar{x}, y, \bar{z})$	$(\bar{x}, \bar{y}, z)$	(y, z, x)	$(y, \bar{z}, \bar{x})$
$\boldsymbol{B}_{\omega}(\varphi)$	1 0 0	1 0 0	−1 0 0	−1 0 0	0 1 0	0 1 0
	0 1 0	0 −1 0	0 1 0	0 −1 0	0 0 1	0 0 −1
	0 0 1	0 0 −1	0 0 −1	0 0 1	1 0 0	−1 0 0
h_i	$h_7 - 3_{\bar{1}11}$	$h_8 - 3_{1\bar{1}1}$	$h_9 - 3_{111}$	$h_{10} - 3^2_{\bar{1}11}$	$h_{11} - 3^2_{11\bar{1}}$	$h_{12} - 3^2_{\bar{1}11}$
	$(\bar{y}, z, \bar{x})$	$(\bar{y}, \bar{z}, x)$	(z, x, y)	$(z, \bar{x}, \bar{y})$	$(\bar{z}, x, \bar{y})$	$(\bar{z}, \bar{x}, y)$
$\boldsymbol{B}_{\omega}(\varphi)$	0 −1 0	0 −1 0	0 0 1	0 0 1	0 0 −1	0 0 −1
	0 0 1	0 0 −1	1 0 0	−1 0 0	1 0 0	−1 0 0
	−1 0 0	1 0 0	0 1 0	0 −1 0	0 −1 0	0 1 0
h_i	$h_{13} - 2_{\bar{x}y}$	$h_{14} - 4_z$	$h_{15} - 4^3_z$	$h_{16} - 2_{xy}$	$h_{17} - 2_{\bar{y}z}$	$h_{18} - 2_{yz}$
	$(\bar{y}, \bar{x}, \bar{z})$	$(\bar{y}, x, z)$	$(y, \bar{x}, z)$	$(y, x, \bar{z})$	$(\bar{x}, \bar{z}, \bar{y})$	$(\bar{x}, z, y)$
$\boldsymbol{B}_{\omega}(\varphi)$	0 −1 0	0 −1 0	0 1 0	0 1 0	−1 0 0	−1 0 0
	−1 0 0	1 0 0	−1 0 0	1 0 0	0 0 −1	0 0 1
	0 0 −1	0 0 1	0 0 1	0 0 −1	0 −1 0	0 1 0
h_i	$h_{19} - 4_x$	$h_{20} - 4^3_x$	$h_{21} - 2_{\bar{x}z}$	$h_{22} - 4^3_y$	$h_{23} - 2_{xz}$	$h_{24} - 4_y$
	$(x, \bar{z}, y)$	$(x, z, \bar{y})$	$(\bar{z}, \bar{y}, \bar{x})$	$(\bar{z}, y, x)$	$(z, \bar{y}, x)$	$(z, y, \bar{x})$
$\boldsymbol{B}_{\omega}(\varphi)$	1 0 0	1 0 0	0 0 −1	0 0 −1	0 0 1	0 0 1
	0 0 −1	0 0 1	0 −1 0	0 1 0	0 −1 0	0 1 0
	0 1 0	0 −1 0	−1 0 0	1 0 0	1 0 0	−1 0 0
H_i	$H_1 - 1$	$H_2 - 6_z$	$H_3 - 3_z$	$H_4 - 2_z$	$H_5 - 3^{-1}_z$	$H_6 - 6^{-1}_z$
	(x, y, z)	$(x-y, x, z)$	$(\bar{y}, x-y, z)$	$(\bar{x}, \bar{y}, z)$	$(y-x, \bar{x}, z)$	$(y, y-x, z)$
$\boldsymbol{B}_{\omega}(\varphi)$	1 0 0	1 −1 0	0 −1 0	−1 0 0	−1 1 0	0 1 0
	0 1 0	1 0 0	1 −1 0	0 −1 0	−1 0 0	−1 1 0
	0 0 1	0 0 1	0 0 1	0 0 1	0 0 1	0 0 1
H_i	$H_7 - 2_y$	$H_8 - 2_{\bar{x}y}$	$H_9 - 2_x$	$H_{10} - 2_{210}$	$H_{11} - 2_{xy}$	$H_{12} - 2_{120}$
	$(\bar{x}, y-x, \bar{z})$	$(\bar{y}, \bar{x}, \bar{z})$	$(x-y, \bar{y}, \bar{z})$	$(x, x-y, \bar{z})$	$(y, x, \bar{z})$	$(y-x, y, \bar{z})$
$\boldsymbol{B}_{\omega}(\varphi)$	−1 0 0	0 −1 0	1 −1 0	1 0 0	0 1 0	−1 1 0
	−1 1 0	−1 0 0	0 −1 0	1 −1 0	1 0 0	0 1 0
	0 0 −1	0 0 −1	0 0 −1	0 0 −1	0 0 −1	0 0 −1

In international notation a rotation is given by a rotation axis and by up to three indices indicating the direction of the axis or a direction symbol [*uvw*]. The superscript on the rotation axis denotes the number of elementary revolutions, i.e. the rotation angle; the subscript shows the direction of the rotation axis in the selected system of coordinates. For example, the rotation h_{13} through 180° about the direction $[\bar{1}10]$ corresponds to the element $2_{\bar{x}y}$, which means rotation through 180° about the bisector of the angle formed by the coordinate axes $-x$ and $+y$.

Translation Groups. The three-dimensional regularly filled space of a crystal is referred to as the crystal lattice, i.e. a regular periodically repeated system of points determined by the three shortest nonparallel translations. If the vector basis $\boldsymbol{t}_1$, $\boldsymbol{t}_2$, $\boldsymbol{t}_3$ is known, the position of any lattice site $\boldsymbol{r}_n$ is determined by the vector

$$\boldsymbol{r}_n = n_1\boldsymbol{t}_1 + n_2\boldsymbol{t}_2 + n_3\boldsymbol{t}_3 \,, \tag{2.8}$$

where n_1, n_2, and n_3 are integers; $[[n_1n_2n_3]]$ is the site symbol. The parallelepiped constructed using the three shortest translations of a crystal forms a primitive cell, which uniquely characterizes the whole translation group, because it contains all three generators $\boldsymbol{t}_1$, $\boldsymbol{t}_2$, $\boldsymbol{t}_3$. In addition to the linearly independent translation vectors $\boldsymbol{t}_1$, $\boldsymbol{t}_2$, $\boldsymbol{t}_3$, all vectors $\boldsymbol{t} = u_1\boldsymbol{t}_1 + u_2\boldsymbol{t}_2 + u_3\boldsymbol{t}_3$ ($u_{1,2,3}$ are arbitrary integers) are translation vectors of the crystal. The set of all translation vectors of a crystal is called the lattice of translation vectors or the vector lattice. If every integral linear combination $\boldsymbol{t} = u_1\boldsymbol{a}_1 + ... + u_n\boldsymbol{a}_n$ is a lattice vector than a basis of n vectors $\boldsymbol{a}_1, ...,$ $\boldsymbol{a}_n$ is a crystallographic basis the of n-dimensional vector lattice. Such a crystallographic basis is a primitive basis if every lattice vector $\boldsymbol{t}$ can be obtained as an integral linear combination of the basis vectors. Nonprimitive crystallographic bases are used conventionally for the description of centered lattices.

The crystal lattice is described by six quantities in a chosen system of coordinates: three non-parallel axial translations *a, b* and *c*, and three axial angles α, β and γ. There are seven geometrically different cells whose figure symmetry is given by point groups $\bar{1}$, $2/m$, mmm, $\bar{3}m$, $4/mmm$, $6/mmm$ and $m3m$, which correspond to seven crystal systems. An important aspect of a lattice is its unit cell. A unit cell of the vector lattice is the set of all vectors $x_1\boldsymbol{a}_1 + ... + x_n\boldsymbol{a}_n$ with $0 \le x_i < 1$. The term "unit cell" is usually applied to real crystal structures. In the general case, a unit cell includes atoms of different species, which are not interconnected by translations or any other symmetry elements. If the primitive unit cell does not have the symmetry of the crystal lattice than the crystal lattice is described by a so-called centered cell. The most important classification which takes into account both the lattice point-group symmetry and the centering mode was suggested by Bravais in 1866. The resulting classes are called Bravais lattices. Thus, Bravais lattices (or systems of translations) can be: (1) primitive *P* if they are assigned by translations ***a***, ***b***, and ***c***; (2) centered *C* (*A*, *B*) if they are assigned by the set of translations ***a***, ***b***, ***c***, and $(\boldsymbol{a} + \boldsymbol{b})/2$ (or $(\boldsymbol{b} + \boldsymbol{c})/2$ or $(\boldsymbol{a} + \boldsymbol{c})/2$); (3) body-centered *I* with the set of translations ***a***, ***b***, ***c***, and $(\boldsymbol{a} + \boldsymbol{b} + \boldsymbol{c})/2$; and (4) face-centered *F* with the set of translations ***a***, ***b***, ***c***, $(\boldsymbol{a} + \boldsymbol{b})/2$, and $(\boldsymbol{b} + \boldsymbol{c})/2$. As a result, crystals have one out of 14 different independent lattices (Table 2.4).

Table 2.4. Conventional cells of the three-dimensional Bravais crystal lattices

Crystal system	Parameters of unit cell	Lattices and their basis: primitive *P* Basis: 0 0 0	centred *C* (*A*, *B*) Basic: 0 0 0; 1/2 1/2 0	body centred *I* Basic: 0 0 0; 1/2 1/2 1/2	face centred *F* Basic: 0 0 0; 1/2 1/2 0; 1/2 0 1/2; 0 1/2 1/2
Triclinic	a, b, c; α, β, γ	*P*1 (C_1^1) 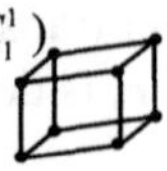			
Mono clinic	a, b, c; $\alpha = \gamma = 90° \neq \beta$	*P*2/*m* (C_{2h}^1) 	*C*2/*m* (C_{2h}^3) 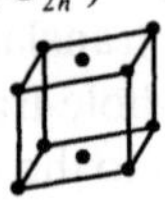		
Ortho rhombic	a, b, c; $\alpha = \beta = \gamma = 90°$	*Pmmm* (D_{2h}^1) 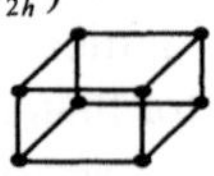	*Cmmm* (D_{2h}^{19}) 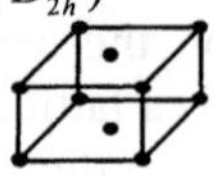	*Immm* (D_{2h}^{25}) 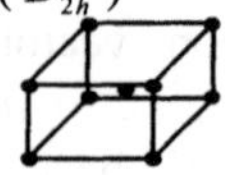	*Fmmm* (D_{2h}^{33}) 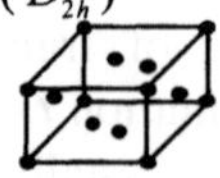
Trigonal (rhombo hedral)	$a = b = c$; $\alpha = \beta = \gamma \neq 90°$	$R\bar{3}m$ (D_{3d}^5) 			
Tetra gonal	$a = b \neq c$; $\alpha = \beta = \gamma = 90°$	*P*4/*mmm* (D_{4h}^1) 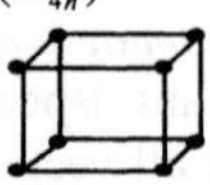		*I*4/*mmm* (D_{4h}^{17}) 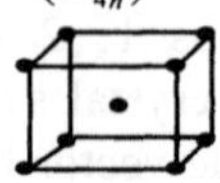	
Cubic	$a = b = c$; $\alpha = \beta = \gamma = 90°$	$Pm\bar{3}m$ (O_h^1) 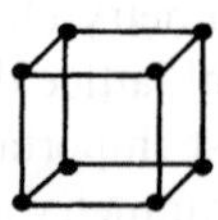		$Im\bar{3}m$ (O_h^9) 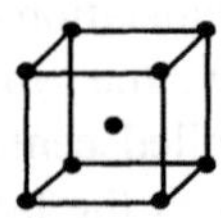	$Fm\bar{3}m$ (O_h^5) 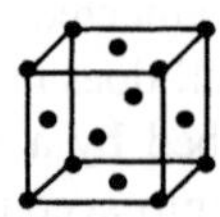
Hexa gonal	$a = b = c$; $\alpha = \beta = 90°$ $\gamma = 120°$	*P*6/*mmm* (D_{6h}^1) 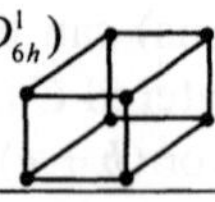			

The choice of the conventional lattice bases should satisfy the following conditions:
1. The lattice basis should correspond to the crystal symmetry and have one of the seven point groups of symmetry depending on the system to which the crystal belongs.
2. If the first condition is met, the number of right angles in the cell should be a maximum.
3. If the first and second conditions are met, the cell should have a minimum volume.

The Seitz cell is often used in solid-state physics to describe a crystal. In this case, some point of the lattice, which is the center of symmetry, is taken as the center of the cell. Connecting vectors are drawn from the chosen center to the nearest equivalent sites and then planes are constructed, which are perpendicular to and bisect these vectors. The volume bounded by these planes is called the Seitz cell.

Space Groups of Symmetry. The whole set of symmetry elements of a crystal forms its space group *G*. A space group can include the following symmetry elements:
1. Translations to the vector $\boldsymbol{t}_n = n_1\boldsymbol{t}_1 + n_2\boldsymbol{t}_2 + n_3\boldsymbol{t}_3$.
2. Rotations about symmetry axes of the order $n = 1, 2, 3, 4, 6$.
3. Inversion rotations $\bar{1}$, m, $\bar{3}$, $\bar{4}$, $\bar{6}$.
4. Screw rotations, which means a combination of transformations of rotations and translations along the rotation axis, labeled as 2_1, 3_1, 4_1, 4_2, 4_3, 6_1, 6_2, 6_3, 6_4, 6_5. The relation of the subscript to the order of the rotation axis indicates the translation value in fractions of the lattice constant in this direction.
5. Glide planes imply a combination of transformations including symmetry planes and translations in these planes. Glide reflections, which are directed along edges by the vectors ***a***, ***b***, and ***c***, are designated *a*, *b*, and *c*, respectively. Their length is equal to half the lattice constant in the translation direction. Glide planes, whose translations are directed along the diagonal of the cell, are equal to one-half or one-quarter of the diagonal length. They are designated *n* and *d* respectively.

Space groups, which do not contain screw rotations and glide planes, are called symmorphic.

Thus, a space group has symmetry elements of both the point groups and the group of translations, including also a combination of elements comprising these groups. Taking all symmetry elements possible in crystals, one can construct 230 independent space (Fedorov) groups. All these groups are defined and comprehensively covered in the literature [7, 14–18].

The full or short international notation of a space group indicates, like for a point group, generating symmetry elements only, i.e. generators. The symbol of a space group starts with the type of centering of the Bravais lattice and ends with the crystal system of the group. For example, the full and short symbols of a space group with a structure of a *B*1 (NaCl) type have the form $F\frac{4}{m}\bar{3}\frac{2}{m} = Fm\bar{3}m\ (O_h^5)$. The Schoenflies designation for this space group is given in brackets.

The notion of position multiplicity is significant for a characterization of space groups. If some atom in a crystal occupies an arbitrary point with coordinates x, y, and z, the operation of all (except full translations) symmetry elements of the crystal can give other points hosting within one cell the same crystallographically equivalent atoms. This set of points is called a position or the asymmetric unit. The number of points, which are produced from one arbitrary point under the action of symmetry elements, is referred to as the position multiplicity. Any crystalline structure can be described if one specifies its space group and the positions occupied by atoms, providing numerical values of the parameters in fractions of the cell edges, the edge length, and axial angles.

Space groups are defined differently in terms of the Seitz notation. In this case all rotation symmetry elements are placed at the origin of coordinates and the element of the space group G is labeled $g = \{h|\boldsymbol{t}_h + \boldsymbol{t}\}$. Here h is the revolving part of the element, $\boldsymbol{t}_h$ denotes the accompanying nontrivial translation, and $\boldsymbol{t}$ is an arbitrary full translation. The revolving parts h of the elements $\{h|\boldsymbol{t}_h + \boldsymbol{t}\}$ represent revolving elements h_1–h_{48} of the cubic point group $m\bar{3}m$ (O_h) or H_1–H_{24} of the hexagonal point group $6/mmm$ (D_{6h}). In accordance with Seitz, the general structure of any space group has the form

$$\{h_1|\boldsymbol{t}_1\},\ \{h_k|\boldsymbol{t}_k\},\ \dots,\ \{h_p|\boldsymbol{t}_p\};\ \ \{h_1|\boldsymbol{t}_1+\boldsymbol{t}\},\ \{h_k|\boldsymbol{t}_k+\boldsymbol{t}\},\ \dots,\ \{h_p|\boldsymbol{t}_p+\boldsymbol{t}\};\ \dots$$

In the initial (zero) block of elements all accompanying translations $\boldsymbol{t}_h$ are nontrivial (are smaller than the crystal lattice constant). In the next block, accompanying translations $\boldsymbol{t}_h + \boldsymbol{t}$ differ from those of the zero block: the full translation $\boldsymbol{t}$ (the lattice constant) is added. If the lattice type (Bravais lattice) is known, elements of the zero block need only be written explicitly to fully characterize a space group. All 230 space groups are described in this form in [16, 17].

In terms of the Seitz notation, net translations are designated $\{E|\boldsymbol{t}\}$ or $\{h_1|\boldsymbol{t}\}$. For symmorphic groups $\boldsymbol{t}_h = 0$ and therefore their revolving elements are elements of the type $\{h_k|0\}$.

2.2 Reciprocal Lattice and Wave Vectors

Reciprocal Space and Reciprocal Lattice. All physical quantities, which characterize properties of the crystal structure, possess the lattice periodicity. Among these quantities is, for example, the probability that an atom is located at some site of the lattice or the charge density formed by electrons. A crystal represents a system of interacting particles, namely nuclei and electrons. The Hamiltonian describing this system should include the kinetic energy of all particles, the potential energy of all particles in external fields, and the energy for the interparticle interaction in the crystal. If the adiabatic and single-electron approximations are introduced, the Hamiltonian of the crystal expands to Hamiltonians of noninteracting particles.

Let us consider an electron, which has no effect on the crystal state. The Hamiltonian $\hat{H}$ describing the action of the lattice field on this electron is of the form

$$\hat{H} = \hat{T} + \hat{U}(\boldsymbol{r}), \tag{2.9}$$

where $\hat{T}$ is the kinetic energy operator and $\hat{U}(\boldsymbol{r})$ is the potential energy operator. The potential field $\hat{U}(\boldsymbol{r})$ possesses the symmetry of the crystal space group, i.e. $\hat{U}(\boldsymbol{r})$ is invariant under any transformation of the symmetry g_i of this crystal. The same is true of the kinetic energy operator, which commutates both with transformations of the translation group and point transformations of the space group of the crystal. Considering that kinetic and potential energy operators $\hat{T}$ and $\hat{U}$ commutate with symmetry transformations of the space group G, the Hamiltonian $\hat{H}$ possesses this property too.

The Schrödinger equation for one electron in the crystal field $\hat{U}(\boldsymbol{r})$ is of the form

$$\hat{H}\psi(\boldsymbol{r}) = E\psi(\boldsymbol{r}) . \tag{2.10}$$

Using symmetry properties of the lattice field $\hat{U}(\boldsymbol{r})$, it is possible to determine the main properties of the electron energy spectrum E and wave functions $\psi(\boldsymbol{r})$. Any element $\hat{T}(\boldsymbol{t})$ of a translation group corresponds to a one-dimensional matrix of the form $\exp(-i\boldsymbol{kt})$. Therefore the resulting effect of the translation operator $\hat{T}(\boldsymbol{t})$ on the wave function $\psi(\boldsymbol{r})$ is

$$\hat{T}(\boldsymbol{t})\psi(\boldsymbol{r}) = \psi(\boldsymbol{r}+\boldsymbol{t}) = \exp(-i\boldsymbol{kt})\psi(\boldsymbol{r}) . \tag{2.11}$$

From (2.11) it is obvious that the wave function $\psi(\boldsymbol{r})$ can be written as

$$\psi_k(\boldsymbol{r}) = \exp(i\boldsymbol{kr})u_k(\boldsymbol{r}) , \tag{2.12}$$

where $u_k(\boldsymbol{r} + \boldsymbol{t}) \approx u_k(\boldsymbol{r})$ stands for an arbitrary periodic function, which is invariant under the translation. Thus, the function $\psi_k(\boldsymbol{r})$ represents a plane wave $\exp(i\boldsymbol{kr})$ with an amplitude $u_k(\boldsymbol{r})$. The vector $\boldsymbol{k}$ of the plane wave propagation is called the wave vector. The equalities (2.11) and (2.12) are different formulations of the Bloch theorem. In accordance with the Bloch theorem, eigenstates ψ of the one-electron Hamiltonian $\hat{H}$ for all sites $\boldsymbol{r}$ of the lattice may be chosen such that their wave functions have the form of a plane wave multiplied by the function with the lattice periodicity, in other words, such that equality (2.12) is fulfilled. In the general case, the functions $\psi_k(\boldsymbol{r})$ are different for different vectors $\boldsymbol{k}$. Therefore the eigenvalues of the energy E also depend on $\boldsymbol{k}$ and (2.10) can be written as

$$\hat{H}\psi_k(\boldsymbol{r}) = E(\boldsymbol{k})\psi_k(\boldsymbol{r}) . \tag{2.13}$$

This expression shows that the energy is a function of points $\boldsymbol{k}$ in some space. Let us consider properties of this space.

Periodicity of the potential field $\hat{U}(\boldsymbol{r})$ means that

$$\hat{U}(\boldsymbol{r}+\boldsymbol{t}) = \hat{U}(\boldsymbol{r}) \tag{2.14}$$

at any integer translation $\boldsymbol{t} = n_1\boldsymbol{a}_1 + n_2\boldsymbol{a}_2 + n_3\boldsymbol{a}_3$ ($\boldsymbol{a}_1$, $\boldsymbol{a}_2$, and $\boldsymbol{a}_3$ being basis lattice constants). The periodic function $\hat{U}(\boldsymbol{r})$ can be expanded into three-dimensional Fourier series as

$$\hat{U}(\boldsymbol{r}+\boldsymbol{t}) = \sum_k U_k \exp(i\boldsymbol{kr})\exp(i\boldsymbol{kt}), \tag{2.15}$$

where the summation is taken over all values of the vector $\boldsymbol{k}$. The function $\hat{U}$ has the required translation periodicity only when the scalar product $\boldsymbol{kt}$ is a multiple of 2π, because in this case $\boldsymbol{kt} = 2\pi m$ (m being an integer) and $\exp(i\boldsymbol{kt}) = \exp(i2\pi m) = 1$. Taking in sequence the lattice constants $\boldsymbol{a}_1$, $\boldsymbol{a}_2$, and $\boldsymbol{a}_3$ as $\boldsymbol{t}$, we obtain three equations

$$\boldsymbol{a}_1\boldsymbol{k} = 2\pi m_1, \quad \boldsymbol{a}_2\boldsymbol{k} = 2\pi m_2, \text{ and } \quad \boldsymbol{a}_3\boldsymbol{k} = 2\pi m_3 \tag{2.16}$$

whose solution has the form

$$\boldsymbol{k} = \boldsymbol{b}_1 m_1 + \boldsymbol{b}_2 m_2 + \boldsymbol{b}_3 m_3 . \tag{2.17}$$

The vectors $\boldsymbol{b}_i$ are determined from the vectors $\boldsymbol{a}_i$ by the formula

$$\boldsymbol{b}_i = 2\pi \frac{\boldsymbol{a}_j \times \boldsymbol{a}_k}{\boldsymbol{a}_1 \cdot (\boldsymbol{a}_2 \times \boldsymbol{a}_3)}, \tag{2.18}$$

where $i, j, k = 1, 2, 3$. In crystallography the factor 2π in the relation (2.18) is usually omitted.

The geometrical product $\boldsymbol{a}_1 \cdot (\boldsymbol{a}_2 \times \boldsymbol{a}_3)$ is the volume of the primitive cell constructed using vectors $\boldsymbol{a}_1$, $\boldsymbol{a}_2$, and $\boldsymbol{a}_3$, while the product $\boldsymbol{a}_j \times \boldsymbol{a}_k$ corresponds to the surface area of three faces of this cell. Hence, the vectors $\boldsymbol{b}_i$ have dimensions of reciprocal length. As distinct from the covariant vectors $\boldsymbol{a}_1$, $\boldsymbol{a}_2$ and $\boldsymbol{a}_3$, the vectors $\boldsymbol{b}_1$, $\boldsymbol{b}_2$ and $\boldsymbol{b}_3$ are called contravariant basis vectors or basis vectors of the reciprocal lattice. The reciprocal lattice is a lattice in Fourier space. The wave vector $\boldsymbol{k}$ has period $2\pi\boldsymbol{b}_i$ in the reciprocal space. As a result, the energy $E(\boldsymbol{k})$ is periodic in the wave vector, since points $\boldsymbol{k}$ and $\boldsymbol{k}' = \boldsymbol{k} + 2\pi\boldsymbol{b}_i$ are equivalent.

Thus, each crystalline structure has two lattices: a real (direct) crystal lattice and a reciprocal lattice, which are related to one an other by the expression (2.18). The translation periods of a primitive cell of a direct crystal lattice should be used to construct the reciprocal lattice. From (2.18) it follows that the vectors of the reciprocal lattice are perpendicular to the coordinate planes of the crystal lattice. If the Kronecker symbol δ_{ik} is used, the relation between the vectors of the direct and reciprocal lattices can be written as

$$\frac{1}{2\pi}\boldsymbol{a}_i\boldsymbol{b}_k = \delta_{ik} = \begin{cases} 1 \text{ when } i = k \\ 0 \text{ when } i \neq k \end{cases} \quad (i, k = 1, 2, 3). \tag{2.19}$$

The reciprocal-lattice parameters a^*, b^*, c^*, α^*, β^*, and γ^* are related to the parameters of the direct crystal lattice as follows:

$$\left.\begin{array}{ll} a^* = (\sin\alpha)/a\omega, & \cos\alpha^* = (\cos\beta\cos\gamma - \cos\alpha)/\sin\beta\sin\gamma \\ b^* = (\sin\beta)/b\omega, & \cos\beta^* = (\cos\alpha\cos\gamma - \cos\beta)/\sin\alpha\sin\gamma \\ c^* = (\sin\gamma)/c\omega, & \cos\gamma^* = (\cos\alpha\cos\beta - \cos\gamma)/\sin\alpha\sin\beta \end{array}\right\} \tag{2.20}$$

where $\omega = \sqrt{1 - \cos^2\alpha - \cos^2\beta - \cos^2\gamma + 2\cos\alpha\cos\beta\cos\gamma}$.

The crystal lattice in the direct space can be viewed as a set of plane nodal frameworks (*hkl*) with interplanar distance d_{hkl}. The radius-vectors of the reciprocal space $\boldsymbol{g}_{hkl} = \boldsymbol{a}^*h + \boldsymbol{b}^*k + \boldsymbol{c}^*l$ are perpendicular to the planes (*hkl*) and they are equal in absolute value to $|\boldsymbol{g}_{hkl}| = 1/d_{hkl}$. Therefore, to present a crystal in reciprocal space, it suffices to assign a common point 000 for the origin of vectors and their end points $\boldsymbol{g}_{hkl}$, which will form a system of regularly repeated nodes in the space, i.e. the reciprocal lattice of the crystal. Thus, each family of parallel planes of a crystal is given in reciprocal space by one node.

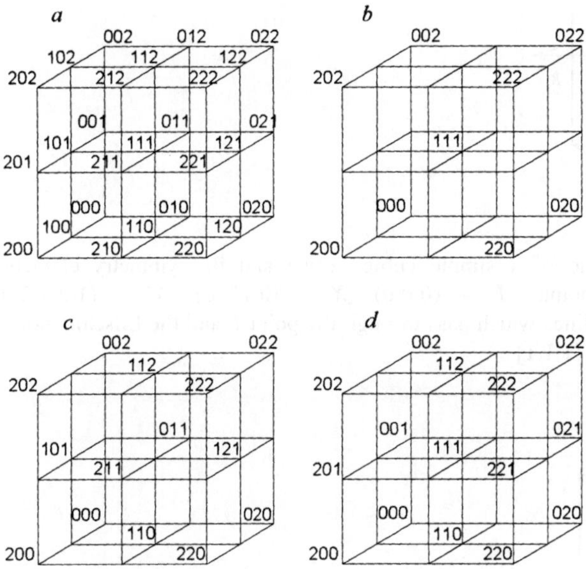

Fig. 2.1. Reciprocal lattices of crystals with primitive (*a*), face-centered (*b*), body-centered (*c*), and *C*-centered (*d*) cubic lattices

The reciprocal lattice of the primitive lattice is primitive (Fig. 2.1*a*). A face-centered lattice (*F*) corresponds to a reciprocal body-centered lattice (*I*) of the same symmetry (Fig. 2.1*b*). A direct body-centered lattice (*I*) corresponds to a reciprocal face-centered lattice (*F*) (Fig. 2.1*c*). A direct centered lattice (*C*) has, as its counterpart, a reciprocal centered lattice (*C*) of the same symmetry (Fig. 2.1*d*).

The Brillouin Zone and its Symmetrical Points. Special significance in solid-state theory is attached to the Seitz cell in the reciprocal lattice, which is called the first Brillouin zone. The first Brillouin zone is the unit cell with the smallest volume possible. It is a convex polyhedron and is fully bounded by at most 14 planes, which bisect the reciprocal-lattice vectors perpendicular to these planes. These vectors are drawn from the node chosen as the origin of coordinates. In the general case, the shape of the Brillouin zone depends on the relationship between the parameters of the unit cell. The first Brillouin zone is a cube in a simple cubic lattice, a right-angled hexagonal prism in a hexagonal lattice, a right-angled rhombododecahedron in a body-centered cubic lattice, and a truncated octahedron in a face-centered cubic lattice (Figs. 2.2–2.5).

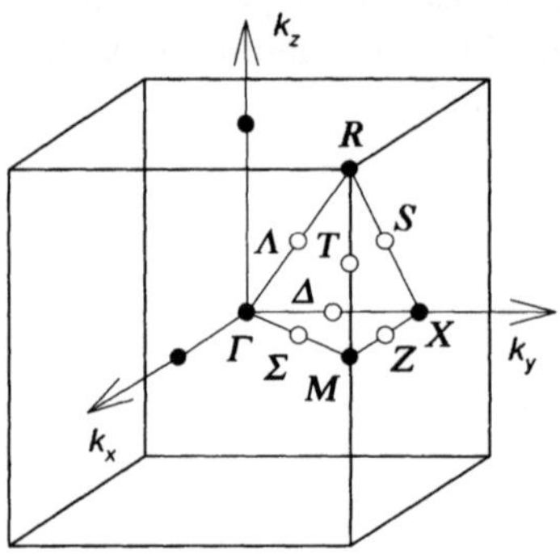

Fig. 2.2. First Brillouin zone of a simple cubic lattice and its symmetry elements. Coordinates of Lifschitz points: Γ = (0 0 0), X = (0 1/2 0), M = (1/2 1/2 0), R = (1/2 1/2 1/2). Symmetry lines which pass through the point Γ and the Lifschitz points: Δ = [0 1 0], Σ = [1 1 0], Λ = [1 1 1]

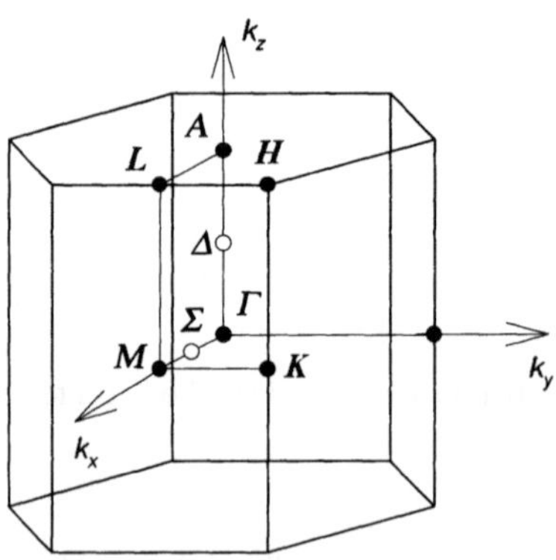

Fig. 2.3. First Brillouin zone of a hexagonal lattice and its symmetry elements. Coordinates of Lifschitz points: Γ = (0 0 0), M = (1/2 0 0), K = (1/3 1/3 0), L = (1/2 0 1/2), A = (0 0 1/2), H = (1/3 1/3 1/2). Symmetry lines which pass through the point Γ and the Lifschitz points: Δ = [0 0 1], Σ = [1 0 0]

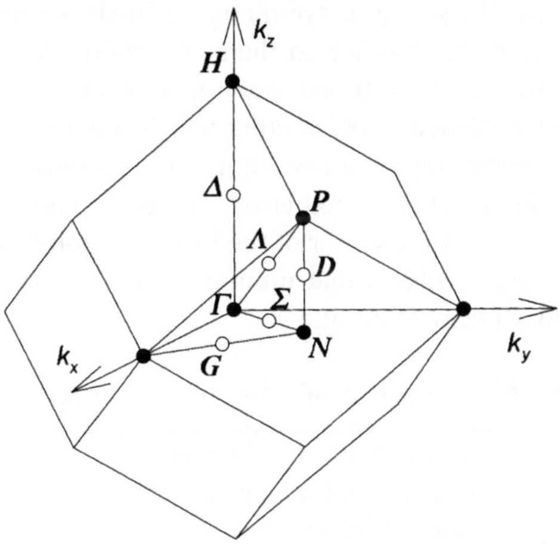

Fig. 2.4. First Brillouin zone of a body-centered cubic lattice and its symmetry elements. Coordinates of Lifschitz points: $\Gamma = (0\ 0\ 0)$, $N = (1/2\ 1/2\ 0)$, $P = (1/2\ 1/2\ 1/2)$, $H = (0\ 0\ 1)$. Symmetry lines which pass through the point Γ and the Lifschitz points: $\Delta = [0\ 0\ 1]$, $\Sigma = [1\ 1\ 0]$, $\Lambda = [1\ 1\ 1]$

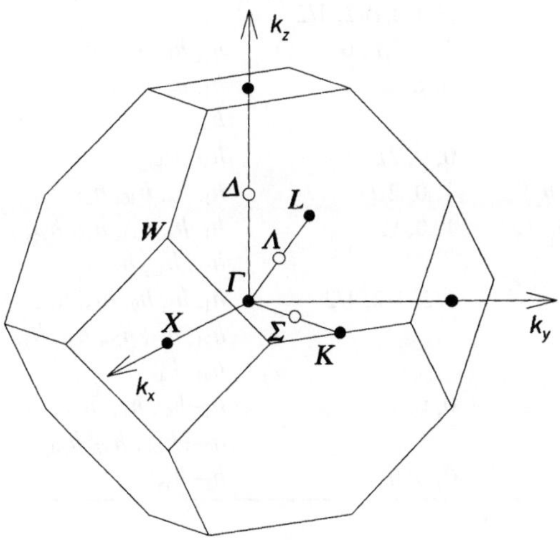

Fig. 2.5. First Brillouin zone of a face-centered cubic lattice and its symmetry elements. Coordinates of Lifschitz points: $\Gamma = (0\ 0\ 0)$, $W = (1\ 0\ 1/2)$, $L = (1/2\ 1/2\ 1/2)$, $X = (1\ 0\ 0)$. Symmetry lines which pass through the point Γ and the Lifschitz points: $\Delta = [0\ 0\ 1]$, $\Sigma = [1\ 1\ 0]$, $\Lambda = [1\ 1\ 1]$

The symmetry of the reciprocal lattice allows the classification all points in the Brillouin zone. The so-called symmetrical (Lifschitz or symmetry-defined) points are of special significance. These points have attached to them a certain set of symmetry elements, which determine their identity transformation or transfer to equivalent positions. This set of symmetry elements (or, in other words, the group of the wave vector $\boldsymbol{k}$, because this vector corresponds to any point in the reciprocal space) necessarily includes axes, which intersect one another or symmetry planes, or the inversion points. Thus, the group of the wave vector of the Lifschitz point in the Brillouin zone must contain symmetry elements that mutually intersect at a single point. Such groups are said to have a central point.

Table 2.5. Groups $G_{\boldsymbol{k}_s^{(1)}}$ of the wave vectors $\boldsymbol{k}_s^{(1)}$ of the Brillouin zone of a fcc lattice

Star $\{\boldsymbol{k}_s\}$	[1] Wave vector $\boldsymbol{k}_s^{(1)}$ as a combination of basis vectors $\boldsymbol{b}_1, \boldsymbol{b}_2, \boldsymbol{b}_3$ of a reciprocal lattice	Coordinates of vector $\boldsymbol{k}_s^{(1)}$ in the reciprocal lattice (in units $2\pi/a$)	[2] Group of the vector $\boldsymbol{k}_s^{(1)}$
$\{\boldsymbol{k}_0\}$	$\boldsymbol{k}_0^{(1)} = \mu_1\boldsymbol{b}_1 + \mu_2\boldsymbol{b}_2 + \mu_3\boldsymbol{b}_3$	$\mu_2 + \mu_3 - \mu_1$, $\mu_1 + \mu_3 - \mu_2$, $\mu_1 + \mu_2 - \mu_3$	h_1
$\{\boldsymbol{k}_1\}$	$\boldsymbol{k}_1^{(1)} = \mu_1(\boldsymbol{b}_1 + \boldsymbol{b}_3) + \mu_2(\boldsymbol{b}_2 + \boldsymbol{b}_3)$	$2\mu_2$, $2\mu_1$, 0	h_1, h_{28}
$\{\boldsymbol{k}_2\}$	$\boldsymbol{k}_2^{(1)} = \mu_1(\boldsymbol{b}_1 + \boldsymbol{b}_2) + \mu_2\boldsymbol{b}_3$	$\mu_2, \mu_2, 2\mu_1 - \mu_2$	h_1, h_{37}
$\{\boldsymbol{k}_3\}$	$\boldsymbol{k}_3^{(1)} = (\boldsymbol{b}_1 + \boldsymbol{b}_2 + \boldsymbol{b}_3)/2 + \mu(\boldsymbol{b}_1 - \boldsymbol{b}_2)$	$(1 - 4\mu)/2$, $(1 + 4\mu)/2$, 1/2	h_1, h_{13}
$\{\boldsymbol{k}_4\}$	$\boldsymbol{k}_4^{(1)} = \mu(\boldsymbol{b}_1 + \boldsymbol{b}_2 + 2\boldsymbol{b}_3)$	2μ, 2μ, 0	$h_1, h_{16}, h_{28}, h_{37}$
$\{\boldsymbol{k}_5\}$	$\boldsymbol{k}_5^{(1)} = \mu(\boldsymbol{b}_1 + \boldsymbol{b}_2 + \boldsymbol{b}_3)$	μ, μ, μ	$h_1, h_5, h_9, h_{37}, h_{41}, h_{45}$
$\{\boldsymbol{k}_6\}$	$\boldsymbol{k}_6^{(1)} = \mu(\boldsymbol{b}_1 + \boldsymbol{b}_2)$	0, 0, 2μ	h_{37}, h_{40}
$\{\boldsymbol{k}_7\}$	$\boldsymbol{k}_7^{(1)} = (\boldsymbol{b}_1 + \boldsymbol{b}_3)/2 + \mu(\boldsymbol{b}_1 + \boldsymbol{b}_2)$	1, 0, 2μ	h_1, h_4, h_{26}, h_{27}
$\{\boldsymbol{k}_8\}$	$\boldsymbol{k}_8^{(1)} = (\boldsymbol{b}_1 + \boldsymbol{b}_2)/4 + (\boldsymbol{b}_2 + \boldsymbol{b}_3)/2$	1, 0, ½	$h_1, h_4, h_{13}, h_{16}, h_{26}, h_{27}, h_{38}, h_{39}$
$\{\boldsymbol{k}_9\}$	$\boldsymbol{k}_9^{(1)} = (\boldsymbol{b}_1 + \boldsymbol{b}_2 + \boldsymbol{b}_3)/2$	1/2, 1/2, 1/2	$h_1, h_5, h_9, h_{13}, h_{17}, h_{21}, h_{25}, h_{29}, h_{33}, h_{37}, h_{41}, h_{45}$
$\{\boldsymbol{k}_{10}\}$	$\boldsymbol{k}_{10}^{(1)} = (\boldsymbol{b}_1 + \boldsymbol{b}_2)/2$	0, 0, 1	h_1–h_4, h_{13}–h_{16}, h_{25}–h_{28}, h_{37}–h_{40}
$\{\boldsymbol{k}_{11}\}$	$\boldsymbol{k}_{11}^{(1)} = 0$	0, 0, 0	h_1–h_{48}

[1] $\boldsymbol{b}_1 = (-2\pi/a, 2\pi/a, 2\pi/a)$, $\boldsymbol{b}_2 = (2\pi/a, -2\pi/a, 2\pi/a)$, $\boldsymbol{b}_3 = (2\pi/a, 2\pi/a, -2\pi/a)$.

[2] The group of vector $\boldsymbol{k}_s^{(1)}$ includes only generators (rotations) of class $m\bar{3}m$ (O_h) which transform the vector $\boldsymbol{k}_s^{(1)}$ into itself or an equivalent vector ($\boldsymbol{k}_s^{(1)} + \boldsymbol{b}$). The group of vector $\boldsymbol{k}_s^{(1)}$ does not include the generators which transform the vector $\boldsymbol{k}_s^{(1)}$ into the vector ($-\boldsymbol{k}_s^{(1)} + \boldsymbol{b}$). Non-included generators may be found by multiplication of these generators (in the table) by inversion.

Wave vectors $\boldsymbol{k}$ of the form $\boldsymbol{k} = \mu_1\boldsymbol{b}_1 + \mu_2\boldsymbol{b}_2 + \mu_3\boldsymbol{b}_3$, which correspond to Lifschitz points, have only fixed values of the parameter μ (in most cases, the parameter μ of Lifschitz points is equal to 0 or $\pm 1/2$ and, occasionally, 1/3 or 1/4). Vectors of non-Lifschitz points are written using some current parameter μ. The Baukart–Wigner–Smolukhovski notation is used in the literature to label qualitatively different points in the Brillouin zone: Γ, X, M, K, L, H, Σ, Δ, etc. (see Figs. 2.2–2.5). Another method consists in conditional numbering of points [16, 17]: the position of each point in the zone is determined both by its coordinates in the reciprocal lattice and by the corresponding wave vectors. Table 2.5 illustrates the description of qualitatively different points in the Brillouin zone of a fcc lattice [16, 17]. In addition to symmetrical points, symmetrical lines and planes are distinguished in the Brillouin zone.

The Wave-Vector Star. The concept of the wave-vector star is broadly presented in [19]. Given below are only those data and propositions that are necessary in the context of this book and will be used in the discussion of structural phase transformations.

Any point in the Brillouin zone (of the common or particular position) is described by the wave vector $\boldsymbol{k}$. The number of qualitatively different points is limited and small. In the case of symmorphic groups, the reciprocal lattice has all revolving symmetry elements of the space group G of the direct lattice. These symmetry elements will make the given vector $\boldsymbol{k}$ move to equivalent (if the vector $\boldsymbol{k}$ occupies a symmetric position) or nonequivalent positions. The set of all nonequivalent vectors, which are produced from the given vector $\boldsymbol{k}$ under the action of revolving elements of the group G, is called the irreducible star of the wave vector and is labeled $\{\boldsymbol{k}\}$. Individual vectors comprising this set are called star rays. Obviously, the number of star rays cannot be greater than the number of elements in the point group of a crystal. For points of the common type the number of star rays equals the order of the group G. If the point $\boldsymbol{k}$ is located on a symmetry element, the operation of this element does not change the vector $\boldsymbol{k}$. Consequently, the number of star rays may be smaller than the order of the group G. For example, if the vector $\boldsymbol{k}$ is directed along a symmetry axis, it does not change for any rotation about this axis. The vector $\boldsymbol{k} = \boldsymbol{b}_i/2$ ($\boldsymbol{b}_i$ being one of the basis vectors of the reciprocal lattice) transforms to the equivalent vector $-\boldsymbol{k} = -\boldsymbol{b}_i/2 = \boldsymbol{k} - \boldsymbol{b}_i$ during inversion, etc. The set of elements g of the group G, which do not change the vector $\boldsymbol{k}$ or transfer it to an equivalent position, is referred to as the wave-vector group (small group) G_k. The small group G_k represents a subgroup of the group G and includes all net translations of the group G.

To obtain all rays of a star, it suffices that the first ray $\boldsymbol{k}^{(1)}$ is subjected to representative elements g_L, which are obtained from an expansion of the group G in terms of its subgroup G_k into adjacent classes, i.e.

$$\boldsymbol{k}^{(j)} = g_L\boldsymbol{k}^{(1)} . \tag{2.21}$$

Since $g_L = \{h_L|\boldsymbol{t}_L\}$ and the vector $\boldsymbol{k}$ is acted upon only by the revolving part h_L of the element g_L, (2.21) can be rewritten as

$$\boldsymbol{k}^{(j)} = h_L \boldsymbol{k}^{(1)} . \tag{2.22}$$

The number of rays of the star $\{\boldsymbol{k}\}$ is equal to the index of the subgroup G_k relative the group G (the index is the relation between the orders of the group and the subgroup). For any crystallographic space group G there cannot be more than 48, which is the maximum number of symmetry elements in the point group of a crystal.

Rays of Lifschitz stars are designated as main vectors of the reciprocal lattice using numerically fixed parameters (for example, rays of the stars $\{\boldsymbol{k}_8\}$, $\{\boldsymbol{k}_9\}$, $\{\boldsymbol{k}_{10}\}$, and $\{\boldsymbol{k}_{11}\}$ in the Brillouin zone of a fcc lattice, see Table 2.5). In the case of non-Lifschitz stars of common type, $\boldsymbol{k}_0^{(1)} = \mu_1\boldsymbol{b}_1 + \mu_2\boldsymbol{b}_2 + \mu_3\boldsymbol{b}_3$, values for the parameter μ_i at vectors $\boldsymbol{b}_i$ are not determined by any symmetry condition. All stars, which have the same formulation of rays and similar symmetry, but differ in the numerical values for the parameter μ_i, are assumed to be a single star. Lines and planes, rather than symmetrical points (as in the case of Lifschitz stars), will correspond to vectors of such stars in the Brillouin zone. Any arbitrary point in the volume of the Brillouin zone may correspond to a star of the most common type. The group of the wave vector G_k contains only one element h_1, namely the identity transformation E, for points of common type.

Handbooks [16, 17] list all qualitatively different stars in Brillouin zones of the 14 Bravais lattices. It should be remembered that the Brillouin zone may contain qualitatively similar stars, all rays of which differ from rays of a star [16, 17] by the whole vector of the reciprocal lattice. Put another way, the stars are equivalent. For example, the star $\{\boldsymbol{k}_1\}$ in the Brillouin zone of a fcc crystal is determined by the first ray $\boldsymbol{k}_1^{(1)} = \mu_1(\boldsymbol{b}_1 + \boldsymbol{b}_3) + \mu_2(\boldsymbol{b}_2 + \boldsymbol{b}_3)$, whose group is made up of the elements h_1 and h_{28} (see Table 2.5). The full equivalent to this star will be a star of similar symmetry defined by the ray $\boldsymbol{k}_1^{(1)} = \mu_1\boldsymbol{b}_1 + \mu_2\boldsymbol{b}_2 + (\mu_1 + \mu_2 - 1)\boldsymbol{b}_3$ having the same group h_1 and h_{28}. In a fcc lattice, any ray entering the Brillouin zone, with one coordinate being unity and two other coordinates being nonzero, e.g. $(x\; y\; 1)$, is equivalent to a ray of one of the stars listed in Table 2.5.

Lifschitz stars, their rays and representative elements in Brillouin zones of the 14 Bravais lattices are tabulated elsewhere [19]. In [22] the tables contain additionally Lifschitz points in Brillouin zones and coordinates of these points in the reciprocal lattice.

This book focuses on order–disorder phase transitions in the fcc sublattice of strongly nonstoichiometric compounds. Therefore, for the reader's convenience, all odd rays of the stars from $\{\boldsymbol{k}_1\}$ to $\{\boldsymbol{k}_8\}$ and all rays of Lifschitz stars $\{\boldsymbol{k}_9\}$ and $\{\boldsymbol{k}_{10}\}$ in the Brillouin zone of a fcc lattice are listed in Table 2.6. Each ray is labeled with its own number (superscript) and the number of the star, to which the ray belongs (subscript). For example, $\boldsymbol{k}_3^{(15)}$ denotes the fifteenth ray of the star $\{\boldsymbol{k}_3\}$. Rays of the common-type star $\{\boldsymbol{k}_0\}$ are omitted in Table 2.6, because the numbers of the 48 rays of this star coincide with the numbers of the representative elements h_L, which are used to obtain each ray from the first ray $\boldsymbol{k}_0^{(1)}$. The representative element for an even ray of the stars from $\{\boldsymbol{k}_1\}$ to $\{\boldsymbol{k}_8\}$ is $h_{24+k} = h_k h_{25}$ where the element h_k corresponds to the previous odd ray.

Table 2.6. List of the rays of the wave vectors of the Brillouin zone of a fcc lattice

Odd rays $\mathbf{k}_s^{(2n-1)}$ of the star $\{\mathbf{k}_s\}$	Coordinates of odd rays $\mathbf{k}_s^{(j)}$ (in units of $2\pi/a$)	h_L	Interval of variation of μ
1	2	3	4
	Star $\{\mathbf{k}_1\}$		
$\mathbf{k}_1^{(1)} = \mu_1(\mathbf{b}_1+\mathbf{b}_3)+\mu_2(\mathbf{b}_2+\mathbf{b}_3)$	$2\mu_2,\ 2\mu_1,\ 0$	h_1	$0<\mu_2<3/8$
$\mathbf{k}_1^{(3)} = -\mu_1(\mathbf{b}_1+\mathbf{b}_3)+\mu_2(\mathbf{b}_2+\mathbf{b}_3)$	$2\mu_2,\ -2\mu_1,\ 0$	h_2	$0<\mu_1<\mu_2$
$\mathbf{k}_1^{(5)} = \mu_1(\mathbf{b}_2+\mathbf{b}_3)+\mu_2(\mathbf{b}_1+\mathbf{b}_2)$	$2\mu_1,\ 0,\ 2\mu_2$	h_5	or
$\mathbf{k}_1^{(7)} = \mu_1(\mathbf{b}_2+\mathbf{b}_3)-\mu_2(\mathbf{b}_1+\mathbf{b}_2)$	$2\mu_1,\ 0,\ -2\mu_2$	h_6	$3/8<\mu_2<1/2$
$\mathbf{k}_1^{(9)} = \mu_1(\mathbf{b}_1+\mathbf{b}_2)+\mu_2(\mathbf{b}_1+\mathbf{b}_3)$	$0,\ 2\mu_2,\ 2\mu_1$	h_9	$0<\mu_1<3/4-\mu_2$
$\mathbf{k}_1^{(11)} = -\mu_1(\mathbf{b}_1+\mathbf{b}_2)+\mu_2(\mathbf{b}_1+\mathbf{b}_3)$	$0,\ 2\mu_2,\ -2\mu_1$	h_{11}	
$\mathbf{k}_1^{(13)} = \mu_1(\mathbf{b}_2+\mathbf{b}_3)-\mu_2(\mathbf{b}_1+\mathbf{b}_3)$	$2\mu_1,\ -2\mu_2,\ 0$	h_{15}	
$\mathbf{k}_1^{(15)} = \mu_1(\mathbf{b}_2+\mathbf{b}_3)+\mu_2(\mathbf{b}_1+\mathbf{b}_3)$	$2\mu_1,\ 2\mu_2,\ 0$	h_{16}	
$\mathbf{k}_1^{(17)} = \mu_1(\mathbf{b}_1+\mathbf{b}_2)-\mu_2(\mathbf{b}_2+\mathbf{b}_3)$	$-2\mu_2,\ 0,\ 2\mu_1$	h_{18}	
$\mathbf{k}_1^{(19)} = \mu_1(\mathbf{b}_1+\mathbf{b}_2)+\mu_2(\mathbf{b}_2+\mathbf{b}_3)$	$2\mu_2,\ 0,\ 2\mu_1$	h_{19}	
$\mathbf{k}_1^{(21)} = \mu_1(\mathbf{b}_1+\mathbf{b}_3)+\mu_2(\mathbf{b}_1+\mathbf{b}_2)$	$0,\ 2\mu_1,\ 2\mu_2$	h_{22}	
$\mathbf{k}_1^{(23)} = -\mu_1(\mathbf{b}_1+\mathbf{b}_3)+\mu_2(\mathbf{b}_1+\mathbf{b}_2)$	$0,\ -2\mu_1,\ 2\mu_2$	h_{23}	
	Star $\{\mathbf{k}_2\}$		
$\mathbf{k}_2^{(1)} = \mu_1(\mathbf{b}_1+\mathbf{b}_2)+\mu_2\mathbf{b}_3$	$\mu_2,\ \mu_2,\ 2\mu_1-\mu_2$	h_1	$0<\mu_2<1/4$
$\mathbf{k}_2^{(3)} = -\mu_1(\mathbf{b}_1+\mathbf{b}_2)+\mu_2\mathbf{b}_2$	$\mu_2,\ -\mu_2,\ \mu_2-2\mu_1$	h_2	$0<\mu_1<1/2+\mu_2/2$
$\mathbf{k}_2^{(5)} = -\mu_1(\mathbf{b}_1+\mathbf{b}_2)+\mu_2\mathbf{b}_1$	$-\mu_2,\ \mu_2,\ \mu_2-2\mu_1$	h_3	or
$\mathbf{k}_2^{(7)} = \mu_1(\mathbf{b}_1+\mathbf{b}_2)-\mu_2(\mathbf{b}_1+\mathbf{b}_2+\mathbf{b}_3)$	$-\mu_2,\ -\mu_2,\ 2\mu_1-\mu_2$	h_4	$1/4<\mu_2<1/2$
$\mathbf{k}_2^{(9)} = \mu_1(\mathbf{b}_1+\mathbf{b}_3)+\mu_2\mathbf{b}_2$	$\mu_2,\ 2\mu_1-\mu_2,\ \mu_2$	h_5	$\mu_2<\mu_1<3/4-\mu_2/2$
$\mathbf{k}_2^{(11)} = -\mu_1(\mathbf{b}_1+\mathbf{b}_3)+\mu_2\mathbf{b}_3$	$\mu_2,\ \mu_2-2\mu_1,\ -\mu_2$	h_6	or
$\mathbf{k}_2^{(13)} = \mu_1(\mathbf{b}_1+\mathbf{b}_3)-\mu_2(\mathbf{b}_1+\mathbf{b}_2+\mathbf{b}_3)$	$-\mu_2,\ 2\mu_1-\mu_2,\ -\mu_2$	h_7	$0<\mu_2<1/2$
$\mathbf{k}_2^{(15)} = -\mu_1(\mathbf{b}_1+\mathbf{b}_3)+\mu_2\mathbf{b}_1$	$-\mu_2,\ \mu_2-2\mu_1,\ \mu_2$	h_8	$0<\mu_1<\mu_2$
$\mathbf{k}_2^{(17)} = \mu_1(\mathbf{b}_2+\mathbf{b}_3)+\mu_2\mathbf{b}_1$	$2\mu_1-\mu_2,\ \mu_2,\ \mu_2$	h_9	or
$\mathbf{k}_2^{(19)} = \mu_1(\mathbf{b}_2+\mathbf{b}_3)-\mu_2(\mathbf{b}_1+\mathbf{b}_2+\mathbf{b}_3)$	$2\mu_1-\mu_2,\ -\mu_2,\ -\mu_2$	h_{10}	$1/2<\mu_2<3/4$
$\mathbf{k}_2^{(21)} = -\mu_1(\mathbf{b}_2+\mathbf{b}_3)+\mu_2\mathbf{b}_3$	$\mu_2-2\mu_1,\ \mu_2,\ -\mu_2$	h_{11}	$0<\mu_1<3/4-\mu_2/2$
$\mathbf{k}_2^{(23)} = -\mu_1(\mathbf{b}_2+\mathbf{b}_3)+\mu_2\mathbf{b}_2$	$\mu_2-2\mu_1,\ -\mu_2,\ -\mu_2$	h_{12}	
	Star $\{\mathbf{k}_3\}$		
$\mathbf{k}_3^{(1)} = (\mathbf{b}_1+\mathbf{b}_2+\mathbf{b}_3)/2+\mu(\mathbf{b}_1-\mathbf{b}_2)$	$(1-4\mu)/2,\ (1+4\mu)/2,\ 1/2$	h_1	$0<\mu<1/4$
$\mathbf{k}_3^{(3)} = -\mathbf{b}_1/2-\mu(\mathbf{b}_1+\mathbf{b}_2+2\mathbf{b}_3)$	$(1-4\mu)/2,\ -(1+4\mu)/2,\ -1/2$	h_2	
$\mathbf{k}_5^{(3)} = -\mathbf{b}_2/2+\mu(\mathbf{b}_1+\mathbf{b}_2+2\mathbf{b}_3)$	$(4\mu-1)/2,\ (1+4\mu)/2,\ -1/2$	h_3	
$\mathbf{k}_3^{(7)} = -\mathbf{b}_3/2+\mu(\mathbf{b}_2-\mathbf{b}_1)$	$(4\mu-1)/2,\ -(1+4\mu)/2,\ 1/2$	h_4	
$\mathbf{k}_3^{(9)} = (\mathbf{b}_1+\mathbf{b}_2+\mathbf{b}_3)/2+\mu(\mathbf{b}_3-\mathbf{b}_1)$	$(1+4\mu)/2,\ 1/2,\ (1-4\mu)/2$	h_5	
$\mathbf{k}_3^{(11)} = -\mathbf{b}_1/2+\mu(\mathbf{b}_1+\mathbf{b}_2+2\mathbf{b}_3)$	$(1+4\mu)/2,\ -1/2,\ (4\mu-1)/2$	h_6	
$\mathbf{k}_3^{(13)} = -\mathbf{b}_2/2+\mu(\mathbf{b}_1-\mathbf{b}_3)$	$-(1+4\mu)/2,\ 1/2,\ (4\mu-1)/2$	h_7	
$\mathbf{k}_3^{(15)} = -\mathbf{b}_3/2-\mu(\mathbf{b}_1+\mathbf{b}_2+\mathbf{b}_3)$	$-(1+4\mu)/2,\ -1/2,\ (1-4\mu)/2$	h_8	
$\mathbf{k}_3^{(17)} = (\mathbf{b}_1+\mathbf{b}_2+\mathbf{b}_3)/2+\mu(\mathbf{b}_2-\mathbf{b}_3)$	$1/2,\ (1-4\mu)/2,\ (1+4\mu)/2$	h_9	
$\mathbf{k}_3^{(19)} = -\mathbf{b}_1/2+\mu(\mathbf{b}_3-\mathbf{b}_2)$	$1/2,\ (4\mu-1)/2,\ -(1+4\mu)/2$	h_{10}	
$\mathbf{k}_3^{(21)} = -\mathbf{b}_2/2-\mu(2\mathbf{b}_1+\mathbf{b}_2+\mathbf{b}_3)$	$-1/2,\ (1-4\mu)/2,\ -(1+4\mu)/2$	h_{11}	
$\mathbf{k}_3^{(23)} = -\mathbf{b}_3/2+\mu(2\mathbf{b}_1+\mathbf{b}_2+\mathbf{b}_3)$	$-1/2,\ (4\mu-1)/2,\ (1+4\mu)/2$	h_{12}	

Table 2.6 (continued)

1	2	3	4
	Star $\{\boldsymbol{k}_4\}$		
$\boldsymbol{k}_4^{(1)}=\mu(\boldsymbol{b}_1+\boldsymbol{b}_2+2\boldsymbol{b}_3)$	$2\mu,\ 2\mu,\ 0$	h_1	$0<\mu_2\leq 3/8$
$\boldsymbol{k}_4^{(3)}=\mu(\boldsymbol{b}_2-\boldsymbol{b}_1)$	$2\mu,-2\mu,\ 0$	h_2	
$\boldsymbol{k}_4^{(5)}=\mu(\boldsymbol{b}_1+2\boldsymbol{b}_2+\boldsymbol{b}_3)$	$2\mu,\ 0,\ 2\mu$	h_5	
$\boldsymbol{k}_4^{(7)}=\mu(\boldsymbol{b}_3-\boldsymbol{b}_1)$	$2\mu,\ 0,-2\mu$	h_6	
$\boldsymbol{k}_4^{(9)}=\mu(2\boldsymbol{b}_1+\boldsymbol{b}_2+\boldsymbol{b}_3)$	$0,\ 2\mu,\ 2\mu$	h_9	
$\boldsymbol{k}_4^{(11)}=\mu(\boldsymbol{b}_3-\boldsymbol{b}_2)$	$0,\ 2\mu,-2\mu$	h_{11}	
	Star $\{\boldsymbol{k}_5\}$		
$\boldsymbol{k}_5^{(1)}=\mu(\boldsymbol{b}_1+\boldsymbol{b}_2+\boldsymbol{b}_3)$	$\mu,\ \mu,\ \mu$	h_1	$0<\mu<1/2$
$\boldsymbol{k}_5^{(3)}=-\mu\boldsymbol{b}_1$	$\mu,-\mu,-\mu$	h_2	
$\boldsymbol{k}_5^{(5)}=-\mu\boldsymbol{b}_2$	$-\mu,\ \mu,-\mu$	h_3	
$\boldsymbol{k}_5^{(7)}=-\mu\boldsymbol{b}_3$	$-\mu,-\mu,\ \mu$	h_4	
	Star $\{\boldsymbol{k}_6\}$		
$\boldsymbol{k}_6^{(1)}=\mu(\boldsymbol{b}_1+\boldsymbol{b}_2)$	$0,\ 0,\ 2\mu$	h_1	$0<\mu<1/2$
$\boldsymbol{k}_6^{(3)}=\mu(\boldsymbol{b}_1+\boldsymbol{b}_3)$	$0,\ 2\mu,\ 0$	h_5	
$\boldsymbol{k}_6^{(5)}=\mu(\boldsymbol{b}_2+\boldsymbol{b}_3)$	$2\mu,\ 0,\ 0$	h_9	
	Star $\{\boldsymbol{k}_7\}$		
$\boldsymbol{k}_7^{(1)}=(\boldsymbol{b}_1+\boldsymbol{b}_3)/2+\mu(\boldsymbol{b}_1+\boldsymbol{b}_2)$	$0,\ 1,\ 2\mu$	h_1	$0<\mu<1/4$
$\boldsymbol{k}_7^{(3)}=(\boldsymbol{b}_2+\boldsymbol{b}_3)/2+\mu(\boldsymbol{b}_1+\boldsymbol{b}_3)$	$1,\ 2\mu,\ 0$	h_5	
$\boldsymbol{k}_7^{(5)}=(\boldsymbol{b}_1+\boldsymbol{b}_2)/2+\mu(\boldsymbol{b}_2+\boldsymbol{b}_3)$	$2\mu,\ 0,\ 1$	h_9	
$\boldsymbol{k}_7^{(7)}=-(\boldsymbol{b}_2+\boldsymbol{b}_3)/2-\mu(\boldsymbol{b}_1+\boldsymbol{b}_2)$	$-1,\ 0,-2\mu$	h_{13}	
$\boldsymbol{k}_7^{(9)}=-(\boldsymbol{b}_1+\boldsymbol{b}_2)/2-\mu(\boldsymbol{b}_1+\boldsymbol{b}_3)$	$0,-2\mu,-1$	h_{17}	
$\boldsymbol{k}_7^{(11)}=-(\boldsymbol{b}_1+\boldsymbol{b}_3)/2-\mu(\boldsymbol{b}_2+\boldsymbol{b}_3)$	$-2\mu,-1,\ 0$	h_{21}	
	Star $\{\boldsymbol{k}_8\}$		
$\boldsymbol{k}_8^{(1)}=(\boldsymbol{b}_1+\boldsymbol{b}_2)/4+(\boldsymbol{b}_2+\boldsymbol{b}_3)/2$	$1,\ 0,\ 1/2$	h_1	
$\boldsymbol{k}_8^{(3)}=(\boldsymbol{b}_1+\boldsymbol{b}_3)/4+(\boldsymbol{b}_1+\boldsymbol{b}_2)/2$	$0,\ 1/2,\ 1$	h_5	
$\boldsymbol{k}_8^{(5)}=(\boldsymbol{b}_2+\boldsymbol{b}_3)/4+(\boldsymbol{b}_1+\boldsymbol{b}_3)/2$	$1/2,\ 1,\ 0$	h_9	
	Star $\{\boldsymbol{k}_9\}$: all rays are given		
$\boldsymbol{k}_9^{(1)}=(\boldsymbol{b}_1+\boldsymbol{b}_2+\boldsymbol{b}_3)/2$	$1/2,\ 1/2,\ 1/2$	h_1	
$\boldsymbol{k}_9^{(2)}=\boldsymbol{b}_1/2$	$-1/2,\ 1/2,\ 1/2$	h_{26}	
$\boldsymbol{k}_9^{(3)}=\boldsymbol{b}_2/2$	$1/2,-1/2,\ 1/2$	h_{27}	
$\boldsymbol{k}_9^{(4)}=\boldsymbol{b}_3/2$	$1/2,\ 1/2,-1/2$	h_{28}	
	Star $\{\boldsymbol{k}_{10}\}$: all rays are given		
$\boldsymbol{k}_{10}^{(1)}=(\boldsymbol{b}_1+\boldsymbol{b}_2)/2$	$0,\ 0,\ 1$	h_1	
$\boldsymbol{k}_{10}^{(2)}=(\boldsymbol{b}_1+\boldsymbol{b}_3)/2$	$0,\ 1,\ 0$	h_5	
$\boldsymbol{k}_{10}^{(3)}=(\boldsymbol{b}_2+\boldsymbol{b}_3)/2$	$1,\ 0,\ 0$	h_9	
	Star $\{\boldsymbol{k}_{11}\}$		
$\boldsymbol{k}_{11}^{(1)}=0$	$0,\ 0,\ 0$	h_1	

The variation limits of the current parameters μ in Table 2.6 were found on condition that the numbers of rays of non-Lifschitz stars were uniquely determined. Rays of all non-Lifschitz stars, which have negative values of the current parameters μ_i, are equivalent to rays of the same stars having positive values, but which are equal in absolute value for the parameters μ_i. However, mutually equivalent rays are numbered differently. To ensure that equivalent wave vectors are not assigned different numbers of rays of qualitatively similar stars whose current parameters are equal in magnitude but different in sign, the variation of μ in Table 2.6 is limited to positive values. If the current parameters μ_i of a wave vector have a negative value, one should look in Table 2.6 for the equivalent vector with positive values of μ_i.

2.3 Elements of Diffraction Theory

Diffraction Spectrum. The main method used to examine the structure of crystals is diffraction of waves interacting with atoms. The length of these waves should be comparable to the interatomic spacing. The most popular method is X-ray diffraction. Neutron diffraction is widely used too. The electron diffraction method is a rare guest in structural studies and serves primarily to analyze surfaces and films.

Diffraction studies rely on changes of intensity of scattered radiation. Oscillations of the electric field of an incident electromagnetic wave cause a displacement of electrons in an atom. The electrons vibrate at a frequency equal to the oscillation frequency of the electromagnetic wave and act as a source of radiation having the same frequency. The absorption and release of radiation by the electrons of an atom represent scattering. Only the atomic nuclei interact in the case of neutron diffraction.

Let us consider the incidence of a wave having wavelength λ on a family of parallel atomic planes (hkl) in a crystal with the interplanar distance d_{hkl}. The difference in the path of the beams reflected from adjacent atomic planes is $2d_{hkl}\sin\theta$, where θ is the Bragg angle. For elastic scattering incident and scattered radiation waves have equal length. The radiation reflected from adjacent atomic planes can be amplified or attenuated by interference. The amplification, i.e. the interference maximum, is observed for scattering from adjacent planes providing the same phase angle. In this case the difference in the beam paths should be an integral number of wavelengths, i.e.

$$2d_{hkl}\sin\theta = n\lambda\,. \tag{2.23}$$

This condition is known as the Wulff–Bragg law and results from the periodicity of the space lattice. It is not connected with the atomic arrangement of a cell, which determines the relative intensity of diffracted radiation. The relation (2.23) is used to determine interplanar distances and then to calculate parameters of the unit cell of a crystal according to the general formula

$$\frac{1}{d_{hkl}^2}=\frac{1}{\omega^2}\left[\left(\frac{h\sin\alpha}{a}\right)^2+\left(\frac{k\sin\beta}{b}\right)^2+\left(\frac{l\sin\gamma}{c}\right)^2+\frac{2hk(\cos\alpha\cos\beta-\cos\gamma)}{ab}\right.$$
$$\left.+\frac{2kl(\cos\beta\cos\gamma-\cos\alpha)}{bc}+\frac{2hl(\cos\alpha\cos\gamma-\cos\beta)}{ac}\right], \tag{2.24}$$

where ω is given in (2.20) and $1/d_{hkl} = |\mathbf{g}_{hkl}|$ is the interplanar distance in the reciprocal lattice. The characteristic behavior for the interplanar distances versus the parameters of the unit cell of crystals for various systems can be found in numerous monographs and reference books [e.g., 26–29]. To solve the inverse problem, i.e. to construct a diffraction spectrum, (2.23) is used in the form

$$\theta=\arcsin\left(\frac{\lambda}{2}\frac{n}{d_{hkl}}\right). \tag{2.25}$$

The expression (2.25) suggests that each family of planes (*hkl*) in a crystal lattice corresponds to a set of interference peaks, which are observed at certain values of the angle θ. The whole set of different-intensity reflections from various planes (*hkl*) forms a diffraction spectrum (a diffraction pattern).

If the indices of all lines of the diffraction spectrum of a powder are determined, it is possible to calculate dimensions and to find the shape of the unit cell. Indexing is simple if the type of structure or the crystal system of the material is known. For example, the ratios $\sin^2\theta_i/\sin^2\theta_1$ equal to $(h_i^2+k_i^2+l_i^2)/(h_1^2+k_1^2+l_1^2)$ are considered for crystals of the cubic system. Subscripts denote ordinal numbers of lines in a spectrum. The set of interference indices is known for crystals with any type of structure in the cubic system. Consequently, one can select a square sum of indices for the first line such that the square sum of indices for the other lines will correspond to reflections (*hkl*) allowed for a given type of structure. The same method is used for indexing of crystals of hexagonal and tetragonal systems.

If the crystal system of a material is unknown, a consecutive series of $1/d_i^2$ values is considered starting from small angles θ. These values are equal to the square of interplanar distances in the reciprocal lattice, $|\mathbf{g}_{hkl}|^2$. It was shown in Sect. 2.2 that one node of the reciprocal lattice corresponds to each family of planes in the crystal lattice. This means that each reflection in a diffraction spectrum corresponds to a vector in the reciprocal lattice. Therefore, three values corresponding to axial translations of a primitive cell of the reciprocal lattice are selected from the set of $1/d_i^2$ values. If the cell so determined does not have the lattice symmetry, it is transformed into a unit cell. This technique is described in detail in [26]. An example of the practical application of the reciprocal lattice for indexing the diffraction spectrum of an ordered phase and for determining the symmetry of an ordered phase is given in [23].

Thus, the Bragg angles, at which a crystal reflects incident radiation, depend on interplanar distances, i.e. on the dimensions of the unit cell. Put another way, the directions for all reflections from a crystal are determined only by the values of axial translations and axial angles of the unit cell and are independent of the particular species of atoms present in the crystal. The relative intensity of different reflections *hkl* depends on how the particular atoms are located in the unit cell. The simultaneous analysis of the position and the relative intensities of reflections allows one to solve the crystal structure.

Structural Amplitude. The theory of X-ray and neutron scattering has been comprehensively elaborated and described in detail in many monographs (for example, [29–31]). Therefore we shall introduce the concept of the structural amplitude or of the structure factor without rigorous reasoning. It is this quantity which determines the intensity of diffraction reflections.

Let us consider scattering of radiation in a kinematic approximation, i.e. under the assumption of single scattering and in the absence of the interaction between incident and scattered waves. Further absorption of incident and scattered waves in a crystal is assumed to be negligibly small.

Let the distribution of some physical quantity in a crystal be described by the periodic potential $U(\boldsymbol{r})$, which depends on the atomic coordinates. This physical quantity is the electron density for X-ray radiation and it is the nuclear force field for neutron diffraction. The scattering amplitude can be found by expanding $U(\boldsymbol{r})$ in a Fourier series. On the assumption of single scattering, we shall consider only the first term of the expansion normalized to the unit volume of a crystal:

$$U(\boldsymbol{q}) = (1/V)\int U(\boldsymbol{r})\exp(-i\boldsymbol{qr})d^3\boldsymbol{r} \ . \tag{2.26}$$

The value of $U(\boldsymbol{r})$ depends on the diffraction vector $\boldsymbol{q} = \boldsymbol{k}_2 - \boldsymbol{k}_1$, where $\boldsymbol{k}_1$ and $\boldsymbol{k}_2$ denote the wave vectors of the incident and of the scattered radiation respectively. The intensity of the diffracted radiation for elastic scattering is

$$I(\boldsymbol{q}) = |U(\boldsymbol{q})|^2 \ . \tag{2.27}$$

Consider a crystal with *mnp* unit cells, each containing *s* atoms. The position of the *j*-th atom of a cell is determined by the vector

$$\boldsymbol{r}_j = x_j\,\boldsymbol{a} + y_j\,\boldsymbol{b} + z_j\,\boldsymbol{c} \tag{2.28}$$

drawn from the node

$$\boldsymbol{r}_{mnp} = m\boldsymbol{a} + n\boldsymbol{b} + p\boldsymbol{c} \ , \tag{2.29}$$

which belongs to the unit cell numbered *mnp*. The position of the *j*-th node of the cell *mnp* is determined by the vector $\boldsymbol{r}_j + \boldsymbol{r}_{mnp}$. In the general case the arbitrary periodic potential $U(\boldsymbol{r})$ can be presented as a superposition of potentials of individual atoms $u_j(\boldsymbol{r} - \boldsymbol{r}_j - \boldsymbol{r}_{mnp})$, i.e.

$$U(\boldsymbol{r}) = \sum_{mnp}\sum_{j=1}^{s} u_j(\boldsymbol{r} - \boldsymbol{r}_j - \boldsymbol{r}_{mnp}) \ . \tag{2.30}$$

In the formula (2.30) the first summation is made over all atoms of the basis, while the second summation is taken with respect to all unit cells of the crystal.

Considering (2.26) and (2.30), the formula (2.27) for the scattering intensity can be rewritten as

$$I(\boldsymbol{q}) = \left| \frac{1}{V} \int \sum_{mnp} \sum_{j} \{ u_j(\boldsymbol{r} - \boldsymbol{r}_j - \boldsymbol{r}_{mnp}) \exp(-i\boldsymbol{q}\boldsymbol{r}) d^3\boldsymbol{r} \} \right|^2$$

$$= \left| \frac{1}{V} \sum_{mnp} \sum_{j} \{ \exp[-i\boldsymbol{q}(\boldsymbol{r}_j + \boldsymbol{r}_{mnp})] \times \int u_j(\boldsymbol{r}') \exp(-i\boldsymbol{q}\boldsymbol{r}') d^3\boldsymbol{r}' \} \right|^2 , \qquad (2.31)$$

where $\boldsymbol{r}' = \boldsymbol{r} - \boldsymbol{r}_j - \boldsymbol{r}_{mnp}$ and $\int u_j(\boldsymbol{r}') \exp(-i\boldsymbol{q}\boldsymbol{r}') d^3\boldsymbol{r}' = f_j$ is the atomic scattering factor (form factor). Write the expression (2.31) in the form

$$I(\boldsymbol{q}) = \left| \frac{1}{V} \sum_{mnp} \exp(-i\boldsymbol{q}\boldsymbol{r}_{mnp}) \sum_{j} f_j \exp(-i\boldsymbol{q}\boldsymbol{r}_j) \right|^2 . \qquad (2.32)$$

The first factor $(1/V)\sum_{mnp} \exp(-i\boldsymbol{q}\boldsymbol{r}_{mnp})$ is nonzero only when the diffraction vector $\boldsymbol{q}$ is equal to an arbitrary vector of the reciprocal lattice:

$$\boldsymbol{q} = h\boldsymbol{b}_1 + k\boldsymbol{b}_2 + l\boldsymbol{b}_3 . \qquad (2.33)$$

Here $\boldsymbol{b}_{1,2,3}$ denotes the main vectors of the reciprocal lattice; h, k, and l are integers. For the vectors $\boldsymbol{q}$ that satisfy the condition (2.33), the first factor in the formula (2.32) is unity. The sum $\sum_j f_j \exp(-i\boldsymbol{q}\boldsymbol{r}_j)$ is called the structural amplitude or the geometric structure factor. For some reflection hkl, the structural amplitude F_{hkl} is

$$F_{hkl} = \sum_j f_j \exp[-i2\pi(x_j h + y_j k + z_j l)] , \qquad (2.34)$$

where x_j, y_j and z_j are the coordinates of the j-th atom of the basis. In the general case the structural amplitude is a complex quantity. Therefore in practical calculations the structural factor F^2, which is equal to the square of the absolute value of the structural amplitude (or the product of the structural amplitude and the complex conjugate F^*_{hkl}), i.e.

$$F^2_{hkl} = |F_{hkl}|^2 = F_{hkl} F^*_{hkl} \qquad (2.35)$$

is used. The trigonometric form of the structural factor is

$$F^2_{hkl} = \left[\sum_j f_j \cos 2\pi(x_j h + y_j k + z_j l) \right]^2$$

$$+ \left[\sum_j f_j \sin 2\pi(x_j h + y_j k + z_j l) \right]^2 . \qquad (2.36)$$

The atomic scattering factor f_j in expression (2.34) is a measure of the scattering capacity of the j-th species atom. X-rays are scattered most significantly from the electrons of the atoms. Therefore the potential $u(\boldsymbol{r})$ representing an individual atom cannot be thought of as being concentrated at the nucleus. In the case of X-ray radiation, the atomic scattering factor is the Fourier representation of the electron density $n_e(\boldsymbol{r})$ in an atom and depends on the imparted momentum $\hbar\boldsymbol{q}$. In other words, in the case of X-ray radiation, the atomic scattering factor depends on the number and distribution of electrons of the scattering atom, and also on the wavelength and the scattering angle. Values for the atomic factors of X-ray radiation scattering from neutral atoms and ions as a function of the value of $(\sin\theta)/\lambda$ are given in [27]. In contrast neutrons interact with atomic nuclei only and therefore the atomic scattering factor for neutrons is constant. The atomic scattering factor of neutrons is usually referred to as the scattering amplitude. Values for the neutron scattering amplitudes of nuclei of various elements have been summarized by the Committee for Neutron Diffraction at the International Union of Crystallographers and they are given in many books (for example, [32, 33]).

In the general case, the measured integral intensity of scattered radiation is determined from the formula

$$I = KI_0 F_{hkl}^2 PLf_T, \qquad (2.37)$$

where K is the instrumental constant; I_0 the incident radiation intensity; P the multiplicity factor; L the Lorentz factor; and f_T the thermal (temperature) factor. All factors, except F^2, in (2.37) are nonzero and positive for any kind of radiation. The structural amplitude, which may be to zero for some values of hkl, can eliminate some reflections allowed by the space lattice.

We now analyze how a crystalline structure influences the structural amplitude for some materials in the cubic system. We shall consider the structure types $A2$, $B2$, $A1$, $B1$, and $C1$. The multiplicity factors P for reflections from atomic planes (hkl) of these structures are given in Table 2.7.

1. Materials with the $A2$ structure (like tungsten W) have a bcc lattice (space group $Im\bar{3}m$ (O_h^9)). Tungsten, alkali metals, vanadium, niobium, tantalum, and chromium crystallize in this structure. The lattice basis contains two atoms with the coordinates 000 and 1/2 1/2 1/2. In accordance with (2.34) and with the coordinates of the atoms in the basis, the structural amplitude is

$$F = f\{1 + \exp[-i\pi(h + k + l)]\}, \qquad (2.38)$$

where f is the atomic scattering factor of the metal. Table 2.7 gives F^2 values for reflections from various planes (hkl). From Table 2.7 it is seen that reflections from the planes (100), (111), (210), (300), etc. are absent in the diffraction spectrum. This is explained as follows. A bcc lattice contains parallel equivalent planes (100), (200), (300), ... or (111), (211), (311), ..., etc. The planes (100) and (300) delimit an elementary cube and reflections from these planes are observed if the reflected beams are out of phase by 2π. However, the intermediate plane (200) is located halfway between these planes and the beam reflected from the plane (200) is shifted in phase relative the beam reflected from the plane (100) by π. As a result, the reflection (100) is quenched or forbidden. Other reflections can be treated analogously. Obviously, reflection (100) is suppressed in the bcc lattice of metals having the *A*2 structure because planes (100) and (200) include an equal number of the same atoms occupying positions of similar type, i.e. because these planes are equivalent.

2. Quenching does not take place in the *B*2 structure (CsCl) of the space group $Pm\bar{3}m$ (O_h^1). Cesium halides, compounds like β-brass (for example, CuZn), and many other alloys have a structure of *B*2 type. In this structure atoms of each species form a simple cubic lattice. The structure type *B*2 differs from the structure type *A*2 by alternation of planes, which are composed of atoms of species 1 and of atoms of species 2, in the direction [100]. Scattering factors of these atoms are f_1 and f_2. Planes formed by different atoms are nonequivalent and therefore the diffraction spectrum includes all reflections (Table 2.7). The lattice basis has two distinct atoms positioned at the coordinates 0 0 0 and 1/2 1/2 1/2, respectively. Hence, the structural amplitude is

$$F = f_1 + f_2\exp[-i\pi(h + k + l)]. \tag{2.39}$$

3. The structure type *A*1 has a fcc lattice and the space group is $Fm\bar{3}m$ (O_h^5). This structure is typical for copper and for noble metals. The lattice basis contains four identical atoms with coordinates 0 0 0, 1/2 1/2 0, 1/2 0 1/2, and 0 1/2 1/2. The structural amplitude is

$$F = f\{1 + \exp[-i\pi(h + k)] + \exp[-i\pi(h + l)] + \exp[-i\pi(k + l)]. \tag{2.40}$$

4. The structure type *B*1 (NaCl) is composed of two fcc lattices, which are displaced relative to one another by a half-period. The space group is $Fm\bar{3}m$ (O_h^5). This structure occurs frequently in nature. In particular, strongly nonstoichiometric monocarbides MC_y of titanium, zirconium, hafnium, vanadium, niobium and tantalum, mononitrides MN_y of titanium, zirconium, hafnium, vanadium and niobium, and monoxides MO_y of titanium and vanadium, which will be discussed in this book, have this structure. The lattice basis contains four metal atoms with coordinates 0 0 0, 1/2 1/2 0, 1/2 0 1/2, 0 1/2 1/2, and four nonmetal atoms with coordinates 1/2 1/2 1/2, 0 0 1/2, 0 1/2 0, 1/2 0 0. If the atomic scattering factors of the metal and nonmetal atoms are f_1 and f_2, the structural amplitude is

$$F = f_1\{1 + \exp[-i\pi(h + k)] + \exp[-i\pi(h + l)] + \exp[-i\pi(k + l)]\}$$
$$+ f_2\{\exp[-i\pi(h + k + l)] + \exp(-i\pi h) + \exp(-i\pi k) + \exp(-i\pi l)]\}. \tag{2.41}$$

F^2 values for reflections from atomic planes (*hkl*) are listed in Table 2.7.

5. The structure type *C*1 (like fluorite CaF_2) has the space group $Fm\bar{3}m$ (O_h^5). It is characteristic for some fluorides and oxides. In a fluorite-like structure the metal atoms form a fcc sublattice, while nonmetal atoms form a simple cubic lattice. Each nonmetal atom is located at the center of a tetrahedron made up of four metal atoms. The lattice basis contains four metal atoms with coordinates 0 0 0, 1/2 1/2 0, 1/2 0 1/2, 0 1/2 1/2, and eight nonmetal atoms with the

coordinates 1/4 1/4 1/4, 3/4 3/4 1/4, 3/4 1/4 3/4, 1/4 3/4 3/4, 3/4 3/4 3/4, 1/4 1/4 3/4, 1/4 3/4 1/4, 3/4 1/4 1/4. The expression describing the structural amplitude of the *C*1 structure is rather cumbersome and therefore it is omitted. Values of the square of the structural amplitude F^2 for various reflections (*hkl*) are given in Table 2.7.

In summary crystals having different symmetry also have different sets of diffraction reflections. Diffraction spectra of crystals, which have conventional lattices of similar symmetry but different types of structures, contain the same reflections. However, the defined ratio for the reflection intensities is characteristic for each type of structure. Thus, the set of diffraction reflections and their intensity ratio represent a unique characteristic of the crystalline structure.

Table 2.7. Multiplicity factors *P* and structure factors F^2 for some cubic structures

		F^2 for the structures				
hkℓ	*P*	*A*1 (Cu)	*B*1 (NaCl)	*C*1 (CaF_2)	*A*2 (W)	*B*2 (CsCl)
100	6	0	0	0	0	$(f_1-f_2)^2$
110	12	0	0	0	$4f^2$	$(f_1+f_2)^2$
111	8	$16f^2$	$16(f_1-f_2)^2$	$16f^2$	0	$(f_1-f_2)^2$
200	6	$16f^2$	$16(f_1+f_2)^2$	$16(f_1-2f_2)^2$	$4f^2$	$(f_1+f_2)^2$
210	24	0	0	0	0	$(f_1-f_2)^2$
211	24	0	0	0	$4f^2$	$(f_1+f_2)^2$
220	12	$16f^2$	$16(f_1+f_2)^2$	$16(f_1+2f_2)^2$	$4f^2$	$(f_1+f_2)^2$
221	24	0	0	0	0	$(f_1-f_2)^2$
300	6	0	0	0	0	$(f_1-f_2)^2$
310	24	0	0	0	$4f^2$	$(f_1+f_2)^2$
311	24	$16f^2$	$16(f_1-f_2)^2$	$16f^2$	0	$(f_1-f_2)^2$
222	8	$16f^2$	$16(f_1+f_2)^2$	$16(f_1-2f_2)^2$	$4f^2$	$(f_1+f_2)^2$
320	24	0	0	0	0	$(f_1-f_2)^2$
321	48	0	0	0	$4f^2$	$(f_1+f_2)^2$
400	6	$16f^2$	$16(f_1+f_2)^2$	$16(f_1+2f_2)^2$	$4f^2$	$(f_1+f_2)^2$
322	24	0	0	0	0	$(f_1-f_2)^2$
410	24	0	0	0	0	$(f_1-f_2)^2$
330	12	0	0	0	$4f^2$	$(f_1+f_2)^2$
411	24	0	0	0	$4f^2$	$(f_1+f_2)^2$
331	24	$16f^2$	$16(f_1-f_2)^2$	$16f^2$	0	$(f_1-f_2)^2$
420	24	$16f^2$	$16(f_1+f_2)^2$	$16(f_1-2f_2)^2$	$4f^2$	$(f_1+f_2)^2$
421	48	0	0	0	0	$(f_1-f_2)^2$
332	24	0	0	0	$4f^2$	$(f_1+f_2)^2$
422	24	$16f^2$	$16(f_1+f_2)^2$	$16(f_1+2f_2)^2$	$4f^2$	$(f_1+f_2)^2$
500	6	0	0	0	0	$(f_1-f_2)^2$
430	24	0	0	0	0	$(f_1-f_2)^2$
510	24	0	0	0	$4f^2$	$(f_1+f_2)^2$

Effect of Ordering on Diffraction Spectra. Structural order–disorder phase transitions are connected with the redistribution of atoms at crystal lattice sites. Ordering alters the crystal symmetry, because some symmetry elements present in the disordered crystal may disappear in an ordered crystal. In consequence ordering has a direct effect on the diffraction spectrum. Since the translational symmetry changes during ordering, the sizes of the main translation vectors of an ordered crystal increase compared to vectors in a disordered crystal and the main translations of the reciprocal lattice will decrease correspondingly.

Ordering will thus lead to the appearance of a small-scale reciprocal lattice which is inscribed in the reciprocal lattice of the disordered crystal. The positions of the reflections correspond to structural vectors of the reciprocal lattice and to structural reflections in diffraction spectra. Other reflections which may appear in the reciprocal lattice of an ordered crystal are called superstructure reflections.

In a disordered solid solution (alloy) all parallel atomic planes have an equal average composition and they will scatter radiation in the same way.

When a solution (alloy) becomes ordered, atoms are redistributed and the number of atoms of some species may become different in individual planes, i.e. adjacent planes become nonequivalent. In consequence the amplitudes of radiation scattered by these planes are no longer equal in magnitude and superstructure reflections appear in the diffraction spectrum. In general the intensity of superstructure reflections is weaker than the intensity of the structural reflections. This can be illustrated taking a CuZn alloy of an equiatomic composition as an example. If this alloy is disordered, copper and zinc atoms are distributed over all sites of the bcc lattice with an equal probability and scattering is determined by the average atomic scattering factor $f = (f_1 + f_2)/2$. Thus the scattering amplitude is

$$F = [(f_1 + f_2)/2]\{1 + \exp[-i\pi(h + k + l)]\}. \tag{2.42}$$

The structure factor F^2 is zero and $(f_1 + f_2)^2$ for reflections from the planes (100) and (110), respectively. Obviously the reflection (110) is the first structural reflection of the disordered alloy. In the ordered state the CuZn alloy has the $B2$ structure and the structural amplitude is now given by (2.39). The structure factor F^2 of the ordered alloy is $(f_1 - f_2)^2$ and $(f_1 + f_2)^2$ for reflections for the planes (100) and (110), respectively. However the intensity of the first superstructure reflection (100) will be much smaller than the intensity of the structure reflection (110). The intensity of superstructure reflections is the smaller, the closer the values of the atomic scattering factors f_1 and f_2. If $f_1 = f_2$ then superstructure reflections will not be observed at all in the diffraction spectrum of an ordered alloy having the β-brass structure.

Neutron diffraction has a remarkable feature for studying ordering of solid solutions. While for X-ray radiation atomic scattering factors are positive for all elements, some elements (for example, Ti and Mn) have negative scattering amplitudes in the case of neutron scattering. This can result in an unusual intensity ratio of structural and superstructural lines as compared to X-ray scattering. Let the concentration of the second component in the solution be c. If f_1 and f_2 are different, the intensity of the superstructure reflection becomes proportional to $|f_1 - f_2|^2$ and may become larger than the intensity of the structural reflection which is proportional to $|f_1 - c(f_1 - f_2)|^2$.

The scattering of various types of waves by the crystal lattice of an ordered solution was discussed in detail in [31, 34]. A point of significance, for the purpose of this book, is that the intensity of superstructure reflections is proportional to η^2 (η being the long-range order parameter). Thus it becomes possible to determine not only the crystalline structure of an ordered phase but also the degree of long-range order from a diffraction spectrum. The relationship between the intensity of superstructure reflections and the parameters of the long-range order will be detailed in Chap. 5.

References

1 E. S. Fedorov: Elements of the study of figures, Zapiski Imper. S.-Petersburg. Mineralog. Obshestva, Second Seriya, Part 21, 1–289 (1885) (in Russian)

2 E. S. Fedorov: Studies in analytical crystallography. Second scetch, Gornyi Zhurnal **1**, 395–425 (1886) (in Russian)

3 E. S. Fedorov: Symmetry of regular systems of figures, Zapiski Imper. S.-Petersburg. Mineralog. Obshestva, Second Seriya, Part 28, 1–147 (1891) (in Russian). (Engl. Transl.: The Symmetry of Regular Systems of Figures, in E. S. Fedorov: *Symmetry of Crystals*, ACA Monograph No. 7, Amer. Crystallographic Association, New York 1971, pp. 50–131).

4 E. S. Fedorov: Theorie der Kristallstruktur. Einleitung. Regelmässige Punktsysteme, Z. Kristallogr. Min. **24**, 209–252 (1895)

5 E. S. Fedorov: Theorie der Kristallstruktur. I. Mögliche Structurarten (mit übersichtlicher graphischer Darstellung der Symmorphen Structurarten), Z. Kristallogr. Min. **25**, 113–124 (1895)

6 E. S. Fedorov: Reguläre Plan- und Raumverteilung, Abhandlungen der K. Bayer Akademie der Wissenschaft, Math.-Phys. Cl., (München) **XX**, 465–588 (1899–1900)

7 E. S. Fedorov: *Symmetry and Structure of Crystals* (Nauka, Moscow 1949) 630 pp. (in Russian)

8 E. S. Fedorov: *Tiling of Plane and Space* (Nauka, Leningrad 1979) 272 pp. (in Russian)

9 E. S. Fedorov: *Symmetry of Crystals*, ACA Monograph No. 7 (Amer. Crystallographic Association, New York 1971) 315 pp.

10 A. Schönflies: Über reguläre Gebietstheilungen des Raumes, Nachrichten der Gesellschaft der Wissenschaft zu Goettingen No 9, 223–237 (1889)

11 A. Schönflies: Über Gruppen von Transformationen des Raumes in sich, Mathemat. Annalen. **34**, 172–203 (1889)

12 A. Schoenflies: *Krystallsysteme und Krystallstructur* (Teubner, Leipzig 1891) 638 SS.

13 A. Schoenflies: *Theorie der Kristallstruktur* (Gebrüder Borntraeger, Berlin 1923) 556 SS.

14 T. Hahn (Ed.): *International Tables for X-ray Crystallography* **A**, *Space group symmetry* 4th revised edn (Kluwer Academic Publishers, Dordrecht-Boston-London 1995) 878 pp.
15 G. B. Bokii: *Introduction to Crystal Chemistry* (US Atomic Energy Commission, Office of Technical Information, Oak Ridge, Tennessee 1960) 528 pp.
16 O. V. Kovalev: *Irreducible Representations of the Space Groups* (Gordon & Breach, New York 1965) 210 pp.
17 O. V. Kovalev: *Irreducible and Induced Representations and Corepresentation of Fedorov Groups* (Nauka, Moscow 1986) 368 pp. (in Russian)
18 W. Ludwig, C. Falter: *Symmetries in Physics* (Springer, Berlin-Heidelberg 1988) 461 pp.
19 Yu. A. Izyumov, V. N. Naish, R. P. Ozerov: *Neutron Diffraction of Magnetic Materials* (Consultants Bureau, New York 1981) 339 pp.
20 M. Wagner: *Gruppentheoretische Methoden in der Physik* (Spriger, Berlin-Heidelberg 1998) 461 SS.
21 R. L. Flurry: *Symmetry Groups: Theory and Chemical Applications* (Prentice - Hall Inc., Englewood Cliffs 1980) 356 pp.
22 Yu. A. Izyumov, V. N. Syromyatnikov: *Phase Transitions and Symmetry of Crystals* (Nauka, Moscow 1984) 248 pp. (in Russian)
23 A. I. Gusev, A. A. Rempel: *Structural Phase Transitions in Nonstoichiometric Compounds* (Nauka, Moscow 1988) 308 pp. (in Russian)
24 D. Shechtman, I. Blech, D. Gratias, J. W. Chan: Metallic phase with long-range orientational order and no translational symmetry, Phys. Rev. Lett. **53**, 1951–1953 (1984)
25 R. D. Field, H. L. Fraser: Precipitates possessing icosahedral symmetry in a rapidly solidified Al-Mn alloy, Mater. Sci. Eng. **68**, L17–L21 (1984)
26 L. V. Azaroff, M. J. Buerger: *The Powder Method in X-ray Crystallography* (McGraw-Hill, New York-London 1958) 342 pp.
27 A. J. C. Wilson (Ed.): *International Tables for X-ray Crystallography* **C**, *Mathematical, Physical and Chemical Tables* (Kluwer Academic Publishers, Dordrecht-Boston-London 1992) 883 pp.
28 C. Giacovazzo, H. L. Monaco, D. Viterbo, F. Scordari, G. Gilli, G. Zanotti, M. Catti: *Fundamentals of Crystallography* (Oxford University Press, Oxford 1992) 654 pp.
29 M. von Laue: *Röntgenstrahlinterferenzen* (Akademische Verlagsgesellschaft Becker & Erler Kom.-Ges., Leipzig 1941) 367 pp.
30 R. W. James: *The Optical Principles of the Diffraction of X-rays* (G. Bell and Sons Ltd., London 1954) 624 pp.
31 M. A. Krivoglaz: *Theory of X-ray and Thermal-Neutron Scattering by Real Crystals* (Plenum Press, New York 1969) 405 pp.
32 M. Bee: *Quasielastic Neutron Scattering: Principles and Applications in Solid State Chemistry, Biology, and Materials Science* (Adam Hilger, Bristol – Philadelphia 1988) 437 pp.
33 Yu. Z. Nozik, R. P. Ozerov, K. Hennig: *Structural Neutron Diffraction* (Atomizdat, Moscow 1979) 344 pp. (in Russian)
34 M. A. Krivoglaz, A. A. Smirnov: *The Theory of Order–Disorder in Alloys* (American Elsevier, New York 1964) 427 pp.

3. Strongly Nonstoichiometric Compounds

This chapter gives the general characteristics of strongly nonstoichiometric compounds which are interstitial phases and it describes their crystal structures and homogeneity intervals. Also the principal methods of preparing nonstoichiometric carbides and nitrides are described. The position of the lower boundary of the homogeneity interval is discussed in detail. The last section presents a systematic review of experimental data on ordering in strongly nonstoichiometric compounds and their solid solutions.

In nonstoichiometric compounds the chemical composition does not coincide with the concentration of crystal lattice sites for different species of the compound. Nonstoichiometry is possible only in two- and multiatomic substances. Strongly nonstoichiometric compounds are substances with structural vacancies and homogeneity intervals such that the concentration of vacancies is sufficient for their interaction [1, 2]. A homogeneity interval is a concentration region within which the crystal structure of a compound or a substitutional solid solution remains unchanged if the composition is changed. The term "strongly nonstoichiometric compound" denotes a particular group of compounds, and it should be regarded as an appropriate term. It applies also to a stoichiometric compound if found within a homogeneity interval. To our knowledge, a similar word combination, "grossly nonstoichiometric phases", was used by Anderson [3] in his discussion of nonstoichiometry in chalcogenides and sulfides; however it had not been broadly accepted then.

The group of strongly nonstoichiometric interstitial compounds includes carbides, nitrides and oxides MX_y and M_2X_y (X = C, N, O) of groups IV and V transition metals M, and related ternary compounds with extended homogeneity intervals.

The history of the investigation of these truly unique compounds may be divided into three periods.

In the first period (early 20th century to 1940), carbides and nitrides were regarded as ordinary chemical compounds of stoichiometric composition. The majority of publications in that period were dedicated to defining the types of crystal structure and unit cell parameters of carbides and nitrides obtained by different methods. The composition of these compounds was assumed to be stoichiometric. Towards the end of the first period (1930–1940s), investigations of the thermodynamics and of various physical properties of carbides and nitrides developed on a broad scale. It was at that time that mass defiencies were noted in results reported in some of the papers concerned with properties of one and the same carbide or nitride. The discrepancies observed failed to be attributed to measurement errors or the use of different techniques. Detailed crystal chemistry investigations showed that the observed discrepancies were caused by differences in the concentration of interstitial atoms (carbon and nitrogen) in one and the same compound. It was found that, unlike other compounds, the composition of carbides and nitrides of groups IV and V transition metals may vary over a wide range, with the crystal structure remaining unchanged.

The second period (1940–1970s) of the investigation of strongly nonstoichiometric compounds of interstitial phase type is linked to studies of composition–property relationships. Interest in these compounds and properties increased in the 1960s when thousands of papers dealing with the influence of composition on crystal, chemical, thermodynamic, thermophysical, electric, magnetic, optical and other properties of these materials appeared. The principal qualitative concepts of the nature of chemical bonds in nonstoichiometric compounds were formulated during the same period. In the 1970s, the purely fundamental investigations of nonstoichiometric compounds decreased noticeably in number giving way to a growing amount of applied research. The faulty idea of nonstoichiometric compounds being completely investigated was widely spread among the majority of researchers.

For a long time after detecting extensive homogeneity intervals in nonstoichiometric compounds it was believed that nonmetal interstitial atoms and structural vacancies (vacant interstitial sites on the metallic sublattice) are randomly distributed in the lattice. But crystallographic investigations made in 1967 and in subsequent years showed that under certain conditions interstitial atoms and vacancies may redistribute on the crystal lattice sites to form ordered structures of various types. These works marked the beginning of the third period in the investigation of nonstoichiometric compounds. This third period is connected with investigations of ordering phenomena and its influence on properties of strongly nonstoichiometric compounds. It has become clear today that the ordered state is the thermodynamic equilibrium state of nonstoichiometric compounds at temperatures below 1000 K, while the disordered state at these temperatures is metastable.

The investigations of the early 1980s concerned with properties of strongly nonstoichiometric compounds in relation to the distribution of interstitial atoms and vacancies showed that changes relating to ordering are comparable to changes in the same properties caused by a variation of the composition of the nonstoichiometric compound within the homogeneity interval of the ordered phase. Thus, changes in composition of strongly nonstoichiometric compounds and the redistribution of atoms and vacancies in their crystal lattice present two equal factors determining the properties of those compounds.

3.1 Preparation Techniques and Degree of Homogeneity

Normally, polycrystalline carbide and nitride samples are synthesized using techniques from powder metallurgy. There are two principal ways to synthesize carbides and nitrides: either directly from elements, or from compounds containing these elements.

The chemically purest samples of MC_y carbides which also have a predefined concentration of carbon are produced by solid-state sintering under vacuum of metal powders and carbon black:

$$M + yC \rightarrow MC_y \,. \tag{3.1}$$

Prior to charge blending, carbon black is calcinated for 4 h at 700–900 K at a pressure lower than 1 Pa to remove moisture. Metal powders are blended with calcinated carbon in the necessary ratio, thoroughly mixed and then pressed to pellets under a pressure of 2500–2700 kg cm^{-2}.

The temperature conditions of solid-state vacuum synthesis of titanium, zirconium, hafnium, vanadium, niobium and tantalum carbides by sintering of metal powders and carbon [4] are illustrated in Fig. 3.1.

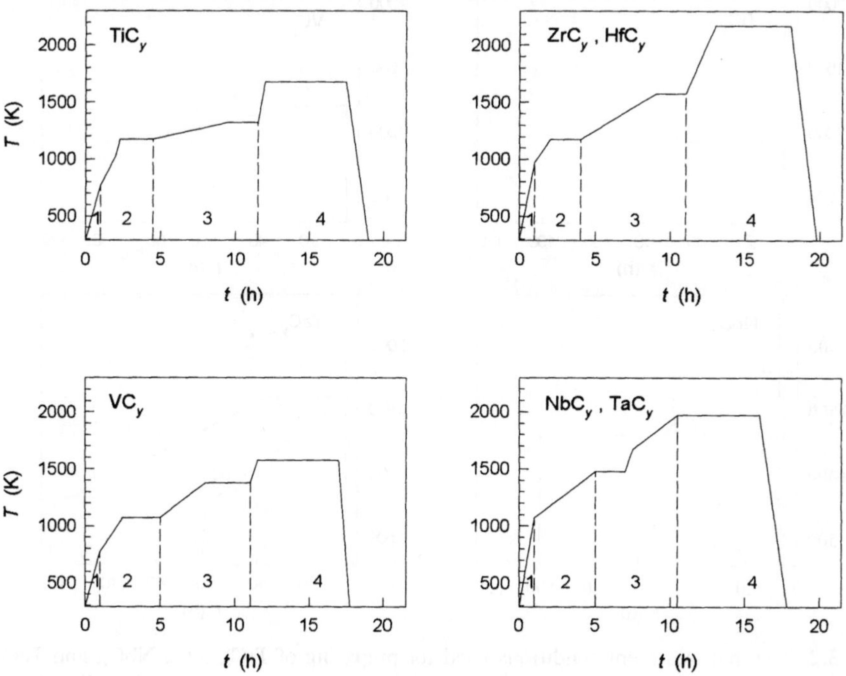

Fig. 3.1. Temperature conditions of the solid-state vacuum synthesis of carbides TiC_y, ZrC_y, HfC_y, VC_y, NbC_y and TaC_y [4]: (1) outgassing, (2) preliminary sintering, (3) chemical interaction between metal and carbon and formation of carbide, (4) homogenization of carbide

The synthesis route comprises four stages: The first heating stage lasts 1 h and provides outgassing of the charge. During the second stage the charge is pre-sintered. During the third or main stage the metal and carbon interact chemically to form the carbide. During this stage continuous linear or constant heating of specimens is needed and also increasing temperature is needed for a change of the rate of chemical interaction. The fourth stage provides the homogenization of the compound. Quick cooling (quenching) after the fourth stage of sintering produces nonstoichiometric carbides in a disordered state. The duration and temperature of the homogenization stage depend on whether the nonstoichiometric carbide evaporates congruently or incongruently. A further important parameter is the evaporation rate of the carbide. The conditions are chosen such that the carbide composition remains unchanged during homogenization. Nonstoichiometric carbides synthesized by the above method have the highest degree of homogeneity [4] and contain minimum impurities to render them suitable for precision research.

Synthesized samples may be subject to additional thermal treatments to produce nonstoichiometric carbides with different degrees of order. Ordered carbides are normally obtained by long-term annealing below 1300–1500 K and then the temperature is decreased slowly down to 300 K (or higher, see Fig. 3.2).

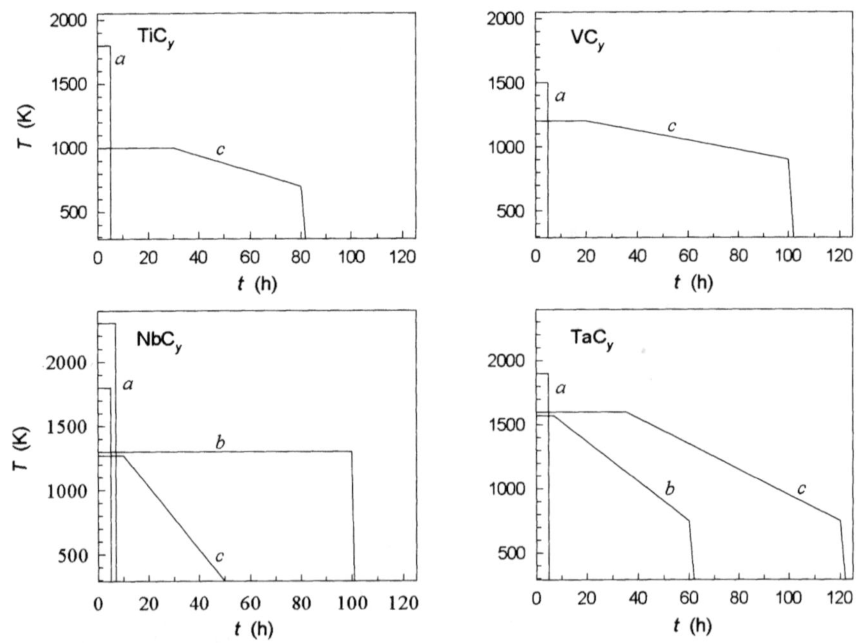

Fig. 3.2. Thermal treatment conditions used for preparing of TiC_y, VC_y, NbC_y, and TaC_y carbides in disordered and ordered states [4]: (*a*) quenching for preparing disordered carbides; (*b*), (*c*) annealing for preparing nonstoichiometric carbides with different degrees of long-range order

Figure 3.2 presents typical thermal treatment sequences intended to produce ordered nonstoichiometric titanium, vanadium, niobium and tantalum carbides. A procedure along the path *a*, characterized by rapid quenching from 1850–2300 K to 300 K at a rate of 2000 K min^{-1}, is used to obtain a disordered state which is metastable at room temperature. Paths *b* and *c* produce an equilibrium state with long-range order of carbon atoms and structural vacancies. These routes need long-time (up to 100 h) annealing and a slow decrease of temperature from 1000–1600 K (depending on the species of metal atoms) to 600–800 K. A typical cooling rate is 0.05–0.20 K min^{-1}. A route along *c*, characterized by a very low cooling rate, provides nonstoichiometric carbides with a higher degree of long-range order than the regime *b*.

Fundamental research into carbide and nitride systems often employs the method of solid-state diffusion [5] to produce the entire spectrum of phases formed in the M–X system. This method is time-consuming but ensures a reasonably accurate knowledge of the chemical and phase composition.

Modifications of the solid-state method are arc melting in vacuum or in an inert atmosphere and hot extrusion.

Carbides can also be synthesized by sintering of metal hydrides and carbon

$$MH + yC \rightarrow MC_y + H_2\uparrow \qquad (3.2)$$

or by interaction between metal and a carbonizing gas

$$M + C_nH_m \rightarrow MC_y + H_2\uparrow, \qquad (3.3)$$

$$M + CO \rightarrow MC_y + CO_2\uparrow. \qquad (3.4)$$

Synthesis of carbides and nitrides by precipitation reactions from the gas phase is based on the interaction of chlorides or carbonyls of the transition metals with a hydrogen–hydrocarbon or hydrogen–nitrogen mixture:

$$MCl_4 + C_nH_m + H_2 \rightarrow MC_y + HCl\uparrow + C_pH_q\uparrow, \qquad (3.5)$$

$$MCl_4 + N_2 + H_2 \rightarrow MN_y + HCl\uparrow. \qquad (3.6)$$

Nitrides are produced from reactions of metal powders or metal hydrides in a nitrogen or ammonia atmosphere:

$$M\ (MH) + N_2 \rightarrow MN_y + (H_2\uparrow), \qquad (3.7)$$

$$M\ (MH) + NH_3 \rightarrow MN_y + H_2\uparrow. \qquad (3.8)$$

The composition and homogeneity of synthesized nitrides depend on the synthesis temperature, the pressure of the nitrogen, and the reaction time.

Synthesis of carbides MC_y and nitrides MN_y by thermal decomposition involves various precursors [6] including carbonyls $M(CO)_n$, amides $M(NR_2)_4$, and metal-organic compounds like $(C_5H_5)_2MX_n$, or $(C_2H_5)_3Ti(CH_3)_2$, or polymers (e.g. $[(C_6H_4O_2)_2Ti]_n$).

The method defined as self-propagating high-temperature synthesis (SHS) has been used frequently. SHS is a fast process of solid-state combustion of reagents (a metal and carbon for carbides or a metal in a nitrogen environment in the case of nitrides) at temperatures between 2500 and 3000 K [7].

The method of mechanical synthesis of carbides through intense grinding of metal and carbon powders has been developed during the last decade [8].

Carbothermal reduction of transition metal oxides in an inert or in a reducing atmosphere is the cheapest and most commonly used commercial method for producing carbides with a near-stoichiometric composition:

$$MO + C \rightarrow MC_y + CO\uparrow. \qquad (3.9)$$

Usually this reaction is realized with excess carbon. In analogy the carbothermal reduction of oxides in a nitrogen atmosphere is used to synthesize nitrides:

$$MO + C + N_2 \rightarrow MN_y + CO\uparrow. \qquad (3.10)$$

A detailed description of various methods used to synthesize carbides and nitrides of transition metals can be found in [9–11].

Physical chemistry and solid-state chemistry commonly use the notion of homogeneity, which is a qualitative characteristic of the extent to which a multi-component system or a compound is compositionally homogeneous. However it is crucial to consider homogeneity, which can provide the basis for an objective comparison of compounds produced by different methods. Methods of analytical chemistry fail in assessing the degree of homogeneity of a compound. Thus, no attempts were made until recently to estimate homogeneity on a quantitative level. This is especially significant for crystalline compounds where the composition can considerably deviate from stoichiometry. It has been demonstrated [12] that the degree of homogeneity can be assessed quantitatively by diffraction methods for all nonstoichiometric compounds and substitutional solid solutions (alloys).

Inhomogeneity is a fluctuation of the concentration c in a volume V of a solid solution or a nonstoichiometric compound. Concentration fluctuations at points $\boldsymbol{r}$ of a crystal are characterized by a continuous fluctuation distribution function $\delta c(\boldsymbol{r})$ expandable in a Fourier series $\delta c(\boldsymbol{r}) = \sum_k c_k \exp(-i\boldsymbol{kr})$, where $c_k = \frac{1}{V}\int \delta c(\boldsymbol{r})\exp(i\boldsymbol{kr})dV$ and $\boldsymbol{k}$ is a fluctuation wave vector. Since fluctuations yield only a concentration redistribution between different parts of a crystal rather than to the concentration variation, the integral $\int \delta c(\boldsymbol{r})dV = 0$. In a crystal the concentration fluctuations produce inhomogeneous static displacements.

Structural defects are determined from line broadening of Bragg reflections in diffraction experiments on powder specimens. All types of defects displace atoms from their lattice sites and therefore affect the structural factor which determines the intensity of Bragg reflections and diffuse scattering. The well-known expression for the intensity of Bragg reflections from a defective crystal was derived in [13]. In accordance with this expression, all defects are divided conditionally into two groups. Defects in the first group only weaken the intensity of diffraction line and do not cause line broadening whereas defects in the second group yield a broadening of reflections. Broadening can be due to the small size D of crystallites or stacking faults. In this latter case the broadening is proportional to $\sec\theta$, whereas broadening caused by lattice microdeformations or by randomly distributed dislocations is proportional to $\tan\theta$.

Nonstoichiometric compounds MX_y can have an additional origin for line broadening, relating to a nonuniform composition over the volume of the compounds.

Let us visualize a powder specimen as a set of crystallites with different compositions y. Assume the crystallites are arranged in space statistically. In this case reflection profiles will widen in relation to the fluctuations of the number of defects in different crystallites. Since reflections from different crystallites merge, the resulting intensity distribution will be smeared out in the neighborhood of reciprocal lattice points.

Let us consider a nonstoichiometric carbide MC_{y_0} with the basic structure $B1$. If the carbon concentration distribution y for the carbide is described by the normal (Gaussian) distribution

$$g(y) = [1/(\Delta y\sqrt{2\pi})]\exp[-(y-y_0)/(2\Delta y^2)], \qquad (3.11)$$

where y_0 is the mean relative concentration of carbon and Δy is the half-width of the distribution function $g(y)$, the MC_{y_0} carbide composition is preset to within $y = y_0 \pm \Delta y$ and the quantity Δy is the degree of carbide inhomogeneity. The usual normalization conditions $\int_{-\infty}^{\infty} g(y)dy = 1$ and $\int_{-\infty}^{\infty} yg(y)dy = y_0$ are met in this case. Let us proceed from the distribution in y to the distribution in the diffraction angle θ using the Wulff–Bragg equation written for first-order diffraction reflections in terms of the lattice constant $a_{B1}(y) = d_{hkl}(y)(h^2+k^2+l^2)^{1/2}$ as $\theta_0 = \arcsin\{[\lambda(h^2+k^2+l^2)^{1/2}]/2a_{B1}(y)\}$ (d_{hkl} is the interplanar distance for the reflection (hkl) with diffraction angle θ_0, $a_{B1}(y)$ is the lattice constant as a function of the MC_y composition, and λ denotes the wavelength). The distribution (3.11) takes the form

$$g(\theta) = \left(1/\Delta y\sqrt{2\pi}\right)\exp[-(\theta-\theta_0)^2/(2\theta_y^2)], \qquad (3.12)$$

where $\theta_y = \dfrac{|a'_{B1}(y)|_{y=y_0}|}{K(\theta_0)}\Delta y$ and $K(\theta_0) = \dfrac{\pi\lambda\sqrt{h^2+k^2+l^2}}{360}\dfrac{\cos\theta_0}{\sin^2\theta_0}$.

The experimental integrated intensity of the reflection $\int_{-\infty}^{\infty} R(\theta'-\theta)g(\theta')d\theta'$ is a convolution of the theoretical distribution function $g(\theta)$ and instrumental resolution function

$$R(\theta) = \left(1/\theta_R\sqrt{2\pi}\right)\exp[-(\theta-\theta_0)^2/(2\theta_R^2)]. \qquad (3.13)$$

It has the form

$$I(\theta) = [1/\sqrt{2\pi(\theta_R^2+\theta_y^2)}]\exp\{-(\theta-\theta_0)^2/[2(\theta_R^2+\theta_y^2)]\}. \qquad (3.14)$$

From (3.14) it is clear that the half-width θ_{exp} of the experimental function can be found from the equation

$$\theta_{\rm exp}^2 = \theta_{\rm R}^2 + \theta_y^2 = \theta_{\rm R}^2 + \left[\frac{\left| a'_{B1}(y)\,|_{y=y_0} \right|}{K(\theta_0)} \Delta y \right]^2 \qquad (3.15)$$

and the line broadening β is $\beta = \dfrac{\left| a'_{B1}(y)\,|_{y=y_0} \right|}{K(\theta_0)} \Delta y$ where $a'_{B1}(y) = da_{B1}(y)/dy$.

Considering the angular dependence of the coefficient $K(\theta_0)$, it follows that the broadening due to inhomogeneity Δy varies as $(\sin^2\theta)/\cos\theta$. Finitesize broadening and deformation broadening are proportional to $\sec\theta$ and $\tan\theta$, respectively. Therefore different angular dependencies allow one to distinguish between the three types of line broadening.

If size and deformation broadening are absent, the inhomogeneity Δy can be determined from the broadening of reflections as

$$\Delta y = \frac{K(\theta_0)}{\left| a'_{B1}(y)\,|_{y=y_0} \right|} \sqrt{\theta_{\rm exp}^2 - \theta_{\rm R}^2}\,. \qquad (3.16)$$

We shall refer to the quantity $(1 - \Delta y)$ as the degree of homogeneity. Clearly, the closer $(1 - \Delta y)$ is to unity, the more homogeneous is the crystal of a nonstoichiometric compound. The half-width $\theta_{\rm exp}$ of the experimental function and the half-width $\theta_{\rm R}$ of the resolution function in (3.15) and (3.16) are expressed in terms of the angles θ. For a Gaussian distribution, the half-width $\theta_{\rm R}$ expressed as angles θ is related to the full width at half maximum $\mathrm{FWHM_R}(2\theta) \equiv \Delta(2\theta_{\rm R})$, measured in of angles 2θ, by the known expression $\theta_{\rm R}(\theta) = \Delta(2\theta_{\rm R})/(2 \times 2.35)$. This relation is readily derived by integrating the Gaussian distribution.

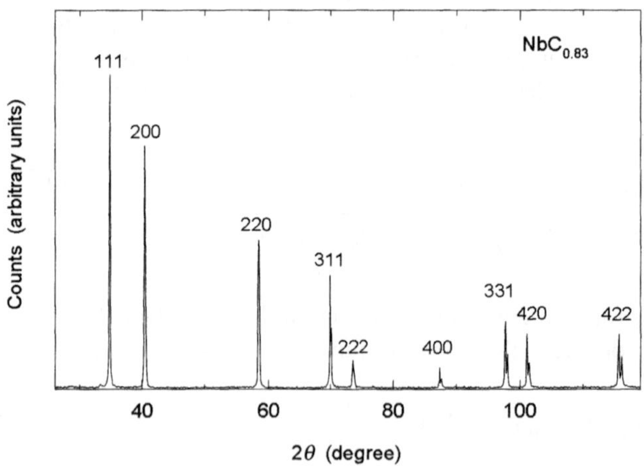

Fig. 3.3. X-ray diffraction spectrum of disordered nonstoichiometric niobium carbide $NbC_{0.83}$. A high degree of carbide homogeneity is confirmed by splitting of the Cu$K\alpha_{1,2}$ doubles, which is already observed for the reflection $(311)_{B1}$

As an example, let us evaluate the degree of homogeneity of disordered niobium carbide $NbC_{0.83}$ produced by the solid-state sintering method as shown in Fig. 3.1. Structural investigation by X-ray the powder methods suggests a high degree of homogeneity of the synthesized carbides: $CuK\alpha_{1,2}$ doublets split for lines having small Miller indices. As a typical example, Fig. 3.3 depicts an X-ray diffraction spectrum of disordered carbide $NbC_{0.83}$ where doublets split starting from the $(311)_{B1}$ line. The diffraction spectrum was measured using a Siemens D-500 automatic diffractometer with $CuK\alpha$-radiation in the Bragg–Brentano geometry. The half-width θ_R of the instrumental resolution function was determined in special experiments with a single crystal of hexagonal silicon carbide 6H-SiC and stoichiometric tungsten carbide WC. The half-width θ_R is 0.0301° in $CuK\alpha_1$ radiation at the diffraction angle $\theta = 35°$. In accordance with [4], the first derivative of the concentration dependence of the lattice constant with respect to y at the point y = 0.83 is $a'_{B1}(y)|_{y=0.83}$ = 0.01655 nm for disordered niobium carbide NbC_y. The experimental full width at half maximum $\Delta(2\theta_{exp})$ equals 0.15° for the first split doublet (311) in $CuK\alpha_1$ radiation. Therefore the experimental half-width $\theta_{exp}(\theta) = \Delta(2\theta_{exp})/(2 \times 2.35)$ of the reflection $(311)\alpha_1$ equals 0.0319° in angles θ. (The half-width measured in angles 2θ is $\theta_{exp}(2\theta) = \Delta(2\theta_{exp})/2.35$.) The coefficient $K(\theta_0)$ = 0.636 nm rad^{-1} (0.0111 nm degree^{-1}) for the reflection $(311)\alpha_1$ with diffraction angle θ_0 = 35°. Considering the $K(\theta_0)$, $a'_{B1}(y)|_{y=0.83}$, θ_{exp}, and θ_R values, the inhomogeneity Δy determined from broadening of the reflection $(311)\alpha_1$ equals 0.007, i.e. the degree of homogeneity of synthesized nonstoichiometric carbide $NbC_{0.83}$ is 0.993 and approaches unity. Estimates by other reflections give the same degree of homogeneity $(1 - \Delta y)$ = 0.993 for niobium carbide $NbC_{0.83}$. It should be noted that other methods fail to produce nonstoichiometric carbides MC_y with such low inhomogeneity. The quantity Δy was estimated at 0.015–0.025 or even higher for carbides produced by carbothermal reduction or self-propagating high-temperature synthesis.

Thus, the routes shown in Fig. 3.1 allow the production of high-purity homogeneous carbides TiC_y, ZrC_y, HfC_y, VC_y, NbC_y and TaC_y in the form of bulky samples or powders with the carbon concentration preset within ±0.005–0.007 or better. Powdered nonstoichiometric carbides have a narrow grain size distribution: the grain-size analysis shows that 80% of all particles are between 2 and 3 μm in size. Bulky samples of disordered nonstoichiometric carbides MC_y have grains 1–2 μm in size at $y \approx 1.0$ and up to 10–15 μm in size at $y \leq 0.70$.

In the X-ray experiment with $CuK\alpha_{1,2}$ radiation, homogeneity of a nonstoichiometric compound is confirmed by the splitting of diffraction reflections. Lines α_1 and α_2 are split when the distance $\Delta(2\theta(\alpha_1,\alpha_2))$ between doublets α_1 and α_2 is larger than the double half-width of the experimental function (Fig. 3.4), i.e. $\Delta(2\theta(\alpha_1,\alpha_2)) \geq 2\theta_{exp}(2\theta) = \Delta(2\theta_{exp})/1.175$. Hence,

$$\Delta(2\theta_{exp}) \leq 1.175\Delta(2\theta(\alpha_1,\alpha_2)). \tag{3.17}$$

Considering (3.15), $\theta_{exp} = \Delta(2\theta_{exp})/4.7$, and $\theta_R = \Delta(2\theta_R)/4.7$, inequality (3.17) can be presented as a condition of doublet splitting in the X-ray pattern in the form

$$[\Delta(2\theta_R)]^2 + \left[4.7 \times \frac{a'_{B1}(y)|_{y=y_0}}{K(\theta_0)} \Delta y\right]^2 \le [1.175\Delta(2\theta(\alpha_1,\alpha_2))]^2 \ . \tag{3.18}$$

Inequality (3.18) permits the assessment of the degree of homogeneity Δy of a compound MX_y from the distance $\Delta(2\theta(\alpha_1,\alpha_2))$ between doublets α_1 and α_2 of the first split reflection. Figure 3.4 shows the narrowing of the profiles for the lines α_1 and α_2 and the separation of the reflection $(311)_{B1}$ into doublets α_1 and α_2 as the degree of homogeneity of carbide $NbC_{0.83}$ increases (Δy decreases).

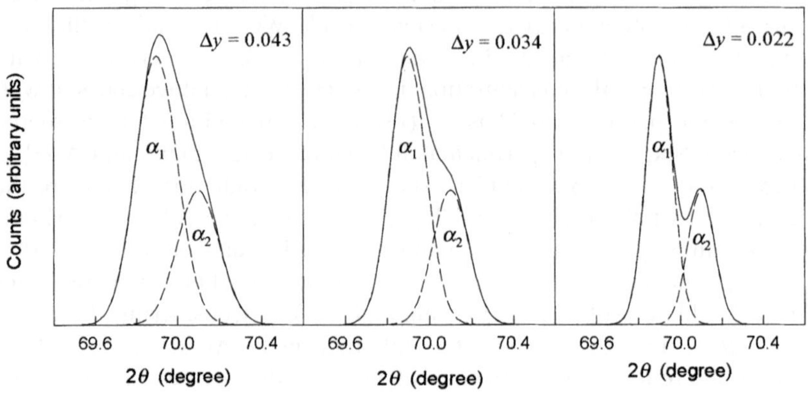

Fig. 3.4. Splitting of the diffraction reflection $(311)_{B1}$ into doublets α_1 and α_2 depending on the degree of homogeneity Δy of the nonstoichiometric carbide $NbC_{0.83}$ (a smaller value of Δy corresponds to a more homogeneous carbide)

The above analysis of X-ray diffraction reflections of nonstoichiometric compounds makes it possible to find their degree of homogeneity. The quantitative homogeneity index proposed in [4, 12] allows the adjustment and optimization of the synthesis conditions to produce strongly nonstoichiometric carbides with a high degree of homogeneity. The degree of homogeneity is useful in characterizing both disordered and ordered nonstoichiometric carbides since it allows one to distinguish between inhomogeneous and two-phase samples.

3.2 Crystal Structure and the Homogeneity Interval

Compounds of groups IV and V transition metals with carbon, nitrogen and oxygen have a similar simple structure and broad homogeneity intervals. Compounds of this group are referred to in the literature as interstitial phases, interstitial compounds or alloys, or compounds with variable composition [14–20]. The most appropriate name for them is strongly nonstoichiometric compounds of interstitial phase type or strongly nonstoichiometric compounds [1, 2, 21–23]. This name provides a comprehensive and true characteristic of fundamental features of the compounds under discussion.

The term "interstitial phase" was first used by Hagg [14] while discussing the structure of transition-metal carbides, nitrides, hydrides and borides. Hagg applied this term only to substances where "atoms H, B, C or N are located in a simple metallic lattice". Indeed, a characteristic structural feature of substances discussed by Hagg was the presence of a face-centered cubic (fcc) or hexagonal close-packed (hcp) lattice of the metal, while nonmetal atoms were located at the centers of octahedral interstitial sites of the metallic lattice. However, the symmetry of the metallic lattice in carbides and nitrides differs from the symmetry of the crystal lattice of transition metals, i.e. the metal crystal structure is altered upon formation of carbides and nitrides (thorium being an exception). Hcp transition metals of group IV (Ti, Zr, Hf) form carbides and nitrides with an fcc metallic sublattice. Transition metals with a body-centered cubic (bcc) structure (vanadium, niobium, tantalum, chrome, molybdenum, tungsten) form carbides and nitrides with an fcc or hcp metallic sublattice. Alteration of the metal crystal structure in carbides or nitrides suggests vigorous interactions between metal and nonmetal atoms. Therefore the term "interstitial phase" does not quite fit well these substances. The point is that only limited interstitial solid solutions are true interstitial phases.

According to [20–24], only carbides, nitrides and lower oxides of transition metals MX_y with the *B*1 structure, hexagonal carbides and nitrides M_2X_y with the *L*'3 (W_2C) structure, and some ternary compounds akin to these (carbosilicides $M_5Si_3C_x$, siliconitrides $M_5Si_3N_x$ and silicoborides $M_5Si_3B_x$ with the $D8_8$ (Mn_5Si_3) structure, and aluminidonitrides M_2AlN_x and $M_3Al_2N_x$ with structures of the Cr_2AlC and *A*13 (β-Mn) types respectively) are strongly nonstoichiometric interstitial compounds in the full sense of the word. All these compounds have considerable homogeneity intervals, and an element of their structure is a regular or distorted octahedron comprising six atoms of a transition metal with the interstitial atom or vacancy at the center of octahedron.

Transition-metal hydrides, borides and silicides should not be included in the group of strongly nonstoichiometric compounds. The point is that stability of nitrides and their maximum hydrogen concentration strongly depend on pressure and temperature. The type of chemical bonds in hydrides with one and the same hydrogen concentration can vary with external conditions. Clearly, the difference between transition-metal hydrides and solid solutions of hydrogen in transition metals is blurred. Borides and especially silicides have virtually no homogeneity intervals. Moreover, direct B–B and Si–Si bonds are significant in these compounds. Direct interactions between nonmetal atoms are negligible in nonstoichiometric carbides, nitrides and oxides.

The structure and properties of highly nonstoichiometric compounds have been the focus of hundreds of original research works and a large number of reviews and monographs [1–3, 14–23, 25–42]. Books on carbides and nitrides, particularly those published before the years 1975–1980 [15–17, 25–29] or setting out the ideas of the 1970s [36, 37], are obsolete and are only of historic interest. Nonetheless the literature provides ample evidence of nonstoichiometric compounds. Therefore this section will just touch upon the main features of their crystal structure.

Hagg [14] proposed empirical rules for constructing crystal structures of nonstoichiometric interstitial compounds. Exceptions to these rules are now known but their basic postulates are valid. In accordance with [14], nonstoichiometric interstitial compounds are formed if the atomic radii of the metal R_M and nonmetal R_X meet the condition $0.41 < R_X/R_M < 0.59$. When this condition is fulfilled, nonmetal atoms are located at the largest interstitial sites of the metallic lattice, which are a little smaller than the interstitial nonmetal atoms. The change of symmetry and a slight expansion of the metallic lattice in carbide, nitride or oxide ensure stability of the structure. Table 3.1 lists R_X/R_M values for carbon, nitrogen, oxygen and transition metals of groups IV–VI. If $R_X/R_M > 0.59$, compounds with a more complex structure are formed, which have no homogeneity intervals: for example, $R_C/R_{Cr} = 0.609$ and chromium carbides $Cr_{23}C_6$, Cr_7C_3, and Cr_3C_2 have no homogeneity intervals.

A characteristic feature of nonstoichiometric compounds is a seemingly independent metallic sublattice that serves as a matrix for atoms occupying its intersite free spaces and forming the nonmetallic sublattice. Vacant interstitial sites (structural vacancies □) are analogs of interstitial atoms to a first approximation. Structural vacancies and interstitial atoms form a substitutional solution in the nonmetallic sublattice.

Table 3.1. Atomic radii for transition metals of IV, V and VI groups, carbon, nitrogen and oxygen, and radius ratios, R_X/R_M , for non-metal X (X = C, N, O) and metal M

Element	[1] R (nm)	R_C/R_M	R_N/R_M	R_O/R_M
Ti	0.1467	0.526	0.504	0.450
Zr	0.1597	0.483	0.463	0.414
Hf	0.1585	0.486	0.467	0.417
Th	0.1796	0.430	0.412	0.367
V	0.1338	0.576	0.553	0.493
Nb	0.1456	0.530	0.508	0.453
Ta	0.1457	0.529	0.508	0.453
Cr	0.1267	0.609	0.584	0.521
Mo	0.1386	0.556	0.534	0.476
W	0.1394	0.553	0.531	0.473
U	0.1530	0.505	0.484	0.431
C	0.0772	–	–	–
N	0.0740	–	–	–
O	0.0660	–	–	–

[1] Atomic radii R are given for coordination number 12.

A high concentration of structural vacancies is the most significant property of strongly nonstoichiometric compounds. Under certain conditions the presence of structural vacancies in nonstoichiometric interstitial compounds such as carbides, nitrides and oxides may lead to ordering. The concentration and ordered or disordered distribution of structural vacancies strongly affect the properties of these compounds. References [21–23, 38–42] sum up a considerable body of experimental and theoretical data concerning the effect of the atomic and vacancy distribution on the structure and properties of nonstoichiometric compounds.

Strongly nonstoichiometric interstitial compounds and their solid solutions are the hardest and most high-melting materials of all known compounds. Moreover, they are superconductors with superconducting critical temperature of up to 18 K. They are also radiation resistant thanks to the high concentration of structural vacancies [43, 44]. Table 3.2 presents temperature of melting, microhardness and superconducting critical temperature of typical nonstoichiometric compounds.

Table 3.2. Maximal values of the temperature of melting T_m, microhardness H_V (at 300 K) and superconducting critical temperature T_c which are reached in the homogeneity intervals of strongly nonstoichiometric disordered carbides and nitrides MX_y

MX_y	Structure	$T_m \pm 50$ (K)	$H_V \pm 0.2$ (GPa)	$T_c \pm 0.1$ (K)
TiC_y	*B*1 (NaCl)	3340 ($TiC_{0.78}$) [29,33]	[1] 27.9 ($TiC_{0.96}$) [45] [2] 28.6 ($TiC_{0.97}$) [45] [3] 27.7 ($TiC_{0.98}$) [45]	< 0.1 [28]
ZrC_y	*B*1 (NaCl)	3710 ($ZrC_{0.82}$) [29,33]	[1] 25.6 ($ZrC_{0.96}$) [45] [3] 25.8 ($ZrC_{0.96}$) [45]	< 0.1 [28]
HfC_y	*B*1 (NaCl)	4200 ($HfC_{0.94}$) [29,33]	[1] 25.5 ($HfC_{0.98}$) [45] [3] 24.6 ($HfC_{0.98}$) [45]	< 1.7 [28]
VC_y	*B*1 (NaCl)	3020 ($VC_{0.88}$) [29,33]	[1] 31.0 ($VC_{0.88}$) [46]	< 0.1 [28]
NbC_y	*B*1 (NaCl)	3880 ($NbC_{0.80}$)[29,33]	[1] 33.0 ($NbC_{0.77}$) [46]	11.1 ($NbC_{1.00}$) [28]
TaC_y	*B*1 (NaCl)	4250 ($TaC_{0.88}$) [29,33]	[1] 28.9 ($TaC_{0.81}$) [46] [2] 29.0 ($TaC_{0.85}$) [46]	10.5 ($TaC_{1.00}$) [28]
TiN_y	*B*1 (NaCl)	3320 ($TiN_{1.00}$) [47]	23.0 ($TiN_{0.50}$) [47] 17.0 ($TiN_{1.00}$) [47]	5.8 ($TiN_{1.00}$) [47]
ZrN_y	*B*1 (NaCl)	3250 ($ZrN_{1.00}$) [47]	15.0 ($ZrN_{1.00}$) [47]	8.7 ($ZrN_{0.85}$) [47] 10.5 ($ZrN_{1.00}$) [47]
HfN_y	*B*1 (NaCl)	3600 ($HfN_{1.06}$) [47]	13.9 ($HfN_{0.94}$) [47] 15.9 ($HfN_{1.00}$) [47]	8.7 ($HfN_{0.85}$) [47] 6.9 ($HfN_{1.00}$) [47]
VN_y	*B*1 (NaCl)	2620 ($VN_{1.00}$) [47]	13.0 ($VN_{0.70}$) [47] 5.7 ($VN_{1.00}$) [47]	2.7 ($VN_{0.76}$) [47] 8.9 ($VN_{1.00}$) [47]
NbN_y	*B*1 (NaCl)	decomposed	13.0 ($NbN_{0.84}$) [47] 11.0 ($NbN_{1.00}$) [47]	13.8 ($NbN_{0.84}$) [47] 17.2 ($NbN_{1.00}$) [47]
TaN_y	CoSn	decomposed	32.0 ($TaN_{1.00}$) [47]	8.9 ($TaN_{0.94}$) [47]

[1, 2, 3] Microhardness for samples which are prepared by gas carburization of metal, hot pressing and arc melting, respectively.

These compounds are used in state-of-the-art technology for the production of heavy-duty tool materials serviceable at high temperatures, in aggressive environments, under high loads, etc. Strongly nonstoichiometric compounds are the subject of close study not only due to their practical significance but also as convenient model objects to retrieve information on the interrelation between crystal and electron structures, composition and properties.

Cellular Disorder in Nonstoichiometric Compounds. Generally, disorder always disturbs crystal regularity. However, crystal lattice symmetry is preserved in disordered interstitial solid solutions thanks to the probability of filling of any site $\boldsymbol{r}$ with an atom of one species or another.

In ideal solutions, which disregard the interatomic interaction, the substitution of atoms of one species by atoms of another species does not lead to distortion of the crystal lattice. This model describes an ideal substitutional disorder when probabilities of sites being filled with atoms of particular species coincide with their concentration.

In the case of real solid solutions, this model is realized only in the low-concentration region and only for some elements whose atoms have similar electronic structures and sizes. Replacement of an atom of species A by an atom of species B brings about considerable lattice distortions and static atomic displacements from ideal positions in most real solid solutions. The statistical disorder model overlooks atomic displacements and, therefore, one needs to introduce the notion of cellular disorder.

In solid-state physics, each atom is correlated to a Seitz cell (see Sect. 2.1). By introducing the notion of cellular disorder, i.e. picturing a crystal as a set of Seitz cells, it is possible to bypass the problem of lattice distortions in the solid solution and atomic displacements. In the first approximation, the cells can be taken to be similar, although atoms in these cells are located with some displacement $\boldsymbol{u}_n = \boldsymbol{R}_n - \boldsymbol{r}_n$ relative to the cell center ($\boldsymbol{R}_n$ being the vector defining the location of the n-th atom in a real lattice taking account of displacement, and $\boldsymbol{r}_n$ the location of the atom in an ideal lattice, i.e. at the cell center.) If atoms of unlike species are arranged irregularly in a substitutional solution, one can speak about the cellular disorder in this solution. In this case the probability that the site $\mathbf{r}_n$ is occupied with an atom of a given species can be written as

$$P(\boldsymbol{r}_n) = \int_V n(\boldsymbol{r} - \boldsymbol{r}_n)\delta(\boldsymbol{r} - \boldsymbol{R}_n)d\boldsymbol{r}\,, \tag{3.19}$$

where $n(\boldsymbol{r})$ is the atomic distribution function of the crystal and $\delta(\boldsymbol{r} - \boldsymbol{R}_n)$ is the Dirac delta-function ($\delta(\boldsymbol{r} - \boldsymbol{R}_n) = 0$ at $\boldsymbol{r} \neq \boldsymbol{R}_n$). The integration in (3.19) is taken over the volume V of a Seitz cell.

The notion of cellular disorder can be applied to nonstoichiometric interstitial compounds, because their nonmetallic sublattice contains a substitutional solid solution between interstitial atoms and structural vacancies.

Indeed, the number of positions in the nonmetallic sublattice of nonstoichiometric compounds is larger than that of interstitial atoms. Under certain conditions, interstitial atoms will be distributed randomly over available positions and will occupy sites of the nonmetallic sublattice with equal probability. This means that all sites of the nonmetallic sublattice are crystallographically equivalent in a disordered nonstoichiometric compound.

Most nonstoichiometric compounds with broad homogeneity intervals have a high-symmetry structure of $B1$ type, which is preserved upon disordered distribution of interstitial atoms and vacancies. A disordered distribution causes a disturbance of the symmetry in the local environment of each individual atom. Distortion of the local symmetry leads to static atomic displacements $\boldsymbol{u}_n = \boldsymbol{R}_n - \boldsymbol{r}_n$. As distinct from dynamic displacements, which are due to thermal atomic oscillations, static displacements are a consequence of an asymmetric angular distribution of bonds between adjacent atoms and the different energies of these bonds.

Cubic symmetry of disordered nonstoichiometric compounds with the $B1$-type structure permits only a spherically symmetric distribution of static atomic displacements, otherwise the crystallographic equivalence of various crystal sites would be disturbed. A symmetric distribution of static displacements in disordered nonstoichiometric compounds is provided by the random distribution of vacancies. At present static displacements have been determined for most transition-metal carbides and nitrides by X-ray and neutron diffraction methods. The results of research into static atomic displacements are summarized in [21, 38, 39] and will be comprehensively covered in Chap. 4.

Clearly, the notion of cellular disorder applies to disordered nonstoichiometric compounds. It should be remembered however that an ideal cellular disorder is rare because of correlations between atoms located at neighboring lattice sites.

The ordered state is the thermodynamically equilibrium one at 300 K for nonstoichiometric compounds with a high concentration of structural vacancies. This state is normally achieved thanks to long-term annealing or other special heat treatment procedures providing diffusion redistribution of atoms over the lattice sites. On the contrary, the state with cellular disorder is realized easily enough: for example, when compounds produced by high-temperature sintering are cooled from the synthesis temperature to 300 K at a rate of 25 K min^{-1} or higher.

Thus, it is disordered nonstoichiometric compounds that are produced by high-temperature synthesis. This fact aroused a widely spread erroneous belief that a disordered (statistical) distribution of interstitial atoms and vacancies is the main if not the only possible structural state of these compounds, while the ordered state is something of a rarity. Actually, the disordered state of nonstoichiometric compounds is in thermodynamic equilibrium only at a temperature above 1200–1500 K. At a temperature below 1000 K, only the ordered state is in thermodynamic equilibrium, while the disordered state is metastable. This raises the question as to the validity of numerous theoretical estimates of the electron energy spectrum of disordered nonstoichiometric compounds, which were made neglecting thermal excitation (that is, for a temperature of 0 K). Only the ordered state is in equilibrium at 0 K.

The Structure and Homogeneity Intervals. Table 3.3 presents data on the crystal structure and homogeneity intervals of disordered carbides, nitrides and oxides belonging to the group of strongly nonstoichiometric interstitial compounds. Values of N (the number of formula units in a unit cell) for binary nonstoichiometric compounds with different types of structure are given in Table 3.3 also.

Table 3.3. Homogeneity intervals and crystal structure of binary disordered strongly nonstoichiometric interstitial compounds

Stoichio metric compo sition	Homogeneity interval at 1700 K	Crystal structure					
		space group	structure type	N	compo sition	lattice constants (nm)	
						a	c
1	2	3	4	5	6	7	8
TiC	$TiC_{0.48}$–$TiC_{1.00}$	$Fm\bar{3}m$	$B1$ (NaCl)	4	$TiC_{0.48}$	0.4296	–
					$TiC_{1.00}$	0.4326	–
ZrC	$ZrC_{0.60}$–$ZrC_{0.98}$	$Fm\bar{3}m$	$B1$ (NaCl)	4	$ZrC_{0.63}$	0.4694	–
					$ZrC_{0.98}$	0.4698	–
HfC	$HfC_{0.56}$–$HfC_{1.00}$	$Fm\bar{3}m$	$B1$ (NaCl)	4	$HfC_{0.60}$	0.4602	–
					$HfC_{1.00}$	0.4641	–
ThC	$ThC_{0.72}$–$ThC_{1.00}$ (at 300 K)	$Fm\bar{3}m$	$B1$ (NaCl)	4	$ThC_{0.78}$	0.5317	–
					$ThC_{1.00}$	0.5342	–
VC	$VC_{0.65}$–$VC_{0.875}$	$Fm\bar{3}m$	$B1$ (NaCl)	4	$VC_{0.65}$	0.4128	–
					$VC_{0.87}$	0.4164	–
V_2C	$VC_{0.42}$–$VC_{0.50}$	$P6_3/mmc$	$L'3$ (W_2C)	2	$VC_{0.50}$	0.2902	0.4577
NbC	$NbC_{0.70}$–$NbC_{1.00}$	$Fm\bar{3}m$	$B1$ (NaCl)	4	$NbC_{0.70}$	0.4425	–
					$NbC_{1.00}$	0.4470	–
Nb_2C	$NbC_{0.35}$–$NbC_{0.50}$	$P6_3/mmc$	$L'3$ (W_2C)	2	$NbC_{0.50}$	0.3127	0.4972
TaC	$TaC_{0.71}$–$TaC_{1.00}$	$Fm\bar{3}m$	$B1$ (NaCl)	4	$TaC_{0.71}$	0.4406	–
					$TaC_{0.99}$	0.4456	–
Ta_2C	$TaC_{0.38}$–$TaC_{0.50}$	$P6_3/mmc$	$L'3$ (W_2C)	2	$TaC_{0.50}$	0.3106	0.4945
[1] γ-MoC	$MoC_{0.64}$–$MoC_{0.98}$	$P\bar{6}m2$	B_h (WC)	2	$MoC_{0.67}$	0.2898	0.2809
WC	$WC_{0.98}$–$WC_{1.02}$	$P\bar{6}m2$	B_h (WC)	2	$WC_{1.00}$	0.2906	0.2837
[2] W_2C	$WC_{0.36}$–$WC_{0.52}$	$P6_3/mmc$	$L'3$ (W_2C)	2	$WC_{0.45}$	0.2992	0.4722
UC	Very narrow	$Fm\bar{3}m$	$B1$ (NaCl)	4	$UC_{1.00}$	0.4961	–
[3] TiN	$TiN_{0.38}$–$TiN_{1.00}$	$Fm\bar{3}m$	$B1$ (NaCl)	4	$TiN_{0.38}$	0.4210	–
					$TiN_{1.00}$	0.4244	–
ZrN	$ZrN_{0.55}$–$ZrN_{1.00}$	$Fm\bar{3}m$	$B1$ (NaCl)	4	$ZrN_{0.55}$	0.4565	–
					$ZrN_{0.99}$	0.4566	–
HfN	$HfN_{0.74}$–$HfN_{1.00}$	$Fm\bar{3}m$	$B1$ (NaCl)	4	$HfN_{0.74}$	0.4518	–
					$HfN_{1.00}$	0.4524	–
ThN	Very narrow	$Fm\bar{3}m$	$B1$ (NaCl)	4	$ThN_{1.00}$	0.5155	–

(*Continued*)

Table 3.3. (continued)

1	2	3	4	5	6	7	8
VN	$VN_{0.72}$–$VN_{1.00}$	$Fm\bar{3}m$	*B*1 (NaCl)	4	$VN_{0.72}$	0.4060	–
					$VN_{1.00}$	0.4134	–
V_2N	$VN_{0.48}$–$VN_{0.50}$	*P*312	ε-Fe_2N ?		$VN_{0.50}$	0.2839	0.4560
[4] δ-NbN	$NbN_{0.88}$–$NbN_{0.98}$	$Fm\bar{3}m$	*B*1 (NaCl)	4	$NbN_{0.88}$	0.4383	–
					$NbN_{0.98}$	0.4391	–
Nb_2N	$NbN_{0.40}$–$NbN_{0.50}$	$P6_3/mmc$	*L*'3 (W_2C)	2	$NbN_{0.50}$	0.3056	0.4957
[5] TaN	$TaN_{0.98}$–$TaN_{1.00}$	*P*6/*mmm*	*B*35 (CoSn)	3	$TaN_{1.00}$	0.5191	0.2910
Ta_2N	$TaN_{0.30}$–$TaN_{0.50}$	$P6_3/mmc$	*L*'3 (W_2C)	2	$TaN_{0.41}$	0.3048	0.4915
CrN	[6] Narrow	$Fm\bar{3}m$	*B*1 (NaCl)	4	CrN	0.4149	–
Cr_2N	$CrN_{0.42}$–$CrN_{0.50}$	–	hexagonal		$CrN_{0.50}$	0.4759	0.4438
MoN	[6] Narrow	–	hexagonal		MoN	0.5725	0.5608
Mo_2N	$MoN_{0.42}$–$MoN_{0.50}$	$Fm\bar{3}m$	*B*1 (NaCl)	4	$MoN_{0.50}$	0.4165	–
WN	[6] Narrow	$P\bar{6}m2$	B_h (WC)	2	WN	0.2893	0.2826
W_2N	[6] Narrow	$Fm\bar{3}m$	*B*1 (NaCl)	4	$WN_{0.50}$	0.4126	–
UN	Very narrow	$Fm\bar{3}m$	*B*1 (NaCl)	4	$UN_{1.00}$	0.4889	–
[7] TiO	$TiO_{0.80}$–$TiO_{1.25}$	$Fm\bar{3}m$	*B*1 (NaCl)	4	$TiO_{0.80}$	0.4192	–
					$TiO_{1.25}$	0.4169	–
[8] VO	$VO_{0.85}$–$VO_{1.23}$	$Fm\bar{3}m$	*B*1 (NaCl)	4	$VO_{0.85}$	0.4024	–
					$VO_{1.23}$	0.4130	–
[9] NbO	[6] Narrow	$Fm\bar{3}m$	*B*1 (NaCl)	4	NbO	0.440	–
TaO	[6] Narrow	$Fm\bar{3}m$	*B*1 (NaCl)	4	TaO	0.439	–

[1] Molybdenum carbide is assumed to have the composition Mo_3C_2 . There are several modifications of MoC_y carbide with different sequences of packing of metal atom layers.
[2] A cubic modification of W_2C carbide exists at a temperature above 2800 K.
[3] Cubic nitride $Ti_{0.96}N$ with vacancies in metallic sublattice exists in the fine-grained state.
[4] There are several related modification of NbN nitride including hexagonal ones.
[5] Series of polytypes with a composition-dependent crystal structure.
[6] The boundaries of the homogeneity interval are unknown.
[7] Cubic "stoichiometric" titanium monoxide TiO contains about 16.7 at.% of vacancies in titanium and oxygen sublattices simultaneously and has a real composition $Ti_{0.833}O_{0.833}$.
[8] Cubic vanadium monoxide VO contains vacancies in metal and oxygen sublattices simultaneously.
[9] It is proposed that niobium monoxide NbO has no homogeneity interval and is an individual phase with primitive cubic lattice in which every fourth niobium atom and every fourth oxygen atom are absent.

The structure of nonstoichiometric compounds can be analyzed by a variety of diffraction methods capable of identifying the lattice symmetry. However, the concentration of structural vacancies in the nonmetallic and metallic sublattices of a nonstoichiometric compound need be known for detailed characterization of its structure. Vacancies may occur either in one of the sublattices or in both at a time.

Most nonstoichiometric carbides and nitrides contain vacancies in the nonmetallic sublattice only. The presence of structural vacancies in the nonmetallic and metallic sublattices is typical of cubic titanium and vanadium oxides, oxycarbides and oxynitrides. Superstoichiometric nitrides formed in thin films and ultrafine powders contain structural vacancies in the metallic sublattice.

In order to determine the vacancy concentration or the degree of occupancy of the sublattice, one needs to know not only the chemical composition of nonstoichiometric compound MX_y but its density as well. Consider a compound $A_{p_1}B_{p_2}...S_{p_j}$ with the density d, the unit cell volume V, and the mass fraction of the j-th element equal to c_j. If unit cell contains N formula units of the compound, the mass of the j-th element in one formula unit of the compound is $m_j = Vc_jd/N$. If atoms of the j-th element occupy all sites of their sublattice, the mass of that element in one formula unit is $p_jA_jm_u$, where p_j is the number of atoms of the j-th element in one stoichiometric formula unit (p_j is integer always and $p_j \equiv 1$ for the compounds MX_y), A_j is the atomic mass of the j-th element and $m_u = 1.66 \times 10^{-27}$ kg is the atomic mass constant. The degree to which the j-th element fills sites of its sublattice can be presented as the ratio of m_j to $p_jA_jm_u$, i. e.

$$n_j = Vc_jd/(Np_jA_jm_u) \,. \tag{3.20}$$

The true density d is determined experimentally by the most reliable pycnometric method. Precision measurements of the density of fine powders eliminate the effect of micropores, voids and cracks on the density d. The relative concentration of vacancies in the j-th sublattice is $c_{\square j} = 1 - n_j$.

Transition metals of groups IV and V form carbides with an fcc or hcp metallic sublattice having the highest melting point. The stoichiometric carbide $MC_{1.0}$ normally corresponds to the upper boundary of the homogeneity interval of cubic monocarbides.

Group IV metals (titanium, zirconium, hafnium) form only monocarbides TiC_y, ZrC_y, and HfC_y with the $B1$ (NaCl) structure. These monocarbides have the widest homogeneity intervals. For example, titanium carbide has a homogeneity interval from $TiC_{1.00}$ to $TiC_{0.48}$, i.e. the carbon sublattice contains over 50 at.% of structural vacancies at the lower boundary of the homogeneity interval.

Thorium carbide ThC_y is of special significance among cubic monocarbides $M^{(IV)}C$. Metallic α-Th has an fcc lattice and the metal structure remains unchanged when carbide ThC_y with the $B1$ structure is formed. Octahedral interstitial sites of the thorium sublattice are so large that they can host two carbon atoms each. In addition to the carbide ThC_y, the Th–C system includes monoclinic dicarbide α-ThC_2 at a temperature from 300 to 1500 K, tetragonal dicarbide β-ThC_2 in the temperature interval from 1500 to 1700 K, and cubic dicarbide γ-ThC_2 with a KCN-type structure at $T > 1700$ K. Cubic thorium monocarbide ThC_y has a very broad homogeneity interval from $ThC_{0.01}$ to $ThC_{1.00}$ at a temperature from 1400 to 1700 K. The lower and upper boundaries of the homogeneity interval of ThC_y shift towards larger carbon concentrations at higher temperatures. For example, cubic carbide ThC_y has a homogeneity interval from $ThC_{0.25}$ to $ThC_{1.86}$ at 2200 K.

Group V transition metals (vanadium, niobium, tantalum) form, along with the cubic carbides VC_y, NbC_y, and TaC_y, lower carbides V_2C_y, Nb_2C_y, and Ta_2C_y with an hcp structure of $L'3$ (W_2C) type. Homogeneity intervals of monocarbides of group V transition metals are much narrower than those of monocarbides of group IV metals, but are still broad enough. Stoichiometric cubic vanadium carbide does not exist under normal conditions. The upper boundary of the homogeneity interval of this compound is the carbide $VC_{0.87}$ containing 13 at.% structural vacancies in the carbon sublattice. Lower carbides M_2C_y have narrow homogeneity intervals.

Carbides of group VI metals, except MoC and W_2C, have no homogeneity intervals. Structural data for molybdenum carbides are conflicting. Five structures of carbide MoC are known. A particular modification of carbide MoC depends on its synthesis method and heat treatment conditions. For example, fcc carbide MoC can be produced at a temperature of up to 2800 K and pressure from 4 to 7 GPa [48] or in a thin film by combined evaporation of molybdenum and carbon at 1500 K [49]. The instability of molybdenum carbide is confirmed by the fact that its structure can be easily changed by the introduction of alloying additions. For example, small additions of NbC in MoC stabilize the $B1$-type structure, while WC stabilizes the simple hexagonal structure of WC type. The high-temperature modification β-Mo_2C has a disordered structure of $L'3$ (W_2C) type. Cubic uranium monocarbide has the stoichiometric composition $UC_{1.0}$ in the temperature interval from 300 to 1700 K and possesses no homogeneity interval. Octahedral interstitial sites of the uranium sublattice may host two carbon atoms each at $T > 1700$ K. Uranium carbide has the homogeneity interval $UC_{1.00}$–$UC_{1.10}$ at 2100 K.

Transition-metal nitrides are close to carbides in structure. Titanium, zirconium and hafnium nitrides MN_y have $B1$-type structures and broad homogeneity intervals. Thorium nitride ThN_y with a $B1$-type structure has a very narrow homogeneity interval near the stoichiometric composition $ThN_{1.00}$ at $T < 1700$ K. The homogeneity interval of ThN_y widens with increasing temperature to cover the range $ThN_{0.87}$–$ThN_{1.07}$ at 2300 K. Group V transition metals form lower hexagonal nitrides M_2N_y in addition to cubic mononitrides MN_y. Tantalum nitride TaN_y has a CoSn-type hexagonal structure below 1800 K. A cubic phase TaN_y ($y \approx 1.0$) with a $B1$-type structure is observed in bulk samples only at $T > 1800$ K and can occur in thin films at room temperature.

Nitrides CrN, Mo_2N, and W_2N have a $B1$-type structure, while Cr_2N, MoN, and WN have a hexagonal structure. Group VI transition metals are unstable and decompose at low temperatures. For example, cubic chromium nitride CrN, which has virtually no homogeneity interval, transforms to a tetragonal antiferromagnetic compound below 280 K. Group VI metal nitrides all have very narrow homogeneity intervals whose exact boundaries are unknown. For example, cubic nitride UN_y with the $B1$ structure has a very narrow homogeneity interval near the stoichiometric composition at a temperature from 1100 to 2500 K. At $T < 1000$ K the homogeneity interval is so small that it cannot be identified.

A characteristic representative of nonstoichiometric monoxides is TiO with the *B*1-type structure. This compound can deviate significantly from the stoichiometric composition with the formation of structural vacancies in oxygen and titanium sublattices. The question of whether or not zirconium and hafnium monoxides exist has yet to be answered conclusively. Vanadium monoxide VO has, like TiO, a structure of *B*1 type and contains a large number of vacancies in both sublattices. Cubic monoxides NbO and TaO have narrow homogeneity intervals and are formed upon oxidation of thin metallic films. These monoxides are observed in bulky samples only in the presence of higher oxides.

Thus, nonstoichiometric carbides, nitrides and oxides of transition metals generally have cubic or hexagonal structures. These structures can be visualized as successively alternating layers of atoms of unlike species. For example, in monocarbides and mononitrides having a *B*1-type structure planes containing sites of the metallic or nonmetallic sublattice only alternate perpendicular to the direction $[111]_{B1}$ (or perpendicular to equivalent directions). These planes are spaced by $\sqrt{3}a/6$, while the distance between nearest atoms in a single plane equals $\sqrt{2}a/2$ (a is the crystal lattice constant). The alternation of metallic atomic planes provides a sequence of ABCABC... type, i.e. the closest cubic packing. Layers formed by the nonmetallic sublattice sites alternate in the same sequence XYZXYZ... The general alternation sequence of atomic layers in the direction $[111]_{B1}$ in a *B*1-type cubic structure has the form AXBYCZ AXBYCZ....

Nonmetal atoms statistically fill octahedral interstitial sites of the metallic sublattice in nonstoichiometric compounds MX_y with a *B*1-type structure (Fig. 3.5).

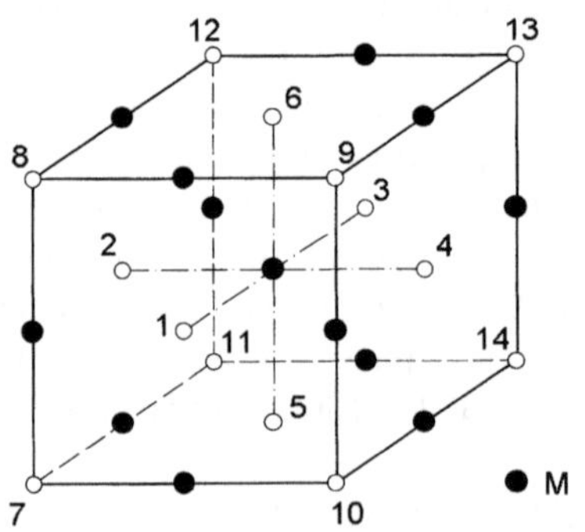

Fig. 3.5. Structure type *B*1 (NaCl) and the nearest environment of the metal atom M (•). Sites of the nonmetallic sublattice are numbered. The first coordination sphere of the metal atom is formed by six sites (1–6) of the nonmetallic sublattice, which are located at the centers of the faces. The third coordination sphere is formed by eight sites (from the 7–14) of the nonmetallic sublattice, which are located at the vertices. The second coordination sphere of the metal atom comprises twelve metal atoms located in the middle of the edges

Any site of the nonmetallic sublattice has an octahedral environment of six nearest sites of the metallic sublattice, whereas any metal atom has an octahedral environment of six nearest sites of the nonmetallic sublattice. Table 3.4 gives the distribution of metallic and nonmetallic sublattice sites over coordination spheres for a $B1$-type structure.

Table 3.4. $B1$ (NaCl) type crystal structure: Distribution in the coordination spheres (CS) of metallic (M) and nonmetallic (X) sublattices sites and coordination number z_j (the nonmetallic sublattice site is the center of the coordination spheres)

CS	[1] Coordinates of the site which belongs to the j-th CS				Sublattice sites	Number of metallic or nonmetallic CS		z_j	[2] Relative radius of the j-th CS
J	hkl	x/a	y/a	z/a		X	M		R_j/a
0	000	0	0	0	X	0	–	1	0
1	100	1/2	0	0	M	–	1		1/2
2	110	1/2	1/2	0	X	1	–		√2/2
3	111	1/2	1/2	1/2	M	–	2		√3/2
4	200	1	0	0	X	2	–		1
5	210	1	1/2	0	M	–	3		√5/2
6	211	1	1/2	1/2	X	3	–		√6/2
7	220	1	1	0	X	4	–		√2
8	221	1	1	1/2	M	–	4	30	3/2
	300	3/2	0	0					
9	310	3/2	1/2	0	X	5	–	24	√10/2
10	311	3/2	1/2	1/2	M	–	5	24	√11/2
11	222	1	1	1	X	6	–	8	√3
12	320	3/2	1	0	M	–	6	24	√13/2
13	321	3/2	1	1/2	X	7	–	48	√14/2
14	400	2	0	0	X	8	–	6	2
15	322	3/2	1	1	M	–	7	48	√17/2
	410	2	1/2	0					
16	330	3/2	3/2	0	X	9	–	36	√18/2
	411	2	1/2	1/2					
17	331	3/2	3/2	1/2	M	–	8	24	√19/2
18	420	2	1	0	X	10	–	24	√5
19	421	2	1	1/2	M	–	9	48	√21/2
20	332	3/2	3/2	1	X	11	–	24	√22/2
21	422	2	1	1	X	12	–	24	√6

[1] The coordinates of the rest of the coordination sphere sites can be obtained with the aid of symmetry operations of the point group $m\bar{3}m$ (O_h).

[2] The radius of j-th coordination sphere is equal to $R_j = (a_{B1})\sqrt{h^2 + k^2 + l^2}$, where a_{B1} is a basic $B1$-type lattice constant.

A site of the nonmetallic sublattice is located at the center of the spheres. The *B*1-type structure is invariant with respect to mutual replacement of sites of one sublattice by those of the other. Therefore it suffices to replace M by X and X by M in Table 3.4 to view the sequence of coordination spheres constructed about a site in the metallic sublattice.

The majority of lower carbides and nitrides M_2X_y have a structure of *L*'3 type (Fig. 3.6). Nonmetal atoms in this structure statistically fill half of all octahedral interstitial sites of the metallic sublattice. Layers of metal atoms alternate with layers of the nonmetallic sublattice sites in the direction of the *c*-axis following the general sequence AXBXAXBX...

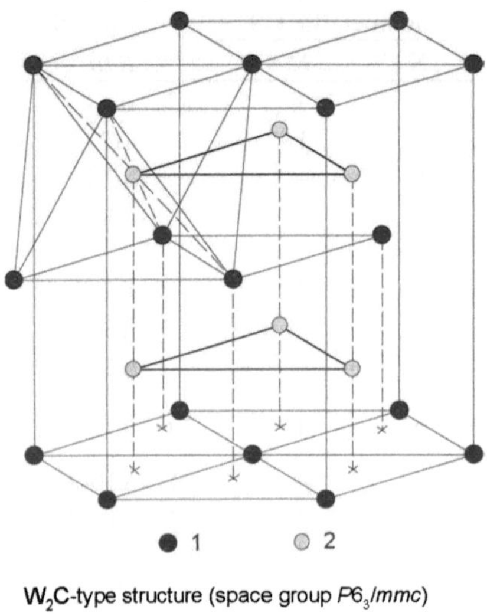

Fig. 3.6. Hexagonal close-packed structure type *L*'3 (W_2C): (1) metal atoms; (2) nonmetallic sublattice sites, one-half of which are statistically occupied with nonmetal atoms

There are many ternary nonstoichiometric compounds in addition to nonstoichiometric binary interstitial compounds. They are formed in ternary systems M–Si–B, M–Si–C, M–Si–N, M–Al–C, and M–Al–N (M is a transition metal). A detailed description of ternary nonstoichiometric interstitial compounds and ternary interstitial compounds having homogeneity intervals but not possessing stoichiometry can be found in [24]. Structural characteristics of ternary compounds entering the group of strongly nonstoichiometric interstitial compounds are given in Table 3.5. Phase equilibria in ternary transition-metal systems M–X–X' and M–Al–X (X = B, C, N, Si) have received much less study than those in binary systems M–X. Therefore some ternary nonstoichiometric compounds may not be found in Table 3.5.

Table 3.5. Ternary disordered strongly nonstoichiometric interstitial compounds [24]

Compound	Symmetry and structural type	Space group	[1] N	Lattice constants (nm)	
				a	c
[2] $Zr_5Si_3(B)$	Hexagonal, type $D8_8$ (Mn_5Si_3)	$P6_3/mcm$	2	0.7788	0.5558
$Ti_5Si_3C_y$	Hexagonal, type $D8_8$ (Mn_5Si_3)	$P6_3/mcm$	2	–	–
$Zr_5Si_3C_y$	Hexagonal, type $D8_8$ (Mn_5Si_3)	$P6_3/mcm$	2	0.7911	0.5569
$V_5Si_3C_y$	Hexagonal, type $D8_8$ (Mn_5Si_3)	$P6_3/mcm$	2	0.7135	0.4842
$Nb_5Si_3C_y$	Hexagonal, type $D8_8$ (Mn_5Si_3)	$P6_3/mcm$	2	0.7536	0.5249
$Ta_5Si_3C_y$	Hexagonal, type $D8_8$ (Mn_5Si_3)	$P6_3/mcm$	2	0.7494	0.5242
$Cr_5Si_3C_y$	Hexagonal, type $D8_8$ (Mn_5Si_3)	$P6_3/mcm$	2	0.6982	0.4731
$Mo_5Si_3C_y$	Hexagonal, type $D8_8$ (Mn_5Si_3)	$P6_3/mcm$	2	0.7292	0.5043
$Mn_5Si_3C_y$	Hexagonal, type $D8_8$ (Mn_5Si_3)	$P6_3/mcm$	2	0.6915	0.4823
$Ti_5Si_3N_y$ $0 \le y \le 1.0$	Hexagonal, solid solution of N in Ti_5Si_3 , type $D8_8$ (Mn_5Si_3)	$P6_3/mcm$	2	–	–
$Zr_5Si_3N_y$ $0.1 \le y \le 0.3$	Hexagonal, type $D8_8$ (Mn_5Si_3)	$P6_3/mcm$	2	0.7949	0.5573
$Hf_5Si_3N_y$ $y \le 0.1$	Hexagonal, type $D8_8$ (Mn_5Si_3)	$P6_3/mcm$	2	0.7801	0.5523
$V_5Si_3N_y$ $0.2 \le y \le 1.0$	Hexagonal, type $D8_8$ (Mn_5Si_3)	$P6_3/mcm$	2	0.7121	0.4847
$Nb_5Si_3N_y$ $y \le 0.3$	Hexagonal, type $D8_8$ (Mn_5Si_3)	$P6_3/mcm$	2	0.7517	0.5249– –0.5310
$Ta_5Si_3N_y$ $0.3 \le y \le 1.2$	Hexagonal, type $D8_8$ (Mn_5Si_3)	$P6_3/mcm$	2	0.7501	0.5229– –0.5246
Ti_3AlC_y	Cubic, type $E2_1$ ($CaTiO_3$)	$Pm\bar{3}m$	1	0.4156	–
Mn_3AlC_y	Cubic, type $E2_1$ ($CaTiO_3$)	$Pm\bar{3}m$	1	0.3869	–
Fe_3AlC_y	Cubic, type $E2_1$ ($CaTiO_3$)	$Pm3m$	1	0.3728– –0.3788	–
Co_3AlC_y	Cubic, type $E2_1$ ($CaTiO_3$)	$Pm\bar{3}m$	1	0.3697	–
Ni_3AlC_y	Cubic, solid solution of C in Ni_3Al, type $L1_2$ (Cu_3Au)	$Pm3m$	1	0.3583– –0.3617	–
$Zr_2Al_3C_{5-y}$	Hexagonal	$P31c$	2	0.3345	0.7734
$ZrAlC_{2-y}$	Hexagonal	$P6_3/mmc$	6	0.3345	2.7610
$ZrAlC_{2-y}$	Hexagonal, type $D8_8$ (Mn_5Si_3)	$P6_3/mcm$	2	0.7734	0.5245
Ti_2AlN_y	Hexagonal, type Cr_2AlC	$P6_3/mmc$	6	0.2991– –0.3040	1.361– –1.369
$Zr_5Al_3N_y$ $0.1 \le y \le 0.9$	Hexagonal, type $D8_8$ (Mn_5Si_3)	$P6_3/mcm$	2	0.8170– –0.8240	0.5655– –0.5694
$Hf_5Al_3N_y$	Hexagonal, type $D8_8$ (Mn_5Si_3)	$P6_3/mcm$	2	0.8062	0.5603
$Ta_5Al_3N_y$	Tetragonal, type $D8_\ell$ (Cr_5B_3)	$I4/mcm$	4	–	–
Nb_3Al_2N	Cubic, type $A13$ (β-Mn)	$P4_132$	4	~0.7031	–

[1] N is the number of formula units per unit cell.

[2] The ternary compound $Zr_5Si_3(B)$ may contain from 5 to 8 at.% of boron.

The largest number of nonstoichiometric ternary interstitial compounds was found in the systems M–Si–C and M–Si–N. The hexagonal structure of $D8_8$ type (Mn_5Si_3) is the most frequent structure of nonstoichiometric ternary compounds.

Strongly nonstoichiometric binary and ternary interstitial compounds belong to the class of octahedral structures. The main structural element of nonstoichiometric binary and ternary compounds is an ideal or distorted octahedron comprising six atoms of transition metal M. The octahedral interstitial site may either be occupied by an interstitial nonmetal atom B, C, N or be vacant. The possibility of octahedral interstitial sites being completely or partially filled determines narrow or broad homogeneity intervals.

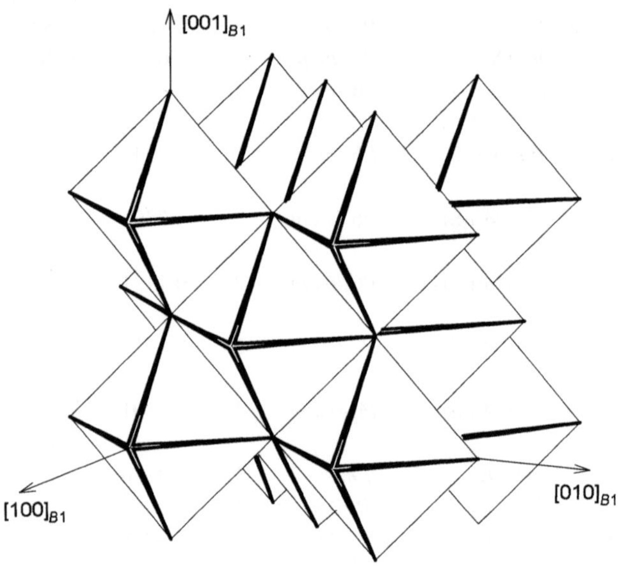

Fig. 3.7. Structure type *B*1 (NaCl) of nonstoichiometric compounds MX_y pictured as a set of octahedra XM_6

Octahedrons $\square M_6$ or XM_6 in cubic carbides and nitrides MX_y with a *B*1-type structure are linked with all twelve edges (Fig. 3.7) to form the closest packing. In hexagonal lower carbides and nitrides M_2X_y with an *L*'3-type structure octahedral groups of metal atoms are linked by opposite faces in the direction of the *c*-axis and share edges in planes perpendicular to the *c*-axis (Fig. 3.6).

Structures of ternary nonstoichiometric compounds differ in the method by which vacant $\square M_6$ or filled XM_6 octahedrons are bound together. Octahedrons XM_6 are interconnected by opposite faces in hexagonal compounds $M_5Si_3X_y$ and $M_5Al_3X_y$ having a $D8_8$ (Mn_5Si_3)-type structure (Fig. 3.8).

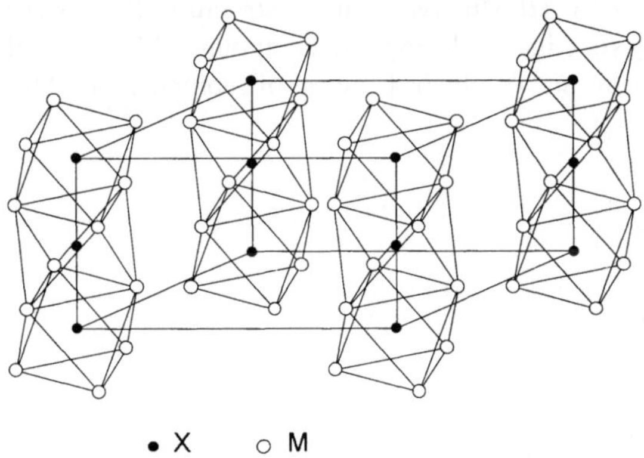

M_5Si_3X (type Mn_5Si_3, space group $P6_3/mcm$)

Fig. 3.8. Crystal structure type $D8_8$ (Mn_5Si_3) of ternary hexagonal compounds $M_5Si_3X_y$ and $M_5Al_3X_y$ (space group $P6_3/mcm$). The figure shows only atoms of a transition metal M at sites 6(*g*) and nonmetal interstitial atoms X at sites 2(*b*) that form octahedra XM_6. Octahedra XM_6 are interconnected by two edges and form columns going along the [001] axis

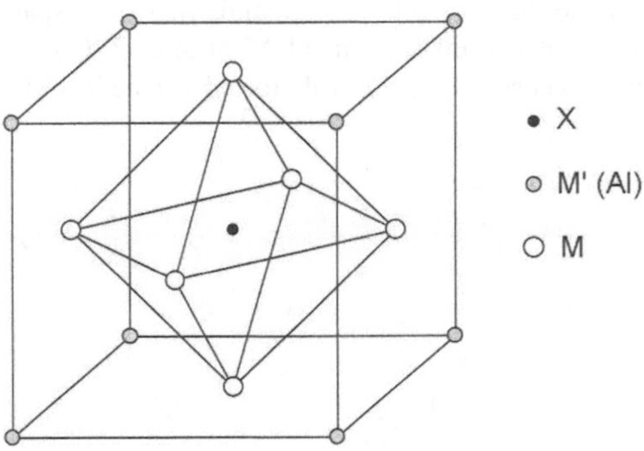

$M_3M'X$ (type $CaTiO_3$, space group $Pm\bar{3}m$)

Fig. 3.9. Unit cell of ternary compounds M_3AlX_y with a cubic (space group $Pm\bar{3}m$) structure type $E2_1$ ($CaTiO_3$ perovskite type)

Octahedra are interconnected by all six vertices in compounds M_3AlX_y with an $E2_1$-type cubic structure ($CaTiO_3$ perovskite type) (Fig. 3.9).

Distorted octahedra XM_6 are also bound together by all six vertices in compounds $M_3Al_2X_y$ with an *A*13 (β-Mn)-type cubic structure (Fig. 3.10). Octahedra XM_6 are connected by six edges in hexagonal compounds M_2AlX_y with a Cr_2AlC-type structure to form an octahedral layer of composition M_2X (Fig. 3.11).

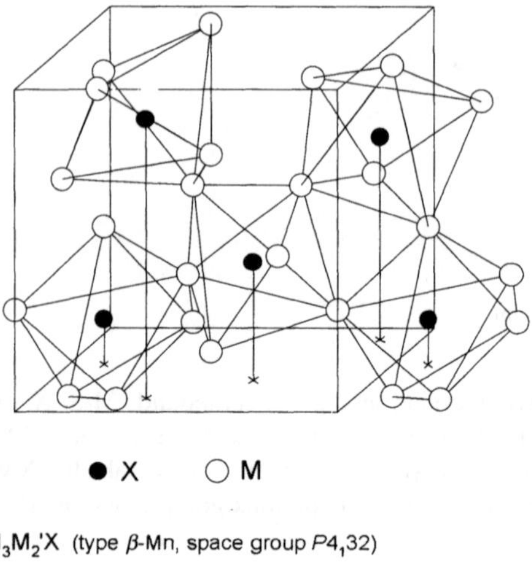

Fig. 3.10. Crystal structure type *A*13 (β-Mn) of ternary cubic compounds $M_3Al_2X_y$ (space group $P4_132$). The figure shows only atoms of a transition metal M at sites 12(*d*) and nonmetal interstitial atoms X at sites 4(*a*). These atoms form distorted octahedra XM_6 interconnected by all the vertices

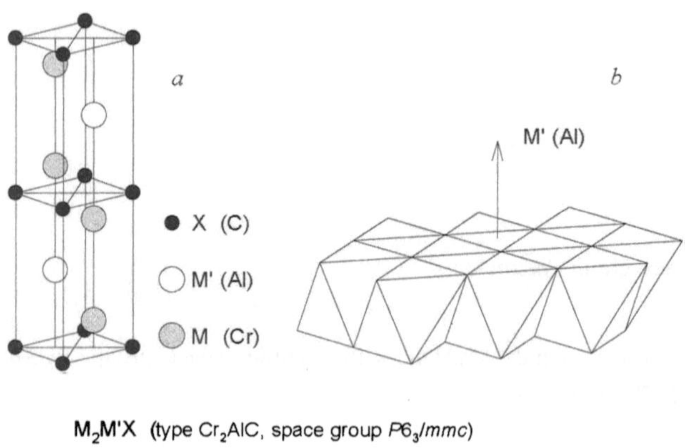

Fig. 3.11. Crystal structure type Cr_2AlC of ternary hexagonal compounds $M_2M'X_y$ (space group $P6_3/mmc$): (*a*) unit cell; (*b*) layer of octahedra XM_6 interconnected by six edges. Atoms of nontransition metal M' (Al) are located over tetrahedral voids

The structure of nonstoichiometric ternary interstitial compounds is looser than that of nonstoichiometric binary compounds. Indeed, the volume V/n per atom (V being the volume of a unit cell and n the number of atoms in the unit cell) in ternary compounds is larger than its counterpart in binary compounds. Atoms of the third element complicate and loosen the crystal structure of a ternary compound as compared with an analogous binary compound of a transition metal.

Nonstoichiometric binary and ternary compounds can be in a disordered or ordered state. The phenomenon of ordering is well understood for nonstoichiometric binary interstitial compounds [21–23, 40–42], while nothing is known so far about ordering in ternary nonstoichiometric compounds.

Chemical Bonds in Nonstoichiometric Compounds. Band and cluster methods based on a single electron approximation are currently used for calculating the structure of a solid.

Methods for estimating the energy bands focus on specific features of the energy spectrum obtained taking account of translational symmetry. A solid is viewed as an infinite regular crystal. Band states are largely delocalized and describe the motion of an electron distributed over all unit cells of the infinite crystal. Therefore band methods are reasonably efficient in describing properties determined by delocalized electrons. On the other hand, certain difficulties may be involved with band methods describing local properties. Thus, the band approach has the drawback that any deviation from the ideal crystal model or disturbance of the translational periodicity, i.e. the presence of impurities, defects, interfaces etc., largely complicates calculations of energy bands. However, crystals with some disturbance of translational symmetry can be adequately described using the band approach and Green's functions.

A single-electron state in a solid can be modeled by the cluster approximation. In this case a single-electron orbital is resolved with respect to the basis of atomic orbitals localized on atoms of a cluster. Cluster methods describe the electronic structure as a set of discrete energy levels of a relatively small group of atoms selected as a cluster in the crystal. These methods are simple enough and are applicable to compounds of various compositions with different crystal structures. Cluster methods are used for estimating electronic states in disordered solids containing defects of various types. A drawback of cluster methods is that crystals with wide bands are difficult to describe by these methods. Moreover, the application of cluster methods has been a problem since one needs to determine boundary conditions or take into account the cluster environment in the crystal lattice.

In accordance with modern ideas, a combined covalent-metallic-ionic type of chemical bond is found in nonstoichiometric interstitial compounds MX_y [18, 22, 30, 34, 35]. This is in line with specific feature of nonstoichiometric compounds such as the combination of main parameters of metals (a simple structure and high thermal and electric conductivity decreasing with temperature) and covalent compounds (high hardness and low plasticity). The band structure of transition-metal carbides and nitrides was comprehensively reviewed in [50].

Researchers [30, 34, 35] have analyzed the results of numerous calculations performed by band and cluster methods and have postulated regularities of the structure of the valence band in nonstoichiometric compounds MX_y, with a *B*1-type structure. In accordance with [30, 34, 35], the valence band includes three bands: a low-energy 2*s*(X) band containing small contributions from *s*-, *p*- and *d*-states of the metal; the main valence-binding band formed by strong mixing of 2*p*(X) and *d*(M) wave functions; and a partially filled high-energy conduction band formed mostly by *d*(M) functions with an admixture of 2*p*(X), *p*(M), and *s*(M) functions. In the series from carbides to nitrides and oxides, the low-energy 2*s*(X) band, the main hybridized 2*p*(X)–*d*(M) band, and the delocalized high-energy *d*,*s*(M) band become narrower and shift to the region of lower energies. The redistribution of some atomic states in nonstoichiometric compounds causes a partial charge transfer between metal and nonmetal atoms. This accounts for the ionic component of chemical bonds. The results of X-ray emission and photoelectron spectroscopy and calculation data suggest the electron transfer from metal to nonmetal. The transferred charge increases a little in the carbide–nitride–oxide series to produce a larger ionic component of chemical bonds.

Researchers in [51, 52] were the first to report considerable changes in electron and energy spectra of carbides and the emergence of local peaks in those spectra caused by the formation of vacancies in the carbon sublattice. The appearance of vacancies in the nonmetallic sublattice of the compounds at hand leads to an increase in the width and occupancy of the *d*-band of the metal, narrowing of the 2*p*(X) band, and diminishing of the atomic charge. An additional vacancy peak of the density of electronic states arises simultaneously in the conduction band below the Fermi level. Generally, these changes in the electron energy spectrum can be interpreted as the increase in the metal–metal interaction combined with diminishing of covalent and ionic components of metal–nonmetal bonds.

The effect of structural vacancies on the electronic structure of titanium carbide and nitride was discussed in sufficient detail in [53, 54]. Researchers [55, 56] have performed a comprehensive analysis of the electronic spectrum variation with increasing concentration of structural vacancies in titanium, vanadium, zirconium and niobium carbides.

Specific features of the electronic structure and chemical bonds in nonstoichiometric ternary interstitial compounds were discussed in [57].

A considerable advance has been made recently in studying the electronic structure of disordered nonstoichiometric interstitial compounds, yet there is room for further research. The point is that all calculations of the electronic structure of disordered nonstoichiometric compounds were made for the ground state, i.e. for the state at 0 K. However, the disordered state of nonstoichiometric compounds is stable only at a high temperature $T > 1300$ K, whereas ordered phases of nonstoichiometric compounds are in thermodynamic equilibrium at low temperatures. Clearly, calculations of the ground state of the electronic structure should take into account the degree of order in the distribution of atoms and vacancies over the crystal lattice of a nonstoichiometric compound. However, ordering of nonstoichiometric compounds is rarely considered in calculations of the electron energy spectrum. The reader is referred to the estimates of the electronic structure of ordered zirconium nitride Zr_3N_4 ($Zr_{0.75}N$) with a defective metallic sublattice [58] and ordered titanium carbide Ti_4C_3 ($TiC_{0.75}$) and nitride Ti_4N_3 ($TiN_{0.75}$) [53, 54]. It should be noted that these ordered phases do not exist in reality. Also, an attempt was made [59, 60] to compute the ordering energy of transition-metal carbides and nitrides proceeding from the electronic structure of these compounds in the disordered state. Calculated energies of paired interactions in the nonmetallic sublattice were used to predict a possible type of ordering.

Thus, the equilibrium, i.e. ordered, state of nonstoichiometric compounds should be considered when computing the ground state of their electronic subsystem. Calculations of the electronic structure of disordered nonstoichiometric compounds should allow for thermal excitation, i.e. for $T > 0$ K.

In the final analysis, quantum-chemical calculations of the electronic structure of a solid should provide a theoretical explanation of various macroscopic properties proceeding from atomic and electronic considerations. Generally, this problem is solved in two steps. The first step involves determining the electron energy spectrum of a compound in the adiabatic approximation when nuclei (ion cores) are assumed to be immobile. In order to analyze equilibrium properties at the second step, one needs to find the statistical sum of all admissible positions of nuclei and the thermodynamic potential of the crystal as a function of independent thermodynamic variables. Today only the first part of the problem is solved by various methods of quantum chemistry of nonstoichiometric compounds. As a result, the derived information is limited and incomplete.

Let us also discuss the applicability of the adiabatic approximation to nonstoichiometric compounds. The behavior is adiabatic if the electron spectrum is free of excitations with energies approaching the nuclear oscillation energy $\hbar\omega$ ($\omega = \langle v_{nucl}\rangle/\Delta R$ being the nuclear oscillation frequency, $\langle v_{nucl}\rangle$ the mean oscillation rate, and ΔR the nuclear displacement), i.e.

$$\hbar\omega = \hbar\langle v_{nucl}\rangle/\Delta R \ll \Delta E_e , \tag{3.21}$$

where ΔE_e is the excitation energy or the energy gap between the energy of the outer (valence) electrons in the ground state and the energy of the first excited level. This criterion is not fulfilled for materials with metallic conductivity, including nonstoichiometric carbides and nitrides, because electron transitions with a vanishingly small excitation energy may occur near the Fermi surface and the energy spectrum has no gap, i.e. $\Delta E_e = 0$. This means that electron–phonon excitation, which leads to electron renormalization at the Fermi level, should be taken into account when describing nonstoichiometric compounds in order to refine the electron energy spectrum in a static lattice. Earlier calculations of the electronic structure of nonstoichiometric compounds disregard this fact.

Thus, the Hamiltonian of a system, which is used in the description of nonstoichiometric compounds by methods of quantum chemistry, should include not only the kinetic energy of electrons and potential energies of electron–electron and electron–nuclear (ionic cores) interactions, but also the kinetic energy of nuclei (cores) and the potential energy of nuclear interaction. Also, thermal excitation of the system should be taken into account by a special method. The last requirement is of particular importance for disordered nonstoichiometric compounds.

3.3 Structural Stability Boundaries

The majority of disordered nonstoichiometric carbides and nitrides MX_y of groups IV and V transition metals have structures of $B1$ type and broad homogeneity intervals (see Table 3.3). Metal atoms in these compounds form a metallic fcc sublattice whose octahedral interstitial sites host nonmetal interstitial atoms. Interstitial atoms can occupy all or just part of octahedral interstitial sites depending on the composition of the compound $MX_y \equiv MX_{1-z}$. Thus, octahedral interstitial sites are positions in the nonmetallic fcc sublattice whose sites can host interstitial atoms or vacancies. A disordered nonstoichiometric compound has a homogeneity interval if the type of its crystal structure is preserved when the concentration of structural vacancies changes.

The upper boundary of the homogeneity interval of nonstoichiometric compounds is normally a compound of stoichiometric composition ($MX_{1.0}$, $M_5Si_3X_{1.0}$, etc.) where all octahedral interstitial sites of the crystal structure are filled with interstitial atoms. An exception is cubic vanadium carbide for which carbide $VC_{0.875}$ is the upper boundary of the homogeneity interval. The lower boundary of the homogeneity interval is peculiar to each nonstoichiometric compound. The concentration of vacancies in cubic binary carbides and nitrides near the lower boundary of their homogeneity interval is 30–50 at.% or higher.

The change in the composition of disordered cubic carbides and nitrides MX_y from the upper to the lower boundary of their homogeneity interval, i.e. the growth of the concentration of structural vacancies, is accompanied by the decrease in the lattice constant a_{B1} (a weak maximum in the dependence $a_{B1}(y)$ is observed for carbides TiC_y, ZrC_y, and HfC_y at $y > 0.9$).

Each metal atom in carbides with a $B1$-type structure has an octahedral environment of six sites of the nonmetallic sublattice, while each site of the nonmetallic sublattice is surrounded by six metal atoms. If one or more structural vacancies are present in the nearest neighborhood of a metal atom, this atom is displaced statically because of the combined asymmetric effect of the nearest neighbors. Let us discuss the direction in which atoms can be displaced to provide for the experimentally observed decrease in the lattice constant of carbides MC_y with growing concentration of structural vacancies.

If metal atoms shift towards a vacancy, compression of vacant octahedral interstitial sites $\square M_6$ will be opposed by M–C interactions in adjacent occupied octahedra CM_6. If the concentration of vacant interstitial sites, which have a smaller linear size than occupied octahedral interstitial sites, increases, the lattice constant a_{B1} will shrink provided vacancy-induced static displacements of metal atoms decrease monotonically and tend asymptotically to zero with increasing distance from the vacancy. The lattice constant will decrease even if vacancy-induced disturbances extend to the first coordination sphere only. However, in this case it is impossible to explain the weak maximum in $a_{B1}(y)$ relating the lattice constant to the composition of titanium, zirconium and hafnium carbides.

If metal atoms nearest to a vacancy shift away from this vacancy, metal atoms forming the next coordination sphere of the vacancy should be displaced in the opposite direction, i.e. toward the vacancy, so that the lattice constant decreases. So, a vacancy-induced field of disturbances should extend to at least two coordination spheres of metal atoms. In this case attenuation of disturbances with distance from the vacancy represents Friedel oscillations. In accordance with estimates [58–62], the effective disturbance radius in nonstoichiometric cubic carbides MC_y is larger than the unit cell constant and the disturbance covers more than two coordination spheres. As long as the concentration of vacancies is small and vacancy-induced disturbance regions in the lattice do not overlap, the lattice constant a_{B1} will increase with growing concentration of vacancies. When the disturbance regions overlap, static atomic displacements induced by neighboring vacancies are mutually compensated and the lattice constant decreases. Consequently, a maximum should be observed in the dependence $a_{B1}(y)$ for disordered nonstoichiometric compounds MX_y. The position of this peak depends on the effective radius of the disturbance around a vacancy. Probably, vacancy-induced disturbances have a longer range and cover more coordination spheres in group V transition-metal carbides than in group IV transition-metal carbides. As a result, disturbance regions in carbides $M^{(V)}C_y$ overlap at a small concentration of vacancies $c_{\square} = 1 - y < 0.01$ and the maximum in the dependence $a_{B1}(y)$ is unobservable. The effective disturbance radius in carbides $M^{(IV)}C_y$ is smaller and the maximum in the dependence $a_{B1}(y)$ is observed at the vacancy concentration $c_{\square} = 1 - y \approx 0.05$-$0.07$. There is experimental evidence [61–67] that atomic displacements in nonstoichiometric carbides oscillate in direction and magnitude and metal atoms of the first coordination sphere shift away from the vacancy. It will be shown in Chapter 11 that displacement of atoms away from a vacancy in the first coordination sphere also accounts for the increase in the basic lattice constant of nonstoichiometric carbides upon ordering.

Thus, structural vacancies bring about static distortions of the crystal lattice. The resistance of a $B1$-type structure to the formation of structural vacancies is probably due to the fact that occupied octahedral groups XM_6 preserve the system of metal atom packing and resist stresses arising around a vacancy.

For quite a long time the literature presented only qualitative speculations on factors responsible for the location of the lower boundary of the homogeneity interval in nonstoichiometric compounds. This gap was bridged by work [22, 67–70] concerned with a quantitative method for estimating a limiting concentration of structural vacancies corresponding to the lower boundary of the homogeneity interval in nonstoichiometric compounds.

Condition of Stability Loss by the Crystal Lattice upon Formation of Structural Vacancies. Consider the disordered nonstoichiometric compound $MX_{1-z} \equiv MX_y$ with a $B1$-type structure. Nonmetal atoms and vacancies are located in crystallographically equivalent positions in the disordered state and form a substitutional solution in the nonmetallic lattice. Ample experimental data suggest that the lattice constant exhibits a nonlinear dependence on the composition of nonstoichiometric compounds (for example, carbides MC_y) and a_{B1} decreases on approaching the lower boundary of the homogeneity interval (Fig. 3.12).

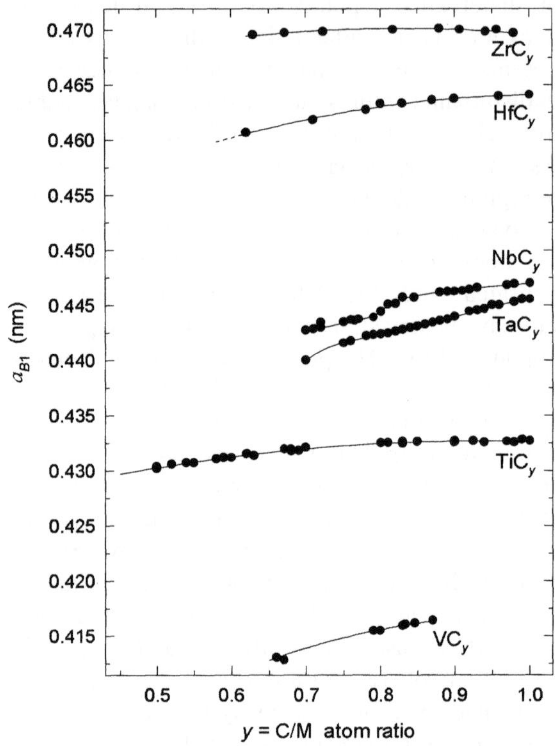

Fig. 3.12. Variation of the crystal lattice constant a_{B1} in homogeneity intervals of disordered nonstoichiometric carbides TiC_y, ZrC_y, HfC_y, VC_y, NbC_y and TaC_y

The lattice constant measured by diffraction methods represents a statistical average over sizes of a multitude of unit cells. Therefore in the case of a compound MX_y with a $B1$-type structure containing $z = 1 - y$ vacancies the lattice constant a_{B1} can be presented as

$$a_{B1}(y) = 2[R^*_M + yR^*_X + (1-y)R_\square], \qquad (3.22)$$

where R_M^* and R_X^* are metal and nonmetal atomic radii in the compound concerned, and $R_\Box$ is the vacancy radius (the radius of a vacant octahedral interstitial site). The increase in the overlap of disturbance regions with growing concentration of vacancies is equivalent to the decrease in the vacancy radius $R_\Box$ when $y \to y_c$ (y_c is the relative concentration of nonmetal atoms at the lower boundary of the homogeneity interval in compound MX_y). If it is assumed that the metal atomic radius R_M^* remains unchanged in the homogeneity interval of compound MX_y, the observed decrease in the carbide lattice constant (see Fig. 3.12) is due to smaller $R_\Box$.

Suppose that the lower boundary of the homogeneity interval corresponds to some minimum radius of a vacant octahedral interstitial site $R_\Box^{min}$, when local displacements of metal atoms around a vacancy cause changes in its nearest environment, i.e.

$$R_\Box^{min} = \inf\{R_\Box : z_c - z\}, \tag{3.23}$$

where $z_c = 1 - y_c$ is the critical concentration of structural vacancies corresponding to the lower boundary of the homogeneity interval. Obviously, for the whole crystal to lose stability, the concentration of vacancies with a minimum admissible radius $R_\Box^{min}$ should be such that a continuous (infinite) vacancy cluster extending throughout the crystal is formed in the nonmetallic sublattice. In this case, local displacements of metal atoms capable of altering the structure cover the whole volume of the crystal and render it unstable.

Thus, it is possible to define necessary and sufficient conditions for the loss of stability by the crystal lattice of a nonstoichiometric compound upon formation of structural vacancies, i.e. the lower boundary of the homogeneity interval. The necessary condition of stability loss reduces to the appearance of structural vacancies having the smallest admissible size $R_\Box^{min}$. The sufficient condition of stability loss is that the concentration of vacancies with the smallest size $R_\Box^{min}$ should be sufficient for the formation of an infinite vacancy cluster in the crystal.

Thus, to determine the lower boundary of the homogeneity interval for a nonstoichiometric compound, one needs to find the $R_\Box^{min}$ value and calculate the vacancy concentration sufficient for the formation of an infinite vacancy cluster.

Crossing of the lower boundary of the homogeneity interval of a nonstoichiometric compound is followed by the appearance of a new phase with a different crystal structure, i.e. results from bifurcation instability of a disordered atomic lattice. Critical phenomena in disordered systems are efficiently described by percolation theory [71–73]. In our case, the system that can be described in terms of percolation theory is a disordered solution of interstitial atoms and vacancies in the nonmetallic sublattice of a nonstoichiometric compound.

Consider a nonmetallic sublattice as a graph G with a multitude of vertices v, which are sites of the nonmetallic sublattice. The state of this system can be described by a combination of two numbers +1 and –1. The value –1 corresponds to a vacant site and the value +1 refers to an occupied site. In this case, the configuration space of the whole system can be presented as $S = \prod_{v}\{-1, 1\}$ [73].

For disordered compound $MX_y \equiv MX_{1-z}$ each site of the nonmetallic sublattice is free (vacant) with some probability $p = 1 - y \equiv z$, the states of all sites being independent. We shall refer to vacancies that are nearest neighbors in the nonmetallic sublattice as bound vacancies and a set of bound vacancies as a conducting cluster $W(v)$ of site v. Put another way, the conducting cluster $W(v)$ of site v is the totality of all sites and edges of the graph G that can be reached from site v through vacant sites only. They are isolated from one another when the vacancy concentration is small ($z << 1$). Indeed, the probability of a given vacancy being isolated in a nonmetallic fcc sublattice with coordination number 12 is $P_1 = (1 - p)^{12} = (1 - z)^{12}$ and is almost unity at $z << 1$. The probability that a chosen vacancy is not isolated is $P = 1 - (1 - z)^{12}$ and approaches zero at $z << 1$.

The probability that vacancies are isolated from one another decreases rapidly with increasing vacancy content ($0 < z < 1$) and vacancy clusters $W(v)$ having different sizes appear in the nonmetallic sublattice. As soon as some concentration z_c^* is reached, a single cluster going throughout the crystal lattice is formed in the system. It is called an infinite cluster. As the vacancy concentration increases further ($z_c > z_c^*$), the system acquires [74, 75] two numerical intervals of the parameter $p = z$ with clearly different behavior of the system: when $p < p_c = z_c^*$, the system is free of vacancy clusters; when $p >> p_c = z_c^*$, one infinite vacancy cluster is present with probability 1. Correspondingly, the critical probability or the percolation threshold is the quantity

$$p_c = \sup\{p: \theta(p) = 0\}: \theta(p) = 0 \text{ when } p < p_c, \text{ and } \theta(p) \geq 0 \text{ when } p \geq p_c, \quad (3.24)$$

where $\theta(p)$ is the limit of a probability of an infinite cluster being formed at a given probability p and $v \to \infty$.

Estimates of the percolation threshold in the problem of tangent spheres showed [76] that percolation occurs when part of the system volume occupied by all spheres described around free sites is equal to or exceeds some critical value f_c independent of the crystal lattice type. Experimental determination of the percolation threshold in the problem of hard spheres showed that part of the critical volume f_c is also independent of the radius ratio of the spheres filling the lattice space [72]. For three-dimensional systems $f_c = 0.16$ [72] and the critical volume is f_cV, where V is the whole volume of the system. Consequently, an infinite cluster is formed if $V_p \geq f_cV$ (V_p being the volume of spheres described about free sites). For nonstoichiometric compounds MX_y, the disordered system, where an infinite cluster can appear, is the nonmetallic sublattice. Since $V = V_M + V_X$ (V_M and V_X denote the volume of the metallic and nonmetallic sublattices respectively), then

$$V_p \geq f_cV = f_c(V_M + V_X) = f_cV_M + f_cV_X. \quad (3.25)$$

Hence, the inequality

$$V_{pX} \geq f_cV_X \quad (3.26)$$

should hold for the nonmetallic sublattice (V_{pX} being the total volume of vacancies in the nonmetallic sublattice). The volume of the nonmetallic sublattice in a crystal with a $B1$-type structure comprising $N >> 1$ unit cells with the lattice constant a (3.22) is

$$V_X = N[a(y)]^3[a(y) - 2R_M^*]^3 / \{[a(y) - 2R_M^*]^3 + (2R_M^*)^3\}. \quad (3.27)$$

As mentioned in the foregoing, the lattice of nonstoichiometric compound $MX_y \equiv MX_{1-z}$ loses stability if an infinite cluster is formed by vacancies having the smallest admissible size. In this case

$$V_{pX} \geq z_c nNV_\square^{min} \equiv (1 - y)nNV_\square^{min}, \quad (3.28)$$

where n is the number of formula units per one unit cell ($n = 4$ for a structure of $B1$ type) and $V_\square^{min} = 4\pi(R_\square^{min})^3/3$ is the minimum admissible volume of a vacancy. Considering (3.26)–(3.28), the condition of stability loss by the crystal lattice of disordered nonstoichiometric compound $MX_y \equiv MX_{1-z}$ can be written as

$$(1 - y_c)nV_\square^{min} \geq f_c[a(y)]^3[a(y) - 2R_M^*]^3 / \{[a(y) - 2R_M^*]^3 + (2R_M^*)^3\}. \quad (3.29)$$

Inequality (3.29) can be solved with respect to $z_c = 1 - y_c$, which is the critical concentration of structural vacancies corresponding to the lower boundary of the homogeneity interval of the nonstoichiometric compound $MX_y \equiv MX_{1-z}$.

Calculation of the Lower Boundary of the Homogeneity Interval for Cubic Carbides. Let us calculate the lower boundary of homogeneity intervals for cubic carbides MC_y of groups IV and V transition metals. Write the lattice constant variation with the carbide composition as a polynomial $a(y) = \sum_{k=0}^{m} a_k y^k$. Considering the accuracy of experimental measurements, $a(y)$ can be described by a second-order polynomial. The coefficients of the polynomials in Table 3.6 were determined [67, 77–79] from the experimental dependencies of the lattice constants of TiC_y, VC_y, NbC_y, and TaC_y. The coefficients of the polynomials $a(y)$ describing the dependence of lattice constants of disordered cubic carbides ZrC_y and HfC_y were calculated from the data of [80, 81]. The lattice constant $a_0 = 2(R_M^* + R_C^*)$ for the defect-free carbide $MC_{1.0}$ ($z = 0$). Experimental a_0 values deviate from $2(R_M + R_C)$, where R_M and R_C are atomic radii of transition metals and carbon for coordination number 12 (see Table 3.1). The discrepancy is due to the change in the sizes of metal and carbon atoms upon carbide formation. Therefore $z_c = 1 - y_c$ was calculated using atomic radius values normalized to a_0: $R_M^* = a_0 R_M / [2(R_M + R_C)]$ and $R_C^* = a_0 R_C / [2(R_M + R_C)]$ (see Table 3.6).

The minimum admissible size of vacancy, $R_\square^{min}$, needs be known to calculate the vacancy concentration at the lower boundary of the homogeneity interval. The size $R_\square^{min}$ depends on the phase with which the nonstoichiometric compound is in equilibrium at the lower boundary of the homogeneity interval. Carbides of group IV transition metals are in equilibrium with bcc metals at the lower boundary of the homogeneity interval. An exception is hafnium carbide, which is in equilibrium with hcp α-Hf. Octahedral interstitial sites in the structure of metals are much smaller than the least possible size ($(\sqrt{2} - 1)R_M^*$) of octahedral interstitial sites of the fcc metallic sublattice of carbides.

From crystal chemistry it is known [82] that this size of octahedral interstitial site limits the stability of a $B1$-type structure, which is typical of carbides. Therefore one may think that the minimum admissible size of a vacant octahedral interstitial site in carbides of group IV transition metals is $R_{\square}^{\min} \approx 0.414\, R_{\mathrm{M}}^{*}$.

Nonstoichiometric cubic carbides MC_y of group V transition metals are in equilibrium with lower carbides M_2C_y having a $L'3$-type structure at the lower boundary of the homogeneity interval. Carbon atoms statistically occupy one-half of the octahedral interstitial sites of the hcp metallic sublattice, while the rest of octahedral interstitial sites are vacant in this structure. Clearly, the size of a vacant octahedral interstitial site, $R_{\square}^{\mathrm{hex}}$, in the lower carbide M_2C represents the minimum admissible size of vacancies, $R_{\square}^{\min}$, in the cubic carbide MC_y. Therefore $R_{\square}^{\min}$ can be found as follows. The diagonal of an octahedron formed by six metal atoms in the hexagonal lattice of the lower carbide M_2C is $d = [(16a^2 + 3c^2)/12]^{1/2}$, where a and c are constants of the unit cell of carbide the M_2C. The octahedron size is an average of the sizes of octahedral interstitial sites in the hcp metal sublattice filled with and free of carbon atoms. The number of occupied and vacant interstitial sites in a $L'3$-type structure is equal and, therefore, $d = \frac{1}{2}[2(R_{\mathrm{M}}^{*} + R_{\mathrm{C}}^{*}) + 2(R_{\mathrm{M}}^{*} + R_{\square}^{\mathrm{hex}})]$. The quantity $2(R_{\mathrm{M}}^{*} + R_{\mathrm{C}}^{*})$ is the size of an occupied octahedral interstitial site and equals the lattice constant a_0 of stoichiometric cubic carbide $MC_{1.0}$, while $R_{\square}^{\min} = R_{\square}^{\mathrm{hex}}$. Considering this, $R_{\square}^{\min} = R_{\square}^{\mathrm{hex}} = d - 0.5a_0 - R_{\mathrm{M}}^{*}$ for nonstoichiometric carbides of group V transition metals. Table 3.6 gives $R_{\square}^{\min}$ values used to calculate the critical concentration of structural vacancies $z_c = 1 - y_c$ in carbides of groups IV and V transition metals.

Table 3.6. The critical structural vacancy concentration $z_c = 1 - y_c$ and the position of the lower boundary of the homogeneity interval for nonstoichiometric carbides MC_y with a $B1$-type structure

MC_y	Coefficients of the polynomial $a_{B1}(y) = a_0 + a_1 y + a_2 y^2$			R_{M}^{*} (nm)	$R_{\square}^{\min}$ (nm)	z_c	Lower boundary of the homogeneity interval	
	a_0 (nm)	a_1 (nm)	a_2 (nm)				calculation	experiment
TiC_y	0.4216	0.0237	−0.0127	0.142	0.059	0.462	$TiC_{0.54}$	$TiC_{0.48}$ [21], $TiC_{0.49}$ [25]
ZrC_y	0.4564	0.4213	−0.0202	0.158	0.066	0.355	$ZrC_{0.64}$	$ZrC_{0.62}$ [80]
HfC_y	0.4412	0.0449	−0.0220	0.156	0.065	0.360	$HfC_{0.64}$	$HfC_{0.56}$ [22], $HfC_{0.60}$ [25]
VC_y	0.3829	0.0690	−0.0348	0.132	0.066	0.343	$HfC_{0.64}$	$VC_{0.65}$ [22]
NbC_y	0.4093	0.0701	−0.0322	0.146	0.068	0.301	$NbC_{0.70}$	$NbC_{0.72}$ [25], $NbC_{0.70}$ [22]
TaC_y	0.4213	0.0342	−0.0099	0.146	0.067	0.329	$TaC_{0.67}$	$TaC_{0.74}$ [28], $TaC_{0.71}$ [22]

Examination of the calculated values of the critical concentration of structural vacancies, z_c, shows that titanium carbide TiC_y deviates most from carbon stoichiometry. The lower boundary of the homogeneity interval corresponds to a smaller concentration of structural vacancies in the nonmetallic sublattice of carbides of group V transition metals than in the similar sublattice of carbides of group IV transition metals. In general, the lower boundary of the homogeneity interval calculated for nonstoichiometric carbides MC_y with a *B*1-type structure is in good agreement with experimental data (see Table 3.6).

The minimum concentration of free sites sufficient for the formation of an infinite cluster in an fcc lattice is $z_c^* = 0.2$ [72]. Calculations for carbides give $z_c > z_c^* = 0.2$. Consequently, an infinite cluster of vacancies with $R_\square > R_\square^{min}$ is first formed in carbides with growing concentration of vacancies. The size of vacancies diminishes to $R_\square^{min}$ as their concentration increases further. An infinite cluster formed by vacancies with $R_\square > R_\square^{min}$ causes instability of the lattice and determines the lower boundary of the homogeneity interval.

Thus, the use of some postulates of percolation theory suggest a method for calculating the limits of structural stability of strongly nonstoichiometric interstitial compounds. This method can be employed to calculate regions of interstitial solid solutions. To this end, one needs to know only the crystal volume versus the number of interstitial atoms, i.e. the concentration dependence of lattice constants of the disordered interstitial phase, and also the radius of the matrix atoms. Proceeding from the explanation [69] of the lattice instability in nonstoichiometric compounds, later it will be possible to work out a method for calculating the limits of structural instability, which will be based on first principles and will not require experimental data. However, it should be remembered that only thermodynamic calculations provide an absolutely logical and consistent determination of the homogeneity interval boundaries of a nonstoichiometric compound.

3.4 Ordering in Strongly Nonstoichiometric Carbides, Nitrides and Oxides

Ordering as Atomic Redistribution. The notion of ordering is a synonym of a disorder–order structural phase transformation. Ordering as a phase transformation results from atomic redistribution over lattice sites of a substitutional solid solution. Mutually substitutional components of a solid solution are distributed randomly over lattice sites in the disordered state and the probability that any lattice site is occupied with an atom of a given species coincides with the concentration of this species in the solution. A decrease in temperature is followed by a disorder–order phase transformation. As a result, lattice sites of the disordered solution may separate into several sublattices differing in the probabilities of their sites being filled with atoms of a given species. Ordered phases, which appear after such separation, are called superstructures.

Ordering is not an exclusive property of substitutional solid solutions. It can occur in interstitial solid solutions if the number of interstitial positions is bigger than the number of atoms occupying these positions. Vacant positions and interstitial atoms act as the components of a substitutional solution, while the solvent atoms form a static atomic host structure in which interstitial atoms and vacancies rearrange. Interstitial atoms and structural vacancies are regarded as components of a binary substitutional solution in the nonmetallic sublattice of nonstoichiometric compounds. In the simplest case ordering in the nonmetallic sublattice formed by octahedral interstitial sites of the metallic sublattice creates two new sublattices. The sites of the first superstructural sublattice are occupied with interstitial atoms, while those of the second sublattice are vacant. It is reasonable to speak of a vacancy sublattice when nonstoichiometric compounds are ordered. Ordering of the nonmetallic sublattice is possible in substitutional solid solutions, for example, carbonitrides MC_yN_{1-y}, oxycarbides MC_yO_{1-y}, etc. In these solutions, the sites of one superstructural sublattice are occupied mostly by nonmetal atoms of the first species, while those of the second sublattice host nonmetal atoms of the second species.

Ordering in the nonmetallic sublattice of nonstoichiometric compounds is accompanied by a decrease in symmetry of the crystal space group. Some symmetry transformations in a disordered nonmetallic sublattice are not symmetry elements of the ordered crystal, as these sites are crystallographically nonequivalent.

For a long time X-ray structural analysis was used to determine the positions of interstitial atoms in the lattice of nonstoichiometric compounds. However, interstitial atoms have small scattering amplitudes for X-ray radiation compared to the metal atoms and, consequently, it is not always possible to deduce the positions of interstitial atoms from X-ray scattering. This led many researchers to believe that the interstitial atoms in the nonmetallic sublattice of nonstoichiometric compounds are distributed randomly under all conditions. High-sensitivity diffraction techniques and, particularly, neutron diffraction revealed that under certain conditions interstitial atoms may have an ordered distribution in nonstoichiometric compounds. In neutron diffraction the scattering power of many transition metals and of light interstitial elements are comparable. Ordering of interstitial atoms was also detected by electron diffraction and diverse methods of physical and chemical analysis.

Crystal Structures of Nonstoichiometric Compounds in the Ordered State. Ordering of interstitial atoms is quite common in nonstoichiometric carbides, nitrides and oxides of group IV and V transition metals and has been found experimentally in most nonstoichiometric compounds. However, this feature is frequently ignored in the literature. Among 2090 papers referred to in [36], only 21 measurements were performed between 1981 and 1985. No references were made to studies carried out after 1985. Issues of short- and long-range order and its connection with nonstoichiometry were completely ignored. From the scientific viewpoint, the book [36] satisfies the level of knowledge existing between 1960 and 1965. The handbook [37], published in 1996, gives the following short note about ordering in nonstoichiometric transition metal carbides: "When the concentration of carbon-atom vacancies is high, long-range ordering in their arrangement within the metal lattice is usually observed. The effect of this ordering on the structure and physical properties of the carbide may be considerable in some cases but the mechanisms of formation, the structural and bonding arrangement are still not well defined".

The only books [21–23] in the scientific literature describing disorder–order transitions in strongly nonstoichiometric compounds are published in Russian and are inaccessible to readers outside the former USSR.

Starting from crystallographic considerations, it was shown in [83] that ordered structures corresponding to stoichiometric compositions M_2C, M_3C_2, M_8C_5, M_4C_3, M_5C_3, M_5C_4, M_6C_5, and M_8C_7 can be formed in nonstoichiometric carbides. An estimate of the ordering in strongly nonstoichiometric cubic compounds MX_y [22, 84, 85] by the order parameter functional method [21–23, 86–88] showed that only superstructures M_2X, M_3X_2 and M_6X_5 can appear if thermodynamic limitations are taken into account. If the upper boundary of the homogeneity interval occupies a special position, as is the case with cubic vanadium carbide VC_y, the ordered cubic phase M_8X_7 can also be formed. Experimental studies from 1980 to 2000 adduced reliable evidence for the existence of ordered phases of M_2X and M_6X_5 types. In recent years experiments brought further evidence for the existence of phases of M_3X_2 type.

In [89] a pioneering review was made of ordered structures of transition-metal carbides and nitrides. The atomic ordering of nonstoichiometric compounds was also discussed later in [18, 20, 28, 29, 90–94]. The most complete, comprehensive and detailed analysis of ordering of nonstoichiometric compounds is presented in the reviews [2, 21–23, 40–42].

Self-sufficient rhombohedral ζ-phases M_4C_{3-x} are present in systems V–C, Nb–C, and Ta–C in addition to nonstoichiometric carbides M_2C_y and MC_y [95]. It should be noted that carbides ζ-M_4C_{3-x} are not ordered phases of cubic monocarbides MC_y. Phase equilibria in systems V–C, Nb–C, and Ta–C involving phases ζ-V_4C_{3-x}, ζ-Nb_4C_{3-x}, and ζ-Ta_4C_{3-x} were closely studied in [96].

Table 3.7 summarizes data on crystal structures of ordered phases formed in M–C systems.

The known data on the structure of ordered phases of nonstoichiometric compounds are far from complete and even conflicting in some cases.

Occasionally it seems that some papers concerned with ordering in nonstoichiometric carbides report, with reference to earlier research, phases that do not exist. Among them are superstructures types M_8X_5 (Ti_8C_5) and M_4X_3 (Nb_4C_3 and Ta_4C_3), which were presumed in cubic nonstoichiometric compounds MX_y, but could not be confirmed. It was suggested that the ordered phase Ti_2C has a trigonal structure (space group $P3_121$) with unit cell constants $a = b = a_{B1}/\sqrt{2}$ ($\mathbf{a}$ $= \{1/2\ 1/2\ 0\}_{B1}$) and $c = 3\sqrt{2} \times a_{B1}$ ($\mathbf{c} = \{2\ 2\ 2\}_{B1}$), which is likely to be erroneous.

Table 3.7. Ordered phases of strongly nonstoichiometric carbides of transition metals of IV and V groups

Ordered phase				
composition	region of existence of the ordered phase or nonmetal content (at.%)	lattice type or symmetry	space group	[1] unit cell parameters (lattice constants in nm)
1	2	3	4	5
Ti_2C	$TiC_{0.52}$–$TiC_{0.71}$	cubic	$Fd\bar{3}m$	$a = 2a_0$
Ti_2C	$TiC_{0.58}$–$TiC_{0.63}$	trigonal	$R\bar{3}m$	$a = b = c = (a_0\sqrt{6})/2$
Ti_3C_2	$TiC_{0.64}$–$TiC_{0.68}$		$C222_1$	$a = a_0\sqrt{2}$, $b = 3a_0\sqrt{2}$, $c = 2a_0$
Ti_2C	$TiC_{0.49}$–$TiC_{0.54}$	cubic	$Fd\bar{3}m$	$a = 2a_0$
Ti_2C	$TiC_{0.55}$–$TiC_{0.59}$	trigonal	$R\bar{3}m$	$a = b = c = (a_0\sqrt{6})/2$
Ti_3C_2	$TiC_{0.63}$–$TiC_{0.67}$	orthorhombic	$C222_1$	$a = a_0\sqrt{2}$, $b = 3a_0\sqrt{2}$, $c = 2a_0$
Ti_6C_5		hexagonal		$a = b = 0.529 = (a_0\sqrt{6})/2$, $c = 0.996 = (4a_0\sqrt{3})/3$
Ti_2C	$TiC_{0.47}$–$TiC_{0.54}$			
Ti_3C_2	$TiC_{0.62}$–$TiC_{0.70}$			
Ti_6C_5	$TiC_{0.80}$–$TiC_{0.90}$			
δ'-ZrC_y	39.0–43.0	trigonal		$a = 0.663 \approx (a_0\sqrt{2})/2$, $b = 1.491 \approx 2a_0\sqrt{3}$
Zr_2C	$ZrC_{0.63}$–$ZrC_{0.74}$	cubic		$a = 2a_0$
$ZrC_{0.75}$	$ZrC_{0.69}$–$ZrC_{0.77}$	cubic ?		$a = 2a_0$
Zr_2C	$ZrC_{0.47}$–$ZrC_{0.54}$			
Zr_3C_2	$ZrC_{0.60}$–$ZrC_{0.71}$			
Zr_6C_5	$ZrC_{0.76}$–$ZrC_{0.92}$			

Disordered basic phase		References	Remarks
composition	lattice type		
6	7	8	9
TiC_y	*B*1 (NaCl)	97, 98	Degree of ordering of carbon sublattice is small. $T_{trans} \approx$ 950-1030 K. Assumption [97] about existence of Ti_2C phase up to 2170 K is erroneous
TiC_y	*B*1 (NaCl)	94, 99	Calculation. $T_{trans} <$ 900 K. Stated homogeneity interval corresponds to 750 K
TiC_y	*B*1 (NaCl)	62, 94	Calculation. $T_{trans} <$ 720 K. Stated homogeneity interval corresponds to 650 K
TiC_y	*B*1 (NaCl)	100	$T_{trans} <$ 1000 K. Stated homogeneity interval corresponds to ~600 K
TiC_y	*B*1 (NaCl)	100	$T_{trans} <$ 1010 K. Stated homogeneity interval corresponds to ~600 K
TiC_y	*B*1 (NaCl)	100	$T_{trans} <$ 990 K. Stated homogeneity interval corresponds to ~600 K
TiC_y	*B*1 (NaCl)	101	Phase is found in thin films
TiC_y	*B*1 (NaCl)	1, 2, 102	Calculation. $T_{trans} <$ 950 K. Stated homogeneity interval corresponds to 600 K
TiC_y	*B*1 (NaCl)	1, 2, 102	Calculation. $T_{trans} <$ 930 K. Stated homogeneity interval corresponds to 600 K
TiC_y	*B*1 (NaCl)	1, 2, 102	Calculation. $T_{trans} <$ 920 K. Stated homogeneity interval corresponds to 600 K
ZrC_y	*B*1 (NaCl)	92	
ZrC_y	*B*1 (NaCl)	98	$T_{trans} \approx$ 1170 K
ZrC_y	*B*1 (NaCl)	103	This phase is isomorphic to ordered phase $ThC_{0.76}$ [104]. $T_{trans} \approx$ 1330-1350 K
ZrC_y	*B*1 (NaCl)	1, 2, 22, 84, 85, 105, 106	Calculation. $T_{trans} <$ 1180 K. Stated homogeneity interval corresponds to 600 K
ZrC_y	*B*1 (NaCl)	1, 2, 22, 84, 85, 105, 106	Calculation. $T_{trans} <$ 1130 K. Stated homogeneity interval corresponds to 600 K
ZrC_y	*B*1 (NaCl)	1, 2, 22, 84, 85, 105, 106	Calculation. $T_{trans} <$ 1130 K. Stated homogeneity interval corresponds to 600 K

(*Continued*)

Table 3.7 (continued)

1	2	3	4	5
Hf_3C_2	$HfC_{0.67}$ - $HfC_{0.72}$			
Hf_6C_5	$HfC_{0.77}$ - $HfC_{0.90}$			
$ThC_{0.75}$	$ThC_{0.74}$ - $ThC_{0.76}$	cubic ?		$a = 2a_0$
β_{rh}-V_2C	~33.0 ($VC_{0.5}$)	orthorhombic, type ζ-Fe_2N	*Pbcn*	$a = 0.4567 = c_0$, $b = 0.5744 = 2a_0$, $c = 0.5026 \approx a_0\sqrt{3}$
β'-V_2C	~33.0 ($VC_{0.5}$)	hexagonal		$a = 0.4997 \approx a_0\sqrt{3}$, $c = 0.4568 = c_0$
V_3C_2	$VC_{0.67}$–$VC_{0.72}$			
V_6C_5	~45.5 ($VC_{0.83}$)	trigonal	$P3_1$, $P3_2$	$a = 0.509$, $c = 1.44$
V_6C_5	~45.5 ($VC_{0.83}$)	monoclinic	*C*2	$a = 0.509$, $b = 1.018$, $c = 0.882$, $\gamma = 109.8°$
V_6C_5	~45.5 ($VC_{0.83}$)	monoclinic	*C*2/*m*	
δ'-VC_y	~45.5 ($VC_{0.83}$)	trigonal		$a = 0.5885 \approx a_0\sqrt{2}$, $c = 1.443 \approx 2a_0\sqrt{3}$
δ^*_M-VC_y	45.0–47.0	monoclinic		$a = (a_0\sqrt{3})/2$, $b = a_0\sqrt{6}$, $c = 8a_0\sqrt{2}$, $\gamma = 109.47°$
V_8C_7	~46.6 ($VC_{0.87}$)	cubic	$P4_332$	$a = 0.8332 \approx 2a_0$
Nb_2C	~33.3 ($NbC_{0.5}$)	orthorhombic, type ζ-Fe_2N	*Pnma*	$a = 1.236$, $b = 1.089$, $c = 0.496$
Nb_2C	~33.3 ($NbC_{0.5}$)	hexagonal, type ε-Fe_2N		$a = a_0\sqrt{3}$, $c = c_0$
Nb_6C_5	$NbC_{0.81}$–$NbC_{0.88}$	monoclinic	*C*2/*m*	$a = c = 0.54605 = (a_0\sqrt{6})/2$, $b = 0.94579 = (a_0\sqrt{18})/2$ $\beta = 109.47°$
Nb_3C_2	$NbC_{0.67}$–$NbC_{0.71}$			
Nb_6C_5	$NbC_{0.78}$–$NbC_{0.90}$			
Nb_6C_5	$NbC_{0.75}$–$NbC_{0.84}$	trigonal	$P3_1$, $P3_2$	
Ta_2C	~33.3 ($TaC_{0.5}$)	trigonal	$P\bar{3}m1$	
Ta_6C_5	$TaC_{0.79}$–$TaC_{0.89}$	M_6C_5 type		
Ta_6C_5	$TaC_{0.81}$–$TaC_{0.89}$			

[1] a_0, b_0, c_0 stand for the disordered basic phase lattice constants.

6	7	8	9
HfC_y	*B*1 (NaCl)	1, 2, 22, 81, 106–109	Calculation. T_{trans} < 780 K. Stated homogeneity interval corresponds to 500 K
HfC_y	*B*1 (NaCl)	1, 2, 22, 81, 106–109	Calculation. T_{trans} < 800 K. Stated homogeneity interval corresponds to 500 K
ThC_y	*B*1 (NaCl)	104	T_{trans} = 1020-1070 K
V_2C_y	*L*′3 (W_2C)	91, 92	T_{trans} < 1100 K
V_2C_y	*L*′3 (W_2C)	91, 92	Ordered phase is produced by quenching from 1800 K
VC_y	*B*1 (NaCl)	77, 110	Calculation. T_{trans} < 1140 K
VC_y	*B*1 (NaCl)	77, 110–112	
VC_y	*B*1 (NaCl)	113, 114	T_{trans} < 1420 K
VC_y	*B*1 (NaCl)	115	Structure of the Nb_6C_5 type [116–119]
VC_y	*B*1 (NaCl)	92	T_{trans} < 1300 K
VC_y	*B*1 (NaCl)	92	T_{trans} < 1500 K
VC_y	*B*1 (NaCl)	112, 120–124	First-kind transition. T_{trans} < 1500 K
Nb_2C_y	*L*′3 (W_2C)	125	Low-temperature ordered phase
Nb_2C_y	*L*′3 (W_2C)	125	High-temperature ordered phase
NbC_y	*B*1 (NaCl)	116–119, 126, 127	First-kind transition. For $NbC_{0.83}$ carbide T_{trans} = 1304 K and ΔH_{trans} = 2.2 kJ mol^{-1}
NbC_y	*B*1 (NaCl)	1, 2, 22, 84, 85, 105, 106	Calculation. T_{trans} < 760 K. Stated homogeneity interval corresponds to 400 K
NbC_y	*B*1 (NaCl)	1, 2, 22, 84, 85, 105, 106	Calculation. T_{trans} < 1580 K. Stated homogeneity interval corresponds to 600 K
NbC_y	*B*1 (NaCl)	113, 128, 129	Trigonal phase Nb_6C_5 is similar to V_6C_5 phase [111]. T_{trans} = 1313 K
Ta_2C_y	*L*′3 (W_2C)	130	Carbon atoms fill in an ordered fashion half the octahedral interstitial sites
TaC_y	*B*1 (NaCl)	79, 131–134	Incommensurate phase of the M_6C_5 type. First-kind transition. T_{trans} ≈ 1110 K
TaC_y	*B*1 (NaCl)	1, 2, 22, 84, 85, 105, 106	Calculation. T_{trans} < 1400 K. Stated homogeneity interval corresponds to 800 K

In accordance with [97, 98], the ordered cubic phase Ti_2C, which is formed in titanium carbide $TiC_{0.52}$–$TiC_{0.71}$ during slow cooling, has double the lattice parameter as compared to the disordered carbide. Goretsky [97] erroneously supposed that this phase exists and is stable even at a temperature below 2170 K. Actually (see, for example [66, 67, 94, 98, 100] and many other experimental studies), the ordered phase Ti_2C is formed as a result of slow cooling from 1300 to 800 K and its disorder–order transition temperature is about 1000 K. Ordering proceeds by the phase transition channel including one of the rays of the star $\{\boldsymbol{k}_9\}$ and meets the Landau criterion [135] for phase transitions of the second kind. However, a study of the ordering kinetics of carbide $TiC_{0.60}$ [136] showed that the formation of the superstructure Ti_2C is a phase transition of the first kind. The disorder–order transition temperature is ~1030 K, while the order–disorder transition point is about 1080 K. It was found [94] that the trigonal ordered phase (space group $R\bar{3}m$) Ti_2C is formed as a result of ordering of nonstoichiometric titanium carbide in the $TiC_{0.58}$–$TiC_{0.63}$ range, while a rhombic superstructure Ti_3C_2 (space group $C222_1$) can appear at a higher carbon concentration.

A detailed structural and electrokinetic study of disorder–order phase transformations in the carbide TiC_y ($0.5 < y < 1.0$) [100] revealed the formation of cubic (space group $Fd\bar{3}m$) and trigonal (space group $R\bar{3}m$) superstructures of Ti_2C type and of a rhombic superstructure (space group $C222_1$) Ti_3C_2. In accordance with [100, 137, 138], the transformations TiC_y–Ti_2C are weak phase transitions of the first kind with $T_{trans} \approx 980$ K. The rhombic ordered phase Ti_3C_2 is formed by the mechanism of a phase transition of the first kind at a temperature about 1000 K in the range $TiC_{0.63}$–$TiC_{0.67}$ [100, 137, 138]. The ordered phase Ti_6C_5 with hexagonal symmetry was found in thin Ti–C films [101]. The films were produced by electron beam sputtering of sintered carbide $TiC_{0.9}$, magnetron sputtering of titanium and carbon, and laser sputtering of titanium and graphite or carbide $TiC_{0.9}$. In accordance with [101], the disordered carbide $TiC_{0.83}$ corresponds to the ordered phase Ti_6C_5. However, lattice parameters of the hexagonal ordered phase, which were determined in [101], correspond to disordered carbide TiC_y with basic cubic lattice constant $a_0 = 0.4319$ nm. More reliable data [67] suggest that the carbide $TiC_{0.68\text{-}0.70}$ has this lattice constant, while disordered carbide $TiC_{0.83}$ has $a_0 = 0.4326$ nm. The low value of the lattice constant of cubic titanium carbide in films may be due to an oxygen impurity.

The lower boundary of the homogeneity interval of zirconium carbide ZrC_y is not known with certainty. It may fall within the interval $ZrC_{0.55}$–$ZrC_{0.60}$. Neutron diffraction analysis [98] showed that the ordered phase Zr_2C of nonstoichiometric zirconium carbide ZrC_y has a cubic unit cell and is formed near the lower boundary of the homogeneity interval of ZrC_y at the carbon concentration y from 0.63 to 0.74. The lattice of ordered carbide Zr_2C is assumed to have trigonal symmetry [92] if one considers distortions. Ordering of carbide ZrC_y produced by arc melting and annealed at 1870 K for 6 hours was studied in [103]. Measurements of electric resistivity of carbide $ZrC_{0.51}$ revealed a surge in the $\rho(T)$ plot at about 1130 K, but X-ray diffraction patterns of carbide $ZrC_{0.51}$ did not contain superstructure reflections. It was argued [103] that the sample had two phases (zirconium carbide and metallic zirconium). The observed kink was attributed to the transformation α-Zr $\rightarrow$ β-Zr. The $\rho(T)$ curve of carbide $ZrC_{0.70}$ exhibited an abrupt change at a temperature between 1330 and 1350 K, which was confirmed by the DTA method. The X-ray diffraction pattern of carbide $ZrC_{0.70}$ annealed at 1220 K for 30 hours contained superstructure reflections. In accordance with [103], a cubic ordered phase, whose unit cell constant is twice as large as that of the disordered carbide, is formed in the carbide $ZrC_{0.70}$. The space group of the superstructure was not identified. The ordered phase of zirconium carbide was assumed to have the structure of thorium carbide $ThC_{0.76}$ type [104]. It is formed in the range $ZrC_{0.69}$–$ZrC_{0.77}$ and has a stoichiometric composition corresponding to the carbide $ZrC_{0.75}$. The statement on the stoichiometric composition $ZrC_{0.75}$ seems doubtful.

Direct structural evidence for ordering in nonstoichiometric hafnium carbide HfC_y is lacking. Measurements of the magnetic susceptibility [81, 108, 109] revealed abnormal behavior for carbides $HfC_{0.71}$, $HfC_{0.78}$ and $HfC_{0.83}$ at a temperature between 870 and 930 K likely associated with the formation of superstructural short-range order in HfC_y assumed to correspond to the ordered phase Hf_3C_2 in $HfC_{0.71}$ and $HfC_{0.78}$, and to the ordered phase Hf_6C_5 in $HfC_{0.83}$.

Trigonal (space group $P3_1$) [77, 111, 112] and monoclinic (space group $C2$) [113, 114] ordered phases of V_6C_5 type, and also the cubic (space group $P4_332$) ordered phase V_8C_7 [112, 120–124] were identified in vanadium carbide VC_y. Also [115] fine precipitates of monoclinic (space group $C2/m$) ordered carbide V_6C_5 in high-vanadium cast iron were identified. This ordered phase is similar in structure to ordered monoclinic niobium carbide Nb_6C_5 (space group $C2/m$) [116–119].

Ordering in nonstoichiometric niobium monocarbide NbC_y was studied by electron [113, 139] and neutron diffraction [116–119, 126, 128, 129]. It was found that annealing at a temperature below 1300 K caused the formation of the ordered phase Nb_6C_5 over a broad composition range around $NbC_{0.83}$. A trigonal (space group $P3_1$) structure analogous to V_6C_5 [111] was proposed [113, 128, 129, 139] for describing the observed superstructure reflections. X-ray examination of the ordered phase of niobium monocarbide revealed only one ordered phase Nb_6C_5. This phase was initially believed [140, 141] to have trigonal symmetry, but later the same authors [127] came to the conclusion that the phase Nb_6C_5 was monoclinic. The most comprehensive study of ordering in nonstoichiometric niobium carbide NbC_y was performed by [116–119, 126].

Samples of niobium carbide NbC_y of various compositions within the homogeneity interval of the basic cubic phase were produced by solid-state sintering of niobium powders and carbon black. The synthesis conditions are shown in Fig. 3.1. In order to produce niobium carbide with different degrees of order, synthesized samples underwent thermal treatment along four routes with different annealing temperatures, annealing times, and cooling rates (Fig. 3.2).

The structure of the ordered phase of niobium carbide was determined by neutron and X-ray-diffraction. Long-term annealing of NbC_y samples ($0.81 \leq y \leq 0.88$) led to the appearance of superstructure lines in neutron diffraction and X-ray patterns (Figs. 3.13 and 3.14). As seen from Fig. 3.13, neutron diffraction patterns for the annealed samples $NbC_{0.81}$, $NbC_{0.83}$, $NbC_{0.845}$, and $NbC_{0.88}$ contain a similar system of superstructure reflections. The analysis of the intensity and the position of the superstructure reflections showed that they correspond to the ordered phase Nb_6C_5 with monoclinic (space group *C*2/*m*) symmetry.

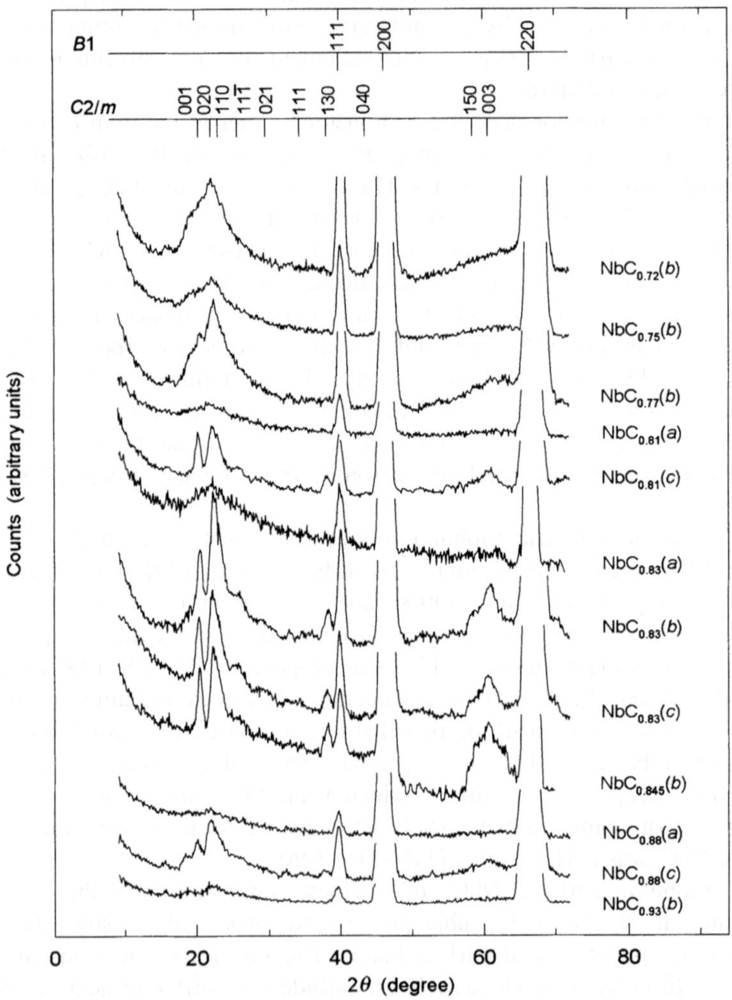

Fig. 3.13. Neutron diffraction patterns of niobium carbide NbC_y with different degrees of ordering upon thermal treatment in routes (*a*), (*b*) and (*c*) shown in Fig. 3.2. The neutron wavelength is $\lambda = 0.1694$ nm [119, 126]

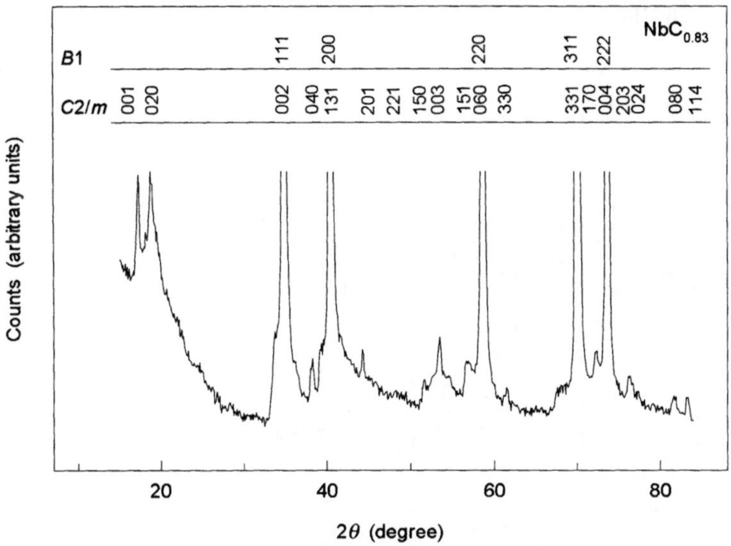

Fig. 3.14. X-ray pattern of ordered niobium carbide $NbC_{0.83}$ [119]

The structure of the ordered phase of niobium carbide was determined by neutron and X-ray-diffraction. Long-term annealing of NbC_y samples ($0.81 \leq y \leq 0.88$) led to the appearance of superstructure lines in neutron diffraction and X-ray patterns (Figs. 3.13 and 3.14). As seen from Fig. 3.13, neutron diffraction patterns for the annealed samples $NbC_{0.81}$, $NbC_{0.83}$, $NbC_{0.845}$, and $NbC_{0.88}$ contain a similar system of superstructure reflections.

The analysis of the intensity and the position of the superstructure reflections showed that they correspond to the ordered phase Nb_6C_5 with monoclinic (space group *C2/m*) symmetry. Researchers [21, 126] calculated the parameters of the unit cell parameters of the ordered phase Nb_6C_5 using Ito's method [142]. The monoclinic unit cell of the ordered niobium carbide Nb_6C_5 was found to belong to space group *C2/m* with the following parameters: $a = c = 0.54605$ nm, $b = 0.94579$ nm, $\alpha = \gamma = 90°$, and $\beta = 109.471°$. This unit cell is shown in Sect. 5.4.4 and comprises two formula units Nb_6C_5, i.e. 12 niobium atoms, 10 carbon atoms and 2 vacancies. A primitive translational triclinic cell in the vacancy sublattice with parameters $a = b = c = 0.54605$ nm, $\alpha = \beta = 99.6°$ and $\gamma = 120°$ can be selected in the ordered structure of niobium carbide. This cell comprises six niobium atoms, five carbon atoms and one vacancy (Fig. 3.15).

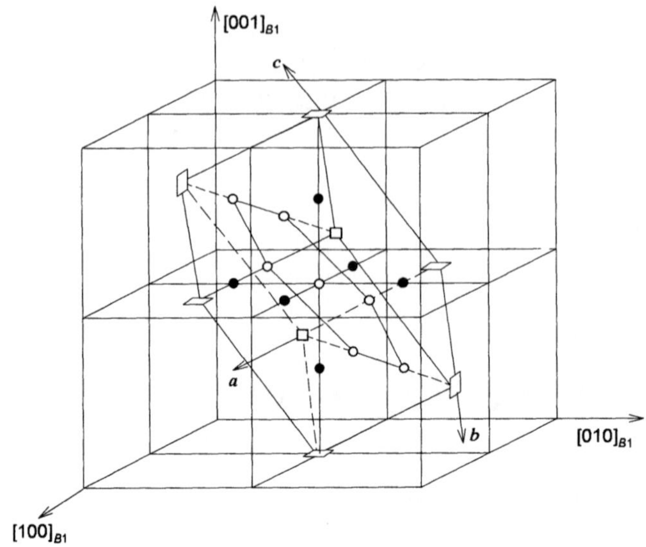

Fig. 3.15. Primitive translational cell with respect to the vacancy sublattice of monoclinic (space group *C*2/*m*) ordered phase Nb_6C_5: (○) interstitial atoms, (•) niobium atoms, (□) vacancies [21, 126]

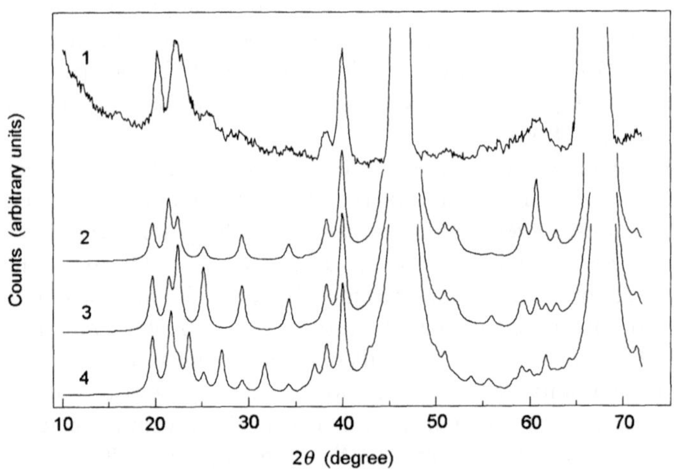

Fig. 3.16. Comparison of experimental (1) (data of [126]) and theoretical (2, 3, 4) neutron diffraction patterns (λ = 0.1694 nm) of ordered phase Nb_6C_5 with the following structure: (2) monoclinic (space group *C*2/*m*) structure allowing for atomic displacements [117, 126], (3) trigonal (space group $P3_1$) structure type V_6C_5 [111], (4) monoclinic (space group *C*2) structure type V_6C_5 [113, 114]

Calculations of theoretical neutron diffraction spectra for the monoclinic structure [116–119, 126] in agreement with earlier descriptions of the ordered phase M_6C_5 [111, 113, 114] shows that the monoclinic (space group *C*2/*m*) structure (Fig. 3.16) fits the observed data best (with *R*-factor equal to 0.71).

The structure of the ordered carbide Nb_6C_5 identified in [116–119, 126] is very similar to the structure of V_6C_5 [111, 113, 114]. Figure 3.17 shows correlations between these structures. If one considers the nonmetallic sublattice only (the symmetry of the metallic sublattice remains virtually unchanged during the ordering), complete filled planes containing carbon atoms only will alternate with partially filled (defective) planes containing carbon atoms and vacancies along the direction $[1\bar{1}1]_{B1}$. The vacancies occupy one third of the carbon positions and they form regular hexagons in the defective planes. Figure 3.17 presents the mutual arrangement of the defective planes responsible for three different structures of M_6C_5 type.

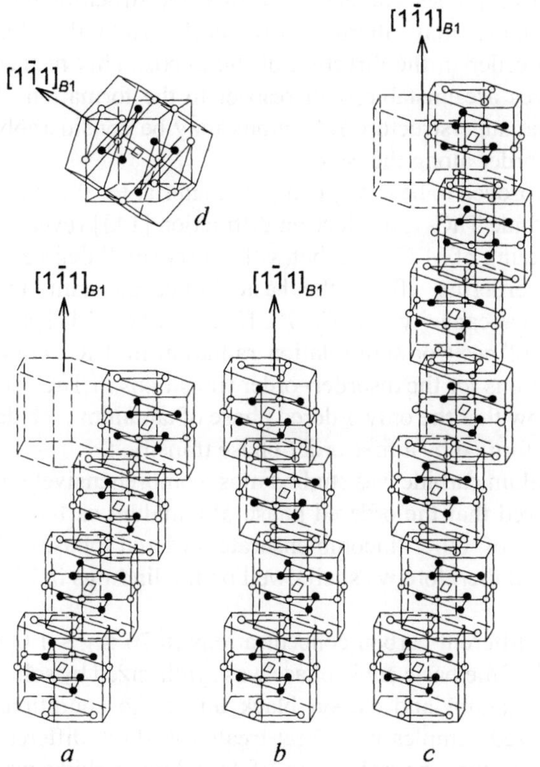

Fig. 3.17. Models of related structures of ordered carbide phases of M_6C_5 type: (*a*) monoclinic (space group *C*2/*m*) structure M_6C_5, (*b*) trigonal (space group $P3_1$) structure M_6C_5, (*c*) monoclinic (space group *C*2) structure M_6C_5, (*d*) position of quasimolecule $M_6C_5\square$ in a lattice type *B*1, (○) carbon atoms, (•) metal atoms, (□) vacancies

A monoclinic (space group *C*2/*m*) structure is formed when atoms and vacancies of equidistant defective planes are successively displaced three times relative to one another in the direction $\{1/2\ -1/2\ 1\}_{B1}$ (Fig. 3.17*a*). A trigonal structure of V_6C_5 type appears upon successive helical displacement along the vectors $\{1/2\ -1/2\ 1\}_{B1}$, $\{1/2\ -1\ 1/2\}_{B1}$ and $\{1\ -1/2\ 1/2\}_{B1}$ [111] (Fig. 3.17*b*). A monoclinic (space group *C*2) superstructure proposed for the low-temperature ordered phase V_6C_5 [113, 114] is formed when atoms and vacancies of defective planes are displaced alternately along vectors $\{1/2\ -1/2\ 1\}_{B1}$ and $\{1/2\ -1\ 1/2\}_{B1}$ (Fig. 3.17*c*). As seen from Fig. 3.17, all the superstructures of M_6C_5 type can be visualized as a set of specially arranged quasimolecules $M_6C_5\square$ (Fig. 3.17*d*) differing only in the mutual arrangement of the defective planes. These superstructures have identical short-range order and they are similar up to the third coordination sphere with respect to the metal atom. Consequently, they have nearly equal formation energies, and a "mixed" sequence in the mutual arrangement of partially defective planes characteristic of each structure may easily develop. As a result, planes may alternate irregularly. Actually, this corresponds to a disturbance of order in the direction of the *c*-axis. This further implies that these superstructures are unstable with respect to the formation of planar antiphase domains. Some superstructure reflections may be considerably attenuated due to this disturbed order along the *c*-axis.

Cubic tantalum carbide TaC_y proved to be a very complex material for the study of ordering. Investigation of carbide $TaC_{0.83}$ by electron diffraction [143] revealed a diffuse band indicating an ordering of M_6C_5 type, but with a very small degree of order. There is evidence that ordering affects the basic lattice constant, the magnetic susceptibility and the heat capacity of TaC_y [78, 131, 132, 144–147], and the angular correlation spectra of positron annihilation radiation in TaC_y [148, 149]. Thermodynamical calculations for the disorder–order transitions in TaC_y [1, 2, 84, 85, 88, 105, 106, 150] show that the only ordered phase of tantalum carbide is a superstructure of M_6C_5 type formed by a first-order phase transition.

The structure of ordered tantalum carbide was studied most comprehensively in refs. [79,133,134]. It was observed that the ordered phase of tantalum carbide is incommensurate. Any evidence of other incommensurate ordered phases of carbides or nitrides is lacking, and therefore we shall dwell on the findings in [79, 133, 134] more closely.

Tantalum carbide TaC_y with different carbon concentrations ($0.70 \le y \le 1.00$) within the homogeneity interval of the basic cubic phase was synthesized by solid-state sintering of powders of tantalum and carbon black under the conditions shown in Fig. 3.1. The synthesized samples were heat treated in three different ways (Fig. 3.2). The structure of the ordered phase of tantalum carbide was examined by neutron diffraction ($\lambda = 0.1694$ nm) with step-by-step scanning of $\Delta(2\theta) = 0.1°$ in the 2θ angle range from 12° to 90°. Long counting times were used to allow for the observation of weak superstructure reflections. The background intensity was about 5000 pulses. The neutron diffraction spectra contained higher harmonic reflections from radiation with wavelength $\lambda/2$ and $\lambda/3$ corresponding to structural reflections $(200)_{B1}$, $(220)_{B1}$ and $(222)_{B1}$. The intensity of these reflections accounts for about 0.5% of the intensity of the main reflections.

Figure 3.18 presents typical neutron diffraction spectra of nonstoichiometric carbide TaC_y. Additional reflections are observed along with the reflection $(111)_{B1}$. The intensity of the reflection $(111)_{B1}$ is small and approaches the intensity of superstructure reflections, because it varies as $(f_{Ta} - f_C)^2$, while atomic scattering amplitudes of tantalum (f_{Ta}) and carbon (f_C) are nearly equal ($f_{Ta} = 0.70 \times 10^{-12}$ and $f_C = 0.665 \times 10^{-12}$ cm). Superstructure reflections were observed in the spectra of annealed tantalum carbide in the interval of $TaC_{0.79}$–$TaC_{0.89}$. The intensity seems to be determined by the conditions of the heat treatment. Weak superstructure reflections were also observed for quenched carbide $TaC_{0.83}$ (Fig. 3.18). Probably, the quenching rate (200 K min^{-1}) was smaller than the ordering rate and it was insufficient for tantalum carbide to reach a fully developed disordered state. Superstructure reflections are much broader than the basic reflection $(111)_{B1}$. Their intensity quickly diminishes with increasing diffraction angle 2θ, and they are virtually unobservable beyond $2\theta > 50°$.

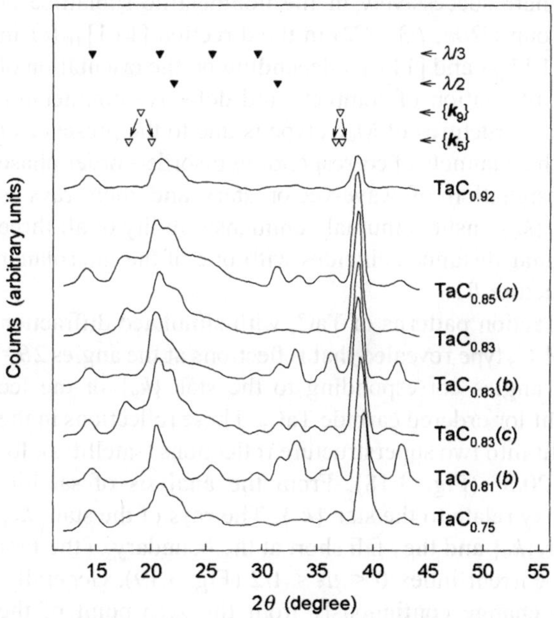

Fig. 3.18. Neutron diffraction patterns of tantalum carbide TaC_y after thermal treatment in ways (*a*), (*b*) and (*c*) shown in Fig. 3.2. The neutron wavelength is $\lambda = 0.1694$ nm; ▼ denotes positions of reflections due to diffraction with wavelengths $\lambda/2$ and $\lambda/3$; ∇ stands for positions of superstructure reflections corresponding to stars $\{\boldsymbol{k}_9\}$ and $\{\boldsymbol{k}_5\}$. Arrows show splitting of reflections corresponding to the star $\{\boldsymbol{k}_9\}$ and are observed in spectra of commensurate superstructures of type M_6C_5, into experimental superstructure reflections corresponding to the star $\{\boldsymbol{k}_5\}$ [79]

This suggests that the domains of the ordered phase of tantalum carbide are much smaller than the grains of disordered carbide with a basic structure of $B1$ type. From this it may be inferred that a structure with ideal long-range order does not develop in TaC_y even after long-time annealing, although the ordering itself seems to be quick.

The angular range $2\theta = 19–25°$ corresponds to wave vectors limited to the first Brillouin zone in the neutron diffraction pattern of tantalum carbide. The superstructure reflections present in this zone suggest that static concentration waves with wave vectors restricted to the first Brillouin zone exist in the crystal. In accordance with [1, 21, 22], superstructures of M_6C_5 type with space groups $C2/m$, $P3_1$ and $C2$ are formed in nonstoichiometric monocarbides of group V transition metals upon ordering. The position of superstructure reflections for a basic cubic lattice with constant $a_{B1} = 0.4428$ nm corresponding to the carbide $TaC_{0.83}$ were calculated for those superstructures. Comparison of the observed positions shows that the superstructure of tantalum carbide differs from known superstructures of M_6C_5 type.

For an interpretation the following model has been proposed. Complete and partially defective planes alternate successively in the nonmetallic sublattice of M_6C_5 superstructures (space group $C2/m$, $P3_1$, $C2$) in the direction $[1\bar{1}1]_{B1}$ (or in equivalent directions $[111]_{B1}$, $[\bar{1}11]_{B1}$ and $[11\bar{1}]_{B1}$ depending on the orientation of the unit cell) (Fig. 3.17). The alternation of complete and defective nonmetallic atomic planes $(111)_{B1}$ in ordered structures of M_6C_5 type is due to the presence of the ray $\boldsymbol{k}_9^{(3)}$ of the star $\{\boldsymbol{k}_9\}$ in the channels of corresponding disorder–order phase transitions [21, 151]. The designation of wave-vector stars and their rays is described in Chap. 2. The star $\{\boldsymbol{k}_9\}$ ensures mutual commensurability of all three structures, because the interplanar distance coincides with one of the interplanar distances of the basic cubic structure $B1$.

Comparison of neutron diffraction patterns of TaC_y with simulated diffraction spectra for superstructures of M_6C_5 type revealed that reflections at the angles $2\theta =$ 19.0°, 37.0°, 59.0° and other angles corresponding to the star $\{\boldsymbol{k}_9\}$ of the fcc nonmetallic sublattice are absent for ordered carbide TaC_y. These reflections in the diffraction patterns of TaC_y split into two superstructure reflections (satellites), for example at $2\theta = 18.0°$ and 20.5° (Fig. 3.18). From the analysis of satellite positions it follows that they may relate to the star $\{\boldsymbol{k}_5\}$. The rays of the star $\{\boldsymbol{k}_5\}$ are collinear to those of the star $\{\boldsymbol{k}_9\}$ and they fall short at the boundary of the first Brillouin zone. They have the current index $0 < \mu_5 < 1/2$ (Fig. 3.19). Generally, the location of rays $\boldsymbol{k}_5^{(j)}$ can change continuously from the zero point of the reciprocal space (the center of the Brillouin zone) to the L point at the zone boundary of the first Brillouin zone. Points L correspond to the rays of the star $\{\boldsymbol{k}_9\}$. The presence of star $\{\boldsymbol{k}_5\}$ rays in the phase transition channel actually means that the tantalum carbide superstructure is incommensurate. The numerical value of the index μ_5 can be deduced from experimental neutron diffraction patterns (Fig. 3.18). The ray $\boldsymbol{k}_9^{(3)}$ is collinear to the ray $\boldsymbol{k}_5^{(6)} = \{\mu_5, -\mu_5, \mu_5\}$. The modulus of the diffraction vector $\boldsymbol{q}$, which coincides with the wave vector $\boldsymbol{k}_5^{(6)}$, is $|\boldsymbol{q}| = |\boldsymbol{k}_5^{(6)}| = \sqrt{3}\mu_5$. The relation $q = (2a_{B1}\sin\theta)/\lambda$ yields

$$\mu_5 = (2a_{B1}\sin\theta_5)/\sqrt{3}\lambda, \tag{3.30}$$

where $\theta_5 \approx 9.00$–$9.02°$ is the angle corresponding to the superstructure reflection due to a wave vector with length $|\boldsymbol{k}_5^{(6)}|$. Calculations showed that $\mu_5 \approx 0.473$ irrespective of the composition of the tantalum carbide (within the accuracy of the diffraction experiment). The reciprocal superstructure vector $\boldsymbol{k}_5^{(5)} = -\boldsymbol{k}_5^{(6)}$ corresponding to the superstructure vector $\boldsymbol{k}_5^{(6)}$ is present in the first Brillouin zone. The vector $\boldsymbol{k}_5^{(5)}$ is not equivalent to the vector $\boldsymbol{k}_5^{(6)}$ and therefore it is included in the phase transition channel along with the vector $\boldsymbol{k}_5^{(6)}$. Thus, the disorder–order phase transition channel associated with the formation of an incommensurate ordered phase in nonstoichiometric tantalum carbide includes rays $\boldsymbol{k}_5^{(6)} \approx 0.473\boldsymbol{b}_2$ and $\boldsymbol{k}_5^{(5)} = -\boldsymbol{k}_5^{(6)}$ (Fig. 3.19).

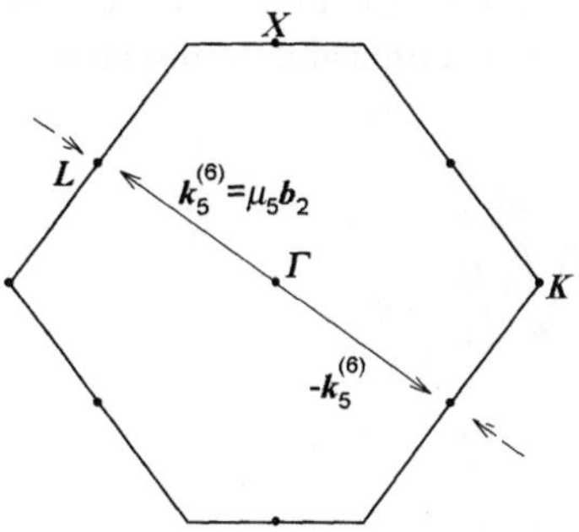

Fig. 3.19. Positions of superstructure vectors $\boldsymbol{k}_5^{(6)}$ and $-\boldsymbol{k}_5^{(6)}$ in the cross-section (100) of the first Brillouin zone. Dotted superstructure vectors are formed by nearest sites of the reciprocal lattice. Two superstructure vectors $\boldsymbol{k}_5^{(j)}$ with $\mu_5 \approx 0.473$ near the L point cause the appearance of two similar superstructure reflections in diffraction spectra at $2\theta \approx 18.0°$ and $20.5°$ (Γ, K, L and X are special symmetry points of the first Brillouin zone)

Let us consider functions that describe the probability of a carbon atom to be on a site of the nonmetallic plane $(11\bar{1})_{B1}$, i.e. the degree of occupancy $P^{(111)}$ by carbon atoms of nonmetallic planes $(111)_{B1}$ in carbide MC_y having any degree of long-range order η. For commensurate superstructures of M_6C_5 type, if one considers their corresponding distribution functions [11, 30, 141], the degree of occupancy $P_{com}^{(11\bar{1})}$ can be written as

$$P_{com}^{(11\bar{1})} = y - (\eta_9/6)\cos(2\pi\mu_9 z), \tag{3.31}$$

where z is the ordinal number of the nonmetallic plane $(11\bar{1})_{B1}$, η_9 is the long-range order parameter corresponding to the star $\{\boldsymbol{k}_9\}$, and $\mu_9 = 1/2$. In the case of an incommensurate ordered structure approaching M_6C_5 type, the degree of occupancy of nonmetallic planes $(11\bar{1})_{B1}$ by C atoms in carbide MC_y has the form

$$P_{\text{inc}}^{(1\bar{1}1)} = y - (\eta_5/6)\cos(2\pi\mu_5 z). \tag{3.32}$$

From equations (3.31) and (3.32) it follows that the maximum and minimum degrees of occupancy of nonmetallic planes $(1\bar{1}1)_{B1}$ by carbon atoms are $P_{\text{max}}^{(1\bar{1}1)} = y + \eta/6$ and $P_{\text{min}}^{(1\bar{1}1)} = y - \eta/6$, where η is η_9 or η_5 for commensurate and incommensurate superstructures respectively. Figure 3.20 presents functions characterizing the degree of occupancy of nonmetallic planes $(1\bar{1}1)_{B1}$ in commensurate and incommensurate ordered structures of M_6C_5 type. It is noted that these functions are physically meaningful for $(1\bar{1}1)_{B1}$ planes only.

Only nonmetallic atomic planes $(1\bar{1}1)_{B1}$ with maximum $P_{\text{max}}^{(1\bar{1}1)}$ and minimum $P_{\text{min}}^{(1\bar{1}1)}$ degrees of occupancy alternate in the direction $[1\bar{1}1]_{B1}$ in commensurate superstructures of M_6C_5 type (Fig. 3.20*a*). Let us label these planes as type 1 and type 2 planes, respectively. The maxima and minima of the function $P^{(1\bar{1}1)}$ do not coincide with nonmetallic planes $(1\bar{1}1)_{B1}$ for the incommensurate superstructure (Fig. 3.20*b*) and the degree of occupancy of these planes by carbon atoms is intermediate between $P_{\text{max}}^{(1\bar{1}1)}$ and $P_{\text{min}}^{(1\bar{1}1)}$ (Fig. 3.20*c*). In other words, the concentration wave corresponding to the star $\{\boldsymbol{k}_5\}$ has maxima and minima that do not coincide with nonmetallic planes $(1\bar{1}1)_{B1}$.

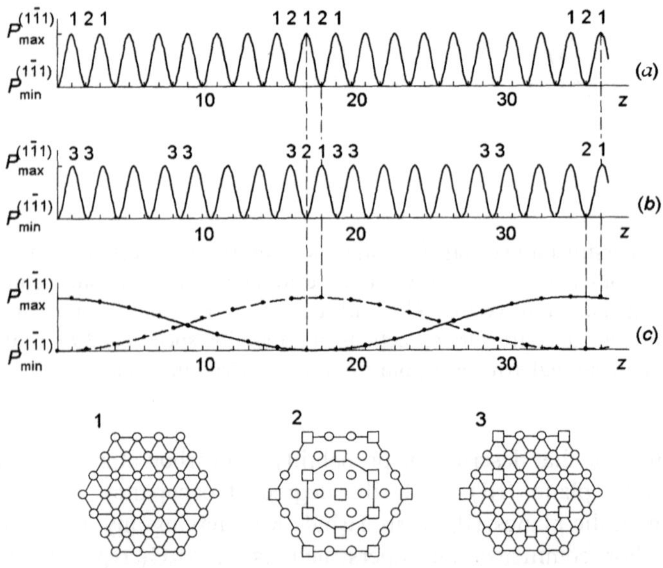

Fig. 3.20. Occupancy $P^{(1\bar{1}1)}$ of nonmetal atomic planes $(1\bar{1}1)_{B1}$ with carbon atoms in the direction $[1\bar{1}1]_{B1}$: (*a*) commensurate superstructure type M_6C_5; (*b*) incommensurate superstructure of tantalum carbide similar to type M_6C_5; (*c*) occupancy $P^{(1\bar{1}1)}$ of odd (solid line) and even (dashed line) nonmetal planes $[1\bar{1}1]_{B1}$ in incommensurate ordered carbide TaC_y; (1) complete nonmetal plane $[1\bar{1}1]_{B1}$ with the maximum occupancy $P_{\text{max}}^{(1\bar{1}1)} = y + \eta/6$; (2) defective ordered nonmetal plane $[1\bar{1}1]_{B1}$ with the minimum occupancy $P_{\text{min}}^{(1\bar{1}1)} = y - \eta/6$; (3) defective nonmetal plane $[1\bar{1}1]_{B1}$ with an intermediate occupancy between $P_{\text{max}}^{(1\bar{1}1)}$ and $P_{\text{min}}^{(1\bar{1}1)}$, (○) carbon atoms, (□) vacancies

As seen from Fig. 3.20*c*, every 18th nonmetallic plane $(1\bar{1}1)_{B1}$ in incommensurate tantalum carbide has a peak value for the occupancy of carbon atoms (i.e. the 18th, 37th, 54th, 73rd, etc.). In contrast in a commensurate superstructure the nonmetallic planes $(1\bar{1}1)_{B1}$ with a maximum value for the occupancy (type 1 planes) occur every other plane (i.e. the 1st, 3rd, 5th, etc.) (Fig. 3.20*a*). Thus, the translation period in the direction $[1\bar{1}1]_{B1}$ increases by a factor of about 18 from 0.5112 nm in the commensurate phase of the M_6C_5 type to 8.9–9.1 nm in incommensurately ordered tantalum carbide. The exact value of the translation period depends on the TaC_y composition and the conditions of the heat treatment.

The mismatch between the maxima and minima of the concentration wave and the nonmetallic planes $(1\bar{1}1)_{B1}$ implies that the probabilities of carbon and vacancy positions being filled in ordered tantalum carbide are largely different from 1 and 0, respectively. Thus, the degree of long-range order in ordered tantalum carbide does not even come close to the maximum values possible. If the long-range order parameters $\eta_9 = \eta_5 = \eta$ are equal, the difference between the degrees of occupancy of nonmetallic planes $(1\bar{1}1)_{B1}$ in incommensurate and commensurate superstructures is

$$P_{\text{inc}}^{(1\bar{1}1)} - P_{\text{com}}^{(1\bar{1}1)} = (\eta/3)\sin[\pi z(\mu_9 + \mu_5)]\sin[\pi z(\mu_9 - \mu_5)]\,. \qquad (3.33)$$

Let us estimate the number of the plane $(1\bar{1}1)_{B1}$ in the nonmetallic sublattice for which the probabilities $P_{\text{inc}}^{(1\bar{1}1)}$ and $P_{\text{com}}^{(1\bar{1}1)}$ will be equal. The condition is $P_{\text{inc}}^{(1\bar{1}1)} - P_{\text{com}}^{(1\bar{1}1)} = 0$. Considering numerical values of μ_9 and μ_5, it follows from (3.33) that $z \approx 37m$ (m = 0, 1, 2, ... being an integer). Thus, the occupancy of nonmetallic planes $(1\bar{1}1)_{B1}$ with carbon atoms in incommensurate and commensurate superstructures will coincide in every 37th plane (Fig. 3.20*c*).

Structural incommensurability means that the ordered phase of TaC_y does not have an exact stoichiometric composition. This, among other things, explains the existence of a weak correlation between the intensity of superstructure reflections and the composition of carbide TaC_y (see Fig. 3.18) as pointed out in [79, 134]. However, this does not imply that any ordered nonstoichiometric phase is incommensurate.

The distribution pattern of carbon atoms and vacancies in ordered TaC_y can also be explained in an alternative way. An ordered structure of nonstoichiometric tantalum carbide can be visualized as sequences $[(CD)_nC]_\infty$ or $[(CD)_nD]_\infty$ of alternating complete C and ordered defective D nonmetallic planes $(1\bar{1}1)_{B1}$. This description represents a long-periodic structure. The parameter μ_5 should vary with the composition of carbide TaC_y having a long range-periodic structure $[(CD)_nC]_\infty$ or $[(CD)_nD]_\infty$. However, such a correlation was not observed within the accuracy of the diffraction experiment.

A common feature of ordering in lower hexagonal carbides M_2C_y of group V transition metals is the fact that ordered nonmetal atomic planes have different degrees for the occupancy of carbon atoms. These planes alternate successively in ordered phases perpendicular to the *c*-axis of the initial structure of *L*'3 type.

Table 3.8. Ordered phases of strongly nonstoichiometric nitrides and oxides of transition metals of IV and V groups

Ordered phase				
composition	region of existence of the ordered phase or nonmetal content (at.%)	lattice type or symmetry	space group	[1] unit cell parameters (lattice constants in nm)
1	2	3	4	5
δ'-Ti_2N	~33.0–33.3	tetragonal	$I4_1/amd$	$a = 0.4144 \approx a_0$, $c = 0.8787 \approx 2a_0$
Ti_2N	$TiN_{0.45}$–$TiN_{0.50}$	tetragonal	$I4_1/amd$	$a = 0.4226 = a_0$, $c = 0.8452 = 2a_0$
δ'-Ti_2N	$TiN_{0.50}$–$TiN_{0.51}$	tetragonal	$I4_1/amd$	
Ti_2N	$TiN_{0.52}$–$TiN_{0.57}$			
Ti_3N_2	$TiN_{0.58}$–$TiN_{0.72}$			
Ti_6N_5	$TiN_{0.77}$–$TiN_{0.90}$			
Ti_6N_5	44.7–45.6	ortho rhombic ?		$a = b = (a_0\sqrt{6})/2$, $c = (a_0\sqrt{18})/2$, $\beta = 109.47°$
V_2N	~33.3	hexagonal, type ε-Fe_2N		
V_9N_4	~33.0	hexagonal	$P6_322$	
δ'-VN_y	44.0–47.0	tetragonal		$a = c = 0.815 \approx 2a_0$
$V_{32}N_{26}$	~$VN_{0.78}$	tetragonal	$P4_2/nmc$	
Nb_4N_3	~43.0	tetragonal	$I4/mmm$	$a = 0.4382 \approx a_0$, $c = 0.8632 \approx 2a_0$
$TiO_{1.0}$ (Ti_5O_5)	$TiO_{0.9}$–$TiO_{1.1}$	monoclinic	$A2/m$, Am	$a \approx a_0\sqrt{2}$, $b \approx a_0\sqrt{5}$, $c \approx a_0$, $\gamma \approx 107.53°$
$Ti_{90}O_{90}$	~50.0	cubic	$Fm\bar{3}m$	$a = 1.254 \approx 3a_0$
Ti_5O_6 ($TiO_{1.20}$)	~54.5	ortho-rhombic	$Immm$, $Imm2$	$a = a_0/\sqrt{2}$, $b = 3a_0/\sqrt{2}$, $c = a_0$
Ti_4O_5	~55.5	tetragonal	$I4/m$	$a = a_0\sqrt{10}/2$, $c = a_0$
Ti_4O_5 ?	~54.1	cubic	$Pn\bar{3}m$	$a = 5a_0$
Ti_5O_5	$TiO_{0.9}$–$TiO_{1.1}$	monoclinic	$C2/m$ or $A12/m1$	$a \approx a_0\sqrt{2}$, $b \approx a_0$, $c \approx a_0\sqrt{5}$, $\gamma \approx 108.43°$
$V_{52}O_{64}$	$VO_{1.20}$–$VO_{1.30}$	tetragonal	$I4_1/amd$	$a = 1.172 \approx (2\sqrt{2})a_0$ $c = 0.8245 \approx 2a_0$
$VO_{1.17}$	$VO_{1.16}$–$VO_{1.20}$	tetragonal	$I4_1/a$	$a = 2.608$, $c = 0.830$
$V_{204}O_{256}$	~56.0	tetragonal		$a = 1.6623 \approx 4a_0$, $c = 1.6515 \approx 4a_0$
$Nb_{0.75}O_{0.75}$ (Nb_3O_3)	~50.0	cubic primitive		$a = 0.421$

[1] a_0, b_0, c_0 stand for the disordered basic phase lattice constants.

Disordered basic phase		Refe rences	Remarks
compo sition	lattice type		
6	7	8	9
TiN_y	*B*1 (NaCl)	152–156	$T_{trans} < 900$ K
TiN_y	*B*1 (NaCl)	157	Annealing temperature 800 K
TiN_y	*B*1 (NaCl)	158–161	Metastable phase [162]. Temperature interval of stability from 800 to 1180 K
TiN_y	*B*1 (NaCl)	102, 107	Calculation. $T_{trans} < 1070$ K
TiN_y	*B*1 (NaCl)	102, 107	Calculation. $T_{trans} < 1060$ K
TiN_y	*B*1 (NaCl)	102, 107	Calculation. $T_{trans} < 1060$ K
TiN_y	*B*1 (NaCl)	163	Calculation. $T_{trans} < 600$ K. Unit cell [163] coincide completely with monoclinic (space group *C*2/*m*) unit cell for Nb_6C_5 phase [119]
V_2N_y	*L′*3 (W_2C)	139	
V_9N_4	*L′*3(W_2C)	164	Phase is stable below 900 K
VN_y	*B*1 (NaCl)	92	Phase is stable below 800 K
VN_y	*B*1 (NaCl)	164	First-kind phase transition. $T_{trans} = 790$ K
NbN_y	*B*1 (NaCl)	165, 166	Phase is stable below 1700 K
TiO_y	*B*1 (NaCl)	167–170	Phase is stable below 1250 K
TiO_y	*B*1 (NaCl)	171	Phase is stable in interval 1250-1520 K
TiO_y	*B*1 (NaCl)	171	Phase is stable below 1090 K
TiO_y	*B*1 (NaCl)	170, 171	Phase is stable below 990 K
TiO_y	*B*1 (NaCl)	172	
TiO_y	*B*1 (NaCl)	173	Sublattices of the ideal monoclinic Ti_5O_5 phase each contain ~16.7 at. % vacancies
VO_y	*B*1 (NaCl)	174	Annealing temperature 870-1070 K
VO_y	*B*1 (NaCl)	174	Prolonged annealing at 370, 1070 и 870 K
VO_y	*B*1 (NaCl)	175	Phase is stable below 1170 K
NbO	*B*1 (NaCl)	176, 177	One Nb atom and one O atom are absent in a unit cell with *B*1 structure

Table 3.8 summarizes data for the structure of ordered phases of nonstoichiometric nitrides and cubic oxides of group IV and V transition metals.

The ordering phenomena for nitrides are less clear than the ordering of carbides. Ordered phases were detected in cubic titanium, vanadium, and niobium (TiN_y, NbN_y, VN_y) nitrides, and also in lower hexagonal vanadium nitride V_2N_y. Data on the ordering of cubic zirconium and hafnium nitrides are lacking.

A tetragonal (space group $I4_1/amd$) ordered phase Ti_2N (or δ'-Ti_2N) with a narrow homogeneity interval was observed in titanium nitride TiN_y [152–158]. In accordance with [159, 160, 162], the phase δ'-Ti_2N is metastable and exists only in a narrow temperature interval from 900 to 1180 K.

The phase diagram for the Ti–N system taking into account ordering in nitride TiN_y [102, 107] was calculated by the order parameter functional method. Calculations showed that ordered phases of Ti_2N, Ti_3N_2, and Ti_6N_5 types can appear in various concentration intervals at temperatures below 1060 K. The formation of an ordered phase of Ti_6N_5 type at a temperature below 600 K is also confirmed by calculations [163] made by the Monte Carlo and cluster variation methods. In accordance with [163], the superstructure Ti_6N_5 has rhombic symmetry. The unit cell parameters and the arrangement of nonmetal atoms and vacancies in this superstructure are similar to those in the monoclinic (space group $C2/m$) superstructure Nb_6C_5 [116–119, 126].

In accordance with [164], cubic mononitride VN_y and lower hexagonal nitride V_9N_4, rather than V_2N_y as commonly assumed (see, for example, [16, 28, 178]), are stable in the V–N system above 900 K. Ordering of cubic vanadium carbide $VN_{0.78}$ is accompanied by the appearance of the tetragonal (space group $P4_2/nmc$) phase $V_{32}N_{26}$ and it takes place as a phase transition of the first kind below 793 K [164].

Ordered phases of nonstoichiometric cubic titanium, vanadium and niobium mononitrides have tetragonal structures and they differ considerably from the corresponding ordered carbides. The vacancies are located at sites of a body-centered tetragonal lattice in the unit cells of ordered titanium nitride Ti_2N (space group $I4_1/amd$) and niobium nitride Nb_4N_3 (space group $I4/mmm$).

Depending on the oxygen concentration and the conditions of heat treatment, a variety of ordered phases with cubic, tetragonal, rhombic or monoclinic lattices are formed upon ordering in cubic titanium (TiO_y) and vanadium (VO_y) monoxides (Table 3.8). Diverse superstructures of cubic titanium and vanadium monoxides are due to the presence of structural vacancies both in the metallic and oxygen sublattices. Superstructures of cubic titanium monoxide are described in detail in [179]. The authors in ref. [173] show that the metallic and nonmetallic sublattices of the ideal monoclinic (space group $C2/m$) Ti_5O_5 superstructure each contain ~16.7 at.% vacancies implying that the vacancy content in the titanium and oxygen sublattices of the disordered titanium monoxide $TiO_{1.0}$ is also equal to 16.7 at.% rather than to 15–16 at.%, as was assumed earlier [167, 180].

First principle calculations by the local density approximation were made in [181] on the electronic structure of the low-temperature ordered phase of titanium monoxide. The results indicated that the superstructure contains 16.7 at.% vacancies in each sublattice, that the composition corresponds to Ti_5O_5, and that the structure is monoclinic. Comparative calculations for superstructures Ti_3O_3 with 25 at.% vacancies and Ti_7O_7 with 12.5 at.% vacancies showed that these structures are less stable than the structure of Ti_5O_5.

Cubic niobium monoxide $Nb_{0.75}O_{0.75}$ with a primitive lattice can either be considered as an ordered phase of cubic oxide NbO (with the *B*1 structure) or as an independent phase. In the former case, the cubic unit cell contains 25 at.% vacancies each in the niobium and oxygen sublattices, i.e. one niobium atom and one oxygen atom is absent at the vertex and at the center of the cubic cell of *B*1 type. However, niobium monoxide has no homogeneity interval and it can therefore be considered as an independent phase with a unique structure of NbO type, which contains free sites, but no vacancies.

Few reports are available on the ordering in substitutional solid solutions based on transition-metal carbides, nitrides, and oxides. A neutron diffraction investigation of titanium and zirconium carbonitrides, MC_xN_y, [182, 183] has shown that ordering is absent in low-vacancy carbonitrides (with $x + y \geq 0.75$). In titanium carbonitride with a large vacancy content $(x + y < 0.74)$ an ordered cubic (space group $Fd\bar{3}m$) phase arises where the lattice constant is doubled compared to the disordered carbonitride. The observed ordering is due to filling all 16(*c*) and some 16(*d*) positions by carbon and nitrogen atoms whereas structural vacancies are in 16(*d*) positions only. No separation into a carbon sublattice and a nitrogen sublattice occurs. According to [184], a low-temperature trigonal (space group $R\bar{3}m$) ordered phase and a high-temperature cubic (space group $Fd\bar{3}m$) ordered phase form in titanium carbonitrides TiC_xN_y with $x + y \approx 0.63$. Increasing the nitrogen content in TiC_xN_y leads to a lowering of the disorder–order transition temperature [183, 184]. Indirect evidence of ordering in titanium carbonitrides $Ti(C_xN_{1-x})_{0.82}$ is provided by discontinuities in the heat capacity at about 1000 K [185]. Apparently, these effects are due to ordering of nonmetal interstitial atoms and vacancies in $Ti(C_xN_{1-x})_{0.82}$ following M_6X_5-type ordered phases.

The authors in ref. [186] suggest ordering of titanium carbonitrides TiC_xN_y containing 58–65 at.% Ti and less than 20 at.% N. They observed a second cubic phase after a carbonitride solid solution was annealed at 770 K. The lattice constant of the second phase is smaller as compared to the initial disordered phase. The second phase is thought to be an ordered carbonitride phase [186]. This idea is questionable since the main reflections and the additional reflections have nearly equal intensity in the X-ray patterns. Most probably, the solid solutions are inhomogeneous and the two observed cubic phases are titanium carbonitrides of very similar compositions.

When ordering takes place in titanium oxycarbide, the hypothetic cubic ordered phase $TiC_{0.5}O_{0.5}$ is characterized by the presence of a carbon sublattice and an oxygen sublattice [187, 188]. However, no information is given in [187, 188] concerning the concentration of structural vacancies in the nonmetal and metal sublattices of oxycarbides TiC_xO_y $(x + y \approx 1.0)$ of different compositions. Cubic titanium monoxide $TiO_{1.0}$ actually has the composition $Ti_{0.83}O_{0.83}$ and contains structural vacancies in both sublattices. Therefore oxycarbide TiC_xO_y with a large oxygen content ($y > 0.4$) contains vacancies in the metal and the nonmetal sublattices. In addition, titanium monoxide $TiO_{1.0}$ forms an ordered monoclinic phase Ti_5O_5 [173]. This fact was overlooked by the authors of [187, 188]. According to [189], the ordered oxycarbide $Ti_2(C_xO_y)$ has a cubic unit cell (space group $Fd\bar{3}m$) with a doubled lattice constant $a = 2a_{B1}$ and it forms in a composition interval from $TiC_{0.5}O_{0.2}$ to $TiC_{0.1}O_{0.6}$.

In $Ti_xV_{1-x}C_{0.5}$ solid solutions with $0.2 \le x \le 0.3$, the ordered anti-$CaCl_2$-type phase has been observed [190]. The authors of [191] have found that the intrusion of hydrogen in nonstoichiometric titanium carbide TiC_y ($0.55 \le y \le 0.71$) does not alter the type of cubic Ti_2C superstructure.

References

1 A. I. Gusev, A. A. Rempel: Phase diagrams of metal-carbon and metal-nitrogen systems and ordering in strongly nonstoichiometric carbides and nitrides, Phys. Stat. Sol. (a) **163**, 273–304 (1997)

2 A. I. Gusev, A. A. Rempel: Atomic ordering and phase equilibria in strongly nonstoichiometric carbides and nitrides, in *Materials Science of Carbides, Nitrides and Borides*, Y. G. Gogotsi, R. A. Andrievski (Eds.) (Kluwer Academic Publishers, Dordrecht 1999) pp. 47–64

3 J. S. Anderson: The thermodynamics and theory of nonstoichiometric compounds, in *Problems of Nonstoichiometry*, A. Rabenau (Ed.) (North Holland Publ., Amsterdam-London 1970) pp. 1–76; Defect chemistry and non-stoichiometric compounds, in *Modern Aspects of Solid State Chemistry*, C. N. R. Rao (Ed.) (Plenum Press, New York 1970) pp. 29–105

4 A. A. Rempel, A. I. Gusev: Preparation of disordered and ordered highly nonstoichiometric carbides and evaluation of their homogeneity, Fiz. Tverd. Tela **42**, 1243–1249 (2000) (in Russian). (Engl. transl.: Physics of the Solid State **42**, 1280–1286 (2000))

5 W. Lengauer: The temperature gradient diffusion couple technique: an application of solid-solid phase reactions for phase diagram imaging, J. Solid State Chem. **91**, 279–285 (1991)

6 N. T. Kuznetsov: Precursors for carbide, nitride and boride synthesis, in *Materials Science of Carbides, Nitrides and Borides*, Y. G. Gogotsi, R. A. Andrievski (Eds.) (Kluwer Academic Publishers, Dordrecht 1999) pp. 223–246

7 A. G. Merzhanov, S. Yu. Sharivker: Self-propagating high-temperature synthesis of carbides, nitrides and borides, in *Materials Science of Carbides, Nitrides and Borides*, Y. G. Gogotsi, R. A. Andrievski (Eds.) (Kluwer Academic Publishers, Dordrecht 1999) pp. 205–222

8 M. S. El-Eskandarany, M. Omori, T. Kamiyama, T. J. Konno, K. Sumiyama, T. Hirai, K. Suzuki: Mechanically induced carbonization for formation of nanocrystalline TiC alloy, Sci. Reports of Res. Inst. Tohoku Univ. (Sendai, Japan) **43**, 181–193 (1997)

9 W. Weimer (Ed.): *Carbide, Nitride and Boride Materials Synthesis and Processing* (Chapman & Hall, London 1997) 671 pp.

10 Y. G. Gogotsi, R. A. Andrievski (Eds.): *Materials Science of Carbides, Nitrides and Borides* (Kluwer Academic Publishers, Dordrecht 1999) 360 pp.
11 L.-M. Berger, M. Hermann, A. I. Gusev, A. A. Rempel: Verfahren zur Herstellung nichtstöchiometrischer Carbide definierter Zusammensetzung. Offenlegungsshrift DE 198 07 589 A 1. Int. Cl.[6]: C 01 B 31/30 (C 04 B 35/36). Bundesrepublik Deutschland: Deutsches Patentamt: Anmeldetag 23.02.1998, Offenlegungstag 10.09.1998. 3 pp.
12 A. A. Rempel, S. V. Rempel, A. I. Gusev: Quantitative assessment of homogeneity of nonstoichiometric compounds, Doklady Akademii Nauk **369**, 486–490 (1999) (in Russian). (Engl. transl.: Doklady Phys. Chem. **369**, 321–325 (1999))
13 M. A. Krivoglaz: *Theory of X-ray and Thermal-Neutron Scattering by Real Crystals* (Plenum Press, New York 1969) 405 pp.
14 G. Hägg: Gezetsmässigkeiten im Kristallbau bei Hydriden, Boriden, Karbiden und Nitriden der Übergangselemente, Z. Phys. Chem. **12**, 33–56 (1931)
15 B. F. Ormont (Ed.): *Compounds with Variable Composition* (Khimiya, Leningrad 1969) 520 pp. (in Russian)
16 H. J. Goldschmidt: *Interstitial Alloys* (Plenum Press, New York 1967) 820 pp.
17 R. A. Andrievski, Ya. S. Umanskii: *Interstitial Phases* (Nauka, Moscow 1977) 240 pp. (in Russian)
18 G. P. Shveikin, S. I. Alyamovskii, Yu. G. Zainulin, A. I. Gusev, V. A. Gubanov, E. Z. Kurmaev: *Compounds of Variable Composition and Their Solid Solutions* (Ural Scientific Centre, Sverdlovsk 1984) 292 pp. (in Russian)
19 A. I. Gusev, S. I. Alyamovskii, Yu. G. Zainulin, G. P. Shveikin: Structural vacancies in compounds of variable composition, Uspekhi Khimii **55**, 2067–2085 (1986) (in Russian). (Engl. transl.: Russ. Chem. Rev. **55**, 1175–1185 (1986))
20 A. I. Gusev, A. A. Rempel: *Thermodynamics of the Structural Vacancies in Nonstoichiometric Interstitial Phases* (Ural Scientific Centre, Sverdlovsk 1987) 114 pp. (in Russian)
21 A. I. Gusev, A. A. Rempel: *Structural Phase Transitions in Nonstoichiometric Compounds* (Nauka, Moscow 1988) 308 pp. (in Russian)
22 A. I. Gusev: *Physical Chemistry of Nonstoichiometric Refractory Compounds* (Nauka, Moscow 1991) 286 pp. (in Russian)
23 A. A. Rempel: *Effects of Ordering in Nonstoichiometric Interstitial Compounds* (Nauka, Ekaterinburg 1992) 232 pp. (in Russian)
24 A. I. Gusev: Phase equilibria in M-X-X' and M-Al-X ternary systems (M - transition metal; X, X' = B, C, N, Si) and crystal chemistry of related ternary compounds, Uspekhi Khimii **65**, 407–451 (1996) (in Russian). (Engl. transl.: Russ. Chem. Rev. **65**, 379–419 (1996))
25 R. B. Kotel'nikov, S. N. Bashlykov, Z. G. Galiakbarov, A.I. Kashtanov: *Fundamentals of Refractory Elements and Compounds* / Handbook (Metallurgiya, Moscow 1969) 376 pp. (in Russian)
26 E. Storms: *The Refractory Carbides* (Academic Press, New York and London 1967) 300 pp.
27 G. V. Samsonov, G. Sh. Upadhaya, V. S. Heshpor: *Physical Material Science of Carbides* (Naukova Dumka, Kiev 1974) 456 pp. (in Russian)
28 L. Toth: *Transition Metal Carbides and Nitrides* (Academic Press, New York and London 1971) 280 pp.
29 R. A. Andrievski, A. G. Lanin, G. A. Rymashevskii: *Strength of High-Melting Compounds* (Metallurgiya, Moscow 1974) 232 pp. (in Russian)

30 A. L. Ivanovskii, V. A. Gubanov, E. Z. Kurmaev, G. P. Shveikin: Electronic structure and chemical bonding in nonstoichiometric refractory compounds on a base of subgroup IVa, Va transition metals, Uspekhi Khimii **52**, 704–742 (1983) (in Russian)
31 T. Ya. Kosolapova (Ed.): *Properties, Production and Application of Refractory Compounds* / Handbook (Metallurgiya, Moscow 1986) 928 pp. (in Russian)
32 H. Holleck: *Binäre und Ternäre Carbid- und Nitridsysteme der Übergangsmetalle* (Gebrüder Borntraeger, Berlin-Stuttgart 1984) 295 SS.
33 R. A. Andrievski, I. I. Spivak: *Strength of Refractory Compounds and Materials* (Metallurgiya, Chelyabinsk 1989) 368 pp. (in Russian)
34 A. L. Ivanovskii, V. P. Zhukov, V. A. Gubanov: *Electronic Structure of High-Melting Transition Metals Carbides and Nitrides* (Nauka, Moscow 1990) 224 pp. (in Russian)
35 V. A. Gubanov, A. L. Ivanovsky, V. P. Zhukov: *Electronic Structure of Refractory Carbides and Nitrides* (Cambridge University Press, Cambridge 1994) 256 pp.
36 G. S. Upadhyaya: *Nature and Properties of Refractory Carbides* (Nova Science Publishers, New York 1996) 545 pp.
37 H. O. Pierson: *Handbook of Refractory Carbides and Nitrides: Properties, Characteristics, Processing and Applications* (Noyes Publications, Westwood 1996) 318 pp.
38 A. I. Gusev: Short-range order and local displacements of the atoms in nonstoichiometric compounds, Uspekhi Khimii **57**, 1595–1621 (1988) (in Russian). (Engl. transl.: Russ. Chem. Rev. **57**, 913–928 (1988))
39 A. I. Gusev: Short-range order in nonstoichiometric transition metal carbides, nitrides and oxides, Phys. Stat. Sol. (b) **156**, 11–40 (1989)
40 A. I. Gusev: Disorder and long-range order in nonstoichiometric interstitial compounds: transition metal carbides, nitrides and oxides, Phys. Stat. Sol. (b) **163**, 17–54 (1991)
41 A. I. Gusev, A. A. Rempel: Superstructures of non-stoichiometric interstitial compounds and the distribution functions of interstitial atoms, Phys. Stat. Sol. (a) **135**, 15–58 (1993)
42 A. A. Rempel: Atomic and vacancy ordering in nonstoichiometric carbides, Uspekhi Fiz. Nauk **166**, 33-62 (1996) (in Russian). (Engl. transl.: Physics – Uspekhi **39**, 31–56 (1996))
43 A. I. Gusev, A. A. Rempel, G. P. Shveikin: Radiation damage resistance of materials and nonstoichiometry, Doklady Akademii Nauk **357**, 490–494 (1997) (in Russian). (Engl. transl.: Doklady Phys. Chem. **357**, 373–376 (1997))
44 A. I. Gusev, A. A. Rempel, G. P. Shveikin: Nonstoichiometry and radiation damage resistance of machinery materials, Problems of Atomic Science and Technique. Seriya "Material Science and New Materials" No 1(54), 10–19 (1997) (in Russian)
45 L. Ramqvist: Variation of lattice parameter and hardness with carbon content of group 4b metal carbides, Jernkontorets Annaler **152**, 517–523 (1968)
46 L. Ramqvist: Variation of hardness, resistivity and lattice parameter with carbon content of group 5b metal carbides, Jernkontorets Annaler **152**, 467–475 (1968)
47 P. Ettmayer, W. Lengauer: Nitrides: transition metal solid state chemistry, in *Encyclopedia of Inorganic Chemistry* (Wiley, New York 1994) pp. 2498–2515
48 E. V. Clougherty, K. H. Lothrop, J. A. Kafalas: New phase formed by high pressure treatment: face-centered cubic molybdenum monocarbide, Nature **191**, 1194–1195 (1961)

49 G. Kaye: Face-centered cubic molybdenum monocarbide, Nature **195**, 1195–1196 (1962)

50 K. Schwarz: Band structure and chemical bonding in transition metal carbides and nitrides, Critical Reviews in the Solid State and Materials Science **13**, 211–257 (1987)

51 P. V. Geld, V. A. Tskhai, A. S. Borukhovich, L. B. Dubrovskaya, I. I. Matveenko: Conduction band of IVa and Va subgroup transition metal carbides (I), Phys. Stat. Sol. (b) **42**, 85–93 (1970)

52 A. S. Borukhovich, P. V. Geld, V. A. Tskhai, L. B. Dubrovskaya, I. I. Matveenko: Conduction band of IVa and Va subgroup transition metal carbides (II), Phys. Stat. Sol. (b) **45**, 179–187 (1971)

53 J. Redinger, R. Eibler, P. Herzig, A. Neckel, R. Podloucky, E. Wimmer: Vacancy induced changes in the electronic structure of titanium carbide, J. Phys. Chem. Solids **46**, 383–398 (1985); **47**, 387–393 (1986) .

54 P. Herzig, J. Redinger, R. Eibler, A. Neckel: Vacancy induced changes in the electronic structure of titanium nitride, J. Sol. State Chem. **70**, 281–294 (1987)

55 P. Marksteiner, P. Weinberger, A. Neckel, R. Zeller, P. H. Dederichs: Electronic structure of substoichiometric carbides and nitrides of titanium and vanadium, Phys. Rev. B **33**, 812–822 (1986)

56 P. Marksteiner, P. Weinberger, A. Neckel, R. Zeller, P. H. Dederichs: Electronic structure of substoichiometric carbides and nitrides of zirconium and niobium, Phys. Rev. B **33**, 6709–6718 (1986)

57 A. L. Ivanovsky, A. I. Gusev, G. P. Shveikin: *Quantum Chemistry in Materials Science. Ternary Carbides and Nitrides of Transition Metals and Elements of Sungroup IIIb and IVb* (Ural Division of the Russian Academy of Sciences, Ekaterinburg 1996) 340 pp. (in Russian)

58 K. Schwarz, A. R. Williams, J. J. Cuomo, J. M. E. Harper, H.T. G. Hentzel: Zirconium nitride – a new material for Josephson junctions, Phys. Rev. B **32**, 8312–8316 (1985)

59 J. P. Landesman, P. Turchi, F. Ducastelle, G. Treglia: Electronic structure and stability of vacancy-ordered compounds of transition metal carbides and nitrides, in *Materials Research Society*, Symposium Proceedings (Elsevier, Amsterdam 1984) **21**, pp. 363–368

60 F. Ducastelle: Electronic structure and ordering in transition alloys and compounds, in *Materials Research Society*, Symposium Proceedings (Elsevier, Amsterdam 1984) **21**, pp. 375-380

61 V. Moisy-Maurice, C. H. de Novion, A. N. Christensen, W. Just: Elastic diffuse neutron scattering study of the defect structure of $TiC_{0.76}$ and $NbC_{0.73}$, Solid State Commun. **39**, 661–665 (1981)

62 V. Moisy-Maurice: *Structure atomique des carbures non-stoechiometriques de metaux de transition*, Rapport CEA-R-5127 (Commissariat à l'Energie Atomique, Gif-sur-Yvette (France) 1981) 184 pp.

63 T. H. Metzger, J. Peisl, R. Kaufmann: X-ray determination of local atomic displacements around carbon vacancies in NbC_{1-c} single crystals, J. Phys. F: Metal Phys. **13**, 1103–1113 (1983)

64 R. Kaufmann, O. Meyer: Ion channeling in NbC_{1-c}: determination of local atomic displacements around carbon vacancies, Phys. Rev. B **28**, 6216–6226 (1983)

65 T. Priem, B. Beuneu, C. H. de Novion, J. Chevrier, F. Livet, A. Finel, S. Lefevbre: Short-range order and static displacements in non-stoichiometric transition metal carbides and nitrides, Physica B **156–157**, 47–49 (1989)

66 T. Priem: *Etude de l'ordre à courte distance dans les carbures et nitrures non-stoechiometriques de metaux de transition par diffusion diffuse de neutrons*, Rapport CEA-R-5499 (Commissariat à l'Energie Atomique, Centre d'Etudes Nucléaires de Saclay, Gif-sur-Yvette (France) 1989) 162 pp.

67 L. V. Zueva, A. I. Gusev: Effect of nonstoichiometry and ordering on the period of the basis structure of cubic titanium carbide, Fiz. Tverd. Tela **41**, 1134–1141 (1999) (in Russian). (Engl. transl.: Physics of the Solid State **41**, 1032–1038 (1999)

68 A. I. Gusev, A. A. Rempel, G. P. Shveikin: Structural stability and the boundaries of a homogeneity region of nonstoichiometric casrbides, Doklady Akademii Nauk SSSR **298**, 890–894 (1988) (in Russian). (Engl. transl.: Doklady Phys. Chem. **298**, 121–125 (1988))

69 A. I. Gusev: Structural stability boundaries for nonstoichiometric compounds, Phys. Stat. Sol. (a) **111**, 443–450 (1989)

70 A. I. Gusev, A. A. Rempel: The boundaries of the structural stability of non-stoichiometric carbides, Zh. Neorgan. Khimii **34**, 556–561 (1989) (in Russian). (Engl. transl.: Russ. J. Inorg. Chem. **34**, 309–312 (1989))

71 J. W. Essam: Percolation theory, Rep. Progr. Phys. **43**, 883–912 (1980)

72 L. Efros: *Physics and Geometry of Disorder* (Nauka, Moscow 1982) 175 pp. (in Russian)

73 H. Kesten: *Percolation Theory for Mathematicians* (Birkhäuser, Boston-Basel-Stuttgart 1982) 423 pp.

74 T. E. Harris: Lower bound for the critical probability in a certain percolation process, Proc. Cambridge Philosoph. Soc. **56**, 13–20 (1960)

75 M. E. Fisher: Critical probabilities for cluster size and percolation problems, J. Math. Phys. **2**, 620–627 (1961)

76 H. Sher, R. Zallen: Critical density in percolation processes, J. Chem. Phys. **53**, 3759–3761 (1970)

77 V. N. Lipatnikov, A. I. Gusev, P. Ettmayer, W. Lengauer: Phase transformation in non-stoichiometric vanadium carbide, J. Phys.: Condens. Matter **11**, 163–184 (1999)

78 V. N. Lipatnikov, A. A. Rempel, A. I. Gusev: Effect of nonstoichiometry and ordering on the basal structural parameters of the monocarbides of niobium and tantalum, Izv. AN SSSR. Neorgan. Materialy **26**, 2522–2526 (1990) (in Russian). (Engl. transl.: Inorganic Materials **26**, 2171–2175 (1990))

79 A. I. Gusev, A. A. Rempel, V. N. Lipatnikov: Incommensurate ordered phase in non-stoichiometric tantalum carbide, J. Phys.: Condens. Matter **8**, 8277–8293 (1996)

80 R. V. Sara: The system zirconium–carbon, J. Amer. Ceram. Soc. **48**, 243–246 (1968)

81 A. N. Zyryanova, A. I. Gusev: Magnetic susceptibility and ordering in non-stoichiometric hafnium carbide, Zh. Fiz. Khimii **72**, 2234–2242 (1998) (in Russian). (Engl. transl.: Russ. J. Phys. Chem **72**, 2034–2041 (1998))

82 G. B. Bokii: *Introduction to Crystal Chemistry* (US Atomic Energy Commission, Office of Technical Information, Oak Ridge, Tennessee 1960) 528 pp.

83 E. Parthe, K. Yvon: Crystal chemistry of the close packed transition metal carbides. Proposal for the notation of the different crystal structures, Acta Crystallogr. B **26**, 153–163 (1970)

84 A. I. Gusev: Phase equilibria and phase diagrams of systems with atomic ordering, Doklady Akad. Nauk SSSR **313**, 887–893 (1990) (in Russian). (Engl. transl.: Doklady Phys. Chem. **313**, 615–620 (1990))

85 A. I. Gusev: Phase diagrams for ordering systems in order–parameter functional method, Fiz. Tverd. Tela **32**, 2752–2760 (1990) (in Russian). (Engl. transl.: Sov. Physics - Solid State **32**, 1595–1599 (1990))
86 A. I. Gusev, A. A. Rempel: Order parameter functional method in the theory of atomic ordering, Phys. Stat. Sol. (b) **131**, 43–51 (1985)
87 A. I. Gusev, A. A. Rempel: Calculating the energy parameters for CV and OPF methods, Phys. Stat. Sol. (b) **140**, 335–346 (1987)
88 A. I. Gusev: Atomic ordering and the order parameter functional method, Philosoph. Mag. B **60**, 307–324 (1989)
89 H. Nowotny, F. Benesovsky: Ordered structures of transition metal carbides and nitrides, in *Metallovedenie*: *Physical Metallurgy. Structure and Properties of Steels and Alloys*, N. V. Ageev (Ed.) (Nauka, Moscow 1971) pp. 315–323 (in Russian)
90 C. H. de Novion, V. Maurice: Order and disorder in carbides and nitrides, J. Physique Colloq. (France) **38**, C7-211–C7-220 (1977)
91 B. V. Khaenko: Phase transitions in M_2X-type carbides and nitrides of metals of groups V and VI, Poroshkovaya Metallurgiya No 1, 71–74 (1978) (in Russian)
92 B.V. Khaenho: Order in cubic carbides and nitrides of transition metals of groups IV and V, Izv. AN SSSR. Neorgan. Materialy **15**, 1952–1960 (1979) (in Russian). (Engl. transl.: Inorg. Mater. **15**, 1535–1543 (1979))
93 C. H. de Novion, J. P. Landesman: Order and disorder in transition metal carbides and nitrides: experimental and theoretical aspects, Pure and Appl. Chem. **57**, 1391–1402 (1985)
94 C. H. de Novion, B. Beuneu, T. Priem, N. Lorenzelli, A. Finel: Defect structures and order-disorder transformations in transition metal carbides and nitrides, in *The Physics and Chemistry of Carbides, Nitrides and Borides*, R. Freer (Ed.) (Kluwer Acad. Publ., Netherlands 1990) pp. 329–355
95 K. Yvon, E. Parthe: Crystal structure of the ζ-V, Nb and Ta carbides, Acta Crystallogr. B **26**, 149–152 (1970)
96 H. Wiesenberger, W. Lengauer, P. Ettmayer: Reactive diffusion and phase equilibria in the V–C, Nb–C, Ta–C and Ta–N systems, Acta Mater. **46**, 651–666 (1998)
97 H. Goretzki: Neutron diffraction studies of titanium-carbon and zirconium-carbon alloys, Phys. Stat. Sol. **20**, K141–K143 (1967)
98 I. Karimov, V. T. Em, I. Khidirov, I. S. Latergaus: Neutron diffraction investigation of ordering in titanium and zirconium carbides, Izv. AN Uzb. SSR. Seriya Fiz. Mat. Nauk No 4, 81–83 (1979) (in Russian)
99 N. Lorenzelli, R. Caudron, J. P. Landesman, C. H. de Novion: Influence of the ordering of carbon vacancies on the electronic properties of $TiC_{0.625}$, Solid State Commun. **59**, 765–769 (1986)
100 V. N. Lipatnikov, A. Kottar, L. V. Zueva, A. I. Gusev: Disorder–order phase transformations and electrical resistivity of a nonstoichiometric titanium carbide, Fiz. Tverd. Tela **40**, 1332–1340 (1998) (in Russian). (Engl. transl.: Physics of the Solid State **40**, 1211–1218 (1998))
101 N. V. Dzhalabadze, B. G. Eristavi, N. I. Maisuradze, E. R. Kuteliya: Structure of crystalline phases in Ti–C thin films, Fiz. Metall. Metalloved. **86**, 85–92 (1998) (in Russian). (Engl. transl.: Phys. Met. Metallogr. **86**, 59–64 (1998))
102 A. I. Gusev, A. A. Rempel: Phase diagrams for Ti–C and Ti–N systems and atomic ordering of nonstoichiometric titanium carbide and nitride, Doklady Akademii Nauk

332, 717–721 (1993) (in Russian). (Engl. transl.: Doklady Chemistry **332**, 456–461 (1993))
103 N. Obata, N. Nakasawa: Superlattice formation in zirconium-carbon system, J. Nucl. Mater. **60**, 39–42 (1976)
104 R. Lorenzelli, I. de Dieuleveult: Existence d'une surstructure dans le carbure de thorium sous-stoechiometrique: $ThC_{0.76}$, J. Nucl. Mater. **29**, 349–353 (1969)
105 A. I. Gusev, A. A. Rempel: Thermodynamic model of atomic ordering. Phase diagrams of ordered systems, Zh. Fiz. Khimii **65**, 625–633 (1991) (in Russian). (Engl. transl.: Russ. J. Phys. Chem. **65**, 330–334 (1991))
106 A. I. Gusev, A. A. Rempel: Calculation of phase diagrams of interstitial compounds, J. Phys. Chem. Solids **55**, 299–304 (1994)
107 A. I. Gusev: Phase diagrams of ordered nonstoichiometric hafnium carbide and titanium nitride, Doklady Akademii Nauk **322**, 918–923 (1992) (in Russian). (Engl. transl.: Doklady Phys. Chem. **322**, 84–88 (1992))
108 A. N. Zyryanova, S. Z. Nazarova, A. I. Gusev: Magnetic susceptibility anomaly as evidence for ordering of nonstoichiometric hafnium carbide HfC_y , Doklady Akademii Nauk **359**, 348–353 (1998) (in Russian). (Engl. transl.: Doklady Phys. Chem. **359**, 91–96 (1998))
109 A. I. Gusev, A. N. Zyryanova: Atomic-vacancy ordering and magnetic susceptibility of nonstoichiometric hafnium carbide, Pis'ma v ZhETF **69**, 296–301 (1999) (in Russian). (Engl. transl.: JETP Letters **69**, 324–329 (1999))
110 V. N. Lipatnikov, A. I. Gusev, P. Ettmaier, W. Lengauer: Order-disorder phase transformations and specific heat of nonstoichiometric vanadium carbide, Fiz. Tverd. Tela **41**, 529–536 (1999) (in Russian). (Engl. transl.: Physics of the Solid State **41**, 474–480 (1999))
111 J. D. Venables, D. Kahn, R. G. Lye: Structure of ordered compound V_6C_5, Philosoph. Mag. **18**, 177–192 (1968)
112 V. N. Lipatnikov, W. Lengauer, P. Ettmayer, E. Keil, G. Groboth, E. Kny: Effects of vacancy ordering on structure and properties of vanadium carbide, J. Alloys Comp. **261**, 192–197 (1997)
113 J. Billingham, P. S. Bell, M. H. Lewis: Vacancy short-range order in substoichiometric transition metal carbides and nitrides with the NaCl structure. Electron diffraction studies of short-range ordered compounds, Acta Crystallogr. A **28**, 602–606 (1972)
114 J. Billingham, P. S. Bell, M. H. Lewis: Superlattice with monoclinic symmetry based on compounds V_6C_5 , Philosoph. Mag. **25**, 661–671 (1972)
115 R. Kesri, S. Hamar-Thibault: Structures ordonnées à longue distance dans les carbures MC dans les fontes, Acta Met. **36**, 149–166 (1988)
116 A. A. Rempel, A. I. Gusev, V. G. Zubkov, G. P. Shveikin: Structure of ordered niobium carbide Nb_6C_5, Doklady AN SSSR **275**, 883–887 (1984) (in Russian). (Engl. transl.: Sov. Physics Doklady **29**, 257–259 (1984))
117 A. I. Gusev, A. A. Rempel: Ordering in the carbon sublattice of nonstoichiometric niobium carbide, Fiz. Tverd. Tela **26**, 3622–3627 (1984) (in Russian). (Engl. transl.: Sov. Physics – Solid State **26**, 2178–2181 (1984))
118 A. A. Rempel, A. I. Gusev: Order-disorder phase transition in nonstoichiometric niobium carbide, Kristallografiya **30**, 1112–1115 (1985) (in Russian). (Engl. transl.: Sov. Physics – Crystallography **30**, 648–650 (1985))
119 A. I. Gusev, A. A. Rempel: Order–disorder phase transition channel in niobium carbide, Phys. Stat. Sol. (a) **93**, 71–80 (1986)

120 C. H. de Novion, N. Lorenzelli, P. Costa: Superlattice structure in vanadium carbide VC_{1-x}, Compt. Rend. Acad. Sci. Paris B **263**, 775–778 (1966)
121 S. I. Alyamovskii, P. V. Geld, G. P. Shveikin, E. N. Schetnikov: New phase in the vanadium – carbon system, Zh. Neorgan. Khimii **13**, 895–897 (1968) (in Russian). (Engl. transl.: Russ. J. Inorg. Chem. **13**, 472–473 (1968))
122 M. P. Arbuzov, B. V. Khaenko, V. G. Fak, Yu. F. Nosachev: Study of ordering process in alloys on a base of vanadium monocarbide, Ukrain. Fiz. Zh. **22**, 291–297 (1977) (in Russian)
123 T. Athanassiadis, N. Lorenzelli, C. H. de Novion: Diffraction studies of the order–disorder transformation in V_8C_7, Ann. Chim. France **12**, 129–142 (1987)
124 D. Rafaja, W. Lengauer, P. Ettmayer, V. N. Lipatnikov: Rietveld analysis of the ordering in V_8C_7, J. Alloys Comp. **269**, 60–62 (1998)
125 E. Rudy, C. E. Brukl: Lower temperature modification of the Nb_2C and V_2C, J. Amer. Ceram. Soc. **50**, 265–268 (1967)
126 A. A. Rempel, A. I. Gusev: *Ordering in Nonstoichiometric Niobium Monocarbide* (Urals Scientific Centre, Sverdlovsk 1983) 68 pp. (in Russian)
127 B. V. Khaenko, O. P. Sivak: Structure of order of niobium monocarbide, Kristallografiya **35**, 1110–1115 (1990) (in Russian)
128 J. P. Landesman, A. N. Christensen, C. H. de Novion, N. Lorenzelli, P. Convert: Order–disorder transition and structure of the ordered vacancy compound Nb_6C_5: powder neutron diffraction studies, J. Phys. C: Solid State Phys. **18**, 809–823 (1985)
129 A. N. Christensen: Vacancy order in Nb_6C_5, Acta Chem. Scand. A **39**, 803–804 (1985)
130 A. L. Bowman, T. C. Wallace, J. L. Yarnell, R. G. Wenzel, E. K. Storms: The crystal structures of V_2C and Ta_2C, Acta crystallogr. **19**, 6–9 (1965)
131 V. N. Lipatnikov, A. A. Rempel, A. I. Gusev: The influence of ordering of the vacancies on the magnetic susceptibility of tantalum carbide, Zh. Neorgan. Khimii **33**, 1860–1863 (1988) (in Russian). (Engl. transl.: Russ. J. Inorg. Chem. **33**, 1058–1059 (1988))
132 A. I. Gusev, A. A. Rempel, V. N. Lipatnikov: Magnetic susceptibility and atomic ordering in tantalum carbide, Phys. Stat. Sol. (a) **106**, 459–466 (1988)
133 A. A. Rempel, V. N. Lipatnikov, A. I. Gusev: The superstructure in nonstoichiometric tantalum carbide, Doklady AN SSSR **310**, 878–882 (1990) (in Russian). (Engl. transl.: Sov. Physics Doklady **35**, 103–106 (1990))
134 A. I. Gusev, A. A. Rempel, V. N. Lipatnikov: Incommensurate superlattice and superconductivity in tantalum carbide, Fiz. Tverd. Tela **33**, 2298–2305 (1991) (in Russian). (Engl. transl.: Sov. Physics – Solid State **33**, 1295–1299 (1991))
135 L. D. Landau: Theory of phase transitions, in *Collected Papers*, **1** (Nauka, Moscow 1969) pp. 234–261 (in Russian)
136 A. Sh. Remeev, I. Karimov: Neutron diffraction study of kinetics of structural phase transition, Izv. AN Uzb. SSR. Seriya Fiz. Mat. Nauk No 2, 87–88 (1986) (in Russian)
137 V. N. Lipatnikov, A. I. Gusev: High-temperature heat capacity and order–disorder phase transformations in nonstoichiometric titanium carbide, Phys. Stat. Sol. (b) **212**, R11–R12 (1999)
138 V. N. Lipatnikov, A. I. Gusev: Effect of ordering on the structure and specific heat of nonstoichiometric titanium carbide, Pis'ma v ZhETF **69**, 631–637 (1999) (in Russian). (Engl. transl.: JETP Letters **69**, 669–675 (1999))

139 M. H. Lewis, J. Billingham, P. S. Bell: Non-stoichiometry in ceramic compounds, in *Solid State Chemistry*, Proc. of 5th Intern. Mater. Res. Symp. (NBS Special Publ. 364). (NBS Publ., Berkley, California 1972) pp. 1084–1114
140 M. P. Arbuzov, B. V. Khaenko, O. P. Sivak: Superstructures of order in niobium monocarbide, Doklady AN Ukr. SSR. Seriya A, No 10, 86–88 (1984) (in Russian)
141 B. V. Khaenko, O. P. Sivak, V. S. Sinel'nikova: X-ray study of ordered modification of niobium monocarbide, Izv. AN SSSR. Neorgan. Materialy **20**, 1825–1828 (1984) (in Russian)
142 L. V. Azaroff, M. J. Buerger: *The Powder Method in X-ray Crystallography* (McGraw-Hill, New York-London 1958) 342 pp.
143 J. D. Venables, M. H. Meyerhoff: Ordering effect in NbC and TaC, in *Solid State Chemistry*, Proc. of 5th Intern. Mater. Res. Symp. (NBS Special Publ. 364). (NBS Publ., Berkley, California 1972) pp. 583-590
144 V. N. Lipatnikov, A. I. Gusev, A. A. Rempel, G. P. Shveikin: Effect of structural transition on magnetic susceptibility of tantalum carbide, Doklady AN SSSR **297**, 849–853 (1987) (in Russian). (Engl. transl.: Sov. Physics Doklady **32**, 988–990 (1987))
145 V. N. Lipatnikov, A. A. Rempel, A. I. Gusev: Specific heat of tantalum carbide in states with different degrees of order, Fiz. Tverd. Tela **31**, 285–287 (1989) (in Russian). (Engl. transl.: Sov. Physics. – Solid State **31**, 1818–1819 (1989))
146 A. I. Gusev, A. A. Rempel, V. N. Lipatnikov: Thermodynamic model of atomic ordering. 5. Specific heat of non-stoichiometric compounds in different structural states, Zh. Fiz. Khimii **64**, 2343–2348 (1990) (in Russian). (Engl. transl.: Russ. J. Phys. Chem. **64**, 1265–1267 (1990))
147 A. I. Gusev, A. A. Rempel, V. N. Lipatnikov: Heat capacity of niobium and tantalum carbides NbC_y and TaC_y in disordered and ordered states below 300 K, Phys. Stat. Sol. (b) **194**, 467–482 (1996)
148 A. A. Rempel, A. A. Druzhkov, V. N. Lipatnikov, A. I. Gusev, S. M. Klotsman, G. P. Shveikin: Angular correlation of annihilation radiation in nonstoichiometric tantalum carbide, Doklady AN SSSR **300**, 90–95 (1988) (in Russian). (Engl. transl.: Sov. Physics Doklady **33**, 357–359 (1988))
149 A. A. Rempel, A. A. Druzhkov, A. I. Gusev: Positron annihilation in tantalum and its carbide, Fiz. Metall. Metalloved. **68**, 271–279 (1989) (in Russian). (Engl. transl.: Phys. Met. Metallogr. **68**, 59–68 (1989))
150 A. I. Gusev, A. A. Rempel: Thermodynamic model of atomic ordering. Phase diagrams of ordered systems, Zh. Fiz. Khimii **65**, 625–633 (1991) (in Russian). (Engl. transl.: Russ. J. Phys. Chem. **65**, 330–334 (1991))
151 A. I. Gusev, A. A. Rempel: Vacancy distribution in ordered Me_6C_5-type carbides, J. Phys. C: Solid State Phys. **20**, 5011–5025 (1987)
152 B. Holmberg: Structural studies on the titanium–nitrogen system, Acta Chem. Scand. **16**, 1255-1261 (1962)
153 C. Lobier, J. P. Marcon: Etude et structure d'une nouvelle phase du sous-nitrure de titane Ti_2N, Compt. Rend. Acad. Sci. Paris. Ser. C **268**, 1132–1135 (1969)
154 S. Nagakura, T. Kusunoki: Structure of titanium nitride TiN_x studied by electron diffraction and microscopy, J. Appl. Crystallogr. **10**, 52–56 (1977)
155 M. P. Arbuzov, S. Ya. Golub, B. V. Khaenko: X-ray study of titanium nitrides. Izv. AN SSSR. Neorgan. Materialy **13**, 1779–1782 (1977) (in Russian)

156 A. N. Christensen, A. Alamo, J. P. Landesman: Structure of vacancy-ordered titanium heminitride δ'-Ti_2N by powder neutron diffraction, Acta Crystallogr. C **41**, 1009–1011 (1985)

157 I. Khidirov, I. Karimov, V. T. Em: Order in titanium nitride and nitridohydride, Doklady AN Uzb. SSR, No 10, 21–22 (1980) (in Russian)

158 D. Sundararaman, A. L. E. Terrance, V. Seetharaman, V. S. Raghunathan: Electron microscopy study of microstructural changes in a Ti-1.6 at.% nitrogen alloy, Trans. Japan. Inst. Metals **24**, 510–513 (1983)

159 W. Lengauer, P. Ettmayer: Thermal decomposition of ε-Ti_2N and δ'-$TiN_{0.50}$ investigated by high-temperature x-ray diffraction, High Temp. – High Pressure **19**, 673–676 (1987)

160 W. Lengauer, P. Ettmayer: Recent advances in the field of transition-metal refractory nitrides, High Temp. -- High Pressure **22**, 13–24 (1990)

161 E. Etchessahar, Sohn Yong-Un, M. Harmelin, J. Debuigne: The Ti-N system: kinetic, calorimetric, structural and metallurgical investigations of the δ'-$TiN_{0.51}$ phase, J. Less-Common Metals **167**, 261–281 (1991)

162 W. Lengauer: The titanium–nitrogen system: a study of phase reactions in the subnitride region by means of diffusion couples, Acta Metall. Mater. **39**, 2985–2995 (1991)

163 T. Priem, B. Beuneu, C. H. de Novion, A. Finel, F. Livet: Short-range order, atomic displacements and effective interatomic ordering energies in $TiN_{0.82}$, J. Physique France **50**, 2217–2242 (1989)

164 T. Onozuka: Nitrogen ordering in vanadium nitrides (V_9N_4 and $V_{32}N_{26}$) studied by neutron and electron diffraction, Trans. Japan. Inst. Metals **23**, 315–319 (1982)

165 N. Terao: Structure des nitrures de niobium, Japan. J. Appl. Phys. **4**, 353–367 (1965)

166 A.Yu. Chervyakov, V.A.Somenkov, Ya.S.Umanskii: Neutron diffraction investigation of ordered phase of niobium nitride, Izv. Vyssh. Ucheb. Zaved. Tsvetnaya Metallurgiya No 5, 140–144 (1971) (in Russian)

167 D. Watanabe, J. R. Castles, A. Jostsons, A. S. Malin: Ordered structure of titanium oxide, Nature **210**, 934–936 (1966)

168 D .Watanabe, J. R. Castles, A. Jostsons, A. S. Malin: The ordered structure of TiO, Acta Crystallogr. **23**, 307–313 (1967)

169 E. Hilti: Röntgenographische Untersuchung der Titanmonoxid-Tieftemperatur-Modifikation, Naturwissenschaften **55**, 131–132 (1968)

170 D. Watanabe, O. Terasaki, A. Jostsons, J. R. Castles: Electron microscopy study on the structure of low temperature modification of titanium monoxide phase, in *The Chemistry of Extended Defects in Non-Metallic Solids*, L. Eyring, M. O'Keeffe (Eds.) (North-Holland Publ., Amsterdam-London 1970) pp. 238–257

171 E. Hilti: Neue Phasen in System Titan-Sauerstoff, Naturwissenschaften **55**, 130–131 (1968)

172 M. P. Arbuzov, B. V. Khaenko, E. T. Kachkovskaya: Crystal-geometric characteristics of ordered modification of titanium monoxide with oxygen atomic content about 54 % , Doklady Akad. Nauk Ukrain. SSR. Seriya A. Fiz.-Mat. i Tekhn. Nauki, No 12, 71–74 (1981) (in Russian)

173 A. A. Valeeva, A. A. Rempel, A. I. Gusev: Two-sublattice ordering in titanium monoxide, Pis'ma v ZhETF **71**, 675–681 (2000) (in Russian). (Engl. transl.: JETP Letters **71**, 460–464 (2000))

174 B. Andersson, J. Gionnes: Ordered phases in the monoxide region of the vanadium–oxygen system, Acta Chem. Scand. **24**, 2250-2252 (1970)
175 S. Westman, C. Nordmark: Phase analysis studies on the vanadium–oxygen system within $VO_{0.25}$–$VO_{1.5}$ region at 800 °C, Acta Chem. Scand. **14**, 465–470 (1960)
176 G. Brauer: Die Oxide Niobs, Z. Anorg. Allgem. Chem. **248**, 1–31 (1941)
177 V. V. Klechkovskaya: Oxidation of thin films of niobium at low temperatures, Kristallografiya **15**, 358–361 (1970) (in Russian)
178 O. N. Carlson, J. F. Smith, R. H. Nafziger: The vanadium–nitrogen system: a review, Metall. Trans. A **17**, 1647–1656 (1986)
179 B.-O. Marinder: Ordered, defective structures of titanium monoxide, the anti-anatase structure and related structures described through an open packing of cube units (die packing), Chem. Scripta. **18**, 169–175 (1981)
180 S. Andersson, B. Collen, U. Kuylenstierna, A. Magneli: Phase analysis studies on the titanium–oxygen system, Acta Chem. Scand. **11**, 1641–1647 (1957)
181 C. Leung, M. Weinert, P. B. Allen, R. M. Wentzcovitch: First-principles study of titanium oxides, Phys. Rev. B **54**, 7857–7864 (1996)
182 I. Karimov, V. T. Em, V. F. Petrunin, I. S. Latergaus, V. S. Polishuk: Neutron diffraction study of titanium carbonitride, Izv. AN SSSR. Neorgan. Materialy **12**, 1492–1494 (1976) (in Russian)
183 V. T. Em, I. Karimov, I. S. Latergaus: Influence of nitrogen on characteristics of order-disorder phase transition in TiC_x, Metallofizika **9**, 113–114 (1987) (in Russian)
184 V. T. Em, M. Yu. Tashmetov: The structure of the ordered phase in rocksalt type titanium carbide, carbidenitride, and carbidehydride, Phys. Stat. Sol. (b) **198**, 571–575 (1996)
185 W. Lengauer, S. Binder, K. Aigner, P. Ettmayer, A. Guillou, J. Debuigne: Solid state properties of group IVb carbonitrides, J. Alloys Comp. **217**, 137–147 (1995)
186 M. P. Arbuzov, S. Ya. Golub, B. V. Khaenko: Study of the phases of the Ti-TiC-TiN system, Izv. AN SSSR. Neorgan. Materialy **14**, 1442–1445 (1978) (in Russian)
187 V. G. Zubkov, I. I. Matveenko, L. B. Dubrovskaya, G. D. Bogomolov, P. V. Geld: Neutron diffraction study of structure of titanium oxycarbides, Doklady Akad. Nauk **191**, 323–325 (1970) (in Russian)
188 I. I. Matveenko, L. B. Dubrovskaya, G. D. Bogomolov, V. G. Zubkov, P. V. Geld: Electrical conductivity of titanium oxycarbides, Izv. AN SSSR. Neorgan. Materialy **6**, 1190–1191 (1970) (in Russian)
189 J. Vicens, J.-L. Chermant: Precipitation du titane dans le système titane–carbone–oxygène, Phys. Stat. Sol. (a) **7**, 217–225 (1971)
190 A. Emiraliev, I. Karimov, F. Faizullaev, M. Patiev: Formation of anti-$CaCl_2$-type superstructure in Ti-V-C system, Kristallografiya **33**, 778–780 (1978) (in Russian)
191 I. S. Latergaus, I. Karimov, V. S. Presman, V. T. Em: Order-disorder phase transition in titanium carbohydrides, Izv. AN SSSR. Neorgan. Materialy **23**, 1471–1474 (1987) (in Russian)

4. Short-Range Order

Short-range order characterizes only the radial atomic distribution, i.e. fluctuations of atomic concentrations in various coordination spheres. In contrast long-range order includes in addition the angular distribution and thus it is possible to determine which atomic species occupies a particular site of the crystal lattice. It is very important to consider short-range order when long-range order is absent i.e. when the mutual arrangement of atoms is correlated. The correlations found relate to the difference in the interaction energies of like and unlike atoms.

The structure of a solid solution, i.e. the mutual arrangement of atoms of different species on crystal lattice sites, is described by short-range order parameters.

4.1 Short-Range Order Parameters and Two-Particle Correlations

The short-range order is described by the short range order parameter α or by the correlation parameter $\varepsilon(R_j)$. They can be determined using several methods [1–11].

Let us consider a binary solid solution AB, where the atomic concentration of the component A is c, while the total number of atoms equals N. Assume that the coordinate number of the j-th coordination sphere is z_j. In this case the number of atoms of species A is Nc and the total number of atoms in the j-th coordination sphere of any atomic species A is Nz_jc. Denote the number of atoms of species A and B in the j-th coordination spheres as $N^{(j)}_{(A)A}$ and $N^{(j)}_{(A)B}$. Then the relative number of atoms of a particular species in the j-th coordination sphere becomes $n^{(j)}_{(A)A} = N^{(j)}_{(A)A} / Nz_jc$ and $n^{(j)}_{(A)B} = N^{(j)}_{(A)B} / Nz_jc$. If the atoms are distributed statistically, the number of atoms of species B located in the j-th coordination sphere relative to the number of species A atoms is $Nz_jc(1-c)$. If any short-range order is present, i.e. if the mutual arrangement of atoms is correlated, then $N^{(j)}_{(A)B} \neq Nz_jc(1-c)$. Therefore a quantity, which characterizes the relative deviation of the atomic distribution in the j-th coordination sphere from the statistical value,

$$\alpha_j = [Nz_jc(1-c) - N^{(j)}_{(A)B}] / Nz_jc(1-c) = 1 - n^{(j)}_{(A)B} / (1-c) \tag{4.1}$$

is introduced as the short-range order parameter. This short-range order parameter was first proposed by Cowley [6] and Warren [7]. An analogous expression

$$\alpha_j = 1 - n^{(j)}_{(\mathrm{B})\mathrm{A}} / c \tag{4.2}$$

is obtained if an atom of species B is taken as the central atom. The expressions (4.1) and (4.2) give the same value for the short-range order parameter. If the atoms are distributed statistically, $n^{(j)}_{(\mathrm{A})\mathrm{B}}$ and $n^{(j)}_{(\mathrm{B})\mathrm{A}}$ calculated by averaging over all j-th coordination spheres around the chosen central atoms A and B, respectively, coincide with the relative concentrations $(1 - c)$ and c and therefore $\alpha_j = 0$. If correlations are present and if the j-th coordination sphere of an atom of a particular species hosts mostly atoms of the other species, i.e. $n^{(j)}_{(\mathrm{A})\mathrm{B}} > 1 - c$ and $n^{(j)}_{(\mathrm{B})\mathrm{A}} > c$, the short-range order parameter becomes negative: $\alpha_j < 0$. If the relative fraction of atoms of the other species in the j-th coordination sphere of an atom of a particular species is less than in the case of the statistical distribution, i.e. $n^{(j)}_{(\mathrm{A})\mathrm{B}} < 1 - c$ and $n^{(j)}_{(\mathrm{B})\mathrm{A}} < c$, the short-range order parameter is positive: $\alpha_j > 0$. Thus, the sign of the short-range order parameter shows whether the environment in the j-th coordination sphere is dominated by unlike ($\alpha_j < 0$) or like ($\alpha_j > 0$) atoms.

If the short-range order parameter α_j is known from formulas (4.1) and (4.2), it is possible to determine $n^{(j)}_{(\mathrm{A})\mathrm{B}}$ and $n^{(j)}_{(\mathrm{B})\mathrm{A}}$, which in essence represent conditional probabilities, i.e. probabilities that atoms of a particular species are located in the j-th coordination sphere on condition that atoms of a given species are at the centre of this coordination sphere:

$$n^{(j)}_{(\mathrm{A})\mathrm{B}} = (1 - c)(1 - \alpha_j); \quad n^{(j)}_{(\mathrm{B})\mathrm{A}} = c(1 - \alpha_j). \tag{4.3}$$

Since $(n^{(j)}_{(\mathrm{A})\mathrm{B}} + n^{(j)}_{(\mathrm{A})\mathrm{A}}) = 1$ and $(n^{(j)}_{(\mathrm{B})\mathrm{A}} + n^{(j)}_{(\mathrm{B})\mathrm{B}}) = 1$ and considering (4.3) we obtain

$$n^{(j)}_{(\mathrm{A})\mathrm{A}} = c + \alpha_j(1 - c); \quad n^{(j)}_{(\mathrm{B})\mathrm{B}} = 1 - c + c\alpha_j\,. \tag{4.4}$$

In some cases the short-range order parameter is more convenient to use in a slightly different form. Take a binary solid solution AB containing N particles and consider the j-th coordination sphere with the coordination number z_j. In the case of a random distribution the total number of particle pairs AA, AB, BA and BB in the j-th coordination sphere is equal to their sum, i.e. $N^{(j)} = N^{(j)}_{\mathrm{AA}} + N^{(j)}_{\mathrm{AB}} + N^{(j)}_{\mathrm{BA}} + N^{(j)}_{\mathrm{BB}}$. If atoms are distributed statistically, the number of pairs of like and unlike atoms is determined by a binomial distribution and

$$\begin{aligned} N^{(j)} &= \tfrac{1}{2}[Nz_j c^2 + 2Nz_j c(1-c) + Nz_j(1-c)^2] \\ &= \tfrac{1}{2}Nz_j[c + (1-c)]^2 = \tfrac{1}{2}Nz_j\,. \end{aligned} \tag{4.5}$$

If any short-range order is absent, the j-th coordination sphere contains an equal number of unlike atomic pairs: $N^{\mathrm{bin}(j)}_{\mathrm{AB}} = N^{\mathrm{bin}(j)}_{\mathrm{BA}} = \tfrac{1}{2}Nz_j c(1-c)$. In the presence of short-range order $N^{\mathrm{bin}(j)}_{\mathrm{AB}} \neq N^{\mathrm{bin}(j)}_{\mathrm{BA}}$. Therefore a quantity, which characterizes the relative deviation of $N^{(j)}_{\mathrm{AB}}$ from $N^{\mathrm{bin}(j)}_{\mathrm{AB}}$, i.e.

$$\alpha_j = (N^{\mathrm{bin}(j)}_{\mathrm{AB}} - N^{(j)}_{\mathrm{AB}}) / N^{\mathrm{bin}(j)}_{\mathrm{AB}} = 1 - (N^{(j)}_{\mathrm{AB}} / N^{\mathrm{bin}(j)}_{\mathrm{AB}})\,, \tag{4.6}$$

is taken as the short-range order parameter.

Expression (4.6) can be more conveniently written in terms of probabilities $P^{\mathrm{bin}}_{\mathrm{AB}} = N^{\mathrm{bin}(j)}_{\mathrm{AB}} / N^{(j)} = Nz_j c(1-c) / Nz_j = c(1-c)$ and $P^{(j)}_{\mathrm{AB}} = N^{(j)}_{\mathrm{AB}} / N^{(j)}$ of unlike-atom pairs AB:

$$\alpha_j = 1 - P_{AB}^{(j)} / P_{AB}^{bin} . \quad (4.7)$$

It should be emphasized that (4.1), (4.2) and (4.7) give the same value of the short-range order parameter when a solid solution having a preset composition and structural state is described. Also note that (4.7) was derived considering atomic pairs AB and BA separately, although the number and probabilities of these pairs are equal ($N_{AB}^{(j)} = N_{BA}^{(j)}$, $P_{AB}^{(j)} = P_{BA}^{(j)}$) irrespective of the structural state of a solid solution. The coefficient $\lambda_{AB} = 2$ is a multiplicity factor, i.e. it denotes the number of equivalent configurations of a pair formed by unlike atoms. When the short-range order parameter is calculated from (4.7), one can use both probabilities P_{AB} neglecting multiplicity and full probabilities. In the latter case coefficients should be considered both in the numerator and the denominator of the second term, i.e.

$$\alpha_j = 1 - \lambda_{AB} P_{AB}^{(j)} / \lambda_{AB} P_{AB}^{bin} . \quad (4.8)$$

Expression (4.8) is identical to formula (4.7). However, in some instances it is more convenient for practical calculations.

It is on relatively rare occasions that the short-range order parameter as introduced by Bethe [5] is used:

$$\alpha_j^* = (P_{AB}^{(j)} - P_{AB}^{bin}) / (P_{AB}^{ord(j)} - P_{AB}^{bin}) , \quad (4.9)$$

where $P_{AB}^{ord(j)}$ is the probability that pairs AB, for which the distance between atoms A and B is equal to the radius of the j-th coordination sphere, are formed in a solid solution with the largest possible long-range order. Under this definition, the short-range order parameter α_j^* changes from zero in a disordered solid solution ($P_{AB}^{(j)} = P_{AB}^{bin}$) to unity in an ordered solid solution ($P_{AB}^{(j)} = P_{AB}^{ord(j)}$).

In addition to short-range order parameters, correlation parameters are frequently used to describe the structural state of a partially ordered solid solution. These correlation parameters are defined [1] as the difference between probabilities $P_{AB}^{(1)(2)}$ of pairs AB, whose atoms are located in different and nonequivalent sublattices (for example, first and second sublattices) and as products like $P_A^{(1)} P_B^{(2)}$:

$$\varepsilon_{AB}^{(1)(2)}(R_j) = P_{AB}^{(1)(2)} - P_A^{(1)} P_B^{(2)} . \quad (4.10)$$

Probabilities $P_A^{(1)}$ and $P_B^{(2)}$ are uniquely determined in terms of long-range order parameters. The product of these probabilities determines the short-range order parameter (4.7) if atoms A and B are distributed statistically on sublattices 1 and 2. Obviously, this situation can hardly occur in a real partially ordered solid solution and therefore the correlation parameter (4.10) describes more accurately the structural state of a solid solution. In the general case correlation parameters may be positive and negative. In a fully ordered crystal all correlation parameters are zero, because the atomic distribution in this state is exactly described by long-range order parameters.

In the absence of long-range order, short-range order is often described by correlation parameters, which are defined as the difference of probabilities that pairs AB are located in the j-th coordination sphere of a solid solution having a preset structural state in the j-th coordination sphere of a disordered solid solution:

$$\varepsilon_{AB}(R_j) = P_{AB}^{(j)} - P_{AB}^{bin} = P_{AB}^{(j)} - c(1-c)\,. \qquad (4.11)$$

If one considers pairs AB and BA to be equivalent, paired correlations are equal to each other: $\varepsilon_{AB}(R_j) = \varepsilon_{BA}(R_j)$. In the absence of long-range order, the correlation function (4.11) tends to zero with increasing radius R of the coordination sphere.

Formulas (4.7) and (4.11) suggest that in the absence of long-range order the short-range order parameter is related to the paired correlation as

$$\varepsilon_{AB}(R_j) = -c(1-c)\alpha_j\,. \qquad (4.12)$$

Analogously,

$$\varepsilon_{AA}(R_j) = \varepsilon_{BB}(R_j) = c(1-c)\alpha_j\,. \qquad (4.13)$$

Effect of Short-Range Order on Diffuse Scattering Intensity. Imperfections in a crystal, which will appear in the state of thermodynamic equilibrium, can be viewed as fluctuations of some internal parameters. Consequently, it is possible to describe X-ray or neutron scattering on statistical inhomogeneities of a crystal as scattering at fluctuation waves of the composition or order parameters or on waves of static atomic displacements produced by fluctuation waves. Using the method of fluctuation waves, which was developed by Krivoglaz [8], it is possible to take into account in describing the diffuse scattering intensity the effect of correlations and short-range order in the mutual arrangement of atoms.

In the general case, the scattering intensity of a crystal can be presented as the sum

$$I(\boldsymbol{q}) = I_0(\boldsymbol{q})\exp(-2M) + I_{DAD}(\boldsymbol{q})[1 - \exp(2\pi i \boldsymbol{q}\boldsymbol{u}_j)] + I_D(\boldsymbol{q})\,, \qquad (4.14)$$

where $I_0(\boldsymbol{q})$ is the theoretical intensity of structural reflections in the absence of atomic displacements; $\boldsymbol{q}$ denotes the diffraction vector ($|\boldsymbol{q}| = (2\sin\theta)/\lambda$); $\exp(-2M)$ is a factor allowing for attenuation of structural reflections as a result of static and dynamic (thermal) atomic displacements; $I_{DAD}(\boldsymbol{q})$ denotes the intensity of the diffuse background due to displacement $\boldsymbol{u}_j$ of atoms from crystal lattice sites; and $I_D(\boldsymbol{q})$ is the diffuse scattering intensity resulting from the difference between atomic scattering factors and correlations (short-range order) in the mutual arrangement of atoms.

Let us consider in more detail the third term in (4.14), i.e. the quantity $I_D(\boldsymbol{q})$, which takes into account the effect of short-range order on the diffuse scattering intensity.

We take the occupancy numbers $c(\boldsymbol{r})$ to characterize the atomic arrangement in a crystal of a binary solid solution AB. The occupancy number $c(\boldsymbol{r})$ is 1 if a species A atom occupies a site $\boldsymbol{r}$ and $c(\boldsymbol{r}) = 0$ if the site $\boldsymbol{r}$ hosts an atom of the other species. The atomic scattering factor, which corresponds to some site $\boldsymbol{r}$, can be presented as a superposition of atomic factors f_A and f_B at that site, i.e.

$$f(\boldsymbol{r}) = f_A c(\boldsymbol{r}) + f_B[1 - c(\boldsymbol{r})] . \tag{4.15}$$

The scattering factor (4.15) at the site $\boldsymbol{r}$ includes constant and variable parts. The constant part is the mean atomic scattering factor equal to $\langle f \rangle = \sum_i f_i c_i$ (in this case $\langle f \rangle = f_A c_A + f_B c_B$ or, considering $c_A = c$ and $c_B = 1 - c$, it becomes $\langle f \rangle = c(f_A - f_B) + f_B$). The variable part of the scattering ability of the site $\boldsymbol{r}$, i.e. fluctuation, is

$$\Delta f(\boldsymbol{r}) = (f_A - f_B)[c(\boldsymbol{r}) - c] . \tag{4.16}$$

In accordance with [8], the diffuse scattering intensity due to fluctuation of the scattering ability of N sites can be written as

$$I_D(\boldsymbol{q}) = \sum_{j=0}^{N}\sum_{n=0}^{N} \Delta f(\boldsymbol{r}_n)\Delta f(\boldsymbol{r}_{n+j}) \exp[2\pi i \boldsymbol{q}(\boldsymbol{r}_{n+j} - \boldsymbol{r}_n)] . \tag{4.17}$$

Let us take an arbitrary site $\boldsymbol{r}_n$ as the original site relative to which we shall consider correlations in the atomic arrangement and assume $\boldsymbol{r}_n = 0$. The vector $(\boldsymbol{r}_{n+j} - \boldsymbol{r}_n)$, which determines the position of any lattice site relative the original site $\boldsymbol{r}_n$, will be denoted as $\boldsymbol{R}_j$ (when $\boldsymbol{r}_n = 0$, the vector $\boldsymbol{R}_j = \boldsymbol{r}_{n+j}$). We now introduce the mean value

$$\left\langle \Delta f(\boldsymbol{r}_n)\Delta f(\boldsymbol{r}_{n+j}) \right\rangle = \frac{1}{N}\sum_{n=0}^{N} \Delta f(\boldsymbol{r}_n)\Delta f(\boldsymbol{r}_{n+j}) . \tag{4.18}$$

Using the above nomenclature, expression (4.17) takes the form

$$I_D(\boldsymbol{q}) = N\sum_{j=0}^{N} \left\langle \Delta f(0)\Delta f(\boldsymbol{R}_j) \right\rangle \exp(2\pi i \boldsymbol{q}\boldsymbol{R}_j) \tag{4.19}$$

and substituting expression (4.16) for the fluctuations,

$$I_D(\boldsymbol{q}) = (f_A - f_B)^2 N\sum_{j=0}^{N} \left\langle [c(0) - c][c(\boldsymbol{R}_j) - c] \right\rangle \exp(2\pi i \boldsymbol{q}\boldsymbol{R}_j) . \tag{4.20}$$

Equation (4.20) can be broken up into two terms. The first term evaluates the condition $j = 0$, while the second term is a sum taken over j from 1 to N (note that vector $\boldsymbol{R}_j = 0$ for $j = 0$):

$$I_D(\boldsymbol{q}) = (f_A - f_B)^2 N \left\langle [c(0) - c]^2 \right\rangle$$
$$+ (f_A - f_B)^2 N\sum_{j=1}^{N} \left\langle [c(0) - c][c(\boldsymbol{R}_j) - c] \right\rangle \exp(2\pi i \boldsymbol{q}\boldsymbol{R}_j) . \tag{4.21}$$

Taking into account $\langle c(\boldsymbol{r}) \rangle = c$ and the identity $c^2(\boldsymbol{r}) \equiv c(\boldsymbol{r})$, the quantity $\langle [c(0) - c]^2 \rangle = \langle c^2(0) \rangle - 2c\langle c(0) \rangle + c^2 = c - 2c^2 + c^2 = c(1 - c)$ and the first term in (4.21) can be written as the Laue formula in the form

$$I_{DL}(\boldsymbol{q}) = Nc(1 - c)(f_A - f_B)^2 . \tag{4.22}$$

This term describes the Laue background, i.e. the contribution to diffuse scattering, which is independent of the degree of short-range order and which results from the difference of atomic scattering factors.

Let us now consider in more detail the second term in formula (4.21). The factors $[c(0) - c]$ and $[c(\boldsymbol{R}_j) - c]$ represent the excess probabilities that sites $\boldsymbol{r} = 0$ and $\boldsymbol{r} = \boldsymbol{R}_j$ are occupied by atoms of species A and therefore their product average is the correlation parameter $\varepsilon_{AA}(R_j)$. Indeed, when $\boldsymbol{R}_j = 0$,

$$\langle[c(0) - c][c(\boldsymbol{R}_j) - c]\rangle = \langle c(0)c(\boldsymbol{R}_j)\rangle - c\langle c(\boldsymbol{R}_j)\rangle - c\langle c(0)\rangle + c^2 \tag{4.23}$$

or, considering that $c(\boldsymbol{r}) = c$,

$$\langle[c(0) - c][c(\boldsymbol{R}_j) - c]\rangle = \langle c(0)c(\boldsymbol{R}_j)\rangle - c^2 = \varepsilon_{AA}(\boldsymbol{R}_j) \ . \tag{4.24}$$

In accordance with (4.24), the diffuse scattering intensity, which is due to short-range order, becomes

$$I_{\rm DSRO}(\boldsymbol{q}) = (f_{\rm A} - f_{\rm B})^2 N \sum_{j=1}^{N} \varepsilon_{\rm AA}(\boldsymbol{R}_j) \exp(2\pi i \boldsymbol{q}\boldsymbol{R}_j) \ . \tag{4.25}$$

Alternatively by considering (4.13) and substituting the short-range order parameter α_j for the correlation parameter $\varepsilon_{AA}(R_j)$ it can be written

$$I_{\rm DSRO}(\boldsymbol{q}) = c(1-c)(f_{\rm A} - f_{\rm B})^2 N \sum_{j=1}^{N} \alpha_j \exp(2\pi i \boldsymbol{q}\boldsymbol{R}_j) \ . \tag{4.26}$$

Mutually replaceable atoms in a nonideal solid solution have different dimensions and therefore atoms are displaced from their exact positions, distorting the lattice. Long-range elastic forces play a significant role in the presence of distortions. Consequently, different atomic radii in nonideal solutions affect the diffuse scattering intensity both directly and indirectly. The direct effect is connected with scattering from distortions. The indirect effect is related to the appearance of an elastic energy, which affects the probability of fluctuations and, hence, diffuse scattering. As a result of different atomic dimensions, expression (4.26) includes additionally an exponential factor $\exp(-\beta_j)$, which attenuates the diffuse scattering intensity $I_{\rm DSRO}$ from short-range order. It is very important to take into account distortions when lattice parameters strongly depend on the concentration and when the atomic scattering factors $f_{\rm A}$ and $f_{\rm B}$ are nearly equal. The effect of distortions on the diffuse scattering intensity was comprehensively studied by Krivoglaz (see paragraphs 18 and 26 in the monograph [8]).

In the case of a single crystal the diffuse scattering intensity due to short-range order will be present according to expression (4.26) if the static displacements of the atoms are neglected.

If the effect of the static displacements is to be neglected, the qualitative character of the short-range order can be determined from an analysis of the diffuse scattering intensity $I_{\rm D} = I_{\rm DL} + I_{\rm DSRO}$. The diffuse scattering from short-range order modulates the Laue background (Fig. 4.1).

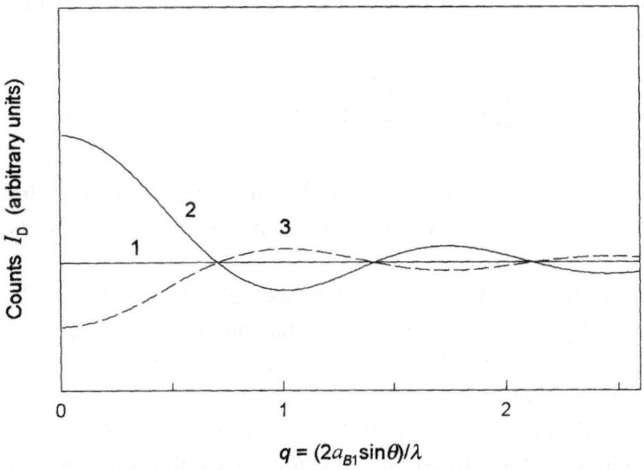

Fig. 4.1. Effect of the structural state of the solid solution on the diffuse scattering intensity $I_D = f(2\pi q)$ of X-ray radiation and thermal neutrons: (1) diffuse scattering by a fully disordered ($\alpha_j = 0$) solid solution is due to the Laue background only; (2) diffuse scattering intensity in the case of short-range decomposition ($\alpha_j > 0$); (3) diffuse scattering in the case of short-range order ($\alpha_j < 0$). $q = (2a_{B1}\sin\theta)/\lambda$ is the dimensionless reduced diffraction vector

If the nearest neighborhood of an atom of a particular species in a solid solution contains more atoms of the same species than in the case of a statistical distribution, the short-range order parameter is positive: $\alpha_j > 0$. Therefore, when $\theta = 0$, the short-range redistribution makes a maximum positive contribution to I_D, because at $\theta = 0$ the exponential factor $\exp(2\pi i\boldsymbol{qR}_j) = \cos[4\pi R_j(\sin\theta)/\lambda] = 1$. When θ and R_j increase, the exponential factor quickly reaches negative values and then grows again. However, the short-range order parameter α_j tends to zero with increasing R_j and therefore the absolute contribution to the diffuse scattering intensity decreases (Fig. 4.1, curve 2). If the nearest neighbors of an atom of a given species are predominantly the atoms of the other species ($\alpha_j < 0$), the contribution of the short-range order to the intensity I_D will be opposite in sign to the contribution made at $\alpha_j > 0$. The change of I_D during short-range ordering ($\alpha_j < 0$) is shown in Fig. 4.1 (curve 3).

In the case of polycrystals, the diffuse scattering intensity due to short-range order is given by an expression corresponding to (4.26):

$$I_{\mathrm{DSRO}}(\boldsymbol{q}) = c(1-c)(f_{\mathrm{A}} - f_{\mathrm{B}})^2 N \sum_{j=1}^{\infty} \alpha_j z_j \frac{\sin(2\pi q R_j)}{2\pi q R_j}. \qquad (4.27)$$

The intensity of the Laue background, I_{DL}, for a polycrystalline binary solid solution is described by (4.22).

Short-range order parameters α_j can be found from X-ray or neutron diffuse scattering curves. The function

$$Q(q) = \frac{I_{\mathrm{DSRO}} + I_{\mathrm{DL}}}{c(1-c)(f_{\mathrm{A}} - f_{\mathrm{B}})^2 N} - 1 = \sum_{j=1}^{\infty} \alpha_j z_j \frac{\sin(2\pi q R_j)}{2\pi q R_j} \tag{4.28}$$

is usually considered for convenience of calculations. From (4.28) it follows that all stray types of scattering, including diffuse scattering at static displacements, are excluded.

Parameters for α_j can be found most easily from measured values of $Q(q)$ by the least-squares methods, i.e. by solving a system of equations relating unknown $x_j = \alpha_j z_j$ to preset values of R_j and $Q(q_j)$, where $q_j = |\boldsymbol{q}_j|$ is the diffraction vector at all measured points of the curve $Q(q)$.

Another method for calculating α_j considers a reduction to the Fourier transform of the function $Q(q)$. Let us introduce the ideal distribution function $U(R)$, which describes deviations from the concentration of species B atoms in some coordination sphere of an atom A from the concentration of species B atoms in the same coordination sphere for a statistical distribution. This deviation is $\Delta n_j = \alpha_j z_j$. Alternatively

$$\Delta n_j = \int_{R_j-\delta}^{R_j+\delta} U(R)dR \tag{4.29}$$

and, therefore,

$$\alpha_j z_j = \int_{R_j-\delta}^{R_j+\delta} U(R)dR \ . \tag{4.30}$$

Substituting (4.30) into (4.28) consider the function $qQ(q)$ and change from summation to integration. Then

$$2\pi q Q(q) = \int_0^{\infty} \frac{U(R)}{R} \sin(2\pi q R)dR \ . \tag{4.31}$$

The Fourier transform gives some function $f(R)$:

$$\frac{2}{\pi} f(R) = \frac{U(R)}{R} = 2\pi \int_0^{\infty} qQ(q)\sin(2\pi q R)dq \ . \tag{4.32}$$

Hence,

$$U(R) = 2R\, f(R)/\pi \ . \tag{4.33}$$

Considering (4.30), we obtain

$$\alpha_j z_j = \frac{2R_j}{\pi} \int_{R_j-\delta}^{R_j+\delta} f(R)dR \ . \tag{4.34}$$

Thus, calculation of short-range order parameters by this method is reduced to the following: The curve $U(R)$ is constructed using measured values of $Q(q)$. The area under this curve near R_j values gives, in accordance with (4.34), values of $\alpha_j z_j$. Obviously, the curve $U(R)$ should have peaks near R_j only. A drawback of this method is that it does not permit resolving peaks with nearly equal radii R_j and R_{j+1} of adjacent coordination spheres. Consequently, the corresponding short-range order parameters α_j and α_{j+1} cannot be calculated.

4.2 Short-Range Order in Strongly Nonstoichiometric Compounds

The most important experimental methods for the examination of short-range order in strongly nonstoichiometric interstitial compounds are nuclear magnetic resonance (NMR), neutron diffuse scattering (distribution of interstitial atoms), and X-ray diffuse scattering (distribution of metal atoms). None of these methods can be universally applied to any nonstoichiometric compound to determine the distribution of atoms of all species. The NMR method is the least applicable to nonstoichiometric compounds. Its limitations are due to physical reasons: magnetic resonance signals can be generated only by nuclei which have a magnetic moment, i.e. by nuclei with a nonzero spin I (nuclei possess nonzero spin if the number of protons or the number of neutrons or the number of protons and neutrons are odd). Therefore the NMR method can be used for analysis of nonstoichiometric compounds in cases like vanadium ($I = 7/2$ for ^{51}V), niobium ($I = 9/2$ for ^{93}Nb), or tantalum ($I = 7/2$ for ^{181}Ta) whose most common isotopes have nuclei with a nonzero spin $I \neq 0$. In contrast, NMR cannot be used for the analysis of nonstoichiometric compounds of titanium, zirconium, and hafnium, because the most common isotopes of these metals (^{48}Ti, ^{46}Ti, ^{92}Zr, ^{94}Zr, ^{180}Hf, ^{176}Hf, ^{174}Hf) have nuclei without a magnetic moment ($I = 0$). Still, the NMR method is in principle applicable to the study of ^{49}Ti ($I = 7/2$), ^{91}Zr ($I = 5/2$) and ^{179}Hf ($I = 9/2$) isotopes. However, no such NMR investigations of nonstoichiometric compounds are known to us. Referring to interstitial atoms (carbon, nitrogen, and oxygen), the nuclei of the most frequent isotopes ^{12}C and ^{16}O possess zero spin and therefore magnetic resonance cannot be observed. The most common nitrogen isotope ^{14}N has a magnetic moment ($I = 1$), but until recently magnetic resonance measurements of ^{14}N in nonstoichiometric compounds was not performed. NMR studies of nuclei of the carbon isotope ^{13}C with spin $I = 1/2$ show great promise. They could considerably expand our knowledge of the structure of nonstoichiometric carbides.

In instances when the NMR method can be applied, it proves to be a very powerful technique, as this method is highly sensitive to small changes in the environmental symmetry of atoms.

4.2.1 NMR Studies of Short-Range Order

The NMR method allows one to determine the mutual arrangement of interstitial atoms and structural vacancies in the nonmetal sublattice of nonstoichiometric compounds, i.e. it permits one to analyse the short-range order in these compounds. In addition, some specific features of the electronic structure of nonstoichiometric compounds can be traced by NMR.

Let us introduce some notation for the convenience of discussion. The nearest neighborhood of a metal atom in carbide MC_y or nitride MN_y having the *B*1 structure includes three coordination spheres (see Fig. 3.5). The first coordination sphere is formed by six occupied or vacant sites of the nonmetal sublattice. The next coordination sphere comprises twelve metal atoms. The second nonmetal coordination sphere is formed by eight sites of the nonmetal sublattice. When a metal atom has such a neighborhood, its position can be described conveniently using a subscript and a superscript, showing the number of vacancies in the first and second nonmetal coordination spheres, respectively. For example, the position of a metal atom M_0^2 means that its first coordination sphere is complete (is free of vacancies), while the second nonmetal coordination sphere has two vacancies.

Compounds which contain vanadium have received the most extensive studies among nonstoichiometric compounds. Researchers [12] have analyzed samples of vanadium carbide VC_y (y = 0.66, 0.70, 0.80, 0.84, and 0.875) over the whole homogeneity interval of the cubic phase. NMR spectra of ^{51}V in carbides $VC_{0.875}$, $VC_{0.84}$ and $VC_{0.70}$ exhibited several resonance lines whose intensity changed with the VC_y composition. Comparison of the total intensity of resonance lines in metallic vanadium and vanadium carbides showed that the NMR method detects nearly all vanadium nuclei in carbide VC_y irrespective of the carbide composition. The presence of vacancies on the carbon sublattice of carbide VC_y has a marked effect on the relative intensity of observed resonance lines: when the carbon concentration decreases, the intensity of two lines reduces quickly while the intensity of the third line increases and one additional weak line appeares in the spectrum of carbide $VC_{0.70}$.

All lines and their satellites in NMR spectra of ^{51}V in VC_y have been correlated [12] to the positions of vanadium atoms having particular vacancy configurations in the first coordination sphere. Such an identification proved to be possible because of a splitting of satellites whose central lines do not separate. The electric field gradient at a vanadium nucleus should be much less with the first coordination sphere being free of vacancies as compared to a situation where the nearest neighborhood includes one or several vacancies. Starting from these considerations, observed lines have been correlated to positions V_0 (without vacancies in the first coordination sphere), V_1, V_2, and V_3.

An expression has been derived in ref. [12] relating the number of vacancies n in the nearest neighborhood of a vanadium atom and the Knight shift jump K:

$$K_n - K_{n-1} \approx nK_0 , \tag{4.35}$$

where $K_0 \approx 0.07$–0.09%, i.e. the Knight shift jump is proportional to the number of vacancies in the first coordination sphere of a vanadium atom in VC_y.

To determine the distribution of vacancies in vanadium carbide, researchers [12] have compared the experimental probability distribution $P(n)$ of vanadium atom positions with the random probability distribution, which has been calculated from the formula

$$P(n) = C_6^n (1-y)^n y^{6-n} . \tag{4.36}$$

The experimental distribution proved to be different from the statistical distribution in all samples of vanadium carbide VC_y. For example, when the carbon concentration decreased, the probability of positions V_0 with a complete (without vacancies) environment in the first coordination sphere diminished much faster than expected for a random distribution and positions V_3 occurred much more rarely than in the case of a random distribution.

Froidevaux and Rossier [12] explained the small number of positions V_3 and the absence of positions of vanadium atoms with four or more vacancies in the first coordination sphere by a repulsion between adjacent vacancies. So, in accordance with [12], a short-range order exists to a certain degree in VC_y carbide.

The NMR spectrum of ^{51}V in $VC_{0.875}$ differs markedly from the spectra of vanadium carbide with different carbon concentrations. Specifically, the intensity ratio of the lines corresponding to positions V_0 and V_1 is 1/3. This ratio can be explained by assuming that the relative intensity of each line is proportional to the relative number of the corresponding lines. In this case three-quarters of all vanadium atoms in carbide $VC_{0.875}$ have one vacancy in the first coordination sphere, while the rest of the vanadium atoms have a complete environment. This distribution of vacancies suggests the presence of a superstructure of type V_8C_7 with the space group $P4_332$ or $P4_132$ [13]. However, it is suggested in ref. [12] that this model of vacancy distribution does not describe a real lattice completely accurately. Indeed, this superstructure with a helicoidal arrangement of vacancies includes positions V_0^1 and V_1^1 only, whereas the analysis of satellites suggests the existence of three types of positions V_1, which could not be resolved by Froidevaux and Rossier [12].

An NMR study of an annealed single crystal of carbide $VC_{0.84}$ [14, 15] gave interesting results. The NMR spectra of ^{51}V in $VC_{0.84}$ could be resolved into lines corresponding to vanadium atom positions V_0 and V_1. Analyzing the satellites, it was possible to decompose the intense line V_1 into lines correspond-ing to positions V_1^0, V_1^1 and V_1^2 , i.e. without vacancies and with one and two vacancies in the second coordination sphere, respectively. The intensity ratio of lines V_1^1 and V_1^2 in the spectra was 2:1. Therefore one might think that the number of positions V_1^1 is twice the number of positions V_1^2 in the present carbide. The high intensity of the lines corresponding to positions V_1^1 and V_1^2 allows one to consider them as positions of an ordered structure of vanadium carbide.

Venables and Kahn [14] examined the carbide $VC_{0.84}$ using electron diffraction. They proposed a structure with trigonal symmetry for the ordered vanadium carbide V_6C_5. Knowing from the NMR data the number ratio for the positions V_1^1 and V_1^2, the authors could describe the distribution of carbon atoms and vacancies in a trigonal unit cell, i.e. they could determine the V_6C_5 structure. The trigonal superstructure of V_6C_5 with the space group $P3_1$ can be viewed as spiral chains made up of quasimolecules $V_6C_5\square$ (see Fig. 3.17*b*). The presence of weak lines corresponding to positions V_0 and V_1^0 was attributed to a deviation from the ideal composition of the ordered phase V_6C_5 ($VC_{0.83}$). It is worth noting that those lines can also be explained by a deviation in the degree of long-range order from the maximum value of the sample.

An effort was made to obtain quantitative data about short-range order in the carbide VC_y by fitting NMR spectra of ^{51}V [16, 17]. Experimental NMR spectra of ^{51}V in carbides $VC_{0.75}$, $VC_{0.84}$, $VC_{0.85}$, $VC_{0.86}$, and $VC_{0.87}$ are analogous to those described in [12] and include resonance lines corresponding to positions V_0, V_1, and V_2. If it is assumed that the complex shape of the line spectrum is due to the presence of nuclei having different local environments (different positions of metal atoms) in the crystal and the intensity of the line is proportional to the number of positions of the given type, then the theoretical intensity can be written as

$$I_{\text{theor}}(\nu) = \sum_i \lambda_i P_i I_i(\nu)\,, \tag{4.37}$$

where P_i is the probability of a cluster with the *i*-th configuration of the nearest neighborhood resonating the core by nonmetal atoms; λ_i is the multiplicity of the *i*-th configuration of the basis cluster; and $I_i(\nu)$ is the mathematical model of an ideal spectral line corresponding to the *i*-th configuration of the nearest neighborhood of the resonating core (in [16] the lines were described by the Gaussian distribution; the basis cluster was an octahedron comprising six sites of the nonmetal sublattice with a metal atom at the center). The probabilities P_i were expressed as an expansion in terms of paired and multiparticle correlation parameters in accordance with [18]. Correlation parameters were found by fitting a functional of the type $\sum[I_{\text{theor}}(\nu) - I_{\text{exp}}(\nu)]^2$. NMR spectra of ^{51}V in carbides VC_y were fitted in [16, 17] taking into account correlation parameters starting from paired parameters and ending with six-fold correlation parameters. The spectra were described best when paired and triple correlations were considered. The spectra changed little when higher-order correlation moments were used.

In accordance with the results obtained, paired correlations ε_{21}, which describe the distribution of carbon atoms (or vacancies) at two adjacent sites of an octahedron located in the first coordination sphere one relative the other, are negative in the entire homogeneity interval of vanadium carbide. This means that in the whole homogeneity interval of VC_y the probability of finding vacancies that represent nearest neighbors is smaller than the probability for the statistical distribution. In accordance with (4.13), the short-range order parameter for the first coordination sphere is $\alpha_1 = \varepsilon_{21}/[y(1-y)]$ and equals nearly -0.1. As the degree of nonstoichiometry increases, this parameter grows slowly and passes the minimum near the composition $VC_{0.83}$.

It can be stated that the calculations of correlations and the short-range order parameter for vanadium carbide agree well with the data [12, 14, 15], suggesting the presence of short-range order is caused by a repulsion between the vacancies in the carbide VC_y. However, in ref. [17] it is stated erroneously that the appearance of ordered phases of vanadium carbide is due to a certain type of short-range order. It was shown [19, 20] that long-range order may not form in nonstoichiometric compounds with the *B*1 structure even at a maximum short-range order in the first and second coordination spheres. Therefore the presence of short-range order in strongly nonstoichiometric carbides is a necessary but insufficient condition for the appearance of long-range order. The relationship between short- and long-range order for nonstoichiometric compounds and solid solutions having different basic structures will be discussed comprehensively in Chap. 6.

The effect of the composition of VN_y on NMR spectra of ^{51}V was analyzed in [21]. Samples had compositions covering the entire homogeneity interval from $VN_{0.98}$ to $VN_{0.706}$ and the measurements performed between 93 and 300 K showed that the shape and position of resonance lines in the spectra were independent of temperature. The total intensity of the NMR response normalized to the vanadium concentration proved to be constant at any nitrogen concentration. This means that in nitride VN_y, as in carbide VC_y [12], all ^{51}V nuclei contribute to the resonance absorption line. This allows for a quantitative analysis of the spectra.

In a magnetic field of 0.39 T the NMR spectra of vanadium nitrides with different nitrogen concentrations represent a single line whose shape and position is independent of the VN_y composition. As the magnetic field is increased, the resonance line becomes asymmetrically distorted where the character of the distortion depends on the VN_y composition. The distortion was explained in [21] by the fact that the observed lines were due to a superposition of several lines from ^{51}V nuclei with different Knight shifts *K*. Vanadium atoms with a different nearest neighborhood in nitride VN_y are responsible for the appearance of those lines.

The resonance peak in the spectra was resolved into two lines corresponding to a complete site of vanadium atoms V_0 and positions V_n with one or more vacancies in the first coordination sphere ($n \geq 1$ denotes the number of vacancies in the nearest neighborhood of a vanadium atom). The comparison of the experimental intensity of the line corresponding to positions V_n with the theoretical intensity calculated for a random distribution by the formula

$$I_{\text{theor}} = I_0(1 - y^x) \tag{4.38}$$

shows that $x = 6$ for a small vacancy concentration . Thus the observed effects are determined mainly by interactions in the first coordination sphere of a vacancy. In formula (4.38) I_0 stands for the integral intensity of the spectrum and x denotes the number of V atoms within the effective range of a vacancy.

When the vacancy concentration is high, experimental points deviate systematically from the curve for a random distribution. This means that nitrogen atoms and vacancies are distributed with a certain degree of order rather than statistically in the nonmetal sublattice of vanadium nitride. Researchers [21] hold to the opinion that the vacancy distribution (the presence of just two types of positions V_1 and V_2), which they observed near a nitride composition $VN_{0.75}$, corresponds closely to the ordered phase V_4N_3 having a structure of Ni_3V type.

Short-range order in vanadium nitride was analyzed quantitatively [17] by the scheme used for vanadium carbide. Fitting of NMR spectra showed that parameters of paired and triple correlations changed monotonically with the VN_y composition. In accordance with calculations, the parameter ε_{21} is negative and the parameter ε_{22} is positive in the whole homogeneity interval of nitride VN_y. The correlation parameter ε_{21} decreases and ε_{22} increases with a growing number of vacancies in nitrides. In nitride VN_y a vanadium atom has an octahedral environment of six sites in the nonmetal sublattice. The sign and variation of the correlation parameters ε_{21} and ε_{22} with concentration suggest that vanadium nitride shows pronounced short-range order and the vacancies are located preferentially at opposite vertices of the octahedron.

Cubic vanadium monoxide VO_y was studied by NMR methods of wide lines and by spin echo at temperatures between 1.4 and 300 K [22, 23]. NMR spectra of ^{51}V in vanadium oxide exhibit only one line whose width increased with the oxygen concentration of VO_y. The measurements also show that the Knight shift is independent of temperature and the VO_y composition.

The NMR method was also used to examine the short-range order in solid solutions of nonstoichiometric vanadium compounds, mainly in oxycarbides and oxynitrides [17, 24]. NMR spectra of ^{51}V in oxycarbides VC_xO_y represent a set of four lines whose width and position depend little on the oxycarbide composition. From a comparison of the integral intensity of NMR spectra of carbide $VC_{0.87}$ and oxycarbide VC_xO_y ($0.75 \le x + y \le 0.94$) it was inferred [17] that all vanadium nuclei contribute to the spectrum of oxycarbide. Oxycarbides VC_xO_y ($y \ge 0.4$) have vacancies both in the nonmetal and metal sublattices. However, the study [17] did not reveal any effect of vanadium vacancies on the integral intensity of the spectra.

The analysis of NMR spectra allowed the correlation of spectrum lines to nonequivalent positions of vanadium atoms. It was noted that the line intensity depends on the oxygen concentration of the measured samples. In accordance with [17], all observed lines correspond to completed environments of the vanadium atoms by carbon and oxygen atoms. However, the number ratio could be different for the latter species, for example, $[VC_5O_1]$, $[VC_4O_2]$, $[VC_1O_5]$, and $[VC_3O_3]$. The intensities of the lines in the spectrum were found to be proportional to the probabilities of the corresponding positions. Comparison of the probabilities, which were determined experimentally and calculated for a statistical distribution of carbon and oxygen atoms, showed that nonmetal atoms had a nearly random distribution in oxycarbides $VC_{0.61}O_{0.14}$ and $VC_{0.67}O_{0.15}$. In contrast the distribution was found to be far from random at higher oxygen concentrations.

NMR spectra of ^{51}V in oxynitride VN_xO_y exhibit [17] three lines, two of which are correlated (by analogy with VC_xO_y) with two positions $[VN_5O_1]$ and $[VN_3O_3]$.

The analysis of NMR spectra of ^{51}V in solid solutions of VC_xO_y, VN_xO_y and VC_xN_y [17] showed that the replacement of a nonmetal atom in the nearest environment by an atom of another species causes an increase in the Knight shift compared to the shift achieved with a homogeneous complete environment. The Knight shift is largest when the number of atoms of different species is equal in the first coordination sphere of the vanadium atom. It is worth noting that the Knight shift also increases in vanadium carbide, nitride and oxide when the vanadium atom has an increased number of vacancies in its nearest environment. All these data suggest that the introduction of any impurity defect in the nearest environment of the vanadium atom in VC_y, VN_y, VO_y, VC_xO_y, VN_xO_y or VC_xN_y should increase the Knight shift compared to the shift achieved with a homogeneous complete environment.

There are only a few NMR studies on ^{93}Nb dealing with an examination of nonstoichiometric niobium compounds. Still, ref. [25], concerned with the phase composition of niobium nitride, deserves mentioning. It was observed that each phase (λ-, δ- or ε-phase) of niobium nitride produces its particular lines in the NMR spectrum. Therefore it is possible to determine the phases present in a sample by analyzing the NMR spectrum. In [25] it is also noted that quadrupolar interactions are stronger in niobium nitride than in vanadium nitride.

Froidevaux and Rossier [12] obtained NMR spectra of ^{93}Nb in carbide NbC_y. All spectra exhibit only one line whose intensity decreases quickly when going from $NbC_{0.94}$ to $NbC_{0.90}$. This decrease in intensity appears to be smaller at lower carbon concentrations implying that many positions of niobium atoms become unobservable upon deviation from stoichiometry. The reason is that large gradients of the electric field at some positions of the niobium atoms provoke a quadrupole broadening of second order.

The width of the observed line changes somewhat with the magnetic field intensity, but the variation also depends on temperature. The linewidth increases slowly at a temperature of about 80 K and decreases at 300 K as the magnetic field grows.

To explain these results, it is assumed in ref. [12] that the electric field gradient has some critical value equal to q_0. The positions of the niobium atoms with gradient $q > q_0$ do not contribute to the observed line. The number of positions having a large gradient increases with the vacancy concentration and, consequently, the intensity $I(y)$ of the measured signal drops. It is suggested in [12] that the signal intensity for carbide NbC_y is given by $I(y) = I_0 y^m$, where m is the number of niobium atom positions whose signal cannot be measured. The m value was determined from experimental data of the intensity of the observed lines in spectra of two samples $NbC_{0.98}$ and $NbC_{0.94}$ having a similar composition. In accordance with [12], $m = 14$. Therefore, the six niobium atoms forming the nearest neighborhood of a vacancy and the eight niobium atoms comprising the next coordination sphere do not contribute to the resonance line. Furthermore it is assumed [12] that the line, which is observed in the NMR spectra of ^{93}Nb in carbide NbC_y, relates to Nb atom positions with a complete neighborhood up to the third coordination sphere. This conclusion can hardly be valid if one considers that the number of excluded positions was determined by a linear extrapolation from two experimental points in close proximity, whereas the whole set of experimental data $\lg I(y) = f(\lg y)$ cannot be described by a linear function. A considerable temperature dependence of the Knight shift at ^{93}Nb nucleii in niobium carbide was detected too [12].

Comparison of the second-order quadrupole widening of NMR spectra of vanadium and niobium carbides shows that this widening is much larger for NbC_y. Therefore it may be argued that the electric field gradient in carbide NbC_y is much larger than in VC_y. In accordance with estimates [12], the valence contribution to the electric field gradient in niobium carbide is nearly 3.5 times higher than in vanadium carbide.

Since quadrupolar interactions in niobium carbide and nitride are stronger than those in vanadium carbide and nitride [12, 25], it is reasonable to assume that quadrupolar interactions in the carbides and nitrides are determined the dominantly by the metal atoms.

NMR and Short-Range Order in Nonstoichiometric Niobium Carbide. Short-range order in the carbide NbC_y was studied extensively by high-sensitivity pulse NMR spectroscopy [26–30]. The results from these studies present special significance, because in addition to the NMR studies, the structure of the same NbC_y samples was examined by X-ray and neutron diffraction.

Samples of the nonstoichiometric niobium carbides $NbC_{0.995}$, $NbC_{0.88}$, $NbC_{0.845}$, $NbC_{0.83}$, $NbC_{0.81}$ and $NbC_{0.75}$ were prepared by solid-state sintering of niobium and carbon powders in vacuum. The sintering conditions are shown in Fig. 3.1. The samples underwent additional thermal treatment (quenching, annealing) in order to produce niobium carbide NbC_y with different degrees of ordering. Thermal treatment conditions are given in Fig. 3.2.

Thermal treatment has a considerable effect on the phase composition of niobium carbide: annealing at a temperature above 1300 K and subsequent quenching (regime *a*) produced disordered samples; annealing at a temperature of 1300 K and subsequent slow cooling (regime *c*) produced niobium carbide in the ordered state. Samples which contained both ordered and disordered phases were obtained under regime *b*. In this text, the conditions for thermal treatment regimes will be specified in brackets after the niobium carbide formula, for example, $NbC_{0.88}(c)$.

All superstructure reflections, which were observed in neutron diffraction patterns of NbC_y samples, are due to the monoclinic (space group $C2/m$) ordered phase Nb_6C_5 [26, 31–33].

NMR absorption spectra of ^{93}Nb nuclei of powder samples of carbide NbC_y were registered at frequencies of 22.0, 21.5 and 14.01 MHz, and at a temperature of 300 K using a Bruker SXP 4-100 pulse spectrometer. The "spin echo" signal was integrated simultaneously with the magnetic field scan according to Clark's method [34]. The first pulse was 1.16 μs long, the delay time was about 100 μs, and the pulse burst time was 0.1 s. To improve the signal-to-noise ratio, the "echo" signal was multiply stored.

NMR spectra of ^{93}Nb nuclei in nonstoichiometric niobium carbide NbC_y with different carbon content are shown in Fig. 4.2. The effect of ordering on the NMR spectra of carbide $NbC_{0.83}$ is shown in Fig. 4.3. The line 0 is observed in most spectra. The position of this line coincides with the position of the line observed by Froidevaux and Rossier [12] in NMR spectra of niobium carbide. The intensity of this line decreases and it even vanishes when the vacancy concentration of the carbon sublattice increases. However, the line 0 appears again (see Fig. 4.2) in spectrum IV corresponding to carbide $NbC_{0.75}(c)$. Since the line 0 decays slower than the line [12] in the series from $NbC_{0.95}$ to $NbC_{0.845}$ and widens (Fig. 4.4), it cannot be related to the positions of niobium atoms with a complete environment in the first and second coordination spheres, as this was done in [12]. One can assume that the line 0 in spectra from I to IV (Fig. 4.2) is due to both the complete positions of niobium atoms Nb_0^0 (Fig. 4.5*a*) and the positions of Nb atoms having vacancies in the second coordination sphere only, for example, Nb_0^1 (Fig. 4.5*b*), Nb_0^2.

When the vacancy concentration of niobium carbide increases, the line 0 first widens and then narrows abruptly in carbide $NbC_{0.75}(c)$ (see Fig. 4.4). One may think that this relates to niobium atoms appearing in positions with a very small electric field gradient q. The fact that the line 0 has an equal width in the spectra of $NbC_{0.995}$ and $NbC_{0.75}(c)$ is indicative of q being nearly zero. An almost zero electric field gradient may be true of the position Nb_0^8. This position of niobium atoms is free of vacancies in the first coordination sphere and does not contain atoms in the second nonmetal coordination sphere. Since the intensity of the line 0 in the NMR spectrum of carbide $NbC_{0.75}(b)$ is relatively small, it is likely that the cubic ordered phase Nb_4C_3 with positions Nb_0^8 is formed in this carbide. Moreover, neutron diffraction of annealed niobium carbide $NbC_{0.75}$ does not reveal any superstructure reflections corresponding to the ordered phase Nb_4C_3 with space group $Pm\bar{3}m$. The line 0 may appear in the NMR spectrum of carbide $NbC_{0.75}$ if this carbide contains some amount of the ζ-Nb_4C_3 phase [35], which is not an ordered phase. The ζ-Nb_4C_3 phase is free of structural vacancies and, consequently, the electric field gradient at ^{93}Nb nuclei may be small. However, that phase was not detected by diffraction methods either[26–33]. Thus, the line 0 present in spectrum IV of carbide $NbC_{0.75}(b)$ [26–29] could not unambiguously be identified.

Figure 4.3 presents NMR spectra of ^{93}Nb nuclei in niobium carbide $NbC_{0.83}$ with a different degree of ordering. The spectra were taken at frequencies 21.5 and 14.01 MHz. One wide peak is observed on each side of the line 0 in spectrum I corresponding to carbide $NbC_{0.83}(a)$. Spectra II and III, which correspond to $NbC_{0.83}(b)$ and $NbC_{0.83}(c)$, exhibit five peaks in addition to the line 0. Spectra IV and V were registered for the same samples as spectra I and II, but at a lower frequency. Spectra IV and V are wider than spectra I and II. The number of peaks in spectrum V is as high as eight.

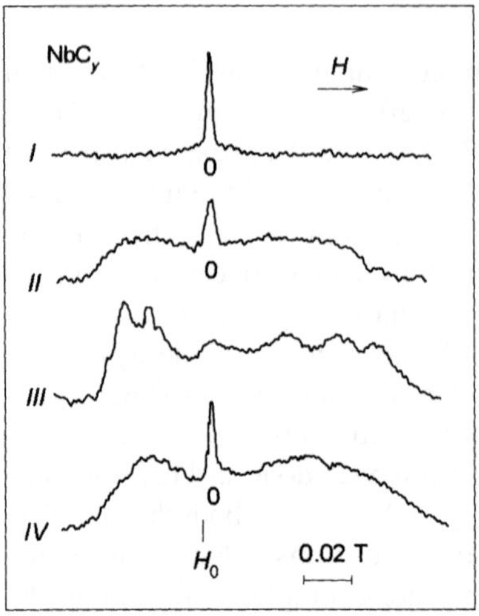

Fig. 4.2. Variation of the shape of the [93]Nb NMR spectra depending on the composition of niobium carbide NbC_y (ν_0 = 21.5 mHz, H_0 = 2.069 T) [29]: I – $NbC_{0.995}$, II – $NbC_{0.88}$(b), III – $NbC_{0.845}$(b), IV – $NbC_{0.75}$(c)

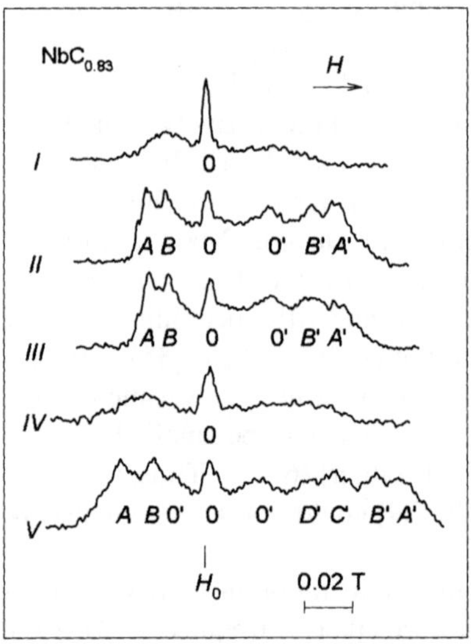

Fig. 4.3. Effect of ordering on NMR spectra of [93]Nb in carbide $NbC_{0.83}$ produced under different thermal treatment regimes [29]: I, IV – route a; II, V – route b; III – route c (I–III, ν_0 = 21.5 mHz, H_0 = 2.069 T; IV, V, ν_0 = 14.01 mHz, H_0 = 1.346 T). Thermal treatment routes are shown in Fig. 3.2

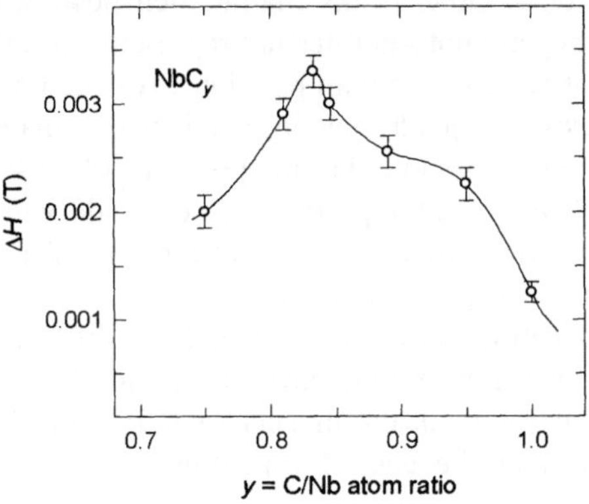

Fig. 4.4. Width of line 0 in NMR spectra depending on the composition of carbide NbC_y ($\nu_0 = 21.5$ mHz, $H_0 = 2.069$ T) [29]

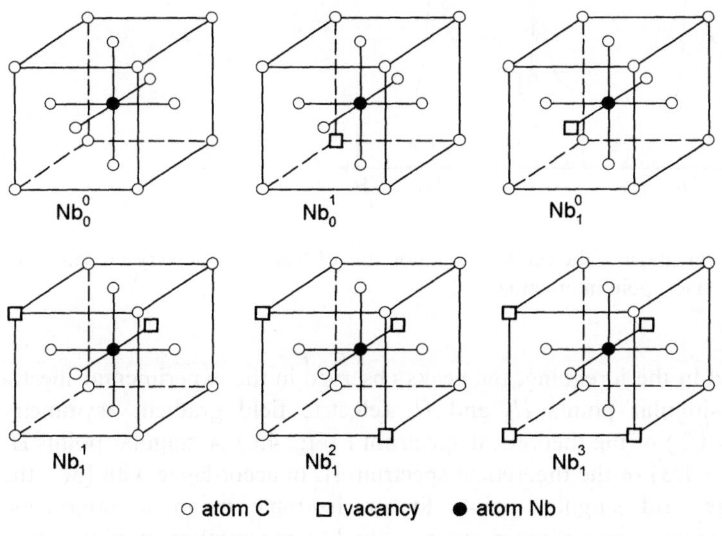

Fig. 4.5. Most probable positions of niobium atoms in niobium carbide NbC_y [29]

Considering the relatively low intensity (about 4%) of the line 0 (spectrum II, Fig. 4.3) related to the positions of niobium atoms Nb_0 and a nearly equal number of positions Nb_0 and Nb_2 in carbide $NbC_{0.83}$, one may assume that the peaks in spectrum II are due mainly to positions Nb_1 (positions Nb_2 are unobservable owing to the second-order quadrupole effect).

The analysis of the frequency dependence of the peaks in the NMR spectra (spectra II and V, Fig. 4.3) shows that only the peaks 0' may be satellites. Their central transition line may be the line 0 only. Other peaks change their positions depending on the frequency. Therefore they are not satellites but represent central transition lines corresponding to different types of niobium positions. A widening of NMR spectra at low frequencies suggests that quadrupole interactions dominate over magnetic interactions. Therefore one may assume that the positions of peaks *A*, *B*, *A*', *B*' (spectrum II) and *A*, *B*, *A*', *B*', *C*', *D*' (spectrum V) coincide with singular points of split lines of the central transition. Satellites of those lines are not observed in accordance with the first-order quadrupole effect.

The positions of peaks in NMR spectra, which were found at different frequencies, change in a complicated manner as a result of asymmetry and of large values for the electric field gradient. Therefore peaks in spectra I, III, and V (Fig. 4.3), which relate to different lines of the central transition, cannot be resolved without doing calculations.

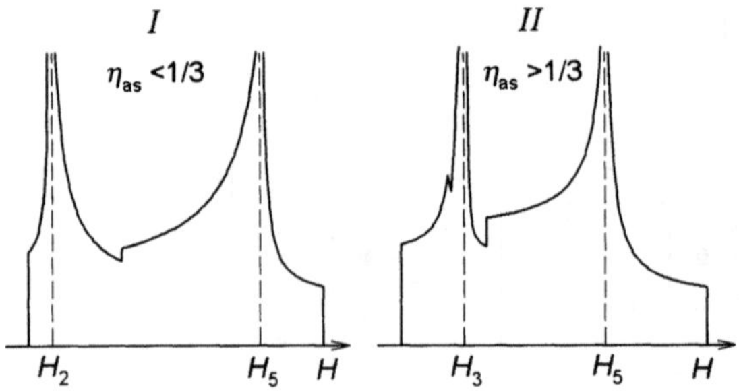

Fig. 4.6. Theoretical shape of the central transition line of NMR in a polycrystal caused by the second-order quadrupole interaction [30]

As mentioned in the foregoing, the peaks observed in the experimental spectra coincide with singular points H_2 and H_5 (electric field gradient asymmetry parameter $\eta_{as} < 1/3$) of the theoretical spectrum I (Fig. 4.6) or singular points H_3 and H_5 (for $\eta_{as} > 1/3$) of the theoretical spectrum II. In accordance with [36], the positions of the said singular points for an isotropic magnetic interaction (magnetic interaction tensor components are equal to one another, $\sigma_1 = \sigma_2 = \sigma_3 = \sigma$) can be given by formulas which are rearranged [27, 29] for the field scan method:

$$\left.\begin{aligned} H_2 &= H_0 - F\frac{(3-\eta_{as})^2}{144H_0} - \sigma H_0 \\ H_5 &= H_0 + F\frac{1-\eta_{as}}{9H_0} - \sigma H_0 \end{aligned}\right\} \text{ if } 0 < \eta_{as} < 1/3 , \qquad (4.39)$$

$$\left.\begin{aligned} H_3 &= H_0 - F\frac{1-\eta_{as}^2}{18H_0} - \sigma H_0 \\ H_5 &= H_0 + F\frac{1-\eta_{as}}{9H_0} - \sigma H_0 \end{aligned}\right\} \text{ if } 1/3 < \eta_{as} < 1 \,, \qquad (4.40)$$

where $F = (\nu_Q/\gamma_{Nb})^2[I(I-1) - 3/4]$; ν_Q is the quadrupole coupling frequency; γ_{Nb} is the gyromagnetic ratio for the niobium atomic core; H is the magnetic field intensity; and $I = 9/2$ is the ^{93}Nb core spin.

Table 4.1. Hyperfine interaction parameters of ^{93}Nb NMR spectra in disordered and ordered niobium carbide $NbC_{0.83}$ [29,30]

Niobium atom position	Label of position (Fig. 4.5)	Vacancy position (Fig. 3.5)	Model calculation results			
			$\nu_Q/\lvert q_C - q_\square\rvert$ (MHz)	η_{as}^{calc}	P_i^{disord}	P_i^{ord}
Nb_0^0	*a*	–	0	0	0.09	0
Nb_0^1	*b*	11	0.2	0	0.13	0
Nb_0^2	–	9, 11	0.4	0	0.01	0
Nb_0^2	–	11, 12	0.2	1.0	0.04	0
Nb_0^2	–	11, 13	0.2	1.0	0.04	0
Nb_1^0	*c*	1	1.05	0	0.10	0
Nb_1^1	*d*	1, 11	1.05	0.2	0.07	0.666
Nb_1^2	*e*	1, 11, 13	1.05	0.4	0.007	0.333
Nb_1^3	*f*	1, 11, 12, 13	1.05	0.6	0.001	0
Nb_2^0	–	1, 3	2.1	0	0.01	0

Niobium atom position	Singular points in central lines in ^{93}Nb NMR spectra at ν_0 (MHz)		Experimental results			
	$\nu_0 = 21.5$	$\nu_0 = 14.01$	ν_Q (MHz)	η_{as}^{exp}	$\lvert q_C - q_\square\rvert$	$K \pm 0.05$ (%)
Nb_0^0	0	0	0	–	–	–0.11
Nb_0^1	0, 0'	0, 0'	0.3 ± 0.07	0	1.5	–0.11
Nb_0^2	0	0	–	–	–	–0.11
Nb_0^2	0	0	–	–	–	–0.11
Nb_0^2	0	0	–	–	–	–0.11
Nb_1^0	*A*, *A*'	*A*, *A*'	1.9 ± 0.3	0	1.8	–0.25
Nb_1^1	*A*, *B*'	*A*, *B*'	2.0 ± 0.3	0.2 ± 0.05	1.9	–0.14
Nb_1^2	*B*, 0'	*B*, *C*'	2.1 ± 0.3	0.4 ± 0.05	2.0	0.25
Nb_1^3	*B*, 0'	*B*, *D*'	2.2 ± 0.3	0.5 ± 0.05	2.1	0.50
Nb_2^0	–	–	–	–	–	–

It follows from (4.39) and (4.40) that the positions of the peaks in NMR spectra are determined by three unknowns: ν_Q, η_{as} and σ. To find these unknowns, it is sufficient to know the positions of three peaks: for example (see Fig. 4.3), the positions of two peaks in spectrum V and the position of one peak in spectrum II measured at a different frequency. Solving the system of three equations and using the values found for ν_Q, η_{as} and σ, it is possible to calculate the position of the fourth peak. If the spectra contain such a peak, both pairs of peaks refer to the same central transition line. The next three peaks need to be chosen in case the calculated position does not coincide with any peak in the experimental spectrum.

This procedure was performed for all possible combinations of peaks in spectra II and V (Fig. 4.3). It was shown that in addition to the line 0, four more central transition lines are contained in these spectra. The positions of the singular points of the central transition lines and the corresponding parameters for the hyperfine interactions ν_Q, η_{as} and $K \equiv \sigma$ are given in Table 4.1 (the isotropic shift σ of the NMR line is equal to the Knight shift K in niobium carbide since it represents a compound with a metallic type of conductivity).

To correlate the lines observed in spectra II and V (Fig. 4.3) to certain positions of niobium atoms, the electric field gradient asymmetry parameter η_{as} was calculated within models with respect to the quadrupole coupling frequency $\nu_Q/|q_C - q_\square|$ and probabilities P_i for different positions of niobium atoms. In accordance with [37], the electric field gradient tensor components in a crystal are

$$q_{ij} = (1-R)q_{ij}^{\text{val}} + (1-\gamma_\infty)q_{ij}^{\text{lat}} , \tag{4.41}$$

where q_{ij}^{val} denotes the electric field gradient due to valence electrons; q_{ij}^{lat} is the electric field gradient due to external charges; and R and γ_∞ stand for the Sternheimer coefficients. The valence contributes further to the electric field gradient if a lattice gradient exists. Therefore one may think that the tensor asymmetry parameter (4.41) coincides with the asymmetry parameter η_{as} of the lattice electric field gradient. In this case, it suffices to calculate the tensor q_{ij}^{lat} to estimate η_{as} and the relative values of the quadrupole interaction constant ν_Q.

In calculating q_{ij}^{lat} for different positions of Nb atoms (see Fig. 4.5), it is assumed that the charges of all carbon atoms and those of all vacancies are equal to q_C and $q_\square$, respectively. All charges are assumed to be localized at sites of an ideal lattice of type $B1$. The Sternheimer coefficient γ_∞ is calculated in accordance with [38]. The electric field gradient tensor components are found from the formula

$$q_{ij}^{\text{lat}} = \sum_N \frac{\partial^2 \varphi_N}{\partial X_i \partial X_j} , \tag{4.42}$$

where φ_N denotes the electric field potential and N is the number of the lattice site.

The rearrangement of the electric field gradient tensor in the form

$$\begin{pmatrix} V_{xx} & 0 & 0 \\ 0 & V_{yy} & 0 \\ 0 & 0 & V_{zz} \end{pmatrix}$$

provides theoretical values of the quadrupole interaction parameters (see Table 4.1):

$$\frac{\nu_Q}{|q_C - q_\square|} = \frac{3e^2 qQ}{2I(2I-1)h}\,; \quad \eta_{as} = (V_{xx} - V_{yy})/V_{zz}\,, \tag{4.43}$$

where Q is the quadrupole momentum of the ^{93}Nb core; $|q_C - q_\square|$ is the absolute difference of the carbon atom and the vacancy charges expressed as units of the electron charge. It should be remembered that the quantity $\nu_Q/|q_C - q_\square|$, which is calculated disregarding the valence contribution, can be used as a qualitative characteristic when interpreting NMR spectra.

In addition to the parameters for the hyperfine magnetic and electric interactions the shape of NMR spectra depends on the relative concentration of niobium atoms occupying different positions in a sample. The probability P_i, that there are niobium atoms at different positions in a crystal, is calculated by the formula

$$P_i = \left\langle \frac{1}{\lambda_i} \sum_{k=1}^{\lambda_i} \prod n(\boldsymbol{r}_{k\nu_C}) \prod [1 - n(\boldsymbol{r}_{k\nu_\square})] \right\rangle, \tag{4.44}$$

where $n(\boldsymbol{r})$ is the distribution function of carbon atoms at a site $\boldsymbol{r}$; and ν_C and $\nu_\square$ denote the nonmetal sublattice sites occupied by carbon atoms C and vacancies □. The products in (4.44) are taken with respect to all position sites i ($i = a, b, \ldots$) of a niobium atom. Summation over k is due to the presence of λ_i different orientations of carbon atoms and vacancies forming the nearest neighborhood of the niobium atom, i.e. its position i. The sign $\langle \ldots \rangle$ means averaging over the microcanonical ensemble. Formula (4.44) represents a particular form of a general expression for the probability P_i obtained in terms of the order parameter functional method (see Sect. 9.1).

In accordance with [30–33], the function $n(\boldsymbol{r})$ describing the distribution of carbon atoms both in ordered and disordered niobium carbides NbC_y depends on three long-range order parameters η_9, η_4, and η_3 (see Sects. 5.2 and 5.4.4, and also formulas (5.33) and (5.72)). Table 4.1 gives calculated probabilities for the positions of niobium atoms in fully disordered P_i^{disord} ($\eta_9 = \eta_4 = \eta_3 = 0$) and fully ordered P_i^{ord} ($\eta_9 = \eta_4 = \eta_3 = 1$) carbide $NbC_{0.83}$.

The large difference between P_i^{disord} and P_i^{ord} suggests that NMR spectra of ordered and disordered phases should be considerably different. For example, the most probable positions of niobium atoms in the disordered phase $NbC_{0.83}$ are Nb_0^0 (Fig. 4.5*a*) and Nb_0^1 (Fig. 4.5*b*) with a complete first coordination sphere. The most probable position in the ordered phase is Nb_1^1 (Fig. 4.5*d*), while positions with a complete first coordination sphere should be absent altogether.

When comparing ν_Q and η_{as} values obtained by model calculations and from experiments, preference was given to the equality of the calculated η_{as}^{calc} and experimental η_{as}^{exp} asymmetry parameters (the small discrepancy may be attributed to the fact that calculations disregard the effect of the coordination sphere formed by niobium atoms). By comparison the experimental values of the hyperfine interaction parameters ν_Q, η_{as} and K corresponding to different positions of niobium atoms (Table 4.1) can be obtained. Considering the environment of niobium atoms up to the second nonmetal coordination sphere, the authors in [29, 30] could explain all peaks observed in NMR spectra of ^{93}Nb nuclei in carbide $NbC_{0.83}$.

As a result of ordering, the intensity of the line 0 (see Fig. 4.3) diminishes and the wide lines of spectrum I split into narrow peaks (spectra II and V). At least two reasons can be invoked for the splitting of lines in NMR spectra of ordered $NbC_{0.83}$(*b, c*). First, peaks in spectra of quenched samples do not split because of the variation in local hyperfine fields corresponding to different types of niobium positions. This variation is small for annealed samples and certain parameters of the hyperfine interactions correspond to equivalent positions of niobium atoms. Second, splitting occurs because the number of different positions of niobium atoms decreases, while the probability is enhanced for positions which are preserved in the carbide upon ordering.

The quadrupole interaction is stronger for positions Nb_1 than for positions Nb_0 with a complete first coordination sphere. Therefore, widening of NMR spectra, which is observed upon ordering, results from the increase in the number of positions Nb_1 compared to the number of positions Nb_0.

Fully ordered niobium carbide $NbC_{0.83}$ can have only two types of niobium atom positions: Nb_1^1 (see Fig. 4.5*d*) and Nb_1^2 (Fig. 4.5*e*). Since spectrum III exhibits peak *A'* and spectrum V includes peaks *A'* and *D'*, the line 0 and its satellites 0' (Fig. 4.3), it may be inferred that a maximum degree of order is not achieved in carbides $NbC_{0.83}$(*b*) and $NbC_{0.83}$(*c*). The similarity of spectrum III (Fig. 4.2), corresponding to carbide $NbC_{0.845}$(*b*), and spectra II and III (Fig. 4.3), corresponding to carbides $NbC_{0.83}$(*b*) and $NbC_{0.83}$(*c*), suggests that the shape of NMR spectra depends more strongly on short-range order than on long-range order.

An example for the variation of NMR spectra as a result of ordering of carbide $NbC_{0.81}$ is shown in Fig. 4.7.

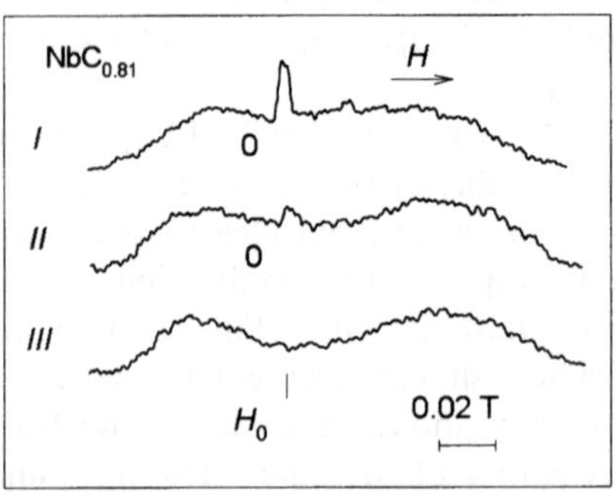

Fig. 4.7. Effect of ordering on NMR spectra of ^{93}Nb in carbide $NbC_{0.81}$ produced under different thermal treatments (ν_0 = 14.01 mHz, H_0 = 1.346 T) [29]: I – route *a*; II – route *b*; III – route *c*. Thermal treatment routes are shown in Fig.3.2

When the degree of order increases, the line 0 in NMR spectra of ^{93}Nb cores in carbide $NbC_{0.81}$ becomes less intense (Fig. 4.7), as in spectra I–III of carbide $NbC_{0.83}$ (Fig. 4.3). From this it follows that ordering in niobium carbide $NbC_{0.81}$ is similar to that observed in the ordered phase Nb_6C_5. Wide lines corresponding to niobium atom positions Nb_1 are also present in the NMR spectra of carbide $NbC_{0.88}$. The fact that the intensity of these lines is enhanced upon annealing is indicative of the presence of some amount of ordered phase Nb_6C_5 in the samples studied. This is in agreement with results of neutron diffraction studies [30–33], which revealed a relatively broad homogeneity interval for the Nb_6C_5 phase.

Analyzing the shape of NMR spectra in different magnetic fields and the variation as a function of the NbC_y composition with respect to the degree of order, it becomes possible [26–30] to resolve the spectra into satellites and central transition lines. Even using basic model calculations for the quadrupolar interaction parameters allowed the correlation of lines in the spectra of ordered $NbC_{0.83}$ samples to niobium atom positions Nb_0^0, Nb_0^1, Nb_1^0, Nb_1^1, Nb_1^2, and Nb_1^3 (Fig. 4.5) and lines in the spectra of disordered $NbC_{0.83}$ samples to positions Nb_0 and Nb_1. Experimental values for quadrupole and magnetic interaction parameters were determined for some positions of the Nb atoms. The data [29, 30] on the quadrupole coupling frequency ν_Q agree well with results [12]. Calculations of the Knight shift K showed that K increases with a growing number of vacancies in the niobium environment in NbC_y as in other similar compounds. Different Knight shifts for various Nb atom positions suggest a different occupancy of their d-orbitals depending on the nearest environment.

It should also be emphasized that the NMR method was not only used [26–29] to obtain information about the nearest environment of the metal atom in the first coordination sphere, but also to gain insight into the environment of the second nonmetal coordination sphere.

NMR studies [26–29] confirmed the existence of the ordered phase Nb_6C_5 with the space group $C2/m$. The mutual arrangement established earlier [31–33] for carbon atoms and vacancies in the nonmetal sublattice of niobium carbide has also been confirmed. Generally the data [26–29] suggest that short-range order in niobium carbide is determined mainly by a repulsive interaction between vacancies in the carbon sublattice.

4.2.2 Diffuse Scattering of Neutrons and Electrons

Diffuse scattering of neutrons and electrons were extensively used to elucidate the short-range order in nonstoichiometric compounds. Experimental data [39] on the diffuse intensity of electron scattering in vanadium carbide $VC_{0.75}$, has been exploited [40, 41] to derive the short-range order parameters α_{hkl} for eight nearest coordination spheres (hkl) of the nonmetal sublattice. A vacancy was taken as the center of the coordination spheres. In the nonstoichiometric compound MX_y the number of carbon atoms n_C and vacancies $n_\square$ in the j-th sphere having the coordination number z_j is

$$n_C = z_j(1 - \alpha_j)y\ ; \qquad n_\square = z_j[1 - (1 - \alpha_j)y]\ . \tag{4.45}$$

Values of the parameter α and the number of carbon atoms n_C and vacancies $n_\square$ in each coordination sphere are given in Table 4.2. For comparison, the same quantities were calculated for the ordered phase V_6C_5 having a monoclinic (space group $C2/m$ or $C2$) or trigonal (space group $P3_1$) superstructure.

Table 4.2. Short-range order parameters α_{hkl} and the number of carbon atoms, n_C, and vacancies, $n_\square$, for the coordination sphere (hkl) of carbide $VC_{0.75}$, ideal ordered carbide V_6C_5 and carbides $NbC_{0.73}$ and $TiC_{0.76}$

(hkl)	z_{hkl}	$VC_{0.75}$ [40]			V_6C_5 [40]		
		α_{hkl}	n_C	$n_\square$	α_{hkl}	n_C	$n_\square$
110	12	−0.178	10.6	1.4	−0.2	12	0
200	6	−0.260	5.7	0.3	−0.2	6	0
211	24	0.176	14.8	9.2	0.2	16	8
220	12	0.008	8.9	3.1	0	10	2
310	24	0.041	17.3	6.7	0	20	4
222	8	−0.171	7.0	1.0	−0.2	8	0
321	48	−0.070	38.5	9.5	−0.1	44	4
400	6	0.143	3.8	2.2	0.2	4	2
(hkl)	z_{hkl}	$NbC_{0.73}$ [43]			$TiC_{0.76}$ [43]		
		α_{hkl}	n_C	$n_\square$	α_{hkl}	n_C	$n_\square$
110	12	−0.095	9.6	2.4	−0.005	9.2	2.8
200	6	−0.275	5.6	0.4	−0.080	4.9	1.1
211	24	0.051	16.6	7.4	0.013	18.0	6.0
220	12	0.072	8.1	3.9	0.006	9.1	2.9
310	24	0.044	16.7	7.3	−0.003	18.3	5.7
222	8	−0.030	6.0	2.0	0.025	5.9	2.1
321	48	−0.020	35.7	12.3	−0.007	36.7	11.3
400	6	0.030	4.2	1.8	0.003	4.5	1.5

[1] Vacancy is a center of coordination spheres. The radius of coordination sphere (hkl) is $R_j = (a_{B1}/2)(h^2 + k^2 + l^2)^{1/2}$ where a_{B1} is the basic lattice constant of a disordered carbide MC_y with $B1$-type structure.

Table 4.2 shows that the distribution of carbon atoms and vacancies in the nonmetal sublattice of carbide $VC_{0.75}$ is far from statistical. For example, the number of vacancies in the first two coordination spheres is much smaller than expected for a random distribution. As the coordination sphere radius increases, α oscillates and its absolute value decreases smoothly. Such distributions of carbon atoms and vacancies can be explained by a repulsion between vacancies. It was argued [40, 41] that an analogous short-range order should be observed in niobium and tantalum carbides. The similarity of short-range order in nonstoichiometric monocarbides of transition metals of the V-th group was also inferred in [42]. De Novion and Maurice [42] hold to the opinion that absolutely disordered (without short-range order) samples of vanadium carbide VC_y with $y > 0.78$ cannot be produced.

The short-range order in $TiC_{0.76}$ and $NbC_{0.73}$ single crystals was analyzed [43] by diffuse neutrons scattering. The intensity distribution was measured at a temperature of 300 K in the $(1\bar{1}0)$ plane of the reciprocal lattice with the diffraction vector $2 < 2\pi q < 40$ nm^{-1}. Prior to the measurements, $TiC_{0.76}$ and $NbC_{0.73}$ single crystals were annealed under vacuum for several days. Diffuse scattering intensity patterns were not found to be periodic, which points to a displacement of atoms from the positions of an ideal undistorted lattice. Using the Fourier transform, the parameters α were determined [43] for eight coordination spheres in the nonmetal sublattice of carbides $TiC_{0.76}$ and $NbC_{0.73}$ (Table 4.2). An examination of Table 4.2 shows that short-range order in $TiC_{0.76}$ is much weaker than in $NbC_{0.73}$, but again the distribution of vacancies depends on the mutual repulsion, as in carbide VC_y. According to ref. [43], the correlations extend at least up to the ninth coordination sphere in niobium carbide. The findings [43] concerning short-range order in niobium carbide are in qualitative agreement with results of Fender [44] who studied carbides $NbC_{0.80}$–$NbC_{0.86}$ by diffuse neutron scattering.

A study of polycrystalline samples of $ZrC_{0.80}$ and $ThC_{0.75}$ by elastic diffuse scattering of neutrons [42] revealed correlations in the distribution of vacancies in three coordination spheres (a vacancy being at the center). In accordance with [42], the intensity distribution of the elastic diffuse scattering are similar for zirconium $ZrC_{0.80}$ and thorium $ThC_{0.75}$ carbides. In carbide $ThC_{0.75}$ the first coordination sphere of a vacancy has few vacancies (fewer vacancies than for a statistical distribution), the second coordination sphere contains no vacancies at all, and the third coordination sphere has a higher vacancy concentration than expected for a statistical distribution ($\alpha_1 = -0.05$; $\alpha_2 = -0.20$; $\alpha_3 = 0.075$). The spectrum of carbide $ZrC_{0.80}$ exhibits a broad peak near the reflection {1/2 1/2 1/2}, which corresponds to the position of the first superstructure line for annealed carbide $ZrC_{0.64}$.

Short-range order in single crystals of $TiC_{0.64}$, $TiC_{0.76}$, $TiN_{0.82}$, $NbC_{0.73}$, and $NbC_{0.83}$ was also studied in [45–51] representing an extension of the investigations in [42, 43, 52]. Christensen used zone melting to produce single crystals. A detailed description of the experimental technique can be found in [53]. Single crystals of titanium and niobium carbides were prepared by zone melting of sintered materials. The samples were melted in an atmosphere of pure helium under a pressure of 2 MPa. The single crystal of titanium nitride was produced by zone melting of a titanium rod at a temperature of 2870 K in an atmosphere of high-purity nitrogen. The nitrogen pressure was again 2 MPa. Single-crystal samples in the form of cylinders about 10 mm in length and in diameter were cut from the primary samples. The unit cell constant a_{B1} of the cubic single crystals $TiC_{0.64}$, $TiC_{0.76}$, $TiN_{0.82}$, $NbC_{0.73}$, and $NbC_{0.83}$ at 300 K was 0.4322, 0.4330, 0.4228, 0.4442, and 0.4453 nm, respectively.

Deriving the short-range order parameters and the atomic displacements from the diffuse scattering intensity represents a complex computational task and the results may have considerable errors. These errors depend on the number of coordination spheres taken into account for fitting, on the size of the zone chosen near a given Bragg reflection (the diffuse scattering intensity depends on this factor), and also on the accuracy of the calibration of the instrument. To reduce these errors, the authors [45–51] have made appropriate studies [48, 50] on estimating statistical calculation errors. In particular, the direct fit of experimental data for a vacancy at the center (000) of coordination spheres gives α_{000}, which differs from the theoretical value equal to 1. This systematic error is due to the undefined value of coefficients in the normalization formula for the elastic scattering section, $d\sigma/d\Omega$. Corrected values (taking $\alpha_{000} = 1$) of α_{hkl} were calculated by the formula

$$\alpha_{hkl}^{\text{corrected}} = \frac{1+(d\sigma/d\Omega)_{\text{inc}}}{\alpha_{000}^{\text{noncorrected}} + (d\sigma/d\Omega)_{\text{inc}}} \times \alpha_{hkl}^{\text{noncorrected}}, \tag{4.46}$$

where $(d\sigma/d\Omega)_{\text{inc}}$ is the incoherent scattering, which gives an additional small contribution to the elastic scattering cross-section. For a nonstoichiometric compound MX_y the quantity of incoherent scattering in Laue units is calculated by the formula $(d\sigma/d\Omega)_{\text{inc}} = (\sigma_{\text{M}}^{\text{inc}} + y\sigma_{\text{X}}^{\text{inc}} + \sigma_{\text{MX}}^{\text{dm}})/[y(1-y)\sigma_{\text{X}}^{\text{coh}}]$, where σ^{coh} and σ^{inc} are the effective coherent and incoherent scattering cross-section, respectively; and σ^{dm} is the diffuse multiple scattering cross-section (1 Laue $= y(1-y)(f_{\text{X}})^2$ and f_{X} is the coherent scattering amplitude of the X atom). The quantity $(d\sigma/d\Omega)_{\text{inc}}$ is equal to 2.56, 3.18, 2.22, 0.27 and 0.37 in (Laue units) or 0.26, 0.26, 0.29, 0.02 and 0.02 (in barns) for $TiC_{0.64}$, $TiC_{0.76}$, $TiN_{0.82}$, $NbC_{0.73}$, and $NbC_{0.83}$ [46].

Table 4.3 gives short-range order parameters for the nonmetal sublattice of nonstoichiometric titanium and niobium carbides and titanium nitride [46] corresponding to the quenched disordered state or a disordered state being in equilibrium at the measurement temperature T, which is much higher than the order–disorder transition temperature T_{trans} for these compounds. For comparison, Table 4.3 presents short-range order parameters for the rhombic (space group $C222_1$) superstructure M_3X_2 and superstructures of M_6X_5 type.

The data in Table 4.3 suggest the following conclusions:

- As the vacancy concentration in titanium carbide increases the correlations extend to higher and higher coordination spheres. Short-range order is almost absent in the fourth and subsequent coordination spheres in carbide $TiC_{0.76}$. Correlations are present up to the eighth coordination sphere (400) in $TiC_{0.64}$.
- Short-range order parameters are negative in the first (110) and second (200) coordination sphere and they are positive in the third (211) coordination sphere under all conditions studied. The sign and the relative value of parameters α_{110}, α_{200} and α_{211} of carbides $TiC_{0.76}$, $NbC_{0.73}$ and $NbC_{0.83}$ and nitride $TiN_{0.82}$ relate mainly to the M_6X_5 superstructure. The same parameters for $TiC_{0.64}$ approximate a rhombic (space group $C222_1$) with a superstructure type M_3X_2 (see Sect. 6.2).
- The absolute value for all short-range order parameters for the nitride $TiN_{0.82}$ decrease with increasing temperature.

Table 4.3. Short-range order parameters α_{hkl} for nonmetallic sublattice of carbides $TiC_{0.64}$, $TiC_{0.76}$, $NbC_{0.73}$, $NbC_{0.83}$ and nitride $TiN_{0.82}$ [46]

[1]CS (*hkl*)	M_3X_2 (space group $C222_1$)	$TiC_{0.64}$ [2]1170 K	$TiC_{0.76}$ [2]1170 K	$NbC_{0.73}$ [3]quenching	[2]1470 K	$NbC_{0.83}$ [2]1470 K	M_6X_5
000	1	1.000	0.934	1.000	1.000	1.000	1
110	−1/8	−0.055	−0.047	−0.085	−0.087	−0.063	−1/5
200	−1/2	−0.276	−0.122	−0.190	−0.196	−0.166	−1/5
211	1/8	0.037	0.035	0.088	0.072	0.094	1/5
220	1/4	0.088	0.018	0.039	0.018	0.005	0
310	1/8	0.021	−0.003	0.035	0.029	0.026	0
222	−1/8	−0.039	0.008	−0.044	−0.029	−0.019	−1/5
321	−3/16	−0.019	−0.001	−0.038	−0.024	−0.030	−1/10
400	0	0.072	0.009	0.010	0.010	0.009	1/5
330	0	0.001	0.010	0.009	−0.003	0.005	2/15
411	0	−0.014	−0.007	−0.022	−0.028	−0.035	2/15
420	0	−0.029	0.010	0.008	0.001	−0.001	−1/5
332	1/4	0.010	−0.002	0.031	0.006	0.002	0
422	0	0.016	−0.001	0.018	0.013	0.005	2/5

[1]CS (*hkl*)	M_3X_2 (space group $C222_1$)	$TiN_{0.82}$ [3]quenching	[2]970 K	[2]1070 K	[2]1170 K	M_6X_5
000	1	1.000	1.000	1.000	1.000	1
110	−1/8	−0.107	−0.101	−0.096	−0.091	−1/5
200	−1/2	−0.114	−0.114	−0.102	−0.096	−1/5
211	1/8	0.036	0.045	0.040	0.034	1/5
220	1/4	−0.002	0.018	0.013	0.013	0
310	1/8	0.004	0.019	0.016	0.014	0
222	−1/8	−0.041	−0.028	−0.023	−0.018	−1/5
321	−3/16	−0.025	−0.018	−0.015	−0.013	−1/10
400	0	−0.002	0.003	0.000	−0.005	1/5
330	0	0.012	−0.002	0.000	0.003	2/15
411	0	−0.019	−0.010	−0.008	−0.005	2/15
420	0	−0.005	0.003	0.004	0.004	−1/5
332	1/4	0.002	0.013	0.010	0.008	0
422	0	0.004	0.010	0.008	0.005	2/5

[1] Coordination sphere (*hkl*) of nonmetallic sublattice.
[2] Neutron diffraction measurements of disordered samples are performed at the given temperature T which is somewhat larger than the transition temperature T_{trans}.
[3] Neutron diffraction measurements of a quenched disordered sample are performed at 300 K.

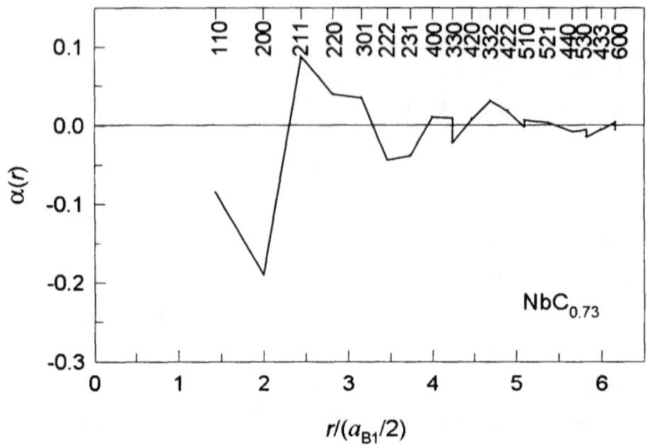

Fig. 4.8. Variation of short-range order parameters α_{hkl} of disordered carbide $NbC_{0.73}$ with increasing radius $r = (a_{B1}/2)(h^2 + k^2 + l^2)^{1/2}$ of the coordination sphere (hkl) [50]

Figure 4.8 shows the oscillation of short-range order parameters α_{hkl} for quenched disordered nonstoichiometric niobium carbide $NbC_{0.73}$ with increasing radius $r = (a_{B1}/2)(h^2 + k^2 + l^2)^{1/2}$ of the (hkl) coordination sphere.

Investigations [45–51] by diffuse neutron scattering show that a pronounced short-range order is preserved in the nearest coordination spheres even at temperatures 200–300 K higher than the order–disorder transition temperature. Clearly, short-range interactions are very significant for nonstoichiometric compounds. However, one cannot assert that short-range order in strongly nonstoichiometric compounds is exclusively due to the repulsion between vacancies. This type of short-range order dominates in nonstoichiometric compounds $MX_y\square_{1-y}$ with a relative concentration of structural vacancies $(1 - y) \leq 0.3$. If the vacancy concentration is larger (for example, in titanium carbide $TiC_{0.5}$–$TiC_{0.7}$), vacancies cannot be isolated from one another. In this case, short-range order is probably due to the formation of vacancy clusters which contain two, three or more vacancies. These vacancy clusters are spaced at large distances.

Electron microscopy and electron diffraction were used to examine short- and long-range orders in the nonmetal sublattice of carbide solid solutions $Ti_{0.95}V_{0.05}C_{0.21}$, $Ti_{0.95}V_{0.05}C_{0.55}$, $Ti_{0.75}V_{0.25}C_{0.21}$ and $Ti_{0.75}V_{0.25}C_{0.55}$ [54]. Samples of the solid solutions were produced by arc melting and subsequent quenching from the melt. They contained two phases: β-phase (bcc solid solution of carbon in the β-(Ti, V) melt) and cubic $B1$-type δ-phase. Samples of $Ti_{0.95}V_{0.05}C_{0.21}$, $Ti_{0.95}V_{0.05}C_{0.55}$, $Ti_{0.75}V_{0.25}C_{0.21}$ and $Ti_{0.75}V_{0.25}C_{0.55}$ contained 52.0, 97.2, 51.2 and 88.0 mol.% δ-phase with the formulas $Ti_{0.997}V_{0.003}C_{0.493}$, $Ti_{0.984}V_{0.016}C_{0.570}$, $Ti_{0.991}V_{0.009}C_{0.519}$ and $Ti_{0.918}V_{0.082}C_{0.679}$ respectively. Precipitates of the δ-phase were in the shape of dendrites with a size up to 10 μm. They were embedded in a matrix of the β-(Ti, V) alloy. An examination of electron diffraction in areas which supposedly contained only the δ-phase revealed additional periodic diffuse scattering and diffraction reflections other than those of the pure δ-phase. Both effects are due to short-range order formed in cubic carbide TiC_y. All extra diffraction reflections are superstructure reflections and correspond to the diffraction vector {1/2 1/2 1/2}. This provides strong evidence that an ordered phase of the Ti_2C type is formed in titanium carbide. In this study the small dimensions of the ordered phase made it impossible to determine the space group ($R\bar{3}m$ or $Fd\bar{3}m$). The short-range order parameters for eight coordination spheres were determined from the diffuse scattering intensity of the nonmetal sublattice [54] of the cubic carbide phases based on titanium carbide (Table 4.4). The parameters for the short-range order, α_{hkl}, correlate well with theoretical values for α calculated for the ideallly ordered phase Ti_2C. The sign of the parameter α indicates that C–□ pairs dominate in the first and second coordination spheres, while C–C and □–□ pairs prevail in the third coordination sphere of the nonmetal sublattice.

Table 4.4. Short-range order parameters α_{hkl} for the ordered Ti_2C-type phase which forms in cubic carbides $Ti_{0.997}V_{0.003}C_{0.49}$, $Ti_{0.991}V_{0.009}C_{0.52}$ and $Ti_{0.985}V_{0.015}C_{0.57}$ [54]

Coordination sphere (*hkl*)	α_{exper}	α^*	α^{**}	α_{theor}
110	−0.183	−0.139	−0.117	0
200	−0.254	−0.147	−0.216	−1
211	0.189	0.083	0.070	0
220	−0.026	−0.009	0.046	1
310	0.050	0.013	0.011	0
222	−0.128	−0.025	0.052	−1
321	−0.089	−0.013	−0.011	0
400	0.171	0.019	0.034	1

α_{exper} experimental short-range order parameter values.

α^* short-range order parameters corrected with allowance for the width of diffraction reflection (according to [54], parameters of correlational short-range order);

α^{**} short-range order parameters corrected with allowance for the irregular distribution of intensity of diffraction reflection (according to [54], parameters α^{**} include correlational and superstructural short-range order simultaneously);

α_{theor} theoretical short-range order parameters for ideal ordered phase Ti_2C (space group $R\bar{3}m$ or $Fd\bar{3}m$).

If corrections can be established for the width of broad diffraction reflections and the nonuniform distribution of the intensity then the short-range order parameters can be determined much more accurately. To establish them ideas about superstructural and correlational short-range orders as advanced in [55, 56] have been followed [54]. They are based on the fact that short-range order parameters α^*, which were determined from the width of the reflections, characterize correlational short-range order only, while parameters α^{**}, which are calculated with corrections for a nonuniform distribution of intensity, include both superstructural and correlational components. It is noted that the superstructure parameter α_{000} is identically equal to 0 for cubic and trigonal ordered phases Ti_2C, whereas in [54] the parameter α_{000} is negative both without (α_{exper}) and with (α^*, α^{**}) the corrections introduced (see Table 4.4). These discrepancies may be attributed to factors, which were overlooked in [54] like the scattering on vanadium atoms that took place in the studied diffraction regions or the static displacement of atoms.

Diffuse scattering of electrons was used to examine short-range order in low hexagonal carbides of vanadium V_2C_y and niobium Nb_2C_y [57]. The study showed that short- and long-range orders in these carbides are connected with a regular arrangement of carbon atoms and structural vacancies in rows C–□–C–□-... parallel to the *c*-axis. Comparing experimental and calculated values of the diffuse scattering intensity, various possible types of order in carbides V_2C and Nb_2C were proposed [57].

4.2.3 Short-Range Order and Magnetic Susceptibility

The magnetic susceptibility method was originally used for nonstoichiometric niobium carbide to analyze atomic distributions and to evaluate short-range order parameters [11, 26, 58, 59]. So far, nonstoichiometric carbides of titanium, hafnium and tantalum have been studied by this method [60–69]. These studies will be comprehensively covered in Sect. 11.5. Here we shall just briefly discuss some results of relevance for short-range order and a comparison with elastic diffuse neutron scattering.

Measurements of the magnetic susceptibility χ of niobium carbide with different compositions performed at temperatures between 300 and 1300 K showed that the χ value depends on the conditions of thermal treatments for NbC_y with $0.81 \le y \le 0.88$. The χ value of NbC_y samples quenched from temperatures considerably higher than the disorder–order transition decreased abruptly down to the value of the magnetic susceptibility of ordered samples of the same composition. After $\chi(T)$ was measured superstructure reflections appeared in neutron diffraction patterns of NbC_y samples. This means that ordering took place during the susceptibility measurements. It is also found that the susceptibility of samples containing both disordered and ordered phases was almost equal to the susceptibility of NbC_y samples of the same composition and being single ordered phased only. Hence it appears, that the susceptibility is independent of the concentration of the ordered phase. The formation of long-range order is followed by the appearance of short-range order. Consequently, the independence of χ on the concentration of the ordered phase and the change of χ upon thermal treatment suggest that the magnetic susceptibility is determined by the appearance of short-range order rather than long-range order.

A crystal of a nonstoichiometric compound MX_y with the *B*1 structure can be represented as a set of noninteracting octahedral clusters, which include six sites of the nonmetal sublattice forming the first coordination sphere of a metal atom located at the center of an octahedral cluster. In this case, the magnetic susceptibility χ of the compound can be given as the susceptibility χ_i and the probability P_i of separate clusters, $\chi(y,T) = \Sigma\chi_i(T)\lambda_i P_i(y)$, where λ_i is the multiplicity (the number of equivalent i-configurations of a cluster, in which n sites out of six sites are vacant). The susceptibility χ_i of clusters was determined [11] from susceptibility measurements of a disordered compound on the assumption that the probabilities P_i of separate clusters in the compound can be described by a binomial distribution $P_i(y) = P_i^{\text{bin}}(y) = y^{(6-n)}(1-y)^n$. Probabilities of different clusters, $P_i^{\text{ord}}(y)$, in ordered carbides were determined using calculated susceptibility values χ_i of separate clusters and experimental results on the susceptibility of ordered carbides. Octahedral clusters include sites of the nonmetal sublattice, located relative to positions in the first and second coordination spheres. Therefore they can be used to determine the short-range order in these coordination spheres from known probabilities $P_i^{\text{ord}}(y)$.

Short-range order parameters α_1 and α_2 for the first and second coordination spheres of the nonmetal sublattice can be found from the formula

$$\alpha_j = 1 - (P_{\text{X}-\square}^{\text{ord}(j)} / P_{\text{X}-\square}^{\text{bin}}), \tag{4.47}$$

where $P_{\text{X}-\square}^{\text{bin}} = \lambda_{\text{X}-\square} y(1-y)$ stands for the binomial probability that an "interstitial atom–vacancy" pair X–□ ($\lambda_{\text{X}-\square} = 2$) is formed. This binomial probability corresponds to a disordered state of compound MX_y and depends on its composition only. The probabilities of X–□ pairs being formed in any i-th coordination sphere of the nonmetal sublattice of the ordered compound are given by the expression

$$P_{\text{X}-\square}^{\text{ord}(j)} = \sum_i n_i^{(j)} \lambda_i P_i^{\text{ord}}(y), \tag{4.48}$$

where $n_i^{(j)}$ is the relative fraction of "interstitial atom–vacancy" pairs in the j-th coordination sphere of a cluster having the i-th configuration.

The comparison of calculated short-range order parameters α_1 and α_2 in ordered carbides TiC_y, HfC_y, NbC_y and TaC_y with theoretical values of α_1 and α_2 in ideal ordered phases M_2C, M_3C_2 and M_6C_5 provides support for the formation of the above phases (Table 4.5). Since the values for α_1 and α_2, found from magnetic susceptibility, differ from the values of these parameters for an ideally ordered phase, this indicate that complete ordering was not achieved in the carbides studied.

It should be remembered that the data [11, 26, 58–69] concerning short-range order in ordered nonstoichiometric carbides obtained from measurements of the magnetic susceptibility can only be regarded as rather rough estimates for the short-range order parameters.

Table 4.5. Short-range order parameters α in ordered nonstoichiometric carbides TiC_y, HfC_y, NbC_y and TaC_y (magnetic susceptibility method) [11, 58–69]

Carbide MC_y	Experimental short-range order parameters		Type and space group of ordered phase	Theoretical short-range order parameters for ideal ordered phase	
	α_1 ±0.005	α_2 ±0.005		α_1	α_2
$NbC_{0.81}$	−0.153	−0.137	Nb_6C_5 ($C2/m$)	−0.2	−0.2
$NbC_{0.83}$	−0.132	−0.126	Nb_6C_5 ($C2/m$)	−0.2	−0.2
$NbC_{0.88}$	−0.064	−0.040	Nb_6C_5 ($C2/m$)	−0.2	−0.2
$TaC_{0.82}$	−0.080	−0.086	Ta_6C_5	−0.2	−0.2
$TaC_{0.83}$	−0.063	−0.080	Ta_6C_5	−0.2	−0.2
$TaC_{0.85}$	−0.059	−0.043	Ta_6C_5	−0.2	−0.2
$HfC_{0.71}$	−0.207	−0.001	M_3C_2 ($P2$, $Immm$)	−0.25	0
$HfC_{0.78}$	−0.123	−0.001	M_3C_2 ($P2$, $Immm$)	−0.25	0
$HfC_{0.83}$	−0.079	−0.023	M_6C_5 ($C2/m$, $P3_1$, $C2$)	−0.2	−0.2
$TiC_{0.50}$	0	−0.793	Ti_2C ($R\bar{3}m$, $Fd\bar{3}m$)	0	−1.0
$TiC_{0.55}$	0	−0.700	Ti_2C ($R\bar{3}m$, $Fd\bar{3}m$)	0	−1.0
$TiC_{0.59}$	0.004	−0.668	Ti_2C ($R\bar{3}m$, $Fd\bar{3}m$)	0	−1.0
$TiC_{0.63}$	−0.055	−0.406	Ti_3C_2 ($C222_1$)	−0.125	−0.5
$TiC_{0.68}$	−0.053	−0.394	Ti_3C_2 ($C222_1$)	−0.125	−0.5
$TiC_{0.69}$	−0.011	−0.441	Ti_3C_2 ($C222_1$)	−0.125	−0.5
$TiC_{0.81}$	−0.093	−0.096	Ti_6C_5 ($C2/m$, $P3_1$, $C2$)	−0.2	−0.2

Indeed, the assumption that ideal disorder is realized in quenched disordered nonstoichiometric compounds and, consequently, that the probabilities of clusters in a disordered compound can be described by a binomial distribution is not quite correct. Studies of nonstoichiometric carbides and nitrides by elastic diffuse neutron scattering showed that these compounds preserve to some degree a short-range order at least in the three nearest coordination spheres even at a temperature 200 K higher than the order–disorder transition point (see Table 4.3). To quantify the degree of accuracy to which the magnetic susceptibility method estimates short-range order parameters, one needs to perform an experiment on the magnetic susceptibility and the elastic diffuse neutron scattering using the same samples of nonstoichiometric carbide in disordered and ordered states. Short-range order parameters of disordered carbide determined by neutron diffraction should be used for calculating the probabilities of clusters P_i in the disordered compound and, subsequently, estimating the susceptibility χ_i of separate clusters. This procedure would considerably enhance the accuracy of determining the probabilities of clusters and short-range order parameters in ordered carbides. A comparison of the short-range order parameters for ordered carbides from magnetic susceptibility and from the diffuse neutron scattering experiment would show decisively to which degree measurements of magnetic susceptibility can be applied for a quantitave analysis of short-range order of the distribution of interstitial atoms and vacancies.

Because no such experiment has been performed so far, we can only compare data on short-range order obtained for different samples. It shows that the results for the short-range order of ordered titanium and niobium carbides from the magnetic susceptibility agree well with the data on short-range order obtained by diffuse neutron scattering [42–44, 46, 47, 50–52] and NMR [12, 27–29] .

4.2.4 Short-Range Order in Carbide Solid Solutions

X-ray diffuse scattering has been used frequently to analyze short-range order in the metal sublattice of solid solutions formed by transition-metal carbides. A central issue for the short-range order in solid solutions $M^{(I)}C$–$M^{(II)}C$ concerns the nonrandom distribution of metal atoms on the metal sublattice sites and the effect of carbon on the correlations for the arrangement of the metal atoms.

Umanskii and Fadeyeva were the first [70–72] to analyze the local distribution of atoms on the metal sublattice of equimolar solid solutions NbC–TaC and ZrC–HfC. It could be assumed that the absence of vacancies in the carbon sublattice and similar atomic radii for the substituted metals (Nb and Ta, Zr and Hf, respectively) might lead to solid solutions with lattices free of static distortions. Short-range order parameters for the three nearest coordination spheres of the metal sublattice were determined from diffuse scattering intensities $I_D = I_{DL} + I_{DSRO}$ (intensities I_{DL} and I_{DSRO} are given by (4.22) and (4.27) respectively). Short-range order parameters of the first and second coordination spheres proved to be positive and the values for α_1 and α_2 increase as the quenching or annealing temperature is decreased (Table 4.6). Positive values of α_1 and α_2 suggest that in the solid solutions studied the first and second coordination spheres of a metal atom of some species contain more atoms of the same species than in the case of the statistical distribution. Put another way, the metal sublattice of solid solutions of NbC–TaC or of ZrC–HfC shows some typ of short-range separation. This is also observed for the metallic alloys Nb–Ta [73]. The short-range order parameter α_3 is slightly negative indicating that correlations in the atomic arrangement in the metal sublattice apply only to the first two nearest coordination spheres. Additional annealings of samples at a low temperature and introduction of 1 mol.% WC enhances correlations in the arrangement of metal atoms.

X-ray diffuse scattering in solid solutions $Ti_{0.5}Nb_{0.5}C$, $Ti_{0.5}Ta_{0.5}C$, $Ti_{0.5}W_{0.5}C$, $V_{0.5}Nb_{0.5}C$, and $V_{0.5}W_{0.5}C$ and metallic alloys $Ti_{0.5}Ta_{0.5}$, $V_{0.64}Nb_{0.36}$ and $V_{0.65}W_{0.35}$ was studied in [74, 75]. The short-range order parameter α_1 proved to be positive for $Ti_{0.5}Ta_{0.5}$ and negative for $V_{0.64}Nb_{0.36}$ and $V_{0.65}W_{0.35}$. The parameter α_1 was negative for carbide solid solutions $Ti_{0.5}Nb_{0.5}C$, $Ti_{0.5}Ta_{0.5}C$, $Ti_{0.5}W_{0.5}C$, and $V_{0.5}Nb_{0.5}C$; $\alpha_1 > 0$ for $V_{0.5}W_{0.5}C$. The sign of the parameter α_1 in carbide solid solutions $Ti_{0.5}Nb_{0.5}C$, $Ti_{0.5}Ta_{0.5}C$, and $V_{0.5}W_{0.5}C$ is opposite to that in the alloys Nb_xTi_{1-x} [76], $Ti_{0.5}Ta_{0.5}$, and $V_{0.65}W_{0.35}$.

Table 4.6. Short-range order parameters α in a metallic sublattice of carbide solid solutions and alloys

Composition	Heat treatment	Short-range order parameters			References
		α_1	α_2	α_3	
$Nb_{0.5}Ta_{0.5}C$	annealing, 2000 K	0.058	0.025	0	70
$Nb_{0.5}Ta_{0.5}C$	annealing, 1800 K	0.089	0.049	0	70
$Nb_{0.5}Ta_{0.5}C$	annealing, 1500 K	0.120	0.071	0	70
$Nb_{0.5}Ta_{0.5}C$ + 1 mol.% WC	annealing, 1800 K	0.11	0.10	0.02	72
$Nb_{0.5}Ta_{0.5}C$ + 1 mol.% WC	annealing, 1500 K	0.14	0.13	0.04	72
$Nb_{0.5}Ta_{0.5}$	quenching, 1600 K	0.13	0.09	0.02	73
$Nb_{0.5}Ta_{0.5}$	annealing, 1400 K	0.16	0.08	−0.04	73
$Zr_{0.5}Hf_{0.5}C$	quenching, 2200 K	0.06	0	0	71
$Zr_{0.5}Hf_{0.5}C$	quenching, 1800 K	0.09	0	0	71
$Nb_{0.5}Ti_{0.5}C$	quenching, 2200 K	−0.06	–	–	9, 74, 75
$Nb_{0.75}Ti_{0.25}$	quenching, 1500 K	0.066	0.103	–	76
$Nb_{0.5}Ti_{0.5}$	quenching, 1500 K	0.063	0.047	–	76
$Nb_{0.25}Ti_{0.75}$	quenching, 1500 K	0.091	0.017	–	76
$Ti_{0.5}Ta_{0.5}C$	quenching, 2200 K	−0.06	–	–	9, 74, 75
$Ti_{0.5}Ta_{0.5}$	quenching, 1500 K	0.19	–	–	9, 74, 75
$Ti_{0.5}Zr_{0.5}C$	quenching	0.07	–	–	9
$Ti_{0.6}Hf_{0.4}C$	quenching	0.03	–	–	9
$V_{0.5}Nb_{0.5}C$	quenching, 2100 K	−0.01	–	–	9, 74, 75
$V_{0.64}Nb_{0.36}$	quenching	−0.23	–	–	9, 74, 75
$V_{0.5}W_{0.5}C$	quenching, 2100 K	0.03	–	–	9, 74, 75
$V_{0.65}W_{0.35}$	quenching	−0.10	–	–	9, 74, 75
$Ti_{0.5}W_{0.5}C$	quenching, 2200 K	−0.06	–	–	9, 74, 75
$Ti_{0.61}W_{0.39}C_{0.98}$	quenching, 2400 K	−0.09	0.03	0	77
$Ti_{0.77}W_{0.23}C_{0.98}$	quenching, 2400 K	−0.07	0	0	77
$Ti_{0.84}W_{0.16}C_{0.98}$	quenching, 2400 K	−0.06	0.02	0.01	77
$Ti_{0.91}W_{0.09}C_{0.98}$	quenching, 2400 K	−0.06	−0.02	−0.02	77
$Ti_{0.58}W_{0.42}C_{0.94}$	quenching, 2400 K	−0.06	0.04	0	77
$Ti_{0.58}W_{0.42}C_{0.86}$	quenching, 400 K	0.05	0.16	0.04	77
$Ti_{0.59}W_{0.41}C_{0.78}$	quenching, 2400 K	0.07	0.15	0.03	77

Short-range order in solid solutions of TiC–WC was studied in [77–79] by X-ray diffuse scattering on samples with different compositions and treated for various thermal conditions (sintering at various temperatures, quenching, cooling in furnace, etc.). All samples of solid solutions TiC–WC with a complete (without vacancies) carbon sublattice showed considerable short-range order in the first coordination sphere ($\alpha_1 < 0$) while the parameters α_2 and α_3 for the second and third coordination spheres approached zero (Table 4.6) demonstrating that correlations in the atomic arrangement of titanium and tungsten take place at short distances only. Further the short-range order parameters always increase after additional low-temperature annealings. A pure statistical distribution of metal atoms has never been observed.

If the tungsten concentration in solid solutions TiC–WC is increased, the degree of short-range order in the first coordination sphere is enhanced. The carbon concentration has a great effect on the distribution of metal atoms. When the carbon concentration deviates from stoichiometry, short-range order parameters of all the three coordination spheres become positive (see Table 4.6). Specific features of the X-ray diffuse scattering intensity of nonstoichiometric solid solutions $Ti_xW_{1-x}C_y$ is indicative [77] of two types of short-range order – with $\alpha_1 < 0$ and $\alpha_1 > 0$ – present in the solid solution structure. It can be assumed that in addition to segregations of atoms of some species, regions of short-range order with a structure typical of a stoichiometric solid solution are preserved in nonstoichiometric solid solution $Ti_xW_{1-x}C_y$.

It should be remembered however, that maxima of diffuse scattering may also result from small impurities of interstitial atoms in transition metals, for example, a single crystal of niobium doped with a controlled amount of nitrogen of 0.0015 atoms per niobium atom to form an interstitial solid solution of $NbN_{0.0015}$. Even such a small impurity can be observed in favourable cases by neutron scattering. No such peak was present in the spectrum of pure niobium. Analogous results were obtained for the solid solution of oxygen in niobium $NbO_{0.0015}$ [80].

4.3 Local Atomic Displacements

Nonstoichiometric compounds in a disordered state are characterized by a random distribution of interstitial atoms and structural vacancies at the nonmetal sublattice sites and, therefore, they do not possess the translational symmetry of a stoichiometric crystal. However, a random distribution of interstitial atoms at the nonmetal sublattice sites means that probabilities of the sublattice sites being occupied are equal and, consequently, all sites of the nonmetal sublattice are crystallographically equivalent. In other words, the translational symmetry applies to some lattice of occupation probabilities in disordered nonstoichiometric compounds.

Most strongly nonstoichiometric compounds with wide homogeneity intervals have a high-symmetry structure $B1$ (see Sect. 3.2), which is preserved for a disordered distribution of interstitial atoms and vacancies. As a result of the disordered distribution, the symmetry of the local environment of each individual atom changes. In turn, distortion of the local symmetry leads to static displacements of both metal and interstitial atoms. Static displacements mean deviations of atoms from positions corresponding to the ideal basic structure of a compound, i.e. some quantity $\boldsymbol{u}_j = \boldsymbol{R}_j - \boldsymbol{r}_j$, where $\boldsymbol{R}_j$ and $\boldsymbol{r}_j$ denote radii-vectors that determine the location of the j-th atom in real and in ideal lattices. As distinct from dynamic displacements, which are due to thermal oscillations of atoms, static displacements arise from an asymmetric angular distribution of bonds between adjacent atoms and different energies of these bonds.

The cubic symmetry of disordered nonstoichiometric compounds allows only a symmetric distribution of static atomic displacements, since otherwise the crystallographic equivalence of crystal sites would be disturbed. A random location of vacancies ensures a symmetric distribution of static displacements in disordered nonstoichiometric compounds. Considering a radial distribution of static displacements and using the analogy with dynamic displacements, one can pass on to root-mean-square (rms) static displacements.

The crystal symmetry decreases upon ordering and, simultaneously, the symmetry of the local atomic environment changes. We assume that the local environment symmetry of an ordered stoichiometric phase coincides with the crystal symmetry. Since the local environment symmetry is not cubic in the ordered state, static atomic displacements necessarily occur upon ordering. The values for the static displacements are different for disordered and ordered states of a crystal, because the local atomic environments in these states are not the same.

If the values and directions of the static displacements of each atom in nonstoichiometric compounds having different structural states are known, it is possible to determine equilibrium positions from which the atoms are displaced dynamically. In other words, the crystal structure can be established more accurately and fully. Comparing the values of the static displacements in different structural states of a crystal, one can deduce possible mechanisms of order–disorder transitions and analyze changes in the interparticle interactions upon ordering.

Determination of Static Displacements. X-ray and thermal neutron diffraction are powerful techniques to determine atomic displacements in crystals. Elastic scattering of neutrons and scattering of X-rays on static irregularities of a crystal are described essentially by the same formulas and can be considered in terms of a unified theory [8, 81]. However, different theoretical parameters may provide considerable methodical advantages.

The X-ray scattering amplitudes on different atoms depend on the number of electrons and they grow monotonically with the ordinal number of an element. Amplitudes of neutron scattering for different atoms change irregularly. When applied to nonstoichiometric carbides and nitrides of transition metals, this means that in the case of X-ray scattering the main contribution to the intensity of diffraction peaks is due to atoms of transition metals, whereas information about the nonmetal sublattice is more difficult to extract.

In neutron diffraction, the neutron scattering amplitudes for the different components of nonstoichiometric compounds may have comparable values (with the exception of titanium and vanadium). For example, the coherent scattering amplitudes for zirconium, hafnium, niobium, tantalum, carbon and nitrogen are 0.71×10^{-12}, 0.78×10^{-12}, 0.71×10^{-12}, 0.70×10^{-12}, 0.665×10^{-12} and 0.94×10^{-12} cm [82]. Therefore, both metal and interstitial atoms contribute to the intensity of diffraction reflections in the case of thermal neutron scattering. In addition the neutron diffraction method provides better accuracy in the determination of displacements on powder samples than the X-ray method, because the neutron scattering amplitude is independent of the scattering angle and does not require corrections for extinction for powder samples. Moreover, most materials of concern here absorb neutrons much less than X-rays. It seems that neutron diffraction is preferable for the determination of displacements of atoms of both species in nonstoichiometric compounds.

Both dynamic (thermal) and static displacements of atoms from sites in an ideal crystal lattice attenuate the intensity of diffraction reflections. In the absence of ordering, these two factors decrease the intensity according to the same law [8, 83]. Indeed, dynamic displacements make atomic positions dependent on time. However, in the first aproximation considering that the scattered radiation frequency is several orders of magnitude higher than the thermal oscillation frequency of atoms, scattering may be assumed to take place on immobile atoms occupying some instant positions. Hence it follows that differences between dynamic and static displacements do not matter much in diffraction experiments and one can take into account their superposition only. Consequently, the spectrum intensity depends on the rms value of the sum of the dynamic u_{dyn} and static u_{st} displacements

$$\left\langle u_{\Sigma}^{2}\right\rangle = \left\langle u_{\text{st}}^{2}\right\rangle + \left\langle u_{\text{dyn}}^{2}\right\rangle . \qquad (4.49)$$

Still dynamic and static displacements can be separated, e.g. from the temperature dependence of the diffraction intensities.

The kinematic theory treats the scattering intensity of a crystal as the sum of contributions from separate small coherently scattering blocks. The amplitude of a wave scattered by such a block is small compared to the amplitude of an incident wave. Therefore the complex interference interaction between incident and scattered waves can be disregarded. If the dimensions of a crystal are large and its structure is regular over the whole volume, the intensity scattered by such a perfect crystal should be calculated taking into account the interaction between incident and scattered waves. The kinematic theory provides a sufficiently accurate description of the intensity of diffraction reflections for most crystalline materials [10,82–84]. In the case of a large ideal crystal, the intensity distribution in reciprocal lattice space is, from kinematic theory, described by narrow peaks localized at the reciprocal lattice sites. By changing the value and the direction of the diffraction vector $q = (2\sin\theta)/\lambda$, it is possible to probe large regions in the reciprocal space and determine the distribution of the scattered radiation intensity.

Crystal irregularities alter the scattered intensity distribution, shift the maxima of the diffraction peaks, change the integral intensities and yield a smooth distribution of diffuse scattering intensity in between the Bragg peaks (see expression (4.14)). In addition, new reflections may appear in some cases.

The intensity of diffraction lines in a neutron data set for a polycrystalline sample can be written as

$$I_{\text{exper}} = KLPF^2, \tag{4.50}$$

where K is an instrument constant; $L = (\sin\theta\sin2\theta)^{-1}$ is the Lorentz factor accounting for the time required for a reciprocal lattice site to pass a Ewald sphere; F^2 is the structure factor; and P denotes the multiplicity factor, which is equal to the number of reciprocal lattice vectors of equal length. The multiplicity factor is introduced because crystallographic planes having reciprocal lattice vectors of equal length contribute, in accordance with the Wulff–Bragg condition (2.23), to the same intensity maximum. Taking into account displacements of atoms from positions of an ideal lattice, the structure amplitude F can be presented in a form well known from structural analysis:

$$F_{hkl} = \sum_j f_j \exp[-i2\pi(x_j h + y_j k + z_j l)]\exp(\pi i\boldsymbol{q}\boldsymbol{u}_j), \tag{4.51}$$

where x_j, y_j, and z_j denote coordinates of the j-th atom of the basis, and f_j is the atomic scattering factor of the j-th atom. The last factor in (4.51) is the Debye–Waller factor. The summation in (4.51) is taken over all sites of the unit cell. The factor $\exp(\pi i\boldsymbol{q}\boldsymbol{u}_j)$ is averaged over all pairs of atoms at a distance $\boldsymbol{R}_j$. It assumes the form

$$\langle\exp(\pi i\boldsymbol{q}\boldsymbol{u}_j)\rangle = \exp\{-8\pi^2\langle u^2{}_{q,\Sigma}\rangle\,[(\sin\theta_{hkl})/\lambda]^2\}\ , \tag{4.52}$$

where $\langle u^2{}_{q,\Sigma}\rangle$ is the mean square of the projection of the displacements on the direction of the diffraction vector $\boldsymbol{q}$. In the Debye approximation, oscillations of each atom are isotropic and are independent of adjacent atoms. For a crystal of cubic symmetry $\langle u^2{}_{q,\Sigma}\rangle = \langle u_\Sigma^2\rangle/3$, where $\langle u_\Sigma^2\rangle$ is the mean square value of the atomic displacements. Consequently, the Debye–Waller factor $\langle\exp(\pi i\boldsymbol{q}\boldsymbol{u}_j)\rangle$ for cubic crystals can be written as $\langle\exp(\pi i\boldsymbol{q}\boldsymbol{u}_j)\rangle = \exp(-M_j)$, where M has the form

$$M = \frac{8\pi^2}{3\lambda^2}\langle u_\Sigma^2\rangle\sin^2(\theta_{hkl})\,. \tag{4.53}$$

Other formulas are used to calculate M for lower symmetry crystals, which lead to more complicated formulas. This is due to the fact that the lower symmetry crystals do not preserve isotropy of the dynamic and static displacements, as the atoms can occupy crystallographically nonequivalent positions.

In the case of a nonstoichiometric compound MX_y with the $B1$ structure expression (4.51) can be rearranged as:

$$F = 4[f_M\exp(-M_M) \pm yf_X\exp(-M_X)]\ , \tag{4.54}$$

where M_M and M_X are obtained from (4.53) and $\langle u_\Sigma^2\rangle$ is replaced by $\langle u_{\Sigma M}^2\rangle$ and $\langle u_{\Sigma X}^2\rangle$ respectively. The + and – signs refer to reflections from planes with even and odd Miller indices.

Expressions (4.50), (4.53) and (4.54) allow the determination of the rms dislacements of atoms of each species by solving a system of equations written for even and odd Miller indices. This separation can be done in practice only if the scattering amplitudes of the metal and the nonmetal atoms are considerably different. Otherwise the structural amplitudes F and experimental intensities of reflections from planes with odd Miller indices are so weak that the intensities of these reflections can be determined with an accuracy insufficient for calculating the displacements. In this case, nonstoichiometric compounds MX_y whose components have similar atomic factors f_M and f_X are analyzed using the rms displacement of the atomic complex, i.e. the quantity $\langle u^2_{\Sigma MX_y} \rangle$ averaged over both sublattices.

In accordance with [8], if $|M_M - M_X| \ll 1$ and if the structural amplitudes f_M and f_X are similar, the structural amplitude can be written including the attenuation factor as

$$F = 4(f_M \pm y f_X)\exp(-M_{MX_y}). \tag{4.55}$$

Substituting (4.55) in (4.50), we obtain

$$I_{exper} = 16KLP(f_M \pm y f_X)^2 \exp(-2M_{MX_y}). \tag{4.56}$$

In (4.56) only the last factor depends on displacements, whereas the product of the other factors represents the theoretical intensity I_0 of a diffraction reflection in the absence of displacements, i.e.

$$I_{exper}(hkl) = I_0 \exp\left\{-\frac{16\pi^2}{3}\frac{\sin^2(\theta_{hkl})}{\lambda^2}\langle u^2_{\Sigma MX_y}\rangle\right\}. \tag{4.57}$$

From (4.57) it follows that the rms of the sum of the displacement of the atomic complex (or the effective rms displacement) $\langle u^2_{\Sigma MX_y} \rangle$ is the slope of the following equation:

$$\ln\frac{I_{exper}(hkl)}{I_0} = -\frac{16\pi^2}{3}\frac{\sin^2(\theta_{hkl})}{\lambda^2}\langle u^2_{\Sigma MX_y}\rangle. \tag{4.58}$$

From plots of $\ln[I_{exper}(hkl)/I_0]$ versus $[(\sin\theta)/\lambda]^2$ for several experimentally observed reflections (hkl), it is possible to find the $\langle u^2_{\Sigma MX_y} \rangle$ value from the slope of the straight line. Considering the inaccurate values of I_{exper}, it is wise to use the least squares method for calculation of displacements.

Atomic Displacements in Disordered Carbides and Nitrides. Static displacements have been determined for several carbides and nitrides. Almost all of the studies were concerned with disordered compounds.

Effective rms displacements $\langle u^2_{\Sigma MX_y} \rangle$ in the homogeneity interval of titanium, zirconium and hafnium carbides were determined at 300 K by X-ray diffraction [85]. Thermal oscillations of atoms are found to be small at room temperature and dynamic distortions of the lattice can be neglected. Therefore it is thought that $\langle u^2_{\Sigma MX_y} \rangle \approx \langle u^2_{st,MX_y} \rangle$.

The most important static displacements have been observed in titanium carbide. Displacements decrease in going to zirconium and hafnium carbides. A comparison of "crystal lattice constant–composition" and "static displacements–composition" curves shows that an abrupt change of the carbide lattice constant is accompanied by greater static displacements.

Neutron diffraction was used [86] to determine the static displacements in titanium, zirconium, vanadium, niobium and tantalum carbides. Effective rms displacements $\langle u^2_{\Sigma \mathrm{MC}_y} \rangle$ were found from (4.58) using experimental data for the intensity of various diffraction reflections. The static displacements are small in near-stoichiometric carbides since the lattice is almost free of vacancies. Therefore $\langle u^2_{\Sigma \mathrm{MC}} \rangle = \langle u^2_{\mathrm{dyn,MC}} \rangle$ was assumed for stoichiometric carbides $MC_{1.0}$. This assumption is supported by the fact that $\langle u^2_{\Sigma \mathrm{MC}} \rangle$ values agree fairly well with calculated $\langle u^2_{\mathrm{dyn,MC}} \rangle$ values [87] using elastic constants in the Debye approximation (in accordance with [87], $\langle u^2_{\mathrm{dyn,MC}} \rangle$ equals 0.0081, 0.0069 and 0.0072 nm for TiC, ZrC and NbC respectively). Assuming a weak dependence of the Debye temperature on the carbide composition, it is supposed in [86] that the dynamic distortions of the lattice change little with decreasing carbon concentration and their dependence on the MC_y composition can be neglected to a first approximation. Hence, the static distortions can be estimated from the formula

$$\langle u^2_{\mathrm{st,MC}_y} \rangle = \langle u^2_{\Sigma \mathrm{MC}_y} \rangle - \langle u^2_{\mathrm{dyn,MC}} \rangle = \langle u^2_{\Sigma \mathrm{MC}_y} \rangle - \langle u^2_{\Sigma \mathrm{MC}} \rangle . \tag{4.59}$$

Values of displacements determined in [86] for titanium carbide proved to be much smaller than those [85] found by X-ray diffraction. The results [85, 86] show better agreement in the case of zirconium carbide ZrC_y (Table 4.7).

Neutron diffraction was used [88] to determine rms displacements $\langle u^2_{\Sigma \mathrm{MN}} \rangle$, $\langle u^2_{\Sigma \mathrm{M}} \rangle$ and $\langle u^2_{\Sigma \mathrm{N}} \rangle$ in the near-stoichiometric nitrides $TiN_{0.98}$, $ZrN_{1.00}$ and $HfN_{0.98}$ (Table 4.7). Values [88] of effective rms displacements in titanium and zirconium nitrides agree well with those of dynamic displacements calculated in ref. [89] for these compounds in the Debye approximation using elastic constants ($\langle u^2_{\mathrm{dyn,MN}} \rangle^{1/2}$ is 0.0083 and 0.0080 nm for TiN and ZrN respectively).

Dynamic displacements of titanium and carbon atoms and static displacements of titanium atoms in carbide $TiC_{0.967}$ (see Table 4.7) were determined in [90]. The intensity of Bragg reflections was measured at temperatures from 293 to 1173 K. In accordance with [90], dynamic displacements of titanium and carbon atoms are equal in magnitude when the temperature changes between 293 and 673 K. Static displacements of titanium atoms, $\langle u^2_{\mathrm{st,Ti}} \rangle^{1/2}$, in the first coordination sphere of a carbon vacancy were estimated [90] at 0.0095 ± 0.0005 nm. Titanium atoms are displaced towards the vacancy.

Dynamic displacements $\langle u^2_{\mathrm{dyn,MX}} \rangle$, $\langle u^2_{\mathrm{dyn,M}} \rangle$ and $\langle u^2_{\mathrm{dyn,X}} \rangle$ in titanium carbide $TiC_{0.96}$, titanium nitride $TiN_{0.98}$ and titanium carbonitride $TiC_{0.48}N_{0.51}$ were determined in [91] from neutron diffraction. Using $B = (8\pi^2/3)\langle u^2 \rangle$ values [91], we calculated the rms displacements $\langle u^2 \rangle$ (see Table 4.7). The displacement values for titanium carbide and nitride agree well with neutron diffraction data [86, 88]. The study of titanium carbide does not reveal any discrepancy between the intensity of diffraction reflections and the intensity calculated in terms of the model of random displacements of atoms from positions in an ideal lattice [91].

Experimental data show that static distortions of the lattice are enhanced (see Table 4.7) with increasing vacancy concentration in the carbon sublattice of carbides TiC_y, ZrC_y and HfC_y.

Table 4.7. Mean-square displacements $\langle u^2 \rangle^{1/2}$ (nm) in transition-metal nonstoichiometric compounds MX_y at a temperature of 300 K

Compound	$\langle u^2_{\Sigma MX_y} \rangle^{1/2}$	$\langle u^2_{st,MX_y} \rangle^{1/2}$	$\langle u^2_{dyn,MX_y} \rangle^{1/2}$	$\langle u^2_{dyn,M} \rangle^{1/2}$	$\langle u^2_{dyn,X} \rangle^{1/2}$	References
$TiC_{0.82}$	0.0160	0.0160	–	–	–	85
$TiC_{0.66}$	0.0184	0.0184	–	–	–	85
$TiC_{0.50}$	0.0199	0.0199	–	–	–	85
$TiC_{0.96}$	0.0084	–	–	–	–	86
$TiC_{0.86}$	0.0104	0.0065	–	–	–	86
$TiC_{0.81}$	0.0119	0.0087	–	–	–	86
$TiC_{0.65}$	0.0142	0.0118	–	–	–	86
$TiC_{0.967}$	–	–	–	0.0044	0.0045	90
$TiC_{0.96}$	–	–	0.0103	0.0084	0.0109	91
$TiC_{0.94}$	–	–	–	0.0049	0.0058	92
$ZrC_{0.80}$	0.0066	0.0066	–	–	–	85
$ZrC_{0.70}$	0.0120	0.0120	–	–	–	85
$ZrC_{0.60}$	0.0138	0.0138	–	–	–	85
$ZrC_{0.97}$	0.0075	–	–	–	–	86
$ZrC_{0.93}$	0.0101	0.0067	–	–	–	86
$ZrC_{0.87}$	0.0110	0.0080	–	–	–	86
$ZrC_{0.67}$	0.0142	0.0120	–	–	–	86
$ZrC_{0.63}$	0.0147	0.0126	–	–	–	86
$HfC_{0.90}$	0.0080	0.0080	–	–	–	85
$HfC_{0.70}$	0.0091	0.0091	–	–	–	85
$HfC_{0.60}$	0.0095	0.0095	–	–	–	85
$VC_{0.845}$	0.0096	0.0045	–	–	–	86
$VC_{0.814}$	0.0106	0.0063	–	–	–	86
$NbC_{1.0}$	0.0081	–	–	–	–	86
$NbC_{0.93}$	–	–	–	0.0048	–	93
$NbC_{0.88}$	–	–	–	0.0046	–	93
$NbC_{0.98}$	–	–	–	0.0051	0.0058	94
$NbC_{0.89}$	–	–	–	0.0049	0.0058	94
$TaC_{1.00}$	0.0091	–	–	–	–	86
$TiN_{0.98}$	–	–	0.0089	0.0078	0.0100	88
$TiN_{0.98}$	–	–	0.0105	0.0081	0.0096	91
$TiN_{0.99}$	–	–	–	0.0054	0.0055	92
$ZrN_{1.0}$	–	–	0.0085	0.0061	0.0112	88
$HfN_{0.98}$	–	–	0.0081	0.0049	0.0126	88
$TiC_{0.48}N_{0.51}$	–	–	0.0107	–	–	91

Local displacements of atoms located around a carbon vacancy in niobium carbide NbC_y ($0.88 \le y \le 0.95$) were determined in a study [93] concerned with an X-ray examination of single crystals. Monocrystalline samples of NbC_y with $y \le 0.93$ were produced by zone melting of pre-sintered carbides with different carbon concentrations. Monocrystalline samples with a high carbon concentration were also obtained by sputtering of a thin layer of carbon on the surface of a $NbC_{0.93}$ single crystal and by subsequent diffusion of carbon at a temperature of 1770 K. Four samples $NbC_{0.88}$, $NbC_{0.89}$, $NbC_{0.93}$ and $NbC_{0.95}$ were analyzed.

To separate the dynamic from the static displacements, spectra of carbides $NbC_{0.88}$ and $NbC_{0.93}$ were registered at two temperatures of 299 and 479 K. It was possible to determine the mean (for the given temperature interval) Debye–Waller thermal factor equal to $2M_{NbC_y} = \ln(I_{T_2} / I_{T_1})$. In accordance with [93], the effective dynamic displacements $\langle u^2_{dyn,NbC_y} \rangle \approx \langle u^2_{dyn,Nb} \rangle$ decrease with a growing number of vacancies in the carbon sublattice of carbide NbC_y and can be described by the linear dependence

$$\langle u^2_{dyn,NbC_y} \rangle \approx \langle u^2_{dyn,Nb} \rangle = [0.27 - 0.5(1-y)] \times 10^{-4} \ (nm^2) . \tag{4.60}$$

Since X-ray diffraction provides information about the metal sublattice only, the observed changes in the diffraction spectra were explained by displacements of six and eight niobium atoms in the first and second metal coordination spheres of a vacancy respectively. Displacements of metal atoms in subsequent coordination spheres and displacements of carbon atoms were disregarded. It was assumed also that vacancies were spaced far from one another and therefore only one nearest vacancy affected the positions of niobium atoms. It is worth noting that the last assumption holds only for near-stoichiometric carbides NbC_y whereas it is not valid for the carbides $NbC_{0.88}$, $NbC_{0.89}$ and $NbC_{0.93}$ studied in [93].

Static displacements of niobium atoms in the first and second coordination spheres of a vacancy were $\langle u^2_{st,Nb} \rangle^{1/2}_1 \approx 0.0125$ and $\langle u^2_{st,Nb} \rangle^{1/2}_2 \approx 0.0005$–$0.001$ nm respectively. The directions of static displacements have also been determined: Nb atoms in the first coordination sphere move away from the vacancy, while those in the second coordination sphere are displaced towards the vacancy. Although displacements of Nb atoms in the second coordination sphere, $\langle u^2_{st,Nb} \rangle^{1/2}_2$, are an order of magnitude smaller than the displacements in the first coordination sphere, $\langle u^2_{st,Nb} \rangle^{1/2}_1$, the displacements $\langle u^2_{st,Nb} \rangle^{1/2}_2$ compensate those in the first coordination sphere and lead to a decrease in the lattice constant of carbide NbC_y as the vacancy content increases. This interpretation of experimental data is questionable since the work under discussion does not report the coordinates of the carbon atoms nor does it present full information about the positions of niobium atoms.

Kaufmann and Meyer [94] determined the static displacements of atoms around a carbon vacancy in carbide NbC_y ($0.82 \le y \le 0.98$). Single crystals of niobium carbide were examined by the method of ion channeling at a temperature of 5, 78 and 295 K. The experimental data were analyzed using the Monte Carlo method. The dynamic displacements $\langle u^2_{dyn,Nb} \rangle^{1/2}$ and $\langle u^2_{dyn,C} \rangle^{1/2}$ of Nb and C atoms in

$NbC_{0.98}$ were 0.0051 and 0.0058 nm at 295 K and changed insignificantly on transition to carbide $NbC_{0.89}$ (see Table 4.7). Calculations of the static displacements $\langle u^2_{st,Nb}\rangle^{1/2}$ and $\langle u^2_{st,C}\rangle^{1/2}$, which were based on experimental data, showed that Nb atoms move away from the vacancy in the first coordination sphere and move towards the vacancy in other coordination spheres. Static displacements of niobium atoms around an isolated vacancy are $\langle u^2_{st,Nb}\rangle^{1/2}_1 = 0.012 \pm 0.0025$ nm in the first coordination sphere and $\langle u^2_{st,Nb}\rangle^{1/2}_2 \leq 0.003 \pm 0.001$ nm in the second coordination sphere. Static displacements also decrease with growing concentration of vacancies in niobium carbide: The displacements of niobium atoms are $\langle u^2_{st,Nb}\rangle^{1/2}_1 = 0.009 \pm 0.0007$ nm and $\langle u^2_{st,Nb}\rangle^{1/2}_2 \leq 0.002 \pm 0.0007$ nm in carbide $NbC_{0.82}$. The displacements in subsequent coordination spheres range between 0.0005 and 0.001 nm. Carbon atoms, which are nearest neighbors of a vacancy, are displaced towards the vacancy.

Static atomic displacements in single crystals of titanium carbide $TiC_{0.94}$ and nitride $TiN_{0.99}$ were determined in [92]. The method of X-ray diffraction with Ag $K\alpha_{1,2}$ radiation was used. Root-mean-square dynamic displacements of Ti and C atoms in $TiC_{0.94}$ were $\langle u^2_{dyn,Ti}\rangle = 0.346 \times 10^{-4}$ and $\langle u^2_{dyn,C}\rangle = 0.319 \times 10^{-4}$ nm^2 as obtained from a rudimentary model. The same model yielded $\langle u^2_{dyn,Ti}\rangle =$ 0.301×10^{-4} and $\langle u^2_{dyn,N}\rangle = 0.325 \times 10^{-4}$ nm^2 for nitride $TiN_{0.99}$. A more sophisticated model, which takes into account nonsphericity of atoms, charge transfer and some further corrections, provided refined values of dynamic displacements: $\langle u^2_{dyn,Ti}\rangle = 0.238 \times 10^{-4}$ and $\langle u^2_{dyn,C}\rangle = 0.336 \times 10^{-4}$ nm^2 for carbide $TiC_{0.94}$, and $\langle u^2_{dyn,Ti}\rangle = 0.294 \times 10^{-4}$ and $\langle u^2_{dyn,N}\rangle = 0.308 \times 10^{-4}$ nm^2 for nitride $TiN_{0.99}$. About 36% all metal atoms in carbide $TiC_{0.94}$ are nearest neighbors of a vacancy and, in accordance with calculations [92], are displaced from the lattice sites for 0.0097 nm in the direction $[100]_{B1}$ ($[010]_{B1}$ or $[001]_{B1}$). Only 6% titanium atoms in nitride $TiN_{0.99}$ have vacancies as their nearest neighbors and no distinct static displacements of titanium atoms were detected.

The effect of structural vacancies on the lattice distortion in nonstoichiometric compounds depends not only on the magnitude but also on the direction of static displacements of atoms around a vacancy. Calculations of atomic displacements around a vacancy in uranium carbide UC, which were performed by the Green function method, showed that nearest metal atoms move away from the vacancy, while carbon atoms move towards the vacancy [95].

Static displacements of atoms around a vacancy in monocrystalline samples of cubic niobium nitride NbN_y ($0.82 \leq y \leq 0.93$) and cubic titanium carbide $TiC_{0.90}$ were found using the channeling method [96]. Measurements were made at a temperature of 5, 78 and 295 K. Static displacements of niobium atoms $\langle u^2_{st,Nb}\rangle^{1/2}$ coincided within the measurement error at all temperatures, were almost independent of the vacancy concentration, and were equal to 0.014 ± 0.002 nm (a weak maximum $\langle u^2_{st,Nb}\rangle^{1/2} \approx 0.015$ nm was observed for nitride $NbN_{0.85}$). Static displacements of nitrogen atoms around a vacancy in niobium nitride did not exceed 0.005 nm. Static displacements of titanium atoms $\langle u^2_{st,Ti}\rangle^{1/2}$ around a vacancy in titanium carbide $TiC_{0.90}$ were 0.004 ± 0.001 nm.

With high accuracy measurement of the intensity of X-ray diffraction from a single crystal of carbide $NbC_{0.72}$, two isotropic temperature parameters (for Nb and C atoms) are distinguished and the values $\langle u^2_{\Sigma Nb} \rangle = 0.41 \times 10^{-4}$ and $\langle u^2_{\Sigma C} \rangle = 1.11 \times 10^{-4}$ nm^2 are found in [97]. These values include dynamic and static atomic displacements. An earlier paper [98] dealing with the carbide $NiC_{0.72}$ reported a slightly larger value of niobium atom displacements $\langle u^2_{\Sigma Nb} \rangle = 0.47 \times 10^{-4}$ nm^2. In accordance with [97], the presence of vacancies in the carbon sublattice leads to considerable static displacements of niobium atoms surrounding a vacancy, while dynamic displacements contribute little to the observed $\langle u^2_{\Sigma Nb} \rangle$ value.

Local Displacements of Atoms in Disordered Niobium Carbide. The dependence of atomic displacements on the composition of disordered niobium carbide NbC_y ($0.75 \le y \le 1.00$) was analyzed [99-101] by neutron diffraction.

The presence of structural vacancies in nonstoichiometric niobium carbide causes a distorted symmetry of the local environment of crystal lattice sites and leads to the appearance of static displacements for the atoms in both the metal matrix and the carbon lattice. A special hardening treatment (see Fig. 3.2), which ensures a statistical distribution of structural vacancies at crystal lattice sites, helps keep the lattice symmetry similar to the lattice symmetry of vacancy-free carbide $NbC_{1.00}$. Thus, disordered nonstoichiometric niobium carbide NbC_y has on the average a high-symmetry lattice with lattice parameters changing with the composition. In this case static distortions can be presented as rms static displacements of atoms in such an averaged lattice.

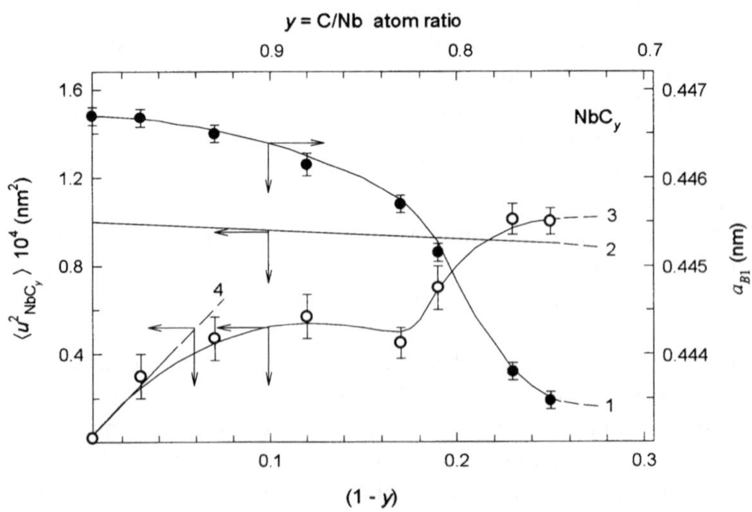

Fig. 4.9. Crystal lattice constant a_{B1} (1) and effective root-mean-square dynamic $\langle u^2_{dyn, NbC_y} \rangle$ (2) and static $\langle u^2_{st, NbC_y} \rangle$ (3) displacements as a function of the composition of disordered niobium carbide NbC_y, and an approximation (4) of the static displacements by (4.62) at low vacancy concentrations [101]

The lattice parameter of the cubic unit cell in disordered carbide NbC_y depends on the compound composition in the nonlinear behavior with an inflection point near $NbC_{0.83}$ (curve 1 in Fig. 4.9).

From neutron diffraction (λ = 0.2643 nm) the rms static and dynamic displacements in the averaged lattice were derived. As the scattering amplitudes of niobium and carbon atoms are similar in magnitude, the effect of the atomic displacements on the intensity of reflections can be accounted for by the Debye–Waller factor averaged for the two atomic species [8, 81, 102].

Table 4.8. Relative intensities of diffraction reflections (*hkl*) in neutron diffraction patterns of disordered niobium carbide NbC_y [100,101]

Carbide	Relative intensities of diffraction reflections (*hkl*)					
NbC_y	200	220	222	400	420	422
$NbC_{1.00}$	590	645	321	209	827	1000
$NbC_{0.97}$	869	903	423	277	1000	892
$NbC_{0.92}$	898	930	436	274	987	1000
$NbC_{0.88}$	1000	997	458	304	995	991
$NbC_{0.83}$	846	927	407	272	1000	969
$NbC_{0.81}$	874	916	406	274	974	1000
$NbC_{0.77}$	947	1000	451	279	986	918
$NbC_{0.75}$	853	894	417	249	939	1000

The integral intensities of the diffraction peaks are given in Table 4.8. Considering the geometry of the diffraction experiment and the fact that disordered carbide NbC_y has a *B*1 structure, the intensity of diffraction peaks with even Miller indices can be written as

$$I_{\text{exper}}(hkl) = K\frac{16P_{hkl}(f_{\text{Nb}} + yf_{\text{C}})^2}{\sin\theta_{hkl}\sin 2\theta_{hkl}}\exp(-2M_{\text{NbC}_y}). \tag{4.61}$$

To a first approximation, the quantity *M* in the Debye–Waller factor can be presented, in accordance with (4.53), as a linear function of the displacements [102]:

$$M = \frac{2\pi^2}{3}q^2\langle u_\Sigma^2\rangle. \tag{4.62}$$

However, a plot of $M_{\text{NbC}_y} = f(q^2)$ curves shows that the dependence is not linear for large q^2. The nonlinearity tends towards higher *M* for all niobium carbide samples studied. The nonlinearity of the function $M = f(q^2)$ may result from the nonadditive character of the dynamic $\langle u_{\text{dyn}}^2\rangle$ and static $\langle u_{\text{st}}^2\rangle$ displacements and on anharmonic oscillations of the atoms. In these cases it may be assumed in accordance with [81] that

$$2M = \tfrac{1}{3}q^2\langle u_{\text{st}}^2\rangle + \tfrac{1}{3}q^2\langle u_{\text{dyn}}^2\rangle + \Delta(\sum_n f_n q^n). \tag{4.63}$$

An analysis of (4.63) shows that in the first case the term that is governed by exponents q^n larger than the second-power exponent is negative at the vacancy concentrations considered. For this reason, the observed increase in M cannot be explained by nonadditivity of dynamic and static displacements. Considering that the dependence $M = f(q^2)$ is also nonlinear for stoichiometric carbide $NbC_{1.00}$, the observed deviations may be attributed to anharmonic dynamic oscillations of atoms in niobium carbide. This agrees with results from investigations of the heat capacity of disordered niobium carbide [103, 104], which revealed significant anharmonic effects at a temperature of 300 K.

The dependence of dynamic displacements on the carbide composition can serve to separate contributions to $\langle u_{\Sigma}^2 \rangle$ from dynamic and static displacements. This was established by Metzger, Peisl, and Kaufmann [93] who analyzed the effect of temperature on the integral intensities of reflections in X-ray patterns of carbide NbC_y. The composition dependence of the dynamic displacements [93] can be presented by (4.60). The scattering capacity of carbon atoms is small in X-ray experiments as compared to the scattering capacity of niobium atoms. Consequently, the relationship (4.60) refers to dynamic atomic displacements in the metal sublattice. An incomplete account for displacements of light carbon atoms results in underestimated values of dynamic displacements.

A superposition model considering both acoustic and optical oscillations was used by ref. [103–105] to examine the low-temperature heat capacity of disordered carbide NbC_y. The distribution function of atomic oscillation frequencies assumes the form

$$g(\omega) = \frac{1}{\sigma}\frac{9\omega^2}{\omega_{\mathrm{D}}^3} + \frac{\sigma - 1}{\sigma} 3\delta(\omega - \omega_{\mathrm{E}}) \tag{4.64}$$

where σ is the number of atoms in a minimum-volume translational cell of the crystal; δ is the Dirac delta function; and ω_{D} and ω_{E} the denote Debye and Einstein frequencies.

Integrating over all frequencies, the following formula can be derived, which allows for the contributions from the acoustic and optical oscillations to the dynamic atomic displacements:

$$\langle u_{\mathrm{dyn}}^2 \rangle = \frac{9\hbar^2}{2k_{\mathrm{B}}m}\left[\frac{1}{4\theta_{\mathrm{D}}} + \frac{T}{\theta_{\mathrm{D}}^2}\Phi\left(\frac{\theta_D}{T}\right) + \frac{1}{6\theta_{\mathrm{E}}}\coth\left(\frac{\theta_{\mathrm{E}}}{2T}\right)\right], \tag{4.65}$$

where $m = (m_{\mathrm{Nb}} + y m_{\mathrm{C}})/(1 + y)$, m_{Nb} and m_{C} stand for the mass of niobium and carbon atoms, θ_{D} and θ_{E} are the Debye and Einstein characteristic temperatures and

$$\Phi(t) = \frac{1}{t}\int_0^t \frac{x\,dx}{\exp x - 1}.$$

The dependence of dynamic atomic displacements on the composition of the disordered carbide NbC_y, which was calculated [101] using (4.65) and values of θ_D and θ_E [103–105], has the form

$$\langle u^2_{dyn} \rangle = [1.06 - 0.36(1-y)] \times 10^{-4} \text{ (nm}^2\text{)}. \tag{4.66}$$

This suggests greater values for the dynamic displacements than determined in [93]. From (4.66), the values of the dynamic displacements decrease with growing concentration of vacancies in niobium carbide (curve 2 in Fig. 4.9). This can be explained by a different contribution from displacements of carbon atoms since their number is less in nonstoichiometric niobium carbide.

Knowing the full rms displacements of atoms $\langle u^2_\Sigma \rangle$ in carbide NbC_y and knowing the concentration dependence of the dynamic displacements (4.66), it is possible to establish the dependence of static displacements $\langle u^2_{st} \rangle$ on the niobium carbide composition (see Fig. 4.9, curve 3). As can be seen from the figure, the static displacements increase with nonstoichiometry in niobium carbide and a local minimum is observed near carbide $NbC_{0.83}$. This minimum is probably due to short-range order in the arrangement of carbon atoms and vacancies, which could not be eliminated even by rapid quenching of the compound. It is suggested that the minimum in $\langle u^2_{st} \rangle$ is observed, in an interval where a monoclinic ordered phase Nb_6C_5 is formed [31–33] after long-term annealing (see Sect. 3.4).

In accordance with [81], when the concentration of defects is low and correlations in the mutual arrangement of atoms are absent, the dependence of the rms static displacements on the concentration c of defects responsible for static displacements is

$$\langle u^2_{st} \rangle = c(1-c)a^2 \frac{M_0}{2\pi^2}, \tag{4.67}$$

where $M_0 = \frac{1}{R}\left(\frac{1}{V}\frac{dV}{dc}\right)^2$, $(1/R)$ is a proportionality factor, a is the unit cell parameter and V is the unit cell volume. Expression (4.67) establishes a relationship between the variation of the medium lattice size with the static atomic displacements, whereas the proportionality factor

$$\frac{1}{R} = M_0\left(\frac{1}{V}\frac{dV}{dc}\right)^{-2}$$

characterizes the response of the crystal to the introduction of defects.

The analysis of (4.67) suggests that R^{-1} is larger when defects lead mainly to local displacements of atoms and have little effect on the lattice parameters. A smaller R^{-1} should be observed when the defects cause a considerable change in the size of the unit cell and, probably also, when defects exert a long-range effect on positions of the atoms in the lattice. R can be considered as an effective interaction radius for the atoms in the crystal.

Defects causing static distortions of the lattice in niobium carbide NbC_y are structural vacancies and therefore $c = (1 - y)/2$. Substituting the cubic cell parameters of NbC_y in (4.67) and performing a numerical differentiation of the cell volume with respect to the defect concentration, it can be shown that (4.67) approximates well the experimentally observed static displacements only in the concentration interval $0.97 < y < 1.00$ (see Fig. 4.9, curve 4). The R^{-1} value determined for this concentration interval is nearly two orders of magnitude larger than the corresponding values calculated [81] for alloys of nontransition metals. This points to larger static atomic displacements than found for alloys and suggests the dominance of strong localized interactions in niobium carbide. These interactions in NbC_y are probably due to strong metal–nonmetal covalent bonds. Delocalized metal bonds are weaker than similar bonds in alloys.

From Fig. 4.9 it is seen that the experimental values of the displacements are much smaller than the corresponding theoretical values at large vacancy concentrations. This is probably due to the fact that (4.67) was derived neglecting the overlap of disturbances from point defects. Such an overlap should necessarily lead to the decrease in $d\langle u_{\mathrm{st}}^2\rangle / dc$ which characterizes the change of static displacements depending on the vacancy concentration.

Static Displacements of Atoms Near a Structural Vacancy. Investigations of the structure of monocrystalline nonstoichiometric carbides and nitrides MX_y by elastic diffuse scattering of neutrons and X-rays allowed the determination of the absolute values and directions of the static displacements of metal and nonmetal atoms near a structural vacancy. The value and the direction of static displacements are directly connected with the change of the lattice constant depending on the composition of the nonstoichiometric compound MX_y and the degree of order in this compound.

The decrease in the concentration of interstitial atoms in nonstoichiometric compounds MX_y is followed by a nonlinear decrease in the constant a_{B1} of the basic cubic lattice with $B1$ structure. Along with the general tendency of the lattice constant a_{B1} to decrease with decreasing y in titanium TiC_y and zirconium ZrC_y carbides, the dependence $a_{B1}(y)$ exhibits a weak maximum in the interval $0.93 < y < 0.99$ (see Fig. 3.12).

In carbides with the structure of $B1$ type each metal atom has an octahedral environment of six sites in the nonmetal sublattice, while each site in the nonmetal sublattice is surrounded by six metal atoms. The presence of one or several structural vacancies in the nearest environment of a metal atom causes static displacements of this atom due to the combined asymmetry effect of its nearest neighbors. Let us discuss the displacement direction of metal atoms necessary to ensure the experimentally observed change of the lattice constant in carbides MC_y with increasing concentration of structural vacancies.

If metal atoms move towards a vacancy, the M–C interactions in adjacent occupied octahedra CM_6 will prevent a compression of vacant octahedral interstices $\square M_6$. The increase in the concentration of vacant interstices, which have a smaller linear size than occupied octahedral interstices, will be accompanied by the decrease in the parameter a_{B1} if vacancy-induced static displacements of metal atoms decrease monotonically and tend asymptotically to zero with increasing distance to the vacancy. The lattice constant will decrease even if vacancy-induced disturbances extend only to the first coordination sphere. It is worth noting that in this case the dependence $a_{B1}(y)$ cannot have a maximum, as observed for carbides TiC_y and ZrC_y.

If the metal atoms next to a vacancy move away from this vacancy, then the metal atoms of the next coordination sphere should be displaced in the opposite direction, i.e. towards the vacancy so that the lattice constant decreases. Thus, the vacancy-induced field of disturbances should extend at least to two coordination spheres of metal atoms. In this case, the attenuation of disturbances with distance represents Friedel oscillations. From estimates [43, 51, 93, 101] it follows that the effective disturbance radius in nonstoichiometric cubic nitrides MC_y is larger than the unit cell parameter and that the disturbance covers more than two coordination spheres (for example, correlations extend to eight coordination spheres in niobium carbide [43]).

In the most general case, the lattice constant of a disordered carbide can decrease with increasing concentration of vacancies when the metal atoms in the first coordination sphere are displaced either away from or towards the vacancy. However, the presence of a maximum in the relation for $a_{B1}(y)$ for carbides TiC_y and ZrC_y implies that metal atoms in the first coordination sphere most probably move away from the vacancy. With this in mind, let us discuss available data concerning static atomic displacements.

It is found in [42] that a partially ordered cubic (space group $Fd\bar{3}m$) phase Zr_2C was formed when zirconium carbide $ZrC_{0.64}$ was annealed at a temperature of 1070 K for 100 h. A neutron diffraction study of partially ordered zirconium carbide $ZrC_{0.64}$ showed that Zr atoms at positions 32(a) were displaced from the vacancy by 0.004 nm. Metal atoms moved towards a carbon atom in monocrystalline titanium $TiC_{0.76}$ and niobium $NbC_{0.73}$ carbides by 0.0030 and 0.0033 nm, respectively [43]. Simultaneously, the carbon atoms moved away from each other: the distance between adjacent carbon atoms in the first and second nonmetal coordination spheres increased by 0.0011 and by 0.0008 nm in titanium carbide $TiC_{0.76}$, and by 0.0006 and 0.0004 nm in niobium carbide $NbC_{0.73}$. Shrinkage of metal–carbon distances means that metal atoms are displaced away from a vacancy.

A method for analyzing experimental data on elastic diffuse scattering of neutrons is proposed by [43, 46]. Not only the short-range order parameters, but also values for the static atomic displacements in various coordination spheres have been determined. In accordance with (4.14), (4.21), (4.22) and (4.25), the intensity of elastic diffuse scattering includes the Laue background I_{DL} and contributions from short-range order I_{DSRO} and atomic displacements I_{DAD}. The contributions due to short-range order and static displacements can be separated directly by fitting the experimental data by the least squares method or the

Sparks–Borie method [106]. In accordance with [43, 46, 50], the intensity I_{DSRO} for the compound MX_y can be written as

$$I_{\mathrm{DSRO}} = \sum_{hkl} \alpha_{hkl} \cos(\pi H_1 h)\cos(\pi H_2 k)\cos(\pi H_3 l), \tag{4.68}$$

where $(h + k + l)$ is an even number for the nonmetal fcc sublattice; α_{hkl} is the short-range order parameter in the coordination sphere (hkl) having radius $\mathbf{R}_{hkl} = h\mathbf{a}/2 + k\mathbf{b}/2 + l\mathbf{c}/2$ ($\mathbf{a}$, $\mathbf{b}$ and $\mathbf{c}$ being fcc cell vectors); H_1, H_2 and H_3 are the coordinates of the reciprocal lattice sites. The intensity I_{DAD} is expressed [43, 46] as the sum of two terms $I_{\mathrm{AD\text{-}1}}$ and $I_{\mathrm{AD\text{-}2}}$. The first term is equal to

$$I_{\mathrm{AD\text{-}1}} = H_1 Q_x(H_1, H_2, H_3) + H_2 Q_x(H_2, H_3, H_1) + H_3 Q_x(H_3, H_1, H_2), \tag{4.69}$$

where

$$Q_x(H_1, H_2, H_3) = \sum_{hkl} \gamma_{hkl} \sin(\pi H_1 h)\sin(\pi H_2 k)\sin(\pi H_3 l),$$

$$\gamma_{hkl} = -\pi[y/(1-y) + \alpha_{hkl}]\langle u^{hkl}_{\mathrm{X\text{-}X}}\rangle \text{ if } (h + k + l) \text{ is an even number,}$$

$$\gamma_{hkl} = -[2\pi/(1-y)](f_{\mathrm{M}}/f_{\mathrm{X}})\langle u^{hkl}_{\mathrm{X\text{-}M}}\rangle \text{ if } (h + k + l) \text{ is an odd number,}$$

$\langle u^{hkl}\rangle$ denotes the relative displacement of two sites spaced from one another by $\mathbf{R}_{hkl}$.

The second term is equal to

$$I_{\mathrm{AD-2}} = \sum_{hkl} \delta_{hkl} H_1^2 \cos(\pi H_1 h)\cos(\pi H_2 k)\cos(\pi H_3 l) + \ldots$$

$$+ \sum_{hkl} \varepsilon_{hkl} H_1 H_2 \sin(\pi H_1 h)\sin(\pi H_2 k)\sin(\pi H_3 l) + \ldots, \tag{4.70}$$

where δ_{hkl} and ε_{hkl} are coefficients related to atomic displacements. The second term $I_{\mathrm{AD\text{-}2}}$ is much smaller than the first term $I_{\mathrm{AD\text{-}1}}$.

In the case of the nonstoichiometric compound $MX_y\square_{1-y}$, the short-range order parameter α_{hkl} in the coordination sphere (hkl) can be determined as probabilities of "vacancy $\square$ – interstitial atom" pairs $P^{hkl}_{\mathrm{X-\square}}$ or $P^{hkl}_{\mathrm{\square-X}}$ from (4.1) or (4.2), which is written in the form

$$\alpha_{hkl} = 1 - \frac{P^{hkl}_{\mathrm{\square-X}}}{y} \equiv 1 - \frac{P^{hkl}_{\mathrm{X-\square}}}{1-y}. \tag{4.71}$$

Therefore $P^{hkl}_{\mathrm{X-\square}} = y(1 - \alpha_{hkl})$. Since $P^{hkl}_{\mathrm{X-\square}} + P^{hkl}_{\mathrm{X-X}} = 1$, then $P^{hkl}_{\mathrm{X-X}} = 1 - P^{hkl}_{\mathrm{X-\square}}$ or, taking into account (4.71), $P^{hkl}_{\mathrm{X-X}} = 1 - (1 - y)(1 - \alpha_{hkl}) = y + (1 - y)\alpha_{hkl}$. The displacements $\langle u^{hkl}_{\mathrm{\square-X}}\rangle$ and $\langle u^{hkl}_{\mathrm{X-X}}\rangle$ are connected with probabilities $P^{hkl}_{\mathrm{\square-X}}$ and $P^{hkl}_{\mathrm{X-X}}$ by the relationship

$$P^{hkl}_{\mathrm{X-X}}\langle u^{hkl}_{\mathrm{X-X}}\rangle + P^{hkl}_{\mathrm{\square-X}}\langle u^{hkl}_{\mathrm{\square-X}}\rangle = 0. \tag{4.72}$$

From (4.69) it follows that $\langle u^{hkl}_{\text{X-X}} \rangle = -\frac{(1-y)}{\pi[y+(1-y)\alpha_{hkl}]}\gamma_{hkl}$. Taking into account $P^{hkl}_{\square\text{-X}}$ and $P^{hkl}_{\text{X-X}}$ values, the expression (4.72) yields a formula for determining rms static displacements in the nonmetal sublattice:

$$\langle u^{hkl}_{\square\text{-X}} \rangle = \frac{1}{2\pi(1-\alpha_{hkl})}\gamma_{hkl}, \tag{4.73}$$

where the sum $(h + k + l)$ is an even number. Displacements $\langle u^{hkl}_{\square\text{-X}} \rangle$ are calculated from formula (4.73) in $a_{B1} \times 10^{-4}/2$ units.

The probabilities of "metal M – vacancy □" and "metal M – interstitial atom X" pairs in a nonstoichiometric compound MX_y are $P^{hkl}_{\text{M-}\square} = 1 - y$ and $P^{hkl}_{\text{M-X}} = y$. These probabilities $P^{hkl}_{\text{M-}\square}$ and $P^{hkl}_{\text{M-X}}$ are related to the rms static displacements $\langle u^{hkl}_{\square\text{-M}} \rangle$ and $\langle u^{hkl}_{\text{X-M}} \rangle$ in the metal sublattice as

$$P^{hkl}_{\text{M-X}}\langle u^{hkl}_{\text{X-M}} \rangle + P^{hkl}_{\text{M-}\square}\langle u^{hkl}_{\square\text{-M}} \rangle = 0. \tag{4.74}$$

From (4.69) it follows that $\langle u^{hkl}_{\text{X-M}} \rangle = -\frac{(1-y)}{2\pi}\frac{f_{\text{X}}}{f_{\text{M}}}\gamma_{hkl}$. Considering further the values for the probabilities $P^{hkl}_{\text{M-}\square}$ and $P^{hkl}_{\text{M-X}}$, (4.74) yields an expression for determining rms static displacements in the metal sublattice of the compound MX_y:

$$\langle u^{hkl}_{\square\text{-M}} \rangle = \frac{yf_{\text{X}}}{2\pi f_{\text{M}}}\gamma_{hkl}, \tag{4.75}$$

where the sum $(h + k + l)$ is an odd number. The displacements $\langle u^{hkl}_{\square\text{-M}} \rangle$ are calculated from (4.73) in $a_{B1} \times 10^{-4}/2$ units.

Values for the coefficients γ_{hkl} and for the static displacements $\langle u^{hkl}_{\square\text{-M}} \rangle$ and $\langle u^{hkl}_{\square\text{-X}} \rangle$, determined in [45–51], are given in Table 4.9. Displacements of metal atoms $\langle u^{hkl}_{\square\text{-M}} \rangle$ were calculated from (4.75) using the following neutron scattering amplitudes f_{M} and f_{X}: $f_{\text{Ti}} = -0.34 \cdot 10^{-12}$, $f_{\text{Nb}} = 0.71 \cdot 10^{-12}$, $f_{\text{C}} = 0.665 \cdot 10^{-12}$ and $f_{\text{N}} = 0.94 \cdot 10^{-12}$ cm. Positive values of displacements mean that the atom moves away from the vacancy, while negative values of displacements indicate that the atom approaches the vacancy.

Table 4.9 shows that metal atoms nearest to a vacancy move away from it, whereas metal atoms in the next coordination sphere approach the vacancy in the carbides $TiC_{0.64}$, $TiC_{0.76}$, $NbC_{0.73}$ and $NbC_{0.83}$ and the nitride $TiN_{0.82}$. The directions and the relative values of the static displacements of metal and nonmetal atoms around a vacancy in the carbides $TiC_{0.64}$, $TiC_{0.76}$, $NbC_{0.73}$ and $NbC_{0.83}$ and the nitride $TiN_{0.82}$ are presented in Figs. 4.10–4.12.

Table 4.9. Coefficients γ_{hkl} and mean-square static displacements $\langle u^{hkl}_{\square-M}\rangle$ and $\langle u^{hkl}_{\square-X}\rangle$ of metal and nonmetal atoms in coordination spheres (*hkl*) of the carbides $TiC_{0.64}$, $TiC_{0.76}$, $NbC_{0.73}$, $NbC_{0.83}$ and the nitride $TiN_{0.82}$ [46] (displacements $\langle u^{hkl}\rangle$ are given in $a_{B1}\times10^{-4}/2$ units)

(*hkl*)	Pair □–M or □–X	$TiC_{0.64}$, a_{B1} = 0.4332 nm, [1] 1170 K γ_{hkl}	$\langle u^{hkl}\rangle$	$TiC_{0.76}$, a_{B1} = 0.4330 nm, [1] 1170 K γ_{hkl}	$\langle u^{hkl}\rangle$	$NbC_{0.73}$, a_{B1} = 0.4442 nm, [2] quenching γ_{hkl}	$\langle u^{hkl}\rangle$	$NbC_{0.73}$, [1] 1470 K γ_{hkl}	$\langle u^{hkl}\rangle$
100	□–M	−0.127	253	−0.098	232	0.146	159	0.151	164
110	□–X	−0.030	−45	−0.025	−38	−0.005	−7	0.001	1
111	□–M	0.024	−48	0.017	−40	−0.047	−51	−0.063	−69
200	□–X	0.034	42	−0.009	−13	−0.045	−60	−0.021	−28
102	□–M	0.026	−52	0.004	−9	−0.007	−8	−0.005	−5
201	□–M	0.004	−8	0	0	−0.015	−16	−0.022	−24
112	□–X	0.008	13	0.006	10	0.002	3	0.002	3
211	□–X	−0.002	−3	−0.002	−3	0.008	14	0.008	14
220	□–X	−0.011	−19	0.001	2	0.008	13	0.012	19
122	□–M	−0.007	14	0.004	−9	0.010	11	0.009	10
212	□–M	0	0	0.002	−5	0.009	10	−0.003	−3
300	□–M	0.028	−56	0.020	−47	−0.037	−40	−0.068	−74
103	□–X	0.006	10	0.002	3	0.007	12	0.006	10
310	□–X	0.003	5	0.002	3	0.010	16	0.013	21

(*hkl*)	Pair □–M or □–X	$NbC_{0.83}$, a_{B1} = 0.4453 nm, [1] 1470 K γ_{hkl}	$\langle u^{hkl}\rangle$	$TiN_{0.82}$, a_{B1} = 0.4228 nm, [2] quenching γ_{hkl}	$\langle u^{hkl}\rangle$	$TiN_{0.82}$, [1] 970 K γ_{hkl}	$\langle u^{hkl}\rangle$	$TiN_{0.82}$, [1] 1170 K γ_{hkl}	$\langle u^{hkl}\rangle$
100	□–M	0.206	255	−0.049	177	−0.057	206	−0.051	184
110	□–X	0.005	7	−0.015	−22	−0.025	−36	−0.027	−39
111	□–M	−0.068	−84	0.021	−76	0.023	−83	0.024	−87
200	□–X	−0.028	−38	0.016	23	0.022	31	0.022	32
102	□–M	0.011	14	0.008	−29	0.011	−40	0.013	−47
201	□–M	−0.020	−25	0.005	−18	−0.007	25	−0.009	32
112	□–X	−0.004	−7	0.001	2	0.001	2	−0.002	−3
211	□–X	0.001	2	−0.007	−12	0.007	12	0.008	13
220	□–X	0.001	2	−0.003	−5	0.004	6	0.002	3
122	□–M	0.010	12	−0.001	4	0	0	0.003	−11
212	□–M	0.008	10	−0.001	4	−0.005	18	−0.003	11
300	□–M	−0.061	−75	0.007	−25	−0.005	18	−0.005	18
103	□–X	0.010	16	−0.002	−3	0.002	3	0.002	3
310	□–X	−0.002	−3	0	0	0.008	13	0.009	15

[1] Neutron diffraction measurements of disordered samples are performed at the given temperature *T*.

[2] Neutron diffraction measurements of quenched samples are performed at 300 K.

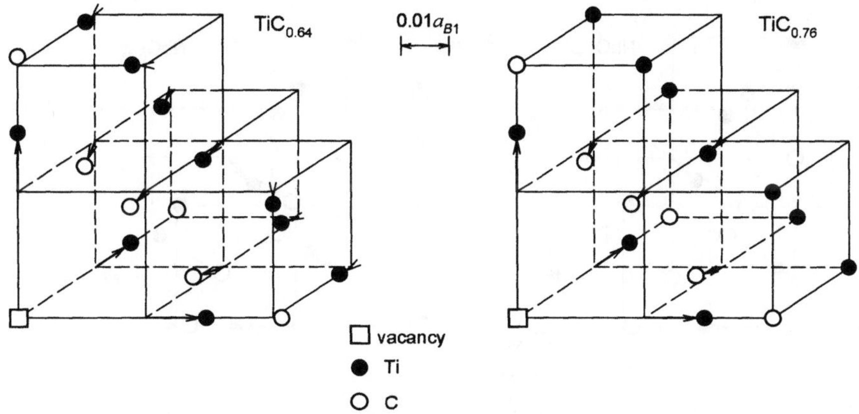

Fig. 4.10. Field of static atomic displacements around a carbon vacancy in single crystals of titanium carbides $TiC_{0.64}$ and $TiC_{0.76}$ with a cubic $B1$ type lattice. This field was determined from measured elastic diffuse neutron scattering at 1170 K [46]. Arrows show the directions of displacements. The displacements are magnified (the displacement scale in $0.01a_{B1}$ is shown at the center top; a_{B1} is the cubic lattice constant)

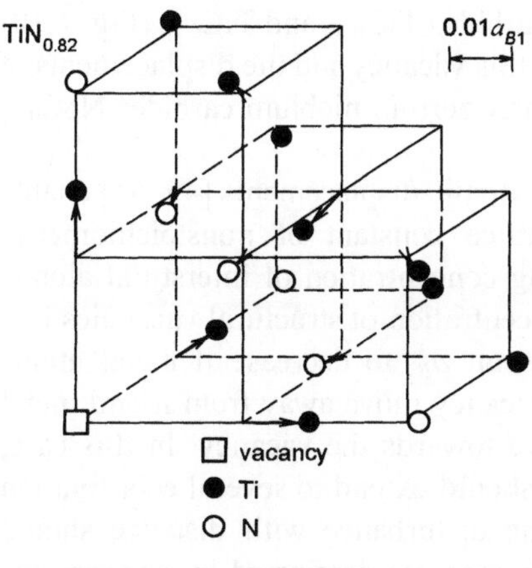

Fig. 4.11. Field of static atomic displacements around a carbon vacancy in a single crystal of titanium nitride $TiN_{0.82}$ with a cubic $B1$ type lattice. This field was determined from measured elastic diffuse scattering of neutrons at 970 K [46, 48]. Arrows show the directions of displacements. The displacements are magnified (the displacement scale in $0.01a_{B1}$ is shown at the center right; a_{B1} is the cubic lattice constant)

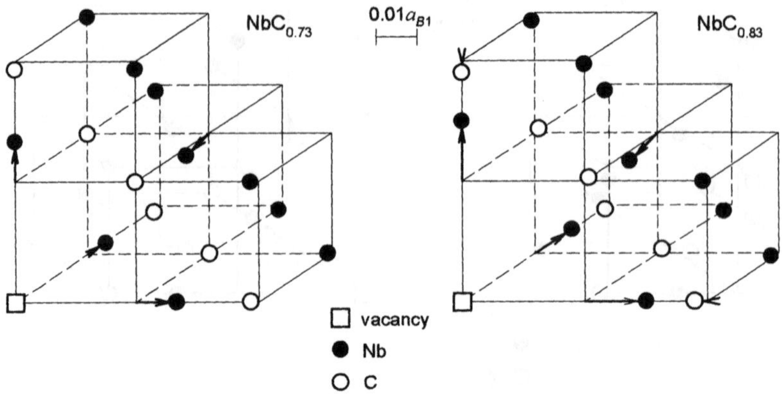

Fig. 4.12. Field of static atomic displacements around a carbon vacancy in single crystals of niobium carbides $NbC_{0.73}$ and $NbC_{0.83}$ with a cubic $B1$ type lattice. This field was determined from measured elastic diffuse scattering of neutrons at 1470 K [46]. Arrows show the directions of displacements. The displacements are magnified (the displacement scale in $0.01a_{B1}$ is shown at the center top; a_{B1} is the cubic lattice constant)

Nonmetal atoms nearest to a vacancy in carbides $TiC_{0.64}$ and $TiC_{0.76}$ (Fig. 4.10) and nitride $TiN_{0.82}$ (Fig. 4.11) move towards this vacancy and the displacements of nonmetal atoms around the vacancy are nearly zero in niobium carbides $NbC_{0.73}$ and $NbC_{0.83}$ (Fig. 4.12).

The values and the directions of the static atomic displacements [45–51] enable us to derive a scheme by which the lattice constant of nonstoichiometric compounds MX_y diminishes with decreasing concentration of interstitial atoms. As mentioned above, the increase in the concentration of structural vacancies in a compound MX_y will cause the lattice constant a_{B1} to decrease if metal atoms forming the first coordination sphere of a vacancy move away from it and metal atoms in the next coordination sphere move towards the vacancy. In this case, vacancy-induced disturbances of the lattice should extend to several coordination spheres and an attenuation of effect of the disturbance with distance should represent Friedel oscillations. These speculations are confirmed by experiments [45–51]. A mechanism for the decrease of the lattice constant in nonstoichiometric compounds MX_y, which takes into account the directions of the atomic displacements around a vacancy, was proposed in [107] and is shown in Fig. 4.13. This scheme was constructed for the $(110)_{B1}$ plane of a nonstoichiometric compound MX_y with the $B1$ structure. The scheme allows for an extension of the displacement field around a vacancy □ out to several coordination spheres and an oscillatory attenuation of the amplitude of the displacement with increasing distance.

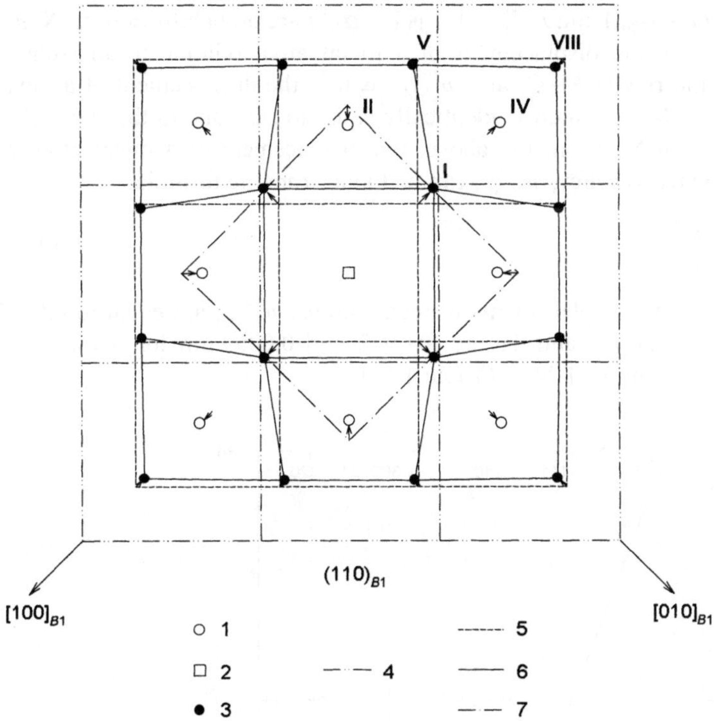

Fig. 4.13. Diagram for the model to explain the decrease of the lattice constant of nonstoichiometric compounds MX_y with the $B1$ structure upon formation of a vacancy □ and static atomic displacements around a vacancy in the $(110)_{B1}$ plane [107]: (1) interstitial atom; (2) vacancy □; (3) metal atom M; (4) defect free ideal lattice; (5) lattice with a vacancy (atomic displacements neglected); (6) real lattice taking account of atomic displacements; (7) unit cell outline; I, II, IV, V and VIII – numbers of coordination spheres (100), (110), (200), (210) and (300) constructed around a vacancy. Arrows show directions of displacements

Figure 4.14 shows the magnitudes and the directions of static displacements of titanium and carbon atoms for 12 coordination spheres of a vacancy in the disordered carbide $TiC_{0.64}$ [46]. The displacements of Ti and C atoms oscillate and their absolute values decrease with distance. Vacancy-induced disturbances extend to a distance nearly equal to two unit cell lattice constants.

The displacements $<u^{hkl}>$ found from diffuse scattering are an average over all atomic configurations of nonstoichiometric compounds MX_y with a large concentration of structural vacancies. To a first approximation, vacancy-induced displacement fields are additive. Therefore the displacements $\langle u^{100}_{\square\text{-M}} \rangle$ of metal atoms in the first coordination sphere can be written as

$$\langle u^{100}_{\square-M} \rangle = P^{200}_{\square-X} u^{100}_{\square-M} + P^{200}_{\square-\square} u^{100(2)}_{\square-M}, \tag{4.76}$$

where $P^{200}_{\square-X} = y(1 - \alpha_{200})$ and $P^{200}_{\square-\square} = 1 - y(1 - \alpha_{200})$ are probabilities of □–X and □–□ pairs; $u^{100}_{\square-M}$ is the displacement of a metal atom relative to an isolated vacancy, i.e. in the row □–M–C; and $u^{100(2)}_{\square-M} \equiv 0$ is the displacement of a metal atom in the row □–M–□, which is identically equal to zero due to the symmetric position of the atom M. From the above, the displacement of a metal atom M relative to an isolated vacancy, $u^{100}_{\square-M}$, can be found from the formula

$$u^{100}_{\square-M} = \frac{\langle u^{100}_{\square-M} \rangle}{y(1-\alpha_{200})}. \tag{4.77}$$

Calculated mean static displacements of metal atoms, $u^{100}_{\square-M}$, are equal to 0.0067 and 0.0059 nm for $TiC_{0.64}$ and $TiC_{0.76}$, 0.0042 and 0.0059 nm for $NbC_{0.73}$ and $NbC_{0.83}$, and 0.0048 nm for $TiN_{0.82}$ (Table 4.10).

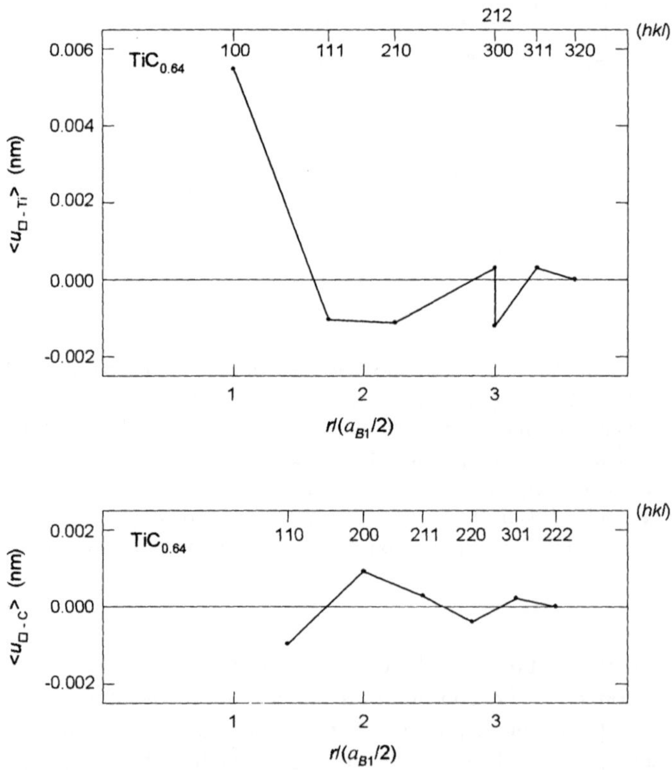

Fig. 4.14. Static displacements of titanium and carbon atoms $\langle u_{\square-Ti} \rangle$ and $\langle u_{\square-C} \rangle$ relative to a vacancy □ in the disordered carbide $TiC_{0.64}$ [46]: positive displacement corresponds to movement of atoms away from the vacancy; negative displacement corresponds to movement of atoms towards the vacancy; *r* is the radius of the coordination sphere (*hkl*)

Table 4.10. Short-range order parameters α_{200}, mean static displacements of metal atoms, $\langle u^{100}_{\square-M} \rangle$ (in $a_{B1} \times 10^{-4}/2$ units), and displacements $u^{100}_{\square-M}$ of metal atoms relative to the isolated vacancy for carbides $TiC_{0.64}$, $TiC_{0.76}$, $NbC_{0.73}$, $NbC_{0.83}$ and nitride $TiN_{0.82}$

Compound MX_y	Lattice constant a_{B1} (nm) at 300 K	α_{200} [46]	$\langle u^{100}_{\square-M} \rangle$ [46]	$\langle u^{100}_{\square-M} \rangle$ (nm) at 300 K	$u^{100}_{\square-M}$ (nm) at 300 K
$TiC_{0.64}$ (1170 K)	0.4322	−0.276	253	0.00547	0.00670
$TiC_{0.76}$ (1170 K)	0.4330	−0.122	232	0.00502	0.00589
$NbC_{0.73}$ (1470 K)	0.4442	−0.196	164	0.00364	0.00417
$NbC_{0.83}$ (1470 K)	0.4453	−0.166	255	0.00567	0.00586
$TiN_{0.82}$ (970 K)	0.4228	−0.114	206	0.00435	0.00476

Values for $u^{100}_{\square-M}$ reported in [46] may be overestimated, because in the calculations α_{110} was used, by mistake, instead of α_{200}. The displacements $u^{100}_{\square-M}$ given in Table 4.10 are average values as compared to literature values: for example, the displacement of Ti atoms from an isolated vacancy is 0.0044 nm in $TiC_{0.90}$ [96], 0.0097 nm in $TiC_{0.94}$ [92], and 0.0080 nm in $TiC_{0.97}$ [108]. The displacement of niobium atoms from a vacancy in the carbide $NbC_{0.83}$, which is 0.0059 nm, is smaller than the displacement u = 0.0090 nm determined for the carbide $NbC_{0.82}$ by ion channeling [94].

In [109] unusual results were obtained concerning the static atomic displacements around a vacancy in ordered cubic (space group $P4_332$) vanadium carbide V_8C_7 ($VC_{0.875}$). Vanadium atoms nearest to a vacancy shift to the center of a vacant octahedron $\square V_6$ for 0.003 to 0.006 nm and the carbon atoms on the first nonmetal coordination sphere of a vacancy approach this vacancy even stronger: the displacement may be as large as 0.05 nm. The dynamic displacements of vanadium atoms V occupying positions 8(*c*) and carbon atoms C occupying positions 12(*d*) are directed mainly along the axis [010]. They are negligibly small compared to the dynamical displacements along other directions. The compression of the vacant octahedra $\square V_6$ is accompanied by the expansion of the occupied octahedra MC_6. In ref. [13], it is argued that the vanadium atoms in the first coordination sphere move away from a vacancy in ordered carbide V_8C_7, i.e. they move in the opposite direction as reported in [109].

Experimental data on short-range order and atomic displacements in strongly nonstoichiometric compounds show that short-range order is preserved at temperatures some 300–400 K higher than the temperature of the equilibrium reversible disorder–order transformation. However, the diffuse scattering may be due not only to the short-range order; it is known that even small impurities of interstitial atoms in transition metals may cause a maximum of diffuse scattering. Therefore one should be careful about the factors responsible for diffuse scattering. Vacancy-induced disturbances of the crystal lattice may cover up to eight or even ten coordination spheres, i.e. they extend to a distance of two basic lattice constants. If disturbances cover such a long distance, then they overlap.

Static displacements increase with the vacancy concentration. However, thanks to the overlapping, the increment in the displacements is smaller each time a vacancy is formed. A complicated dependence of the displacement value on the concentration of structural vacancies and, more so, the presence of short-range order in the high-temperature region should be considered when analyzing the structure of nonstoichiometric compounds or interpreting the phase transformations in these compounds.

References

1 M. A. Krivoglaz, A. A. Smirnov: *The Theory of Order-Disorder in Alloys* (American Elsevier, New York 1964) 427 pp.

2 A. A. Smirnov: *Molecular and Kinetic Theory of Metals* (Nauka, Moscow 1966) 488 pp. (in Russian)

3 A. A. Smirnov: *Theory of Interstitial Alloys* (Nauka, Moscow 1979) 368 pp. (in Russian)

4 A. G. Khachaturian: *Theory of Structural Transformations in Solid* (John Wiley & Sons, New York 1983) 574 pp.

5 H. A. Bethe: Statistical theory of superlattices, Proc. Roy. Soc. London A **150**, 552–575 (1935)

6 J. M. Cowley: An approximate theory of order in alloys, Phys. Rev. **77**, 669-675 (1950)

7 B. E. Warren, B. L. Averbach, B. W. Roberts: Atomic size effect in the X-ray scattering by alloys, J. Appl. Phys. **22**, 1493–1496 (1951)

8 M. A. Krivoglaz: *Theory of X-ray and Thermal-Neutron Scattering by Real Crystals* (Plenum Press, New York 1969) 405 pp.

9 V. I. Iveronova, A. A. Katsnelson: *Short-Range Order in Solid Solutions* (Nauka, Moscow 1977) 256 pp. (in Russian)

10 C. Giacovazzo, H. L. Monaco, D. Viterbo, F. Scordari, G. Gilli, G. Zanotti, M. Catti: *Fundamentals of Crystallography* (Oxford University Press, Oxford 1992) 654 pp.

11 A. I. Gusev, A. A. Rempel: A study of the atomic ordering in the niobium carbide using the magnetic susceptibility method, Phys. Stat.Sol. (a) **84**, 527–534 (1984)

12 C. Froidevaux, D. Rossier: NMR-investigation of the atomic and electronic structure of vanadium and niobium carbides, J. Phys. Chem. Solids **28**, 1197–1209 (1967)

13 C. H. de Novion, N. Lorenzelli, P. Costa: Superlattice structure in vanadium carbide VC_{1-x}, Compt. Rend. Acad. Sci. Paris B. **263**, 775–778 (1966)

14 J. D. Venables, D. Kahn, R. G. Lye: Structure of ordered compound V_6C_5, Philosoph. Mag. **18**, 177–192 (1968)

15 D. Kahn, R. G. Lye: Nuclear magnetic resonance rotation patterns in single-crystal V_6C_5, Bull. Amer. Phys. Soc. **14**, 332–333 (1969)

16 A. V. Dmitriev, V. K. Kapustkin, R. N. Pletnev, V. A. Gubanov, Yu. G. Zainulin: ^{51}V NMR and short-range order in vanadium carbide, Fiz. Tverd. Tela **25**, 66–70 (1983) (in Russian). (Engl. transl.: Sov. Physics – Solid State. **25**, 36-38 (1983))

17 A. V. Dmitriev, R. N. Pletnev, V. A. Gubanov: *Nuclear Magnetic Resonance and Defects in Vanadium Refractory Compounds* (Urals Scientific Centre, Sverdlovsk 1983) 64 pp. (in Russian)

18 F. A. Sidorenko: Significance of cluster figures in crystals with order, Fiz. Tverd. Tela **23**, 3514–3515 (1981) (in Russian)
19 A. A. Rempel, A. I. Gusev: Relation between short-range and long-range order in solid solutions with b.c.c. and f.c.c. structures, Phys. Stat. Sol. (b) **130**, 413–420 (1985)
20 A. A. Rempel, A. I. Gusev: The relationship between short-range and long-range order in ordered alloys, Fiz. Metall. Metalloved. **60**, 847–854 (1985) (in Russian). (Engl. transl.: Phys. Met. Metallogr. **60**, 11–17 (1985))
21 A. F. Shevakin, I. A. Tomilin, V. I. Sarrak: NMR investigation of VN type vanadium nitride, Fizika Metallov i Metallovedenie **32**, 1204–1208 (1971) (in Russian). (Engl. transl.: Phys. Met. Metallogr. **32**, 85–89 (1971))
22 W. W. Warren, A. S. Cossard, M. D. Banus: Onset of magnetism in vanadium oxides: ^{51}V NMR studies of VO, J. Appl. Phys. **41**, 881–882 (1970)
23 S. Takagi, Y. Kitaoka, H. Yasuoka: ^{51}V NMR studies of vanadium monoxide $VO_{1.25}$, J. Phys. Soc. Jap. **49**, 521–527 (1980)
24 A. V. Dmitriev, R. N. Pletnev, V. A. Gubanov, Yu. G. Zainulin, G. P. Shveikin: Hyperfine interactions and chemical bonding in refractory compounds of vanadium, Doklady AN SSSR **275**, 906–909 (1984) (in Russian). (Engl. transl.: Doklady Phys. Chem. **275**, 305–307 (1984))
25 R. A. Bennet, H. O. Hooper: Nuclear magnetic resonance study of the niobium-nitrogen system NbN_x, J. Appl. Phys. **40**, 2441–2444 (1969)
26 A. A. Rempel, A. I. Gusev: *Ordering in Nonstoichiometric Niobium Monocarbide* (Urals Scientific Centre, Sverdlovsk 1983) 68 pp. (in Russian)
27 A. A. Rempel, M. Yu. Belyaev, L. B. Dubrovskaya, G. P. Shveikin: An NMR study of nonstoichiometric niobium monocarbide, Zh. Neorgan. Khimii **29**, 2451–2455 (1984) (in Russian). (Engl. transl.: Russ. J. Inorg. Chem. **29**, 1401–1404 (1984))
28 A. A. Rempel, A. I. Gusev: NMR spectra of ^{93}Nb in ordered niobium carbide, Fiz. Tverd. Tela **25**, 3169–3171 (1983) (in Russian). (Engl. transl.: Sov. Physics – Solid State **25**, 1827–1828 (1983))
29 A. A. Rempel, A. I. Gusev, M. Yu. Belyaev: ^{93}Nb NMR study of an ordered and a disordered non-stoichiometric niobium carbide, J. Phys. C: Solid State Phys. **20**, 5655–5666 (1987)
30 A. I. Gusev, A. A. Rempel: *Structural Phase Transitions in Nonstoichiometric Compounds* (Nauka, Moscow 1988) 308 pp. (in Russian)
31 A. I. Gusev, A. A. Rempel: Ordering in the carbon sublattice of nonstoichiometric niobium carbide, Fiz. Tverd. Tela **26**, 3622–3627 (1984) (in Russian). (Engl. transl.: Sov. Physics – Solid State **26**, 2178–2181))
32 A. A. Rempel, A. I. Gusev: Order-disorder phase transition in nonstoichiometric niobium carbide, Kristallografiya **30**, 1112–1115 (1985) (in Russian). (Engl. transl.: Sov. Physics – Crystallography **30**, 648–650 (1985))
33 A. I. Gusev, A. A. Rempel: Order-disorder phase transition channel in niobium carbide, Phys. Stat. Sol. (a) **93**, 71–80 (1986)
34 G. W. Clark: Pulsed nuclear resonance apparatus, Rev. Sci. Instrum. **35**, 313–333 (1964)
35 E. Parthe, K. Yvon: Crystal chemistry of the close packed transition metal carbides. Proposal for the notation of the different crystal structures, Acta Crystallogr. B **26**, 153–163 (1970)

36 J. F. Baugher, O. S. Taylor, T. Oja, P. J. Bray: Nuclear magnetic resonance powder patterns in the presence of completely asymmetric quadrupole and chemical shift effect, J. Chem. Phys. **50**, 4914–4925 (1969)
37 J. Winter: *Magnetic Resonance in Metals* (Clarendon Press, Oxford 1971) 206 pp.
38 L. G. Makhanek: Antiscreening of nuclear quadrupole moments, Optika i Spektroskopiya **14**, 449–457 (1963) (in Russian)
39 J. Billingham, P. S. Bell, N. H. Lewis: Vacancy short-range order in substoichiometric transition metal carbides and nitrides with the NaCl structure. Electron diffraction studies of short-range ordered compounds, Acta Crystallogr. A **28**, 602–606 (1972)
40 M. Sauvage, E. Parthe: Vacancy short-range order in substoichiometric transition metal carbides and nitrides with the NaCl structure. Numerical calculation of vacancy arrangement, Acta Crystallogr. A **28**, 607–616 (1972)
41 Sauvage, E. Parthe, W. B. Yelon: Neutron diffraction measurement of diffuse scattering in $VC_{0.75}$, Acta Crystallogr. A **30**, 597–599 (1974)
42 C. H. de Novion, V. Maurice: Order and disorder in carbides and nitrides, J. Physique Colloq. (France) **38**, C7-211– C7-220 (1977)
43 V. Moisy-Maurice, C. H. de Novion, A. N. Christensen, W. Just: Elastic diffuse neutron scattering study of the defect structure of $TiC_{0.76}$ and $NbC_{0.73}$, Solid State Commun. **39**, 661–665 (1981)
44 B. E. F. Fender: Diffuse scattering and the study of defect solids, in *Chemical Applications of Thermal Neutron Scattering* (Oxford University Press, London 1973) pp. 250–269.
45 T. Priem, B. Beuneu, C. H. de Novion, R. Caudron, F. Solal, A. N. Christensen: (1/2 1/2 1/2) versus (1 1/2 0) type ordering of nitrogen vacancies in TiN_x, Solid State Commun. **63**, 929–932 (1987)
46 T. Priem: *Etude de l'ordre à courte distance dans les carbures et nitrures non-stoechiometriques de metaux de transition par diffusion diffuse de neutrons*, Rapport CEA-R-5499 (Commissariat à l'Energie Atomique, Centre d'Etudes Nucléaires de Saclay, Gif-sur-Yvette (France) 1989) 162 pp.
47 T. Priem, B. Beuneu, C. H. de Novion, J. Chevrier, F. Livet, A. Finel, S. Lefevbre: Short-range order and static displacements in non-stoichiometric transition metal carbides and nitrides, Physica B **156–157**, 47–49 (1989)
48 T. Priem, B. Beuneu, C. H. de Novion, A. Finel, F. Livet: Short-range order, atomic displacements and effective interatomic ordering energies in $TiN_{0.82}$, J. Physique (France). **50**, 2217–2242 (1989)
49 T. Priem, B. Beuneu, C. H. de Novion: High temperature elastic diffuse neutron scattering study of the defect structure in $TiN_{0.82}$, in *X-ray and Neutron Structure Analysis in Materials Science*, J. Hasek (Ed.), (Plenum Publ. Corp., New York 1989) pp. 149-152
50 B. Beuneu, T. Priem, C. H. de Novion, S. Lefebvre, J. Chevrier, A. N. Christensen: Short-range-order determination in quenched $NbC_{0.73}$ by elastic diffuse neutron scattering, J. Appl. Crystallogr. **23**, 497–508 (1990)
51 C. H. de Novion, B. Beuneu, T. Priem, N. Lorenzelli, A. Finel: Defect structures and order-disorder transformations in transition metal carbides and nitrides, in *The Physics and Chemistry of Carbides, Nitrides and Borides*, R. Freer (Ed.), (Kluwer Acad. Publ., Netherlands 1990) pp. 329–355

52 V. Moisy-Maurice: *Structure atomique des carbures non-stoechiometriques de metaux de transition*, Rapport CEA-R-5127 (Commissariat à l'Energie Atomique, Gif-sur-Yvette (France), 1981) 184 pp.
53 A. N. Christensen: The crystal growth of the transition metal compounds titanium monocarbide, titanium mononitride, and zirconium mononitride by a floating zone technique, J. Cryst. Growth. **33**, 99–104 (1976)
54 J. Bursik, G. C. Weatherly: Ordering of substoichiometric δ-TiC_x phase in Ti-V-C alloys, Phys. Stat. Sol. (a) **174**, 327–335 (1999)
55 A. A. Rempel, A. I. Gusev: Short-range order in ordered alloys and interstitial phases, Fiz. Tverd. Tela **32**, 16–24 (1990) (in Russian). (Engl. transl.: Sov. Physics – Solid State **32**, 8–13 (1990))
56 A. A. Rempel, A. I. Gusev: Short-range order in superstructures, Phys. Stat. Sol. (b) **160**, 389–402 (1990)
57 K. Hiraga, M. Hirabayashi: Long-range and short-range order in interstitial compounds M_2X with special reference to V_2C and Nb_2C, J. Physique Colloq. (France) **38**, C7-224–C7-226 (1977)
58 A. A. Rempel, L. B. Dubrovskaya, A. I. Gusev, G. P. Shveikin: Effect of ordering on the magnetic susceptibility of niobium carbide, Izv. AN SSSR. Neorgan. Materialy **21**, 596–599 (1985) (in Russian). (Engl. transl.: Inorganic Materials **21**, 514–518 (1985))
59 A. I. Gusev, A. A. Rempel: Anomalous changes in the magnetic susceptibility due to ordering of niobium carbide, Fiz. Tverd. Tela **27**, 1528–1530 (1985) (in Russian). (Engl. transl.: Sov. Physics – Solid State **27**, 920–921 (1985))
60 L. V. Zueva, S. Z. Nazarova, V. N. Lipatnikov, A. I. Gusev: Effect of atom-vacancy ordering on magnetic susceptibility of nonstoichiometric titanium carbide, Doklady Akademii Nauk **365**, 82–86 (1999) (in Russian). (Engl. transl.: Doklady Phys. Chem. **365**, 65–69 (1999))
61 L. V. Zueva, V. N. Lipatnikov, A. I. Gusev: Magnetic susceptibility and ordering of titanium carbide, Zh. Fiz. Khimii **73**, 2163–2169 (1999) (in Russian). (Engl. transl.: Russ. J. Phys. Chem. **73**, 1952–1959 (1999))
62 A. N. Zyryanova, S. Z. Nazarova, A. I. Gusev: Magnetic susceptibility anomaly as evidence for ordering of nonstoichiometric hafnium carbide HfC_y , Doklady Akademii Nauk **359**, 348–353 (1998) (in Russian). (Engl. transl.: Doklady Phys. Chem. **359**, 91–96 (1998))
63 A. N. Zyryanova, A. I. Gusev: Magnetic susceptibility and ordering in nonstoichiometric hafnium carbide, Zh. Fiz. Khimii **72**, 2234–2242 (1998) (in Russian). (Engl. transl.: Russ. J. Phys. Chem. **72**, 2034–2041 (1998))
64 A. I. Gusev, A. N. Zyryanova: Atomic-vacancy ordering and magnetic susceptibility of nonstoichiometric hafnium carbide, Pis'ma v ZhETF **69**, 296–301 (1999) (in Russian). (Engl. transl.: JETP Letters **69**, 324–329 (1999))
65 A. I. Gusev, A. N. Zyryanova: Ordering and magnetic susceptibility of non-stoichiometric hafnium carbide, Phys. Stat. Sol. (a) **177**, 417–439 (2000)
66 V. N. Lipatnikov, A. I. Gusev, A. A. Rempel, G. P. Shveikin: Effect of structural transition on magnetic susceptibility of tantalum carbide, Doklady Akad. Nauk SSSR **297**, 849–853 (1987) (in Russian). (Engl. transl.: Sov. Physics Doklady **32**, 988–990 (1987))
67 V. N. Lipatnikov, A. A. Rempel, A. I. Gusev: The influence of ordering of the vacancies on the magnetic susceptibility of tantalum carbide, Zh. Neorgan. Khimii **33**,

1860–1863 (1988) (in Russian). (Engl. transl.: Russ. J. Inorg. Chem. **33**, 1058–1059 (1988))
68 V. N. Lipatnikov, A. I. Gusev: Magnetic susceptibility of a disordered and ordered TaC_y, in *Solid State Chemistry*, P. V. Geld (Ed.) (Ural Polytechnical Institute, Sverdlovsk 1988) pp. 144–149 (in Russian).
69 A. I. Gusev, A. A. Rempel, V. N. Lipatnikov: Magnetic susceptibility and atomic ordering in tantalum carbide, Phys. Stat. Sol. (a) **106**, 459–466 (1988)
70 Ya. S. Umanskii, V. I. Fadeeva: X-ray diffuse scattering by solid solution NbC–TaC, Fiz. Metall. Metalloved. **19**, 473–475 (1965) (in Russian)
71 Ya. S. Umanskii, V. I. Fadeeva: X-ray diffuse scattering by solid solution HfC–ZrC, Fiz. Metall. Metalloved. **20**, 719–722 (1965) (in Russian)
72 Ya. S. Umanskii, V. I. Fadeeva: Influence of WC impurity on short-range decomposition in solid solution TaC–NbC, Izv. AN SSSR. Neorgan. Materialy **2**, 82–86 (1966) (in Russian)
73 Ya. S. Umanskii, V. I. Fadeeva: Features of atomic structure of solid solutions Ta–Nb, Kristallografiya **11**, 196–199 (1966) (in Russian)
74 Ya. S. Umanskii, A. S. Muller: Study of short-range order in refractory compounds. Report 1, Izv. Vyssh. Ucheb. Zaved. Chernaya Metallurgiya No 7, 130–133 (1969) (in Russian)
75 Ya. S. Umanskii, A. S. Muller: Study of short-range order in refractory compounds. Report 2, Izv. Vyssh. Ucheb. Zaved. Chernaya Metallurgiya No 1, 128–131 (1970) (in Russian)
76 P. S. Rudman: X-ray diffuse-scattering study of the Nb–Ti b.c.c. solution, Acta Met. **12**, 1381–1388 (1964)
77 T. B. Gorbacheva, R. V. Rybal'chenko, N. P. Nechaeva, A. A. Katsnelson: Variation of short-range order in solid solutions (Ti,W)C as a function of their composition, Fiz. Metall. Metalloved. **39**, 768–774 (1975) (in Russian)
78 A. A. Katsnelson, T. B. Gorbacheva: Short-range order in solid solution TiC–WC, Fiz. Metall. Metalloved. **32**, 742–746 (1971) (in Russian)
79 A. A. Katsnelson, T. B. Gorbacheva, R. V. Rybal'chenko: Influence of some factors on short-range order in solid solution TiC–WC, Fiz. Metall. Metalloved. **34**, 211–213 (1972) (in Russian)
80 J. M. Rowe, A. Magerl: Neutron diffuse-scattering intensities in niobium, Phys. Rev. B **21**, 1706–1707 (1980)
81 M. A. Krivoglaz: *X-Ray and Neutron Diffraction in Non-Ideal Crystals* (Naukova Dumka, Kiev 1983) 408 pp. (in Russian)
82 M. Bee: *Quasielastic Neutron Scattering: Principles and Applications in Solid State Chemistry, Biology, and Materials Science* (Adam Hilger, Bristol and Philadelphia 1988) 437 pp.
83 B. E. Warren: *X-Ray Diffraction* (Dover Publications, New York 1990) 381 pp.
84 Shih-Lin Chang: *Multiple Diffraction of X-Rays in Crystals* (Springer, Berlin–Heidelberg–New York–Tokyo 1984) 300 pp.
85 I. I. Timofeeva, L. A. Klochkov: Variation of lattice periods and static distortions in homogeneity region of transition metal carbides, in *Refractory Carbides*, G. V. Samsonov (Ed.) (Naukova Dumka, Kiev 1970) pp. 143–147 (in Russian)
86 I. Karimov, V. T. Em, I. Khidirov, F. Faizullaev: Neutron diffraction determination of the Debye temperature and static distortions in carbides of groups IV and V transition metals, Fizika Metallov i Metallovedenie **44**, 184–186 (1977) (in Russian)

87 I. N. Frantsevich, A. B. Lyashchenko: Young's modulus of carbides of some transition metals, Poroshkovaya Metallurgiya No 7, 73–76 (1966) (in Russian)
88 I. Khidirov, I. Karimov, V. T. Em, V. E. Loryan: Neutron diffraction determination of mean-square atomic displacements and the Debye temperature in nitrides of group IV transition metals, Izv. AN Uzb. SSR. Seriya Fiz. Mat. Nauk No 6, 70–73 (1978) (in Russian)
89 I. N. Frantsevich, E. A. Zhurakovskii, A. B. Lyashchenko: Elastic constants and features of electronic structure of some groups of refractory compounds, Izv. AN SSSR. Neorgan. Materialy **3**, 8–12 (1967) (in Russian)
90 P. Capkova, R. Kuzel, J. Sedivy: Thermal vibrations and static displacements of atoms in nonstoichiometric titanium carbide, Phys. Stat. Sol. (a) **76**, 383–390 (1983)
91 V. Valdova, V. Sima, Z. Smetana, P. Capkova: Neutron powder diffraction study of Debye–Waller factor in TiC and TiN compounds and comparison with X-ray results, Phys. Stat. Sol. (a). **88**, 157–163 (1985)
92 A. Dunand, H. D. Flack, K. Yvon: Bonding study of TiC and TiN. I. High-precision x-ray-diffraction determination of the valence-electron density distribution, Debye-Waller temperature factors, and atomic static displacements in $TiC_{0.94}$ and $TiN_{0.99}$, Phys. Rev. B **31**, 2299–2315 (1985)
93 T. H. Metzger, J. Peisl, R. Kaufmann: X-ray determination of local atomic displacements around carbon vacancies in NbC_{1-c} single crystals, J. Phys. F: Metal Phys. **13**, 1103–1113 (1983)
94 R. Kaufmann, O. Meyer: Ion channeling in NbC_{1-c}: determination of local atomic displacements around carbon vacancies, Phys. Rev. B **28**, 6216–6226 (1983)
95 D. Lesueur: Deplacements atomiques autour d'une lacune de carbone dans UC, J. Nuclear Mater. **102**, 87–96 (1981)
96 R. Kaufmann, O. Meyer: Determination of static displacements around non-metal vacancies in NbN_{1-c} and TiC_{1-c} by channeling, Solid State Commun. **51**, 539–543 (1984)
97 M. Morinaga, K. Ohshima, J. Harada, S. Otani: X-ray determination of the atomic displacements in $NbC_{0.72}$, J. Appl. Crystallogr. **19**, 417–419 (1986)
98 K. Ohshima, J. Harada, M. Morinaga, P. Georgopoulos, J. B. Cohen: Distortion-induced scattering due to vacancies in $NbC_{0.72}$, Acta Crystallogr. **44**, 167–176 (1988)
99 A. A. Rempel, A. I. Gusev: Local displacements of atoms in disordered niobium carbide, Fiz. Tverd. Tela **27**, 3678–3680 (1985) (in Russian). (Engl. transl.: Sov. Physics – Solid State **27**, 2216–2217 (1985))
100 A. A. Rempel, A. I. Gusev, Yu. I. Zhdanov: Static crystal-lattice distortion in disordered niobium carbide, Izv. AN SSSR. Neorgan. Materialy **23**, 59–62 (1987) (in Russian). (Engl. transl.: Inorganic Materials **23**, 48–51 (1987))
101 A. I. Gusev, A. A. Rempel: Local static and dynamic atomic displacements in disordered niobium carbide, Phys. Stat. Sol. (b) **154**, 543–459 (1989)
102 R. W. James: *The Optical Principles of the Diffraction of X-rays* (Bell, London 1954) 624 pp.
103 A. A. Rempel, A. I. Gusev: Specific heat of niobium carbide in different structure states, Zh. Fiz. Khimii **62**, 2897–2901 (1988) (in Russian). (Engl. transl.: Russ. J. Phys. Chem. **62**, 1509–1511 (1988))
104 A. A. Rempel, A. I. Gusev: Heat capacity of niobium carbide in different structural states, Phys. Stat. Sol. (a) **113**,353–358 (1989)

105 A. I. Gusev, A. A. Rempel, V. N. Lipatnikov: Heat capacity of niobium and tantalum carbides NbC_y and TaC_y in disordered and ordered states below 300 K, Phys. Stat. Sol. (b) **194**, 467–482 (1996)
106 C. J. Sparks, B. Borie: Methods of analysis for diffuse X-ray scattering modulated by local order and atomic displacements, in *Local Atomic Arrangements Studied by X-ray Diffraction*, J. B. Cohen and J. E. Hillard (Eds), Proc. Symposium held in Chicago, February 15, 1965, Metallurgical Society Conferences **36** (Gordon and Breach Science Publishers, New York 1966) pp. 5–50
107 L. V. Zueva, A. I. Gusev: Effect of nonstoichiometry and ordering on the period of the basis structure of cubic titanium carbide, Fiz. Tverd. Tela **41**, 1134–1141 (1999) (in Russian). (Engl. transl.: Physics of the Solid State **41**, 1032–1038 (1999)
108 V. Moisy-Maurice, C. H. de Novion: An application of Ti-*K* X-ray absorption edges and fine structures to the study of substoichiometric titanium carbide TiC_{1-x}, J. de Physique. France. **49**, 1737–1751 (1988)
109 D. Rafaja, W. Lengauer, P. Ettmayer, V. N. Lipatnikov: Rietveld analysis of the ordering in V_8C_7, J. Alloys Comp. **269**, 60–62 (1998)

5. Long-Range Order

Crystals represent the highest degree of spatial order. From the physical viewpoint, a crystal is an ensemble of an infinite number of atoms that are uniformly packed in regular rows and planes and fill the entire space. In mathematical terms this means that a crystal lattice is invariant under symmetry transformations comprising the space group of the crystal. Symmetry and space groups of crystals were comprehensively discussed in Chap. 2.

Long-range order is often referred to when describing a crystalline solid. What is meant here is a one-to-one correspondence between the value of some physical quantity at an arbitrary point in space and the value of the same quantity at any other point in a crystal including infinitely distant points. This correspondence directly results from the periodicity of the crystal lattice. Full long-range order implies an absolutely regular distribution of atoms at crystal lattice sites.

Any disorder disturbs the crystal symmetry and decreases the degree of long-range order. The simplest type of disorder is realized in ideal binary substitutional solid solutions in which atoms of one species are replaced randomly by atoms of another species without distorting the crystal lattice. Thus, a model of ideal solutions also implies that the occupancy of a site by an atom of a particular species does not depend on which atoms fill neighboring or distant sites. Statistical disorder in a crystalline solid can be realized only at high temperature, when the interatomic interaction forces are small compared to the energy of thermal vibrations and entropy makes a relatively large contribution to the free energy. At low temperature entropy contributes little to the free energy and like and unlike atoms adopt an ordered distribution due to the difference in their interaction energies.

5.1 Long-Range Order Parameters

In a disordered nonstoichiometric compound vacancies and interstitial atoms form a substitutional solution on the nonmetal sublattice. Ordering of interstitial atoms and structural vacancies in the crystal lattice of strongly nonstoichiometric compounds can be characterized by the fraction of sites occupied by atoms and vacancies in nonequivalent sublattices of an ordered crystal. In an ordered solid solution each nonequivalent sublattice is filled mainly by atoms of its "inherent" or "proper" species. Consequently, the translational symmetry of a crystal changes, because the main translation vectors of an ordered solution are several times longer than the main translation vectors of a disordered solution. This distribution of atoms in a solid solution is called long-range order. If translation vectors of the direct lattice increase n times, the main translation vectors of the reciprocal lattice decrease in accordance with (2.18) the same number of times and superstructure sites appear in the reciprocal lattice. These sites correspond to superstructure reflections in a diffraction experiment. For this reason diffraction measurements (X-ray, neutron and electron diffraction) serve as the principal experimental technique for an examination of long-range order.

If long-range order is present, the distribution of atoms of different species in one of the unit cells is replicated in all unit cells of an ordered crystal. The long-range order parameter η is a quantitative measure of long-range order.

The long-range order parameter can be determined by several methods [1–6].

Let us consider a binary solid solution AB with a crystal lattice containing N sites. During ordering the crystal lattice is divided into two nonequivalent sublattices having $N^{(1)}$ and $N^{(2)}$ sites each. In this case the relative fraction of sites of the first sublattice is $\nu = N^{(1)}/N$ and the relative concentration of atoms of species A in the solid solution under consideration is $c = N_A/N$, where N_A denotes the number of lattice sites occupied by atoms A. The relative concentration of atoms B is $c_B = 1 - c$. In an ideal fully ordered solid solution, when $c = \nu$, sites of the first sublattice are occupied only by atoms A and sites of the second sublattice only by atoms B. In the general case, when complete long-range order is not realized or when $c \neq \nu$, there is no unique correspondence between an atomic species and the location of the site occupied by this atom in a particular sublattice. Let $N_A^{(1)}$, $N_B^{(1)}$, $N_A^{(2)}$ and $N_B^{(2)}$ be the number of atoms A and B occupying positions in the first and second sublattices. Then the probability that positions in these sublattices are occupied by atoms of different species can be written as

$$P_\tau^{(i)} = N_\tau^{(i)} / N^{(i)}, \tag{5.1}$$

where $N_\tau^{(i)}$ is the number of atoms of species τ that occupy positions in the i-th sublattice containing $N^{(i)}$ sites. The usual normalization condition

$$\sum_\tau P_\tau^{(i)} = 1 \tag{5.2}$$

is fulfilled for probabilities $P_\tau^{(i)}$ corresponding to one and the same i-th sublattice. The probabilities that positions in different sublattices are occupied by atoms of one and the same species τ are connected to the relative concentration c_τ of atoms of this species by the condition of normalization with respect to composition:

$$\sum_\tau \nu^{(i)} P_\tau^{(i)} = c_\tau \; . \tag{5.3}$$

For a binary solid solution conditions (5.2) and (5.3) can be written as

$$P_A^{(1)} + P_B^{(1)} = 1 ; \quad P_A^{(2)} + P_B^{(2)} = 1 ; \tag{5.4}$$

$$\nu P_A^{(1)} + (1-\nu) P_A^{(2)} = c ; \quad \nu P_B^{(1)} + (1-\nu) P_B^{(2)} = 1 - c \, . \tag{5.5}$$

The long-range order parameter is a quantity showing the extent to which the probability that atoms of a given species occupy sites of their "native" sublattice is greater than the probability that the same atoms fill sites in other sublattices. For a binary solution the long-range order parameter is

$$\eta = P_{A}^{(1)} - P_{A}^{(2)} . \quad (5.6)$$

By finding $P_{A}^{(2)}$ from condition (5.5) and substituting it into (5.6), we obtain a popular formula for calculating the long-range order parameter:

$$\eta = \frac{P_{A}^{(1)} - c}{1-\nu} . \quad (5.7)$$

Bragg and Williams [1, 2] were the first to present the long-range order parameter in this form. Expressions (5.6) and (5.7) suggest that η reaches a minimum value when atoms of all species are distributed statistically over the sites of all sublattices, i.e. $P_{\tau}^{(i)} = c_{\tau}$ for all i and τ. If the degree of long-range order η, the solid solution composition c, and the crystal lattice structure (the fraction of sites in the first sublattice, ν) are known, it is possible from (5.4), (5.5) and (5.7) to find the probabilities that lattice sites are occupied by atoms A and B:

$$P_{A}^{(1)} = c + (1-\nu)/\eta , \quad (5.8)$$

$$p_{A}^{(2)} = c - \nu\eta . \quad (5.9)$$

In accordance with (5.7), the long-range order parameter is proportional to the deviation of the probability $P_{A}^{(1)}$ from its value c in a disordered solid solution. This deviation is zero for disordered solid solutions of any composition. Therefore the parameter η can change from zero to some maximum value η_{max} during ordering. The dependence of η_{max} on the solid solution composition can be found from (5.8) and (5.9).

If the concentration of atoms A is larger than the relative fraction of sites in the first sublattice ($c \geq \nu$), all sites in the first sublattice of an ordered solid solution will be occupied by atoms A only, i.e. $P_{A}^{(1)} = 1$. In this case, it follows from (5.3) that

$$\eta_{max} = (1 - c)/(1 - \nu) \text{ at } c \geq \nu . \quad (5.10)$$

If $c \leq \nu$ and the order is at its maximum, all atoms A will only be located at sites of the first sublattice and the probability that these atoms occupy sites in the second sublattice is zero, i.e. $P_{A}^{(2)} = 0$. Therefore, from (5.9) we have

$$\eta_{max} = c/\nu \text{ at } c \leq \nu . \quad (5.11)$$

Thus, (5.10) and (5.11) determine the dependence of η_{max} on the composition of an ordered solid solution. The limiting value of the long-range order parameter $\eta = 1$ can be achieved only for a stoichiometric solid solution, when $c = \nu$. As examples, Fig. 5.1 shows $\eta_{max}(y)$ for ordered phases of type M_8C_7 ($\nu = 7/8$), M_6C_5 ($\nu = 5/6$), M_3C_2 ($\nu = 2/3$), and M_2C ($\nu = 1/2$), which form in nonstoichiometric carbides MC_y ($y \equiv c$).

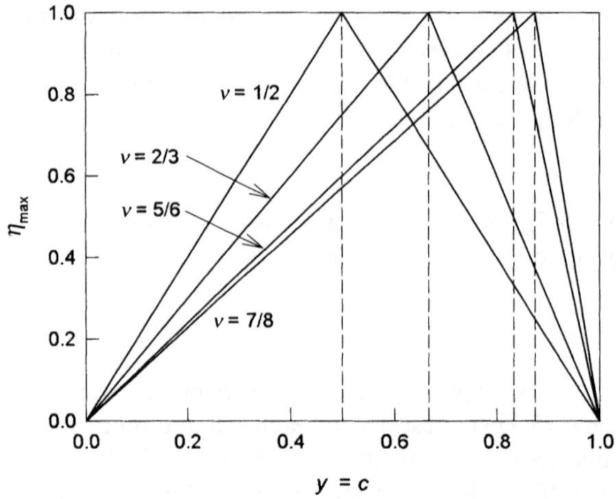

Fig. 5.1. Maximum value of the long-range order parameter η_{max} vs. composition of the nonstoichiometric carbide MC_y with superstructures types M_8C_7 ($\nu = 7/8$), M_6C_5 ($\nu = 5/6$), M_3C_2 ($\nu = 2/3$), and M_2C ($\nu = 1/2$). The dependence is plotted neglecting limited concentration intervals for the ordered phases

Deviations of the degree of long-range order from the maximum value are given by the reduced long-range order parameter $\eta_{red} = \eta/\eta_{max}$. The η_{red} value changes from 0 (in a disordered solid solution) to unity (in a solid solution with the maximum degree of order possible for the given composition) in a solid solution of any composition. Using (5.7) for η and (5.10) or (5.11) for η_{max}, one finds the value of the reduced long-range order parameter:

$$\eta_{red} = \frac{P_A^{(1)} - c}{1 - c} \quad \text{when } c \geq \nu, \tag{5.12}$$

$$\eta_{red} = \frac{\nu(P_A^{(1)} - c)}{c(1 - \nu)} \quad \text{when } c \leq \nu. \tag{5.13}$$

Values of η and η_{red} coincide when $c = \nu$.

An ordered distribution of atoms of two species over two nonequivalent sublattices represents the simplest case of ordering, which is described by one long-range order parameter. Several order parameters are needed to describe the long-range order in systems with more than two species of ordering elements or more than two nonequivalent sublattices. In the general case the number r of long-range order parameters required to describe a system, in which n components are redistributed over p nonequivalent sublattices, is equal to

$$r = (p - 1)(n - 1)\,. \tag{5.14}$$

The most frequent version of ordering in strongly nonstoichiometric interstitial compounds is the distribution of interstitial atoms and structural vacancies on more than two nonequivalent sublattices. The number of long-range order parameters necessary for the description of this ordering is one less than the number of nonequivalent sublattices. Let us consider a typical case of ordering of interstitial atoms and structural vacancies on i nonequivalent sublattices when vacancies form only one nonequivalent sublattice numbered j (the j value is fixed). In this case the long-range order parameters are equal to

$$\eta_{i\neq j} = P_{\mathrm{A}}^{(i)} - P_{\mathrm{A}}^{(j)} \tag{5.15}$$

or, taking into account condition (5.3),

$$\eta_{i\neq j} = (\nu^{(j)} P_{\mathrm{A}}^{(i)} - c + \sum_{i\neq j} \nu^{(i)} P_{A}^{(i)}) / \nu^{(j)} \,. \tag{5.16}$$

If the concentration of vacancies is smaller than the relative fraction of sites in the vacancy sublattice of an ordered structure and ordering is fully developed, then all sites in sublattices $i \neq j$ will be occupied by interstitial atoms. In this case $P_{\mathrm{A}}^{(i\neq j)} = 1$ and the maximum value of the long-range order parameter becomes

$$\eta_{i\neq j}^{\max} = \frac{1-c}{\nu^{(j)}} \quad \text{with } c \geq \sum_{i\neq j} \nu^{(i)} \,. \tag{5.17}$$

The interstitial atoms will be located only at sites of sublattices $i \neq j$ in a fully ordered state if the concentration of vacancies is higher than $\nu^{(j)}$. In this case the maximum value of the long-range order parameter is

$$\eta_{i\neq j}^{\max} = \frac{c}{\sum_{i\neq j} \nu^{(i)}} \quad \text{when } c \leq \sum_{i\neq j} \nu^{(i)} \,. \tag{5.18}$$

The minimum value of the long-range order parameter, $\eta_{i\neq j}^{\min} = 0$, is achieved when atoms and vacancies are distributed statistically on the lattice, i.e. when $P_{\mathrm{A}}^{(i)} = c$. From (5.17) and (5.18) it follows that all long-range order parameters become equal when the order is fully developed.

In the description within the method of static concentration waves [6], the long-range order parameter is introduced through the distribution function $n(\boldsymbol{r})$ of atoms at sites $\boldsymbol{r}$ of an ordered crystal lattice. The function $n(\boldsymbol{r})$ possesses the symmetry of an ordered crystal lattice and represents the probability that an atom of a given species is found at a site $\boldsymbol{r}$. In a disordered crystal the probabilities $n(\boldsymbol{r})$ are similar for all sites and coincide with the fraction of lattice sites occupied by atoms of a given species. For example, the distribution function $n(\boldsymbol{r})$ for atoms A or X in a disordered alloy A_yB_{1-y} or a disordered nonstoichiometric compound $MX_y\square_{1-y}$ is equal to the fraction of lattice sites occupied by these atoms, i.e. $n(\boldsymbol{r}) = y$.

A homogeneous distribution of atoms at crystal lattice sites in a disordered compound is subject to space-periodic modulation during the disorder–order transition. In other words, an ordered phase is formed. The space-periodic modulation of the atomic distribution (deviation of the probability $n(\boldsymbol{r})$ from its value in the case of a statistical distribution) can be represented by the superposition of several plane concentration waves [6]. The vectors of these waves are superstructure vectors $\boldsymbol{k}_s^{(j)}$ of the reciprocal lattice. They are located in the first Brillouin zone of the disordered real space lattice. In terms of the method of static concentration waves [6], the distribution function of atoms of a given species is described by the fraction of sites occupied by these atoms and the superposition of static concentration waves:

$$n(\boldsymbol{r}) = y + \sum_s \sum_{j \in s} \Delta(\boldsymbol{k}_s^{(j)}, \boldsymbol{r}) \ , \tag{5.19}$$

where

$$\Delta(\boldsymbol{k}_s^{(j)}, \boldsymbol{r}) = \tfrac{1}{2}[Q(\boldsymbol{k}_s^{(j)})\exp(i\boldsymbol{k}_s^{(j)}\boldsymbol{r}) + Q^*(\boldsymbol{k}_s^{(j)})\exp(-i\boldsymbol{k}_s^{(j)}\boldsymbol{r})] \tag{5.20}$$

is a standing static plane concentration wave generated by the superstructure vector $\boldsymbol{k}_s^{(j)}$ of the star $\{\boldsymbol{k}_s\}$. In (5.20) $Q(\boldsymbol{k}_s^{(j)})$ denotes the concentration wave amplitude.

Since the distribution function $n(\boldsymbol{r})$ is the probability that an atom occupies a given site, the second term $\sum_s \sum_{j \in s} \Delta(\boldsymbol{k}_s^{(j)}, \boldsymbol{r}) = \Delta_{\boldsymbol{k}}(\boldsymbol{r})$ in (5.19) represents the modulating part of the probability. This term describes the deviation of $n(\boldsymbol{r})$ from the probability that atoms of a given species are found in a disordered crystal, i.e. from the concentration y. Obviously, in alloys or compounds having a certain degree of order the quantity $\Delta_k(\boldsymbol{r})$ is constant in a sublattice, but changes from one sublattice to another. If $n(1)$, $n(2)$, ..., $n(p)$ are the probabilities of finding an atom A (or X) in the first, second, …, p-th sublattices, then

$$\left.\begin{array}{l} \Delta_{\boldsymbol{k}}(1) = n(1) - c, \\ \Delta_{\boldsymbol{k}}(2) = n(2) - c \\ \ldots\ldots\ldots \\ \Delta_{\boldsymbol{k}}(p) = n(p) - c \end{array}\right\} . \tag{5.21}$$

Formulas (5.20) and (5.21) suggest that the deviation $\Delta_k(\boldsymbol{r})$ of the probability $n(\mathbf{r})$ from its average value y is fully determined by the amplitude $Q(\boldsymbol{k}_s^{(j)})$ of the corresponding concentration wave with the wave vector $\boldsymbol{k}_s^{(j)}$ of the star $\{\boldsymbol{k}_s\}$.

Thus, the set of independent amplitudes $Q(\boldsymbol{k}_s^{(j)})$ fully determines an ordered state, where the amplitudes represent the long-range order parameters. However, not the amplitude itself but the quantity η_s, which is proportional to the amplitude, is more conveniently used as the long-range order parameter:

$$Q(\boldsymbol{k}_s^{(j)}) = \eta_s \gamma(\boldsymbol{k}_s^{(j)}) \ , \tag{5.22}$$

where $\gamma(\boldsymbol{k}_s^{(j)}) = \gamma_s \exp(i\varphi_s^{(j)})$ is a coefficient allowing for the crystal symmetry. This coefficient is chosen such that long-range order parameters equal to unity correspond to a fully ordered stoichiometric ($y = \nu$) crystal.

5.2 Disorder-Order Phase Transition Channel[1]

With (5.22), formula (5.20) can be rewritten as

$$\Delta(\boldsymbol{k}_s^{(j)},\boldsymbol{r}) = \tfrac{1}{2}\eta_s[\gamma(\boldsymbol{k}_s^{(j)})\exp(i\boldsymbol{k}_s^{(j)}\boldsymbol{r}) + \gamma^*(\boldsymbol{k}_s^{(j)})\exp(-i\boldsymbol{k}_s^{(j)}\boldsymbol{r})] \;, \tag{5.23}$$

where $\gamma(\boldsymbol{k}_s^{(j)}) = \gamma_s \exp(i\varphi_s^{(j)})$. The physical sense of $\varphi_s^{(j)}$ can be comprehended if a concentration wave is presented in trigonometric form:

$$\Delta(\boldsymbol{k}_s^{(j)},\boldsymbol{r}) = \eta_s\gamma_s \cos(\varphi_s^{(j)} + \boldsymbol{k}_s^{(j)}\boldsymbol{r}) \;, \tag{5.24}$$

where $\eta_s\gamma_s$ and $\varphi_s^{(j)}$ stand for the amplitude and the phase shift of the concentration wave respectively. A concentration wave with the opposite wave vector $-\boldsymbol{k}_s^{(j)}$ has an opposite phase shift equal to $-\varphi_s^{(j)}$.

From (5.19) and (5.23) it follows that the long-range order parameter η_s corresponds to the star $\{\boldsymbol{k}_s\}$ of the wave vector. The distribution function at sites $\boldsymbol{r}$ located in crystallographically equivalent positions has an identical value, because it is invariant under rotation and mirror transformations comprising a point symmetry group of an ordered crystal. The number of values the distribution function can assume at sites of an ordered lattice exceeds the number of long-range order parameters by one.

In accordance with [6], the summation in formula (5.19) is to be taken over all superstructure vectors in the first Brillouin zone of a disordered lattice. However, they include equivalent vectors, which differ only by the structural vector $\boldsymbol{H}$ of the reciprocal lattice. Such vectors are the superstructure vector $\boldsymbol{k}_s^{(j)}$ and the equivalent superstructure vector $\boldsymbol{k}_s^{(j)\mathrm{eq}} = (\boldsymbol{k}_s^{(j)} + \boldsymbol{H})$, which is also located in the first Brillouin zone. It is worth noting that mutually equivalent vectors of the first Brillouin zone always reach the zone boundaries. Translation vectors of the direct and reciprocal lattices, $\boldsymbol{r}$ and $\boldsymbol{H}$, by definition are related by the condition $\boldsymbol{rH} = 2\pi m$ (m being an integer). Therefore, if M equivalent superstructure vectors of the first Brillouin zone are considered instead of one superstructure vector $\boldsymbol{k}_s^{(j)}$, then (5.19) allows for M similar concentration waves $\Delta(\boldsymbol{k}_s^{(j)},\boldsymbol{r})$ whose amplitudes should be summed. However, from the diffraction pattern it is seen that equivalent superstructure vectors $\boldsymbol{k}_s^{(j)}$ and $(\boldsymbol{k}_s^{(j)} + \boldsymbol{H})$ in a three-dimensional reciprocal space correspond to diffraction reflections in different positions, whose intensities are not summed. Indeed, upon ordering, each site of the reciprocal lattice, which is the center of its first Brillouin zone, generates a set of structural sites in the zone. Therefore mutually equivalent superstructure vectors $\boldsymbol{k}_s^{(j)}$ and $(\boldsymbol{k}_s^{(j)} + \boldsymbol{H})$ at the boundary of the Brillouin zone belong to neighboring first Brillouin zones: the site $\boldsymbol{k}_s^{(j)}$ belongs to the first Brillouin zone constructed around a zero structural site $\boldsymbol{H} = 0$ of the reciprocal lattice, while the site $(\boldsymbol{k}_s^{(j)} + \boldsymbol{H})$ belongs to the first Brillouin zone constructed around a nonzero structural site $\boldsymbol{H} \neq 0$. To avoid possible errors, the summation in (5.19) should be taken only over nonequivalent superstructure vectors of the first Brillouin zone.

[1] The content of review article [7] with some changes and additions is given in Sect. 5.2–5.6.

A channel of structural disorder–order phase transition is formed by the whole set of all nonequivalent superstructure vectors $\boldsymbol{k}_s^{(j)}$ of a reciprocal lattice, which refer to a nearest structural site of this reciprocal lattice, are located in the first Brillouin zone and represent rays of stars $\{\boldsymbol{k}_s\}$ associated with the given transformation. This definition of the phase transition channel was formulated by the authors [7] on the basis of earlier studies [8–12]. The term "transition channel" was proposed in [13]. The concept of a phase transition channel is significant for constructing diffraction patterns of ordered crystals and ensures a correct and unique determination of γ_s and $\varphi_s^{(j)}$ in the distribution function $n(\boldsymbol{r})$.

Thus, summation in the distribution function (5.19) is taken over all superstructure vectors comprising a disorder–order phase transition channel.

If ordering takes place in a Bravais lattice, each superstructure vector $\boldsymbol{k}_s^{(j)}$ in the first Brillouin zone has an opposite superstructure vector $-\boldsymbol{k}_s^{(j)}$. The vector $-\boldsymbol{k}_s^{(j)}$ is included in a phase transition channel if it is not equivalent to the vector $\boldsymbol{k}_s^{(j)}$, i.e. if the condition $[\boldsymbol{k}_s^{(j)} - (-\boldsymbol{k}_s^{(j)})] = 2\boldsymbol{k}_s^{(j)} \neq \boldsymbol{H}$ is met. In contrast, the opposite superstructure vector $-\boldsymbol{k}_s^{(j)}$ is not included in a phase transition channel if and only if the double superstructure vector $2\,\boldsymbol{k}_s^{(j)}$ is equal to the structural vector $\boldsymbol{H}$ of the reciprocal lattice. These vectors are the rays of stars $\{\boldsymbol{k}_9\}$ and $\{\boldsymbol{k}_{10}\}$ for ordering in an fcc lattice, the rays of stars $\{\boldsymbol{k}_9\}$ and $\{\boldsymbol{k}_{12}\}$ for ordering in a bcc lattice, and those of the star $\{\boldsymbol{k}_8\}$ when $\mu = 1/2$.

Nonequivalent superstructure vectors $\boldsymbol{k}_s^{(j)}$ and $-\boldsymbol{k}_s^{(j)}$ produce two similar concentration waves. Therefore the terms corresponding to these waves can be included in (5.19) and (5.23) and the function $n(\boldsymbol{r})$ becomes

$$n(\boldsymbol{r}) = y + \tfrac{1}{2}\sum_s \eta_s \sum_{j\in s}[2 - \sum_{\boldsymbol{H}} \delta(2\boldsymbol{k}_s^{(j)} - \boldsymbol{H})]$$
$$\times\,[\gamma(\boldsymbol{k}_s^{(j)})\exp(i\boldsymbol{k}_s^{(j)}\boldsymbol{r}) + \gamma^*(\boldsymbol{k}_s^{(j)})\exp(-i\boldsymbol{k}_s^{(j)}\boldsymbol{r})]\ , \tag{5.25}$$

where

$$\delta(2\boldsymbol{k}_s^{(j)} - \boldsymbol{H}) = \begin{cases} 1, & \text{if } 2\boldsymbol{k}_s^{(j)} = \boldsymbol{H}\,, \\ 0, & \text{if } 2\boldsymbol{k}_s^{(j)} \neq \boldsymbol{H}. \end{cases}$$

It is readily seen that the sum $\sum_{\boldsymbol{H}} \delta(2\boldsymbol{k}_s^{(j)} - \boldsymbol{H})$ is also equal to zero or unity at the whole multitude of sites $\{\boldsymbol{H}\}$ of the reciprocal lattice. The δ-function $\delta(2\boldsymbol{k}_s^{(j)} - \boldsymbol{H})$ allows for the case when the opposite superstructure vector $-\boldsymbol{k}_s^{(j)}$ is equivalent and is not included in a phase transition channel. The summation over s and j in (5.25) means summation over all vectors of a phase transition channel, except opposite vectors $-\boldsymbol{k}_s^{(j)}$, i.e. summation over a truncated phase transition channel.

Let us rearrange the distribution function (5.25) to trigonometric form, which is actually used for a description of specific superstructures:

$$n(\boldsymbol{r}) = y + \sum_s \eta_s \sum_{j\in s}[A_{sj}\cos(\boldsymbol{k}_s^{(j)}\boldsymbol{r}) + B_{sj}\sin(\boldsymbol{k}_s^{(j)}\boldsymbol{r})]\ , \tag{5.26}$$

where

$$A_{sj} = [2 - \sum_{H} \delta(2\mathbf{k}_s^{(j)} - \mathbf{H})]\gamma_s \cos\varphi_s^{(j)}$$

and

$$B_{sj} = -[2 - \sum_{H} \delta(2\mathbf{k}_s^{(j)} - \mathbf{H})]\gamma_s \sin\varphi_s^{(j)} .$$

Knowing numerical values of the coefficients A_{sj} and B_{sj}, i.e. the explicit form of the distribution function $n(\mathbf{r})$, it is possible to find the parameters γ_s and $\varphi_s^{(j)}$:

$$\gamma_s = [2 - \sum_{H} \delta(2\mathbf{k}_s^{(j)} - \mathbf{H})]^{-1} \left|(A_{sj}^2 + B_{sj}^2)^{1/2}\right| , \tag{5.27}$$

$$\varphi_s^{(j)} = -\arctan\left(\frac{B_{sj}}{A_{sj}}\right) + \frac{\pi}{2}\left(1 - \frac{A_{sj}}{|A_{sj}|}\right) + \frac{\pi}{2}\left(1 + \frac{A_{sj}}{|A_{sj}|}\right)\left(1 + \frac{B_{sj}}{|B_{sj}|}\right) . \tag{5.28}$$

A nonmetal fcc sublattice is an ordering Ising lattice in nonstoichiometric compounds with the $B1$ structure. The lattice vectors determining the positions of sites $\mathbf{r}$ in the nonmetal sublattice have the form

$$\mathbf{r} = x_I \mathbf{a}_1 + y_I \mathbf{a}_2 + z_I \mathbf{a}_3 , \tag{5.29}$$

where $\mathbf{a}_1$, $\mathbf{a}_2$, and $\mathbf{a}_3$ are the main translations of the fcc lattice in directions $[100]_{B1}$, $[010]_{B1}$, and $[001]_{B1}$ (with $|\mathbf{a}_1| = |\mathbf{a}_2| = |\mathbf{a}_3| = a$); and x_I, y_I, and z_I are the coordinates (integers and half-integers) of sites in the nonmetal fcc sublattice. The basis vectors of the reciprocal lattice, which correspond to the direct fcc lattice, are equal to $\mathbf{b}_1 = \{\bar{1}11\}$, $\mathbf{b}_2 = \{1\bar{1}1\}$, and $\mathbf{b}_3 = \{11\bar{1}\}$ in $2\pi/a_{B1}$ units.

Below we shall describe a method for calculating the distribution function. As an example, we consider ordering of the M_6C_5 type in niobium carbide NbC_y.

Calculation of the Distribution Function. The ordered monoclinic phase of niobium carbide Nb_6C_5 belongs to the space group $C2/m$ (C_{2h}^3) [10, 11]. The point symmetry group $2/m$ (C_{2h}) includes four symmetry elements h_1 , h_4 , h_{25}, and h_{28} (see Table 2.2). The point group $m\bar{3}m$ (O_h) of the disordered phase of niobium carbide has 48 symmetry elements h_1–h_{48} and therefore the symmetry decreases to 12 upon rotation. A decrease in the translational symmetry corresponds to an increase of the volume of the unit cell and of the number of sites in this cell upon the transition from a disordered state to an ordered state. The symmetry is equal to 3 upon the transformation NbC_y (space group $Fm\bar{3}m$) → Nb_6C_5 (space group $C2/m$). The total decrease of symmetry upon ordering equals the product of the rotational decrease with the decrease in translational symmetry, i.e. $N = n(G)/n(G_D) = 36$, where $n(G)$ and $n(G_D)$ denote the order of the space groups G and G_D of the high-symmetry disordered and low-symmetry (disymmetric) ordered phases.

To calculate the distribution function, one needs to determine the transition channel, i.e. find star vectors $\mathbf{k}_s^{(j)}$ related to the phase transformation. To this end, it is necessary to calculate vectors of the reciprocal lattice from (2.14) using translation vectors of unit cell of the ordered phase. In accordance with [10, 11], unit cell of the ordered niobium carbide phase Nb_6C_5 has translation vectors $\boldsymbol{a} = \{1/2, -1/2, -1\}_{B1}$, $\boldsymbol{b} = \{3/2, 3/2, 0\}_{B1}$, and $\boldsymbol{c} = \{1/2, -1/2, 1\}_{B1}$. With (2.14), the reciprocal lattice vectors become $\boldsymbol{a}^* = \{1/2, -1/2, -1/2\}$, $\boldsymbol{b}^* = \{-1/3, -1/3, 0\}$, and $\boldsymbol{c}^* = \{1/2, -1/2, 0\}$.

Translation of the three superstructural sites of the reciprocal lattice to the vector $\boldsymbol{r} = n_1\boldsymbol{a}^* + n_2\boldsymbol{b}^* + n_3\boldsymbol{c}^*$ shows that the first Brillouin zone of the disordered nonmetallic fcc sublattice of niobium carbide includes eight superstructure vectors of three stars $\{\boldsymbol{k}_9\}$, $\{\boldsymbol{k}_4\}$, and $\{\boldsymbol{k}_3\}$. Six vectors are pairwise equivalent, i.e. differ by a structural vector of the reciprocal lattice. Therefore, among each pair of mutually equivalent vectors, only one vector is included in the transition channel and the order–disorder NbC_y (space group $Fm\bar{3}m$) $\rightarrow$ Nb_6C_5 (space group $C2/m$) transition channel has five nonequivalent vectors $\boldsymbol{k}_9^{(3)} = \boldsymbol{b}_2/2$, $\boldsymbol{k}_4^{(1)} = (\boldsymbol{b}_1 + \boldsymbol{b}_2 + 2\boldsymbol{b}_3)/3$, $\boldsymbol{k}_4^{(2)} = -\boldsymbol{k}_4^{(1)}$, $\boldsymbol{k}_3^{(3)} = -(4\boldsymbol{b}_1 + \boldsymbol{b}_2 + 2\boldsymbol{b}_3)/6$, and $\boldsymbol{k}_3^{(4)} = -\boldsymbol{k}_3^{(3)}$ belonging to three stars $\{\boldsymbol{k}_9\}$, $\{\boldsymbol{k}_4\}$, and $\{\boldsymbol{k}_3\}$. Since each wave-vector star has its corresponding long-range order parameter, the phase transformation in nonstoichiometric niobium carbide is described by three long-range order parameters η_9, η_4, and η_3.

To solve (5.19) taking into account (5.22), it is necessary to find not only the long-range order parameters η_s corresponding to stars $\{\boldsymbol{k}_s\}$ but also the coefficients $\gamma(\boldsymbol{k}_s^{(j)}) = \gamma_s \exp(i\varphi_s^{(j)}) \equiv \gamma_s^{(j)}$.

The translation vector $\boldsymbol{r}$ in the ordering nonmetallic fcc sublattice of niobium carbide has the form (5.29), where x_I, y_I, and z_I are coordinates (integers and half-integers) of sites in the nonmetallic fcc sublattice. Taking into account relations (5.19), (5.22) and (5.29), and considering superstructure vectors that comprise the transition channel, the distribution function for carbon atoms in ordered monoclinic (space group $C2/m$) carbide Nb_6C_5 can be written as

$$\begin{aligned}
n(x_\mathrm{I}, y_\mathrm{I}, z_\mathrm{I}) = y + \tfrac{1}{2}\eta_9\gamma_9 \{&\exp(i\varphi_9^{(3)}) \exp[-i\pi(x_\mathrm{I} - y_\mathrm{I} + z_\mathrm{I})] \\
&+ \exp(-i\varphi_9^{(3)}) \exp[i\pi(x_\mathrm{I} - y_\mathrm{I} + z_\mathrm{I})]\} \\
&+ \tfrac{1}{2}\eta_4\gamma_4 \{\exp(i\varphi_4^{(1)}) \exp[-i4\pi(x_\mathrm{I} + y_\mathrm{I})/3] \\
&+ \exp(-i\varphi_4^{(1)}) \exp[i4\pi(x_\mathrm{I} + y_\mathrm{I})/3] \\
&+ \exp(i\varphi_4^{(2)}) \exp[i4\pi(x_\mathrm{I} + y_\mathrm{I})/3] \\
&+ \exp(-i\varphi_4^{(2)}) \exp[-i4\pi(x_\mathrm{I} + y_\mathrm{I})/3]\} \\
&+ \tfrac{1}{2}\eta_3\gamma_3 \{\exp(i\varphi_3^{(3)}) \exp[-i\pi(x_\mathrm{I} - 5y_\mathrm{I} - 3z_\mathrm{I})/3] \\
&+ \exp(-i\varphi_3^{(3)}) \exp[i\pi(x_\mathrm{I} - 5y_\mathrm{I} - 3z_\mathrm{I})/3] \\
&+ \exp(i\varphi_3^{(4)}) \exp[-i\pi(-x_\mathrm{I} + 5y_\mathrm{I} + 3z_\mathrm{I})/3] \\
&+ \exp(-i\varphi_3^{(4)}) \exp[i\pi(-x_\mathrm{I} + 5y_\mathrm{I} + 3z_\mathrm{I})/3]\}\,. \qquad (5.30)
\end{aligned}$$

Substituting $\gamma_s^{(j)}$ for $\gamma_s \exp(i\varphi_s^{(j)})$ in (5.30) and writing (5.30) in trigonometric notation, we have

$$n(x_{\mathrm{I}}, y_{\mathrm{I}}, z_{\mathrm{I}}) = y + \eta_9 \gamma_9^{(3)} \cos[\pi(x_{\mathrm{I}} - y_{\mathrm{I}} + z_{\mathrm{I}})]$$
$$+\eta_4 \{\gamma_4^{(1)} \cos[4\pi(x_{\mathrm{I}} + y_{\mathrm{I}})/3] + \gamma_4^{(2)} \sin[4\pi(x_{\mathrm{I}} + y_{\mathrm{I}})/3]\}$$
$$+\eta_3 \{\gamma_3^{(3)} \cos[\pi(x_{\mathrm{I}} - 5y_{\mathrm{I}} - 3z_{\mathrm{I}})/3]$$
$$+\gamma_3^{(4)} \sin[\pi(x_{\mathrm{I}} - 5y_{\mathrm{I}} - 3z_{\mathrm{I}})/3]\} \,. \quad (5.31)$$

The coefficients $\gamma_s^{(j)}$ are chosen such that a fully ordered state of the crystal corresponds to long-range order parameters equal to unity, i.e. $\eta_9 = \eta_4 = \eta_3 = 1$. Subject to this condition, the distribution function for carbon atoms will take only two values for the set of all sites in the ordering nonmetal fcc sublattice: $n_1 = 0$ at vacant sites and $n_2 = 1$ at sites occupied by carbon atoms. To determine the coefficients $\gamma_s^{(j)}$, it suffices to calculate values of the distribution function (5.31) at 12 sites of the nonmetal sublattice, which form a unit cell of the ordered phase. Actually, one needs to consider values of the distribution function for six sites only. They include a vacant site in position (*a*), a site in position (*d*) occupied by a carbon atom, and two atoms in each of the positions (*g*) and (*h*) occupied by carbon atoms in a monoclinic (space group *C*2/*m*) superstructure of Nb_6C_5. These sites have coordinates (0 0 0), (0 1/2 1/2), (0 1/3 0), (1/2 1/6 0), (0 1/6 1/2) and (1/2 1/3 1/2) in a monoclinic superstructure. In the basic structure of *B*1 type their coordinates $(x_{\mathrm{I}}, y_{\mathrm{I}}, z_{\mathrm{I}})_{B1}$ are $(0\ 0\ 0)_{B1}$, $(1\ 1/2\ 1/2)_{B1}$, $(1/2\ 1/2\ 0)_{B1}$, $(1/2\ 0\ {-1/2})_{B1}$, $(1/2\ 0\ 1/2)_{B1}$ and $(1\ 0\ 0)_{B1}$. Substituting in sequence the coordinates $(x_{\mathrm{I}}, y_{\mathrm{I}}, z_{\mathrm{I}})_{B1}$ of these sites into (5.31), we obtain a system of six independent equations:

$$\left.\begin{array}{l}
y + \gamma_9^{(3)} + \gamma_4^{(1)} + \gamma_3^{(3)} = 0 \\
y - \gamma_9^{(3)} + \gamma_4^{(1)} - \gamma_3^{(3)} = 1 \\
y + \gamma_9^{(3)} - \gamma_4^{(1)}/2 - \gamma_4^{(2)}\sqrt{3}/2 - \gamma_3^{(3)}/2 - \gamma_3^{(4)}\sqrt{3}/2 = 1 \\
y + \gamma_9^{(3)} - \gamma_4^{(1)}/2 + \gamma_4^{(2)}\sqrt{3}/2 - \gamma_3^{(3)}/2 + \gamma_3^{(4)}\sqrt{3}/2 = 1 \\
y - \gamma_9^{(3)} - \gamma_4^{(1)}/2 + \gamma_4^{(2)}\sqrt{3}/2 + \gamma_3^{(3)}/2 - \gamma_3^{(4)}\sqrt{3}/2 = 1 \\
y - \gamma_9^{(3)} - \gamma_4^{(1)}/2 - \gamma_4^{(2)}\sqrt{3}/2 + \gamma_3^{(3)}/2 + \gamma_3^{(4)}\sqrt{3}/2 = 1
\end{array}\right\} \quad (5.32)$$

This system of equations (5.32) unambiguously determines the composition y of an ideal fully ordered phase of niobium carbide and the coefficients $\gamma_s^{(j)}$. It has the following solution: $y = 5/6$, $\gamma_9^{(3)} = -1/6$, $\gamma_4^{(1)} = -1/3$, $\gamma_4^{(2)} = 0$, $\gamma_3^{(3)} = -1/3$, and $\gamma_3^{(4)} = 0$. Thus, the solution of (5.32) gives a stoichiometric composition of the ordered phase $NbC_{5/6} \equiv Nb_6C_5$, which agrees fairly well with experimental data [10, 11]. In accordance with this solution, the distribution function for carbon atoms in a monoclinic (space group *C*2/*m*) superstructure of Nb_6C_5 type with any degree of order has the form

$$n(x_{\mathrm{I}}, y_{\mathrm{I}}, z_{\mathrm{I}}) = y - (\eta_9/6)\cos[\pi(x_{\mathrm{I}} - y_{\mathrm{I}} + z_{\mathrm{I}})] - (\eta_4/3)\cos[4\pi(x_{\mathrm{I}} + y_{\mathrm{I}})/3]$$
$$- (\eta_3/3)\cos[\pi(x_{\mathrm{I}} - 5y_{\mathrm{I}} - 3z_{\mathrm{I}})/3] \,. \quad (5.33)$$

Calculations by (5.27) and (5.28) show that $\gamma_9 = \gamma_4 = \gamma_3 = 1/6$ and $\varphi_9^{(3)} = \pi$, $\varphi_4^{(1)} = \pi$, $\varphi_4^{(2)} = -\pi$, $\varphi_3^{(3)} = \pi$, $\varphi_3^{(4)} = -\pi$.

Knowing the transition channel and employing the method of static concentration waves, it is possible to determine a general form of the distribution function for any structure even if particular values of long-range order parameters η_s are unknown.

5.3 Transition Channel and Diffraction Reflection Intensity

The scattering intensity at a point $\boldsymbol{q}$ of the reciprocal space (if the geometrical and the Lorentz factors are neglected) is

$$I(\boldsymbol{q}) = F(\boldsymbol{q})F^*(\boldsymbol{q}). \tag{5.34}$$

Taking into account the distribution function $n(\boldsymbol{r})$ and using the Fourier transform, the structural amplitude $F(\boldsymbol{q})$ of a solid solution A_yB_{1-y} can be written as

$$F(\boldsymbol{q}) = \sum_{\nu} f_\nu \sum_{\boldsymbol{r}} n_\nu(\boldsymbol{r}) \exp(i\boldsymbol{q}\boldsymbol{r}), \tag{5.35}$$

where f_ν is the scattering factor of the atomic species ν, $n_A(\boldsymbol{r}) = n(\boldsymbol{r})$ and $n_B(\boldsymbol{r}) = 1 - n(\boldsymbol{r})$. The summation in (5.35) is taken over all atomic species ν and over all sites $\boldsymbol{r}$ of an ordering lattice.

The structural amplitude of a nonstoichiometric compound $MX_y\square_{1-y}$ should take into account an additional contribution from the metal sublattice. Since $f_\square \equiv 0$, the structural amplitude for a compound MX_y is

$$F(\boldsymbol{q}) = f_M \sum_{\boldsymbol{r}_M} \exp(i\boldsymbol{q}\boldsymbol{r}_M) + f_X \sum_{\boldsymbol{r}} n(\boldsymbol{r}) \exp(i\boldsymbol{q}\boldsymbol{r}). \tag{5.36}$$

In (5.36) the summation is taken over all sites $\boldsymbol{r}_M$ of the metal sublattice and over all sites $\boldsymbol{r}$ of the nonmetal sublattice where interstitial atoms X and structural vacancies □ are ordered. Compounds MX_y with the $B1$ structure have an equal number of sites in metal and nonmetal sublattices. In the following we assume that each sublattice contains N sites.

Let the reciprocal lattice have a set of structural sites $\{\boldsymbol{H}\}$ including some site $\boldsymbol{H} = (hkl)$ and all other sites are obtained when the site (hkl) is subject to symmetry operations of the reciprocal lattice of a disordered basic structure. Considering that $\boldsymbol{H}$ is a structural site of the reciprocal lattice and using the property of the sum over primitive lattice sites, we can write

$$\sum_{\boldsymbol{r}} \exp[i\boldsymbol{q}(\boldsymbol{r} - \Delta\boldsymbol{r})] = N_{pr} \exp(-i\Delta\boldsymbol{r}\boldsymbol{H}) \sum_{\boldsymbol{H}} \delta(\boldsymbol{q} - \boldsymbol{H}), \tag{5.37}$$

where N_{pr} is the number of sites in a primitive lattice and $\Delta\boldsymbol{r}$ is the displacement vector of any lattice site relative to the origin of the coordinates. The coordinate axes are parallel to the primitive lattice vectors:

$$\delta(\boldsymbol{q}-\boldsymbol{H})=\begin{cases}1, & \text{if } \boldsymbol{q}=\boldsymbol{H},\\ 0, & \text{if } \boldsymbol{q}\neq\boldsymbol{H}.\end{cases}$$

Note that the last equality is fulfilled for the δ-function when $N_{pr} \to \infty$.

A $B1$ lattice represents eight simple cubic lattices displaced relative to each other. Now let the origin of coordinates be at a site of a simple nonmetal sublattice. In this case, other simple nonmetal sublattices are displaced relative to the first sublattice by vectors $\Delta\boldsymbol{r}$ = (1/2 1/2 0), (0 1/2 1/2), and (1/2 0 1/2). Obviously, in the reciprocal space $\exp(-i\Delta\boldsymbol{r}\boldsymbol{H}) = 1$ for any structural site $\boldsymbol{H} = (hkl)$ of the nonmetal sublattices. Simple cubic metal sublattices are displaced relative to the chosen origin of the coordinates by vectors $\Delta\boldsymbol{r}$ = (1/2 0 0), (0 1/2 0), (0 0 1/2), and (1/2 1/2 1/2). Therefore, $\exp(-i\Delta\boldsymbol{r}\boldsymbol{H}) = \exp(-i\pi(h+k+l)) = \cos[\pi(h+k+l)] = (-1)^{(h+k+l)}$ for all four metal sublattices. In other words, $\exp(-i\Delta\boldsymbol{r}\boldsymbol{H}) = 1$ and $\exp(-i\Delta\boldsymbol{r}\boldsymbol{H}) = -1$ for structural sites (hkl) with even and odd indices respectively.

Let us substitute (5.19) into (5.36) and take the sum over $\boldsymbol{r}$ considering equality (5.37). After various rearrangements the scattering intensity can be written as

$$I(\boldsymbol{q})=I(\boldsymbol{H})\sum_{\boldsymbol{H}}\delta(\boldsymbol{q}-\boldsymbol{H})+\sum_{s}\sum_{j}I(\boldsymbol{k}_s^{(j)})\sum_{\boldsymbol{H}}\delta[\boldsymbol{q}-(\boldsymbol{H}+\boldsymbol{k}_s^{(j)})]\,. \qquad (5.38)$$

Formula (5.38) holds for structure and superstructure reflections that do not overlap in reciprocal space. It holds, in particular, for the $B1$ structure. Summation of the second term, which accounts for the intensity of superstructure reflections, is taken over all superstructure vectors of the phase transition channel.

The intensity of any structural reflection for an fcc solid solution A_yB_{1-y} is

$$I(\boldsymbol{H}) = N^2[yf_A + (1-y)f_B]^2\,. \qquad (5.39)$$

In the case of a nonmetal fcc sublattice of a strongly nonstoichiometric compound MX_y,

$$I(\boldsymbol{H}) = N^2(yf_X)^2\,. \qquad (5.40)$$

Taking into account the contribution from the metal sublattice of a compound MX_y with the $B1$ structure, the intensity of the structure reflection at $\boldsymbol{q} = \boldsymbol{H} = (hkl)$ is

$$I(\boldsymbol{H}) = N^2[(-1)^{(h+k+l)}f_M + yf_X]^2. \qquad (5.41)$$

Let us determine the intensity $I(\boldsymbol{k}_s^{(j)})$ for a single superstructure reflection. Atomic distribution modulations described by nonequivalent superstructure vectors $\boldsymbol{k}_s^{(j)}$ and $-\boldsymbol{k}_s^{(j)}$ are equal to $\Delta(\boldsymbol{k}_s^{(j)},\boldsymbol{r})$ for a direct lattice. Considering that the atomic distribution at a site $\boldsymbol{r}$ of a direct lattice is modulated by a superstructure vector $\boldsymbol{k}_s^{(j)}$ and an unlike nonequivalent vector $-\boldsymbol{k}_s^{(j)}$, the modulation at the site $\boldsymbol{r}$ equals $2\Delta(\boldsymbol{k}_s^{(j)},\boldsymbol{r})$. The structural amplitude corresponding to this modulation has the form

$$F(\boldsymbol{q},\boldsymbol{k}_s^{(j)}) = f_X\sum_{\boldsymbol{r}}[2\Delta(\boldsymbol{k}_s^{(j)},\boldsymbol{r})]\exp(i\boldsymbol{q}\boldsymbol{r})\,. \qquad (5.42)$$

The structure amplitude for a superstructure reflection is obtained by substituting (5.23) into (5.42) and rearranging the last equation with allowance for (5.37):

$$F(\boldsymbol{k}_s^{(j)}) = N f_X \eta_s \gamma_s \exp(i\varphi_s^{(j)}) \,. \tag{5.43}$$

In accordance with (5.34) and (5.43), the intensity of a single superstructure reflection is

$$I(\boldsymbol{k}_s^{(j)}) = N^2 (f_X \eta_s \gamma_s)^2 \,. \tag{5.44}$$

If superstructure vectors $\boldsymbol{k}_s^{(j)}$ and $-\boldsymbol{k}_s^{(j)}$ are equivalent (i.e. when $2\boldsymbol{k}_s^{(j)} = \boldsymbol{H} \neq 0$) and if the vector $-\boldsymbol{k}_s^{(j)}$ is not involved in the phase transition channel, the atomic distribution modulation at a site $\boldsymbol{r}$ is $\Delta(\boldsymbol{k}_s^{(j)}, \boldsymbol{r})$. Therefore the structural amplitude takes the form

$$F(\boldsymbol{q}, \boldsymbol{k}_s^{(j)}) = f_X \sum_{\boldsymbol{r}} \Delta(\boldsymbol{k}_s^{(j)}, \boldsymbol{r}) \exp(i\boldsymbol{q}\boldsymbol{r}) \,. \tag{5.45}$$

After some substitutions and performing rearrangements of (5.45), the expression for the structural amplitude originating from the vector $\boldsymbol{k}_s^{(j)} = \boldsymbol{H}/2$ can be written as

$$F(\boldsymbol{k}_s^{(j)}) = N f_X \eta_s \gamma_s \cos\varphi_s^{(j)} \,. \tag{5.46}$$

Therefore, the intensity of a single superstructure reflection caused by the superstructure vector $\boldsymbol{k}_s^{(j)} = \boldsymbol{H}/2$, which does not have a corresponding unlike vector $-\boldsymbol{k}_s^{(j)}$ in the phase transition channel, is

$$I(\boldsymbol{k}_s^{(j)}) = N^2 f_X^2 \eta_s^2 \gamma_s^2 \cos^2\varphi_s^{(j)} \,. \tag{5.47}$$

For superstructure vectors $\boldsymbol{k}_s^{(j)} = \boldsymbol{H}/2$ the product $\boldsymbol{k}_s^{(j)}\boldsymbol{r} = \boldsymbol{r}\boldsymbol{H}/2 = \pi m$ (m being an integer) and therefore in formula (5.26) the term $B_{sj}\sin(\boldsymbol{k}_s^{(j)}, \boldsymbol{r}) = 0$ and $B_{sj} = 0$. Hence, $\sin\varphi_s^{(j)} = 0$ and $\sin^2\varphi_s^{(j)} = 0$, and $\cos^2\varphi_s^{(j)} = 1$. Consequently, (5.48) takes on the same form as (5.44). The intensity of a single superstructure reflection, which is caused by any vector $\boldsymbol{k}_s^{(j)}$ included in an order–disorder phase transition channel, is described by (5.44).

Although (5.44) for describing the intensity of a superstructure reflection is analogous in form to an expression in ref. [6], they are not equivalent because the definitions of the γ_s values are different. Values determined from (5.27) should be used for a correct calculation of intensities.

From (5.44) it follows that the intensity of superstructure reflections is proportional to η_s^2 and decreases rapidly with decreasing degree of long-range order, and superstructure reflections are absent in a disordered state when $\eta_s = 0$.

The system of superstructure reflections in three-dimensional reciprocal space can be pictured in general as a set of several sublattices, each being identical to the lattice of structural reflections and displaced relative to this lattice by a vector $\boldsymbol{k}_s^{(j)}$. Clearly, the number of superstructure sublattices in reciprocal space is equal to the number of superstructure vectors in a phase transition channel. If sublattices of superstructure reflections do not coincide with one another, then the intensities of single superstructure reflections of one sublattice are equal to $I(\boldsymbol{k}_s^{(j)})$. Superstructure sublattices in reciprocal space can coincide if different crystal sublattices are ordered concurrently in a similar manner (through similar transition channels). This situation is realized, in particular, when titanium atoms and vacancies are ordered in the metal sublattice, while oxygen atoms and vacancies are ordered in the nonmetal sublattice of titanium monoxide. As a result, a monoclinic (space group *C2/m*) superstructure Ti_5O_5 ($Ti_5\blacksquare_1O_5\square_1$) is formed. The intensity of superstructure reflections is determined taking into account the phase difference of static concentration waves (see Sect. 5.4.6) in case superstructure sublattices coincide.

From (5.38), (5.41) and (5.44) it is possible to construct a diffraction pattern for an ordered single crystal having any degree of order. However, polycrystalline samples are used mainly in experimental studies concerned with the structure of strongly nonstoichiometric compounds. This is explained by tedious methods involved in the production of single crystals of ordered phases of refractory nonstoichiometric carbides and nitrides. Formula (5.38) can be rearranged to the form describing diffraction spectra of polycrystals.

If an order–disorder phase transition channel includes wave vectors $\boldsymbol{k}_s^{(j)}$ of the star $\{\boldsymbol{k}_s\}$, a group of superstructural sites $(\boldsymbol{H}+\boldsymbol{k}_s^{(j)})$ corresponds to each structural site $\boldsymbol{H}$ of the reciprocal lattice in diffraction patterns. All wave vectors $\boldsymbol{k}_s^{(j)}$ of one and the same star $\{\boldsymbol{k}_s\}$ give a similar set of superstructure vector moduli $\{|\boldsymbol{H}+\boldsymbol{k}_s^{(j)}|\}$ for a set of sites $\{\boldsymbol{H}\}$ of the reciprocal lattice. Therefore in powder or polycrystal diffraction experiments, equal scattering intensities, which correspond to differently directed but equal in length (modulus) diffraction vectors $\boldsymbol{q}=\boldsymbol{H}+\boldsymbol{k}_s^{(j)}$, are summed both for superstructure and structure reflections. If a phase transition channel includes several (e.g., *m*) rays of the star $\{\boldsymbol{k}_s\}$, the intensity of each superstructure reflection caused by this star increases m_s times. Therefore the diffraction spectrum of an ordered polycrystal can be presented as a function of $q=|\boldsymbol{q}|$:

$$I(q)=I_0(q)+I_{\mathrm{super}}(q)\,, \tag{5.48}$$

where

$$I_0(q)=I(\boldsymbol{H})\sum_{\boldsymbol{H}}P(H)\delta(\boldsymbol{q}-\boldsymbol{H})\,, \tag{5.49}$$

$$I_{\mathrm{super}}(q)=\sum_{s}m_sI(\boldsymbol{k}_s^{(j)})\sum_{\boldsymbol{H}}\boldsymbol{P}(|\boldsymbol{H}+\boldsymbol{k}_s^{(j)}|)\delta[q-(|\boldsymbol{H}+\boldsymbol{k}_s^{(j)}|)]\,. \tag{5.50}$$

In (5.49) and (5.50) $P(H)=P(|\boldsymbol{H}|)$ and $P(|\boldsymbol{H}+\boldsymbol{k}_s^{(j)}|)$ denote multiplicity factors for structure and superstructure reflections of a polycrystal, which correspond to diffraction vectors of the length $q=H=|\boldsymbol{H}|$ and $q=|\boldsymbol{H}+\boldsymbol{k}_s^{(j)}|$. To shorten equation (5.50), we assume

$$\sum_{\boldsymbol{H}}P(|\boldsymbol{H}+\boldsymbol{k}_s^{(j)}|)\delta[\textstyle\sum_{\boldsymbol{H}}P(|q-(|\boldsymbol{H}+\boldsymbol{k}_s^{(j)}|)]=\displaystyle\sum_{\boldsymbol{H}}P(q=|\boldsymbol{H}+\boldsymbol{k}_s^{(j)}|)\,.$$

Then we have

$$I_{\text{super}}(q) = \sum_s m_s I(\boldsymbol{k}_s^{(j)}) \sum_{\boldsymbol{H}} P(q = |\boldsymbol{H} + \boldsymbol{k}_s^{(j)}|) . \tag{5.51}$$

In the following we use (5.48)–(5.51) to calculate diffraction spectra of superstructures formed in strongly nonstoichiometric interstitial compounds MX_y. Diffraction spectra can be calculated correctly from these formulas if one takes into account the following factors. In the presence of a superstructure, a group of superstructural sites $(\boldsymbol{H} + \boldsymbol{k}_s^{(j)})$ corresponds to each structural site $\boldsymbol{H} = (hkl)$ of the reciprocal lattice. Consequently, at $\boldsymbol{q} = \boldsymbol{H}$ each structural reflection in the diffraction spectrum of an ordered polycrystal corresponds to a group of superstructure reflections (when $\{\boldsymbol{q}\} = \{|\boldsymbol{H} + \boldsymbol{k}_s^{(j)}|\}$) located on both sides of the structural reflection. To determine all superstructure reflections in the interval from zero to some limiting value q_b, it is necessary to find all structural sites $\boldsymbol{H}$ of the reciprocal lattice, which give rise to superstructure reflections in the given interval. Such structural sites $\boldsymbol{H} = (h_i k_i l_i)$ of the reciprocal lattice are those for which $(q_b + R_{\max}) \geq |\boldsymbol{H}_i|$, where $R_{\max}$ is the distance from the center to the most distant point of the first Brillouin zone. In an fcc lattice $R_{\max} = \sqrt{5}/2$ for the point $\mathbf{W} = (1\ 0\ 1/2)$. Therefore the interval from 0 to q_b includes superstructure reflections due to all structural sites of the reciprocal lattice from $\boldsymbol{H}_0 = (000)$ to $\boldsymbol{H}_i = (h_i k_i l_i)$ with $|\boldsymbol{H}_i| = (h_i^2 + k_i^2 + l_i^2)^{1/2} \leq q_b + \sqrt{5}/2$. Diffraction spectra of various superstructures, which are discussed in Sect. 5.4 below, were constructed up to $q_b = 3.24$. The interval $0 < q < 3.24$ includes superstructure reflections from structural sites of the reciprocal lattice, for which $(h_i^2 + k_i^2 + l_i^2) < (q_b + \sqrt{5}/2)^2 = 19$, i.e. sites from (000) to (400) … (004) inclusive.

The structure of an ordered phase can be determined readily from a comparison of the experimental diffraction pattern with the calculated diffraction spectra of various superstructures. The following considerations should be taken into account in doing so. Experimental intensities of diffraction reflections of a polycrystalline sample, $I_{\text{exp}}(q) = I(q)L(q) = I_{0\text{exp}}(q) + I_{\text{LRO}}(q)$, include a factor $L(q)$ equal to the product of the Lorentz and geometrical factors. The factor $L(q)$ depends not only on the value of the diffraction vector q, but also on the instrumental geometry. To avoid the dependence of diffraction spectra on the particular measuring technique, one should consider the quantities $I_{\text{LRO}}(q)/L(q) = I_{\text{super}}(q)$ and $I_{0\text{exp}}(q)/L(q) = I_0(q)$ ($I_{\text{LRO}}(q)$ and $I_{0\text{exp}}(q)$ being the measured intensities of superstructure reflections due to the long-range order and structural reflections of the polycrystalline sample).

According to (5.44) the intensity $I(\boldsymbol{k}_s^{(j)})$ depends on the number of lattice sites N, the atomic scattering factor f_X (i.e. the species of interstitial atoms X), and the degree of order η_s. It is more convenient to consider the reduced intensities of superstructure reflections

$$\frac{I_{\text{super}}(q)}{f_X^2 N^2 \eta_s^2} = \frac{I_{\text{LRO}}(q)}{f_X^2 N^2 \eta_s^2 L(q)}$$

which depend only on the superstructure symmetry and type. Therefore all further calculated diffraction patterns will be plotted as reduced intensity $I_{\mathrm{LRO}}(q)/(f_{\mathrm{X}}^2 N^2 \eta_s^2 L(q))$ versus reduced diffraction vector modulus q (q is related to the angle θ as $q = (2a_{B1}\sin\theta)/\lambda$, where a_{B1} denotes the lattice constant of a disordered basic phase having the $B1$ structure and λ is the wavelength).

Calculated diffraction spectra of ideal superstructures $M_{2t}X_{2t-1}$ of nonstoichiometric compounds MX_y are given in Sect. 5.4 below. The calculated diffraction spectra of ideal ordered phases $M_{2t}X_{2t-1}$ contain not only superstructure reflections but also the basic structural reflections in the interval q under consideration. When $q = H$, the reduced intensity of a structural reflection of a polycrystal is

$$I_{0\exp}(q)/(f_{\mathrm{X}}^2 N^2 L(q)) = I_0(q)/(f_{\mathrm{X}}^2 N^2) = [(f_{\mathrm{M}}/f_{\mathrm{X}}) + (-1)^{(h+k+l)}\,y]^2\,P(H)\,.$$

The intensities of structural reflections shown in the diffraction spectra were calculated taking $f_{\mathrm{M}}/f_{\mathrm{X}} = 1.065$ (this value of $f_{\mathrm{M}}/f_{\mathrm{X}}$ is representative for neutron diffraction examination of ZrC_y, NbC_y, and TaC_y carbides) and $y = y_{\mathrm{st}} = (2t-1)/2t$. The intensities of the structural reflection $(111)_{B1}$ and all superstructure reflections in the calculated diffraction spectra are given on the same scale. The intensities of structural reflections $(200)_{B1}$ and $(220)_{B1}$ are much higher than those of superstructure reflections. Therefore only the positions of structural reflections $(200)_{B1}$ and $(220)_{B1}$ are shown in the calculated diffraction spectra.

The diffraction spectra are calculated for ideal superstructures $M_{2t}X_{2t-1}$, i.e. atomic displacements are neglected. Atomic displacements change the intensity ratio of observed reflections slightly. However, if the displacements and in particular those of the metal atoms were very large, additional reflections would appear in the spectra. The intensities of such additional reflections may be comparable with the intensities of superstructure reflections of the ideal ordered phase.

5.4 Distribution Functions and Diffraction for Superstructures of Strongly Nonstoichiometric Compounds

5.4.1 Superstructures of Type M_2X

Trigonal Superstructure M_2X (Space Group $R\bar{3}m$). A unit cell of the trigonal (space group $R\bar{3}m$) superstructure M_2X is shown in Fig. 5.2. The translation vectors of this cell and the coordinates of atoms and vacancies in an ideal trigonal structure M_2X are given in Table 5.1.

Table 5.1. Trigonal M_2X-type superstructure (space group $R\bar{3}m$ (D_{3d}^5)) : $\boldsymbol{a}_{tr} = \frac{1}{2}\langle 1\bar{2}1\rangle_{B1}$, $\boldsymbol{b}_{tr} = \frac{1}{2}\langle 2\bar{1}\bar{1}\rangle_{B1}$, $\boldsymbol{c}_{tr} = \frac{1}{2}\langle \bar{1}\bar{1}2\rangle_{B1}$

Atom	Position and multiplicity	Atomic coordinates in the ideal ordered structure			Values for the distribution function $n(x_I, y_I, z_I)$
		x/a_{tr}	y/b_{tr}	z/c_{tr}	
X1 (vacancy)	1(*a*)	0	0	0	$n_1 = y - \eta_9/2$
X2	1(*b*)	1/2	1/2	1/2	$n_2 = y + \eta_9/2$
M1	2(*c*)	1/4	1/4	1/4	

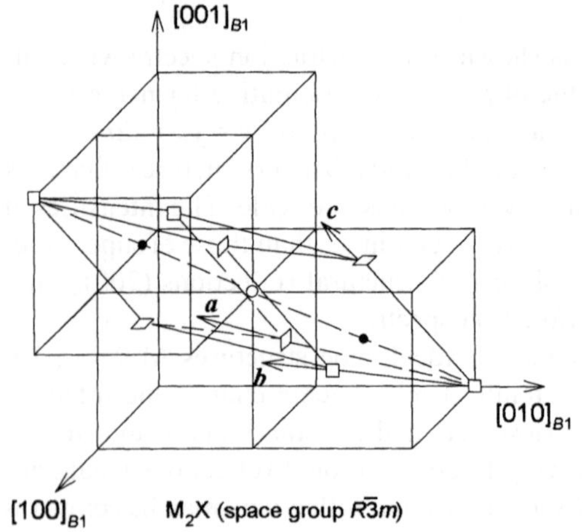

Fig. 5.2. Trigonal (space group $R\bar{3}m$) unit cell of the superstructure M_2X in a lattice with the *B*1 structure: (○) interstitial atoms, (•) metal atoms, (□) vacancies

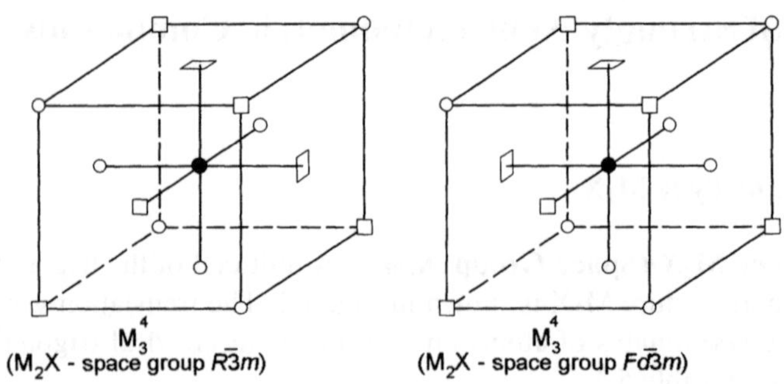

Fig. 5.3. Nearest environment of metal atoms (position M_3^4) in fully ordered trigonal and cubic superstructures type M_2X with space groups $R\bar{3}m$ and $Fd\bar{3}m$: (○) interstitial atoms, (•) metal atoms, (□) vacancies

A trigonal (space group $R\bar{3}m$) ordered phase type M_2X was observed in experiments with titanium carbide [14]. It can form in a nonstoichiometric compound MX_y through a phase transition channel involving one ray $\boldsymbol{k}_9^{(3)} = \boldsymbol{b}_2/2$ of the Lifschitz star $\{\boldsymbol{k}_9\}$, i.e. $m_9 = 1$. (The subscript and the superscript of the wave vector denote numbers of a star and its ray respectively. All numbering is given in accordance with Table 2.6.)

Calculations show that the distribution function describing a trigonal (space group $R\bar{3}m$) superstructure M_2X has the form

$$n(x_I, y_I, z_I) = y - (\eta_9/2)\cos[\pi(x_I - y_I + z_I)] \qquad (5.52)$$

with parameters $\gamma_9 = 1/2$ and $\varphi_9^{(3)} = \pi$. This distribution function also describes a trigonal (space group $R\bar{3}m$) substitutional superstructure AB (type CuPt).

A nonstoichiometric compound MX_y is fully ordered when the long-range order parameter $\eta_9(y)$ is unity and the relative concentration of interstitial atoms in the nonmetal sublattice is $y_{st} = 1/2$.

A complete atomic plane $(\bar{1}11)_{B1}$, in which all sites are occupied by interstitial atoms, and a defect plane $(\bar{1}11)_{B1}$, in which all sites are vacant, alternate in the direction $[1\bar{1}1]_{B1}$ in the nonmetal sublattice of a fully ordered trigonal (space group $R\bar{3}m$) structure M_2X. The metal sublattice of this superstructure should have marked distortions. In a fully ordered ideal trigonal superstructure M_2X all metal atoms have three vacancies in the first coordination sphere and four vacancies in the second coordination sphere (Fig. 5.3*a*), i.e. have a similar environment M_3^4 (the subscript and the superscript denote the number of vacancies in the first and second coordination spheres of a metal atom).

The distribution function (5.52) at all sites of the nonmetal fcc lattice assumes two values, which represent the probabilities that interstitial atoms occupy these sites (Table 5.1).

Cubic Superstructure M_2X (Space Group $Fd\bar{3}m$). A unit cell of the cubic (space group $Fd\bar{3}m$) superstructure M_2X is shown in Fig. 5.4. The translation vectors of this unit cell and the coordinates of atoms and vacancies are given in Table 5.2. A cubic (space group $Fd\bar{3}m$) ordered phase type M_2X was observed in experiments with titanium and zirconium carbides [14–16].

Table 5.2. Cubic M_2X-type superstructure (space group $Fd\bar{3}m$ (O_h^7)): $\boldsymbol{a}_1 = \langle 200\rangle_{B1}$, $\boldsymbol{a}_2 = \langle 020\rangle_{B1}$, $\boldsymbol{a}_3 = \langle 002\rangle_{B1}$

Atom	Position and multiplicity	Atomic coordinates in the ideal ordered structure			Values for the distribution function $n(x_I, y_I, z_I)$
		x/a_1	y/a_2	z/a_3	
X1 (vacancy)	16(*c*)	1/8	1/8	1/8	$n_1 = y - \eta_9/2$
X2	16(*d*)	5/8	5/8	5/8	$n_2 = y + \eta_9/2$
M1	32(*e*)	3/8	3/8	3/8	

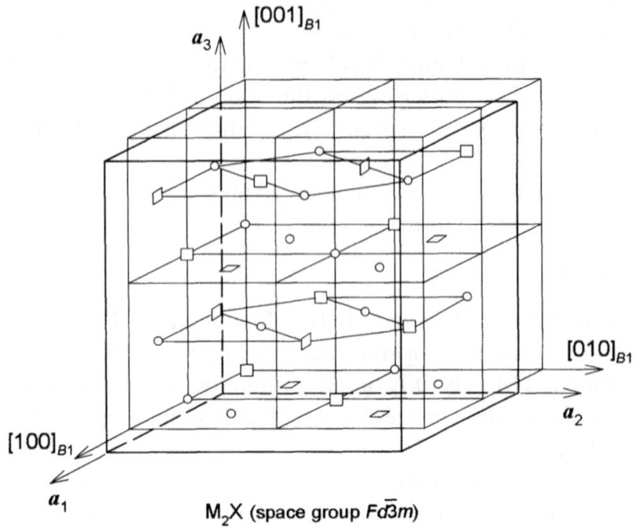

Fig. 5.4. Cubic (space group $Fd\bar{3}m$) unit cell of the superstructure M_2X in a lattice with the $B1$ structure: (○) interstitial atoms, (□) vacancies

The order–disorder phase transition channel MX_y–M_2X (space group $Fd\bar{3}m$) includes all four rays $\boldsymbol{k}_9^{(1)} = (\boldsymbol{b}_1 + \boldsymbol{b}_2 + \boldsymbol{b}_3)/2$, $\boldsymbol{k}_9^{(2)} = \boldsymbol{b}_1/2$, $\boldsymbol{k}_9^{(3)} = \boldsymbol{b}_2/2$, and $\boldsymbol{k}_9^{(4)} = \boldsymbol{b}_3/2$ of the star $\{\boldsymbol{k}_9\}$, i.e. $m_9 = 4$. The cubic superstructure M_2X in hand is described by the following distribution function:

$$n(x_I, y_I, z_I) = y - (\eta_9/4)\{ -\cos[\pi(x_I + y_I + z_I)] + \cos[\pi(-x_I + y_I + z_I)]$$

$$+ \cos[\pi(x_I - y_I + z_I)] + \cos[\pi(x_I + y_I - z_I)]\} \tag{5.53}$$

with parameters $\gamma_9 = 1/4$, $\varphi_9^{(1)} = 0$, and $\varphi_9^{(2)} = \varphi_9^{(3)} = \varphi_9^{(4)} = \pi$. In a fully ordered crystal the relative stoichiometric concentration of interstitial atoms is $y_{st} = 1/2$. The same distribution function describes a substitutional superstructure AB.

The distribution function (5.53) at all sites of the nonmetal basic fcc sublattice assumes the same values n_1 and n_2 (Table 5.2) as the function (5.52) describing a trigonal superstructure M_2X. All metal atoms are located in the position M_3^4 (Fig. 5.3*b*) in a fully ordered ideal cubic superstructure M_2X.

Atomic planes $(111)_{B1}$, in which 1/3 and 2/3 of all sites are vacant, alternate in the nonmetal sublattice of an ideal cubic superstructure M_2X.

Tetragonal Superstructure M_2X (Space Group *P4/mmm*). The location of a tetragonal unit cell of the superstructure M_2X (space group *P4/mmm*) in the type *B*1 lattice is shown in Fig. 5.5. The translation vectors of this tetragonal unit cell and the coordinates of atoms and vacancies are given in Table 5.3.

Table 5.3. Tetragonal M_2X-type superstructure (space group *P*4/*mmm* (D^1_{4h})): $\boldsymbol{a}_t = \frac{1}{2}\langle 1\bar{1}0\rangle_{B1}$, $\boldsymbol{b}_t = \frac{1}{2}\langle 110\rangle_{B1}$, $\boldsymbol{c}_t = \langle 001\rangle_{B1}$

Atom	Position and multiplicity	Atomic coordinates in the ideal ordered structure			Values for the distribution function $n(x_I, y_I, z_I)$
		x/a_t	y/b_t	z/c_t	
X1 (vacancy)	1(*a*)	0	0	0	$n_1 = y - \eta_{10}/2$
X2	1(*d*)	1/2	1/2	1/2	$n_2 = y + \eta_{10}/2$
M1	1(*b*)	0	0	1/2	
M2	1(*c*)	1/2	1/2	0	

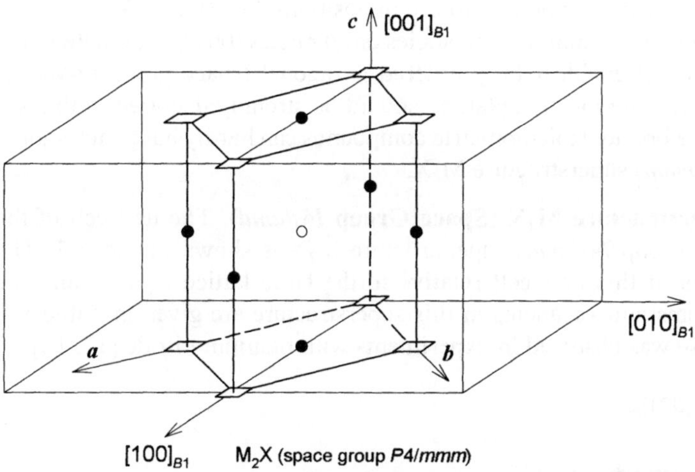

Fig. 5.5. Tetragonal (space group *P*4/*mmm*) unit cell of the superstructure M_2X in a lattice with the *B*1 structure: (○) interstitial atoms, (•) metal atoms, (□) vacancies

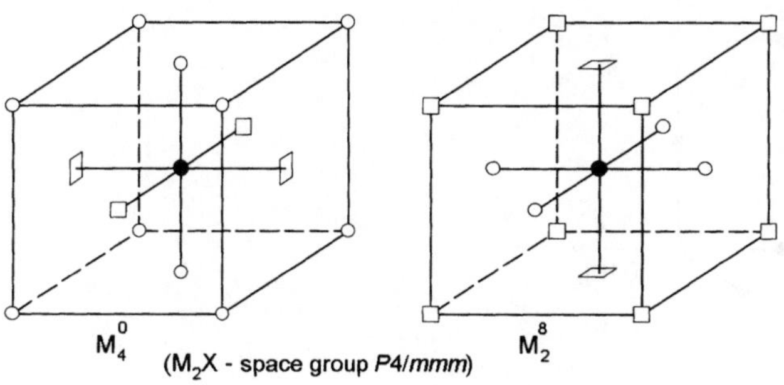

Fig. 5.6. Positions M^0_4 and M^8_2 of metal atoms in an ideal fully ordered tetragonal (space group *P*4/*mmm*) structure M_2X: (○) interstitial atoms, (•) metal atoms, (□) vacancies

The phase transition channel, through which the tetragonal superstructure is formed, includes one ray $\boldsymbol{k}_{10}^{(1)} = (\boldsymbol{b}_1 + \boldsymbol{b}_2)/2$ of the star $\{\boldsymbol{k}_{10}\}$, i.e. $m_{10} = 1$. The distribution function for interstitial atoms has the form

$$n(x_\mathrm{I}, y_\mathrm{I}, z_\mathrm{I}) = y - (\eta_{10}/2)\cos 2\pi z_\mathrm{I} \tag{5.54}$$

with parameters $\gamma_{10} = 1/2$ and $\varphi_{10}^{(1)} = \pi$. The same distribution function describes a tetragonal (space group *P*4/*mmm*) substitutional superstructure AB (type CuAu). In a fully ordered tetragonal phase M_2X, when the long-range order parameter $\eta_{10} = 1$, the relative stoichiometric concentration of interstitial atoms is $y_{st} = 1/2$.

Table 5.3 gives values of the distribution function (5.54) at sites in different positions of the nonmetal sublattice of the superstructure under consideration. In a fully ordered ideal tetragonal superstructure M_2X one-half of all metal atoms occupy positions M_4^0, while the other half are in positions M_2^8 (Fig. 5.6).

Complete and empty (containing vacancies only) planes $(001)_{B1}$ alternate in the nonmetal sublattice of an ideal fully ordered tetragonal (space group *P*4/*mmm*) superstructure M_2X. The metal sublattice should be strongly distorted in this superstructure. Therefore nonstoichiometric compounds can hardly have a tetragonal (space group *P*4/*mmm*) superstructure M_2X.

Tetragonal Superstructure M_2X (Space Group *I*4$_1$/*amd*). The unit cell of the tetragonal (space group $I4_1/amd$) superstructure M_2X is shown in Fig. 5.7. The translation vectors of this unit cell relative to the basic lattice type *B*1 and the coordinates of atoms and vacancies in this superstructure are given in Table 5.4. This ordered phase was observed in experiments with titanium nitride [17, 18].

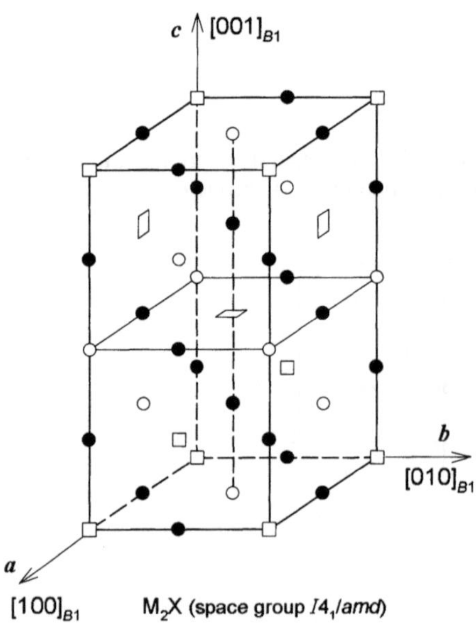

Fig. 5.7. Tetragonal (space group $I4_1/amd$) unit cell of the superstructure M_2X in a lattice with the *B*1 structure: (○) interstitial atoms, (•) metal atoms, (□) vacancies

Table 5.4. Tetragonal M_2X-type superstructure (space group $I4_1/amd$ (D^{19}_{4h})): $\boldsymbol{a}_t = \langle 100 \rangle_{B1}$, $\boldsymbol{b}_t = \langle 010 \rangle_{B1}$, $\boldsymbol{c}_t = \langle 002 \rangle_{B1}$

Atom	Position and multiplicity	Atomic coordinates in the ideal ordered structure			Values for the distribution function $n(x_I, y_I, z_I)$
		x/a_t	y/b_t	z/c_t	
X1 (vacancy)	4 (*a*)	0	0	0	$n_1 = y - \eta_8/2$
X2	4(*b*)	0	0	1/2	$n_2 = y + \eta_8/2$
M1	8(*e*)	0	0	1/4	

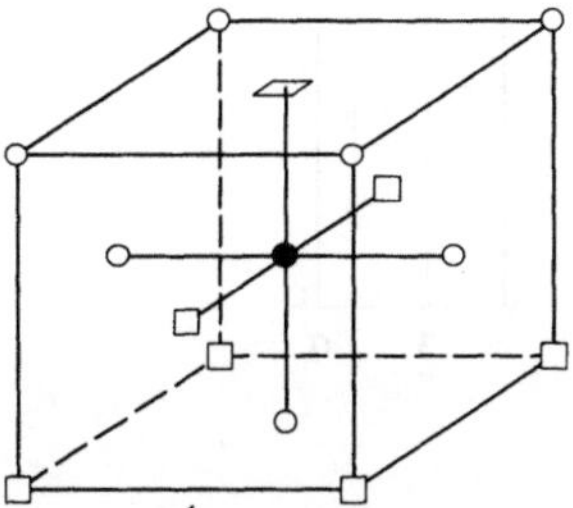

Fig. 5.8. Nearest environment of metal atoms (position M_3^4) in a fully ordered tetragonal (space group $I4_1/amd$) superstructure M_2X: (○) interstitial atoms, (•) metal atoms, (□) vacancies

A tetragonal superstructure M_2X is formed through a phase transition channel involving two rays $\boldsymbol{k}_8^{(1)} = (\boldsymbol{b}_1 + 3\boldsymbol{b}_2 + 2\boldsymbol{b}_3)/4$ and $\boldsymbol{k}_8^{(2)} = -\boldsymbol{k}_8^{(1)}$ of the star $\{\boldsymbol{k}_8\}$, i.e. $m_8 = 2$. The distribution function depends on one long-range order parameter:

$$n(x_I, y_I, z_I) = y - (\eta_8/2)\{\cos[\pi(2x_I + z_I)] + \sin[\pi(2x_I + z_I)]\} \ . \qquad (5.55)$$

In accordance with (5.27) and (5.28), this distribution function has parameters $\gamma_8 = \sqrt{2}/4$, $\varphi_8^{(1)} = 3\pi/4$, and $\varphi_8^{(2)} = -3\pi/4$. As in the previous case, $y_{st} = 1/2$. The same distribution function describes a tetragonal (space group $I4_1/amd$) substitutional superstructure AB. The distribution function (5.55) assumes two values (Table 5.4) at all sites located in different positions of the nonmetal sublattice of the superstructure.

In an ideal ordered tetragonal (space group $I4_1/amd$) superstructure M_2X all metal atoms are located in positions M_3^4 (Fig. 5.8).

Diffraction Spectra of Type M_2X Superstructures. Calculated diffraction spectra of ideal superstructures type M_2X are shown in Fig. 5.9.

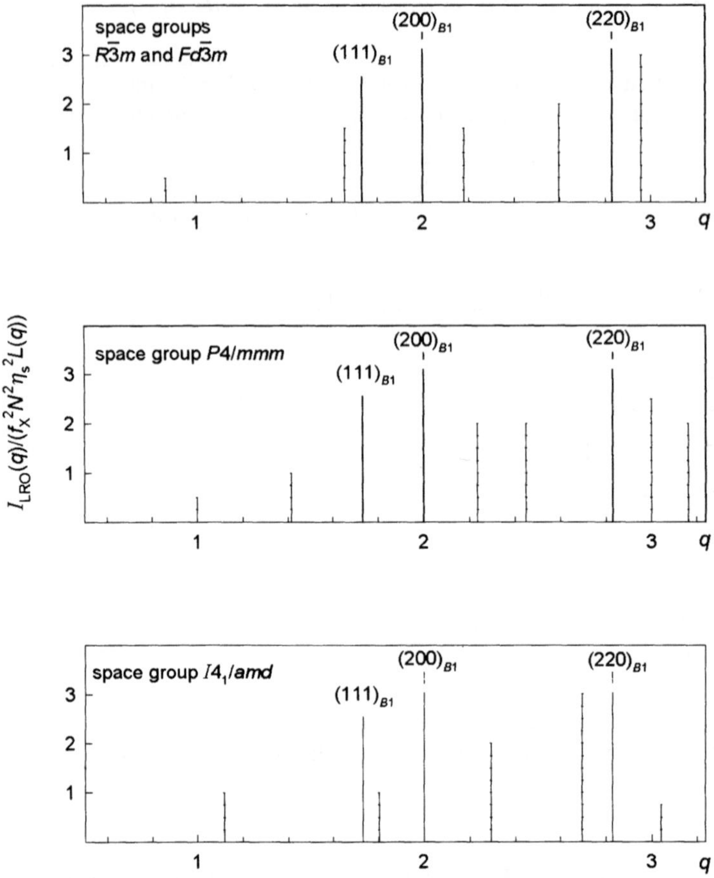

Fig. 5.9. Theoretical diffraction spectra of M_2X-type ideal superstructures plotted as a function of reduced diffraction vector modulus, $q = (2a_{B1}\sin\theta)/\lambda$. The intensities are calculated for a ratio of the atomic scattering factor of $f_M/f_X = 1.065$ which is typical for neutron diffraction. The intensities of superstructure reflections and the structure reflection $(111)_{B1}$ are plotted on the same scale

Given an equal degree of long-range order, the diffraction patterns for an ideal trigonal (space group $R\bar{3}m$) and for a cubic (space group $Fd\bar{3}m$) superstructure of type M_2X coincide. Indeed, considering γ_9 and $\varphi_9^{(1)}$ values for these superstructures, the intensity (5.44) of a superstructure reflection due to any one ray of the star $\{\boldsymbol{k}_9\}$ is

$$I(\boldsymbol{k}_9^{(1)}) = f_X^2 N^2 \eta_9^2 \gamma_{9\text{tr}}^2 = f_X^2 N^2 \eta_9^2 / 4 \tag{5.56}$$

for a trigonal superstructure M_2X. It is four times lower for a cubic superstructure M_2X:

$$I(\boldsymbol{k}_9^{(1)}) = f_X^2 N^2 \eta_9^2 \gamma_9^2 = f_X^2 N^2 \eta_9^2 / 16 \ . \tag{5.57}$$

However, the phase transition channel includes four rays of the star $\{\boldsymbol{k}_9\}$ (i.e. $m_9 = 4$) in the case of a cubic superstructure M_2X, while for a trigonal superstructure of the same type $m_9 = 1$. Considering this, for cubic and trigonal superstructures type M_2X,

$$I_{\text{super}}(q) = (f_X^2 N^2 \eta_9^2 / 4) \sum_{\boldsymbol{H}} P(q = |\boldsymbol{H} + \boldsymbol{k}_9^{(j)}|) \ . \tag{5.58}$$

Therefore, the diffraction spectra of these ideal superstructures coincide both in position and intensity of superstructure reflections. In (5.58) and subsequent analogous formulas $\boldsymbol{k}_s^{(j)}$ stands for an arbitrary superstructure vector of a given star included in a phase transition channel.

In real ordered phases atoms are displaced from the positions in an ideal crystal lattice. The directions of possible displacements of each atom in a unit cell of a given space group depend on the number of free (unfixed) parameters of the position occupied by an atom. In trigonal and cubic superstructures M_2X static displacements of interstitial atoms are forbidden, because any displacement of these atoms changes the superstructure symmetry. However, metal atoms can be displaced along all three axes of each of these superstructures. Different directions of static displacements of atoms in trigonal and cubic superstructures M_2X lead to different intensities of superstructure reflections. If diffraction spectra are processed taking into account static displacements of atoms, it is possible to determine unambiguously the trigonal or cubic symmetry of an ordered phase M_2X.

The intensities of superstructure reflections of an ideal tetragonal (space group *P*4/*mmm*) superstructure M_2X is given by the formula

$$I_{\text{super}}(q) = (f_X^2 N^2 \eta_{10}^2 / 4) \sum_{\boldsymbol{H}} P(q = |\boldsymbol{H} + \boldsymbol{k}_{10}^{(j)}|) \ . \tag{5.59}$$

In the case of an ideal tetragonal (space group $I4_1/amd$) ordered phase M_2X, the intensities of superstructure reflections are

$$I_{\text{super}}(q) = (f_X^2 N^2 \eta_8^2 / 4) \sum_{\boldsymbol{H}} P(q = |\boldsymbol{H} + \boldsymbol{k}_8^{(j)}|) \ . \tag{5.60}$$

Diffraction spectra of tetragonal superstructures type M_2X (see Fig. 5.9) differ from one another and from those of trigonal and cubic superstructures M_2X.

5.4.2 Superstructures of Type M_3X_2

Orthorhombic Superstructure M_3X_2 (Space Group *Immm*). A unit cell of the orthorhombic (space group *Immm*) superstructure M_3X_2 is shown in Fig. 5.10. The translation vectors of this unit cell and the coordinates of atoms and vacancies are given in Table 5.5.

Table 5.5. Orthorhombic M_3X_2-type superstructure (space group *Immm* (D_{2h}^{25})): $\boldsymbol{a}_{\rm rh} = \frac{1}{2}\langle 1\bar{1}0\rangle_{B1}$, $\boldsymbol{b}_{\rm rh} = \frac{3}{2}\langle 110\rangle_{B1}$, $\boldsymbol{c}_{\rm rh} = \langle 001\rangle_{B1}$

Atom	Position and multiplicity	Atomic coordinates in the ideal ordered structure			Values for the distribution function $n(x_I, y_I, z_I)$
		$x/a_{\rm rh}$	$y/b_{\rm rh}$	$z/c_{\rm rh}$	
X1 (vacancy)	2(*a*)	0	0	0	$n_1 = y - 2\eta_4/3$
X2	4(*g*)	0	1/3	0	$n_2 = y + \eta_4/3$
M1	2(*c*)	1/2	1/2	0	
M2	4(*h*)	0	2/3	1/2	

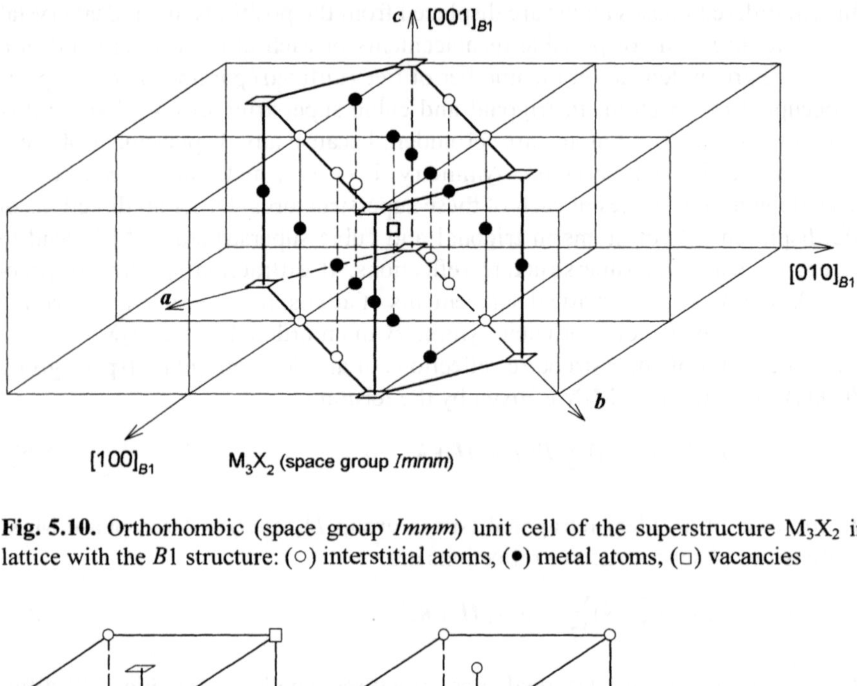

Fig. 5.10. Orthorhombic (space group *Immm*) unit cell of the superstructure M_3X_2 in a lattice with the $B1$ structure: (○) interstitial atoms, (•) metal atoms, (□) vacancies

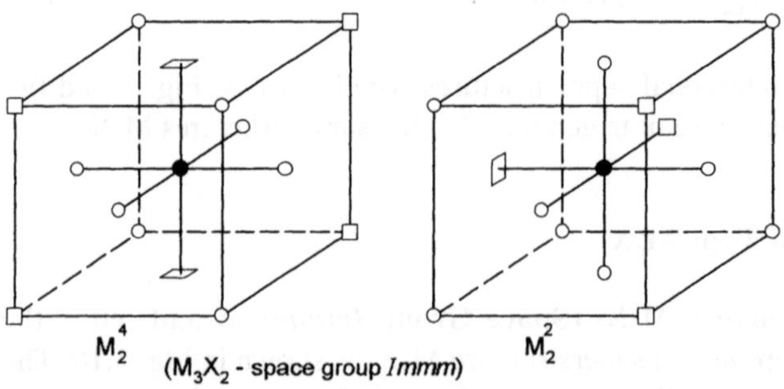

Fig. 5.11. Positions M_2^2 and M_2^4 of metal atoms in an ideal fully ordered orthorhombic (space group *Immm*) structure M_3X_2: (○) interstitial atoms, (•) metal atoms, (□) vacancies

The order–disorder phase transition channel MX_y–M_3X_2 (space group *Immm*) includes two rays $\boldsymbol{k}_4^{(1)} = (\boldsymbol{b}_1 + \boldsymbol{b}_2 + 2\boldsymbol{b}_3)/3$ and $\boldsymbol{k}_4^{(2)} = -\boldsymbol{k}_4^{(1)}$ of the star $\{\boldsymbol{k}_4\}$, i.e. $m_4 = 2$. The current parameter for the rays of the star $\{\boldsymbol{k}_4\}$ is $\mu_4 = 1/3$.

An orthorhombic superstructure M_3X_2 is described by the distribution function

$$n(x_I, y_I, z_I) = y - (2\eta_4/3)\{\cos[4\pi(x_I + y_I)/3] \tag{5.61}$$

with parameters $\gamma_4 = 1/3$, $\varphi_4^{(1)} = \pi$, and $\varphi_4^{(2)} = -\pi$. The same distribution function describes an orthorhombic (space group *Immm*) substitutional superstructure A_2B. In the case of a fully ordered phase M_3X_2, $y_{st} = 2/3$.

As a result of ordering the nonmetal sublattice of a disordered compound MX_y is divided into two nonequivalent sublattices, which differ in probabilities n_1 and n_2 with respect to their sites being occupied by interstitial atoms X (Table 5.5).

In the superstructure M_3X_2 under consideration all nonmetal atomic planes $(1\bar{1}1)_{B1}$ contain sites of the first sublattice, which are surrounded by six sites of the second sublattice. In these planes one-third of all sites belong to the first (vacancy) sublattice (the probability of sites occupied by interstitial atoms X being n_1), while the rest of the sites belong to the second sublattice (the occupancy probability being n_2). A fully ordered ideal rhombic (space group *Immm*) superstructure M_3X_2 has two positions of metal atoms: M_2^4 and M_2^2 (Fig. 5.11). Two-thirds of all metal atoms occupy positions M_2^2 and one-third of atoms are in positions M_2^4.

Trigonal Superstructure M_3X_2 (Space Group $P\bar{3}m1$). A unit cell of the trigonal (space group $P\bar{3}m1$) superstructure M_3X_2 is shown in Fig. 5.12. The translation vectors and the coordinates of atoms and vacancies are given in Table 5.6.

A trigonal (space group $P\bar{3}m1$) superstructure M_3X_2 is formed through a phase transition channel, which includes two rays $\boldsymbol{k}_5^{(5)} = -\boldsymbol{b}_2/3$ and $\boldsymbol{k}_5^{(6)} = -\boldsymbol{k}_5^{(5)}$ of the star $\{\boldsymbol{k}_5\}$, i.e. $m_5 = 2$. The current parameter for the rays $\boldsymbol{k}_5^{(5)}$ and $\boldsymbol{k}_5^{(6)}$ is $\mu_5 = 1/3$.

The distribution function corresponding to this superstructure is of the form

$$n(x_I, y_I, z_I) = y - (2\eta_5/3)\{\cos[2\pi(x_I - y_I + z_I)/3] \tag{5.62}$$

with parameters $\gamma_5 = 1/3$, $\varphi_5^{(1)} = \pi$, and $\varphi_5^{(2)} = -\pi$. As in the previous case, $y_{st} = 2/3$.

Values of the distribution function (5.62) are given in Table 5.6. This function also describes the trigonal (space group $P\bar{3}m1$) substitutional superstructure A_2B.

Table 5.6. Trigonal M_3X_2-type superstructure (space group $P\bar{3}m1$ (D_{3d}^3)): $\boldsymbol{a}_{rh} = \frac{1}{2}\langle 10\bar{1}\rangle_{B1}$, $\boldsymbol{b}_{rh} = \frac{1}{2}\langle 011\rangle_{B1}$, $\boldsymbol{c}_{rh} = \langle 1\bar{1}1\rangle_{B1}$

Atom	Position and multiplicity	Atomic coordinates in the ideal ordered structure			Values for the distribution function $n(x_I, y_I, z_I)$
		x/a_{tr}	y/b_{tr}	z/c_{tr}	
X1 (vacancy)	1(*a*)	0	0	0	$n_1 = y - 2\eta_5/3$
X2	2(*d*)	1/3	2/3	1/3	$n_2 = y + \eta_5/3$
M1	1(*b*)	0	0	1/2	
M2	2(*d*)	1/3	2/3	5/6	

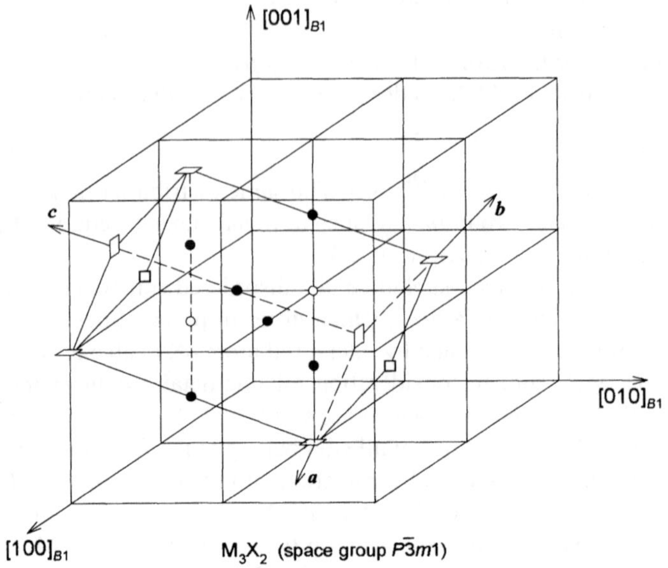

Fig. 5.12. Trigonal (space group $P\bar{3}m1$) unit cell of the superstructure M_3X_2 in a lattice with the $B1$ structure: (○) interstitial atoms, (•) metal atoms, (□) vacancies

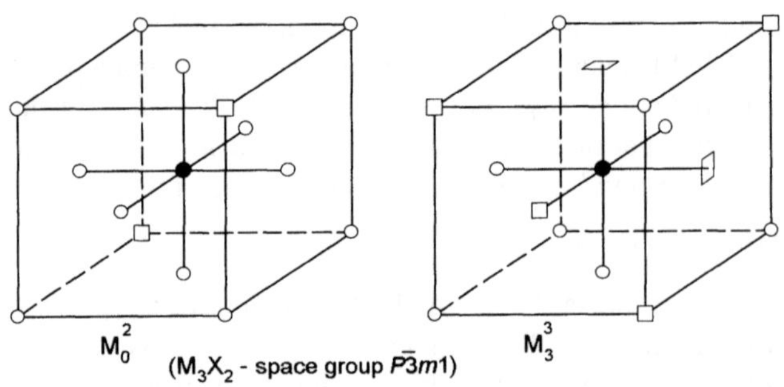

Fig. 5.13. Positions M_3^3 and M_0^2 of metal atoms in an ideal fully ordered trigonal (space group $P\bar{3}m1$) structure M_3X_2: (○) interstitial atoms, (•) metal atoms, (□) vacancies

In a fully ordered trigonal (space group $P\bar{3}m1$) superstructure M_3X_2 two-thirds of all metal atoms occupy positions M_3^3 and the other atoms are in positions M_0^2 (Fig. 5.13). Two complete atomic planes (all sites are occupied by interstitial atoms) and one empty plane (all sites are vacant) alternate in the direction $[1\bar{1}1]_{B1}$ in the nonmetal sublattice of a trigonal (space group $P\bar{3}m1$) superstructure M_3X_2. This superstructure is unlikely to form in nonstoichiometric compounds.

Monoclinic Superstructure M_3X_2 (Space Group *P*2). A unit cell of the monoclinic (space group *P*2) superstructure M_3X_2 is shown in Fig. 5.14. The translation vectors and the coordinates of atoms and vacancies are given in Table 5.7.

A monoclinic (space group *P*2) superstructure M_3X_2 is formed through a phase transition channel, which includes six rays $\boldsymbol{k}_4^{(3)} = (\boldsymbol{b}_2 - \boldsymbol{b}_1)/3$, $\boldsymbol{k}_4^{(4)} = -\boldsymbol{k}_4^{(3)}$, $\boldsymbol{k}_4^{(5)} = (\boldsymbol{b}_1 + 2\boldsymbol{b}_2 + \boldsymbol{b}_1)/3$, $\boldsymbol{k}_4^{(6)} = -\boldsymbol{k}_4^{(5)}$, $\boldsymbol{k}_4^{(9)} = (2\boldsymbol{b}_1 + \boldsymbol{b}_2 + \boldsymbol{b}_3)/3$, and $\boldsymbol{k}_4^{(10)} = -\boldsymbol{k}_4^{(9)}$ of the star $\{\boldsymbol{k}_4\}$, i.e. $m_4 = 6$. In a monoclinic superstructure M_3X_2 the current parameter of the star $\{\boldsymbol{k}_4\}$ is $\mu_4 = 1/3$. This superstructure is described by the function

$$n(x_{\rm I}, y_{\rm I}, z_{\rm I}) = y - (\eta_4/3)\{\cos[4\pi(x_{\rm I} + z_{\rm I})/3] + (\sqrt{3}/3)\sin[4\pi(x_{\rm I} + z_{\rm I})/3]$$
$$+ \cos[4\pi(y_{\rm I} + z_{\rm I})/3] - (\sqrt{3}/3)\sin[4\pi(y_{\rm I} + z_{\rm I})/3]$$
$$+ (2\sqrt{3}/3)\sin[4\pi(x_{\rm I} - y_{\rm I})/3]\} \qquad (5.63)$$

with parameters $\gamma_4 = \sqrt{3}/9$, $\varphi_4^{(3)} = \pi/2$, $\varphi_4^{(4)} = -\pi/2$, $\varphi_4^{(5)} = 5\pi/6$, $\varphi_4^{(6)} = -5\pi/6$, $\varphi_4^{(9)} = 7\pi/6$, and $\varphi_4^{(10)} = -7\pi/6$. For a fully ordered phase M_3X_2, $y_{\rm st} = 2/3$. This distribution function also describes a monoclinic superstructure A_2B.

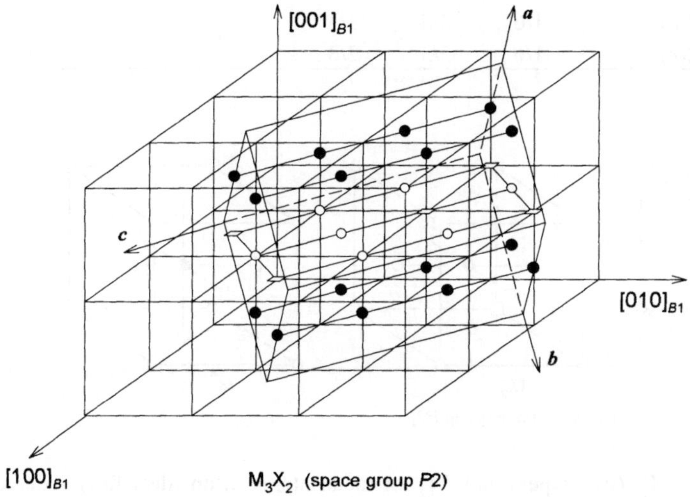

Fig. 5.14. Monoclinic (space group *P*2) unit cell of a superstructure M_3X_2 in a lattice with a *B*1 structure: (○) interstitial atoms, (•) metal atoms, (□) vacancies

At all sites of an fcc nonmetal sublattice the distribution function (5.63) assumes two values n_1 and n_2 (Table 5.7), which are similar to those of the function (5.61) describing the orthorhombic superstructure M_3X_2.

In a fully ordered monoclinic (space group *P*2) structure M_3X_2 ($y = 2/3$, $\eta_4 = 1$) one-third of sites are vacant and the remaining sites are occupied by interstitial atoms in each nonmetal atomic plane $(1\bar{1}1)_{B1}$.

Table 5.7. Monoclinic M_3X_2-type superstructure (space group $P2$ (C_2^1)): $\boldsymbol{a}_m = \frac{1}{2}\langle 1 1 2\rangle_{B1}$, $\boldsymbol{b}_m = \langle 1 1 \bar{1}\rangle_{B1}$, $\mathbf{c}_m = \frac{3}{2}\langle 1 \bar{1} 0\rangle_{B1}$

Atom	Position and multiplicity	Atomic coordinates in the ideal ordered structure			Values for the distribution function $n(x_I, y_I, z_I)$
		x/a_m	y/b_m	z/c_m	
X1 (vacancy)	1(*d*)	1/2	1/2	1/3	$n_1 = y - 2\eta_4/3$
X2 (vacancy)	2(*e*)	1/6	1/6	0	$n_1 = y - 2\eta_4/3$
X3	1(*d*)	1/2	1/2	0	$n_2 = y + \eta_4/3$
X4	1(*d*)	1/2	1/2	2/3	$n_2 = y + \eta_4/3$
X5	2(*e*)	1/6	1/6	1/3	$n_2 = y + \eta_4/3$
X6	2(*e*)	1/6	1/6	2/3	$n_2 = y + \eta_4/3$
M1	1(*c*)	1/2	0	0	
M2	1(*c*)	1/2	0	1/3	
M3	1(*c*)	1/2	0	2/3	
M4	2(*e*)	1/6	2/3	0	
M5	2(*e*)	1/6	2/3	1/3	
M6	2(*e*)	1/6	2/3	2/3	

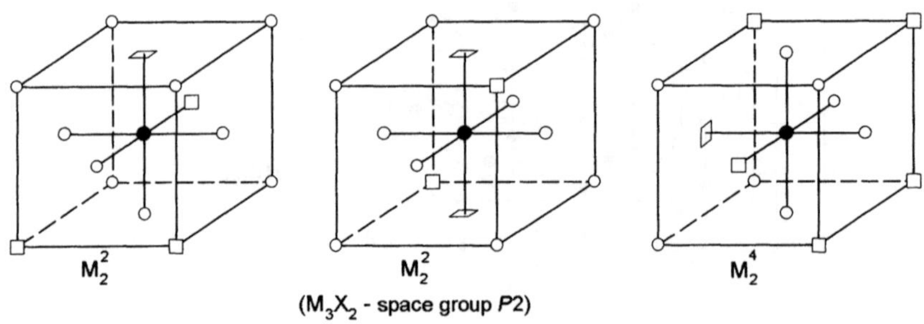

Fig. 5.15. Positions M_2^2 (two types) and M_2^4 of metal atoms in an ideal fully ordered monoclinic (space group $P2$) structure M_3X_2: (○) interstitial atoms, (•) metal atoms, (□) vacancies. One-third of all metal atoms occupy positions M_2^2 with two vacancies in the first coordination sphere spaced $a_{B1}\sqrt{2}/2$ apart; one-third of metal atoms are in positions M_2^2 with two vacancies in the first coordination sphere spaced a_{B1} apart; the remaining metal atoms occupy positions M_2^4

An ideal fully ordered monoclinic superstructure M_3X_2 has three positions of metal atoms: one-third of all metal atoms occupy positions M_2^2 with two vacancies spaced $a_{B1}\sqrt{2}/2$ apart in the first coordination sphere; one-third of metal atoms are in positions M_2^2 with two vacancies spaced a_{B1} apart in the first coordination sphere; the rest of the metal atoms occupy positions M_2^4 (Fig. 5.15).

Orthorhombic Superstructure M_3X_2 (Space Group $C222_1$). An orthorhombic unit cell of the superstructure M_3X_2 of the space group $C222_1$ is shown in Fig. 5.16. The translation vectors of this unit cell and the coordinates of atoms and vacancies are given in Table 5.8. The presence of an ordered orthorhombic (space group $C222_1$) phase M_3X_2 in titanium carbide $TiC_{0.64}$ follows from Monte Carlo calculations [19]. An orthorhombic (space group $C222_1$) ordered phase Ti_3C_2 was observed for nonstoichiometric titanium carbide [14].

The superstructure M_3X_2 is formed through a phase transition channel, which includes two rays $\boldsymbol{k}_4^{(1)} = (\boldsymbol{b}_1 + \boldsymbol{b}_2 + 2\boldsymbol{b}_3)/3$ and $\boldsymbol{k}_4^{(2)} = -\boldsymbol{k}_4^{(1)}$ of the star $\{\boldsymbol{k}_4\}$, and four rays $\boldsymbol{k}_3^{(3)} = -(7\boldsymbol{b}_1 + \boldsymbol{b}_2 + 2\boldsymbol{b}_3)/6$, $\boldsymbol{k}_3^{(4)} = -\boldsymbol{k}_3^{(3)}$, $\boldsymbol{k}_3^{(5)} = (\boldsymbol{b}_1 - 5\boldsymbol{b}_2 + 2\boldsymbol{b}_2)$, and $\boldsymbol{k}_3^{(6)} = -\boldsymbol{k}_3^{(5)}$ of the star $\{\boldsymbol{k}_3\}$. Therefore $m_4 = 2$ and $m_3 = 4$. Current parameters of the stars $\{\boldsymbol{k}_4\}$ and $\{\boldsymbol{k}_3\}$ are equal to $\mu_4 = 1/3$ and $\mu_3 = 1/12$ respectively.

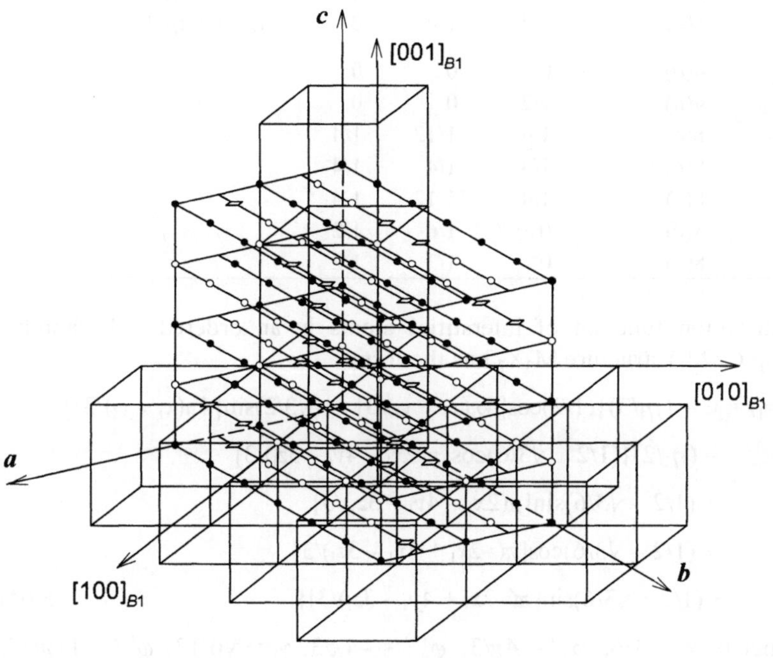

Fig. 5.16. Orthorhombic (space group $C222_1$) unit cell of the superstructure M_3X_2 in a lattice with the $B1$ structure: (○) interstitial atoms, (●) metal atoms, (□) vacancies

Table 5.8. Orthorhombic M_3X_2-type superstructure (space group $C222_1$ (D_2^5)): $\boldsymbol{a}_{rh} = \langle 1\bar{1}0 \rangle_{B1}$, $\boldsymbol{b}_{rh} = \langle 330 \rangle_{B1}$, $\boldsymbol{c}_{rh} = \langle 002 \rangle_{B1}$

Atom	Position and multiplicity	Atomic coordinates in the ideal ordered structure			Values for the distribution function $n(x_I, y_I, z_I)$
		x/a_{rh}	y/b_{rh}	z/c_{rh}	
X1 (vacancy)	4(*b*)	0	1/6	1/4	$n_1 = y - \eta_4/6 - \eta_3/2$
X2 (vacancy)	4(*b*)	0	1/3	1/4	$n_1 = y - \eta_4/6 - \eta_3/2$
X3 (vacancy)	8(*c*)	1/4	5/12	0	$n_1 = y - \eta_4/6 - \eta_3/2$
X4	4(*b*)	0	0	1/4	$n_2 = y + \eta_4/3$
X5	4(*b*)	0	1/2	1/4	$n_2 = y + \eta_4/3$
X6	4(*b*)	0	2/3	1/4	$n_3 = y - \eta_4/6 + \eta_3/2$
X7	4(*b*)	0	5/6	1/4	$n_3 = y - \eta_4/6 + \eta_3/2$
X8	8(*c*)	1/4	1/12	0	$n_3 = y - \eta_4/6 + \eta_3/2$
X9	8(*c*)	1/4	1/4	0	$n_2 = y + \eta_4/3$
M1	4(*a*)	0	0	0	
M2	4(*a*)	1/2	0	0	
M3	8(*c*)	1/4	1/12	1/4	
M4	8(*c*)	1/4	1/4	1/4	
M5	8(*c*)	1/4	5/12	1/4	
M6	8(*c*)	0	1/6	1/2	
M7	8(*c*)	0	1/3	1/2	

The distribution function of interstitial atoms in an ordered orthorhombic (space group $C222_1$) structure M_3X_2 has the form

$$\begin{aligned} n(x_I, y_I, z_I) = y - {} & (\eta_4/3)\{(1/2)\cos[4\pi(x_I + y_I)/3] - (\sqrt{3}/2)\sin[4\pi(x_I + y_I)/3]\} \\ & - (\eta_3/2)\{(1/2 + \sqrt{3}/6)\cos[\pi(2x_I - 4y_I - 3z_I)/3] \\ & + (1/2 - \sqrt{3}/6)\sin[\pi(2x_I - 4y_I - 3z_I)/3] \\ & + (1/2 - \sqrt{3}/6)\cos[\pi(-2x_I + 4y_I - 3z_I)/3] \\ & + (1/2 + \sqrt{3}/6)\sin[\pi(-2x_I + 4y_I - 3z_I)/3]\} \end{aligned} \tag{5.64}$$

with parameters $\gamma_4 = 1/6$, $\varphi_4^{(1)} = 4\pi/3$, $\varphi_4^{(2)} = -4\pi/3$, $\gamma_3 = \sqrt{6}/12$, $\varphi_3^{(3)} = 11\pi/12$, $\varphi_3^{(4)} = -11\pi/12$, $\varphi_3^{(5)} = 7\pi/12$, and $\varphi_3^{(6)} = -7\pi/12$. As in the previous case, $y_{st} = 2/3$. This function also describes an orthorhombic (space group $C222_1$) substitutional superstructure A_2B.

The distribution function (5.64) assumes three values at all sites belonging to different positions in the nonmetal sublattice of the superstructure M_3X_2 under discussion.

In a fully ordered orthorhombic (space group $C222_1$) structure M_3X_2 two-thirds of all metal atoms occupy positions M_2^3, while the remaining atoms are in positions M_2^2 (Fig. 5.17).

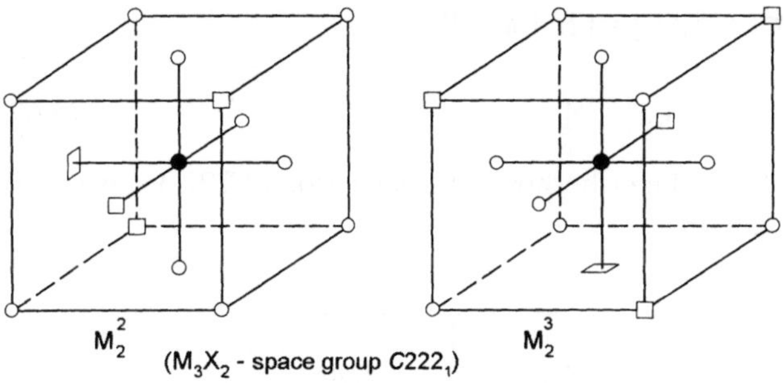

Fig. 5.17. One-third of all metal atoms occupy positions M_2^3 and the remaining metal atoms are in positions M_2^2 in an ideal fully ordered orthorhombic (space group $C222_1$) structure M_3X_2: (○) interstitial atoms, (•) metal atoms, (□) vacancies

Diffraction Spectra of Superstructures of Type M_3X_2. Considering formulas (5.44) and (5.51) and taking into account parameters m_4 and γ_4, the intensities for the superstructure reflections of ideal orthorhombic (space group *Immm*) and of ideal monoclinic (space group *P*2) superstructure type M_3X_2 are

$$I_{\text{super}}(q) = (2f_X^2 N^2 \eta_4^2 / 9) \sum_{\boldsymbol{H}} P(q = |\boldsymbol{H} + \boldsymbol{k}_4^{(j)}|) . \tag{5.65}$$

In other words, the degrees of long-range order η_4 being equal, these superstructures have similar diffraction patterns (Fig. 5.18).

The intensities of superstructure reflections of the ideal ordered phases M_3X_2 (space groups *Immm* and *P*2) will differ slightly because of different directions and values of the atomic displacements that are possible in each of these superstructures. If diffraction spectra are analyzed taking into account static displacements of atoms, it is possible to determine unambiguously the rhombic or monoclinic symmetry of these ordered phases.

In the case of an ideal trigonal (space group $P\bar{3}m1$) superstructure M_3X_2, the intensities of superstructure reflections are

$$I_{\text{super}}(q) = (2f_X^2 N^2 \eta_5^2 / 9) \sum_{\boldsymbol{H}} P(q = |\boldsymbol{H} + \boldsymbol{k}_5^{(j)}|) . \tag{5.66}$$

It should be remembered that the positions of some superstructure reflections coincide with positions of structure reflections in diffraction spectra of an ordered trigonal (space group $P\bar{3}m1$) phase M_3X_2 in a polycrystalline sample (see Fig. 5.18). For example, the structural vector $\boldsymbol{H} = (11\bar{1})$ of the reciprocal lattice has length $q = \sqrt{3}$. The superstructure vectors $(\boldsymbol{H} + \boldsymbol{k}_5^{(6)})$ ($\boldsymbol{H} = (020)$, $(\bar{2}00)$ and $(00\bar{2})$; $\boldsymbol{k}_5^{(6)} = \boldsymbol{b}_2/3$) have the same length. The intensities of coincident structure and superstructure reflections are added up in an experiment.

In the case of an ideal orthorhombic (space group $C222_1$) superstructure M_3X_2 the intensities of superstructure reflections are

$$I_{\text{super}}(q) = f_X^2 N^2 \sum_{H} [(\eta_4^2/18) P(q = |\boldsymbol{H} + \boldsymbol{k}_4^{(j)}|) + (\eta_3^2/6) P(q = |\boldsymbol{H} + \boldsymbol{k}_3^{(j)}|)] \quad (5.67)$$

The diffraction spectrum of the orthorhombic (space group $C222_1$) superstructure M_3X_2 is shown in Fig. 5.18.

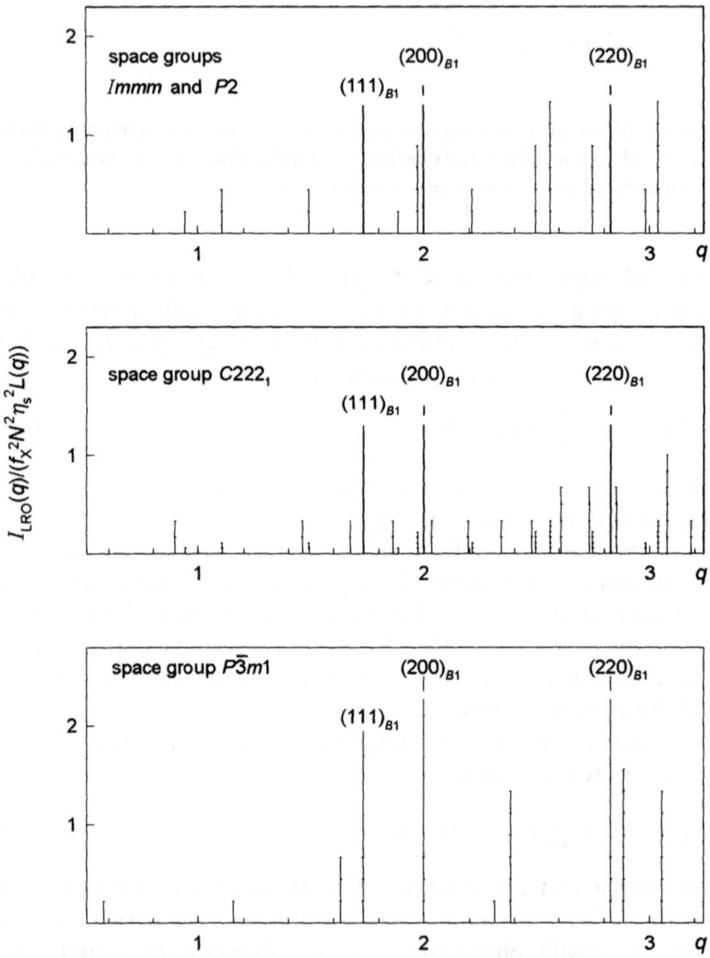

Fig. 5.18. Theoretical diffraction spectra of ideal M_3X_2-type superstructures which are plotted as a function of the reduced diffraction vector modulus, $q = (2a_{B1}\sin\theta)/\lambda$. The reflection intensity is calculated for the ratio of the atomic scattering factor of $f_M/f_X = 1.065$ which is typical for neutron diffraction. The intensities of superstructure reflections and the structure reflection $(111)_{B1}$ are shown on the same scale

5.4.3 Superstructures of Type M_4X_3

Cubic Superstructure M_4X_3 (space group $Pm\bar{3}m$). A unit cell of the cubic (space group $Pm\bar{3}m$) superstructure M_4X_3 is shown in Fig. 5.19. The translation vectors of this unit cell coincide with those of the basic crystal lattice having the $B1$ structure. The coordinates of atoms and vacancies in the ideal cubic superstructure M_4X_3 are given in Table 5.9.

An ordered cubic (space group $Pm\bar{3}m$) phase M_4X_3 is formed through a phase transition channel, which includes all rays $\boldsymbol{k}_{10}^{(1)} = (\boldsymbol{b}_1 + \boldsymbol{b}_2)/2$, $\boldsymbol{k}_{10}^{(2)} = (\boldsymbol{b}_1 + \boldsymbol{b}_3)/2$, and $\boldsymbol{k}_{10}^{(3)} = (\boldsymbol{b}_2 + \boldsymbol{b}_3)/2$ of the star $\{\boldsymbol{k}_{10}\}$. Therefore $m_{10} = 3$.

The distribution function describing the cubic superstructure M_4X_3 has the form

$$(x_{\rm I}, y_{\rm I}, z_{\rm I}) = y - (\eta_{10}/4)(\cos 2\pi x_{\rm I} + \cos 2\pi y_{\rm I} + \cos 2\pi z_{\rm I}) \tag{5.68}$$

with parameters $\gamma_{10} = 1/4$ and $\varphi_{10}^{(1)} = \varphi_{10}^{(2)} = \varphi_{10}^{(3)} = \pi$. The same function (5.68) describes a cubic (space group $Pm\bar{3}m$) substitutional superstructure A_3B (type Cu_3Au).

Table 5.9. Cubic M_4X_3-type superstructure (space group $Pm\bar{3}m$ (O_h^1)): $\boldsymbol{a}_1 = \langle 100 \rangle_{B1}$, $\boldsymbol{a}_2 = \langle 010 \rangle_{B1}$, $\boldsymbol{a}_3 = \langle 001 \rangle_{B1}$

Atom	Position and multiplicity	Atomic coordinates in the ideal ordered structure			Values for the distribution function $n(x_{\rm I}, y_{\rm I}, z_{\rm I})$
		x/a_1	y/a_2	z/a_3	
X1 (vacancy)	1(*a*)	0	0	0	$n_1 = y - 3\eta_{10}/4$
X2	3(*c*)	1/2	0	1/2	$n_2 = y + \eta_{10}/4$
M1	1(*b*)	1/2	1/2	1/2	
M2	3(*d*)	0	1/2	0	

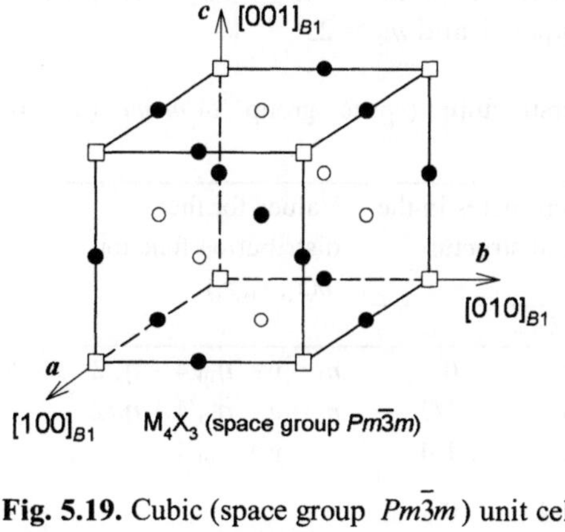

Fig. 5.19. Cubic (space group $Pm\bar{3}m$) unit cell of a superstructure M_4X_3 in a lattice with a $B1$ structure: (○) interstitial atoms, (•) metal atoms, (□) vacancies

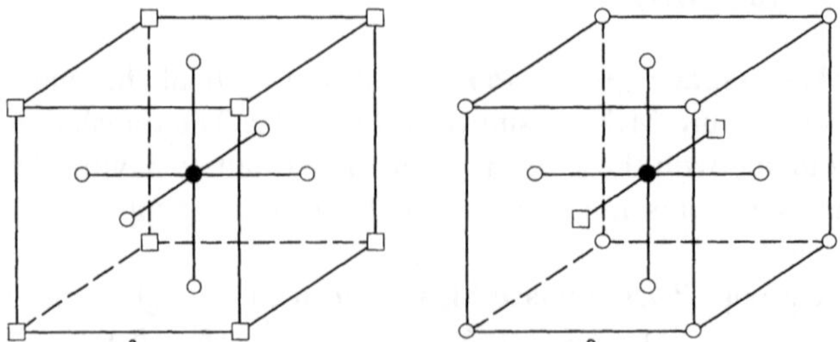

Fig. 5.20. Positions M_2^0 and M_0^8 of metal atoms in an ideal fully ordered cubic (space group $Pm\bar{3}m$) structure M_4X_3: (○) interstitial atoms, (•) metal atoms, (□) vacancies. Three-fourth of all metal atoms occupy positions M_2^0, while the rest of the metal atoms are in positions M_0^8

In the case of a fully ordered phase M_4X_3 the relative stoichiometric concentration of interstitial atoms is $y_{st} = 3/4$.

Values of the distribution function (5.68) at sites belonging to different positions in the nonmetal sublattice are given in Table 5.9.

In an ideal fully ordered cubic (space group $Pm\bar{3}m$) structure M_4X_3 three-quarters of all metal atoms occupy positions M_2^0, while the remaining atoms are in positions M_0^8 (Fig. 5.20).

Tetragonal Superstructure M_4X_3 (Space Group *I4/mmm*). A unit cell of the tetragonal (space group *I4/mmm*) superstructure M_4X_3 is shown in Fig. 5.21. The translation vectors of the unit cell and the coordinates of atoms and vacancies are given in Table 5.10.

The order–disorder transition channel MX_y–M_4X_3 (space group *I4/mmm*) includes one ray $\boldsymbol{k}_{10}^{(1)}$ of the star $\{\boldsymbol{k}_{10}\}$ and two rays $\boldsymbol{k}_8^{(1)} = (\boldsymbol{b}_1 + 3\boldsymbol{b}_2 + 2\boldsymbol{b}_3)/4$ and $\boldsymbol{k}_8^{(2)} = -\boldsymbol{k}_8^{(1)}$ of the star $\{\boldsymbol{k}_8\}$. Therefore $m_{10} = 1$ and $m_8 = 2$.

Table 5.10. Tetragonal M_4X_3-type superstructure (space group *I4/mmm* (D_{4h}^{17})): $\boldsymbol{a}_t = \langle 100 \rangle_{B1}$, $\boldsymbol{b}_t = \langle 010 \rangle_{B1}$, $\boldsymbol{c}_t = \langle 002 \rangle_{B1}$

Atom	Position and multiplicity	Atomic coordinates in the ideal ordered structure			Values for the distribution function $n(x_I, y_I, z_I)$
		x/a_t	y/a_t	z/a_t	
X1 (vacancy)	2(*a*)	0	0	0	$n_1 = y - \eta_{10}/4 - \eta_8/2$
X2	2(*b*)	0	0	1/2	$n_1 = y - \eta_{10}/4 - \eta_8/2$
X3	4(*d*)	0	1/2	1/4	$n_2 = y + \eta_{10}/4$
M1	4(*c*)	0	1/2	0	
M2	4(*e*)	0	0	1/4	

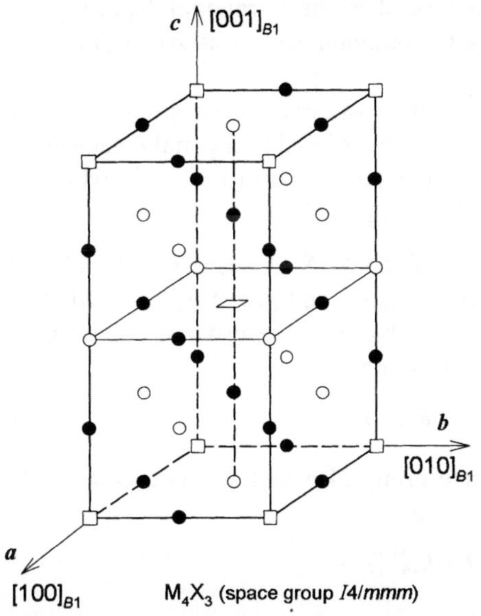

Fig. 5.21. Tetragonal (space group *I4/mmm*) unit cell of the superstructure M_4X_3 in a lattice with the *B*1 structure: (○) interstitial atoms, (•) metal atoms, (□) vacancies

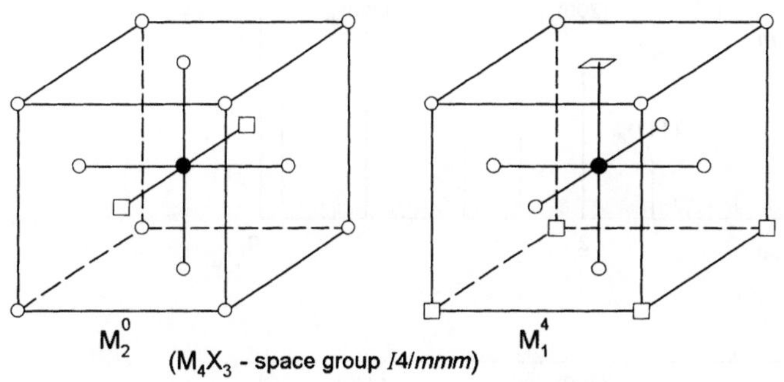

Fig. 5.22. One-half of all metal atoms occupy positions M_2^0, while the remaining metal atoms are in positions M_1^4 in the ideal fully ordered tetragonal (space group *I4/mmm*) structure M_4X_3: (○) interstitial atoms, (•) metal atoms, (□) vacancies

The tetragonal superstructure M_4X_3 is described by the distribution function

$$n(x_I, y_I, z_I) = y - (\eta_{10}/4)\cos 2\pi z_I - (\eta_8/2)\cos[\pi(2x_I + z_I)] , \qquad (5.69)$$

which depends on two long-range order parameters. The parameters of the function (5.69) are $\gamma_{10} = 1/4$, $\varphi_{10}^{(1)} = \pi$, $\gamma_8 = 1/4$, $\varphi_8^{(1)} = \pi$, and $\varphi_8^{(2)} = -\pi$. In the fully ordered phase $y_{st} = 3/4$.

The distribution function (5.69) also describes the tetragonal (space group *I*4/*mmm*) substitutional superstructure A_3B. Compounds such as Al_3Ti and Ni_3V have this structure.

The distribution function (5.69) assumes three values at all sites of an fcc non-metal sublattice (see Table 5.10). In an ideal fully ordered tetragonal (space group *I*4/*mmm*) structure M_4X_3 one-half of metal atoms occupy positions M_2^0, while the other half of these atoms are in positions M_1^4 (Fig. 5.22).

Diffraction Spectra of Superstructures of Type M_4X_3. Considering (5.44) and (5.51) and taking into account the parameters $m_{10} = 3$ and $\gamma_{10} = 1/4$, the intensity of superstructure diffraction reflections of the polycrystalline ordered cubic (space group $Pm\bar{3}m$) phase M_4X_3 is given by the formula

$$I_{\text{super}}(q) = (3 f_X^2 N^2 \eta_{10}^2 / 16) \sum_{\boldsymbol{H}} P(q = |\boldsymbol{H} + \boldsymbol{k}_{10}^{(j)}|) . \tag{5.70}$$

In the case of an ideal tetragonal (space group *I*4/*mmm*) superstructure M_4X_3 the intensities of superstructure reflections are

$$I_{\text{super}}(q) = f_X^2 N^2 \sum_{\boldsymbol{H}} [(\eta_{10}^2 / 16) P(q = |\boldsymbol{H} + \boldsymbol{k}_{10}^{(j)}|)$$

$$+ (\eta_8^2 / 8) P(q = |\boldsymbol{H} + \boldsymbol{k}_8^{(j)}|)] . \tag{5.71}$$

Calculated diffraction spectra of the superstructures M_4X_3 are shown in Fig. 5.23.

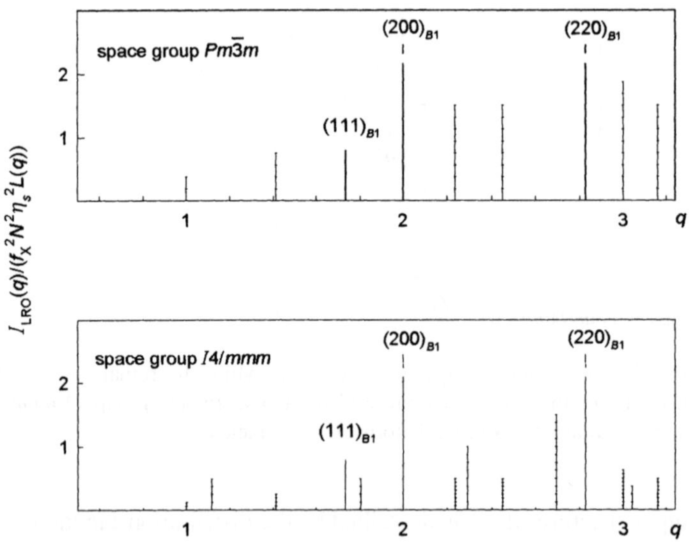

Fig. 5.23. Theoretical diffraction spectra of ideal M_4X_3-type superstructures which are plotted as a function of the reduced diffraction vector modulus, $q = (2a_{B1}\sin\theta)/\lambda$. The reflection intensity is calculated for an atomic scattering factor ratio of $f_M/f_X = 1.065$ which is typical for neutron diffraction. The intensities of superstructure reflections and the structure reflection $(111)_{B1}$ are shown on the same scale

5.4.4 Superstructures of Type M_6X_5

Monoclinic Superstructure M_6X_5 (Space Group *C2/m*). A unit cell of the monoclinic (space group *C2/m*) superstructure M_6X_5 is shown in Fig. 5.24. The translation vectors of this cell and the coordinates of atoms and vacancies are given in Table 5.11. This ordered phase was observed for niobium carbide NbC_y in the interval from $NbC_{0.81}$ to $NbC_{0.88}$ at temperatures below 1300 K [10, 11, 20–23] and for vanadium carbide VC_y [24].

Table 5.11. Monoclinic M_6X_5-type superstructure (space group *C2/m* (C^3_{2h})): $\boldsymbol{a}_m = \frac{1}{2}\langle 1\bar{1}\bar{2}\rangle_{B1}$, $\boldsymbol{b}_m = \frac{3}{2}\langle 110\rangle_{B1}$, $\boldsymbol{c}_m = \frac{1}{2}\langle 1\bar{1}2\rangle_{B1}$

Atom	Position and multiplicity	Atomic coordinates in the ideal ordered structure			Values for the distribution function $n(x_I, y_I, z_I)$
		x/a_m	y/a_m	z/a_m	
X1 (vacancy)	2(*a*)	0	0	0	$n_1 = y - \eta_9/6 - \eta_4/3 - \eta_3/3$
X2	2(*d*)	0	1/2	1/2	$n_2 = y + \eta_9/6 - \eta_4/3 + \eta_3/3$
X3	4(*g*)	0	1/3	0	$n_3 = y - \eta_9/6 + \eta_4/6 + \eta_3/6$
X4	4(*h*)	0	1/6	1/2	$n_4 = y + \eta_9/6 + \eta_4/6 - \eta_3/6$
M1	4(*i*)	1/4	0	3/4	
M2	8(*j*)	1/4	2/3	3/4	

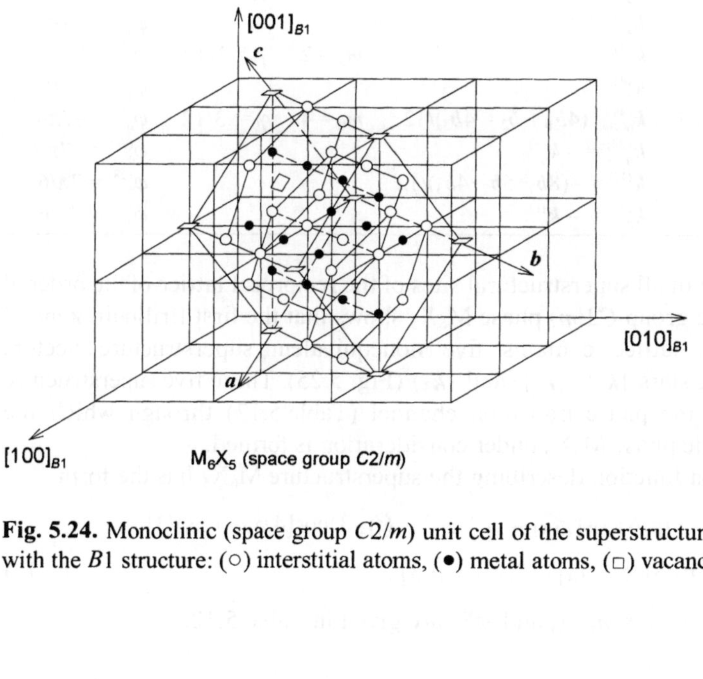

Fig. 5.24. Monoclinic (space group *C2/m*) unit cell of the superstructure M_6X_5 in a lattice with the *B*1 structure: (○) interstitial atoms, (•) metal atoms, (□) vacancies

Table 5.12. MX_y–M_6X_5 disorder–order phase transition channels and parameters of the distribution functions describing the superstructures of M_6X_5 type

Symmetry	Space group	Disorder–order phase-transition channel	m_s	γ_s	$\varphi_s^{(j)}$
Monoclinic	$C2/m$	$\boldsymbol{k}_9^{(3)} = \boldsymbol{b}_2/2$	$m_9 = 1$	$\gamma_9 = 1/6$	$\varphi_9^{(3)} = \pi$
		$\boldsymbol{k}_4^{(1)} = (\boldsymbol{b}_1 + \boldsymbol{b}_2 + 2\boldsymbol{b}_3)/3$	$m_4 = 2$	$\gamma_4 = 1/6$	$\varphi_4^{(1)} = \pi$
		$\boldsymbol{k}_4^{(2)} = -\boldsymbol{k}_4^{(1)}$			$\varphi_4^{(2)} = -\pi$
		$\boldsymbol{k}_3^{(3)} = -(4\boldsymbol{b}_1 + \boldsymbol{b}_2 + 2\boldsymbol{b}_3)/6$	$m_3 = 2$	$\gamma_3 = 1/6$	$\varphi_3^{(3)} = \pi$
		$\boldsymbol{k}_3^{(4)} = -\boldsymbol{k}_3^{(3)}$			$\varphi_3^{(4)} = -\pi$
Trigonal	$P3_1$	$\boldsymbol{k}_9^{(3)}$	$m_9 = 1$	$\gamma_9 = 1/6$	$\varphi_9^{(3)} = \pi$
		$\boldsymbol{k}_4^{(1)}$	$m_4 = 6$	$\gamma_4 = \sqrt{3}/18$	$\varphi_4^{(1)} = 7\pi/6$
		$\boldsymbol{k}_4^{(2)}$			$\varphi_4^{(2)} = -7\pi/6$
		$\boldsymbol{k}_4^{(7)} = (\boldsymbol{b}_3 - \boldsymbol{b}_1)/3$			$\varphi_4^{(7)} = 3\pi/2$
		$\boldsymbol{k}_4^{(8)} = -\boldsymbol{k}_4^{(7)}$			$\varphi_4^{(8)} = -3\pi/2$
		$\boldsymbol{k}_4^{(9)} = (2\boldsymbol{b}_1 + \boldsymbol{b}_2 + \boldsymbol{b}_3)/3$			$\varphi_4^{(9)} = 5\pi/6$
		$\boldsymbol{k}_4^{(10)} = -\boldsymbol{k}_4^{(9)}$			$\varphi_4^{(10)} = -5\pi/6$
		$\boldsymbol{k}_3^{(3)}$	$m_3 = 6$	$\gamma_3 = \sqrt{3}/18$	$\varphi_3^{(3)} = 7\pi/6$
		$\boldsymbol{k}_3^{(4)}$			$\varphi_3^{(4)} = -7\pi/6$
		$\boldsymbol{k}_3^{(9)} = (2\boldsymbol{b}_1 + 3\boldsymbol{b}_2 + 4\boldsymbol{b}_3)/6$			$\varphi_3^{(9)} = \pi/2$
		$\boldsymbol{k}_3^{(10)} = -\boldsymbol{k}_3^{(9)}$			$\varphi_3^{(10)} = -\pi/2$
		$\boldsymbol{k}_3^{(23)} = (2\boldsymbol{b}_1 + \boldsymbol{b}_2 - 2\boldsymbol{b}_3)/6$			$\varphi_3^{(23)} = -5\pi/6$
		$\boldsymbol{k}_3^{(24)} = -\boldsymbol{k}_3^{(23)}$			$\varphi_3^{(24)} = 5\pi/6$
Monoclinic	$C2$	$\boldsymbol{k}_9^{(3)}$	$m_9 = 1$	$\gamma_9 = 1/6$	$\varphi_9^{(3)} = \pi$
		$\boldsymbol{k}_4^{(1)}$	$m_4 = 2$	$\gamma_4 = 1/12$	$\varphi_4^{(1)} = 4\pi/3$
		$\boldsymbol{k}_4^{(2)}$			$\varphi_4^{(2)} = -4\pi/3$
		$\boldsymbol{k}_3^{(3)}$	$m_3 = 2$	$\gamma_3 = 1/12$	$\varphi_3^{(3)} = 4\pi/3$
		$\boldsymbol{k}_3^{(4)}$			$\varphi_3^{(4)} = -4\pi/3$
		$\boldsymbol{k}_0^{(4)} = (4\boldsymbol{b}_1 + \boldsymbol{b}_2 - 4\boldsymbol{b}_3)/12$	$m_0 = 4$	$\gamma_0 = \sqrt{3}/12$	$\varphi_0^{(4)} = -7\pi/6$
		$\boldsymbol{k}_0^{(28)} = -\boldsymbol{k}_0^{(4)}$			$\varphi_0^{(13)} = -7\pi/6$
		$\boldsymbol{k}_0^{(13)} = -(8\boldsymbol{b}_1 + 5\boldsymbol{b}_2 + 4\boldsymbol{b}_3)/12$			$\varphi_0^{(28)} = 7\pi/6$
		$\boldsymbol{k}_0^{(37)} = -\boldsymbol{k}_0^{(13)}$			$\varphi_0^{(37)} = 7\pi/6$

The translation of all superstructural sites of the reciprocal lattice of the ordered monoclinic (space group $C2/m$) phase M_6X_5 shows that the first Brillouin zone of a disordered fcc lattice contains five nonequivalent superstructure vectors belonging to three stars $\{\boldsymbol{k}_9\}$, $\{\boldsymbol{k}_4\}$, and $\{\boldsymbol{k}_3\}$ (Fig. 5.25). These five superstructure vectors comprise the phase transition channel (Table 5.12) through which the ordered monoclinic phase M_6X_5 under consideration is formed.

The distribution function describing the superstructure M_6X_5 has the form

$$n(x_{\rm I}, y_{\rm I}, z_{\rm I}) = y - (\eta_9/6)\cos[\pi(x_{\rm I} - y_{\rm I} + z_{\rm I})] - (\eta_4/3)\cos[4\pi(x_{\rm I} + y_{\rm I})/3] - (\eta_3/3)\cos[\pi(x_{\rm I} - 5y_{\rm I} - 3z_{\rm I})/3]\,. \quad (5.72)$$

Values of the parameters m_s, γ_s, and $\varphi_s^{(j)}$ are given in Table 5.12.

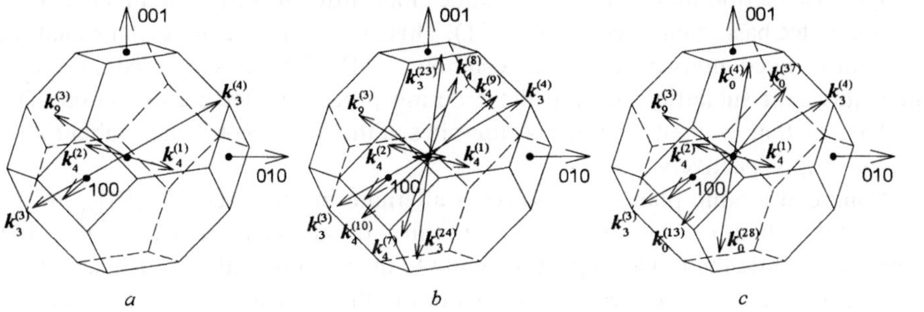

Fig. 5.25. Superstructure vectors of the reciprocal lattice in superstructures type M_6X_5 comprising a disorder–order phase transition channel MX_y–M_6X_5 and the positions of these vectors in the first Brillouin zone of an fcc lattice: (*a*) monoclinic (space group *C*2/*m*) superstructure M_6X_5; (*b*) trigonal (space group $P3_1$) superstructure M_6X_5; (*c*) monoclinic (space group *C*2) superstructure M_6X_5

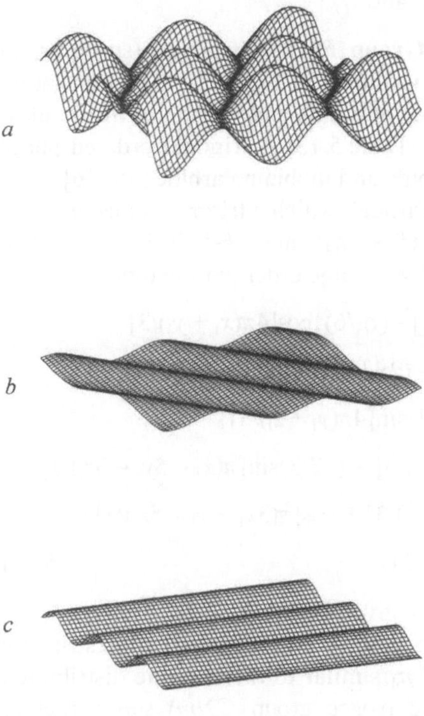

Fig. 5.26. Concentration waves in the defective plane $(1\bar{1}1)_{B1}$ of the monoclinic (space group *C*2/*m*) superstructure M_6X_5: (*a*) $\eta_9 = \eta_4 = \eta_3 = 1$; (*b*) $\eta_9 = 1/3$, $\eta_4 = 0$, $\eta_3 = 1/3$; (*c*) $\eta_9 = \eta_4 = 1/2$, $\eta_3 = 0$

The distribution function (5.72) assumes four different values at all sites of the nonmetal fcc basic sublattice (Table 5.11). This means that the nonmetal sublattice of a disordered nonstoichiometric compound MX_y becomes divided into four nonequivalent sublattices during the ordering process under consideration. The probability that interstitial atoms occupy sites of the first, second, etc., sublattice is n_1, n_2, etc., respectively.

Nonmetal atomic planes of two types alternate in the direction $[1\bar{1}1]_{B1}$ in the nonmetal sublattice. On one type, one-third of all sites belong to the first sublattice, while the remaining sites belong to the third sublattice. Each site of the first sublattice is surrounded by six sites of the third sublattice. On the planes of the second type, one-third of all sites belong to the second sublattice, while the remaining sites are located on the fourth sublattice. Each site of the second sublattice is surrounded by six sites of the fourth sublattice.

As may be inferred from (5.72), this distribution function represents a superposition of three concentration waves related to the three long-range order parameters η_9, η_4, and η_3. Figure 5.26 illustrates the superposition of concentration waves calculated for the plane $(1\bar{1}1)_{B1}$ of the former type taking some reasonable values for the parameters η_9, η_4, and η_3.

Trigonal Superstructure M_6X_5 (Space Group *P*3$_1$). A trigonal (space group $P3_1$) unit cell of the superstructure M_6X_5 is shown in Fig. 5.27. The translation vectors of this unit cell and the coordinates of atoms and vacancies in the ideal trigonal superstructure M_6X_5 are given in Table 5.13. A trigonal ordered phase type M_6X_5 was observed for vanadium carbide and niobium carbide [25, 26].

The order–disorder transition channel, through which a trigonal superstructure M_6X_5 is formed, involves 13 rays of stars $\{\boldsymbol{k}_9\}$, $\{\boldsymbol{k}_4\}$, and $\{\boldsymbol{k}_3\}$ (Table 5.12). The distribution function depends on the three long-range order parameters:

$$\begin{aligned} n(x_\mathrm{I}, y_\mathrm{I}, z_\mathrm{I}) = y &- (\eta_9/6)\cos[\pi(x_\mathrm{I} - y_\mathrm{I} + z_\mathrm{I})] - (\eta_4/6)\{\cos[4\pi(x_\mathrm{I} + y_\mathrm{I})/3] \\ &- (\sqrt{3}/3)\sin[4\pi(x_\mathrm{I} + y_\mathrm{I})/3] - (2\sqrt{3}/3)\sin[4\pi(x_\mathrm{I} - y_\mathrm{I})/3] \\ &+ \cos[4\pi(y_\mathrm{I} + z_\mathrm{I})/3] + (\sqrt{3}/3)\sin[4\pi(y_\mathrm{I} + z_\mathrm{I})/3]\} \\ &- (\eta_3/3)\{\cos[\pi(x_\mathrm{I} - 5y_\mathrm{I} - 3z_\mathrm{I})/3] - (\sqrt{3}/3)\sin[\pi(x_\mathrm{I} - 5y_\mathrm{I} - 3z_\mathrm{I})/3] \\ &+ (2\sqrt{3}/3)\sin[\pi(5x_\mathrm{I} + 3y_\mathrm{I} + z_\mathrm{I})/3] + \cos[\pi(3x_\mathrm{I} + y_\mathrm{I} - 5z_\mathrm{I})/3] \\ &+ (\sqrt{3}/3)\sin[\pi(3x_\mathrm{I} + y_\mathrm{I} - 5z_\mathrm{I})/3]\}. \end{aligned} \tag{5.73}$$

Values of the parameters m_s, γ_s, and $\varphi_s^{(j)}$ are given in Table 5.12.

The distribution function (5.73), which describes the trigonal superstructure M_6X_5, assumes four values n_1, n_2, n_3, and n_4 similar to those of the distribution function (5.72) describing the monoclinic (space group $C2/m$) superstructure M_6X_5. However, the relative location of sites in the four sublattices of the trigonal ordered structure and the monoclinic superstructure is different. Values of the distribution function (5.73) are given in Table 5.13.

Table 5.13. Trigonal M_6X_5-type superstructure (space group $P3_1$ (C_3^2)): $\boldsymbol{a}_{tr} = \frac{1}{2}\langle 2\bar{1}\bar{1}\rangle_{B1}$, $\boldsymbol{b}_{tr} = \frac{1}{2}\langle \bar{1}\bar{1}2\rangle_{B1}$, $\boldsymbol{c}_{tr} = 2\langle 111\rangle_{B1}$

Atom	Position and multiplicity	Atomic coordinates in the ideal ordered structure			Values of the distribution function $n(x_I, y_I, z_I)$
		x/a_{tr}	y/b_{tr}	z/c_{tr}	
X1 (vacancy)	3(*a*)	1/9	8/9	1/6	$n_1 = y - \eta_9/6 - \eta_4/3 - \eta_3/3$
X2	3(*a*)	4/9	5/9	1/6	$n_3 = y - \eta_9/6 + \eta_4/6 + \eta_3/6$
X3	3(*a*)	7/9	2/9	1/6	$n_3 = y - \eta_9/6 + \eta_4/6 + \eta_3/6$
X4	3(*a*)	1/9	5/9	1/3	$n_4 = y + \eta_9/6 + \eta_4/6 - \eta_3/6$
X5	3(*a*)	4/9	2/9	1/3	$n_4 = y + \eta_9/6 + \eta_4/6 - \eta_3/6$
X6	3(*a*)	7/9	8/9	1/3	$n_2 = y + \eta_9/6 - \eta_4/3 + \eta_3/3$
M1	3(*a*)	1/9	5/9	1/12	
M2	3(*a*)	4/9	2/9	1/12	
M3	3(*a*)	7/9	8/9	1/12	
M4	3(*a*)	1/9	2/9	1/4	
M5	3(*a*)	4/9	8/9	1/4	
M6	3(*a*)	7/9	5/9	1/4	

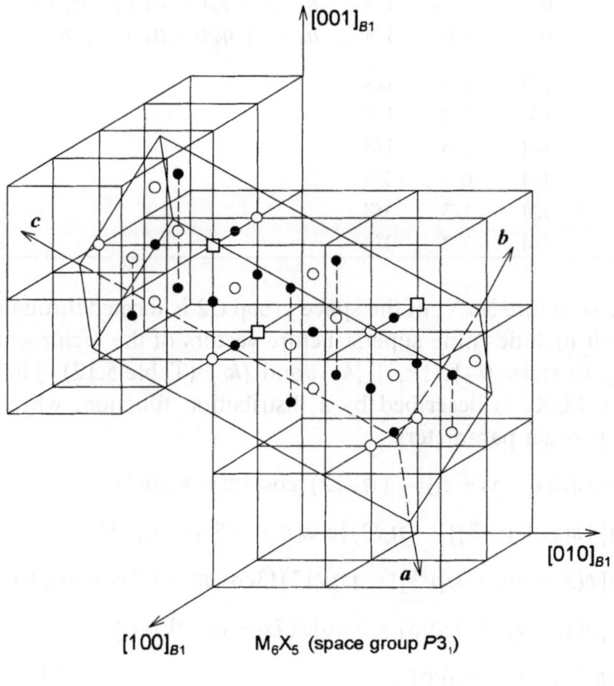

Fig. 5.27. Trigonal (space group $P3_1$) unit cell of the superstructure M_6X_5 in a lattice with the $B1$ structure: (○) interstitial atoms, (•) metal atoms, (□) vacancies

Monoclinic Superstructure M_6X_5 (Space Group *C*2). A unit cell of the monoclinic (space group *C*2) superstructure M_6X_5 is shown in Fig. 5.28. The translation vectors of this unit cell and the coordinates of atoms and vacancies in an ideal monoclinic (space group *C*2) superstructure M_6X_5 are given in Table 5.14. This ordered monoclinic phase type M_6X_5 was observed for carbide VC_y. The V_6C_5 phase is stable at temperatures below 1400 K [27].

Table 5.14. Monoclinic M_6X_5-type superstructure (space group $C2$ (C_2^3)): $\boldsymbol{a}_{\rm m} = \frac{1}{2}\langle 1\bar{1}\bar{2}\rangle_{B1}$, $\boldsymbol{b}_{\rm m} = \frac{3}{2}\langle 110\rangle_{B1}$, $\boldsymbol{c}_{\rm m} = \langle \bar{1}\bar{1}2\rangle_{B1}$

Atom	Position and multiplicity	Atomic coordinates in the ideal ordered structure			Values for the distribution function $n(x_{\rm I}, y_{\rm I}, z_{\rm I})$
		$x/a_{\rm m}$	$y/b_{\rm m}$	$z/c_{\rm m}$	
X1 (vacancy)	2(*a*)	0	0	0	$n_1 = y - \eta_9/6 - \eta_4/12 - \eta_3/12 - \eta_0/2$
X2 (vacancy)	2(*b*)	0	1/3	1/2	$n_1 = y - \eta_9/6 - \eta_4/12 - \eta_3/12 - \eta_0/2$
X3	2(*a*)	0	1/3	0	$n_5 = y - \eta_9/6 - \eta_4/12 - \eta_3/12 + \eta_0/2$
X4	2(*a*)	0	2/3	0	$n_3 = y - \eta_9/6 + \eta_4/6 + \eta_3/6$
X5	2(*b*)	0	0	1/2	$n_5 = y - \eta_9/6 - \eta_4/12 - \eta_3/12 + \eta_0/2$
X6	2(*b*)	0	2/3	1/2	$n_3 = y - \eta_9/6 + \eta_4/6 + \eta_3/6$
X7	4(*c*)	0	1/6	1/4	$n_4 = y + \eta_9/6 + \eta_4/6 - \eta_3/6$
X8	4(*c*)	0	1/2	1/4	$n_2 = y + \eta_9/6 - \eta_4/12 + \eta_3/12$
X9	4(*c*)	0	5/6	1/4	$n_4 = y + \eta_9/6 + \eta_4/6 - \eta_3/6$
M1	4(*c*)	1/4	1/6	1/8	
M2	4(*c*)	1/4	1/2	1/8	
M3	4(*c*)	1/4	5/6	1/8	
M4	4(*c*)	1/4	0	3/8	
M5	4(*c*)	1/4	1/3	3/8	
M6	4(*c*)	1/4	2/3	3/8	

An ordered monoclinic structure M_6X_5 of the space group *C*2 is formed through a transition channel, which includes nine superstructure vectors of the reciprocal lattice. These rays belong to stars $\{\boldsymbol{k}_9\}$, $\{\boldsymbol{k}_4\}$, $\{\boldsymbol{k}_3\}$, and $\{\boldsymbol{k}_0\}$ (Table 5.12). This monoclinic superstructure M_6X_5 is described by a distribution function, which depends on four long-range order parameters:

$$\begin{aligned} n(x_{\rm I}, y_{\rm I}, z_{\rm I}) = y &- (\eta_9/6)\cos[\pi(x_{\rm I} - y_{\rm I} + z_{\rm I})] - (\eta_4/12)\{\cos[4\pi(x_{\rm I} + y_{\rm I})/3] \\ &- (\surd 3)\sin[4\pi(x_{\rm I} + y_{\rm I})/3]\} - (\eta_3/3)\{\cos[\pi(x_{\rm I} - 5y_{\rm I} - 3z_{\rm I})/3] \\ &- (\surd 3)\sin[\pi(x_{\rm I} - 5y_{\rm I} - 3z_{\rm I})/3]\} - (\eta_0/12)\{3\cos[\pi(x_{\rm I} + 7y_{\rm I} + 9z_{\rm I})/6] \\ &- (\surd 3)\sin[\pi(x_{\rm I} + 7y_{\rm I} + 9z_{\rm I})/6] + 3\cos[\pi(7x_{\rm I} + y_{\rm I} - 9z_{\rm I})/6] \\ &- (\surd 3)\sin[\pi(7x_{\rm I} + y_{\rm I} - 9z_{\rm I})/6]\}. \end{aligned} \tag{5.74}$$

Values for the parameters m_s, γ_s, and $\varphi_s^{(j)}$ are given in Table 5.12.

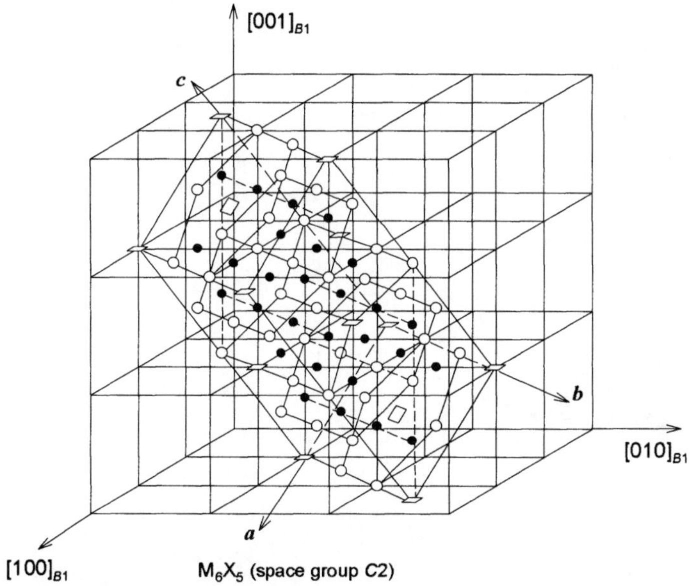

Fig. 5.28. Monoclinic (space group *C*2) unit cell of the superstructure M_6X_5 in a lattice with the *B*1 structure: (○) interstitial atoms, (•) metal atoms, (□) vacancies

The distribution function (5.74) assumes five values at all sites of the nonmetal basic fcc sublattice (Table 5.14). Correspondingly, the nonmetal sublattice of a disordered nonstoichiometric compound is divided during ordering into five nonequivalent sublattices, which differ from one another in the probabilities of the sites occupied by interstitial atoms.

Two types of nonmetal atomic planes alternate in the $[1\bar{1}1]_{B1}$ direction in this superstructure. One type contains an equal number of sites of the first, third and fifth sublattice. In these planes the nearest neighbors of sites of a given sublattice are sites of the other two sublattices only. Planes of the other type comprise sites of the second and fourth sublattice. Each site of the fourth sublattice is surrounded by six sites of the second sublattice.

Figure 5.29 presents the superposition of concentration waves calculated for the nonmetal plane $(1\bar{1}1)_{B1}$ of the above type in the monoclinic superstructure M_6X_5 (space group *C*2). The comparison of Figs. 5.26 and 5.29 shows that the distribution of carbon atoms and vacancies in the superstructure M_6X_5 (space group *C*2) is very different from their distribution in the superstructure M_6X_5 (space group *C*2/*m*).

a

b

c

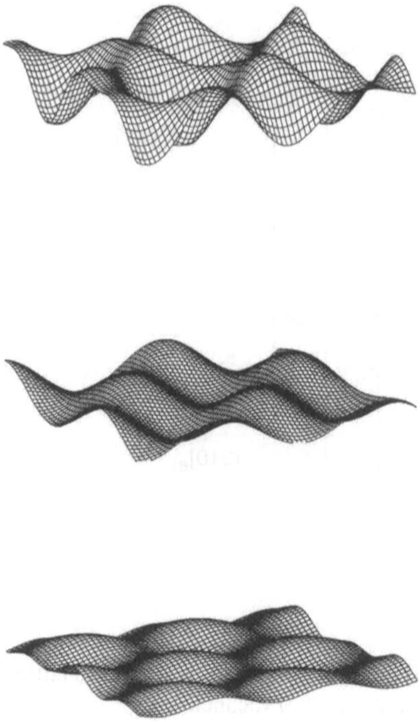

Fig. 5.29. Concentration waves in the defective plane $(1\bar{1}1)_{B1}$ of the monoclinic (space group *C*2) superstructure M_6X_5: (*a*) $\eta_9 = \eta_4 = \eta_3 = \eta_0 = 1$; (*b*) $\eta_9 = 1$, $\eta_4 = \eta_3 = 0$, $\eta_0 = 2/3$; (*c*) $\eta_9 = \eta_0 = 0$, $\eta_4 = \eta_3 = 1/2$

Distribution of Atoms in Superstructures of Type M_6X_5. The concentration of interstitial atoms is $y_{st} = 5/6$ in all fully ordered phases of type M_6X_5. Complete atomic planes $(1\bar{1}1)_{B1}$, in which all sites are occupied by interstitial atoms, and defective planes $(111)_{B1}$, in which one-third of sites are vacant, alternate in the direction $[1\bar{1}1]_{B1}$ in the nonmetal sublattice in these superstructures (Figs. 5.24, 5.27, and 5.28). One-third of all sites are vacant in the defective planes and each vacancy is surrounded by six interstitial atoms forming a regular hexagon. Specific features of superstructures of type M_6X_5 were comprehensively discussed in [12] and are described in Sect. 3.4.

The formation of various ordered structures of M_6X_5 type is associated with different mutual arrangements of defective planes. Figure 5.30 shows the displacement directions of the upper defective planes in the projection on the $(1\bar{1}1)_{B1}$ defective plane. Each sequence of displacements corresponds to an M_6X_5-type superstructure.

The distribution functions (5.72), (5.73) and (5.74) also describe substitutional superstructures of space groups *C*2/*m*, *P*3$_1$, and *C*2.

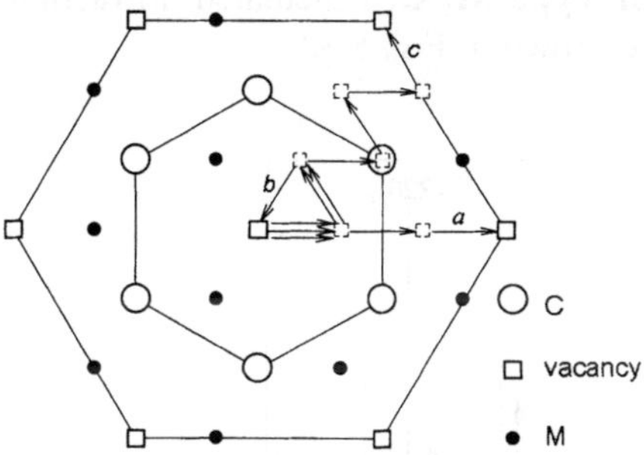

Fig. 5.30. Directions of the relative displacement of $(1\bar{1}1)_{B1}$ defective planes for M_6X_5-type superstructures with space groups *C*2/*m* (*a*), *P*3$_1$ (*b*) and *C*2 (*c*). The $[1\bar{1}1]_{B1}$ direction is perpendicular to the figure plane

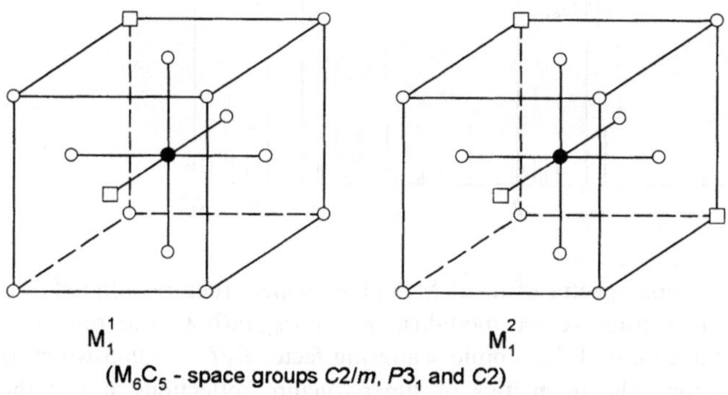

Fig. 5.31. Positions M_1^1 and M_1^2 of metal atoms (•) in ideal fully ordered structures of type M_6X_5 (space groups *C*2/*m*, *P*3$_1$, and *C*2): (o) interstitial atoms, (□) vacancies. Two-thirds of all metal atoms occupy positions M_1^1, and one-third of metal atoms are in positions M_1^2

The metal atoms have a similar environment of nonmetal lattice sites forming the first and second coordination spheres in fully ordered monoclinic (space groups *C*2/*m* and *C*2) and trigonal (space group *P*3$_1$) structures of type M_6X_5. Indeed, the metal atoms of these fully ordered structures are located in two positions only (Fig. 5.31): M_1^1 (a metal atom with one vacancy in the first coordination sphere and one vacancy in the second coordination sphere) and M_1^2 (a metal atom with one vacancy in the first coordination sphere and two vacancies in the second coordination sphere). In the ideal fully ordered structures of type M_6X_5 (space groups *C*2/*m*, *P*3$_1$, and *C*2) two-thirds and one-third of all metal atoms occupy positions M_1^1 and M_1^2 respectively. The presence of these positions was confirmed by NMR studies of short-range order in ordered carbides V_6C_5 [25] and Nb_6C_5 [8, 28–31].

Diffraction Spectra of Superstructures of Type M_6X_5. Calculated diffraction spectra of the ideal superstructures M_6X_5 are shown in Fig. 5.32.

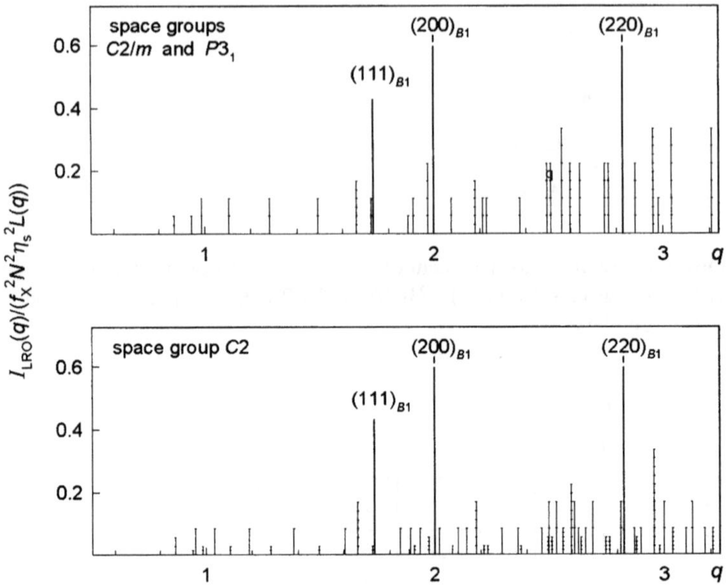

Fig. 5.32. Theoretical diffraction spectra of ideal M_6X_5-type superstructures plotted as a function of the reduced diffraction vector modulus, $q = (2a_{B1}\sin\theta)/\lambda$. The reflection intensities are calculated for a ratio of the atomic scattering factor $f_M/f_X = 1.065$ which is typical for neutron diffraction. The intensities of superstructure reflections and of the structure reflection $(111)_{B1}$ are shown on the same scale

One can see that the ideal monoclinic (space group $C2/m$) and trigonal (space group $P3_1$) superstructures M_6X_5 have similar diffraction patterns. This is quite reasonable since the disorder–order phase transition channels, which correspond to each of these superstructures, include rays of the same stars $\{\boldsymbol{k}_9\}$, $\{\boldsymbol{k}_4\}$, and $\{\boldsymbol{k}_3\}$, while the values of the parameters μ_4 and μ_3, which determine the length of rays $\boldsymbol{k}_4^{(j)}$ and $\boldsymbol{k}_3^{(1)}$, are similar for both superstructures. Hence, diffraction spectra of these ordered polycrystalline phases M_6X_5 appear similar with respect to location of superstructure reflections. From (5.44) and (5.51) it is apparent that given an equal value of η_s, the intensities of superstructure reflections due to the same star $\{\boldsymbol{k}_s\}$ will be different if the values $m_s\gamma_s^2$ corresponding to each of the superstructures are different. But $m_s\gamma_s^2$ values of the same star $\{\boldsymbol{k}_s\}$ are equal for the monoclinic (space group $C2/m$) and trigonal (space group $P3_1$) superstructures M_6X_5. Indeed, considering (5.40) and (5.47) and taking into account the values of the parameters m_s, γ_s and $\varphi_s^{(j)}$ (Table 5.12), the intensities of superstructure reflections for ordered polycrystalline phases M_6X_5 (space groups $C2/m$ and $P3_1$) are described by the formula

$$I_{\text{super}}(q) = f_X^2 N^2 \sum_{H} [(\eta_9^2/36)P(q=|H+k_9^{(j)}|) + (\eta_4^2/18)P(q=|H+k_4^{(j)}|)$$
$$+ (\eta_3^2/18)P(q=|H+k_3^{(j)}|)]\,. \quad (5.75)$$

The diffraction spectra of these ideal superstructures appear similar both in location and intensity of superstructure reflections if the corresponding parameters of long-range order coincide.

However, in real ordered phases M_6X_5 (space groups $C2/m$ and $P3_1$) the intensities of superstructure reflections may fail to agree to some extent because of different directions and values of atomic displacements. The monoclinic or trigonal symmetry of ordered phases of type M_6X_5 can be determined unambiguously if the diffraction spectra are analyzed taking into account the possible static displacements of atoms.

The diffraction patterns of the monoclinic (space group $C2$) superstructure M_6X_5 are more complex than those for the two superstructures M_6X_5 (space groups $C2/m$ and $P3_1$) considered above (Fig. 5.32). In the case of an ideally ordered phase M_6X_5 of the space group $C2$ the intensities of superstructure reflections of a polycrystalline sample is given by the relation

$$I_{\text{super}}(q) = f_X^2 N^2 \sum_{H} [(\eta_9^2/36)P(q=|H+k_9^{(j)}|) + (\eta_4^2/72)P(q=|H+k_4^{(j)}|)$$
$$+ (\eta_3^2/72)P(q=|H+k_3^{(j)}|) + (\eta_0^2/12)P(q=|H+k_0^{(j)}|)]\,. \quad (5.76)$$

Thus, diffraction spectra of this ordered phase include superstructure reflections due to four stars $\{\boldsymbol{k}_9\}$, $\{\boldsymbol{k}_4\}$, $\{\boldsymbol{k}_3\}$, and $\{\boldsymbol{k}_0\}$.

5.4.5 Superstructures of Type M_8X_7

Cubic Superstructure M_8X_7 (Space Group $Fm\bar{3}m$). A cubic (space group $Fm\bar{3}m$) unit cell of the superstructure M_8X_7 is shown in Fig. 5.33. The phase transition channel, through which this superstructure is formed, includes all rays of stars $\{\boldsymbol{k}_{10}\}$ and $\{\boldsymbol{k}_9\}$ (Table 5.15). The translation vectors of this unit cell and the coordinates of atoms and vacancies in the ideally ordered cubic (space group $Fm\bar{3}m$) structure M_8X_7 are given in Table 5.16.

The distribution function of interstitial atoms in the ordered cubic (space group $Fm\bar{3}m$) phase type M_8X_7 has the form

$$n(x_I, y_I, z_I) = y - (\eta_{10}/8)(\cos 2\pi x_I + \cos 2\pi y_I + \cos 2\pi z_I)$$
$$- (\eta_9/8)\{\cos[\pi(x_I + y_I + z_I)] + \cos[\pi(x_I - y_I + z_I)]$$
$$+ \cos[\pi(x_I + y_I - z_I)] + \cos[\pi(-x_I + y_I + z_I)]\}\,. \quad (5.77)$$

Values of the parameters m_s, γ_s, and $\varphi_s^{(j)}$ are given in Table 5.15.

Table 5.15. MX_y–M_8X_7 disorder–order phase transition channels and parameters of the distribution functions describing the superstructures of M_8X_7 type

Symmetry	Space group	Disorder–order phase transition channel	m_s	γ_s	$\varphi_s^{(j)}$
Cubic	$Fm\bar{3}m$	$\boldsymbol{k}_{10}^{(1)} = (\boldsymbol{b}_1 + \boldsymbol{b}_2)/2$	$m_{10} = 3$	$\gamma_{10} = 1/8$	$\varphi_{10}^{(1)} = \pi$
		$\boldsymbol{k}_{10}^{(2)} = (\boldsymbol{b}_1 + \boldsymbol{b}_3)/2$			$\varphi_{10}^{(2)} = \pi$
		$\boldsymbol{k}_{10}^{(3)} = (\boldsymbol{b}_2 + \boldsymbol{b}_3)/2$			$\varphi_{10}^{(3)} = \pi$
		$\boldsymbol{k}_9^{(1)} = (\boldsymbol{b}_1+\boldsymbol{b}_2+\boldsymbol{b}_3)/2$	$m_9 = 4$	$\gamma_9 = 1/8$	$\varphi_9^{(1)} = \pi$
		$\boldsymbol{k}_9^{(2)} = \boldsymbol{b}_1/2$			$\varphi_9^{(2)} = \pi$
		$\boldsymbol{k}_9^{(3)} = \boldsymbol{b}_2/2,\ \boldsymbol{k}_9^{(4)} = \boldsymbol{b}_3/2$			$\varphi_9^{(3)} = \varphi_9^{(4)} = \pi$
Cubic	$P4_332$	$\boldsymbol{k}_9^{(1)}$	$m_9 = 4$	$\gamma_9 = 1/16$	$\varphi_9^{(1)} = 0,$
		$\boldsymbol{k}_9^{(2)}$			$\varphi_9^{(2)} = \pi$
		$\boldsymbol{k}_9^{(3)},\ \boldsymbol{k}_9^{(4)}$			$\varphi_9^{(3)} = \varphi_9^{(4)} = \pi$
		$\boldsymbol{k}_8^{(1)} = (\boldsymbol{b}_1+3\boldsymbol{b}_2+2\boldsymbol{b}_3)/4$	$m_8 = 6$	$\gamma_8 = \sqrt{2}/16$	$\varphi_8^{(1)} = 5\pi/4,$
		$\boldsymbol{k}_8^{(2)} = -\boldsymbol{k}_8^{(1)}$			$\varphi_8^{(2)} = -5\pi/4$
		$\boldsymbol{k}_8^{(3)} = (3\boldsymbol{b}_1+2\boldsymbol{b}_2+\boldsymbol{b}_3)/4$			$\varphi_8^{(3)} = 5\pi/4$
		$\boldsymbol{k}_8^{(4)} = -\boldsymbol{k}_8^{(3)}$			$\varphi_8^{(4)} = -5\pi/4$
		$\boldsymbol{k}_8^{(5)} = (2\boldsymbol{b}_1+\boldsymbol{b}_2+3\boldsymbol{b}_3)/4$			$\varphi_8^{(5)} = 5\pi/4$
		$\boldsymbol{k}_8^{(6)} = -\boldsymbol{k}_8^{(5)}$			$\varphi_8^{(6)} = -5\pi/4$
		$\boldsymbol{k}_4^{(1)} = (\boldsymbol{b}_1+\boldsymbol{b}_2+2\boldsymbol{b}_3)/3$	$m_4 = 12$	$\gamma_4 = 1/16$	$\varphi_4^{(1)} = \pi$
		$\boldsymbol{k}_4^{(2)} = -\boldsymbol{k}_4^{(1)}$			$\varphi_4^{(2)} = -\pi$
		$\boldsymbol{k}_4^{(3)} = (\boldsymbol{b}_2 - \boldsymbol{b}_1)/4$			$\varphi_4^{(3)} = \pi/2$
		$\boldsymbol{k}_4^{(4)} = -\boldsymbol{k}_4^{(3)}$			$\varphi_4^{(4)} = -\pi/2$
		$\boldsymbol{k}_4^{(5)} = (\boldsymbol{b}_1+2\boldsymbol{b}_2+\boldsymbol{b}_3)/4$			$\varphi_4^{(5)} = \pi$
		$\boldsymbol{k}_4^{(6)} = -\boldsymbol{k}_4^{(5)}$			$\varphi_4^{(6)} = -\pi$
		$\boldsymbol{k}_4^{(7)} = (\boldsymbol{b}_3 - \boldsymbol{b}_1)/3$			$\varphi_4^{(7)} = 3\pi/2$
		$\boldsymbol{k}_4^{(8)} = -\boldsymbol{k}_4^{(7)}$			$\varphi_4^{(8)} = -3\pi/2$
		$\boldsymbol{k}_4^{(9)} = (2\boldsymbol{b}_1+\boldsymbol{b}_2+\boldsymbol{b}_3)/4$			$\varphi_4^{(9)} = \pi$
		$\boldsymbol{k}_4^{(10)} = -\boldsymbol{k}_4^{(9)}$			$\varphi_4^{(10)} = -\pi$
		$\boldsymbol{k}_4^{(11)} = (\boldsymbol{b}_3 - \boldsymbol{b}_2)/4$			$\varphi_4^{(11)} = \pi/2$
		$\boldsymbol{k}_4^{(12)} = -\boldsymbol{k}_4^{(11)}$			$\varphi_4^{(12)} = -\pi/2$

Table 5.16. Cubic M_8X_7-type superstructure (space group $Fm\bar{3}m$ (O_h^5)): $\boldsymbol{a}_1 = \langle 200 \rangle_{B1}$, $\boldsymbol{a}_2 = \langle 020 \rangle_{B1}$, $\boldsymbol{a}_3 = \langle 002 \rangle_{B1}$

Atom	Position and multiplicity	Atomic coordinates in the ideal ordered structure			Values for the distribution function $n(x_I, y_I, z_I)$
		x/a_1	y/a_2	z/a_3	
X1 (vacancy)	4(*a*)	0	0	0	$n_1 = y - 3\eta_{10}/8 - \eta_9/2$
X2	4(*b*)	1/2	1/2	1/2	$n_2 = y - 3\eta_{10}/8 + \eta_9/2$
X3	24(*d*)	0	1/4	1/4	$n_2 = y + \eta_{10}/8$
M1	8(*c*)	1/4	1/4	1/4	
M2	24(*e*)	1/4	0	0	

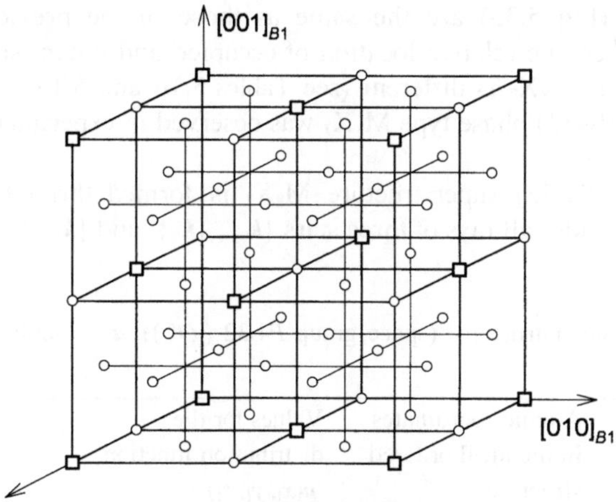

Fig. 5.33. Cubic (space group $Fm\bar{3}m$) unit cell of the superstructure M_8X_7 in a lattice with the *B*1 structure: (○) interstitial atoms, (□) vacancies

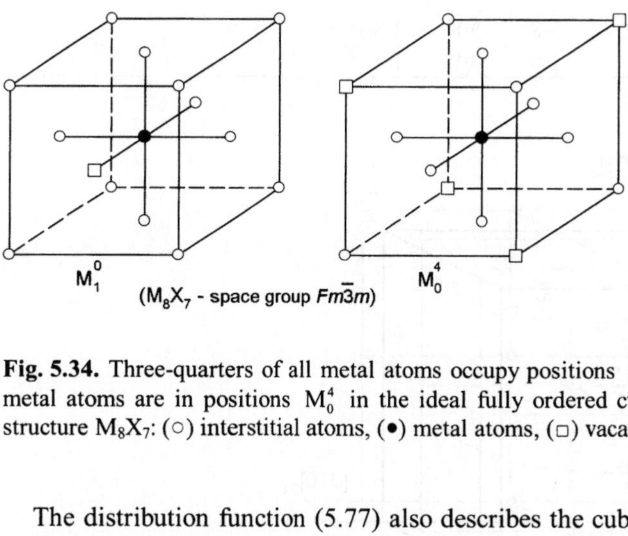

Fig. 5.34. Three-quarters of all metal atoms occupy positions M_1^0 and one-quarter of the metal atoms are in positions M_0^4 in the ideal fully ordered cubic (space group $Fm\bar{3}m$) structure M_8X_7: (○) interstitial atoms, (•) metal atoms, (□) vacancies

The distribution function (5.77) also describes the cubic (space group $Fm\bar{3}m$) substitutional superstructure A_7B (type Pt_7Cu). In the fully ordered cubic phase M_8X_7, $y_{st} = 7/8$.

The distribution function (5.77) depends on two long-range order parameters. The values of this distribution function at sites belonging to different positions of the nonmetal sublattice are given in Table 5.16.

In an ideal ordered cubic (space group $Fm\bar{3}m$) structure M_8X_7 three-quarters and one-quarter of all metal atoms occupy positions M_1^0 and M_0^4 (Fig. 5.34).

Cubic Superstructure M_8X_7 (space group $P4_332$). Translation vectors of a unit cell of this superstructure (Fig. 5.35) are the same as those of the previous superstructure M_8X_7. However, the relative location of occupied and vacant sites in these ordered cubic phases M_8X_7 is different (see Tables 5.16 and 5.17). An ordered cubic (space group $P4_332$) phase type M_8X_7 was observed in experiments with vanadium carbide [32–37].

The cubic (space group $P4_332$) superstructure M_8X_7 is formed through a transition channel, which includes all rays of three stars $\{\boldsymbol{k}_9\}$, $\{\boldsymbol{k}_8\}$, and $\{\boldsymbol{k}_4\}$ ($\mu_4 = 1/4$) (see Table 5.15).

Table 5.17. Cubic M_8X_7-type superstructure (space group $P4_332$ (O^6)): $\boldsymbol{a}_1 = \langle 200 \rangle_{B1}$, $\boldsymbol{a}_2 = \langle 020 \rangle_{B1}$, $\boldsymbol{a}_3 = \langle 002 \rangle_{B1}$

Atom	Position and multiplicity	Atomic coordinates in the ideal ordered structure			Values for the distribution function $n(x_I, y_I, z_I)$
		x/a_1	y/a_2	z/a_3	
X1 (vacancy)	4(*b*)	5/8	5/8	5/8	$n_1 = y - \eta_9/8 - 3\eta_8/8 - 3\eta_4/8$
X2	4(*a*)	1/8	1/8	1/8	$n_2 = y + \eta_9/8 + 3\eta_8/8 - 3\eta_4/8$
X3	12(*d*)	1/8	5/8	5/8	$n_3 = y + \eta_9/8 - \eta_8/8 + \eta_4/8$
X4	12(*d*)	1/8	3/8	7/8	$n_4 = y - \eta_9/8 + \eta_8/8 + \eta_4/8$
M1	8(*c*)	3/8	3/8	3/8	
M2	24(*e*)	1/8	3/8	1/8	

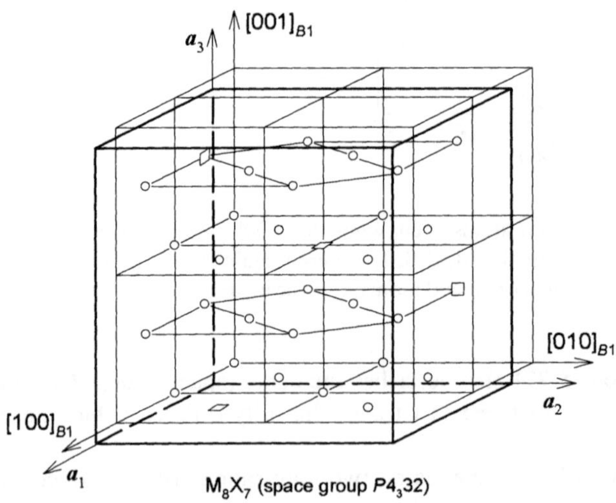

Fig. 5.35. Cubic (space group $P4_332$) unit cell of the superstructure M_8X_7 in a lattice with the $B1$ structure: (○) interstitial atoms, (□) vacancies

The distribution function describing the present superstructure depends on three long-range order parameters:

$$\begin{aligned} n(x_I, y_I, z_I) = y \; - \; &(\eta_9/16)\{\cos[\pi(x_I + y_I + z_I)] - \cos[\pi(x_I - y_I + z_I)] \\ &- \cos[\pi(x_I + y_I - z_I)] - \cos[\pi(-x_I + y_I + z_I)]\} \\ &- (\eta_8/8)\{-\cos[\pi(2x_I + z_I)] + \sin[\pi(2x_I + z_I)] - \cos[\pi(y_I + 2z_I)] \\ &+ \sin[\pi(y_I + 2z_I)] - \cos[\pi(x_I + 2y_I)] + \sin[\pi(x_I + 2y_I)]\} \\ &- (\eta_4/8)\{\cos[\pi(x_I + y_I)] + \cos[\pi(x_I + z_I)] + \cos[\pi(y_I + z_I)] \\ &+ \sin[\pi(x_I - y_I)] - \sin[\pi(x_I - z_I)] + \sin[\pi(y_I - z_I)]\}. \end{aligned} \qquad (5.78)$$

Values for the parameters m_s, γ_s, and $\varphi_s^{(j)}$ of this function are given in Table 5.15. As in the previous case, $y_{st} = 7/8$. The distribution function (5.78) also describes the distribution of atoms of species A in the cubic (space group $P4_332$) substitutional superstructure A_7B.

During ordering, the nonmetal sublattice of a disordered nonstoichiometric compound MX_y is divided into four nonequivalent sublattices. The probabilities that sites of these sublattices are occupied by interstitial atoms X (values of the distribution function (5.78)) are given in Table 5.17.

In an ideal ordered cubic (space group $P4_332$) phase M_8X_7 three-quarters of all metal atoms occupy positions M_1^1, while the rest of the atoms are in positions M_0^1 (Fig. 5.36).

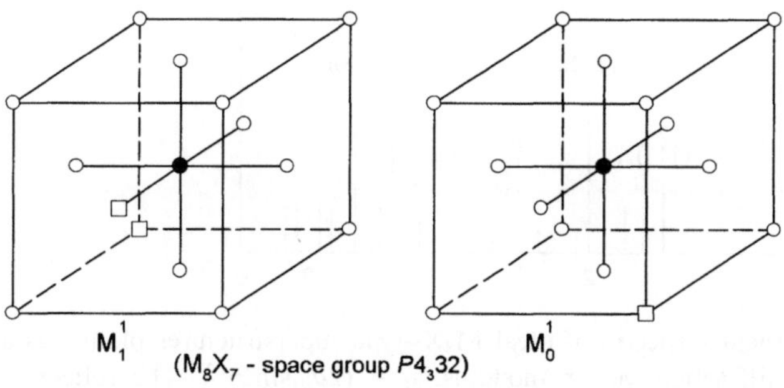

Fig. 5.36. Three-quarters of all metal atoms occupy positions M_1^1 and one-quarter of metal atoms are in positions M_0^1 in the ideal fully ordered cubic (space group $P4_332$) structure M_8X_7: (○) interstitial atoms, (•) metal atoms, (⊔) vacancies

Diffraction Spectra for Superstructures of type M_8X_7. Considering (5.44) and (5.51) and taking into account values for the parameters m_s and γ_s (Table 5.15), the intensities of superstructure reflections in diffraction spectra of the cubic (space group $Fm\bar{3}m$) ordered polycrystalline phase M_8X_7 is described by the formula

$$I_{\text{super}}(q) = f_X^2 N^2 \sum_{\boldsymbol{H}} [(3\eta_{10}^2/64)P(q = |\boldsymbol{H} + \boldsymbol{k}_{10}^{(j)}|)$$
$$+ (\eta_9^2/16)P(q = |\boldsymbol{H} + \boldsymbol{k}_9^{(j)}|)]. \quad (5.79)$$

In the case of the ideal ordered cubic (space group $P4_332$) polycrystalline phase M_8X_7 the intensities of superstructure reflections are

$$I_{\text{super}}(q) = f_X^2 N^2 \sum_{\boldsymbol{H}} [(\eta_9^2/64)P(q = |\boldsymbol{H} + \boldsymbol{k}_9^{(j)}|) + (3\eta_8^2/64)P(q = |\boldsymbol{H} + \boldsymbol{k}_8^{(j)}|)$$
$$+ (3\eta_4^2/64)P(q = |\boldsymbol{H} + \boldsymbol{k}_4^{(j)}|)]. \quad (5.80)$$

Calculated diffraction spectra of ideal superstructures of type M_8X_7 are shown in Fig. 5.37.

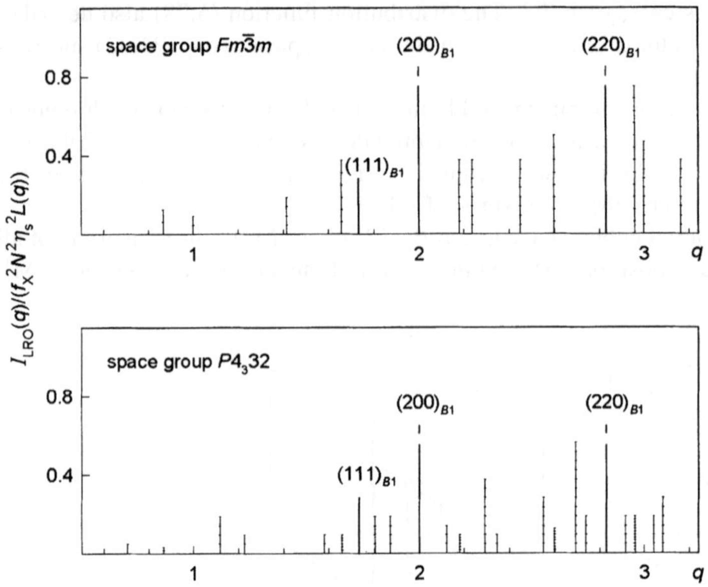

Fig. 5.37. Theoretical diffraction spectra of ideal M_8X_7-type superstructures plotted as a function of the reduced diffraction vector modulus, $q = (2a_{B1}\sin\theta)/\lambda$. The reflection intensity is calculated for a ratio of the atomic scattering factor $f_M/f_X = 1.065$ which is typical for neutron diffraction. The intensities of superstructure reflections and the structural reflection $(111)_{B1}$ are shown on the same scale

5.4.6 Superstructure of Ti_5O_5

Figure 5.38 depicts a unit cell of the monoclinic (space group $C2/m$) superstructure Ti_5O_5, which is formed in titanium monoxide $TiO_y \equiv Ti_xO_z$ containing vacancies both in the titanium and the oxygen sublattices. Translation vectors and coordinates of atoms and vacancies in this unit cell are given in Table 5.18. This superstructure was observed experimentally [38–41].

Table 5.18. Monoclinic Ti_5O_5 ($Ti_5\blacksquare_1O_5\square_1$) superstructure (space group $C2/m$ (C^3_{2h})): $\boldsymbol{a}_m = \langle 10\bar{1}\rangle_{B1}$, $\boldsymbol{b}_m = \langle 010\rangle_{B1}$, $\boldsymbol{c}_m = \langle 102\rangle_{B1}$

Atom	Position and multiplicity	Atomic coordinates in the ideal ordered structure			Values for the distribution function $n_\nu(x_I, y_I, z_I)$ for monoxide Ti_xO_z
		x/a_m	y/a_m	z/a_m	
Ti1 (vacancy)	2(*a*)	0	0	0	$n_{1(Ti)} = x - \eta_{10}/6 - \eta_4/3 - \eta_1/3$
Ti2	2(*d*)	1/2	1/2	1/2	$n_{2(Ti)} = x + \eta_{10}/6 - \eta_4/3 + \eta_1/3$
Ti3	4(*i*)	1/6	0	1/3	$n_{4(Ti)} = x + \eta_{10}/6 + \eta_4/6 - \eta_1/6$
Ti4	4(*i*)	1/3	0	2/3	$n_{3(Ti)} = x - \eta_{10}/6 + \eta_4/6 + \eta_1/6$
O1 (vacancy)	2(*c*)	1/2	0	1/2	$n_{1(O)} = z - \eta_{10}/6 - \eta_4/3 - \eta_1/3$
O2	2(*b*)	0	1/2	0	$n_{2(O)} = z + \eta_{10}/6 - \eta_4/3 + \eta_1/3$
O3	4(*i*)	1/3	0	1/6	$n_{4(O)} = z + \eta_{10}/6 + \eta_4/6 - \eta_1/6$
O4	4(*i*)	1/6	0	5/6	$n_{3(O)} = z - \eta_{10}/6 + \eta_4/6 + \eta_1/6$

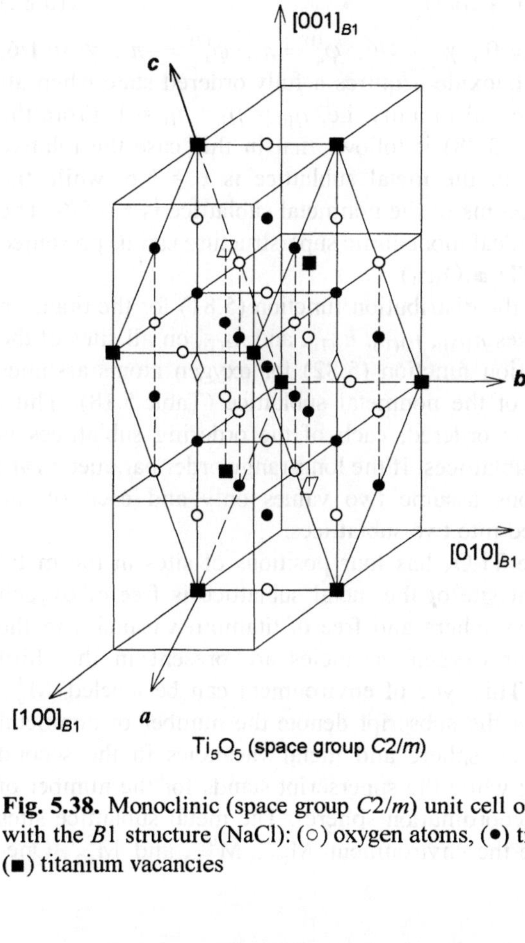

Fig. 5.38. Monoclinic (space group $C2/m$) unit cell of the superstructure Ti_5O_5 in a lattice with the $B1$ structure (NaCl): (○) oxygen atoms, (•) titanium atoms, (□) oxygen vacancies, (■) titanium vacancies

The superstructure Ti_5O_5 is formed through a transition channel including five vectors: the ray $\boldsymbol{k}_{10}^{(1)} = (\boldsymbol{b}_1 + \boldsymbol{b}_2)/2$ of the star $\{\boldsymbol{k}_{10}\}$; two rays $\boldsymbol{k}_4^{(5)} = (\boldsymbol{b}_1 + 2\boldsymbol{b}_2 + \boldsymbol{b}_3)/3$ and $\boldsymbol{k}_4^{(6)} = -\boldsymbol{k}_4^{(5)}$ of the star $\{\boldsymbol{k}_4\}$ with the current parameter $\mu_4 = 1/3$; two rays $\boldsymbol{k}_1^{(7)} = \mu_1^{(1)}(\boldsymbol{b}_2 + \boldsymbol{b}_3) - \mu_1^{(2)}(\boldsymbol{b}_1 + \boldsymbol{b}_2)$ and $\boldsymbol{k}_1^{(8)} = -\boldsymbol{k}_1^{(7)}$ of the star $\{\boldsymbol{k}_1\}$ with current parameters $\mu_1^{(1)} = 1/3$ and $\mu_1^{(2)} = 1/6$.

The distribution function of titanium atoms in a monoclinic (space group $C2/m$) superstructure of titanium monoxide Ti_xO_z has the form

$$n_{Ti}(x_I, y_I, z_I) = x - (\eta_{10}/6)\cos 2\pi z_I - (\eta_4/3)\cos[4\pi(x_I + z_i)/3] - (\eta_1/3)\cos[2\pi(2x_I - z_I)/3] \quad (5.81)$$

with parameters $\gamma_{10} = 1/6$, $\varphi_{10}^{(1)} = \pi$, $\gamma_4 = 1/6$, $\varphi_4^{(1)} = \pi$, $\varphi_4^{(2)} = -\pi$, $\gamma_1 = 1/6$, $\varphi_1^{(1)} = \pi$, $\varphi_1^{(2)} = -\pi$.

The distribution of oxygen atoms in a monoclinic (space group $C2/m$) superstructure of titanium monoxide Ti_xO_z is given by the function

$$n_{O}(x_I, y_I, z_I) = z + (\eta_{10}/6)\cos 2\pi z_I - (\eta_4/3)\cos[4\pi(x_I + z_i)/3] + (\eta_1/3)\cos[2\pi(2x_I - z_I)/3] \quad (5.82)$$

with parameters $\gamma_{10} = 1/6$, $\varphi_{10}^{(1)} = 0$, $\gamma_4 = 1/6$, $\varphi_4^{(1)} = \pi$, $\varphi_4^{(2)} = -\pi$, $\gamma_1 = 1/6$, $\varphi_1^{(1)} = 0$, $\varphi_1^{(2)} = 2\pi$. Titanium monoxide acquires a fully ordered state when all long-range order parameters are equal to unity, i.e. $\eta_{10} = \eta_4 = \eta_1 = 1$. From the distribution functions (5.77) and (5.78) it follows that in this case the relative concentration of titanium atoms in the metal sublattice is $x = 5/6$, while the relative concentration of oxygen atoms in the nonmetal sublattice is $z = 5/6$. The stoichiometric composition of an ideal monoclinic superstructure can be presented as $Ti_{5/6}O_{5/6}$ ($Ti_{0.83}O_{0.83}$) or Ti_5O_5 ($Ti_5■_1O_5□_1$).

In ordered titanium monoxide the distribution function (5.81) for the titanium atoms assumes four different values $n_{1(Ti)}$, $n_{2(Ti)}$, $n_{3(Ti)}$, and $n_{4(Ti)}$ on all sites of the metal sublattice and the distribution function (5.82) for oxygen atoms assumes four similar values on the sites of the nonmetal sublattice (Table 5.18). Thus, when cubic titanium monoxide is ordered, each of the ordering sublattices is divided into four nonequivalent sublattices. If the long-range order parameters are equal, these distribution functions assume two values only and each of the ordering sublattices is only divided into two sublattices.

The monoclinic superstructure Ti_5O_5 has four positions of sites in the metal sublattice (Fig. 5.39). Any vacant site of the metal sublattice is free of oxygen vacancies in the first coordination sphere and free of titanium vacancies in the second coordination sphere. Four oxygen vacancies are present in the third coordination sphere (Fig. 5.39). This type of environment can be labeled $M_{0/0}^{4}$ (the first and second numerals in the subscript denote the number of nonmetal vacancies in the first coordination sphere and metal vacancies in the second coordination sphere, respectively; while the superscript stands for the number of nonmetal vacancies in the third coordination sphere). The metal sublattice sites occupied by titanium atoms have the environment $M_{2/2}^{0}$, $M_{1/3}^{0}$, and $M_{1/2}^{2}$. One-sixth of all sites in the metal sublattice have the environment $M^4_{0/0}$, another one-sixth of the sites are in the environment $M^0_{2/2}$, one-third of the sites occupy positions $M^0_{1/3}$, and one-third of the sites have the environment $M^2_{1/2}$. It is worth noting that all the types of environments are invariant with respect to a replacement of the metal lattice sites by the nonmetal lattice sites, because these sublattices of an ideal ordered monoclinic (space group *C*2/*m*) phase Ti_5O_5 have a similar structure but are displaced relative to each other by {1/2 1/2 0} in the coordinates of the superstructure.

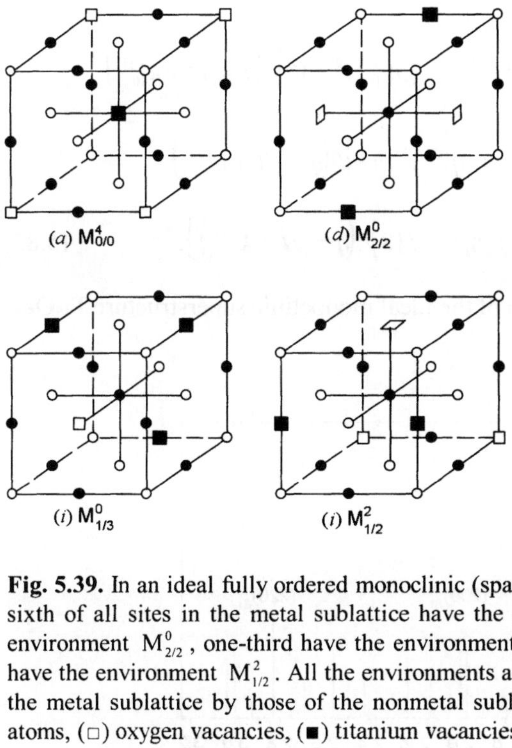

Fig. 5.39. In an ideal fully ordered monoclinic (space group *C*2/*m*) structure of Ti_5O_5 one-sixth of all sites in the metal sublattice have the environment $M^4_{0/0}$, one-sixth have the environment $M^0_{2/2}$, one-third have the environment $M^0_{1/3}$, and one-third of all metal sites have the environment $M^2_{1/2}$. All the environments are invariant to a replacement of sites of the metal sublattice by those of the nonmetal sublattice. (○) oxygen atoms, (•) titanium atoms, (□) oxygen vacancies, (■) titanium vacancies

When calculating intensities of the superstructure reflections of the Ti_5O_5 phase, one should remember that the intensities of the superstructure reflections of star rays $\{\boldsymbol{k}_{10}\}$, $\{\boldsymbol{k}_4\}$, and $\{\boldsymbol{k}_1\}$ of the metal and nonmetal sublattices are pairwise coincident. Therefore the intensity of a superstructure reflection depends on the phase difference between concentration waves of metal and nonmetal atoms and also on the indices (*hkl*) of the structural site $\boldsymbol{H}$, near which the superstructure reflection is formed. For example, the function describing the intensities of superstructure reflections of the star $\{\boldsymbol{k}_4\}$ has the general form

$$I(\{\boldsymbol{k}_4\},\boldsymbol{H}) = N^2 m_4 \gamma_4^2 \sum_{\boldsymbol{H}} \{f_{\text{Ti}}\eta_4^{(\text{Ti})} + [\cos(\varphi_4^{1(\text{Ti})} - \varphi_4^{1(\text{O})})] \times (-1)^{(h+k+l)} f_{\text{O}}\eta_4^{(\text{O})}\}^2$$

$$\times P(q = |\boldsymbol{H} + \boldsymbol{k}_4^{(j)}|) \,.$$

The contributions from stars $\{\boldsymbol{k}_{10}\}$ and $\{\boldsymbol{k}_1\}$ to the diffraction intensities can be written similarly. Taking into account particular values of m_s, γ_s, and φ_s corresponding e.g. to stars $\{\boldsymbol{k}_s\}$ and to titanium and oxygen sublattices, the intensities of superstructure reflections in diffraction spectra of the ordered monoclinic (space group $C2/m$) polycrystalline phase Ti_5O_5 is given by the formula

$$I_{\text{super}}(q) = N^2 \sum_{\boldsymbol{H}} \Big\{ \big[(f_{\text{Ti}}\eta_{10}^{(\text{Ti})} - (-1)^{(h+k+l)} f_{\text{O}}\eta_{10}^{(\text{O})})^2 / 36\big] P\big(q = |\boldsymbol{H} + \boldsymbol{k}_{10}^{(j)}|\big)$$

$$+ \big[(f_{\text{Ti}}\eta_4^{(\text{Ti})} + (-1)^{(h+k+l)} f_{\text{O}}\eta_4^{(\text{O})})^2 / 18\big] P\big(q = |\boldsymbol{H} + \boldsymbol{k}_4^{(j)}|\big)$$

$$+ \big[(f_{\text{Ti}}\eta_1^{(\text{Ti})} - (-1)^{(h+k+l)} f_{\text{O}}\eta_1^{(\text{O})})^2 / 18\big] P\big(q = |\boldsymbol{H} + \boldsymbol{k}_1^{(j)}|\big) \Big\}. \tag{5.83}$$

The calculated diffraction spectrum of the ideal monoclinic superstructure Ti_5O_5 is shown in Fig. 5.40.

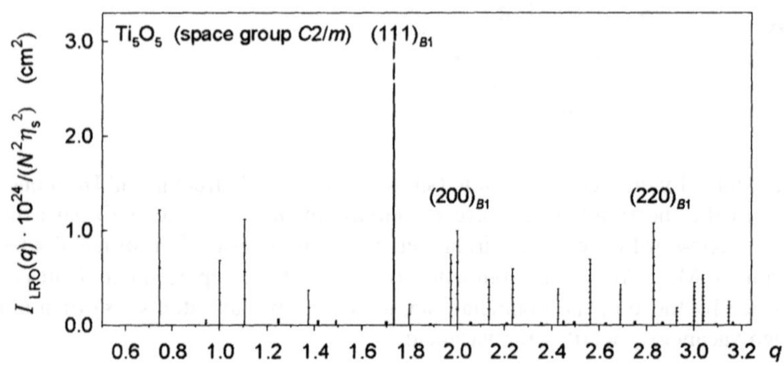

Fig. 5.40. Theoretical neutron diffraction spectrum of an ideal monoclinic (space group $C2/m$) superstructure Ti_5O_5. The spectrum was constructed with atomic scattering factors $f_{\text{Ti}} = -0.33 \times 10^{-12}$ and $f_{\text{O}} = 0.575 \times 10^{-12}$ cm, and the Lorentz factor $L(q) = [\sin\theta \sin 2\theta]^{-1}$, where $\theta = \arcsin(q\lambda/2a_{B1})$, $\lambda = 0.154$ nm, and $a_{B1} = 0.4176$ nm

5.5 Kinds of Disorder–Order Phase Transformations

The term "order" is commonly used for the classification of phase transformations (phase transitions). However, as this book deals with structural ordering in crystals, i.e. with disorder–order transformations we use the term "kind" rather than "order" in the classification of phase transformations.

In accordance with Landau's phenomenological theory of phase transformations [42], a necessary condition for a phase transformations of the second kind is that the coefficient of the cubic term of the expansion of the free energy in the long-range order parameter is identically equal to zero. This relates to the fact that not all changes in the crystal symmetry are possible during phase transformations of the second kind. It is shown by representation theory that an expansion of the thermodynamic potential does not contain third-order terms only if the symmetrical cube of an irreducible representation responsible for the phase transition does not include a unity representation of the symmetry space group of a disordered crystal.

If a phase transformation (transition) of the second kind is accompanied by changes in translational symmetry, the theory-group criterion according to Landau is equivalent to the following necessary condition for a phase transformation of the second kind to occur [6]: a phase transformation of the second kind is possible only if the star vectors associated with the particular transformation do not include three (not necessarily different) vectors whose sum is equal to zero or equal to the superstructure vector $\boldsymbol{H}$ of the reciprocal lattice of a disordered crystal:

$$\left.\begin{array}{r}\boldsymbol{k}_s^{(j_1)}+\boldsymbol{k}_s^{(j_2)}+\boldsymbol{k}_s^{(j_3)} \\ 3\boldsymbol{k}_s^{(j)}\end{array}\right\}\neq 0 \text{ or } \boldsymbol{H}\,. \tag{5.84}$$

The condition (5.84) holds if the space group of an ordered phase represents a symmetry subgroup of a disordered phase.

We consider ordering in a high-symmetry nonmetal fcc sublattice of strongly nonstoichiomertric compounds MX_y with the $B1$ structure therefore all stars $\{\boldsymbol{k}_s\}$ discussed further are the stars of the first Brillouin zone of fcc lattice.

The determination of MX_y–M_2X phase transition channels showed that the formation of any superstructure type M_2X is connected with one of the Lifschitz stars $\{\boldsymbol{k}_{10}\}$, $\{\boldsymbol{k}_9\}$ or $\{\boldsymbol{k}_8\}$. The condition (5.84) is fulfilled for vectors of stars $\{\boldsymbol{k}_9\}$ and $\{\boldsymbol{k}_8\}$ included in the transition channel. This means that disorder–order transitions, which are followed by the formation of superstructures M_2X (space groups $R\bar{3}m$, $Fm\bar{3}m$, and $I4_1/amd$), satisfy the Landau criterion for phase transitions of the second kind. The formation of a tetragonal (space group $P4/mmm$) superstructure M_2X represents a phase transition of the first kind, because it is connected with the star $\{\boldsymbol{k}_{10}\}$, for which the condition (5.84) is not fulfilled.

The formation of cubic structures M_4X_3 (space group $Pm\bar{3}m$) and M_8X_7 (space group $Fm\bar{3}m$), and also of a tetragonal superstructure M_4X_3 (space group $I4/mmm$) is connected with Lifschitz stars too: in the first case, the phase transition channel includes all rays of the star $\{\boldsymbol{k}_{10}\}$ while in the second case all rays of stars

$\{k_{10}\}$ and $\{k_9\}$ contribute and in the third case one ray of the star $\{k_{10}\}$ and two rays of the star $\{k_8\}$ contribute. However, the condition (5.84) does not hold in the presence of the star $\{k_{10}\}$ and therefore the formation of superstructures M_4X_3 (space groups $Pm\bar{3}m$ and *I4/mmm*) and M_8X_7 (space group $Fm\bar{3}m$) can be a phase transition (transformation) of the first kind only.

The formation of superstructure type M_3X_2 is connected with symmetry distortions over one or two non-Lifschitz stars $\{k_5\}$, $\{k_4\}$ or $\{k_3\}$. In addition, $3k_4^{(1)} = (2\bar{2}0)$ for orthorhombic (space groups *Immm* and $C222_1$) superstructures, and $3k_4^{(5)} = (20\bar{2})$ and $3k_5^{(6)} = (111)$ for monoclinic (space group *P*2) and trigonal (space group $P\bar{3}m1$) superstructures. Hence, the condition (5.84) is not fulfilled and ordered phases of type M_3X_2 are formed as a phase transition of the first kind.

In the case of the cubic (space group $P4_332$) superstructure M_8X_7 the phase transition channel includes all rays of Lifschitz stars $\{k_9\}$ and $\{k_8\}$ and the non-Lifschitz star $\{k_4\}$. The sum ($k_4^{(1)} + k_4^{(5)} + k_4^{(9)}$) equals the structural vector (111) of the reciprocal lattice. Obviously, such a superstructure can be formed as a phase transition of the first kind only.

The determination of phase transition channels shows that the formation of superstructures M_6X_5 is connected with several wave-vector stars. A long-range order parameter corresponds to each of these stars. To ascertain the kind of phase transition, one should consider critical order parameters only, which correspond to non-Lifschitz stars. Order parameters corresponding to stars $\{k_4\}$ and $\{k_3\}$ are critical for monoclinic and trigonal (space groups *C*2/*m* and $P3_1$) superstructures M_6X_5. Order parameters corresponding to stars $\{k_4\}$, $\{k_3\}$ and $\{k_0\}$ are critical for a monoclinic (space group *C*2) superstructure M_6X_5. So, critical order parameters for all the three superstructures of M_6X_5 type are more than unity. A simultaneous distortion of symmetry by several irreducible representations suggests that disorder–order transformations MX_y–M_6X_5 are phase transitions of the first kind.

The formation of a monoclinic (space group *C*2/*m*) superstructure Ti_5O_5 is also connected with three wave-vector stars including a Lifschitz star $\{k_{10}\}$. Critical long-range order parameters of this superstructure correspond to non-Lifschitz stars $\{k_4\}$ and $\{k_1\}$. So, critical order parameters (Landau order parameters) are more than unity. A monoclinic superstructure Ti_5O_5 in titanium monoxide TiO_y is formed as a phase transition of the first kind.

Considering superstructure types M_3X_2 (space group $C222_1$), M_6X_5, and M_8X_7 (space group $P4_332$), which were observed in experiments with titanium, vanadium, niobium, and tantalum carbides, the inference on the first kind of disorder–order phase transitions is confirmed by numerous investigations [10, 11, 14, 26, 35, 37, 43–46].

From the above analysis it follows that only superstructures of type M_2X (space groups $R\bar{3}m$, $Fm\bar{3}m$, and $I4_1/amd$) can be formed through a phase transition of the second kind. Superstructures M_2X (space group *P4/mmm*), M_3X_2, M_4X_3, M_6X_5, Ti_5O_5 and M_8X_7 in nonstoichiometric compounds MX_y with the *B*1 structure (or substitutional superstructures A_2B, A_3B, A_5B and A_7B in solid solutions A_yB_{1-y}) can be formed only as phase transitions of the first kind.

5.6 Intervals of Admissible Values of Long-Range Order Parameters

If the long-range order parameters describing a specific superstructure are equal, the nonmetal sublattice of a disordered nonstoichiometric compound MX_y can be broken up during ordering into two nonequivalent sublattices (vacancy and interstitial sublattices). The values of the long-range order parameters being the same, any distribution function degenerates and assumes two values: $n_1^{(d)} = y - (2t-1)\eta/2t$ (the probability for an interstitial atom occuping a site on the vacancy sublattice) and $n_2^{(d)} = y + \eta/2t$ (the probability for an interstitial atom occuping a site on the interstitial sublattice).

An analysis of the distribution functions calculated as an approximation of equal long-range order parameters allows us to establish the relationship between the maximum value for the long-range order parameter and the composition of the nonstoichiometric compound MX_y when a superstructure type $M_{2t}X_{2t-1}$ is formed:

$$\eta^{\max}(y) = \begin{cases} 2t(1-y), & \text{when } y \geq (2t-1)/2t, \\ 2ty/(2t-1), & \text{when } y < (2t-1)/2t. \end{cases} \tag{5.85}$$

As mentioned before, the distribution function $n(\mathbf{r})$ means probability. In general it can assume values between zero and unity. In accordance with calculations, the value of the distribution function depends on the composition of an ordering compound MX_y and on the sublattice of the ordered structure hosting the site $\mathbf{r}$. A comprehensive analysis of values possible for a specific distribution function leads to the significant conclusion that long-range order parameters have an interval of admissible values [8, 12, 47].

In the general case, the dependence of the maximum value of any long-range order parameter on the composition of an ordering compound MX_y is given by (5.85), while the minimum value of long-range order parameters is zero. Therefore the condition

$$0 \leq \eta_s \leq m^*, \tag{5.86}$$

where $m^* = 2t(1-y)$ if $y \geq (2t-1)/2t$ and $m^* = 2ty/(2t-1)$ if $y < (2t-1)/2t$, must be fulfilled for any order parameter describing the superstructures considered above.

The condition (5.86) determines one-dimensional intervals of admissible values of long-range order parameters for superstructures, which are described by one parameter η_s. They include superstructure types M_2X ($t = 1$; space groups $R\bar{3}m$, $Fm\bar{3}m$, $P4/mmm$, and $I4_1/amd$), M_3X_2 ($t = 1.5$; space groups $Immm$, $P\bar{3}m1$, and $P2$), and M_4X_3 ($t = 2$; space group $Pm\bar{3}m$).

However, the condition (5.86) disregards physical restrictions imposed on the values of the distribution function described by several long-range order parameters. Therefore, when determining the interval of admissible values of a particular long-range order parameter, one should be careful that the distribution function values fall within 0 and 1. Considering values assumed by these distribution functions and restrictions imposed, it is possible to determine the interval of admissible values of long-range order parameters for every ordered structure.

An orthorhombic (space group $C222_1$) superstructure M_3X_2 ($t = 1.5$) is described by two order parameters η_4 and η_3. The interval of admissible η values is determined by the conditions

$$\left.\begin{array}{l} -m^* \leq -\eta_4 + 3\eta_3 \leq 2m^* \\ 0 \leq \eta_4 \leq m^* \end{array}\right\}. \tag{5.87}$$

The interval of admissible values of the parameters $\eta_4(y)$ and $\eta_3(y)$ for the superstructure M_3X_2 (space group $C222_1$) is given in Fig. 5.41*a*.

In the case of a tetragonal (space group *I4/mmm*) superstructure M_4X_3 described by two parameters η_{10} and η_8, the interval of admissible values of these long-range order parameters (Fig. 5.41*b*) is determined by the conditions

$$\left.\begin{array}{l} -m^* \leq -\eta_{10} + 2\eta_8 \leq m^* \\ 0 \leq \eta_{10} \leq m^* \end{array}\right\}. \tag{5.88}$$

In the case of a cubic (space group $Fm\bar{3}m$) superstructure M_8X_7 ($t = 4$) described by two parameters η_{10} and η_9, the interval of admissible values of the long-range order parameters is determined, depending on the composition of the compound MX_y, by the conditions

$$\left.\begin{array}{l} -3m^* \leq -3\eta_{10} + 4\eta_9 \leq m^* \\ 0 \leq \eta_{10} \leq m^* \end{array}\right\}. \tag{5.89}$$

The interval of admissible values determined by the conditions (5.89) is shown in Fig. 5.41*c*.

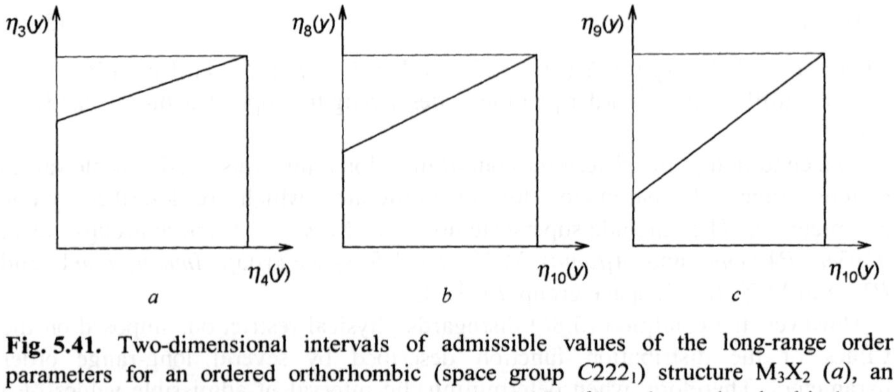

Fig. 5.41. Two-dimensional intervals of admissible values of the long-range order parameters for an ordered orthorhombic (space group $C222_1$) structure M_3X_2 (*a*), an ordered tetragonal (space group *I4/mmm*) structure M_4X_3 (*b*), and an ordered cubic (space group $Fm\bar{3}m$) structure M_8X_7 (*c*)

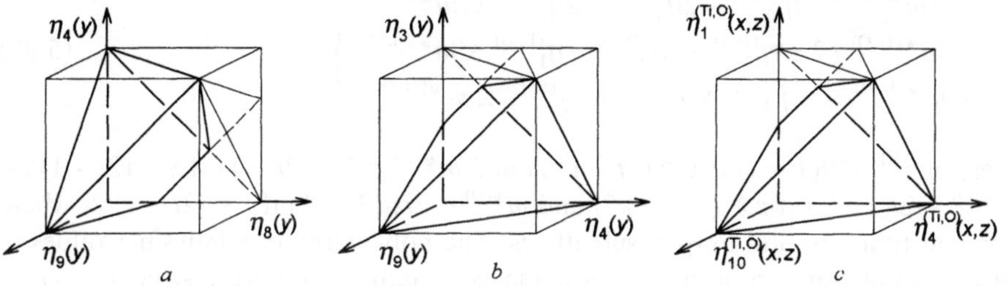

Fig. 5.42. Three-dimensional intervals of admissible values of the long-range order parameters: (*a*) an ordered cubic (space group $P4_332$) structure M_8X_7; (*b*) ordered monoclinic (space group $C2/m$) and trigonal (space group $P3_1$) structures type M_6X_5; (*c*) an ordered monoclinic (space group $C2/m$) structure Ti_5O_5

A cubic (space group $P4_332$) superstructure M_8X_7 ($t = 4$) is described by the distribution function (5.78), which depends on three long-range order parameters η_9, η_8 and η_4. The interval of admissible values of the order parameters in this superstructure is determined by the conditions

$$\left.\begin{array}{l} -3m^* \le \eta_9 + 3\eta_8 - 3\eta_4 \le m^* \\ -m^*/3 \le \eta_9 - \eta_8 + \eta_4 \le m^* \\ -m^* \le -\eta_9 + \eta_8 + \eta_4 \le m^* \end{array}\right\}. \tag{5.90}$$

Figure 5.42*a* presents a polyhedron of admissible values of the order parameters in the cubic (space group $P4_332$) superstructure M_8X_7.

In the case of monoclinic (space group $C2/m$) and trigonal (space group $P3_1$) superstructures M_6X_5 ($t = 3$), which are described by three long-range order parameters η_9, η_4 and η_3, the interval of admissible values is limited by the conditions

$$\left.\begin{array}{l} -2m^* \le \eta_9 - 2\eta_4 + 2\eta_3 \le m^* \\ -m^*/2 \le \eta_9 + \eta_4 - \eta_3 \le m^* \\ -m^* \le -\eta_9 + \eta_4 + \eta_3 \le m^* \end{array}\right\}. \tag{5.91}$$

The intervals of admissible values of the order parameters, which are given by conditions (5.86) and (5.91), represent polyhedra constructed in coordinates $\eta_9(y)$, $\eta_4(y)$ and $\eta_3(y)$ (Fig. 5.42*b*).

The interval of admissible values of the order parameters in any sublattice of a monoclinic (space group $C2/m$) superstructure Ti_5O_5 ($Ti_{2t-1}\blacksquare_1O_{2t-1}\square_1$, where $t = 3$), which appears in titanium monoxide $TiO_y \equiv Ti_xO_z$, is shown in Fig. 5.42*c*. This interval is given by the set of inequalities

$$\left.\begin{aligned} -2m^{*(\mathrm{Ti,\,O})} &\le \eta_{10}^{(\mathrm{Ti,\,O})} - 2\eta_4^{(\mathrm{Ti,\,O})} + 2\eta_1^{(\mathrm{Ti,\,O})} \le m^{*(\mathrm{Ti,\,O})} \\ -m^{*(\mathrm{Ti,\,O})}/2 &\le \eta_{10}^{(\mathrm{Ti,\,O})} + \eta_4^{(\mathrm{Ti,\,O})} - \eta_1^{(\mathrm{Ti,\,O})} \le m^{*(\mathrm{Ti,\,O})} \\ -m^{*(\mathrm{Ti,\,O})} &\le -\eta_{10}^{(\mathrm{Ti,\,O})} + \eta_4^{(\mathrm{Ti,\,O})} + \eta_1^{(\mathrm{Ti,\,O})} \le m^{*(\mathrm{Ti,\,O})} \end{aligned}\right\}. \qquad (5.92)$$

Here $m^{*(\mathrm{Ti})} = 2t(1-x)$ if $x \ge (2t-1)/2t$ and $m^{*(\mathrm{Ti})} = 2tx/(2t-1)$ if $x < (2t-1)/2t$; $m^{*(\mathrm{O})} = 2t(1-z)$ if $z \ge (2t-1)/2t$ and $m^{*(\mathrm{O})} = 2tz/(2t-1)$ if $z < (2t-1)/2t$ where $t = 3$ in titanium and oxygen sublattices. The polyhedron of admissible values of order parameters in a monoclinic (space group $C2/m$) superstructure Ti_5O_5 (Fig. 5.42*c*) coincides in shape with the polyhedron (Fig. 5.42*b*) found for the monoclinic (space group $C2/m$) and trigonal (space group $P3_1$) superstructures of type M_6X_5.

The interval of admissible values of the order parameters in a monoclinic (space group $C2$) superstructure M_6X_5 ($t = 3$) is limited by the conditions

$$\left.\begin{aligned} -m^* &\le 2\eta_9 - \eta_4 + \eta_3 \le 2m^* \\ -m^* &\le \eta_9 + \eta_4 - \eta_3 \le m^* \\ -m^* &\le -\eta_9 + \eta_4 + \eta_3 \le m^* \\ -4m^* &\le -2\eta_9 - \eta_4 - \eta_3 + 6\eta_0 \le 2m^* \end{aligned}\right\}. \qquad (5.93)$$

Figure 5.43 presents three-dimensional sections of a four-dimensional polyhedron of admissible values of the long-range order parameters determined for a monoclinic (space group $C2$) superstructure M_6X_5.

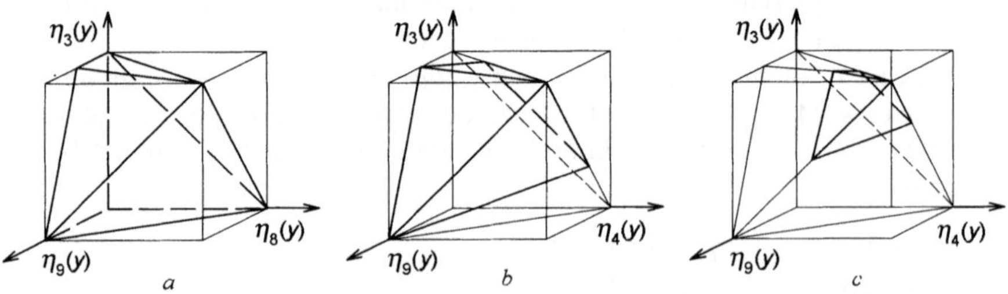

Fig. 5.43. Three-dimensional sections of a four-dimensional interval of the admissible values of the long-range order parameters for an ordered monoclinic (space group $C2$) structure M_6X_5: (*a*) $0 \le \eta_0 \le (1/3)\eta_0^{\max}(y)$; (*b*) $\eta_0 = (2/3)\eta_0^{\max}(y)$; (*c*) $\eta_0 = (5/6)\eta_0^{\max}(y)$

When $0 \le \eta_0 \le \frac{1}{3}\eta_0^{\max}(y)$, the interval of admissible values remains unchanged. This interval shrinks as η_0 increases from $\frac{1}{3}\eta_0^{\max}(y)$ to $\eta_0^{\max}(y)$ (Fig. 5.43*b*,*c*) and degenerates to a point $\eta_9^{\max}(y) = \eta_4^{\max}(y) = \eta_3^{\max}(y) = \eta_0^{\max}(y)$ when $\eta_0^{\max}(y)$ is reached.

References

1 W. L. Bragg, E. J. Williams: The effect of thermal agitation on atomic arrangement in alloys, Proc. Roy. Soc. London A **145**, 699–730 (1934); **151**, 540–566 (1935)

2 E. J. Williams: The effect of thermal agitation on atomic arrangement in alloys-III, Proc. Roy. Soc. London A **152**, 231–252 (1935)

3 M. A. Krivoglaz, A. A. Smirnov: *The Theory of Order-Disorder in Alloys* (American Elsevier, New York 1964) 427 pp.

4 A. A. Smirnov: *Molecular and Kinetic Theory of Metals* (Nauka, Moscow 1966) 488 pp. (in Russian)

5 A. A. Smirnov: *Theory of Interstitial Alloys* (Nauka, Moscow 1979) 368 pp. (in Russian)

6 A. G. Khachaturian: *Theory of Structural Transformations in Solids* (John Wiley & Sons, New York 1983) 574 pp.

7 A. I. Gusev, A. A. Rempel: Superstructures of non-stoichiometric interstitial compounds and the distribution functions of interstitial atoms, Phys. Stat. Sol. (a) **135**, 15–58 (1993)

8 A. I. Gusev, A. A. Rempel: *Structural Phase Transitions in Nonstoichiometric Compounds* (Nauka, Moscow 1988) 308 pp. (in Russian)

9 A. I. Gusev: *Physical Chemistry of Nonstoichiometric Refractory Compounds* (Nauka, Moscow 1991) 286 pp. (in Russian)

10 A. I. Gusev, A. A. Rempel: Ordering in the carbon sublattice of nonstoichiometric niobium carbide, Fiz. Tverd. Tela **26**, 3622–3627 (1984) (in Russian). (Engl. transl.: Sov. Physics - Solid State **26**, 2178–2181 (1984))

11 A. I. Gusev, A. A. Rempel: Order-disorder phase transition channel in niobium carbide, Phys. Stat. Sol. (a) **93**, 71–80 (1986)

12 A. I. Gusev, A. A. Rempel: Vacancy distribution in ordered Me_6C_5-type carbides, J. Phys. C: Solid State Phys. **20**, 5011–5025 (1987)

13 V. E. Naish, V. N. Syromyatnikov: Change of the translation symmetry in the structural phase transitions in the crystals, Kristallografiya **21**, 1085–1092 (1976) (in Russian)

14 V. N. Lipatnikov, A. Kottar, L. V. Zueva, A. I. Gusev: Disorder-order phase transformations and electrical resistivity of a nonstoichiometric titanium carbide, Fiz. Tverd. Tela **40**, 1332–1340 (1998) (in Russian). (Engl. transl.: Phys. Solid State **40**, 1211–1218 (1998))

15 H. Goretzki: Neutron diffraction studies of titanium-carbon and zirconium-carbon alloys, Phys. Stat.Sol. **20**, K141–K143 (1967)

16 I. Karimov, V. T. Em, I. Khidirov, I. S. Latergaus: Neutron diffraction investigation of ordering in titanium and zirconium carbides, Izv. AN Uzb. SSR. Seriya Fiz. Mat. Nauk No 4, 81–83 (1979) (in Russian)

17 M. P. Arbuzov, S. Ya. Golub, B. V. Khaenko: X-ray investigation of titanium nitrides, Izv. AN SSSR. Neorgan. Materialy **13**, 1779–1782 (1977) (in Russian)

18 A. N. Christensen, A. Alamo, J. P. Landesman: Structure of vacancy-ordered titanium heminitride δ'-Ti_2N by powder neutron diffraction, Acta Crystallogr. C **41**, 1009–1011 (1985)

19 C. H. de Novion, B. Beuneu, T. Priem, N. Lorenzelli, A. Finel: Defect structures and order-disorder transformations in transition metal carbides and nitrides, in *The Physics*

and Chemistry of Carbides, Nitrides and Borides, R. Freer (Ed.) (Kluwer Acad. Publ., Netherlands 1990) pp. 329–355
20 A. A. Rempel, A. I. Gusev, V. G. Zubkov, G. P. Shveikin: Structure of ordered niobium carbide Nb_6C_5, Doklady AN SSSR **275**, 883–887 (1984) (in Russian). (Engl. transl.: Sov. Physics Doklady **29**, 257–259 (1984))
21 A. A. Rempel, A. I. Gusev: Order-disorder phase transition in nonstoichiometric niobium carbide, Kristallografiya **30**, 1112–1115 (1985) (in Russian). (Engl. transl.: Sov. Physics – Crystallography **30**, 648–650 (1985))
22 B. V. Khaenko, O. P. Sivak: Displacements of niobium atoms in the superstructure δ-Nb_6C_5 , Doklady AN USSR. Seriya A No 1, 91–94 (1989) (in Russian)
23 B. V. Khaenko, O. P. Sivak: Structure of order in niobium monocarbides, Kristallografiya **35**, 1110–1115 (1990) (in Russian)
24 R. Kesri, S. Hamar-Thibault: Structures ordonnées à longue distance dans les carbures MC dans les fontes, Acta Met. **36**, 149–166 (1988)
25 J. D. Venables, D. Kahn, R. G. Lye: Structure of ordered compound V_6C_5, Philosoph. Mag. **18**, 177–192 (1968)
26 J. P. Landesman, A. N. Christensen, C. H. de Novion, N. Lorenzelli, P. Convert: Order-disorder transition and structure of the ordered vacancy compound Nb_6C_5: powder neutron diffraction studies, J. Phys. C: Solid State Phys. **18**, 809–823 (1985)
27 J. Billingham, P. S. Bell, M. H. Lewis: Superlattice with monoclinic symmetry based on compounds V_6C_5, Philosoph. Mag. **25**, 661–671 (1972)
28 C. Froidevaux, D. Rossier: NMR-investigation on the atomic and electronic structure of vanadium and niobium carbides, J. Phys. Chem. Solids **28**, 1197–1209 (1967)
29 A. A. Rempel, A. I. Gusev: NMR spectra of ^{93}Nb in ordered niobium carbide, Fiz. Tved. Tela **25**, 3169–3171 (1983) (in Russian). (Engl. transl.: Sov. Physics – Solid State **25**, 1827–1828 (1983))
30 A. A. Rempel, M. Yu. Belyaev, L. B. Dubrovskaya, G. P. Shveikin: An NMR study of nonstoichiometric niobium monocarbide, Zh. Neorgan. Khimii **29**, 2451–2455 (1984) (in Russian). (Engl. transl.: Russ. J. Inorg. Chem. **29**, 1401–1404 (1984))
31 A. A. Rempel, A. I. Gusev, M. Yu. Belyaev: ^{93}Nb NMR study of an ordered and a disordered non-stoichiometric niobium carbide, J. Phys. C: Solid State Phys. **20**, 5655–5666 (1987)
32 C. H. de Novion, N. Lorenzelli, P. Costa: Superlattice structure in vanadium carbide VC_{1-x}, Compt. Rend. Acad. Sci. Paris B **263**, 775–778 (1966)
33 S. I. Alyamovskii, P. V. Geld, G. P. Shveikin, E. N. Schetnikov: New phase in the vanadium–carbon system, Zh. Neorgan. Khimii **13**, 895–897 (1968) (in Russian). (Engl. transl.: Russ. J. Inorg. Chem. **13**, 472–473 (1968))
34 A. W. Henfrey, B. E. F. Fender: A neutron diffraction investigation of V_8C_7, Acta Crystallogr. B **26**, 1882–1883 (1970)
35 T. Athanassiadis, N. Lorenzelli, C. H. de Novion: Diffraction studies of the order-disorder transformation in V_8C_7, Ann. Chim. France **12**, 129–142 (1987)
36 D. Rafaja, W. Lengauer, P. Ettmayer, V. N. Lipatnikov: Rietveld analysis of the ordering in V_8C_7, J. Alloys Comp. **269**, 60–62 (1998)
37 V. N. Lipatnikov, A. I. Gusev, P. Ettmayer, W. Lengauer: Phase transformation in non-stoichiometric vanadium carbide, J. Phys.: Condens. Matter **11**, 163–184 (1999)
38 D. Watanabe, J. R. Castles, A. Jostsons, A. S. Malin: Ordered structure of titanium oxide, Nature **210**, 934–936 (1966)

39 D. Watanabe, J. R. Castles, A. Jostsons, A. S. Malin: The ordered structure of TiO, Acta Crystallogr. **23**, 307–313 (1967)
40 E. Hilti: Röntgenographische Untersuchung der Titanmonoxid–Tieftemperatur–Modifikation, Naturwissenschaften **55**, 131–132 (1968)
41 A. A. Valeeva, A. A. Rempel, A. I. Gusev: Two-sublattice ordering in titanium monoxide, Pis'ma v ZhETF **71**, 675–681 (2000) (in Russian). (Engl. ransl.: JETP Letters **71**, 460–464 (2000))
42 L. D. Landau, E. M. Lifschitz: *Statistical Physics. Course of Theoretical Physics* **5** (Pergamon, Oxford 1980) 544 pp.
43 A. I. Gusev, A. A. Rempel, V. N. Lipatnikov: Incommensurate superlattice and superconductivity in tantalum carbide, Fiz. Tverd. Tela **33**, 2298–2305 (1991) (in Russian). (Engl. transl.: Sov. Physics – Solid State **33**, 1295–1299 (1991))
44 A. I. Gusev, A. A. Rempel, V. N. Lipatnikov: Incommensurate ordered phase in nonstoichiometric tantalum carbide, J. Phys.: Condens. Matter **8**, 8277–8293 (1996)
45 V. N. Lipatnikov, A. I. Gusev: Dependence of the resistivity of nonstoichiometric titanium carbide TiC_y on the density and distribution of carbon vacancies, Pis'ma v ZhETF **70**, 284–289 (1999) (in Russian). (Engl. transl.: JETP Letters **70**, 294–300 (1999))
46 V. N. Lipatnikov, A. I. Gusev: High-temperature heat capacity and order-disorder phase transformations in nonstoichiometric titanium carbide, Phys. Stat. Sol. (b) **212**, R11–R12 (1999)
47 A. A. Rempel, V. N. Lipatnikov, A. I. Gusev: Distribution function of carbon atoms and order parameters in ordered carbides of the type M_6C_5, Kristallografiya **32**, 1032–1033 (1987) (in Russian). (Engl. transl.: Sov. Physics – Crystallography **32**, 607–608 (1987))

6. Interrelation of Short- and Long-Range Orders

Short- and long-range orders characterize various types of mutual arrangements of atoms in crystalline solids. Long-range order emerges from the interference of perturbation waves induced by interchangeable elements and covers the whole volume of the crystal. Short-range order characterizes the mutual distribution of atoms of different species depending on the atom at the lattice site chosen. Short-range order vanishes with distance from a chosen atom and normally it covers no more than a few lattice constants.

6.1 Relationship between Short- and Long-Range Orders

Short-range order describes the radial atomic distribution in the lattice of a solid solution and determines only the conditional probability that an atom of a certain species is present at a given site. Long-range order includes both radial and angular distributions and allows us to determine the true probability that a site in some sublattice is filled with an atom of one species or another. However, although short- and long-range orders differ fundamentally, they are still related to a certain extent. If a solid solution exhibits long-range order, it certainly has short-range order too. Short-range order parameters α_j for the j-th coordination sphere are constant quantities, which assume a fixed value for a fully ordered structure (for a fully ordered solid solution, $\alpha_j = \alpha_j^{\text{ord}}$ is constant). Short-range order, which accompanies long-range order, is called superstructural short-range order. If no long-range order is present, short-range order parameters tend to zero with increasing radius of the coordination sphere. This part of the short-range order, which is not connected with long-range order and which is caused by short-range correlations in the atomic arrangement, is termed correlational short-range order [1, 2]. Thus, the presence of short-range order in a solid solution does not necessarily imply that long-range order is present too.

Early theories of order–disorder phase transitions assumed that ordered metal alloys are equivalent to magnetic systems with an antiferromagnetic interaction between nearest spins and could be described in terms of the simple Ising model [3]. This assumption was based on the fact that atoms in ordered alloys were distributed such that the number of pairs of nearest like atoms was as small as possible. However, even the analysis of the Bethe theory [4, 5], which essentially takes into account short-range interactions, clearly showed that the free energy of an alloy determined from short-range interactions does not always permit us to describe long-range order. The same applies to the cluster variation method [6–10], a modern analog of the Bethe–Guggenheim approximation [4, 11]. In particular, it was [12] noted that the ordering in the Cu_3Au alloy was accompanied by a one-dimensional degeneration. In other words, the Au atoms in the first coordination sphere could rearrange themselves with no reduction in the number of unlike atoms (the short-range order in the first coordination sphere was preserved). In this context it can be asked whether the existence of long-range order resulting from minimization of the free energy be inferred from Bethe and Kikuchi's suppositions.

Alternative approximation methods for solving the precise Ising problem (such as the Gorsky–Bragg–Williams theory [13, 14] and the advanced method of static concentration waves [15] in the mean field theory) assume a long-range order in the atomic arrangement. However, another drawback is the difficulty of taking into account correlations (especially multiparticle correlations) in the mutual arrangement of atoms. Thus, the mutual dependence of short- and long-range orders presents a most challenging problem in the context of atomic ordering.

The relations between short- and long-range order can be analyzed quantitatively by the cluster variation method [16–18] used to determine the number of nearest coordination spheres, where the presence of a maximum short-range order is sufficient for the formation of long-range order. In other words, the cluster variation method is all about determining how many short-range order parameters are necessary and sufficient to uniquely determine the presence of long-range order. Let us now consider this question [16–18] in more detail.

In the cluster variation method, the relationship between short- and long-range orders is treated in structural terms, independently of factors responsible for short-range order. However, it is assumed that such factors exist and, under certain conditions, can give rise to maximum short-range order. The method uses the notion of a cluster as being a group of lattice sites in a solid solution. Depending on its size, a cluster may comprise sites shared by several coordination spheres. The choice of a cluster is a fairly complicated task. The maximum cluster size is limited by the unit cell of an ordered structure. To avoid insurmountable calculation difficulties, however, it is normal practice to use a smaller cluster, although still large enough to allow for paired and multiparticle correlations.

In the following a solid solution of an arbitrary composition A_yB_{1-y} is considered as a set of clusters whose sites are occupied by atoms A or B. Under these circumstances there are several possibilities for the arrangements of atoms A and B at the cluster sites, i.e. several different configurations i. Let us label the probability of an i-configured cluster as P_i. Cluster configurations, which coincide after symmetry, are equivalent. Let the number of equivalent configurations of the i-th cluster (configuration multiplicity) be defined as λ_i. In the general case, the multiplicity λ_i is equal to the index of a point symmetry group G_i of an i-configured cluster relative to a point group G_0 of a cluster where all sites are occupied by atoms of one species, i.e. $\lambda_i = n(G_0)/n(G_i)$, where $n(G)$ is the order of the group G. Considering the multiplicity definition, the probability normalization condition will be written as

$$\sum_i \lambda_i P_i = 1 \ . \tag{6.1}$$

For a solid solution A_yB_{1-y}, an equation can also be written to relate the cluster probability P_i to the solid solution composition:

$$\sum_i l_i \lambda_i P_i = y \ , \tag{6.2}$$

where l_i is the fraction of sites occupied by atoms of species A in an i-configured cluster.

For a solid solution A_yB_{1-y}, short-range order in the j-th coordination sphere is characterized by the parameter α_j, which has the form [19]:

$$\alpha_i = 1 - P_{\mathrm{BA}}^{(j)} / P_{\mathrm{BA}}^{\mathrm{bin}} \ . \tag{6.3}$$

Here $P_{\mathrm{BA}}^{(j)}$ is the actual probability for the formation of BA pairs in the j-th coordination sphere; $P_{\mathrm{BA}}^{\mathrm{bin}} = y(1-y)$ is the binomial probability for the formation of BA atomic pairs, which corresponds to a disordered state of a solid solution and which depends on the composition. Let us write the actual probability for the formation of BA pairs in any coordination sphere as the probability and multiplicity of clusters:

$$\lambda_{\mathrm{BA}} P_{\mathrm{BA}}^{(j)} = \sum_i n_i^{(j)} \lambda_i P_i \ , \tag{6.4}$$

where $n_i^{(j)}$ is the relative fraction of A-B pairs in the j-th coordination sphere of an i-configured cluster. Equations (6.3) and (6.4) suggest a functional relationship between the probability P_i and the short-range order parameters α_i for coordination spheres that are accounted for by a chosen cluster. For a fully ordered solid solution, the short-range order parameters in all coordination spheres are constant quantities α_j^{ord}. Therefore, the following restriction is imposed on the probability P_i through (6.3) and (6.4):

$$\sum_i n_i^{(j)} \lambda_i P_i = 2y(1-y)(1-\alpha_j^{\mathrm{ord}}) \ . \tag{6.5}$$

Values of α_j^{ord} are calculated directly from the known structure of an ordered solid solution.

If the coefficients λ_i, l_i, $n_i^{(j)}$ and the quantities y and α_1^{ord} are known, simultaneous solution of (6.1), (6.2), and (6.5) for the first coordination sphere determine the values of the probabilities P_i. Subsequently, the short-range order parameters for remote coordination spheres can be given in terms of the values of P_i with (6.3) and (6.4). From the comparison of calculated α_2, ..., α_j and known short-range order parameters of a fully ordered solid solution, α_2^{ord}, ..., α_j^{ord}, it is possible to ascertain if the existence of a maximum short-range order α_1^{ord} in the first coordination sphere is sufficient for the formation of long-range order. If maximum short-range order in the first coordination sphere is insufficient for the appearance of the long-range order, because $\alpha_2 \neq \alpha_2^{\text{ord}}$, the system of equations (6.1), (6.2), and (6.5) should be extended to include in sequence equation (6.5) written for the second and higher coordination spheres. Such equations should be added until a solution of the system suggests the presence of long-range order in the solid solution. If long-range order is present, the short-range order parameters calculated from (6.3) will coincide with similar parameters for a fully ordered solid solution, i.e. $\alpha_2 = \alpha_2^{\text{ord}}, \ldots, \alpha_j = \alpha_j^{\text{ord}}$. Following this procedure, one can find a minimum number of short-range order parameters for a given ordered structure, which are necessary and sufficient to prove the existence of long-range order.

This method should be checked for binary alloys like β-brass, Cu_3Au or CuAu as an example. These alloys are simple examples for ordering in frequently occurring bcc and fcc basic structures.

Ordered structures with lattices of the β-brass type are observed in CuBe, AgMn, FeAl and other alloys. These structures are formed by the mechanism of a phase transition of the second kind thanks to the continuously varying degree of the long-range order with temperature.

A disordered solid solution or an alloy with a bcc lattice can be visualized as a set of clusters in the form of regular tetrahedral pyramids with atoms A and B varying both in number and mutual arrangement (Fig. 6.1).

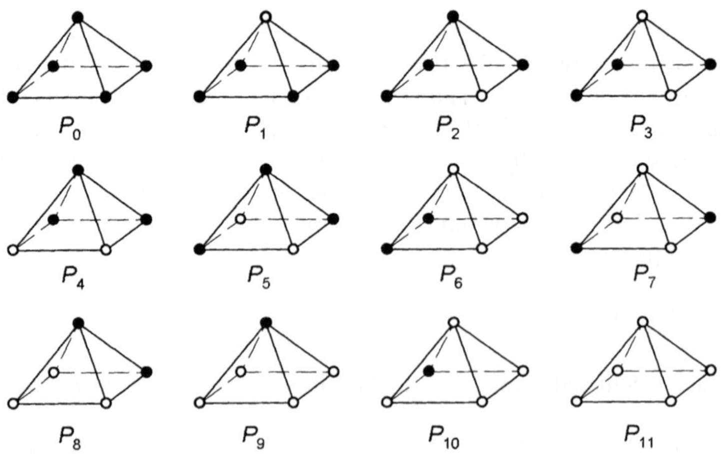

Fig. 6.1. Set of clusters for the description of binary solid solutions A_yB_{1-y} with a bcc basic lattice: (•) A, (○) B

Cluster P_0 has the point symmetry group 4 (C_4) including four symmetry operations h_1, h_4, h_{14}, and h_{15} (see Table 2.3). Such clusters contain atoms from three coordination spheres: relative to any atom at the cluster base, the first coordination sphere includes an atom located at the cluster vertex, the second coordination sphere includes two nearest atoms at the base, the third coordination sphere includes an atom located at the cluster base diagonal. Considering (6.3) and (6.4), the short-range order parameters α_1, α_2 and α_3 can be written as the cluster probabilities P_i. For an ordering alloy of equiatomic composition AB ($y = 0.5$, $P_{\mathrm{BA}}^{\mathrm{bin}} = 0.25$, $\lambda_{\mathrm{BA}} = 2$) these parameters are equal to

$$\left.\begin{aligned} \alpha_1 &= 1 - 2P_1 - 2P_2 - 6P_3 - 4P_4 - 2P_5 - 4P_6 - 2P_7 - 6P_8 - 2P_9 - 2P_{10}\,, \\ \alpha_2 &= 1 - 4P_2 - 4P_3 - 4P_4 - 4P_5 - 4P_6 - 4P_7 - 4P_8 - 4P_{10}\,, \\ \alpha_3 &= 1 - 4P_2 - 4P_3 - 8P_4 - 8P_6 - 4P_8 - 4P_{10}\,. \end{aligned}\right\} \qquad (6.6)$$

For an alloy like β-brass with a maximum degree of long-range order, the short-range order parameters for the first, second and third coordination spheres are $\alpha_1^{\mathrm{ord}} = -1$, $\alpha_2^{\mathrm{ord}} = 1$, and $\alpha_3^{\mathrm{ord}} = 1$. Taking into account $\alpha_1^{\mathrm{ord}} = -1$, the system of equations (6.1), (6.2), and (6.5) is written as

$$\sum_{i=0}^{11} \lambda_i P_i = 1\,, \quad \sum_{i=0}^{11} l_i \lambda_i P_i = 0.5\,, \quad \sum_{i=0}^{11} n_i^{(j)} \lambda_i P_i = 1\,. \qquad (6.7)$$

Values of the coefficients λ_i, l_i and $n_i^{(j)}$ for bcc solid solutions, which are described by clusters in the form of a regular tetrahedral pyramid (Fig. 6.1), are given in Table 6.1.

Table 6.1. Coefficients λ_i, l_i, $n_i^{(1)}$, $n_i^{(2)}$, and $n_i^{(3)}$ corresponding to P_i clusters, which are used for the description of solid solutions with bcc basic structure

Cluster P_i	P_0	P_1	P_2	P_3	P_4	P_5	P_6	P_7	P_8	P_9	P_{10}	P_{11}
λ_i	1	1	4	4	4	2	4	2	4	1	4	1
l_i	1	4/5	4/5	3/5	3/5	3/5	2/5	2/5	2/5	1/5	1/5	0
$n_i^{(1)}$	0	1	1/4	3/4	1/2	1/2	1/2	1/2	3/4	1	1/4	0
$n_i^{(2)}$	0	0	1/2	1/2	1/2	1	1/2	1	1/2	0	1/2	0
$n_i^{(3)}$	0	0	1/2	1/2	1	0	1	0	1/2	0	1/2	0

Considering that the cluster probability is not negative, the system of equations (6.7) has a unique solution: $P_1 = 1/2$, $P_9 = 1/2$, $P_0 = P_2 = P_3 = P_4 = P_5 = P_6 = P_7 = P_8 = P_{10} = P_{11} = 0$. Consequently, at the maximum degree of the short-range order in the first coordination sphere, the parameters α_2 and α_3 for the second and third coordination spheres are, in accordance with (6.6), fixed numbers and coincide with α^{ord}: $\alpha_2 = \alpha_2^{\mathrm{ord}} = 1$ and $\alpha_3 = \alpha_3^{\mathrm{ord}} = 1$. Although short-range order parameters can be calculated only for three coordination spheres, one can state that the maximum degree of the short-range order in the first coordination sphere leads to full long-range order. In this case the only possible form of the ordered structure has a lattice of the β-brass type. This conclusion suggests itself by the fact that clusters P_1 and P_9 whose probability equals 1/2 can be arranged in a bcc lattice in a unique manner only, when atoms A and B have long-range order.

Thus, an order–disorder phase transformation, which takes place in solid solutions and alloys which are ordered like β-brass, can be described by taking into account interactions of atoms A and B with their nearest neighbors only, i.e. interactions within the first coordination sphere.

Let us consider a solid solution with an fcc lattice as a set of octahedron-shaped clusters (Fig. 6.2) with ten nonequivalent configurations of atoms A and B differing in number and mutual arrangement.

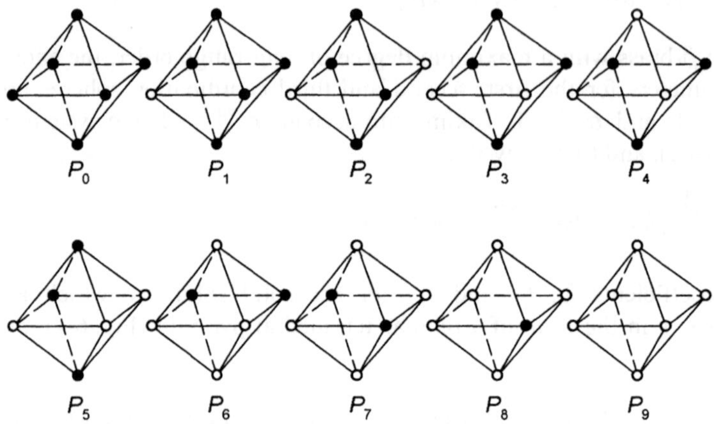

Fig. 6.2. Set of octahedral clusters for the description of binary solid solutions A_yB_{1-y} with an fcc basic lattice: (•) A, (○) B

Cluster P_0 has the point symmetry group $m\bar{3}m$ (O_h) with symmetry elements h_1–h_{48}. Octahedral clusters include atoms located in the first and second coordination spheres. The first coordination sphere includes four nearest sites relative to a randomly chosen octahedron site and the second coordination sphere includes one site located at an octahedron diagonal relative to the chosen site. In accordance with (6.3), short-range order parameters for the first and second coordination spheres in a solid solution A_yB_{1-y} can be written as:

$$\left.\begin{aligned} \alpha_1 &= 1-(P_1+P_2+3P_3+2P_4+4P_5+3P_6+P_7+P_8)/[y(1-y)], \\ \alpha_2 &= 1-(P_1+4P_3+4P_4+2P_5+4P_6+P_8)/[y(1-y)]. \end{aligned}\right\} \quad (6.8)$$

Let us see how the maximum probability $P_{BA}^{(1)\max}$ of the BA pair formation in the first coordination sphere varies with the composition of a solid solution A_yB_{1-y} ($0 \le y \le 1$) having an fcc basic structure. In the general case, when $y < 0.5$, the maximum probability is $P_{BA}^{\max} = P_B^{\max} P_A$, where $P_A = y$, for any coordination sphere. If the concentration of A atoms is small, i.e. $y \to 0$, the probability $P_{\rm B}^{\max} = 1$, and therefore

$$P_{\rm BA}^{\max} = y\,. \tag{6.9}$$

The probability $P_{\rm BA}^{\max}$ at small y values can also be defined as $P_{\rm BA}^{\max} = y(dP_{\rm BA}^{\rm bin} / dy)|_{y=0} = y$, which coincides with (6.9). Let us determine the variation interval of y, for which (6.9) is applicable for the first coordination sphere of a solid solution A_yB_{1-y}.

The maximum admissible degree of short-range order is achieved in fully ordered superstructures. Superstructures of types AB_3, AB and A_3B (for example, $CuAu_3$, CuAu, and Cu_3Au) are formed in an fcc solid solution A_yB_{1-y}. It is also known that superstructures M_8X_7, M_6X_5, M_4X_3, M_3X_2, and M_2X (see Chap. 5) are formed in nonstoichiometric compounds MX_y ($MX_y\square_{1-y}$) having a *B*1 basic structure. Interstitial atoms X and structural vacancies □ are ordered in the nonmetal fcc sublattice of these compounds. Ordered substitutional solid solutions A_7B, A_5B, A_3B, A_2B, and AB are equivalent to these superstructures. Since the structure is invariant to replacement of atoms A by atoms B, ordered solid solutions AB_7 ($A_{1/8}B_{7/8}$ or $y = 1/8$), AB_5 ($y = 1/6$), AB_3 ($y = 1/4$) and AB_2 ($y = 1/3$) should be considered as well. Taking into account all superstructures that may appear in a solid solution A_yB_{1-y}, one needs to find the corresponding $P_{\rm BA}^{(1)\max}$. This can be done using a known distribution of atoms over lattice sites of each superstructure (see Chap. 5) and a set of octahedral clusters (Fig. 6.2) describing an fcc lattice.

The largest values for the probability $\lambda_{\rm BA} P_{\rm BA}^{(1)\max}$ equal to 1/4, 1/3, 1/2, 5/9 and 2/3 are achieved for superstructures of type AB_7 (space groups $Fm\bar{3}m$, $P4_332$), AB_5 (space groups *C*2/*m*, $P3_1$, *C*2), AB_3 (space groups $Pm\bar{3}m$, *I*4/*mmm*), AB_2 (space groups *Immm*, *P*2), and AB (space groups *P*4/*mmm*, $I4_1/amd$) respectively. For superstructures AB_2 (space groups $C222_1$ and $P\bar{3}m1$) the probability $\lambda_{\rm BA} P_{\rm BA}^{(1)\max}$ is equal to 1/2 and 1/3, i.e. it is smaller than $\lambda_{\rm BA} P_{\rm BA}^{(1)\max} = 5/9$ for superstructures AB_2 (space groups *Immm*, *P*2). Similarly, the probability $\lambda_{\rm BA} P_{\rm BA}^{(1)\max}$ equals 1/2 for superstructures AB (space groups $R\bar{3}m$, $Fd\bar{3}m$), i.e. it is smaller than $\lambda_{\rm BA} P_{\rm BA}^{(1)\max} = 2/3$ for superstructures AB (space groups *P*4/*mmm*, $I4_1/amd$). Values of $P_{\rm BA}^{(1)\max}$ can be found from the above mentioned values of $\lambda_{\rm BA} P_{\rm BA}^{(1)\max}$ taking into account that $\lambda_{\rm BA} = 2$.

From the $P_{\rm BA}^{(1)\max}$ values it follows that the relation $P_{\rm BA}^{(1)\max} = y$ holds for superstructures AB_7 ($y = 1/8$), AB_5 ($y = 1/6$), and AB_3 ($y = 1/4$). Therefore it is valid within the interval $0 \le y \le 1/4$. If the probability $P_{\rm BA}^{(1)\max}$ is assumed to be a linear function of y when $y \ge 1/4$ and considering that $P_{\rm BA}^{(1)\max} = 1/4$ at $y = 1/4$ and $P_{\rm BA}^{(1)\max} = 1/3$ at $y = 1/2$, we shall arrive at $P_{\rm BA}^{(1)\max} = (1 + 2y)/6$. Indeed, this relation is valid for superstructures AB_3 (space groups $Pm\bar{3}m$, *I*4/*mmm*), AB_2 (space groups *Immm*, *P*2) and AB (space group *P*4/*mmm*, $I4_1/amd$) at $y = 1/4$, 1/3 and 1/2 respectively.

Since the structure is invariant to a replacement of atoms A by atoms B and vice versa, two more intervals of $P_{\rm BA}^{(1)\max}$ values are obtained if $(1 - y)$ is substituted for y in the $P_{\text{BA}}^{(1)\max}(y)$ relations found. Thus, for fcc solid solutions A_yB_{1-y}, the probability $P_{\text{BA}}^{(1)\max}$ has the form

$$P_{\text{BA}}^{(1)\max} = y, \qquad \text{if } 0 \le y \le 1/4, \tag{6.10}$$

$$P_{\text{BA}}^{(1)\max} = (1+2y)/6, \quad \text{if } 1/4 \le y \le 1/2, \tag{6.11}$$

$$P_{\text{BA}}^{(1)\max} = (3-2y)/6, \quad \text{if } 1/2 \le y \le 3/4, \tag{6.12}$$

$$P_{\text{BA}}^{(1)\max} = (1-y), \qquad \text{if } 3/4 \le y \le 1. \tag{6.13}$$

Similar dependencies of $P_{\text{BA}}^{(1)\max}$ on y are obtained if an fcc lattice is described by a set of clusters in the form of tetrahedra comprising four nearest lattice sites or by clusters including all 14 sites of an unit cell of the fcc lattice. Thus, the shape of $P_{\text{BA}}^{(1)\max}(y)$ is invariant to the shape of the clusters used to describe the fcc lattice.

Let us consider the interval $3/4 \le y \le 1.0$ for which the short-range order parameter $\alpha_1 = 1 - (1/y)$ at the maximum value of the probability $P_{\text{BA}}^{(1)\max}$. Taking into account this α_1 value, the set of equations (6.1), (6.2), and (6.5) takes the form

$$\sum_{i=0}^{9} \lambda_i P_i = 1, \quad \sum_{i=0}^{9} l_i \lambda_i P_i = y, \quad \sum_{i=0}^{9} n_i^{(j)} \lambda_i P_i = 2(1-y). \tag{6.14}$$

Table 6.2 lists values for the coefficients λ_i, l_i and $n_i^{(j)}$ for fcc solid solutions described by octahedron-shaped clusters (see Fig. 6.2).

Table 6.2. Coefficients λ_i, l_i, $n_i^{(1)}$ and $n_i^{(2)}$, corresponding to P_i clusters, which are used for the description of solid solutions with an fcc basic structure

Cluster P_i	P_0	P_1	P_2	P_3	P_4	P_5	P_6	P_7	P_8	P_9
λ_i	1	6	3	12	8	12	12	3	6	1
l_i	1	5/6	2/3	2/3	1/2	1/2	1/3	1/3	1/6	0
$n_i^{(1)}$	0	1/3	2/3	1/2	1/2	2/3	1/2	2/3	1/3	0
$n_i^{(2)}$	0	1/3	0	2/3	1	1/3	2/3	0	1/3	0

Unlike (6.7), equations (6.14) have an infinite number of solutions of type $P_0 = (\lambda_2 P_2 + 6y - 5)$, $P_1 = 1 - y - P_2$ and $P_3 = P_4 = P_5 = P_6 = P_7 = P_8 = P_9 = 0$. Hence, in the case of the maximum degree of the long-range order in the first coordination sphere, the short-range order parameter in the second coordination sphere is in accordance with (6.8) a variable: $\alpha_2 = 1 - P_1/[y(1-y)]$. As a consequence, the atoms are distributed over the lattice with a certain ambiguity, which, in combination with the entropy factor, will prevent an ordered arrangement. This means that short-range order in the first coordination sphere of an fcc lattice is not a sufficient condition for the formation of long-range order. This statement also applies to the Cu_3Au alloy with $y = 3/4$, which is ordered on the basic structure of an fcc lattice. For an ordered Cu_3Au alloy with a maximum degree of short-range order, the cluster formation probabilities $P_0 = 1/4$ and $P_2 = 3/(4\lambda_2)$ are nonzero, which provides a maximum degree for the short-range order in the first coordination sphere. However, this solution clearly shows that a great number of other P_0, P_1 and P_2 values exist, which lead to a maximum degree of short-range order in the first coordination sphere. Thus, the maximum degree of short-range order in the first coordination sphere is a necessary but insufficient condition for the presence of P_0 and P_2 clusters only, which would ensure a formation of an ordered phase with the Cu_3Au structure.

A similar treatment can be given for A_yB_{1-y} with $1/2 \leq y \leq 3/4$, which is appropriate to describe the ordering in CuAu. The parameter α_1^{ord} coincides with a maximum degree of short-range order in the first coordination sphere. A fully ordered alloy includes only clusters P_2 and P_7 with probabilities $P_2 = 1/(2\lambda_2)$ and $P_7 = 1/(2\lambda_7)$. However, α_1 equal to α_1^{ord} can be achieved in alloys having other cluster probabilities, for instance $P_5 = 1/\lambda_5$ (all other probabilities being zero). In the interval $1/2 \leq y \leq 3/4$ the maximum α_1 value is not a sufficient condition for the formation of long-range order. This is because $\alpha_2 = f(P_1)$ even in the second coordination sphere, i.e. α_2 is a variable rather than a constant for a given composition and structure.

A similar treatment of the entire interval of compositions of solid solutions A_yB_{1-y} with an fcc basic structure shows that in fcc solid solutions (unlike in bcc solid solutions) maximum short-range order in the first coordination sphere is not a sufficient condition for the formation of long-range order.

An analysis of the relationship between short- and long-range orders [16–19] suggests that two types of solid solutions exist. The first type includes solid solutions where no long-range order appears in the presence of a maximum degree of short-range order in the first coordination sphere. The second type covers solid solutions where short-range order predetermines the formation of long-range order.

Considering the above features of bcc and fcc solid solutions, it becomes clear why theories [4, 11–14], which allow for nearest neighbor interactions only, describe well the ordering in a bcc lattice and fail in the case of an fcc lattice.

A method developed in ref. [16, 17] permits us to determine the number of short-range order parameters sufficient to describe ordering in fcc solid solutions. To this end, one has to take clusters including atoms from coordination spheres more distant than the second coordination sphere.

As is known, short-range order characterizes a two-particle correlation in atomic arrangements. However, interactions taking place in solid solutions are more complex. The method proposed can be used to analyze not only paired but also multiparticle correlations. In each specific case it allows the determination of whether particular correlations are sufficient for the formation of long-range order. It should be noted that each correlation will entail further equations relating the cluster probability to the probability that groups of three, four or more atoms are formed. The probability of such groups in the ordered structure should be determined beforehand.

The analysis of the relationship between short- and long-range orders suggests one more important deduction concerning the kinds of phase transitions similar to ordering in bcc and fcc basic structures. Depending on the particular reasons responsible for the appearance of long-range order, it can be formulated as a phase transition of the first or second kind. If the structure of an ordered solid solution is such that the presence of short-range order leads to the formation of long-range order, the order–disorder phase transformation can result from a monotonic increase in the degree of the short-range order in the first coordination sphere and, most likely, it will be a transformation of the second kind. The maximum degree of the short-range order in the first coordination sphere in bcc basic structures was shown [16, 17] to be sufficient for the formation of long-range order. This inference is confirmed by numerous studies of alloys having a bcc basic structure, which are usually ordered through phase transition of the second kind [20]. This means that there is no strict functional relationship between short- and long-range orders, i.e. under certain conditions the degree of long-range order can change abruptly and a phase transition of the first kind takes place. Indeed, an analysis of evidence on ordering in solid solutions with an fcc basic structure shows that the overwhelming majority of order–disorder phase transformations in these solutions are of the first kind.

Therefore it is possible to establish a qualitative correlation between short- and long-range orders in a solid solution and the kind of phase transition. If a solid solution can have a crystal lattice with a maximum short-range order and without long-range order, the order–disorder transformation most likely proceeds as a phase transition of the first kind. If the presence of a maximum or lower degree of short-range order in the first coordination sphere automatically leads to the appearance of long-range order, the order–disorder transformation will be a phase transition of the second kind. It is worth noting that the established correlation is in good agreement with the known Landau criterion [21] for phase transitions of the second kind.

A simple analytical method was proposed in ref. [16–18]. It allows one to determine the number of nearest coordination spheres for which short-range order is sufficient for the formation of long-range order. It is applicable to all structures. In [16, 17] the proposed method was used to analyze the ordering in solid solutions with bcc and fcc basic structures as an example. The analysis of the relationship between short- and long-range orders shows that ordering in solid solutions with a bcc basic structure can be described in principle by taking into account short-range order in the first coordination sphere only. In contrast, ordering in solid solutions with an fcc lattice should be described by taking into account short-range order parameters in more than two nearest coordination spheres.

6.2 Short-Range Order in Superstructures

Short-range correlations in superstructures can be due to short-range (e.g. covalent or ionic) or long-range (e.g. conduction electrons, phonons, elastic lattice distortions) interactions. For this reason phase equilibria in alloys and nonstoichiometric compounds should be calculated taking into account possible short-range order both in disordered and ordered phases. However, even most theories based on statistical approaches consider only one type of order. For example, the mean field method [13, 14] considers exclusively long-range order. Most cluster methods (the Bethe theory [4], the cluster variation method [6–10]) feature a precise description of the interactions between the particles in a cluster and an approximate description of further interactions. Cluster methods consider only short-range order.

From a physical point of view, it is not justified to consider only one type of order. However the biggest stumbling block in the statistical analysis of systems where ordering is possible is an apropriate consideration of short- and long-range order. In ref. [1, 2] a method has been proposed which may reduce this obstacle. It is based on a functional relationship between short- and long-range orders and on the determination of the variation interval of short-range order parameters α. Intervals of admissible α values in different coordination spheres of frequent superstructures are useful for the determination of the short-range order from diffuse scattering and from the interpretation of NMR and NGR spectra.

According to [1, 2] we consider a substitutional solid solution A_yB_{1-y}, in which some superstructure is formed. The atomic distribution function $n(\mathbf{r})$ [15] can be used to determine the universal dependence of the short-range order parameter α on the long-range order parameter η. As noted earlier (see Sect. 5.2), the distribution function $n(\mathbf{r})$ is the probability that an atom of a given species is located at site $\mathbf{r}$. It can be presented as the concentration y of atoms of this species and a superposition of static plane waves (formulas (5.19) and (5.20)).

There is a certain probability that lattice sites can be occupied by atoms of both species A and B. The normalization condition is $n(\mathbf{r}) + n_B(\mathbf{r}) = 1$, where $n_B(\mathbf{r})$ is the distribution function for the B atoms.

The probability of filling of any site $\mathbf{r}$ described by the distribution function (5.19) is independent of atomic species occupying other sites. If a site $\mathbf{r}$ is occupied by an A atom, the probability of a site $\mathbf{r}'$ being filled with the same atom, i.e. the conditional probability, equals the product of the distribution function at these sites. This means that the distribution function fails to describe short-range correlations in the atomic arrangement. Hence, the formula derived below for the relation between short- and long-range orders determines only that part of the short-range order which is due directly to the presence of long-range order. This short-range order is called superstructural in contrast to correlational short-range order, which is caused only by short-range correlations in the atomic arrangement.

The short-range order parameter α_j for the j-th coordination sphere can be found from formula (6.3), where the quantity $P_{BA}^{(j)}$ is the probability that BA pairs are formed in a solid solution. The distance between atoms A and B is equal to the radius of the j-th coordination sphere.

Let an ordering solid solution has a superstructure where a unit cell contains M types of nonequivalent positions of the multiplicity g_f ($\sum_{f=1}^{M} g_f = \Phi_s$ being the number of sites in a unit cell of the superstructure). Consider the j-th coordination spheres whose centers are unit cell sites located at nonequivalent positions. Given the same coordination number z_j, each nonequivalent position f has an individual type of neighborhood, which differs by the ratio of atoms of unlike species and their mutual arrangement in the j-th coordination sphere. In this case the probability $P^{(j)}_{f\,\mathrm{BA}}$ that a BA pair is formed in the j-th coordination sphere, whose center is a species B atom located at a site $\boldsymbol{r}_f$ and occupying a position of type f, equals

$$P^{(j)}_{f\,\mathrm{BA}} = \frac{1}{z_j}\sum_{i=1}^{z_j} n_{\mathrm{B}}(\boldsymbol{r}_f)n(\boldsymbol{r}^{(j)}_{if}) = \frac{n_{\mathrm{B}}(\boldsymbol{r}_f)}{z_j}\sum_{i=1}^{z_j} n(\boldsymbol{r}^{(j)}_{if}) \;, \tag{6.15}$$

where $n(\boldsymbol{r}^{(j)}_{if})$ is the distribution function value for species A atoms at sites $\boldsymbol{r}^{(j)}_{if}$ belonging to the j-th coordination sphere. In turn, the probability $P^{(j)}_{\mathrm{BA}}$ is the result of the successive summation and averaging over all sites $\boldsymbol{r}$ included in a unit cell of an ordered phase, allowing for the multiplicity of positions of these sites, i.e.

$$P^{(j)}_{\mathrm{BA}} = \frac{\sum_{f=1}^{M} g_f n_{\mathrm{B}}(\boldsymbol{r}_f)\sum_{i=1}^{z_j} n(\boldsymbol{r}^{(j)}_{if})}{z_j\sum_{f=1}^{M} g_f} = \frac{\sum_{f=1}^{M} g_f[1-n(\boldsymbol{r}_f)]\sum_{i=1}^{z_j} n(\boldsymbol{r}^{(j)}_{if})}{z_j\sum_{f=1}^{M} g_f} \;. \tag{6.16}$$

Considering (6.3) and (6.16), the superstructural short-range order parameter $\alpha_j\{\eta\}$ is

$$\alpha_j\{\eta\} = 1 - \frac{\sum_{f=1}^{M} g_f[1-n(\boldsymbol{r}_f)]\sum_{i=1}^{z_j} n(\boldsymbol{r}^{(j)}_{if})}{y(1-y)z_j\sum_{f=1}^{M} g_f} \;. \tag{6.17}$$

If the distribution function $n(\boldsymbol{r})$ is known, it is easy to find from (6.17) the short-range order parameters $\alpha_j\{\eta\}$ in a solid solution for an arbitrary composition for infinitely remote coordination spheres when long-range order parameters change from zero to a maximum value $\eta_{\max}$. In fact, the determination of the parameters $\alpha_j\{\eta\}$ from (6.17) reduces to the task of computing the average number of atoms in a crystal of a particular species in the j-th coordination sphere. In this respect relationship (6.17) is identical to the Kauli formula [22].

Let us calculate superstructural short-range order parameters for fully ordered alloys of types CuZn (bcc basic structure), CuAu, and Cu_3Au (fcc basic structure), and also for the ordered phases M_2X, M_3X_2, M_4X_3, M_6X_5, and M_8X_7 of strongly nonstoichiometric compounds MX_y with the $B1$ basic structure. The fully ordered alloys and nonstoichiometric compounds have general formulas $M_{2t}X_{2t-1}\square$ and $A_{2t-1}B$, where t = 1, 1.5, 2, 3 or 4. Table 6.3 presents the main structural characteristics of the ordered alloys in question, including superstructure vectors forming an order–disorder transition channel.

Table 6.3. Superstructures of $A_{2t-1}B$-type (t = 1 and 2) in some alloys

Superstructure					Disordered basic phase	
type	[1] disorder–order phase transition channel	symmetry and lattice type	space group	ordered alloys	symmetry and lattice type	space group
AB	$\boldsymbol{k}_{12}^{(1)}$ = {0, 0, 1}	cubic *B*2 (CsCl)	$Pm\bar{3}m$	CuZn AuZn CuPd	bcc *A*2 (W)	$Im\bar{3}m$
AB	$\boldsymbol{k}_{10}^{(1)}$ = {0, 0, 1}	tetragonal $L1_0$ (CuAu)	$P4/mmm$	CuAu, LiBi AlTi, ZnPd FePd, NiPt	fcc *A*1 (Cu)	$Fm\bar{3}m$
A_3B	$\boldsymbol{k}_{10}^{(1)}$ = {0, 0, 1} $\boldsymbol{k}_{10}^{(2)}$ = {0, 1, 0} $\boldsymbol{k}_{10}^{(3)}$ = {1, 0, 0}	cubic $L1_2$ (Cu_3Au)	$Pm\bar{3}m$	Cu_3Au Cr_3Pt, Sn_3U, Zr_3Al,	fcc *A*1 (Cu)	$Fm\bar{3}m$

[1] Superstructure vectors $\boldsymbol{k}_s^{(j)}$ are given in units $2\pi/a$ where a is the lattice constant of the basic disordered phase.

Similar characteristics of superstructures $M_{2t}X_{2t-1}\square$, which appear in non-stoichiometric compounds MX_y with the *B*1 basic structure, are given in Sect. 5.4. If data on order–disorder transition channels are known, the corresponding distribution functions can be calculated from (5.19) and (5.20).

Ordering of AB type (CuZn) in a bcc lattice obeys the distribution function

$$n(\boldsymbol{r}) = y - (\eta_{12}/2)\cos 2\pi z_{\mathrm{I}} \ . \tag{6.18}$$

Ordering of types AB (CuAu) and A_3B (Cu_3Au) in an fcc lattice is given by the distribution functions (5.54) and (5.68) respectively. The distribution functions for the superstructures $M_{2t}X_{2t-1}\square$ of strongly nonstoichiometric interstitial compounds MX_y are discussed in Sect. 5.4; x_{I}, y_{I}, and z_{I} are the coordinates of the sites $\boldsymbol{r}$ in an ordering bcc or fcc lattice in all distribution functions.

Using distribution functions $n(\boldsymbol{r})$ and coordinates of sites $\boldsymbol{r}$ forming a particular coordination sphere, the superstructural short-range order parameters were calculated from (6.17) for the coordination spheres of fully ordered stoichiometric superstructures, i.e. with all long-range order parameters being unity and $y = y_{st} = (2t-1)/2t$. Calculated $\alpha_j(y_{st}, \eta_{max})$ values are given in Tables 6.4, 6.5 and 6.6. The radius of the j-th coordination sphere is $R_j = (a/2)(h^2+k^2+l^2)^{1/2}$ where a is the basic lattice constant.

Table 6.4. Short-range order parameters $\alpha_j(y_{st}, \eta_{max})$ for the superstructure AB ($t = 1$) with bcc basic lattice and for superstructures M_8X_7 and A_7B ($t = 4$) with an fcc basic lattice (z_j is the coordination number)

bcc basic lattice				fcc basic lattice				
Coordination sphere[1]			α_j for super structure AB (space group $Pm\bar{3}m$)	Coordination sphere[1]			α_j for superstructures M_8X_7 and A_7B	
							space group $Fm\bar{3}m$	space group $P4_332$
J	*hkl*	z_j		*j*	*hkl*	z_j		
1	111	8	−1	1	110	12	−1/7	−1/7
2	200	6	1	2	200	6	−1/7	−1/7
3	220	12	1	3	211	24	−1/7	1/7
4	311	24	−1	4	220	12	1	−1/7
5	222	8	1	5	310	24	−1/7	−1/7
6	400	6	1	6	222	8	−1/7	−1/7
7	331	24	−1	7	321	48	−1/7	1/7
8	420	24	1	8	400	6	1	1
9	422	24	1	9	330, 411	36	−1/7	−1/7
10	333, 511	32	−1	10	420	24	−1/7	−1/7
11	440	12	1	11	332	24	−1/7	1/7
12	531	48	−1	12	422	24	1	−1/7
13	442, 600	30	1	13	431, 510	72	−1/7	−1/7
14	620	24	1	14	521	48	−1/7	−1/7
15	533	24	−1	15	440	12	1	1
16	622	24	1	16	433, 530	48	−1/7	−1/7

[1] The vacancy □ is the center of the coordination spheres for superstructures $M_{2t}X_{2t-1}$ of nonstoichiometric interstitial compounds; an atom of species B is the center of the coordination spheres for superstructures $A_{2t-1}B$ of solid solutions (alloys).

Partial coincidence of the short-range order parameters $\alpha_j(y_{st}, \eta_{max})$ takes place for the cubic superstructures of M_8X_7 type (see Table 6.4). In accordance with calculations, also the short-range order parameters $\alpha_j(y_{st}, \eta_{max})$ coincide completely for the superstructures of types AB (space group *P4/mmm*) and A_3B (space group $Pm\bar{3}m$) (see Table 6.5). Fully ordered substitutional solid solutions AB (CuAu) and A_3B (Cu_3Au) have a similar short-range order, because paired interactions do not depend on the composition of these alloys. The same is true for the superstructures M_2X (space group *P4/mmm*) and M_4X_3 (space group $Pm\bar{3}m$) whose nonmetal sublattices are ordered similarly to CuAu and Cu_3Au alloys respectively.

Table 6.5. Short-range order parameters $\alpha_j(y_{st}, \eta_{max})$ for the superstructures M_2X and AB ($t = 1$), and M_4X_3 and A_3B ($t = 2$) with an fcc basic lattice (z_j is the coordination number)

[1] Coordination sphere			M_2X and AB, space groups			M_4X_3 and A_3B, space groups	
j	hkl	z_j	$R\bar{3}m$ and $Fd\bar{3}m$	$I4_1/amd$	$P4/mmm$	$Pm\bar{3}m$	$I4/mmm$
1	110	12	0	−1/3	−1/3	−1/3	−1/3
2	200	6	−1	1/3	1	1	5/9
3	211	24	0	1/3	−1/3	−1/3	1/9
4	220	12	1	−1/3	1	1	1/9
5	310	24	0	−1/3	−1/3	−1/3	−1/3
6	222	8	−1	−1	1	1	−1/3
7	321	48	0	1/3	−1/3	−1/3	1/9
8	400	6	1	1	1	1	1
9	330, 411	36	0	−1/3	−1/3	−1/3	−1/3
10	420	24	−1	1/3	1	1	5/9
11	332	24	0	1/3	−1/3	−1/3	1/9
12	422	24	1	−1/3	1	1	1/9
13	431, 510	72	0	−1/3	−1/3	−1/3	−1/3
14	521	48	0	1/3	−1/3	−1/3	1/9
15	440	12	1	1	1	1	1
16	433, 530	48	0	−1/3	−1/3	−1/3	5/9

[1] The vacancy □ is the center of the coordination spheres for superstructures $M_{2t}X_{2t-1}$ of nonstoichiometric interstitial compounds; an atom of species B is the center of the coordination spheres for superstructures $A_{2t-1}B$ of solid solutions (alloys).

A similar short-range order is observed in fully ordered trigonal and cubic structures M_2X (space groups $R\bar{3}m$ and $Fd\bar{3}m$) (Table 6.5), and also in rhombohedral and monoclinic structures M_3X_2 (space groups *Immm* and *P*2) (see Table 6.6).

The fact that the short-range order parameters $\alpha_j(y_{st}, \eta_{max})$ coincide for three superstructures of type M_6X_5 with different crystal lattices is remarkable (Table 6.6). These superstructures are formed in nonstoichiometric compounds through different transition channels and they have different distribution functions (5.72)–(5.74). Their unit cells belong to the space groups $C2/m$, $P3_1$, and $C2$. An identical short-range order of the superstructures of different symmetry suggests that in addition to strong central interatomic interactions, noncentral long-range forces leading to different symmetries for the ordered phases of type M_6X_5 are significant in the nonstoichiometric compounds MX_y. Even a maximum short-range order seems to be insufficient for the formation of such ordered structures in agreement with the findings in [16, 17].

Table 6.6. Short-range order parameters $\alpha_j(y_{st}, \eta_{max})$ for the superstructures M_3X_2 and A_2B ($t = 1.5$), M_6X_5 and A_5B ($t = 3$), and $Ti_5■_1O_5□_1$ ($t = 3$) with an fcc basic lattice (z_j is the coordination number)

Coordination sphere[1]			M_3X_2 and A_2B, space groups			M_6X_5 and A_5B, space groups	$Ti_5■_1O_5□_1$, space group
J	*hkl*	z_j	*P*2 and *Immm*	$P\bar{3}m1$	$C222_1$	$C2/m$, $P3_1$, $C2$	$C2/m$
1	110	12	−1/4	1/4	−1/8	−1/5	−1/5
2	200	6	0	−1/2	−1/2	−1/5	1/5
3	211	24	1/4	−1/8	1/8	1/5	0
4	220	12	−1/4	1/4	1/4	0	0
5	310	24	0	−1/2	1/8	0	0
6	222	8	1/4	−1/8	−1/8	−1/5	2/5
7	321	48	−1/4	1/4	−3/16	−1/10	−1/10
8	400	6	0	−1/2	0	1/5	1/5
9	330, 411	36	1/4	1/4	0	2/15	1/15
10	420	24	−1/4	1/4	0	−1/5	0
11	332	24	0	−1/2	1/4	0	−1/5
12	422	24	1/4	−1/8	0	2/5	2/5
13	431, 510	72	−1/4	1/4	0	−1/15	−2/15
14	521	48	1/4	−1/8	1/8	0	0
15	440	12	−1/4	1/4	−1/4	0	0
16	433, 600	48	0	−1/2	−1/16	0	−1/10

[1] The vacancy □ is the center of the coordination spheres in superstructures $M_{2t}X_{2t-1}$ of nonstoichiometric interstitial compounds; an atom of species B is the center of the coordination spheres in superstructures $A_{2t-1}B$ of solid solutions (alloys).

In thermodynamic calculations of phase diagrams the interval in which $\alpha_j(y,\eta)$ can vary as a function of the solid solution composition is a crucial parameter. To determine the upper and lower variation limits of $\alpha_j(y,\eta)$, the condition for the normalization of the distribution function at the sites of the j-th coordination sphere to the composition of a solid solution A_yB_{1-y} can be used:

$$\frac{\sum_{f=1}^{M} g_f \sum_{i=1}^{z_j} n(\mathbf{r}_{if}^{(j)})}{z_j \sum_{f=1}^{M} g_f} = y\,. \tag{6.19}$$

The distribution function $n(\mathbf{r})$ may depend on several long-range order parameters. The long-range order parameters $\eta_1 = \eta_2 = \ldots = \eta$ being equal, the distribution function $n(\mathbf{r})$ takes only two values for all sites in an ordered lattice: $n_1 = y - y_{st}\eta$ and $n_2 = y + (1 - y_{st})\eta$ with multiplicity g_1 and g_2 ($y_{st} = g_2/(g_1 + g_2)$ being the stoichiometric composition of the ordered phase). For the superstructures $A_{2t-1}B$ and $M_{2t}X_{2t-1}□$ discussed, the position multiplicity is $g_1 = 1$ and $g_2 = 2t - 1$, while

$y_{st} = (2t-1)/2t$. Taking into account equal parameters for the long-range order, the quantity $(1/z_j)\sum_{i=1}^{z_j} n(\mathbf{r}_{if}^{(j)})$ in (6.17) can be presented as a sum of two terms:

$$(1/z_j)\sum_{i=1}^{z_j} n(\mathbf{r}_{if}^{(j)}) = m_{1f}^{(j)} n_1 + m_{2f}^{(j)} n_2 = y + (m_{2f}^{(j)} - y_{st})\eta \ , \tag{6.20}$$

where $m_{1f}^{(j)}$ and $m_{2f}^{(j)}$ denote the relative number of positions of types 1 and 2 in the j-th coordination sphere whose center is a site of type f.

If the parameters η are equal, the sum over nonequivalent positions f in formula (6.20) reduces to a sum over two positions with multiplicity g_1 and g_2. With the normalization condition (6.19), equation (6.20), and the values *for* n_1 and n_2, formula (6.17) becomes

$$\alpha_j(y,\eta) = -\eta^2(1-y_{st})(m_{21}^{(j)} - y_{st})/[y(1-y)] \ , \tag{6.21}$$

where $m_{21}^{(j)}$ is the relative fraction of sites occupying type 2 positions in the j-th coordination sphere whose center is a site in position 1.

It follows from (6.21) that the maximum values of $\alpha_j(y,\eta)$ are realized when the maximum degree of long-range order is achieved. The maximum value of the long-range order parameter η_{max}, which is possible for a solid solution A_yB_{1-y} of a given composition where a superstructure $A_{2t-1}B$ appears, is determined by (5.85) which yields

$$\eta_{max}(y) = \begin{cases} (1-y)/(1-y_{st}) = 2t(1-y) & \text{if} \quad y_{st} \le y \le 1, \\ y/y_{st} = 2ty/(2t-1) & \text{if} \quad 0 \le y \le y_{st} \, . \end{cases} \tag{6.22}$$

Considering (6.21), (6.22), and $\alpha_j(y_{st},\eta_{max}) = -(m_{21}^{(j)} - y_{st})/y_{st}$, the maximum values of $\alpha_j(y_{st}, \eta_{max})$ become equal to

$$\left.\begin{aligned} -\frac{(1-y)(m_{21}^{(j)} - y_{st})}{y(1-y_{st})} &= \alpha_j(y_{st},\eta_{max})\frac{y_{st}(1-y)}{y(1-y_{st})} \quad \text{if } y_{st} \le y \le 1 \\ -\frac{y(1-y_{st})(m_{21}^{(j)} - y_{st})}{y_{st}^2(1-y)} &= \alpha_j(y_{st},\eta_{max})\frac{y(1-y_{st})}{y_{st}(1-y)} \quad \text{if } 0 \le y \le y_{st} \end{aligned}\right\} . \tag{6.23}$$

For a superstructure of type $A_{2t-1}B$, $y_{st} = (2t-1)/2t$. Therefore (6.21) can be rewritten as

$$\alpha_j(y,\eta) = -(\eta^2/4t^2)(2tm_{21}^{(j)} + 1 - 2t)/[y(1-y)] \, . \tag{6.24}$$

Expressions (6.21) and (6.24) relate the long-range order parameter η to superstructural short-range order parameters α_j for any coordination sphere of ordered solid solutions of an arbitrary composition with different types of ordering in the absence of correlations. It follows that the superstructural short-range order parameter is a quadratic function of the long-range order parameter and equals zero at $\eta = 0$ for any coordination sphere and for any composition. When $\eta = 0$, the short-range order parameter may be negative (if $m_{21}^{(j)} > y_{st}$), zero (if $m_{21}^{(j)} = y_{st}$), or positive (if $m_{21}^{(j)} < y_{st}$). Table 6.7 gives $m_{21}^{(j)}$ values for the superstructures discussed.

Table 6.7. Coefficients $m_{21}^{(j)}$ for the superstructures with bcc and fcc basic lattices

	bcc basic lattice	fcc basic lattice					
j[1]	AB,	M_2X and AB			M_3X_2 and A_2B,		
	space group	space groups			space groups		
	$Pm\bar{3}m$	$R\bar{3}m$ and $Fd\bar{3}m$	$I4_1/amd$	$P4/mmm$	$P2$ and $Immm$	$P\bar{3}m1$	$C222_1$
1	1	1/2	2/3	2/3	5/6	1/2	3/4
2	0	1	1/3	0	2/3	1	1
3	0	1/2	1/3	2/3	1/2	3/4	7/12
4	1	0	2/3	0	5/6	1/2	1/2
5	0	1/2	2/3	2/3	2/3	1	7/12
6	0	1	1	0	1/2	3/4	3/4
7	1	1/2	1/3	2/3	5/6	1/2	19/24
8	0	0	0	0	2/3	1	2/3
9	0	1/2	2/3	2/3	1/2	1/2	2/3
10	1	1	1/3	0	5/6	1/2	2/3
11	0	1/2	1/3	2/3	2/3	1	1/2
12	1	0	2/3	0	1/2	3/4	2/3
13	0	1/2	2/3	2/3	5/6	1/2	2/3
14	0	1/2	1/3	2/3	1/2	3/4	7/12
15	1	0	0	0	5/6	1/2	5/6
16	0	1/2	2/3	2/3	2/3	1	17/24

	fcc basic lattice					
j[1]	M_4X_3 and A_3B,		M_6X_5 and A_5B	$Ti_5■_1O_5□_1$,	M_8X_7 and A_7B,	
	space groups		space groups	space group	space groups	
	$Pm\bar{3}m$	$I4/mmm$	$C2/m$, $P3_1$, $C2$	$C2/m$	$Fm\bar{3}m$	$P4_332$
1	1	1	1	1	1	1
2	0	1/3	1	2/3	1	1
3	1	2/3	2/3	5/6	1	3/4
4	0	2/3	5/6	5/6	0	1
5	1	1	5/6	5/6	1	1
6	0	1	1	½	1	1
7	1	2/3	11/12	11/12	1	3/4
8	0	0	2/3	2/3	0	0
9	1	1	3/18	7/9	1	1
10	0	1/3	1	5/6	1	1
11	1	2/3	5/6	1	1	3/4
12	0	2/3	1/2	½	0	1
13	1	1	8/9	17/18	1	1
14	1	2/3	5/6	5/6	1	1
15	0	0	5/6	5/6	0	0
16	1	1	5/6	11/12	1	1

[1] j is the number of the coordination sphere.

Tables 6.4, 6.5 and 6.6 give $\alpha_j(y_{st}, \eta_{max})$ values for the discussed superstructures. From (6.23) one can easily find the maximum $\alpha_j(y, \eta_{max})$ values for any composition of an ordered alloy or a nonstoichiometric compound. As an example, Fig. 6.3 shows the maximum values of superstructural short-range order parameters $\alpha_j(y, \eta_{max})$ for the first twelve coordination spheres as a function of the composition of a compound MX_y ordered in accordance to the M_6X_5 pattern.

X-ray or neutron diffuse scattering is frequently used to study short-range order in crystals. Formula (6.23) can be used advantageously in analyzing experimental data on short-range order in ordered phases. It allows one to determine the physical limits for the parameters of the function by approximating the intensity of the diffuse scattering. In ref. [23] M_6X_5-ordered carbide $NbC_{0.80}$ was examined by neutron diffuse scattering. A physically erroneous approximation of the data resulted in unreasonable values for the short-range order parameters: $\alpha_5 = -0.194$, $\alpha_6 = 0.228$, and $\alpha_8 = 0.254$. However, for an ordered phase of M_6X_5-type, $\alpha_5(y, \eta) \equiv 0$ at any y values, because $m_{21}^{(5)} = y_{st}$. In the 6th and 8th coordination spheres of ordered carbide $NbC_{0.80}$, short-range order parameters can only vary in the intervals $-0.16 \le \alpha_6(\eta) < 0$ and $0 < \alpha_8(\eta) \le 0.16$. Some physically absurd values (for example, $\alpha_2 = -0.507$!) were also obtained [23] for disordered carbide $NbC_{0.80}$.

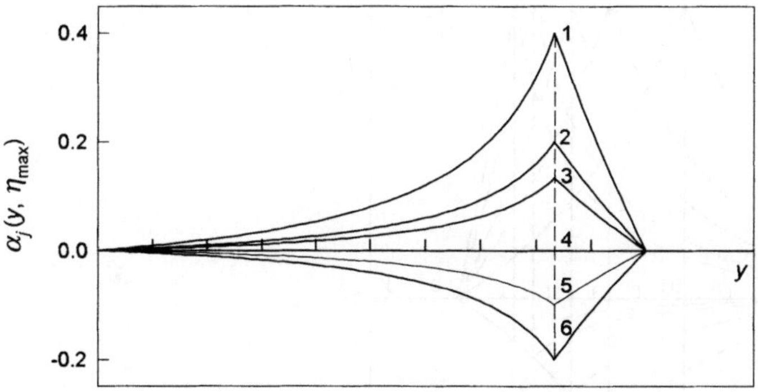

Fig. 6.3. Superstructural short-range order parameter $\alpha_j(y, \eta_{max})$ for the twelve coordination spheres versus the composition of a nonstoichiometric compound MX_y, in which a superstructure of the M_6X_5 type is formed: (1) 12th coordination sphere; (2) 3rd and 8th coordination spheres; (3) 9th coordination sphere; (4) 4th, 5th and 11th coordination spheres; (5) 7th coordination sphere; (6) 1st, 2nd, 6th and 10th coordination spheres

The quantity $m_{21}^{(j)}$ for different values j can only assume the values allowed by the normalization condition (6.19). Substituting (6.20) for the sum $\sum_{i=1}^{z_j} n(\mathbf{r}_{if}^{(j)})$ in (6.19), we obtain a direct normalization condition for $m_{2f}^{(j)}$:

$$(1 - y_{st})m_{21}^{(j)} + y_{st}m_{22}^{(j)} = y_{st} \quad . \tag{6.25}$$

An analysis of (6.25) shows that the relative fraction, $m_{21}^{(j)}$, can change from 0 to 1 when $y_{st} \geq 1/2$ and from 0 to $y_{st}/(1 - y_{st})$ when $y_{st} \leq 1/2$. Substituting these limiting values of $m_{21}^{(j)}$ into equation (6.23), we obtain the lower and upper limits of the short-range order parameter for any coordination sphere depending on the composition of the solid solution and on the type of ordering (Table 6.8).

Table 6.8. Lower (inf(α)) and upper (sup(α)) boundaries of the superstructural short-range order parameter α

Superstructure	$0 \leq y \leq y_{st}$		$y_{st} \leq y \leq 1$	
	inf(α)	sup(α)	inf(α)	Sup(α)
$y_{st} \geq 1/2$	$-\dfrac{y(1-y_{st})^2}{y_{st}^2(1-y)}$	$\dfrac{y(1-y_{st})}{y_{st}(1-y)}$	$-\dfrac{1-y}{y}$	$\dfrac{(1-y)y_{st}}{(1-y_{st})y}$
$y_{st} \leq 1/2$	$-\dfrac{y}{1-y}$	$\dfrac{y(1-y_{st})}{y_{st}(1-y)}$	$-\dfrac{(1-y)y_{st}^2}{(1-y_{st})^2 y}$	$\dfrac{(1-y)y_{st}}{(1-y_{st})y}$

Figure 6.4 displays calculated intervals of admissible values of $\alpha_j(y,\eta)$ for superstructures with y_{st} = 1/2, 2/3, 3/4, 5/6, and 7/8 (t = 1, 1.5, 2, 3, and 4).

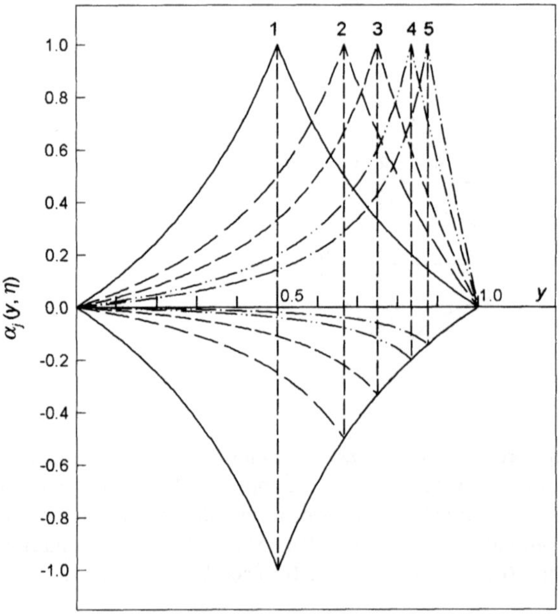

Fig. 6.4. Intervals of admissible $\alpha_j(y,\eta)$ values for different superstructures: 1 – AB or M_2X (y_{st} = 1/2); 2 – A_2B or M_3X_2 (y_{st} = 2/3); 3 – A_3B or M_4X_3 (y_{st} = 3/4); 4 – A_5B or M_6X_5 (y_{st} = 5/6); 5 – A_7B or M_8X_7 (y_{st} = 7/8)

Note that the intervals of $\alpha_j(y,\eta)$ in Fig. 6.4 cover the most extended regions, i.e. they include all the admissible values of $\alpha_j(y,\eta)$ for any basic structures.

Formulas for $\inf(\alpha)$ and $\sup(\alpha)$, which determine the maximum intervals of admissible values of superstructural short-range order parameters, are derived disregarding the specific basic structure that may impose more stringent limitations on $\alpha_j(y,\eta)$. In other words, intervals of $\alpha_j(y,\eta)$ may turn out to be narrower for some specific basic structures responsible for ordering.

As an example we consider short-range order in the first coordination sphere of bcc and fcc basic structures. It was shown in Sect. 6.1 that the interval of negative values of α for the first coordination sphere of a bcc structure coincides with $\inf(\alpha)$ (see Table 6.8). Thus, a bcc basic structure imposes no additional limitations on the interval of admissible values of the short-range order parameter. In contrast, the interval of admissible values for an fcc basic structure is narrower than the maximum interval. This is due to the fact that atoms on the surface of the first coordination sphere of an fcc lattice are spaced from one another by a distance which equals the radius of the sphere. Put another way, the first coordination spheres of different sites in an fcc structure overlap at lattice sites whereas the coordination spheres do not overlap at the lattice sites in a bcc structure.

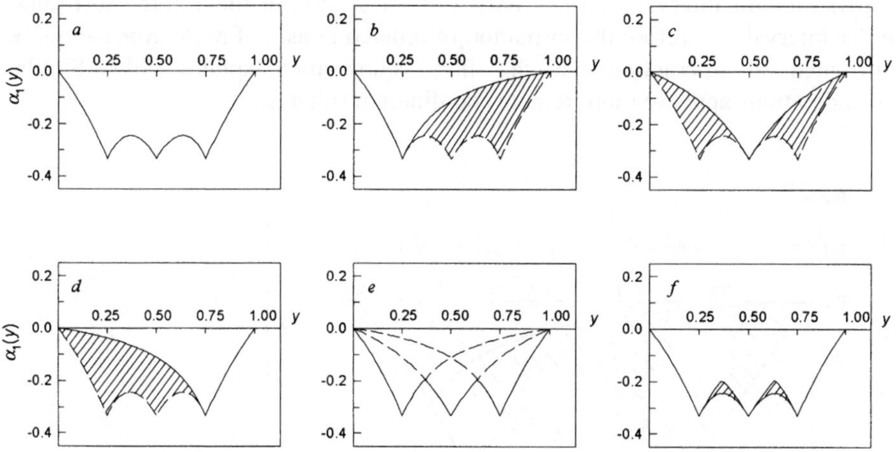

Fig. 6.5. Intervals of admissible values of superstructural and correlational short-range orders in the first coordination sphere of an fcc basic structure (additional intervals only due to short-range correlations are shaded): (*a*) common interval of α_1 for an fcc basic structure; (*b* to *d*) intervals of superstructural $\alpha_1(y,\eta_{max})$ and additional correlational short-range orders for superstructures AB_3 and M_4X, AB and M_2X (space groups *P*4/*mmm* and *I*4$_1$/*amd*), and A_3B (M_4X_3) respectively; (*e*) integral interval of $\alpha_1(y,\eta_{max})$ for superstructures AB_3 (M_4X), AB (M_2X) (space groups *P*4/*mmm* and *I*4$_1$/*amd*), and A_3B (M_4X_3); (*f*) integral interval of $\alpha_j(y,\eta_{max})$ for three superstructures and the interval of a correlational short-range order complementary to the integral interval

Figure 6.5*a* shows the interval of admissible values for the short-range order parameter α_1 for an fcc basic structure. This interval contains all possible α_1 values including those from short-range correlations. A minimum admissible value of $\alpha_1 = -1/3$ is achieved at $y = 1/4$, 1/2, and 3/4. Figures 6.5*b, c, d* illustrate the superposition of superstructural intervals of admissible $\alpha_1(y,\eta_{max})$ values calculated from (6.24) for ordered phases with $y_{st} = 1/4$, 1/2 and 3/4 on the whole interval of admissible α_1 values. The interval in Fig. 6.5*c* corresponds only to superstructures AB or M_2X with space groups $P4/mmm$ and $I4_1/amd$. For superstructures AB or M_2X with space groups $R\bar{3}m$ and $Fd\bar{3}m$, $\alpha_1(y, \eta_{max}) \equiv 0$. Shaded sections of the interval of admissible α_1 values cannot be due to long-range order. They only stem from short-range correlations. These sections represent regions of α_1 values, which complement the interval of a superstructural short-range order. Figure 6.5*e* shows an integral interval of admissible $\alpha_1(y,\eta_{max})$ values for three superstructures ($y_{st} = 1/4$, 1/2, and 3/4). Even this integral interval does not cover the entire interval of admissible α_1 values that can be realized in the presence of correlational short-range order.

Figure 6.6 shows the interval of admissible $\alpha_1(y,\eta_{max})$ values for a superstructure of type M_6X_5 and an additional interval of the parameter α_1. The additional interval for such a superstructure is due exclusively to short-range correlations rather than to the presence of long-range order. It is apparent from Fig. 6.6 that the superstructural interval overlaps only a small section of the whole short-range order interval. Therefore the formation of ordered phases of M_6X_5 type cannot be attributed to interactions in the first coordination sphere only. Similar considerations apply to more remote coordination spheres.

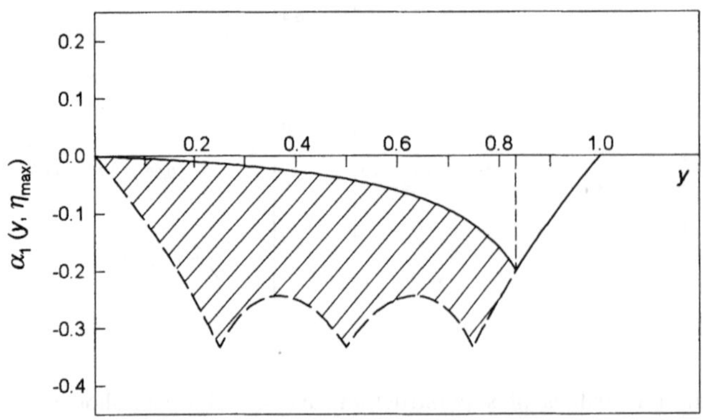

Fig. 6.6. Intervals of admissible values of superstructural $\alpha_1(y,\eta_{max})$ and additional correlational short-range orders in the first coordination sphere for a superstructure of type M_6X_5, which appears in a lattice with an fcc basic structure (the additional interval due to short-range correlations is shaded)

The above analysis suggests that intervals for correlational short-range order exist in addition to intervals for superstructural short-range order. Both intervals of the short-range order should be taken into account to ensure that thermodynamic calculations of ordering in solid solutions are complete.

A further advancement of the theory of atomic ordering seems difficult without considering the contribution of the configuration free energy for both short- and long-range orders. This point was addressed in ref. [1, 2] and Sect. 6.2 outlined a most general approach to that part of the short-range order which depends on the presence of static concentration waves in superstructures. The formulas derived for the calculation of superstructural short-range order can be used to find the probability of various two-particle figures, which can serve, in turn, to describe the configuration energy and entropy of a crystal. This formulation of the free energy is similar to the one used in the order parameter functional method [19, 24-27]. It accounts correctly for the symmetry of an ordered crystal and it describes approximately short-range interactions. As the next step, the theory should take into account an additional correlational short-range order, which is possible in an ordered phase. A combination of correlational and superstructural short-range orders will give the exact probability for the various clusters (atomic groups). It then becomes possible to account accurately for short-range interactions in addition to the proper treatment of the symmetry of an ordered phase. The parameters of correlational and superstructural short-range orders are independent. Their values are subject to considerable limitations, which depend on the magnitude of the superstructural short-range order parameters. This is due to the fact that the sum of those values must not exceed the bounds of the integral interval of the admissible values of the short-range order parameters.

References

1 A. A. Rempel, A. I. Gusev: Short-range order in ordered alloys and interstitial phases, Fiz. Tverd. Tela **32**, 16–24 (1990) (in Russian). (Engl. transl.: Sov. Physics – Solid State **32**, 8–13 (1990)

2 A. A. Rempel, A. I. Gusev: Short-range order in superstructures, Phys. Stat. Sol. (b) **160**, 389–402 (1990)

3 E. Ising: Beitrag zur Theorie des Ferromagnetismus, Z. Phys. **31**, 253–258 (1925)

4 H. A. Bethe: Statistical theory of superlattices, Proc. Roy. Soc. London A **150**, 552–575 (1935)

5 F. C. Nix, W. Shockley: Order–disorder transformations in alloys, Rev. Modern Phys. **10**, 1–71 (1938)

6 R. Kikuchi: Theory of cooperative phenomena, Phys. Rev. **81**, 988–1003 (1951)

7 M. Kurata, R. Kikuchi, T. Watari: Theory of cooperative phenomena. Detailed discussions of the cluster variation method, J. Chem. Phys. **21**, 434–448 (1953)

8 J. Hijmans, J. de Boer: Approximation method for order–disorder problem, Physica **21**, 471–516 (1955)

9 R. Kikuchi, S. G. Brush: Improvement of the cluster-variation method, J. Chem. Phys. **47**, 195–203 (1967)
10 R. Kikuchi: The cluster variation method, J. Physique Colloq. **38**, C7-307–C7-313 (1977)
11 E. A. Guggenheim: Statistical mechanics of regular solutions, Proc. Roy. Soc. London A **148**, 304–312 (1935)
12 R. Kikuchi, H. Sato: Characteristics of superlattice formation in alloys of face centred cubic structure, Acta Met. **22**, 1099–1112 (1974)
13 W. Gorsky: Röntgenographische Üntersuchung von Umwandlungen in der Legierung CuAu, Z. Phys. **50**, 64–81 (1928)
14 W. L. Bragg, E. J. Williams: The effect of thermal agitation on atomic arrangement in alloys, Proc. Roy. Soc. London A **145**, 699–730 (1934)
15 A. G. Khachaturian: *Theory of Structural Transformations in Solids* (John Wiley & Sons, New York 1983) 574 pp.
16 A. A. Rempel, A. I. Gusev: Relation between short-range and long-range order in solid solutions with b.c.c. and f.c.c. structures, Phys. Stat. Sol. (b) **130**, 413–420 (1985)
17 A. A. Rempel, A. I. Gusev: The relationship between short-range and long-range order in ordered alloys, Fiz. Metall. Metalloved. **60**, 847–854 (1985) (in Russian). (Engl. transl.: Phys. Met. Metallogr. **60**, 11–17 (1985))
18 A. A. Rempel, A. I. Gusev: Cluster-variation method of analysis of short-order and long-range order in ordered crystals, in *Solid State Chemistry* **9** P. V. Geld (Ed.) (Urals Polytechnical Institute, Sverdlovsk 1986) pp. 117–124 (in Russian)
19 A. I. Gusev, A. A. Rempel: *Structural Phase Transitions in Nonstoichiometric Compounds* (Nauka, Moscow 1988) 308 pp. (in Russian)
20 N. G. Parsonage, L. A. K. Staveley: *Disorder in Crystals* (Clarendon Press, Oxford 1978) 926 pp.
21 L. D. Landau, E. M. Lifschitz: *Statistical Physics. Course of Theoretical Physics* **5** (Pergamon, Oxford 1980) 544 pp.
22 J. M. Cowley: An approximation theory of order in alloys, Phys. Rev. **77**, 669–675 (1950)
23 D. Ya. Khvatinskaya, I. Karimov, V. S. Presman: Short-range order of carbon atom positions in niobium carbide, Dokl. Akad. Nauk Uzb. SSR No 1, 23–25 (1988) (in Russian)
24 A. I. Gusev, A. A. Rempel: Order parameter functional method in the theory of atomic ordering, Phys. Stat. Sol. (b) **131**, 43–51 (1985)
25 A. I. Gusev, A. A. Rempel: Calculating the energy parameters for CV and OPF methods, Phys. Stat. Sol. (b) **140**, 335–346 (1987)
26 A. I. Gusev: Atomic ordering and the order parameter functional method, Philosoph. Mag. B **60**, 307–324 (1989)
27 A. I. Gusev: *Physical Chemistry of Nonstoichiometric Refractory Compounds* (Nauka, Moscow 1991) 286 pp. (in Russian)

7. Nonstoichiometric Nitrides at High Temperature

Thermodynamic properties of strongly nonstoichiometric compounds MX_y vary considerably with the content of nonmetal interstitial atoms, because homogeneity intervals determine the concentration dependence of integral and partial thermodynamic functions. However, at high temperatures (above 1500 K) the composition of nonstoichiometric compounds, especially nitrides MN_y, may change as a result of incongruent evaporation. Generally, the nitrogen concentration of nitrides decreases due to evaporation, because the equilibrium partial pressure of nitrogen is much higher than that of a metal at a similar temperature. The nitrogen concentration changes, i.e. MN_y evaporates incongruently until the nitride MN_{y_c}, which evaporates congruently (the composition being unchanged) at a given temperature T, is formed. The composition of a congruently evaporating nitride depends both on temperature and nitrogen pressure. Under equilibrium conditions, a complicated functional relationship exists between the nitride composition, nitrogen partial pressure and temperature. A number of experimental investigations were carried out to study this relationship and determine the nitrogen partial pressure as a function of temperature and the composition of nitrides and carbonitrides. Theoretical methods for estimating the nitrogen partial pressure over nonstoichiometric nitrides appeared only in the last decade of the twentieth century.

7.1 Evaporation of Nonstoichiometric Nitrides

Evaporation of such nonstoichiometric compounds as transition metal carbides and nitrides was considered in numerous original studies, whose results are summarized in monographs and reference books [1–6].

Of particular intricacy is the evaporation of compounds with wide homogeneity intervals. Evaporation is pronounced at temperatures above 1200 K, when the disordered state is in thermodynamic equilibrium. Therefore the further discussion refers to disordered nonstoichiometric nitrides.

In the general case evaporation of a nonstoichiometric nitride MN_y (or $MN_y\square_{1-y}$, where $\square$ is a structural vacancy) at a temperature T can be given by the following reactions:

$$MN_y\ (solid) = MN_{y_c}\ (solid) + [(y - y_c)/2]N_2\ (gas),\quad (\text{if}\ \ y > y_c) \tag{7.1}$$

$$MN_{y_c}\,(solid) = M\,(gas) + (y_c/2)N_2(gas)\,, \qquad (7.2a)$$

or

$$MN_{y_c}\,(solid) = M\,(solid) + (y_c/2)N_2(gas)\,, \qquad (7.2b)$$

where y_c is the relative concentration of nitrogen in the nitride evaporating congruently at a temperature *T*.

The equilibrium of a solid–gas heterogeneous system is described by the Gibbs phase rule

$$f = k + 2 - p^*, \qquad (7.3)$$

where f is the number of degrees of freedom (independent external thermodynamic parameters); k is the number of components of the system (the notion "component" is taken here in terms of the thermodynamic theory of phase equilibrium); and p^* is the number of phases in the system.

Reaction (7.1) describes incongruent evaporation resulting in the formation of the nitride MN_{y_c}, whose composition does not change upon subsequent evaporation. Indeed, in accordance with the Gibbs phase rule, the system described by reaction (7.1) has two degrees of freedom. Therefore, in the process of evaporation at a temperature *T*, the composition of the solid phase and the nitrogen pressure p_{N_2} change simultaneously, i.e. evaporation by reaction (7.1) is incongruent. The system described by reaction (7.2*a*) contains one MN_{y_c} component and two (solid and gaseous) phases, i.e. $k = 1$ and $p^* = 2$. Therefore, in accordance with (7.3), reaction (7.2*a*) is monovariant and describes a congruent evaporation during which each temperature in the equilibrium conditions corresponds to a certain vapor pressure, the composition of the solid phase being preserved. The fact that the composition of the solid phase does not change means that the evaporation rate of the nitride components is proportional to their concentration in the nitride and the relative concentrations of metal atoms n_M and nitrogen n_N in the solid and gaseous phases are similar, i.e. $(n_M/n_N)_{gas} = (n_M/n_N)_{solid} = 1/y_c$. Thus, reaction (7.2*a*) meets the kinetic conditions of congruent evaporation. Reaction (7.2*b*) is monovariant too and describes evaporation during which the composition of the initial phase does not change.

It was shown earlier [7–9] that at a given temperature *T* the nonstoichiometric compound MX_y evaporates congruently if its composition corresponds to the minimum of the concentration dependence of the free energy $F(y)$ at this temperature. For condensed systems, if the pressure p is constant (it is usually assumed that $p = 0$), the crystal enthalpy $H = E + pV$ coincides with the internal lattice energy E, and the isobaric potential (Gibbs free energy) $G = E - TS + pV$ coincides with the free energy $F = E - TS$. However, to describe equilibrium in a system, which includes the gaseous phase, one should account for the pressure of this phase using the Gibbs free energy. Hence, from [7–9] it follows that at a temperature *T* the nitride MN_y evaporates congruently when its composition corresponds to the minimum of the concentration dependence of the Gibbs free energy $G(y)$ at this temperature. If the MN_y composition does not correspond to the minimum of the concentration versus $G(y)$ dependence, the composition will change by evaporation until the Gibbs free energy $G(y)$ attains a maximum value. In other words, during evaporation the nonstoichiometric compound MN_y changes their composition up to the MN_{y_c}. After reaching this composition the compound evaporates further congruently. Therefore, equilibrium conditions in (7.1) are established at a certain $y_c(T)$ value, which depends on the nature of the compound and temperature.

If the activities of solid components in (7.1) are assumed to be unity, the equilibrium constant of this reaction takes the form

$$K = (p_{N_2})^{(y-y_c)/2} . \tag{7.4}$$

The equilibrium constant of (7.1) is connected to the variation of the Gibbs free energy $\Delta G(T)$ in this reaction by the known relationship

$$RT \ln K = -\Delta G(T) = \Delta G_f^{\text{init}}(n,T) - \Delta G_f^{\text{fin}}(n,T) , \tag{7.5}$$

where $\Delta G_f^{\text{init}}(n,T)$ and $\Delta G_f^{\text{fin}}(n,T)$ are the free energies of formation of the initial compound evaporating incongruently and the compound evaporating congruently at a given temperature T; $n = (1 - y)$ is the relative concentration of structural vacancies in the nonmetal sublattice of MX_y. As was shown in [9–12], the free energy of formation, $\Delta G_f(n,T)$, of the nonstoichiometric compound $MX_y\square_{1-y}$ or nitride $MN_y\square_{1-y}$ containing structural vacancies □ can be expressed through the free energy of formation, $\Delta G_f(0,T)$, of a defect free compound and the energy of formation, $\Delta G_\square(0,T)$, of a structural vacancy as

$$\Delta G_f(n,T) = \Delta G_f(0,T) + \int_0^n \Delta G\ (n,T)dn - T\Delta S(n) \equiv \Delta G_f^*(n,T) - T\Delta S(n) , \tag{7.6}$$

where $\Delta S(n) = -R[n\ln n + (1-n)\ln(1-n)] \equiv -R[y\ln y + (1-y)\ln(1-y)]$ is the configurational entropy of disordered nitride MN_y. In accordance with [10–13], the formation energy of structural vacancies in carbides and nitrides MX_y is a linear function of the vacancy concentration n, i.e. $\Delta G_\square(n,T) = a(T) + nb(T)$. Therefore, substituting (7.4) and (7.6) into (7.5) and making some manipulations, we obtain a formula for the nitrogen pressure p_{N_2} over the nonstoichiometric nitride MN_y (at $y > y_c$), which evaporates by reaction (7.1):

$$p_{N_2}(y,T) = \frac{y_c^2}{(1-y_c)^2}\left[\frac{(1-y)}{(1-y_c)}\right]^{2(1-y)/(y-y_c)}\left(\frac{y}{y_c}\right)^{2y/(y-y_c)}$$

$$\times \exp\left[-\frac{2\Delta G_\square(n_c = 1-y_c, T)}{RT}\right]\exp\left[\frac{(y-y_c)b(T)}{RT}\right] . \tag{7.7}$$

In the limit $y \to y_c$, formula (7.7) yields the $p_{N_2}(y_c,T)$ value as an arbitrary measurement unit of the nitrogen partial pressure. Indeed, from (7.7) it follows that if $y = y_c$,

$$p_{N_2}(y_c,T)=\frac{y_c^2}{(1-y_c)^2}\exp\left[-\frac{2\Delta G_\square(n_c=1-y_c,T)}{RT}\right]. \tag{7.8}$$

Allowing for (7.8), formula (7.7) can be written differently:

$$p_{N_2}(y,T)=p_{N_2}(y_c,T)\left[\frac{(1-y)}{(1-y_c)}\right]^{2(1-y)/(y-y_c)}\left(\frac{y}{y_c}\right)^{2y/(y-y_c)}$$

$$\times\exp\left[\frac{(y-y_c)b(T)}{RT}\right], \tag{7.9}$$

where $y_c < y \neq 1$.

The nitrogen pressure over the stoichiometric nitride $MN_{1.00}$ ($y = 1$) is described by the formula:

$$p_{N_2}(1,T)=p_{N_2}(y_c,T)y_c^{-2/(1-y_c)}\exp\left[\frac{(1-y_c)b(T)}{RT}\right],\quad \text{when } y=1. \tag{7.10}$$

To determine the nitrogen pressure over the nitride MN_y ($y \le y_c$), whose nitrogen concentration is less than or equal to its counterpart in the congruently evaporating nitride MN_{y_c}, the following reaction should be considered

$$MN_y\,(solid) = MN_{y-x}\,(solid) + (x/2)N_2\,(gas), \tag{7.11}$$

where $x \to 0$. It is easy to see that in this case

$$p_{N_2}(y,T)=\frac{y^2}{(1-y)^2}\exp\left[-\frac{2G_\square(n=1-y,T)}{RT}\right]\quad \text{when } y \le y_c. \tag{7.12}$$

Divide (7.12) by (7.8). Simple rearrangements yield the formula

$$p_{N_2}(y,T)=p_{N_2}(y_c,T)\left[\frac{y(1-y_c)}{y_c(1-y)}\right]^2\exp\left[\frac{(y_c-y)b(T)}{RT}\right]\quad \text{when } y \le y_c. \tag{7.13}$$

Formulas (7.9), (7.10), and (7.13) permit calculating the nitrogen pressure $p_{N_2}(y,T)$ over the nitride MN_y if the arbitrary measurement unit of the equilibrium nitrogen pressure $p_{N_2}(y_c,T)$ is known, i.e. the nitrogen pressure over the nitride MN_{y_c}, which evaporates congruently (the composition being unchanged) at a temperature T.

To determine the nitrogen partial pressure $p_{N_2}(y_c,T)$ of congruently evaporating nitride, let us consider reaction (7.2*a*). If a compound evaporates congruently by reaction (7.2*a*), the solid and gaseous phases share equally metal and nitrogen atoms, i.e. $(n_M/n_N)_{gas} = (n_M/n_N)_{solid} = 1/y_c$. But the gaseous phase contains nitrogen in the molecular state and $n_{N_2} = 0.5n_N$. Therefore $(n_M/n_{N_2})_{gas} = 2(n_M/n_N)_{gas} = 2/y_c$. The partial pressure of atoms of a given species is proportional to their concentration in the gaseous phase, i.e.

$p_M / p_{N_2} = (n_M / n_{N_2})_{gas} = 2 / y_c$. From this it follows that in the case of congruent evaporation by reaction (7.2*a*)

$$p_M = (2/y_c) p_{N_2} . \tag{7.14}$$

The equilibrium constant of reaction (7.2*a*) is

$$K = p_M (p_{N_2})^{y_c/2} \tag{7.15}$$

or, considering (7.14),

$$K = (2/y_c)(p_{N_2})^{(2+y_c)/2} . \tag{7.16}$$

The equilibrium constant of reaction (7.2*b*) takes the form

$$K = a_M (p_{N_2})^{y_c/2} , \tag{7.17}$$

where $a_M = p_M / p^{\circ}_M$ is the activity; p_M is the metal vapor pressure during evaporation of nitride by reaction (7.2*b*); and p°_M is the equilibrium metal vapor pressure over metal M.

The equilibrium constant of reaction (7.2*a*) can be expressed through the Gibbs free energy $\Delta G(T)$ of this reaction as

$$RT \ln K = -\Delta G^{\circ}(T) \equiv \Delta G_f(y_c, T) - \Delta G_f(n_c = 1 - y_c, T) , \tag{7.18}$$

where $\Delta G_f(n_c = 1 - y_c, T)$ is determined by (7.6). Substituting (7.16) into (7.18) and performing some rearrangements, we obtain the formula for the nitrogen pressure over congruently evaporating nitride

$$p_{N_2}(y_c, T) = \frac{y_c^2}{(1-y_c)^2} \left[\frac{(1-y_c)^3}{2y_c} \right]^{2/(2+y_c)} \exp\left[\frac{2\Delta G_f^*(y_c, T)}{(2+y_c)RT} \right] \text{ (atm)}, \tag{7.19}$$

where

$$\Delta G_f^*(y_c, T) \equiv \Delta G_f^*(n_c = 1 - y_c, T) = \Delta G_f(0, T) + \int_0^{n_c} \Delta G_{\square}(n, T) dn$$

is the free energy of formation (disregarding the contribution of configurational entropy). If MN_{y_c} evaporates by reaction (7.2*b*), the nitrogen partial pressure $p_{N_2}(y_c, T)$ over this nitride is determined from the formula

$$p_{N_2}(y_c, T) = \frac{y_c^2}{(1-y_c)^2} \left[\frac{p^{\circ}_M (1-y_c)}{p_M} \right]^{2/y_c} \exp\left[\frac{2\Delta G_f^*(y_c, T)}{y_c RT} \right] \text{ (atm)}. \tag{7.20}$$

When the vacancy formation energy $\Delta G_{\square}(n,T)$ is known, y_c in formulas (7.9), (7.10), (7.13), (7.19), and (7.20) can be found from a self-consistent solution of the equation

$$\ln[y_c/(1 - y_c)] = \Delta G_{\square}(n_c = 1 - y_c, T)/RT. \tag{7.21}$$

The nitrogen pressure $p_{N_2}(y,T)$ over MN_y can be found from (7.9), (7.13), and (7.19), if the nitride evaporates by reactions (7.1) and (7.2*a*). If the nitride evaporates by reactions (7.1) and (7.2*b*), the nitrogen pressure is calculated from (7.9), (7.13), and (7.20).

The disordered carbonitride MC_xN_y ($x \le 1, y \le 1, x+y \le 1$) is a solid solution of carbide and nitride of one and the same metal. As regards the mutual solubility of carbide and nitride, cubic carbonitrides of titanium, zirconium, hafnium, vanadium, and niobium are close to ideal solid solutions with the statistical distribution of carbon and nitrogen atoms in the nonmetal sublattice. The share of nonmetal sublattice sites in MC_xN_y occupied by carbon and nitrogen atoms equals $x + y$. If we consider occupied sites only, the relative share of sites filled with carbon atoms is $x/(x + y)$, whereas the share of sites occupied by nitrogen atoms is $y/(x + y)$. Thus, the relative shares of sites occupied by carbon and nitrogen atoms coincide with the relative concentrations of carbide and nitride in carbonitride. Therefore the composition of carbonitride MC_xN_y can be written as $(MC_{x+y})_{x/(x+y)}(MN_{x+y})_{y/(x+y)}$. The nitrogen partial pressure $p_{N_2}(x,y,T)$ over the carbonitride solid solution approximating the ideal solid solution is proportional to the share of nitride MN_{x+y} in the solid solution and the nitrogen pressure $p_{N_2}(y',T)$ over the nitride and equals

$$p_{N_2}(x,y,T) = \frac{y}{(x+y)} p_{N_2}(y',T)\ , \tag{7.22}$$

where $y' = x + y$ and the pressure $p_{N_2}(y',T)$ is determined from (7.9) and (7.13). Therefore, the nitrogen partial pressure $p_{N_2}(x,y,T)$ over the carbonitride MC_xN_y is

$$p_{N_2}(x,y,T) = p_{N_2}(y_c,T) \frac{y}{(x+y)} \left[\frac{(1-x-y)}{(1-y_c)} \right]^{2(1-x-y)/(x+y-y_c)}$$

$$\times \left(\frac{x+y}{y_c} \right)^{2(x+y)/(x+y-y_c)} \exp\left[\frac{(x+y-y_c)b(T)}{RT} \right] \quad \text{if } x+y \ge y_c \tag{7.23}$$

or

$$p_{N_2}(x,y,T) = p_{N_2}(y_c,T) \frac{y(x+y)(1-y_c)^2}{(1-x-y)^2 y_c^2}$$

$$\times \exp\left[-\frac{2(y_c-x-y)b(T)}{RT} \right] \quad \text{if } x+y \le y_c. \tag{7.24}$$

In the particular case when $x + y = 1$, the equilibrium nitrogen pressure of MC_xN_y is

$$p_{N_2}(x+y,T) = y p_{N_2}(y_c,T) y_c^{-2/(1-y_c)} \exp\left[\frac{(1-y_c)b(T)}{RT} \right] \quad \text{at } x+y = 1. \tag{7.25}$$

Thus, if the nitrogen pressure over the nitride, which evaporates congruently at a given temperature T, is known, we can easily find the nitrogen partial pressure over the carbonitride MC_xN_y or nitride MN_y of any composition.

7.2 Titanium Nitrides and Carbonitrides

Cubic titanium nitride TiN_y and titanium carbonitride TiC_xN_y are extensively used as protective and wear-resistant coatings. The coating composition can be controlled by the varying temperature and nitrogen pressure of the gaseous phase. Although titanium nitride and carbonitride have great practical significance, experimental data on nitrogen partial pressures over these compounds are very scarce.

Evaporation of cubic titanium nitride $TiN_{0.794}$ in a vacuum at temperatures from 2000 to 2250 K was studied by Hoch et al. [14]. In accordance with [14], titanium nitride dissociates during evaporation to titanium and nitrogen, which transform into the gaseous phase. This suggests that TiN_y evaporates by reaction (7.2*a*). The composition of titanium nitride upon evaporation was not determined.

In the homogeneity interval of cubic titanium nitride TiN_y, the nitrogen partial pressure and the coefficients of nitrogen activity at 1273 and 1673 K as a function of the nitride composition were determined by Grieveson [15]. The nitrogen partial pressure was found to be many times higher than the p_{N_2} values determined in [14]. For example, it was found [14] that nitride the $TiN_{0.794}$ had the partial pressure $p_{N_2} = 6.5 \times 10^{-7}$ atm at a temperature of 1987 K, whereas in accordance with [15], the partial pressure of the similar nitride $TiN_{0.792}$ was as high as $p_{N_2} = 8.2 \times 10^{-6}$ atm already at 1273 K, i.e. 13 times larger.

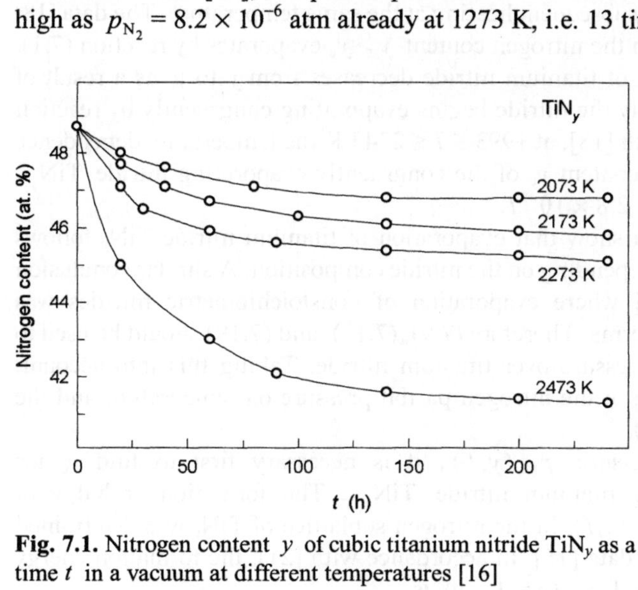

Fig. 7.1. Nitrogen content y of cubic titanium nitride TiN_y as a function of the evaporation time t in a vacuum at different temperatures [16]

Kislyi and Kuzenkova [16] studied the evaporation of titanium nitride at temperatures from 2000 to 2500 K in a vacuum of 0.01 Pa. Initial samples of titanium nitride had the composition $TiN_{0.95}$, while the heating time was 4 hours. The nitrogen concentration decreased rapidly in 1.5–2 hours to the value corresponding to congruently evaporating nitride. During subsequent heating in a vacuum at the same temperature the nitrogen concentration of the nitride remained constant within measurement error (Fig. 7.1).

It was found [16] that under equilibrium conditions each temperature corresponded to definite concentrations of nitrogen and structural vacancies in congruently evaporating nitride. In other words, the y_c value and the temperature T are related by the functional dependence:

T (K)	2023	2073	2123	2173	2223	2273	2473
y_c	0.902	0.884	0.866	0.846	0.825	0.803	0.710

The data [16] suggest that titanium nitride TiN_y evaporates by reaction (7.1) when $y > y_c$ and later, when $y = y_c$, by reaction (7.2*a*). Using the method proposed in [16], researchers [17] examined the evaporation of cubic zirconium nitride ZrN_y in a vacuum and arrived at analogous conclusions about the temperature dependence of the composition of congruently evaporating nitride ZrN_y.

Evaporation of titanium nitride TiN_y in a vacuum in the temperature interval from 1993 to 2243 K was studied in [18]. It was found that the constant composition of titanium nitride samples with any initial nitrogen content is established after heating at a given temperature T in vacuum for 1.0–1.5 h. The chemical composition attained depends on the evaporation temperature only and remains constant during subsequent heating at the same temperature. The data [18] confirmed that TiN_y with the nitrogen content $y > y_c$ evaporates by reaction (7.1), i.e. the nitrogen content of titanium nitride decreases from y to y_c as a result of evaporation. Subsequently the nitride begins evaporating congruently by reaction (7.2*a*). In accordance with [18], at $1993 \leq T \leq 2243$ K the temperature dependence of the relative nitrogen content y_c of the congruently evaporating nitride TiN_{y_c} has the form $y_c = 1.46 - 2.8 \times 10^{-4} T$.

So, experimental data show that evaporation of titanium nitride TiN_y follows reaction (7.1) or (7.2*a*) depending on the nitride composition. A similar conclusion was drawn in [19, 20] where evaporation of nonstoichiometric nitrides was analyzed in theoretical terms. Therefore (7.9), (7.13), and (7.19) should be used to calculate the nitrogen pressure over titanium nitride. Taking this into account, consider the dependence of the nitrogen partial pressure on temperature and the TiN_y composition ($y \leq 1$).

To calculate the pressure $p_{N_2}(y,T)$, it is necessary first to find y_c for congruently evaporating titanium nitride TiN_{y_c}. The formation enthalpy of structural vacancies, $\Delta G_{\square}(n,T)$, in the nitrogen sublattice of TiN_y was determined [21] using experimental data [16]. In accordance with [21], the formation energy of structural vacancies in TiN_y $(n = 1 - y)$ is

$$\Delta G_{\square}(n,T) = 298000 - 136.6T + (33800 + 69.8T)n \quad (\mathrm{J\ mol^{-1}}). \tag{7.26}$$

Values of $y_c(T)$ calculated for TiN_y from (7.20) using the energy $\Delta G_{\square}(n,T)$ (7.26) are given in Table 7.1.

The temperature dependence of the free energy of formation $\Delta G_f(y = 1, T) \equiv \Delta G_f(n = 0, T)$ of the stoichiometric nitride $TiN_{1.0}$ can be found in reference books. However, analogous dependences $\Delta G_f(y < 1, T)$ for nonstoichiometric titanium nitrides TiN_y ($y < 1.0$) are lacking. Let us consider the free energy of nitride formation, $\Delta G_f^*(y,T)$, disregarding the contribution from configurational entropy. Assume as a first approximation that, in the homogeneity interval of titanium nitride, $\Delta G_f^*(y,T)$ changes as the formation enthalpy which is equal to

$$H_{f298}(y) = 3750 - 417900y + 92910y^2 \quad (\mathrm{J\ mol^{-1}}). \tag{7.27}$$

This dependence was derived earlier [22] from experimental data [23]. Since the formation enthalpy changes little with temperature (by less than 1% when the temperature increases from 300 to 1900 K) and the variation of the formation enthalpy can be neglected, then

$$\Delta G_f^*(y, T) = \Delta G_f(y = 1, T) + [\Delta H_{f298}(y) - \Delta H_{f298}(y = 1)] + \Delta\,, \tag{7.28}$$

where $\Delta = \Delta H_{f298}(y=1) - \Delta H^*_{f298}(y=1)$ is a correction which equals the difference between the formation enthalpy of stoichiometric nitride $TiN_{1.0}$ determined from (7.27) and the reference formation enthalpy $\Delta H^*_{f298}(y=1)$ for $TiN_{1.0}$. Substituting Δ into (7.28) in the explicit form, the quantity $\Delta G_f^*(y,T)$ can be written as

$$\Delta G_f^*(y,T) = \Delta G_f(y=1,T) + [\Delta H_{f298}(y) - \Delta H^*_{f298}(y=1)]\,. \tag{7.29}$$

In accordance with [24], $\Delta H^*_{f298}(y=1) = -337860\ \mathrm{J\ mol^{-1}}$. Taking into account this value and the dependence (7.27), we obtain the free energy of formation $\Delta G_f^*(y,T)$ for TiN_y as a function of y and T:

$$\Delta G_f^*(y,T) = \Delta G_f(y = 1,T) + 341610 - 417900y + 92910y^2 \quad (\mathrm{J\ mol^{-1}}). \tag{7.30}$$

Table 7.1 presents the free energies of formation $\Delta G_f^*(y,T)$ for titanium nitride TiN_{y_c} at y_c values calculated from (7.26) and corresponding to different temperatures in the interval from 1400 to 2400 K. Using these values of y_c and $\Delta G_f^*(y_c,T)$, the nitrogen pressure $p_{N_2}(y_c,T)$ over congruently evaporating titanium nitride TiN_{y_c} (Table 7.1) was calculated from (7.19).

The nitrogen equilibrium partial pressure over TiN_y as a function of the nitrogen content of nitride and temperature was calculated from (7.9), (7.10), and (7.13) for the compositions from $TiN_{0.45}$ to $TiN_{1.00}$. The values of y_c and $p_{N_2}(y_c,T)$ were used for computations. Calculated isotherms $p_{N_2} - y$ [21] of cubic titanium nitride at temperatures from 1273 to 2400 K are shown in Fig. 7.2. The lower boundary of the homogeneity interval for cubic titanium nitride TiN_y is given in accordance with [22, 25] and [26].

Table 7.1. Relative nitrogen content y_c, the formation Gibbs free energy $\Delta G_f^*(y_c,T)$ and nitrogen partial pressure $p_{N_2}(y_c,T)$ for cubic titanium nitrides TiN_{y_c} which evaporate congruently at a given temperature T

T (K)	TiN_y		
	y_c	$\Delta G_f^*(y_c,T)$ (kJ mol^{-1})	$p_{N_2}(y_c,T)$ (atm)
1400	0.9999	−189.14	0.000012
1500	0.9994	−179.67	0.000042
1600	0.9975	−169.92	0.000124
1700	0.9914	−159.23	0.000324
1800	0.9766	−146.51	0.000797
1900	0.9505	−131.05	0.001009
2000	0.9151	−112.44	0.004570
2100	0.8742	−92.55	0.010805
2200	0.8310	−71.45	0.024696
2300	0.7877	−50.01	0.054166
2400	0.7453	−28.48	0.113910

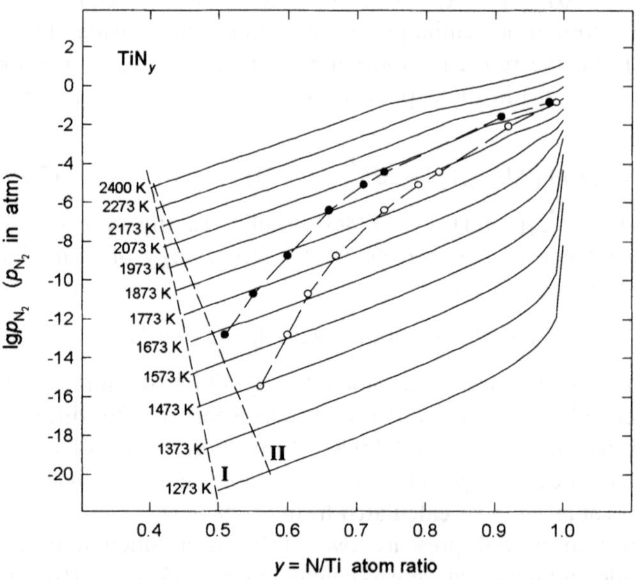

Fig. 7.2. Nitrogen partial pressure p_{N_2} in equilibrium with titanium nitride TiN_y at temperatures from 1273 to 2400 K [21]: solid lines denote calculated isotherms $p_{N_2} - y$; (○) and (●) stand for experimental data [15] measured at temperatures 1273 and 1673 K, respectively; I and II are the positions of the lower boundary of the homogeneity interval of cubic titanium nitride TiN_y as determined in [22, 25] and [26] respectively

Table 7.2. Nitrogen partial pressure $p_{N_2}(y_c, T)$ (atm) over cubic nitride TiN_y at a temperature of 1600, 1800 and 2000 K

y	TiN_y		
	1600 K	1800 K	2000 K
0.43	–	–	9.04×10^{-10}
0.45	–	4.10×10^{-12}	1.61×10^{-9}
0.46	4.42×10^{-15}	5.51×10^{-12}	2.15×10^{-9}
0.50	1.46×10^{-14}	1.78×10^{-11}	6.84×10^{-9}
0.53	3.58×10^{-14}	4.29×10^{-11}	1.63×10^{-8}
0.55	6.51×10^{-14}	7.71×10^{-11}	2.90×10^{-8}
0.58	1.60×10^{-13}	1.87×10^{-10}	6.92×10^{-8}
0.60	2.93×10^{-13}	3.37×10^{-10}	1.24×10^{-7}
0.65	1.34×10^{-12}	1.50×10^{-9}	0.000001
0.69	4.61×10^{-12}	5.05×10^{-9}	0.000002
0.70	6.31×10^{-12}	6.87×10^{-9}	0.000003
0.72	1.19×10^{-11}	1.28×10^{-8}	0.000004
0.75	3.11×10^{-11}	3.30×10^{-8}	0.000011
0.80	1.65×10^{-10}	1.70×10^{-7}	0.000057
0.85	9.90×10^{-10}	0.000001	0.000325
0.90	7.45×10^{-9}	0.000007	0.00233
0.95	9.92×10^{-8}	0.000094	0.0111
0.96	1.97×10^{-7}	0.000185	0.0148
0.97	4.45×10^{-7}	0.000416	0.0204
0.98	0.000001	0.000967	0.0292
0.99	0.000006	0.00194	0.0451
1.00	0.000941	0.00774	0.0896

Calculations showed that the nitrogen pressure over TiN_y dropped quickly with increasing concentration of vacancies. For example, in the interval $0.95 \leq y \leq 0.98$ the pressure p_{N_2} reduced by a factor of 1.5 to 2 each time y decreased by 0.01. The pressure p_{N_2} over the stoichiometric nitride $TiN_{1.0}$ is 8 and 80 times higher than over $TiN_{0.95}$ at a temperature of 2000 K and 1800 K respectively (Table 7.2).

Since systematic experimental studies of the nitrogen pressure over titanium nitride as a function of the composition and temperature are lacking, computational results can be compared with literature data on a rather limited scale. In accordance with [14], the nitrogen partial pressure over $TiN_{0.794}$ at 1987, 2017, and 2058 K is 6.5×10^{-7}, 8.5×10^{-7}, and 14.7×10^{-7} atm respectively. These values are nearly 30–120 times smaller than the computational results. Understated (as compared to the calculated data) values of the equilibrium pressure p_{N_2} can be explained probably by the fact that Hoch et al. used effusion cells with large flow orifices [14]. On the other hand, the data [15] concerning the equilibrium nitrogen pressure for TiN_y at 1273 and 1673 K are obviously overstated (see Fig. 7.2). Moreover, an unusually rapid growth of the pressure p_{N_2} with the nitrogen concentration of TiN_y is doubtful. For example, in accordance with [15], at

1273 K the nitrogen pressure p_{N_2} increases by a factor of 4×10^{14} when the TiN_y composition changes form $TiN_{0.56}$ to $TiN_{0.99}$. As is seen, literature data on the nitrogen pressure over titanium nitride are ambiguous. Our value of the pressure $p_{N_2}(y,T)$ over TiN_y is intermediate between values obtained in [14] and [15].

The calculated values of equilibrium nitrogen pressure over titanium carbonitrides $TiC_{0.25}N_y$ $(0.21 \le y \le 0.75)$, $TiC_{0.50}N_y$ $(0 < y \le 0.50)$, and $TiC_{0.75}N_y$ $(0 < y \le 0.25)$ are given in Table 7.3.

Table 7.3. Nitrogen partial pressure p_{N_2} (atm) over cubic titanium carbonitrides TiC_xN_y ($x = 0.50$ and 0.75) at a temperature of 1600, 1800 and 2000 K

y	$TiC_{0.50}N_y$			$TiC_{0.75}N_y$		
	1600 K	1800 K	2000 K	1600 K	1800 K	2000 K
0	0	0	0	0	0	0
0.05	5.92×10^{-15}	7.01×10^{-12}	2.63×10^{-9}	1.03×10^{-11}	1.06×10^{-8}	0.000004
0.10	4.88×10^{-14}	5.62×10^{-11}	2.06×10^{-8}	1.16×10^{-10}	1.17×10^{-7}	0.000038
0.15	3.09×10^{-13}	3.46×10^{-10}	1.24×10^{-7}	1.24×10^{-9}	0.000001	0.000388
0.20	1.80×10^{-12}	1.96×10^{-9}	6.89×10^{-7}	2.09×10^{-8}	0.000020	0.00233
0.22	3.63×10^{-12}	3.90×10^{-9}	0.000001	1.01×10^{-7}	0.000094	0.00462
0.24	7.30×10^{-12}	7.77×10^{-9}	0.000003	0.000002	0.000470	0.0109
0.25	1.04×10^{-11}	1.10×10^{-8}	0.000004	0.000235	0.00194	0.0224
0.26	1.48×10^{-11}	1.55×10^{-8}	0.000005	–	–	–
0.28	3.01×10^{-11}	3.13×10^{-8}	0.000011	–	–	–
0.30	6.19×10^{-11}	6.38×10^{-8}	0.000021	–	–	–
0.32	1.30×10^{-10}	1.32×10^{-7}	0.000044	–	–	–
0.34	2.76×10^{-10}	2.78×10^{-7}	0.000092	–	–	–
0.35	4.08×10^{-10}	4.08×10^{-7}	0.000134	–	–	–
0.36	6.06×10^{-10}	6.03×10^{-7}	0.000197	–	–	–
0.38	1.38×10^{-9}	0.000001	0.000439	–	–	–
0.40	3.31×10^{-9}	0.000003	0.00103	–	–	–
0.42	8.60×10^{-9}	0.000008	0.00234	–	–	–
0.44	2.54×10^{-8}	0.000024	0.00395	–	–	–
0.45	4.70×10^{-8}	0.000044	0.00524	–	–	–
0.46	9.43×10^{-8}	0.000089	0.00709	–	–	–
0.48	6.23×10^{-7}	0.000473	0.0143	–	–	–
0.50	0.000470	0.00387	0.0448	–	–	–

In accordance with computations (see Table 7.3), the nitrogen pressure for the nonstoichiometric carbonitrides $TiC_{0.50}N_y$ $(y < 0.50)$ and $TiC_{0.75}N_y$ $(y < 0.25)$ rises with temperature more rapidly than for the stoichiometric carbonitrides $TiC_{0.50}N_{0.50}$ and $TiC_{0.75}N_{0.25}$. For example, when the temperature is elevated by 200 K (in the interval from 1600 to 2000 K), the nitrogen partial pressure over stoichiometric carbonitrides increases eight to twelve times, whereas the nitrogen pressure over nonstoichiometric carbonitrides TiC_xN_y (at $x + y = 0.90$) increases by a factor of 350–1000.

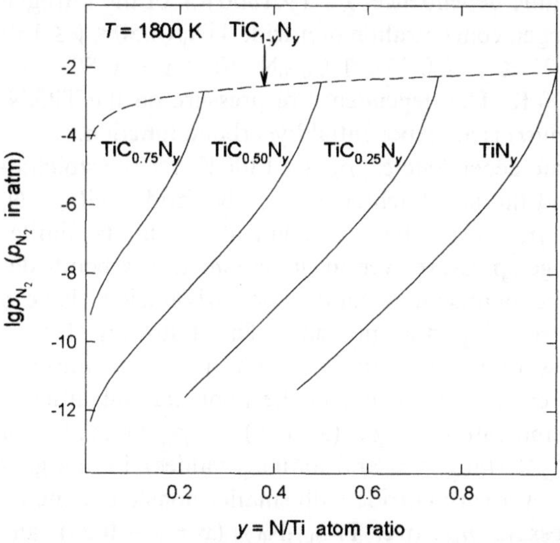

Fig. 7.3. Nitrogen equilibrium pressure p_{N_2} for cubic titanium nitride TiN_y and titanium carbonitrides $TiC_{0.25}N_y$, $TiC_{0.50}N_y$, and $TiC_{0.75}N_y$ at a temperature of 1800 K [19–21]. The dashed line shows the limiting nitrogen partial pressure of stoichiometric carbonitride $TiC_{1-y}N_y$ at 1800 K

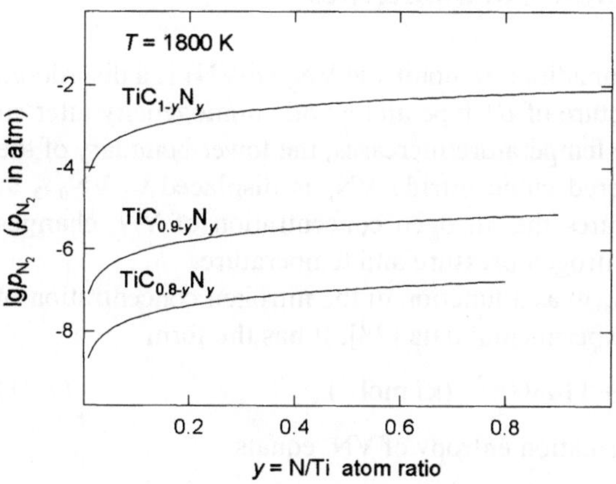

Fig. 7.4. Nitrogen partial pressure p_{N_2} as a function of the composition of stoichiometric $TiC_{1-y}N_y$ and nonstoichiometric $TiC_{0.9-y}N_y$ and $TiC_{0.8-y}N_y$ cubic titanium carbonitrides at a temperature of 1800 K [21]

Figure 7.3 shows the logarithmic dependence $p_{N_2}(y, 1800\,K)$ of the nitrogen equilibrium pressure on the nitrogen concentration of nitride TiN_y ($0.50 \le y \le 1.0$) and carbonitrides $TiC_{0.25}N_y$ ($0.21 \le y \le 0.75$), $TiC_{0.50}N_y$ ($0 < y \le 0.50$), and $TiC_{0.75}N_y$ ($0 < y \le 0.25$) at 1800 K. The dependence of pressure on the TiC_xN_y composition, $p_{N_2}(y)$, (at $x = const$) is an exponential hyperbolic function.

Figure 7.4 presents logarithmic dependencies $p_{N_2}(y, 1800\,K)$ for the stoichiometric carbonitride $TiC_{1-y}N_y$ and the nonstoichiometric carbonitrides $TiC_{0.9-y}N_y$ and $TiC_{0.8-y}N_y$. The pressure p_{N_2} over these carbonitrides exhibits similar qualitative changes, but the nitrogen pressure over nonstoichiometric carbonitrides is smaller. Thus, defects in the nonmetal sublattice of carbonitrides have a considerable effect on the nitrogen partial pressure. The following fact is remarkable. The nitrogen pressure of carbonitrides with a similar nitrogen concentration y, but with different defect content in the nonmetal sublattice is different. Indeed, at a given temperature T, $p_{N_2}(x_1, y, T) > p_{N_2}(x_2, y, T)$ for carbonitrides $MC_{x_1}N_y$ and $MC_{x_2}N_y$ (at $x_1 > x_2$ and similar y values), i.e. a higher nitrogen partial pressure will be over carbonitride with smaller nonstoichiometry. For example, at 2000 K the pressure p_{N_2} over $TiC_{0.75}N_{0.20}$ ($x + y = 0.95$) and $TiC_{0.50}N_{0.20}$ ($x + y = 0.70$) is 2.3×10^{-3} and 6.9×10^{-7} atm respectively.

Analogous calculations of the nitrogen partial pressure p_{N_2} as a function of the composition and temperature were carried out for stoichiometric and nonstoichiometric nitrides and carbonitrides of vanadium and niobium.

7.3 Vanadium Nitrides and Carbonitrides

At temperatures above 1000 K vanadium mononitride VN_y (δ-VN) is a disordered compound having a crystal structure of $B1$ type and a wide homogeneity interval from $VN_{0.69}$ to $VN_{1.00}$. When the temperature increases, the lower boundary of the homogeneity interval of disordered cubic nitride VN_y is displaced to $VN_{0.53}$ at 2000 K [27]. At high temperatures the nitrogen concentration of VN_y changes considerably depending on the nitrogen pressure and temperature.

The formation enthalpy $\Delta H_{f298}(y)$ as a function of the nitrogen concentration of VN_y was calculated [21] using experimental data [28]. It has the form

$$\Delta H_{f298}(y) = -6120 - 309960y + 114800y^2 \quad (\text{kJ mol}^{-1}). \tag{7.31}$$

In the first approximation the formation entropy of VN_y equals

$$\Delta S_{f298}(y) = -20.01 - 66.86y \quad (\text{J mol}^{-1}\,\text{K}^{-1}). \tag{7.32}$$

It follows from (7.31) and (7.32) that $\Delta H_\square(n) = \partial(\Delta H_{f298}(n))/\partial n = 80360 + 229600n$ (J mol^{-1}) and $\Delta S_\square(n) = \partial(\Delta S_{f298}(n))/\partial n = -66.86$ J K^{-1} mol^{-1}. Therefore the energy of formation of structural vacancies, $\Delta G_\square(n,T)$, in the nitrogen sublattice of VN_y is

$$\Delta G_\square(n,T) = 80360 - 66.86T + 229600n \quad (\text{J mol}^{-1}). \tag{7.33}$$

Values of y_c calculated for VN_y from (7.21) taking account of the vacancy formation energy (7.33) are given in Table 7.4.

Table 7.4. Relative nitrogen content y_c, the formation Gibbs free energy $\Delta G_f^*(y_c,T)$ and nitrogen partial pressure $p_{N_2}(y_c,T)$ for cubic vanadium nitrides VN_{y_c} which evaporate congruently at a given temperature T

T (K)	VN_y		
	y_c	$\Delta G_f^*(y_c,T)$ (kJ mol^{-1})	$p_{N_2}(y_c,T)$ (atm)
1000	0.9512	−111.30	0.000061
1100	0.9279	−100.92	0.000264
1200	0.9034	−90.35	0.000908
1300	0.8784	−79.64	0.002609
1400	0.8531	−68.80	0.006518
1500	0.8279	−57.87	0.014547
1600	0.8028	−46.83	0.029668
1700	0.7778	−35.70	0.056137
1800	0.7531	−24.48	0.099871
1900	0.7287	−13.19	0.168560
2000	0.7046	−1.81	0.272320
2100	0.6903	+9.70	0.424140

Table 7.5. Nitrogen partial pressure $p_{N_2}(y_c,T)$ (atm) over cubic nitride VN_y at a temperature of 1600, 1800 and 2000 K

y	VN_y		
	1600 K	1800 K	2000 K
0.55	–	–	0.00100
0.58	–	0.000101	0.00292
0.60	–	0.000220	0.00599
0.65	0.000032	0.00157	0.0365
0.69	0.000181	0.00767	0.158
0.70	0.000280	0.0115	0.230
0.72	0.000679	0.0257	0.363
0.75	0.00260	0.0878	0.639
0.80	0.0260	0.267	1.67
0.85	0.0922	0.789	4.49
0.90	0.326	2.47	12.7
0.95	1.30	8.56	39.6
0.96	1.76	11.2	50.6
0.97	2.43	14.9	65.6
0.98	3.42	20.3	86.4
0.99	5.01	28.4	117.0
1.00	8.28	43.9	172.2

The free energy of formation of vanadium nitride as a function of temperature and the VN_y composition was determined [21] by the same method as for titanium nitride VN_y. Taking into consideration (7.31) and (7.33) and the reference value $\Delta H^{*}_{f298}(y=1) = -217150$ J mol^{-1} [24], the free energy of formation of VN_y takes the form

$$\Delta G_f^*(y,T) = \Delta G_f(y = 1,T) + 211030 - 309960y + 114800y^2 \quad (\text{J mol}^{-1}). \qquad (7.34)$$

Table 7.4 presents free energies of formation, $\Delta G_f^*(y,T)$, of vanadium nitride VN_{y_c} at y_c values calculated from (7.21) and corresponding to different temperatures from 1000 to 2100 K. The nitrogen pressure $p_{N_2}(y_c,T)$ over congruently evaporating titanium nitride VN_{y_c} was found from (7.19) using calculated values of y_c and $\Delta G_f^*(y_c,T)$.

Several researchers [28–37] have studied nonstoichiometry of vanadium nitride VN_y and the "nitrogen partial pressure–composition–temperature" p_{N_2}–y–T relationship. Turkdogan and Fenn [29] plotted an isobar for cubic nitride VN_y at the nitrogen pressure of 1 atm and a temperature from 1173 to 1773 K. The nitrogen dissociation pressure for $VN_{0.72}$ at a temperature between 1573 and 1873 K was measured in [30]. Kozheurov et al. [31, 32] analyzed the dissociation pressure as a function of the temperature and composition of VN_y. Measurements were performed at temperatures between 1573 and 1923 K in the interval of nitrogen pressures from 6.5×10^{-4} to 1 atm. In accordance with [31, 32], stoichiometric vanadium nitride $VN_{1.0}$ cannot be obtained at the nitrogen pressure of 1 atm even if nitride is annealed for several days at a low temperature of 1273 K. Onozuka [33] established the dependence of the VN_y composition on the nitrogen pressure in the interval from 6.6×10^{-4} to 1.0 atm at a temperature of 1573 K. Pompe [34] measured the pressure p_{N_2} in equilibrium with VN_y at temperatures from 1273 to 1773 K. The p_{N_2}–y–T dependence for nitride VN_y at temperatures from 1380 to 1783 K and the nitrogen pressure from 0.1 to 1.0 atm was analyzed in [28, 35]. As is seen from this brief review, the upper limit of the nitrogen equilibrium pressure [28–35] is 1 atm. Lengauer and Ettmayer [36, 37] examined the dependence of the VN_y composition on the nitrogen pressure in the interval from 0.1 to 40 atm at temperatures from 1353 to 1923 K. Generally, literature values of the nitrogen partial pressure p_{N_2} for VN_y vary greatly.

Figure 7.5 shows the nitrogen equilibrium partial pressure p_{N_2} of VN_y ($0.55 \le y \le 1.0$) as a function of temperature and the nitrogen content of nitride. The dependences $p_{N_2}(y,T)$ were determined from (7.9), (7.10), and (7.13) using calculated values y_c and $p_{N_2}(y_c,T)$ (see Table 7.4). When $y = y_c$, i.e. for congruently evaporating nitride VN_{y_c}, the isotherms $p_{N_2}(y_c,T)$ exhibit a break (Fig. 7.5). If the nitrogen concentration y of VN_y is larger or smaller than y_c, nitrides VN_y evaporate by reaction (7.1) or (7.11), while the nitrogen pressure as a function of the composition and temperature is described by different formulas (7.9) or (7.13). As a consequence, isotherms $p_{N_2}(y,T)$ have different slopes at $y < y_c$ and $y > y_c$ and a break at $y = y_c$.

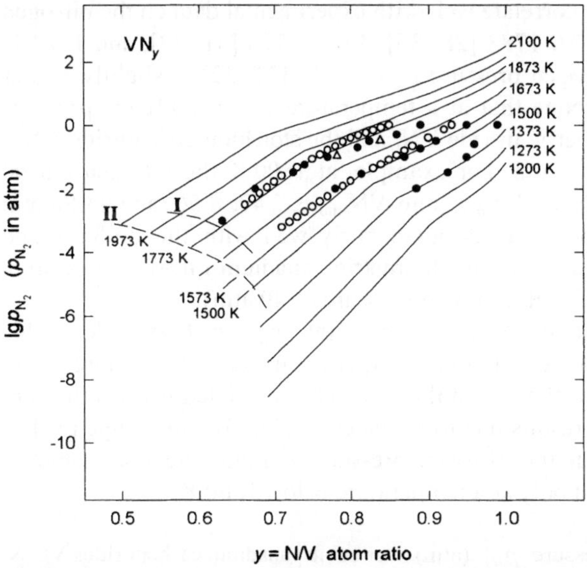

Fig. 7.5. The $p_{N_2} - y$ isotherms for cubic vanadium nitride VN_y at temperatures from 1200 to 2100 K [21]: calculated isotherms are shown by solid lines; (●) data [28, 35] measured at temperatures of 1380, 1577, and 1783 K; (○) data [31, 32] measured at temperatures of 1573 and 1773 K; (Δ) data [36, 37] measured at a temperature of 1723 K; I and II are the positions of the lower boundary of the homogeneity interval of cubic vanadium nitride VN_y as determined in [28] and [31, 32] respectively

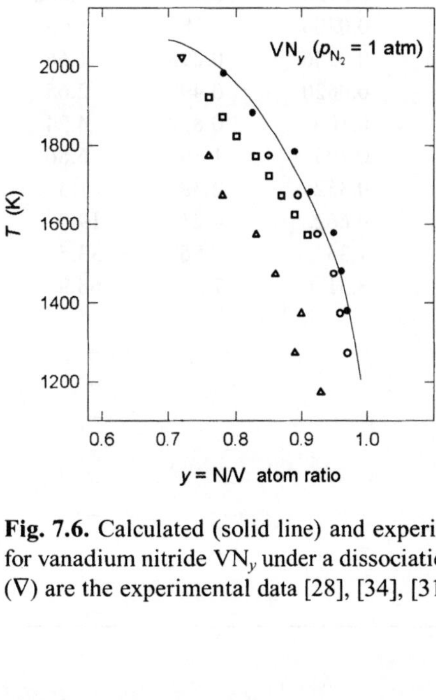

Fig. 7.6. Calculated (solid line) and experimental "nitrogen content–temperature" isobars for vanadium nitride VN_y under a dissociation pressure of 1 atm [21]: (●), (○), (□), (Δ), and (∇) are the experimental data [28], [34], [31, 32], [29], and [30] respectively

Calculated values of p_{N_2} correlate well with experimental data on the nitrogen partial pressure at 1380, 1577, 1783 [28, 35], 1573, 1773 [31, 32], and 1723 K [36, 37] (Fig. 7.5). The nitrogen pressure reported in [31, 32] is slightly higher than the calculated values. Note that at a temperature $T > 1400$ K the pressure p_{N_2} over the near-stoichiometric nitride VN_y and the stoichiometric nitride $VN_{1.0}$ is considerably higher than 1 atm. For example, at 1800 K the nitrogen partial pressure p_{N_2} of nitrides $VN_{0.90}$, $VN_{0.98}$, and $VN_{1.00}$ is 2.47, 8.56, and 43.86 atm respectively (Table 7.5). These values agree fairly well with the findings and conclusions [36] showing that near-stoichiometric vanadium nitride VN_y cannot be obtained in the absence of high nitrogen pressure (≥ 40 atm).

As is seen from Fig. 7.6 displaying isobars (at p_{N_2} = 1 atm) for cubic vanadium nitride VN_y, calculated values and experimental data do not agree well. The data [29] fall well below the rest of the data. The calculated isobar falls very nearly midway between the results due to Kieda et al. [28,35] and Pompe [34]. It follows from Fig. 7.6 that at the nitrogen pressure of 1 atm the stoichiometric nitride $VN_{1.0}$ can be obtained only at a temperature below 1200 K.

Table 7.6. Nitrogen partial pressure p_{N_2} (atm) over cubic vanadium carbonitrides VC_xN_y ($x = 0.40$ and 0.60) at a temperature of 1600, 1800 and 2000 K

y	$VC_{0.40}N_y$			$VC_{0.60}N_y$		
	1600 K	1800 K	2000 K	1600 K	1800 K	2000 K
0.10	–	–	–	0.000040	0.00164	0.0357
0.15	–	–	–	0.000521	0.0176	0.128
0.20	0.000001	0.000073	0.00200	0.00650	0.0668	0.417
0.22	0.000003	0.000171	0.00437	0.0120	0.110	0.662
0.24	0.000008	0.000396	0.00953	0.0206	0.181	1.05
0.26	0.000019	0.000916	0.0207	0.0356	0.298	1.66
0.28	0.000048	0.00212	0.0451	0.0620	0.493	2.65
0.30	0.000120	0.00491	0.0984	0.109	0.822	4.24
0.32	0.000302	0.0114	0.161	0.193	1.39	6.86
0.34	0.000762	0.0267	0.243	0.352	2.38	11.3
0.36	0.00194	0.0546	0.366	0.662	4.21	19.0
0.38	0.00499	0.0853	0.551	1.33	7.86	33.5
0.40	0.0130	0.134	0.833	3.31	17.5	68.9
0.42	0.0229	0.210	1.26	–	–	–
0.44	0.0378	0.332	1.92	–	–	–
0.46	0.0631	0.527	2.94	–	–	–
0.48	0.106	0.846	4.54	–	–	–
0.50	0.181	1.37	7.07	–	–	–
0.52	0.314	2.25	11.2	–	–	–
0.54	0.558	3.78	17.9	–	–	–
0.56	1.03	6.55	29.5	–	–	–
0.58	2.03	12.0	51.1	–	–	–
0.60	4.97	26.3	103.3	–	–	–

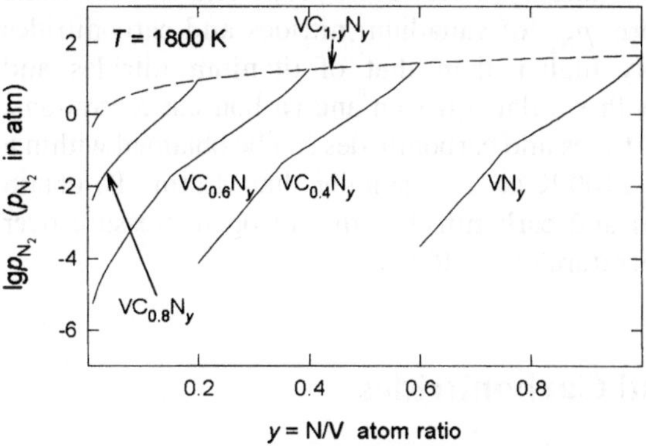

Fig. 7.7. Nitrogen partial pressure p_{N_2} of cubic vanadium nitride VN_y and carbonitrides $VC_{0.4}N_y$, $VC_{0.6}N_y$, and $VC_{0.8}N_y$ at a temperature of 1800 K [21]. The dashed line shows the limiting nitrogen partial pressure p_{N_2} over the stoichiometric carbonitride $VC_{1-y}N_y$ at 1800 K

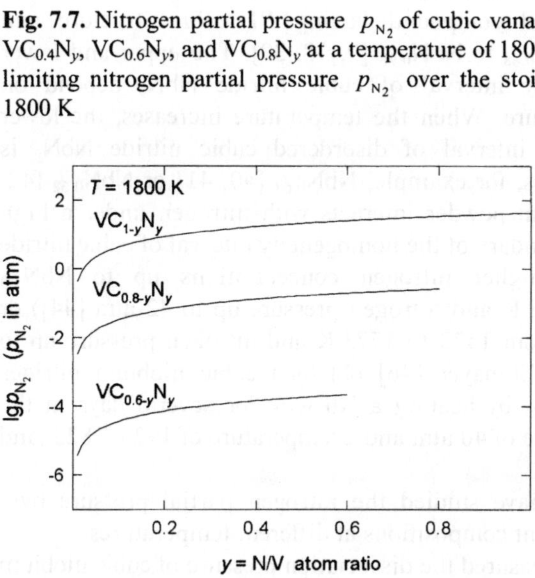

Fig. 7.8. Nitrogen partial pressure p_{N_2} in equilibrium with stoichiometric $VC_{1-y}N_y$ and nonstoichiometric $VC_{0.8-y}N_y$ and $VC_{0.6-y}N_y$ cubic vanadium carbonitrides at a temperature of 1800 K [21]

The nitrogen pressure of cubic vanadium carbonitrides $VC_{0.4}N_y$ $(0.2 \le y \le 0.6)$, $VC_{0.6}N_y$ $(0 < y \le 0.4)$, $VC_{0.8}N_y$ $(0 < y \le 0.2)$, $VC_{1-y}N_y$ $(0 < y \le 1.0)$, $VC_{0.8-y}N_y$ $(0 < y \le 0.8)$, and $VC_{0.6-y}N_y$ $(0 < y \le 0.6)$ as a function of their composition and temperature was calculated from (7.23) and (7.24). The corresponding values are given in Table 7.6 and Figs. 7.7 and 7.8.

The nitrogen pressure p_{N_2} varies as a function of the vanadium carbonitride composition (Figs. 7.7 and 7.8) similarly to p_{N_2} as a function of the titanium carbonitride composition (Figs. 7.3 and 7.4). However, given one and the same temperature, the nitrogen pressure p_{N_2} of vanadium nitrides and carbonitrides VN_y and VC_xN_y is considerably higher than that of titanium nitrides and carbonitrides TiN_y and TiC_xN_y with similar nitrogen and carbon concentrations. Indeed, stoichiometric titanium nitrides and carbonitrides can be obtained within a wide temperature interval up to 2100 K at p_{N_2} smaller than 1 atm. To obtain stoichiometric vanadium nitrides and carbonitrides, the nitrogen pressure over 1 atm is needed even at low temperature ($T < 1400$ K).

7.4 Niobium Nitrides and Carbonitrides

At a temperature $T > 1200$ K and nitrogen pressure of 1 atm, niobium mononitride NbN_y (δ-NbN) is a disordered compound having *B*1 cubic structure and homogeneity interval from $NbN_{0.88}$ to $NbN_{0.98}$ [9, 38, 39]. The upper and lower boundaries of the homogeneity interval of cubic nitride NbN_y depend on temperature and nitrogen pressure. When the temperature increases, the lower boundary of the homogeneity interval of disordered cubic nitride NbN_y is displaced towards lower *y* values, for example, $NbN_{0.68}$ [40, 41] or $NbN_{0.62}$ [42, 43] at 2000 K. When a niobium powder interacts with nitrogen under a high nitrogen pressure, the upper boundary of the homogeneity interval of cubic nitride NbN_y is displaced towards higher nitrogen concentrations up to $NbN_{1.0}$ (temperature from 1613 to 1788 K and nitrogen pressure up to 42 atm [44]) or even to $NbN_{1.06}$ (temperature from 1573 to 1773 K and nitrogen pressure up to 160 atm [45]). Lengauer and Ettmayer [46] obtained cubic niobium nitrides $NbN_{0.924}$, $NbN_{0.953}$, and $NbN_{0.975}$ by heating a Nb wire for several days in the nitrogen atmosphere at a pressure of 40 atm and a temperature of 1923, 1823, and 1723 K respectively.

Many researchers [39–48] have studied the nitrogen partial pressure over niobium nitrides NbN_y of different compositions at different temperatures.

Brauer and Esselborn [39] measured the dissociation pressure of cubic niobium nitrides $NbN_{0.95}$ and $NbN_{0.98}$ and found that at 1673 K the pressure p_{N_2} is equal to 0.28 and 1.0 atm respectively. Five isotherms of the nitrogen equilibrium pressure $p_{N_2}(y,T)$ of cubic nitride NbN_y were constructed for temperatures from 1523 to 1853 K [47]. Measurements were made at the nitrogen pressure from 0.001 to 0.82 atm. The thermogravimetric method was used [40, 41] to analyze the relationship $p_{N_2}-y-T$ in the Nb–N system at temperatures from 1773 to 2023 K and the nitrogen pressure from 0.004 to 1.0 atm. Similar studies [42, 43] were performed later in a wider temperature interval from 1573 to 2073 K. In accordance with [42, 43], at a temperature above 1673 K and nitrogen pressure from 0.001 to 1.0 atm the Nb–N system contains only two phases, β-Nb_2N and δ-NbN. The dependence of the nitrogen partial pressure on the composition of cubic niobium nitride NbN_y was studied [48] at a temperature of 1673, 1773, 1873, and 1973 K.

The scatter of the results [40–43, 47, 48] is insignificant. Given the same temperature and the composition of NbN_y, the p_{N_2} values found by Sudarikov et al. [48] were 3–5 times larger than those reported in [40, 41]. The nitrogen partial pressure over niobium nitride [47] was a little smaller than the value determined in [48]. The values [42, 43] are intermediate between the data of [40, 41] and [48].

To calculate the equilibrium nitrogen pressure p_{N_2} over niobium nitride NbN_y, we need to know the formation energy of structural vacancies $\Delta G_{\square}(n,T)$ in the nitrogen sublattice and the free energy of formation $\Delta G_f{}^*(y,T)$ for NbN_y. Using experimental values of $p_{N_2}(y,T)$ [40, 41, 48], numerical values of y_c, $b(T)$, and $p_{N_2}(y_c,T)$ were calculated [21] by the least-squares fit from (7.9) for temperatures of 1673, 1773, 1873, and 1973 K. The y_c values were used to calculate numerical values of the formation energy of structural vacancies, $\Delta G_{\square}(n,T)$, for the above temperatures from (7.21). In accordance with [21], for cubic niobium nitride NbN_y the least-squares fit gives

$$\Delta G_{\square}(n,T) = (262960 - 169T) + (-425830 + 386T)n \quad (\text{J mol}^{-1}). \tag{7.35}$$

The free energy of formation, $\Delta G_f{}^*(y,T)$, for cubic nitride NbN_y was found from the analysis of numerical values of $p_{N_2}(y_c,T)$ taking account of (7.19) and the calculated energy $\Delta G_{\square}(n,T)$ (7.35). In the temperature interval from 1200 to 2100 K the free energy of formation, $\Delta G_f{}^*(y,T)$, takes the form

$$\begin{aligned}\Delta G_f{}^*(y,T) = &-164955 + 125.1 + (162870 - 217T)y \\ &+ (-212915 + 193T)y^2 \quad (\text{J mol}^{-1}).\end{aligned} \tag{7.36}$$

The y_c values calculated from (7.21) considering the formation energy of structural vacancies (7.35) and estimated values of $\Delta G_f{}^*(y,T)$ are given in Table 7.7. The nitrogen pressure $p_{N_2}(y_c,T)$ over congruently evaporating nitride NbN_y was calculated from (7.19).

Table 7.7. Relative nitrogen content y_c, the formation Gibbs free energy $\Delta G_f^*(y_c,T)$ and nitrogen partial pressure $p_{N_2}(y_c,T)$ for cubic niobium nitrides NbN_{y_c} which evaporate congruently at a given temperature T

T (K)	NbN_y		
	y_c	$\Delta G_f^*(y_c,T)$ (kJ mol^{-1})	$p_{N_2}(y_c,T)$ (atm)
1400	0.9437	−71.79	0.007842
1500	0.8908	−61.40	0.014928
1600	0.8441	−52.07	0.025640
1700	0.8067	−43.53	0.041108
1800	0.7772	−35.53	0.062498
1900	0.7536	−27.90	0.090962
2000	0.7344	−20.52	0.127603
2100	0.7185	−13.33	0.173441

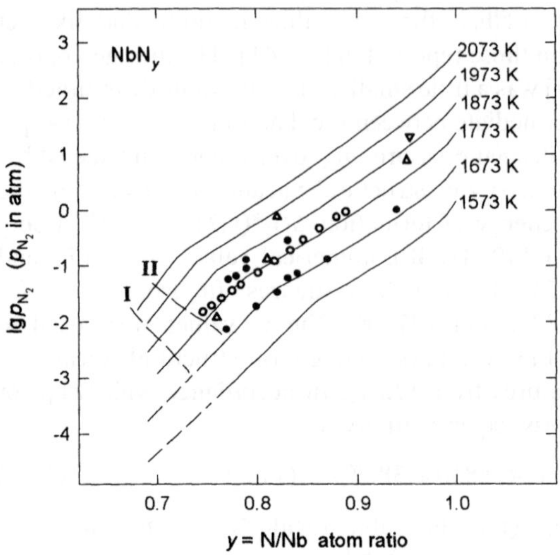

Fig. 7.9. The $p_{N_2} - y$ isotherms for cubic niobium nitride NbN_y at temperatures from 1573 to 2073 K [21]: calculated isotherms are shown by solid lines; (•) data [42, 43] measured at temperatures of 1673 and 1773 K; (○) data [40, 41] measured at a temperature of 1773 K; (∇) data [46] at a temperature of 1823 K; (Δ) data [49] at a temperature of 1743 K; I and II are the positions of the lower boundary of the homogeneity interval of cubic niobium nitride NbN_y as determined in [42, 43] and [40, 41] respectively

Table 7.8. Nitrogen partial pressure $p_{N_2}(y_c, T)$ (atm) over cubic nitride NbN_y at a temperature of 1600, 1800 and 2000 K

y	NbN_y		
	1600 K	1800 K	2000 K
0.69	–	–	0.0130
0.70	–	–	0.0217
0.72	–	0.00435	0.0607
0.75	0.000523	0.0174	0.192
0.80	0.00393	0.108	0.715
0.85	0.0292	0.367	2.76
0.90	0.0925	1.32	11.2
0.95	0.336	5.35	50.6
0.96	0.450	7.25	69.9
0.97	0.612	9.96	97.7
0.98	0.856	14.0	139.2
0.99	1.26	20.3	204.3
1.00	2.14	33.0	328.8

The nitrogen equilibrium pressure p_{N_2} of niobium nitride NbN_y $(0.7 \leq y \leq 1.0)$ is presented in Fig. 7.9 as a function of T and the nitrogen content of nitride. Calculated isotherms $p_{N_2}(y,T)$ correlate well with experimental isotherms measured at 1673K [42, 43] and 1773 K [40–43]. The nitrogen partial pressure p_{N_2} of $NbN_{0.95}$ at 1823 K [46] and $NbN_{0.81}$, $NbN_{0.82}$, and $NbN_{0.94}$ at 1743 K [49] is slightly higher than the calculated p_{N_2}. At a temperature above 1573 K the nitrogen partial pressure of the near-stoichiometric nitride NbN_y exceeds 1 atm (Table 7.8).

The nitrogen partial pressure of cubic niobium carbonitrides $NbC_{0.25}N_y$, $NbC_{0.5}N_y$, $NbC_{0.75}N_y$, $NbC_{1-y}N_y$, $NbC_{0.9-y}N_y$, $NbC_{0.8-y}N_y$, $NbC_{0.7-y}N_y$, $NbC_xN_{0.7}$, $NbC_xN_{0.6}$, $NbC_xN_{0.5}$, and $NbC_xN_{0.4}$ was calculated as a function of the composition and temperature from (7.23), (7.24), and (7.25). The calculated values are given in Table 7.9.

Table 7.9. Nitrogen partial pressure p_{N_2} (atm) over cubic niobium carbonitrides NbC_xN_y at a temperature of 1600 and 1800 K

$NbC_{0.5}N_y$	$NbC_{0.75}N_y$		$NbC_xN_{0.5}$			
	1600 K	1800 K	1600 K	1800 K	1600 K	1800 K
0	–	–	0	0	–	–
0.05	–	–	0.000246	0.00674	–	–
0.10	–	–	0.00344	0.0432	–	–
0.15	–	–	0.0154	0.221	–	–
0.20	0.000021	0.000499	0.0708	1.13	0.000053	0.00125
0.22	0.000049	0.00133	0.139	2.26	0.000112	0.00302
0.24	0.000114	0.00355	0.304	4.93	0.000238	0.00739
0.26	0.000266	0.0095	–	–	0.000512	0.0183
0.28	0.000623	0.0240	–	–	0.00111	0.0429
0.30	0.00147	0.0404	–	–	0.00246	0.0674
0.32	0.00354	0.0683	–	–	0.00553	0.107
0.34	0.00868	0.116	–	–	0.0128	0.170
0.36	0.0153	0.197	–	–	0.0212	0.274
0.38	0.0249	0.339	–	–	0.0328	0.446
0.40	0.0411	0.588	–	–	0.0514	0.736
0.42	0.0691	1.04	–	–	0.0823	1.23
0.44	0.119	1.87	–	–	0.136	2.12
0.46	0.215	3.47	–	–	0.234	3.77
0.48	0.419	6.85	–	–	0.439	7.14
0.50	1.07	16.5	–	–	1.07	16.5

Calculated nitrogen partial pressure p_{N_2} of cubic niobium nitride NbN_y $(0.69 \leq y \leq 1.0)$ and carbonitrides $NbC_{0.25}N_y$ $(0.45 \leq y \leq 0.75)$, $NbC_{0.50}N_y$ $(0.20 \leq y \leq 0.50)$, and $NbC_{0.75}N_y$ $(0 < y \leq 0.25)$ at a temperature of 1873 K is compared in Fig. 7.10 with experimental data [48] for NbN_y, $NbC_{0.25}N_y$, $NbC_{0.49}N_y$, and $NbC_{0.74}N_y$ obtained at the same temperature.

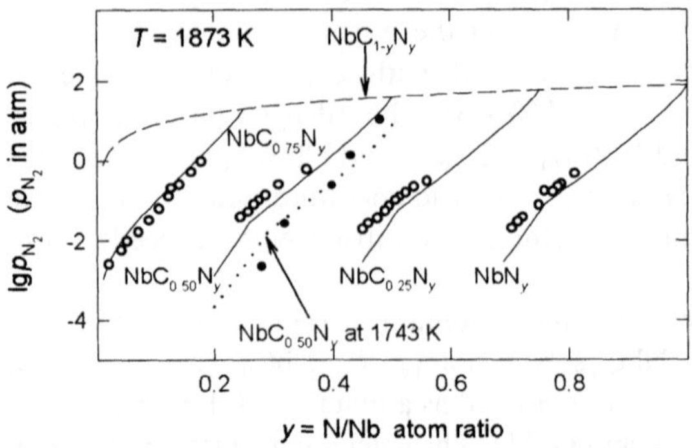

Fig. 7.10. Calculated nitrogen partial pressure p_{N_2} of cubic niobium nitride NbN_y and carbonitrides $NbC_{0.25}N_y$, $NbC_{0.50}N_y$, and $NbC_{0.75}N_y$ at a temperature of 1873 K and carbonitride $NbC_{0.50}N_y$ at a temperature of 1743 K [21]. The nitrogen pressure p_{N_2} over stoichiometric niobium carbonitride $NbC_{1-y}N_y$ at 1873 K is shown by the dashed line; (○) experimental data [48] at a temperature of 1873 K; (●) experimental data [49] at 1743 K

As is seen from Fig. 7.10, experimental values of p_{N_2} for NbN_y, $NbC_{0.25}N_y$, and $NbC_{0.49}N_y$ are nearly 3–10 times higher than the calculated values. The experimental nitrogen partial pressure of $NbC_{0.74}N_y$ is slightly smaller than the calculated pressure. Note that the calculated isotherms of p_{N_2} and experimental isotherms [48] have similar slopes (see Fig. 7.10). The nitrogen partial pressure p_{N_2} of $NbC_{0.50}N_y$ measured at 1743 K by Williams et al. [49] is in good agreement with calculated values at $y = 0.35$ and $y = 0.41$. When $y < 0.35$, p_{N_2} is smaller and at $y > 0.41$ is larger than the calculated values (Fig. 7.10).

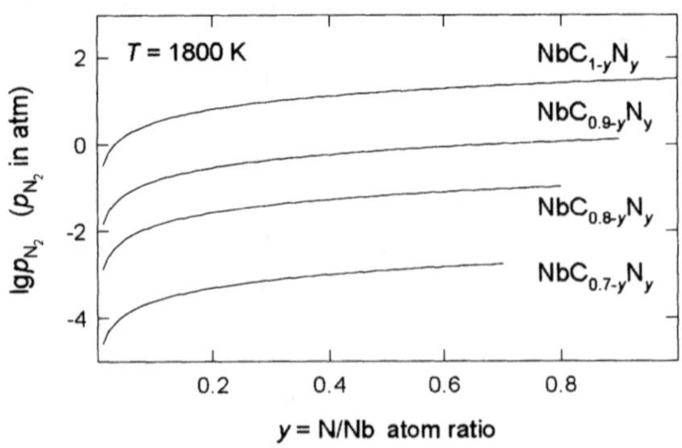

Fig. 7.11. Nitrogen partial pressure p_{N_2} in equilibrium with stoichiometric cubic niobium carbonitride $NbC_{1-y}N_y$ and nonstoichiometric cubic niobium carbonitrides $NbC_{0.9-y}N_y$, $NbC_{0.8-y}N_y$, and $NbC_{0.7-y}N_y$ at a temperature of 1800 K [21]

Figure 7.11 illustrates calculated isotherms $p_{N_2}(y,1800\,K)$ of niobium carbonitrides $NbC_{1-y}N_y$ $(0.01 \le y \le 1.0)$, $NbC_{0.9-y}N_y$ $(0.01 \le y \le 0.9)$, $NbC_{0.8-y}N_y$ $(0.01 \le y \le 0.8)$, and $NbC_{0.7-y}N_y$ $(0.01 \le y \le 0.7)$ with different nonstoichiometry of the nonmetal sublattice. The nitrogen pressure p_{N_2} is reduced when the relative common content of nitrogen and carbon decreases from 1.0 in $NbC_{1-y}N_y$ to 0.7 in $NbC_{0.7-y}N_y$.

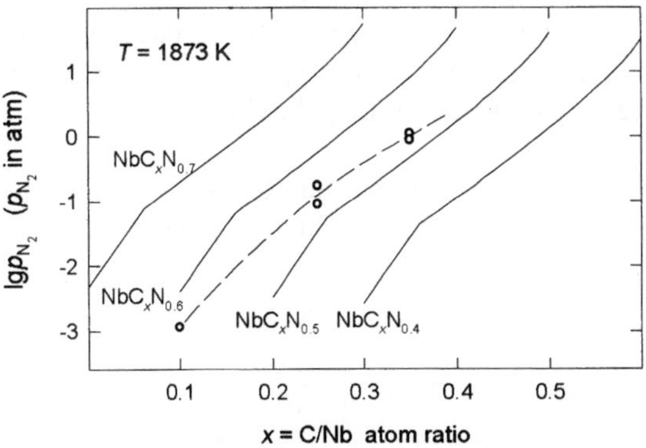

Fig. 7.12. Effect of the carbon content on the nitrogen partial pressure p_{N_2} of cubic niobium carbonitrides $NbC_xN_{0.7}$, $NbC_xN_{0.6}$, $NbC_xN_{0.5}$, and $NbC_xN_{0.4}$ at a temperature of 1873 K [21]: calculated values are shown by the solid line; (○) experimental data [48] at a temperature of 1873 K

The effect of the relative carbon content x on the nitrogen partial pressure $p_{N_2}(x,y,T)$ of cubic carbonitride NbC_xN_y ($y = const$) at a temperature of 1873 K is shown in Fig. 7.12. As follows from the results obtained, an increase in carbon content x of the carbonitride NbC_xN_y (with $y = const$) leads to an increase of the pressure p_{N_2}. A similar dependence of p_{N_2} on the carbon content x of niobium carbonitrides NbC_xN_y was observed in experiments [48, 49].

In general, the effect of nonstoichiometry on the equilibrium nitrogen partial pressure p_{N_2} over transition metal nitrides and carbonitrides boils down to the following.

The nitrogen partial pressure over MN_y ($MN_y\square_{1-y}$ with $y \le 1.0$) and MC_xN_y ($MC_xN_y\square_{1-x-y}$ with $x + y \le 1.0$) drops with increasing concentration of structural vacancies □. The increase in the vacancy concentration affects mostly the pressure p_{N_2} over near-stoichiometric titanium nitrides and carbonitrides.

When the temperature is elevated, the nitrogen pressure over nonstoichiometric nitrides MN_y ($y < 1.0$) and carbonitrides MC_xN_y ($x = const$, $x + y < 1.0$) increases more rapidly than the pressure p_{N_2} over stoichiometric nitrides $MN_{1.0}$ and carbonitrides MC_xN_{1-x} with a similar carbon concentration x. For example, when the temperature is elevated by 200 K (in the interval from 1600 to 2000 K), the nitrogen partial pressure p_{N_2} over stoichiometric vanadium nitride $VN_{1.0}$ and carbonitride $VC_{1-y}N_y$ increases four- to five-fold, whereas the pressure p_{N_2} over nonstoichiometric nitrides VN_y and carbonitrides VC_xN_y increases by a factor of 10–50. The presence of vacancies in the nonmetal sublattice of carbonitrides considerably affects the nitrogen partial pressure. The pressure p_{N_2} drops more rapidly over MC_xN_y (with $x = const$) than over MC_xN_y (with $x + y = const$) when the relative nitrogen content y decreases.

At a given temperature T, two carbonitrides $MC_{x_1}N_y$ and $MC_{x_2}N_y$ with different carbon content x_1 and x_2 and the same nitrogen content y have different nitrogen partial pressures: $p_{N_2}(x_1, y, T) > p_{N_2}(x_2, y, T)$ if $x_1 > x_2$. This means that the higher nitrogen partial pressure p_{N_2} corresponds to carbonitride MC_xN_y with a smaller nonstoichiometry.

The above findings on the nitrogen partial pressure may be used to determine synthesis conditions of stoichiometric and nonstoichiometric nitrides and carbonitrides of transition metals.

References

1 V. V. Fesenko, A. S. Bolgar: *Evaporation of Refractory Compounds* (Metallurgiya, Moscow 1966) 112 pp. (in Russian)

2 R. B. Kotel'nikov, S. N. Bashlykov, Z. G. Galiakbarov, A.I. Kashtanov: *Fundamentals of Refractory Elements and Compounds* (Handbook) (Metallurgiya, Moscow 1969) 376 pp. (in Russian)

3 A. S. Bolgar, A. G. Turchanin, V. V. Fesenko: *Thermodynamic Properties of Carbides* (Naukova Dumka, Kiev 1973) 272 pp. (in Russian)

4 A. S. Bolgar, V. F. Litvinenko: *Thermodynamic Properties of Nitrides* (Naukova Dumka, Kiev 1980) 282 pp. (in Russian)

5 J. P. Hirth, G. M. Pound: *Condensation and Evaporation-Nucleation and Growth Kinetics* (Pergamon Press, Oxford 1963) 192 pp.

6 I. S. Kulikov: *Thermodynamics of Carbides and Nitrides* (Metallurgiya, Chelyabinsk 1988) 320 pp. (in Russian)

7 A. I. Gusev: Thermodynamic conditions for the congruent vaporization of unordered nonstoichiometric compounds, Doklady AN SSSR **311**, 631–635 (1990) (in Russian). (Engl. transl.: Doklady Phys. Chem. **311**, 264–268 (1990))

8 A. I. Gusev: Thermodynamics of the congruent vaporization of nonstoichiometric refractory compounds, Teplofizika Vysokikh Temperatur **29**, 695–701 (1991) (in Russian). (Engl. transl.: High Temp. **29**, 541–546 (1991))

9 A. I. Gusev: *Physical Chemistry of Nonstoichiometric Refractory Compounds* (Nauka, Moscow 1991) 286 pp. (in Russian)

10 A. I. Gusev: Statistical description of structural vacancy formation in compounds of variable composition such as zirconium nitride and carbide, Zh. Fiz. Khimii **53**, 1379–1383 (1979) (in Russian). (Engl. transl.: Russ. J. Phys. Chem. **53**, 780–782 (1979))

11 A. I. Gusev: Energy characteristics of the structural vacancies in titanium and zirconium nitrides, Zh. Neorgan. Khimii **25**, 629–633 (1980) (in Russian). (Engl. transl.: Russ. J. Inorg. Chem. **25**, 345–347 (1980))

12 A. I. Gusev: Characteristics of the formation of structural vacancies in vanadium carbide and thermodynamic properties of defect-free vanadium carbide, Zh. Fiz. Khimii **57**, 1382–1386 (1983) (in Rusian). (Engl. transl.: Russ. J. Phys. Chem. **57**, 837–839 (1983))

13 A. I. Gusev, A. A. Rempel: *Thermodynamics of the Structural Vacancies in Nonstoichiometric Interstitial Phases* (Ural Scientific Centre, Sverdlovsk 1987) 114 pp. (in Russian)

14 M. Hoch, D. P. Dingledy, H. L. Johnston: The vaporization of TiN and ZrN, J. Amer. Chem. Soc. **77**, 304–306 (1955)

15 P. Grieveson: An investigation of the Ti–C–N system, Proc. British Ceram. Soc. No 8, 137–153 (1967)

16 P. S. Kislyi, M. A. Kuzenkova: Variation of structure and properties of titanium nitride at high-temperature heating in vacuum, Izv. AN SSSR. Neorgan. Materialy **5**, 2083–2086 (1969) (in Russian).

17 A. I. Gusev, G. P. Shveikin: Thermodynamic characteristics of vacancies in zirconium nitride, Izv. AN SSSR. Neorgan. Materialy **11**, 653–656 (1975) (in Russian). (Engl. transl.: Inorganic Materials **11**, 561–563 (1975))

18 E. A. Ryklis, A. S. Bolgar, O. P. Kulik, S. A. Shwab, V. V. Fesenko: Evaporation of titanium nitride at high temperatures, Zh. Fiz. Khimii **44**, 1295–1297 (1970) (in Russian)

19 A. I. Gusev: Nitrogen pressure over cubic stoichiometric and nonstoichiometric nitrides and carbonitrides of transition metals, Doklady Akademii Nauk **340**, 758–762 (1995) (in Russian). (Engl. transl.: Physics – Doklady **40**, 60–64 (1995))

20 A. I. Gusev: The influence of composition, nonstoichiometry, and temperature on the partial pressure of nitrogen over transition metal nitrides and carbonitrides, Zh. Fiz. Khimii **70**, 616–621 (1996) (in Russian). (Engl. transl.: Russ. J. Phys. Chem. **70**, 570–575 (1996)

21 A. I. Gusev: Nitrogen partial pressure of stoichiometric and nonstoichiometric titanium, vanadium and niobium nitrides and carbonitrides, Phys. Stat. Sol. (b) **209**, 267–286 (1998)

22 A. I. Gusev: Phase diagrams of ordered nonstoichiometric hafnium carbide and titanium nitride, Doklady Akademii Nauk **322**, 918–923 (1992) (in Russian). (Engl. transl.: Doklady Phys. Chem. **322**, 84–88))

23 M. P. Morozova, M. M. Khernburg: Enthalpy of formation of titanium nitrides as a function of their composition, Zh. Fiz. Khimii **40**, 1125–1128 (1966) (in Russian)

24 I. Barin: *Thermodynamical Data of Pure Substances* (VCH, Weinheim 1995) 1880 pp.

25 A. I. Gusev, A. A. Rempel: Phase diagrams of metal–carbon and metal–nitrogen systems and ordering in strongly nonstoichiometric carbides and nitrides, Phys. Stat. Sol. (a) **163**, 273–304 (1997)

26 W. Lengauer: The titanium–nitrogen system: a study of phase reactions in the subnitride region by means of diffusion couples, Acta Metall. Mater. **39**, 2985–2995 (1991)

27 O. N. Carlson, J. F. Smith, R. H. Nafziger: The vanadium–nitrogen system: review, Metall. Transactions A **17**, 1647–1656 (1986)

28 N. Kieda, N. Mizutani, M. Kato: Non-stoiciometry and phase equilibria of β-V_2N and δ-VN, J. Less-Common Metals **144**, 293–299 (1988)

29 E. T. Turkdogan, E. M. Fenn: A note on non-stoichiometric vanadium nitrides, J. Iron Steel Inst. **181**, 343–344 (1955)

30 G. Brauer, W.-D. Schnell: Zur Kenntnis des Systems Vanadium–Stickstoff und der reinen Vanadiums, J. Less-Common Metals **6**, 326–332 (1964)
31 V. A. Kozheurov, V. M. Zhikharev, V. I. Shishkov, G. V. Gritsishina: Study of thermodynamic properties of vanadium mononitride, Izv. Vyssh. Ucheb. Zaved. Chernaya Metallurgiya No 8, 10–13 (1972) (in Russian)
32 V. A. Kozheurov, V. M. Zhikharev, V. I. Shishkov: Study of thermodynamic properties of vanadium mononitride. Report 2, Izv. Vyssh. Ucheb. Zaved. Chernaya Metallurgiya No 12, 76–78 (1972) (in Russian)
33 T. Onozuka: Vacancy ordering in vanadium nitride (VN_{1-x}), J. Appl. Crystallogr. **11**, 132–136 (1978)
34 R. Pompe: Some thermochemical properties of the system vanadium–nitrogen and vanadium–carbon–nitrogen in the temperature range 1000-1550°C, Thermochim. Acta **57**, 273–281 (1982)
35 N. Kieda, K. Uematsu, N. Mizutani, M. Kato: Nitrogen partial pressure-temperature-composition relations of vanadium mononitride (δ-VN), J. Less-Common Metals **99**, 131–135 (1984)
36 W. Lengauer, P. Ettmayer: Physical and mechanical properties of cubic δ-VN_{1-x}, J. Less-Common Metals **109**, 351–359 (1985)
37 W. Lengauer: Interdependency of composition and T_c of δ-VN_{1-x} and the influence and determination of its nitrogen surface diffusion layers, J. Phys. Chem. Solids **49**, 59–63 (1988)
38 G. Brauer: Nitrides, carbonitrides and oxynitrides of niobium, J. Less-Common Metals **2**, 131–137 (1960)
39 G. Brauer, R. Esselborn: Nitridphasen des Niobs, Z. Anorg. Allgem. Chem. **309**, 154–170 (1961)
40 L. N. Matushina, V. M. Zhikharev: Thermodynamic functions of niobium nitrides. Report 1, Izv. Vyssh. Ucheb. Zaved. Chernaya Metallurgiya No 8, 14–17 (1978) (in Russian)
41 L. N. Matushina, V. M. Zhikharev: Thermodynamic functions of niobium nitrides. Report 2, Izv. Vyssh. Ucheb. Zaved. Chernaya Metallurgiya No 10, 13–15 (1978) (in Russian)
42 G. Brauer, W. Kern: Zersetzungsdrücke und Phasengrenzen von Niobnitriden, Z. Anorg. Allgem. Chem. **507**, 127–141 (1983)
43 G. Brauer, W. Kern: Gezielte Präparation von Niobnitridphasen, Z. Anorg. Allgem. Chem. **512**, 7–12 (1984)
44 G. Brauer, J. Jander, H. Rögener: Zur Struktur und Zusammensetzung von Niob–Stickstoff–verbindungen, Z. Physik **134**, 432–434 (1953)
45 G. Brauer, H. Kirner: Drucksynthese von Niobnitriden und Konstitution von δ-NbN, Z. Anorg. Allgem. Chem. **328**, 34–43 (1964)
46 W. Lengauer, P. Ettmayer: Preparation and properties of compact cubic δ-NbN_{1-x}, Monatsh. Chemie **117**, 275–286 (1986)
47 A. G. Shchurik, I. A. Tomilin: Pressure of dissociation of the NbN-type niobium nitride in a homogeneity region, Zh. Fiz. Khimii **45**, 2049–2052 (1971) (in Russian). (Engl. transl.: Russ. J. Phys. Chem. **45**, 1162–1164 (1971))
48 M. V. Sudarikov, V. M. Zhikharev, A. A. Lykasov: Conditions of equlibrium of a cubic niobium carbonitride with nitrogen, Neorgan. Materialy **32**, 567–570 (1996)
49 M. W. Williams, K. M. Ralls, M. R. Pickus: Superconductivity of cubic niobium carbonitrides, J. Phys. Chem. Solids **28**, 333–341 (1967)

8. Order–Disorder Phase Transformation Theories

Currently there is no such solid-state theory which would allow the prediction of all properties of a substance from its chemical composition. All phase transformation theories designed for the phase description center on the thermodynamic potential and vary in the form of writing this potential in terms of parameters of the theory. Clearly, the extent to which any such theory approaches ideality is determined by the minimum number of phenomenological parameters needed to describe all interparticle interactions.

This chapter provides a brief review of a variety of lines along which phase transformation theory has been advancing. As is known, the phenomenological approach developed in the thermodynamic theory of phase transformations of the second kind addresses general regularities of the process. However, this approach holds only in a narrow temperature interval near the transition point, with the exception of the nearest neighborhood of the transition point itself. Broader temperature and concentration intervals call for approximate statistical methods to describe ordering.

This chapter highlights methods and models proved as the most effective in depicting structural order–disorder phase transformations in nonstoichiometric compounds.

8.1 Thermodynamics of Phase Transformations

Phase Equilibrium Conditions. A phase is a thermodynamically homogeneous part of a system. Depending on the particular external conditions of a system, phases can coexist without mass variation (phase equilibrium) or a transition from one phase state to another (phase transformation) is possible.

Consider a two-phase system as an example. The state of the system is governed by the pressure p, temperature T, and (when the mass m of the system is given) by specific volumes of the first and second phases v_1 and v_2 and the mass of the first phase m_1. Let $f_1(v_1,T)$ and $f_2(v_2,T)$ be the specific free energy of the first and second phases respectively, while $p_1(v_1)$ and $p_2(v_2)$ are the equilibrium pressure in the first and second phases corresponding to specific volumes of these phases. In the equilibrium state, the thermodynamic equilibrium potential Φ_0 of the system will be equal to the sum of the thermodynamic potentials of its two phases:

$$\Phi_0 = m_1[f_1(v_1,T) + v_1p_1(v_1)] + m_2[f_2(v_2,T) + v_2p_2(v_2)]. \tag{8.1}$$

When the pressure changes from $p_1(v_1)$ to p and from $p_2(v_2)$ to p, the total work done by the first and second phases is

$$A = m_1v_1[p_1(v_1) - p] + m_2v_2[p_2(v_2) - p]. \tag{8.2}$$

Subtracting the work (8.2) from (8.1), we obtain the thermodynamic potential $\Phi(p,T,v_1,v_2,m_1)$ of the state of interest as determined with respect to the thermodynamic potential Φ_0 of an arbitrary initial equilibrium state:

$$\Phi(p,T,v_1,v_2,m_1) = m_1(f_1 + pv_1) + m_2(f_2 + pv_2), \tag{8.3}$$

where $f_1 + pv_1 = \varphi_1$ and $f_2 + pv_2 = \varphi_2$ denote specific isobaric–isothermal potentials of the first and second phases. To determine equilibrium conditions, let us find the minimum Φ for constant temperature and pressure on the condition of the first differential being zero:

$$d\Phi = m_1\left(\frac{\partial f_1}{\partial v_1} + p\right)dv_1 + m_2\left(\frac{\partial f_2}{\partial v_2} + p\right)dv_2$$
$$+ (f_1 + pv_1)dm_1 + (f_2 + pv_2)dm_2 = 0. \tag{8.4}$$

Considering $dm_1 = -dm_2$, (8.4) is fulfilled if

$$\frac{\partial f_1}{\partial v_1} = \frac{\partial f_2}{\partial v_2} = -p\,, \tag{8.5}$$

$$f_1 + pv_1 = f_2 + pv_2\,. \tag{8.6}$$

Since $\partial f_1/\partial v_1 = p_1$ and $\partial f_2/\partial v_2 = p_2$, (8.5) expresses the first condition for equilibrium between two phases: the equality of pressures $p_1 = p_2$ in contacting phases. This condition is necessary for mechanical equilibrium of the system. Equation (8.6) describes the second equilibrium condition: the equality of specific thermodynamic potentials of both phases

$$\varphi_1(p, T) = \varphi_2(p, T)\,. \tag{8.7}$$

Condition (8.7) is written similarly to the equality of chemical potentials of equilibrium phases:

$$\mu_1 = \mu_2\,, \tag{8.8}$$

because $(\partial\Phi/\partial m_1)_{m_2} = \mu_1$ and $(\partial\Phi/\partial m_2)_{m_1} = \mu_2$.

Generally for a multiphase system, in which a chemical or phase transformation takes place (which is the same from the thermodynamic viewpoint, i.e. a nonequilibrium process), the following inequality holds:

$$\sum_{i=1}^{n} \mu_i dm_i < 0\,. \tag{8.9}$$

Once equilibrium has been established, the algebraic sum of the effects interchanged by a variety of thermodynamically homogeneous regions (phases) becomes zero to entail the equilibrium condition

$$\sum_{i=1}^{n} \mu_i dm_i = 0 , \tag{8.10}$$

where summation is taken over phase indices.

The equilibrium condition (8.10) for multiphase systems is met if the chemical potentials of a substance are equal in all phases, i.e. $\mu_1 = \mu_2 = \ldots = \mu_n$. In multicomponent systems, the equilibrium condition (8.10) is fulfilled for each component. In other words, when a multiphase multicomponent system is in equilibrium, the chemical potentials of every given component must be equal in all phases. Therefore equilibrium of a multiphase multicomponent system is described by the set of equations

$$\begin{aligned} &\mu_1^{(1)} = \mu_2^{(1)} = \ldots = \mu_n^{(1)} \\ &\mu_1^{(2)} = \mu_2^{(2)} = \ldots = \mu_n^{(2)} \\ &\ldots\ldots\ldots\ldots\ldots\ldots\ldots \\ &\mu_1^{(k)} = \mu_2^{(k)} = \ldots = \mu_n^{(k)} , \end{aligned}$$

where the upper index is the component number and the lower index is the phase number.

Classification of Phase Transformations and Ehrenfest Equations. It has taken great effort to work out a classification of phase transformations. The classification proposed by Ehrenfest [1] has been most popular. According to the definition given in [1], the kind of phase transformation depends on the order of the isobaric–isothermal potential derivative (free Gibbs energy) that is broken first at the transition point.

A sudden change in the specific volume and entropy is typical of phase transformations of the first kind:

$$\Delta v = v_2 - v_1 = \left(\frac{\partial \varphi_2}{\partial p}\right)_T - \left(\frac{\partial \varphi_1}{\partial p}\right)_T \neq 0 , \tag{8.11}$$

$$\Delta S = S_2 - S_1 = \left(\frac{\partial \varphi_2}{\partial T}\right)_p - \left(\frac{\partial \varphi_1}{\partial T}\right)_p \neq 0 . \tag{8.12}$$

A sudden change in entropy is indicative of heat release or absorption and therefore phase transformations of the first kind are characterized by some transition heat ΔH_{trans}. It should be emphasized that the heat capacities of two phases are different for transformations of the first kind. In the case of transformations of the melting type, the heat capacity in the liquid state tends to be higher than that in the solid state. The heat capacity discontinuity may have any sign upon solid-to-solid phase transitions. If ordering proceeds as a transformation of the first kind, the ordered phase (near T_{trans}) normally has a lower heat capacity.

During phase transformations of the second kind, the specific volumes and entropy change continuously. Therefore no transition heat is available. Sudden changes occur at the transition point in the second derivatives of the isobaric–isothermal potential and in quantities they determine, namely, heat capacity

$$C_p = -T\left(\frac{\partial^2 \varphi}{\partial T^2}\right)_p = T\left(\frac{\partial S}{\partial T}\right)_p, \tag{8.13}$$

thermal expansion coefficient

$$\alpha = \frac{1}{v}\left(\frac{\partial^2 \varphi}{\partial T \partial p}\right)_{T,p} = \frac{1}{v}\left(\frac{\partial v}{\partial T}\right)_p, \tag{8.14}$$

and isothermal compressibility

$$\kappa_T = -\frac{1}{v}\left(\frac{\partial^2 \varphi}{\partial p^2}\right)_T = -\frac{1}{v}\left(\frac{\partial v}{\partial p}\right)_T. \tag{8.15}$$

Let us look at the derivation of the Ehrenfest equations describing phase transitions of the second kind in analytical form. If the derivatives do not experience abrupt changes,

$$\Delta\left(\frac{\partial \varphi}{\partial T}\right)_p = 0, \quad \Delta\left(\frac{\partial \varphi}{\partial p}\right)_T = 0. \tag{8.16}$$

If one considers an equilibrium variation of the state of a system, i.e. along the phase equilibrium curve, (8.16) will hold not only for abrupt changes themselves but also for their differentials. Therefore

$$\left.\begin{aligned} &\Delta\left(\frac{\partial^2 \varphi}{\partial p \partial T}\right)_{p,T} dp + \Delta\left(\frac{\partial^2 \varphi}{\partial T^2}\right)_p dT = 0 \\ &\Delta\left(\frac{\partial^2 \varphi}{\partial p^2}\right)_T dp + \Delta\left(\frac{\partial^2 \varphi}{\partial T \partial p}\right)_{T,p} dT = 0 \end{aligned}\right\} \tag{8.17}$$

or

$$\left.\begin{aligned} &\Delta\left(\frac{\partial v}{\partial T}\right)_p dT - \frac{\Delta C_p}{T} dT = 0 \\ &\Delta\left(\frac{\partial v}{\partial p}\right)_T dp + \Delta\left(\frac{\partial v}{\partial T}\right)_p dT = 0 \end{aligned}\right\}. \tag{8.18}$$

The Ehrenfest equations (8.18) have a nontrivial solution if the determinant composed of their coefficients vanishes, i.e.

$$\Delta\left(\frac{\partial v}{\partial p}\right)_T \frac{\Delta C_p}{T} + \left[\Delta\left(\frac{\partial v}{\partial T}\right)_p\right]^2 = 0, \tag{8.19}$$

whence follows an equation relating abrupt changes in physical quantities during phase transformations of the second kind:

$$\Delta C_p = -T\frac{[\Delta(\partial v/\partial T)_p]^2}{\Delta(\partial v/\partial p)_T}. \tag{8.20}$$

Phase transformations of the third kind, etc. can be given a similar consideration. Equation (8.20) is an analog of the Clapeyron–Klausius equation relating abrupt changes in physical quantities during phase transformations of first kind:

$$\Delta H_{\text{trans}} = T\Delta v\frac{\partial p}{\partial T}, \tag{8.21}$$

where the derivative dp/dT is taken along the phase equilibrium curve.

Other phase transformation classifications, which are more comprehensive than the classification proposed by Ehrenfest, are also available. In particular, Pippard [2] suggested that each of Ehrenfest's transitions of the second and third kinds should be broken down into four types depending on whether the quantities dC_p/dT and C_p are finite or infinite on one or both sides of the transition point.

Soviet researchers made a large contribution to the advancement of phase transition theory. This applies particularly to the phenomenological theory of phase transitions, which was developed up to the mid-1960s almost exclusively by the Soviet school of physical thought.

8.2 Phenomenological Theory of Phase Transformations

As was mentioned in Chap. 2, the formation of an ordered phase during structural order–disorder phase transformations supposes a decrease in the system symmetry. The theory of phase transitions of the second kind, which is based on symmetry considerations, was developed by Landau and Lifschitz [3–5].

Landau's phenomenological theory is built around two assumptions that may be outlined as follows. The state of an ordering system can be characterized, in addition to the usual physical parameters (temperature T, pressure p, composition c, etc.), by the long-range order parameter η. In the vicinity of the phase transformation point T_{trans} where the parameter η becomes zero, the thermodynamic potential that describes an ordering system can be expanded in a power series of η. The terms of the series are reasonably small and quite trivial. The above assumption allows finding equations to express the long-range order parameter as a function of temperature and pressure.

A significant feature of the Landau theory is that it relates phase transitions of the second kind to variations in the body symmetry, which take place upon ordering. When describing atomic ordering, crystal lattice symmetry is taken to mean symmetry of the probability density function $n(x, y, z)$ of an atom species occupying the site $\boldsymbol{r}$ with the coordinates (x, y, z). The function $n(x, y, z)$ will be hereinafter referred to as the distribution function. In the case of magnetic ordering, spin density is considered as a characteristic of the system.

The state of a substance can be characterized by generalized coordinates, i.e. by functions of the positions and velocity of atoms and electrons, ion magnetic moments, etc. If fluctuations of generalized coordinates are possible, the thermodynamic potential describing the system may be viewed as a functional of these deviations. This nonequilibrium thermodynamic potential is termed the Landau potential. If a system may include (under various conditions) several phases, the Landau potential describing this system must have several minima at equilibrium values of the generalized coordinates corresponding to each of the phases. Put another way, the Landau potential should be a nonlinear functional of the generalized coordinates. Since sets of nonlinear equations of state are difficult to solve, until recently the nonequilibrium potential method was used only for the description of phase transitions of the second kind or phase transitions of the first kind, which approached those of the second kind.

Let us take a closer look at the concepts of thē Landau phenomenological theory.

Expansion of the Thermodynamic Potential in Terms of Order Parameters. As shown in Chap. 2, the lattice symmetry is a group of coordinate transformations relative to which the distribution function $n(x, y, z)$ is invariant (i.e. the form of the functional dependence of n on x, y, z remains unchanged upon transformations comprising the group G). In total there are 230 types of symmetry of crystalline solid bodies, i.e. 230 space groups.

Let the function $n(\boldsymbol{r})$ be invariant under all symmetry operations of the group G above the transition point and the function $n_{\mathrm{D}}(\boldsymbol{r})$ be invariant under all operations of the group G_{D} below the transition point, while $G_{\mathrm{D}} \subset G$. (Here and henceforth, a high-symmetry disordered phase is assumed to exist at a higher temperature.) As is known from the theory of groups, any arbitrary function can be presented as a linear combination of some functions ψ_1, ψ_2, ... sharing the property of being transformable through each other at all transformations of that group. Such mutually transforming functions are called basis functions.

The most general functions, which mutually transform during rotations and full translations, are Bloch wave functions of the form $\psi_k(\boldsymbol{r}) = \varphi_k(\boldsymbol{r})\exp(i\boldsymbol{kr})$ (see Sect. 2.2). Transformation matrices of basis functions are irreducible representations of the group G, whereas the functions themselves are the basis of irreducible representations. Expansion of the function $n_{\mathrm{D}}(\boldsymbol{r})$ in terms of the full set of basis functions transformed by irreducible representations of the group G has the form

$$n_{\mathrm{D}}(\boldsymbol{r}) = \sum_{\nu}\sum_{i}\eta_i^{(\nu)}\psi_i^{(\nu)}, \tag{8.22}$$

where ν is the number of the irreducible representation, i is the basis function number, and $\eta^{(\nu)}$ is a coefficient having the meaning of the order parameter. The dimensionality (number of components) of the order parameter $\eta^{(\nu)}$ equals the dimensionality of the corresponding irreducible representation $D^{(\nu)}$. If the unit representation in the functions $n_D(\boldsymbol{r})$ and $n(\boldsymbol{r})$ is excluded from the summation, expansion in terms of basis functions will assume the form

$$\delta n(\boldsymbol{r}) = \sum_{\nu}{}' \sum_{i} \eta_i^{(\nu)} \psi_i^{(\nu)} , \tag{8.23}$$

where $n_D(\boldsymbol{r}) = n(\boldsymbol{r}) + \delta n(\boldsymbol{r})$.

The thermodynamic potential Φ of a crystal having a distribution function determined by (8.23) depends on the external thermodynamic parameters and the coefficients $\eta_i^{(\nu)}$. Consequently, it depends on the form of the basis functions $\psi_i^{(\nu)}$. Values of $\eta_i^{(\nu)}$, which are realized depending on pressure and temperature, are determined from the condition of the minimum Φ under thermodynamic equilibrium. The same condition determines the symmetry of the low-symmetry (disymmetrical) phase, because the symmetry group G_D of the disymmetrical phase will include only those irreducible representations whose basis functions have coefficients $\eta_i^{(\nu)}$ other than zero.

All coefficients $\eta_i^{(\nu)}$ are zero above the transition temperature. Since the state of the crystal changes continuously during phase transitions of the second kind, $\delta n(\boldsymbol{r})$ should increase from zero smoothly rather than abruptly when the temperature drops below T_{trans}. Clearly, the coefficients $\eta_i^{(\nu)}$ take arbitrarily small values near the transition point. Therefore expansion of the thermodynamic potential Φ in a power series of $\eta_i^{(\nu)}$ near the transition temperature can be limited to the first few terms. This expansion was shown [6, 7] to be true in the vicinity of the transition temperature but for a narrow interval near the transition point itself where fluctuations of order parameters should be considered. The expansion starts with a term quadratic in η and has the form

$$\Phi(T,p,\eta,c) = \Phi_0(T,p,c) + A(T,p,c)\eta^2 + B(T,p,c)\eta^3 + C(T,p,c)\eta^4 + \ldots \tag{8.24}$$

Equilibrium values of the parameter η are found from the condition of the minimum thermodynamic potential using the following relations:

$$\partial\Phi/\partial\eta = 0 , \tag{8.25}$$

$$\partial^2\Phi/\partial\eta^2 > 0 . \tag{8.26}$$

Clearly, (8.25) holds provided the expansion (8.24) does not contain terms linear in the parameter η. The fact that the expansion (8.24) is free of a term linear in the parameter η can also be understood from other considerations. Since the thermodynamic potential should be invariant with respect to all symmetry operations of group the G, each term of the expansion of Φ in a power series of η should contain only an invariant combination of the quantities $\eta_i^{(\nu)}$ of the corresponding power. However a linear invariant cannot be composed of quantities transformed by an irreducible representation of the group because this representation does not contain a unit representation. This is possible in reducible representations only.

For a crystal to have symmetry G at the equilibrium value $\eta = 0$ above and at the transition temperature T_0, the coefficient A in expansion (8.24) should be positive in accordance with (8.26). For η values to be nonzero below the transition temperature it is necessary that $A < 0$. Therefore $A = 0$ at the transition point. The coefficient A can be assumed to be a linear function of temperature, i.e.

$$A = a(T - T_0) . \tag{8.27}$$

For expansion (8.24), the equation of state (8.25) can be written as

$$2A\eta + 3B\eta^2 + 4C\eta^3 = 0 . \tag{8.28}$$

Now it is possible to find two true nonzero solutions at $A \le 9B^2/32C$:

$$\eta_{1,2} = \left(-3B \pm \sqrt{9B^2 - 32AC}\right)/8C . \tag{8.29}$$

Consider a physically reasonable positive solution for η. In accordance with (8.27), the coefficient A is a linear function of temperature. Therefore the relation $\eta(A)$ can be regarded as the temperature dependence of the order parameter (Fig. 8.1).

At temperatures from T_0 to T_1 corresponding to the interval of A values from $A_0 = 0$ to $A_1 = 9B^2/32C$, the order parameter η varies ambiguously. This points to an unstable state of the system in that temperature interval. The point $A_0 = 0$ corresponds to the boundary of an absolute instability of a high-symmetry disordered phase. The point A_1 determines the boundary of an absolute instability of an ordered phase. In this case $T_1 = T_0 + 9B^2/32aC$. The temperature T_0 is the undercooling temperature of the disordered phase and T_1 is the overheating temperature of the ordered phase. A true phase transition occurs at some temperature $T_0 < T_c < T_1$.

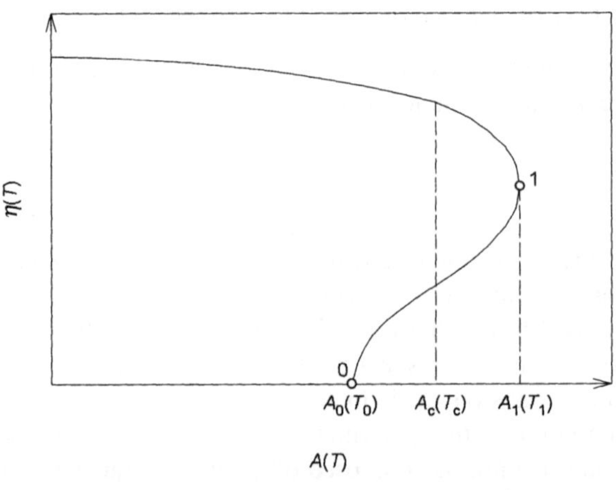

Fig. 8.1. Equilibrium values of the long-range order parameter η for phase transitions of the first kind. Positive solutions to equation (8.28)

When disordered and ordered phases are in equilibrium, their thermodynamic potentials are equal (see Sect. 8.1). Therefore from expansion (8.24) it follows that

$$A\eta^2 + B\eta^3 + C\eta^4 = 0\,. \tag{8.30}$$

Taking into account (8.27), simultaneous solution of (8.28) and (8.30) yields the transition temperature T_c and the abrupt change of the order parameter η_c at the phase transition point:

$$T_c = T_0 - B^2/4aC, \tag{8.31}$$

$$\eta_c = -B/2C. \tag{8.32}$$

The η_c value corresponds to the higher of the two possible order parameter values at $A_c(T_c)$. The state with the smallest η_c value has a higher energy and is metastable.

So, allowing for the cubic term in the expansion of the thermodynamic potential Φ, we arrive at the phase transition of the first kind. It should be noted that the classical Landau theory should be applied to phase transitions of the first kind with great care, because the philosophy of expanding Φ in a power series assumes that η is small near the transition point, whereas the order parameter η changes abruptly during phase transitions of the first kind.

Landau Condition for Phase Transitions of the Second Kind. The state of a crystal changes continuously during phase transitions of the second kind. Therefore the low-temperature phase symmetry can decrease only if some elements of the high-temperature phase symmetry are lost. Considering expansion (8.23), the thermodynamic potential Φ in the vicinity of the transition temperature can be presented as

$$\Phi(T,p,\eta,c) = \Phi_0(T,p,c) + \sum_{v} A^{(v)}(T,p,c) \sum_{i} [\eta_i^{(v)}]^2 + \ldots \tag{8.33}$$

As was shown, the absence of the linear term in expansion (8.33) is due to the condition of minimum Φ (8.25). Moreover, group theory proves that only one quadratic (second-order) invariant $\sum_i [\eta_i^{(v)}]^2$ exists for each irreducible representation of the group G. High-power expansion terms are invariant polynomials of corresponding orders.

Expansion (8.33) is analogous to (8.24) and the above thermodynamic reasoning is applicable to it. As the temperature decreases, one of the coefficients $A^{(v')}$ becomes zero at the transition point. Generally, several coefficients $A^{(v)}$ can go to zero successively one after another, but the transition temperature is the point at which the first coefficient $A^{(v)}$ becomes zero. The fact that only one coefficient of expansion (8.33) vanishes at the transition temperature corresponds to Landau's postulate that the phase transition of the second kind can occur only by one irreducible representation. Since cubic terms present in the thermodynamic potential necessarily entail a phase transition of the first kind, Landau's condition for phase transitions of the second kind entail that the irreducible representation responsible for the phase transition prevents the formation of a third-order invariant. From the theory of representations it is known that third-order quantities comprising the coefficients $\eta_i^{(v)}$ and transformed by some irreducible representation $D^{(v)}$ are themselves transformable as a symmetric cube of that irreducible representation. Hence, the absence of a unit representation $D^{(1)}$ of the group G_D in the irreducible representation is a condition for the absence of third-order invariants in the thermodynamic potential expansion.

Fourth-order terms are always present in the thermodynamic potential expansion, because the direct fourth-order product $[D^{(v)4}] = D^{(v)} \times D^{(v)} \times D^{(v)} \times D^{(v)}$ necessarily contains a unit representation for any irreducible representation $D^{(v)}$. Generally such a product has several different invariants. Considering fourth-order terms, the conditions of the minimum potential Φ (8.24) with respect to η variation have the form

$$A\eta + C\eta^3 = 0, \tag{8.34}$$

$$A + 12C\eta^2 > 0. \tag{8.35}$$

Equation (8.34) has solutions $\eta_1 = 0$ corresponding to a high-temperature disordered state and $\eta_2^2 = -A/4C$ corresponding to a low-temperature ordered state. Since $A < 0$ below the transition temperature, $C > 0$ for phase transitions of the second kind.

Criterion of Superstructure Stability in the Homogeneous State. A necessary condition for stability of superstructures in a homogeneous state was postulated by E. M. Lifschitz [4]. This condition applies to superstructures formed as phase transition of the second kind. The Lifschitz criterion is directly linked to classification properties of irreducible representations. Irreducible representations are defined not only by their number v but also by the wave vector $\boldsymbol{k}$. Therefore the coefficients $A^{(v)}$ in expansion (8.33) are functions not only of the external thermodynamic parameters but also of $\boldsymbol{k}$. If a phase transition is associated with some coefficient $A^{(v)}(\boldsymbol{k})$ vanishing at $\boldsymbol{k} = \boldsymbol{k}_0$, this transition can occur when $A^{(v)}$ as a function of $\boldsymbol{k}$ has a minimum at $\boldsymbol{k} = \boldsymbol{k}_0$, i.e. $(\partial A^{(v)}(\boldsymbol{k})/\partial \boldsymbol{k})_{\boldsymbol{k}=\boldsymbol{k}_0} = 0$. This means that expansion of $A^{(v)}(\boldsymbol{k})$ in terms of $(\boldsymbol{k} - \boldsymbol{k}_0)$ in the vicinity of $\boldsymbol{k}_0$ should not contain linear terms. The $\boldsymbol{k}_0$ value determines the translational symmetry of the function $\delta n(\boldsymbol{r})$, i.e. the lattice periodicity of the low-symmetry phase.

The structure with $\boldsymbol{k} = \boldsymbol{k}_0 + \Delta$ (Δ is a small quantity) differs from the structure with $\boldsymbol{k} = \boldsymbol{k}_0$ by the appearance of an inhomogeneity at distances larger than the lattice constants. Lifschitz [4] suggested a macroscopic description of such inhomogeneities where order parameters are viewed as slowly varying coordinate functions. When the quantities η_i vary in space, the thermodynamic potential of a crystal depends not only on η_i but also on their coordinate derivatives. The stability condition reduces [4, 5] to the absence of linear invariants comprising quantities of the form $[\eta_k(\partial\eta_i/\partial x) - \eta_i\partial\eta_k/\partial x)]$ in the expansion of the thermodynamic potential Φ. Such linear invariants in turn cannot be formed if and only if the group of the vector $\boldsymbol{k}$ contains symmetry elements intersecting at a single point (i.e. contains mutually intersecting axes, intersecting axes and planes, or an inversion). The last requirement imposes major limitations on possible changes of symmetry during phase transitions of the second kind, because only a few vectors $\boldsymbol{k}$ occupying a special position in the reciprocal lattice can have such intrinsic symmetry. Points in the first Brillouin zone, which correspond to these vectors, are usually called Lifschitz points.

Further Advancement of the Phenomenological Theory. In recent decades, the theory of phase transformations has been largely advanced thanks to Dzyaloshinskii's work dealing with theories of weak ferromagnetism [8] and incommensurate phases [9, 10]. In particular, Dzyaloshinskii analyzed modulated phases in crystals and showed that these phases can appear from the solution of a nonlinear equation resulting from minimization of the thermodynamic potential, which includes order parameter gradients (Lifschitz invariants).

Indenbom [11, 12] found a complete solution to the crystallographic problem of specifying all phases between which transitions of the second kind are possible with no impairment of the translational symmetry.

Birman [13] proposed a new scheme for the determination of low-symmetry ordered phases with a given wave vector $\boldsymbol{k}$, which are formed from a high-symmetry disordered phase. The method [13] does not require minimization of the thermodynamic potential and therefore its results are independent of the degree of the thermodynamic potential approximation. The method is based on the selection of admissible subgroups using group theoretical criteria and admissible irreducible representations by which a low-symmetry phase can be formed.

Gufan [14] discussed a complete rational basis of invariants (CRBI), which allows the construction of a polynomial representation of the thermodynamic potential with any degree of approximation. The CRBI is some finite set of independent polynomial invariants. Therefore the problem of complete expansion of the thermodynamic potential reduces to the construction of a CRBI for the symmetry group of a high-symmetry phase.

The aforementioned lines of advancement of the Landau phenomenological theory disregard fluctuations. Therefore deductions of those theories hold only outside a narrow interval near the phase transition point as specified by Levanyuk [6] and Ginzburg [7].

A systematic and expanded presentation of the state-of-the-art phenomenological theory of phase transitions and its lines of advancement is given in the monograph [15].

8.3 Mean Field Approximation

The phenomenological approach developed in the thermodynamic theory of phase transitions of the second kind permits the establishment of general regularities of ordering with no resort to specific model representations. However, this approach is applicable only within a narrow temperature interval near the point of the phase transition of the second kind, when the order parameter changes little. Moreover, until recently the phenomenological theory failed to describe order–disorder phase transitions of the first kind, when the order parameter changes abruptly and immediately assumes a large value on transition to the ordered state. For this reason, various simplified models based on statistical-thermodynamic calculation methods are used to describe ordering in a broad interval of temperatures and concentrations.

The determination of the thermodynamics of a system of numerous interacting particles presents a major problem of the statistical theory of cooperative phenomena in solids and, specifically, atomic ordering theory. In terms of the most popular model, atoms are located at the sites of some rigid crystal lattice. The statistical ordering theory takes into account atomic structure of the crystal, considers the possible atomic distribution over sites of the ordering lattice, and computes (proceeding from some assumed interatomic interactions) the distribution energy. In this case the configuration energy of a crystal is presented as the sum of all paired potentials (energies) of interatomic interactions.

In statistical mechanics, each atom in a system interacts both with the external field and neighboring atoms. The mean field model replaces interatomic interactions by the interaction averaged over all atoms. This model is used most frequently to describe structural phase transitions in a vast variety of systems. Usually interactions between nearest atoms are considered only, but in some instances interactions between more distant atoms are taken into account too.

The subsequent discussion of statistical ordering theory will focus on binary substitutional or interstitial solid solutions. If atoms of two species (A and B) in a substitutional solution or interstitial atoms and structural vacancies in a nonstoichiometric interstitial compound or an interstitial solution can rearrange themselves only over sites of some rigid lattice (Ising lattice), then, in this case, binary substitutional or interstitial solutions and nonstoichiometric interstitial compounds can be described within a single scheme. Indeed, an interstitial solution can be viewed as a binary substitutional solution, which is formed by interstitial atoms and structural vacancies in an ordering sublattice.

Consider an Ising lattice containing N sites each of which can be occupied by an atom of species A or B. Let $c(\boldsymbol{r})$ be the occupancy number equal to 1, if the site $\boldsymbol{r}$ hosts an A species atom, and 0, if the site $\boldsymbol{r}$ is occupied by an atom of the other species. In strongly nonstoichiometric interstitial compounds and interstitial solutions $c(\boldsymbol{r})$ is equal to unity for Ising lattice sites $\boldsymbol{r}$ occupied by interstitial atoms and 0 if sites $\boldsymbol{r}$ are vacant. Let us take the model accounting for interactions between nearest neighbors. If each atom has the first coordination sphere containing z sites, it is exposed to the field

$$H = H_0 + \tfrac{1}{2}\sum_j V_{ij} c(\boldsymbol{r}_i), \tag{8.36}$$

where H_0 is the external field (a constant independent of mutual atomic arrangement) and V_{ij} is the paired interaction potential. The sum in (8.36) is taken with respect to z neighboring sites j. In the mean field approximation, expression (8.36) is replaced by

$$H = H_0 + \frac{1}{2(N-1)} \sum_{j \neq i} V_{ij} c(\boldsymbol{r}_j), \tag{8.37}$$

where the summation and averaging are taken with respect to all $(N-1)$ sites $\boldsymbol{r}_j$ that do not coincide with $\boldsymbol{r}_i$.

The Hamiltonian $\mathcal{H}$ corresponding to (8.37) has the form

$$\mathcal{H} = -\frac{z}{2(N-1)} \sum_{i,j} V_{ij} c(\boldsymbol{r}_i) c(\boldsymbol{r}_j) - H_0 \sum_{i=1}^{N} c(\boldsymbol{r}_i) \tag{8.38}$$

and is called the mean field Hamiltonian. The first sum in (8.38) is taken with respect to all $N(N-1)/2$ possible pairs of sites $\boldsymbol{r}_i$ and $\boldsymbol{r}_j$.

Ising [16] proposed a model where paired interaction potentials are nonzero for nearest neighbors only, while sites form a simple Bravais lattice. At present an exact solution to the Ising problem has been obtained for one-dimensional lattices, for two-dimensional lattices with equiatomic composition, and for few special three-dimensional lattices. Computing the exact statistical sum for most interesting cases of bcc and fcc lattices is currently infeasible. Therefore a variety of approximations are normally used to write the crystal free energy in one form or another.

To determine equilibrium properties of an ordering crystal at a given temperature and preset component concentrations, it is necessary to find the statistical sum

$$Z = \sum_{\kappa} \exp(-E_{\kappa} / k_{\mathrm{B}} T), \tag{8.39}$$

where E_{κ} is the crystal energy in a state κ. In this case the free energy is

$$F = -k_{\mathrm{B}} T \ln Z. \tag{8.40}$$

If κ stands for states with different mutual arrangements of atoms, formula (8.40) expresses the configuration part of the free energy. The configuration energy E_{κ} can be presented as the energies of interaction between pairs of atoms located at different distances ρ_{ℓ} (ρ_{ℓ} is the radius of the l-th coordination sphere). For a binary solution A-B, the configuration energy is

$$E_{\kappa} = -\sum_{\ell} [N_{\mathrm{AA}}^{(\ell)} \varepsilon_{\mathrm{AA}}(\rho_{\ell}) + N_{\mathrm{BB}}^{(\ell)} \varepsilon_{\mathrm{BB}}(\rho_{\ell}) + N_{\mathrm{AB}}^{(\ell)} \varepsilon_{\mathrm{AB}}(\rho_{\ell})], \tag{8.41}$$

where $\varepsilon(\rho_{\ell})$ denotes the energies of corresponding paired interatomic interactions at distances ρ_{ℓ} and $N^{(\ell)}$ is the number of corresponding pairs of atoms located at distances ρ_{ℓ}. In the approximation allowing for the interaction of nearest neighbors only, E_{κ} has the form

$$E_{\kappa} = -(N_{\mathrm{AA}} \varepsilon_{\mathrm{AA}} + N_{\mathrm{BB}} \varepsilon_{\mathrm{BB}} + N_{\mathrm{AB}} \varepsilon_{\mathrm{AB}}). \tag{8.42}$$

All quantities in (8.42) are taken for atoms that are nearest neighbors only. If a crystal comprises N_A atoms of species A and N_B atoms of species B, while the number of pairs of unlike atoms in the first coordination sphere is N_{AB}, then the number of pairs of like atoms will be

$$N_{AA} = (zN_A - N_{AB})/2\,, \tag{8.43}$$

$$N_{BB} = (zN_B - N_{AB})/2\,, \tag{8.44}$$

where z is the coordination number for the first coordination sphere. Considering (8.43) and (8.44), expression (8.42) can be written as

$$E_\kappa = -\tfrac{1}{2}[VN_{AB} + z(N_{AA}\varepsilon_{AA} + N_{BB}\varepsilon_{BB})]\,, \tag{8.45}$$

where $V = 2\varepsilon_{AB} - \varepsilon_{AA} - \varepsilon_{BB}$ is the mixing energy.

Let us now take a closer look at one of the simplest and most popular theories of ordering, namely the Gorsky–Bragg–Williams theory [17–19].

The Gorsky-Bragg-Williams Theory. Efforts to describe critical phenomena in terms of the mean field representation were made independently by Van der Vaals in liquid-state physics and by Weiss in ferromagnetism theory. Atomic ordering takes place at thermodynamic equilibrium. Therefore a correct mean field value can be obtained by minimizing the system's free energy with respect to the order parameter that characterizes the mean field. The order parameter not only characterizes changes in the degree of order during a transition, but also describes most crucial properties of the final ordered state. The thermodynamic approach, which involves minimization of the free energy with respect to order parameters, differs from the first methods used by Van der Vaals and Weiss: it is more general and efficient.

It is this thermodynamic approach that is employed in the Gorsky–Bragg–Williams theory [17–19]. The approximation underlying this ordering theory is that probabilities of species A or B atoms occupying a given lattice site are independent of the atomic configuration at the surrounding sites. They are determined only by the concentration of components and the degree of long-range order. In other words, the long-range order parameter alone is sufficient for an adequate description of the system in that theory at a given composition of solid solution, while short-range correlations of atoms are disregarded. It should be noted that near the transition point the Gorsky–Bragg–Williams approximation yields much the same results as the Landau phenomenological theory of phase transitions. The Landau theory can be viewed as an extension of the mean field theory since it accounts for the order parameter gradient in the expansion of the free energy into a power series of that order.

Let us consider the application of the Gorsky–Bragg–Williams theory to the description of atomic ordering in a binary solid solution. A comprehensive discussion of this issue can be found in the book [20].

Assume that during ordering the crystal lattice of a binary solution A-B breaks down into two nonequivalent sublattices with $N^{(1)}$ and $N^{(2)}$ sites each. The total number of sites in the ordering lattice is $N^{(1)} + N^{(2)} = N$. Since each site can be occupied by one atom only, $N_A + N_B = N$. In this case the relative concentration of sites in the first sublattice is $\nu = N^{(1)}/N$. The probabilities that sites of each sublattice are filled with atoms A and B depend on the ratio between the number of sites in a given sublattice occupied by atoms of a given species and the total number of sites in this sublattice, i.e.

$$\left.\begin{aligned} P_A^{(1)} &= N_A^{(1)} / N^{(1)} \\ P_A^{(2)} &= N_A^{(2)} / N^{(2)} \\ P_B^{(1)} &= N_B^{(1)} / N^{(1)} \\ P_B^{(2)} &= N_B^{(2)} / N^{(2)} \end{aligned}\right\}. \tag{8.46}$$

The usual normalization conditions

$$P_A^{(1)} + P_B^{(1)} = 1, \quad P_A^{(2)} + P_B^{(2)} = 1, \tag{8.47}$$

and the composition normalization condition

$$\nu P_A^{(1)} + (1-\nu) P_A^{(2)} = c \tag{8.48}$$

are met for the probabilities P. Here $c = N_A/N$ denotes the concentration of species A atoms. The concentration of species B atoms is $c_B = 1 - c$.

If the degree of long-range order η is proportional to the deviation of the probability from its value c in a disordered solution, then

$$\eta = (P_A^{(1)} - c)/(1-\nu) \quad \text{if} \quad c \geq \nu, \tag{8.49}$$

$$\eta = (c - P_A^{(2)})/\nu \quad \text{if} \quad c < \nu. \tag{8.50}$$

Consider the ordering of a binary solution A-B in two nonequivalent sublattices having the same number of sites $N/2$ each. Any site of one sublattice is surrounded by sites of the other sublattice only. For example, β-brass CuZn has this structure in the ordered state. The total number of pairs of unlike atoms in crystals with the said structure can be found from probabilities as follows:

$$N_{AB} = \frac{zN}{2}(P_A^{(1)} P_B^{(2)} + P_A^{(2)} P_B^{(1)}). \tag{8.51}$$

Considering (8.47), (8.49) and (8.50), the probabilities will look like

$$\left.\begin{aligned} P_A^{(1)} &= c + (1-\nu)\eta \\ P_A^{(2)} &= c - \nu\eta \\ P_B^{(1)} &= 1 - c - (1-\nu)\eta \\ P_B^{(2)} &= 1 - c + \nu\eta \end{aligned}\right\}. \tag{8.52}$$

Here $\nu = 1/2$. Considering this and expressions (8.42), (8.51) and (8.52), the configuration energy E_κ in the nearest-neighbor approximation can be presented as

$$E_\kappa = -(zN/2)\{c\varepsilon_{AA} + (1-c)\varepsilon_{BB} + V[c(1-c) + (\eta^2/4)]\}. \tag{8.53}$$

In the Gorsky–Bragg–Williams theory, all energies E_κ corresponding to different atomic configurations at lattice sites are assumed equal at a given degree of long-range order. The total number of configurations is equal to the number of rearrangements of atoms A and B over the sites of nonequivalent sublattices, i.e.

$$W = \frac{(N/2)!}{N_A^{(1)}!\,N_B^{(1)}}\,\frac{(N/2)!}{N_A^{(2)}!\,N_B^{(2)}}, \tag{8.54}$$

and the statistical sum, which is replaced by its largest term, is

$$Z = W\exp(-E_\kappa/k_B T). \tag{8.55}$$

Therefore, in conformity with (8.40) and the Stirling approximation $\ln N! = N\ln N - N$, some manipulations give an expression for the configuration part of the free energy:

$$\begin{aligned} F = &-(zN/2)\{c\varepsilon_{AA} + (1-c)\varepsilon_{BB} + V[c(1-c) + (\eta^2/4)]\} \\ &+ (Nk_BT/2)[(c+\eta/2)\ln(c+\eta/2) + (c-\eta/2)\ln(c-\eta/2) \\ &+ (1-c+\eta/2)\ln(1-c+\eta/2) + (1-c-\eta/2)\ln(1-c-\eta/2)]. \end{aligned} \tag{8.56}$$

The value of the long-range order parameter, which is equilibrium for a given temperature and composition, can be found from the condition of the free energy minimum $\partial F/\partial\eta = 0$. Differentiating expression (8.56) with respect to η, we obtain the transcendental equation

$$\ln\left[\frac{(c+\eta/2)(1-c+\eta/2)}{(c-\eta/2)(1-c-\eta/2)}\right] = \frac{zV}{k_B T}\eta\ . \tag{8.57}$$

In the case of phase transitions of the second kind near the transition point, when η is small, only the first summand may be taken into account in the expansion of the left-hand part of (8.57) into a power series of η. In this case the transition temperature is

$$T_c = zc(1-c)V/k_B\ . \tag{8.58}$$

It will be noted that at $T = T_c$ and $\eta = 0$ both the first derivative $\partial F/\partial\eta$ and the second derivative $\partial^2 F/\partial\eta^2$ vanish, as they should at the point of a phase transition of the second kind. It follows from (8.58) that the dependence of the ordering temperature on the crystal composition assumes the form of a parabola whose maximum corresponds to an equiatomic composition.

Expressing the energy V in terms of the ordering temperature T_c by means of relation (8.58) and substituting this energy into (8.57), we obtain an expression for the dependence of the long-range order parameter on the crystal composition and temperature:

$$\ln\left[\frac{(c+\eta/2)(1-c+\eta/2)}{(c-\eta/2)(1-c-\eta/2)}\right] = \frac{T_c}{T}\frac{\eta}{c(1-c)}\ . \tag{8.59}$$

In accordance with (8.59), the long-range order parameter η changes continuously with temperature from zero at $T = T_c$ to $\eta_{max} = 2c$ (at $c \le 1/2$) or $\eta_{max} = 2(1-c)$

(at $c > 1/2$) when T/T_c is nearly zero. Thus, when the model under discussion is used to describe ordering in solid solutions having the structure of β-brass, it leads to a phase transition of the second kind. This statement is also supported by the fact that the configuration part of the entropy $S = -\partial F/\partial T$ changes continuously with temperature. Hence, no thermal effect is observed on such ordering. Heat capacity increases abruptly on transition to the ordered state by the value

$$\Delta C = \tfrac{3}{2} N k_B \frac{c(1-c)}{c^3 + (1-c)^3} . \tag{8.60}$$

Similar considerations can be applied to ordering in solid solutions with different number of sites in nonequivalent sublattices: $N^{(1)} \neq N^{(2)}$.

It is worth noting however that direct application of the Gorsky–Bragg–Williams theory for the description of the ordering in solid solutions with the CuAu structure leads to a qualitatively erroneous conclusion that the order–disorder transition represents a phase transformation of the second kind. But atomic ordering in solid solutions with this structure proceeds as a phase transformation of the first kind. In this case, the approximation, which disregards correlation, proves to be too rough.

Static Concentration Wave Method. A version of the mean field approximation developed by Khachaturian and known as the static concentration wave method [21] is fairly effective in describing order–disorder structural phase transitions in substitutional and interstitial solid solutions.

In terms of the static concentration wave method [21], the atomic distribution in substitutional or interstitial solutions is written using the distribution function $n(\boldsymbol{r})$, which is the probability that an atom of a given species occupies a site $\boldsymbol{r}$ of the Ising lattice. In a disordered crystal, the probabilities $n(\boldsymbol{r})$ are equal for all sites $\boldsymbol{r}$ of the Ising lattice and coincide with the relative fraction of sites occupied by atoms of that species, i.e. $n(\boldsymbol{r}) = c$. The distribution function can assume several values over the whole set of lattice sites during ordering and, consequently, describes several sublattices, which are formed in a disordered crystal lattice upon ordering. The basic concepts of the distribution function and the static concentration wave method were given in Sect. 5.1.

The concept of the distribution function as a superposition of static concentration waves may be introduced if one considers radiation scattering by an ordered crystal.

If thermal fluctuations of the composition and thermal oscillations of atoms are neglected, a disordered solid solution can be viewed as an ideal crystal whose unit cells have equal scattering ability. In this approximation, scattering by a disordered crystal represents a set of structural reflections whose mutual arrangement is described by structural sites of the reciprocal lattice corresponding to the direct lattice of the disordered crystal.

Variation of the translational symmetry of a crystal during ordering leads to formation of a reciprocal lattice inscribed into the reciprocal lattice of the disordered crystal. All lattice sites of this reciprocal lattice coincide with some sites of the reciprocal lattice of the ordered crystal. The other sites in the reciprocal lattice of the ordered crystal, which do not coincide with structural sites, and their corresponding vectors are termed superstructure vectors.

The total intensity of radiation scattering by a crystal can be presented as

$$I(\boldsymbol{q}) = I_0(\boldsymbol{q}) + I_1(\boldsymbol{q})\,, \tag{8.61}$$

where $\boldsymbol{q}$ is the diffraction vector. The first summand in (8.61) is nonzero at structural sites of the reciprocal lattice only and describes the intensity of structural reflections. The second summand is zero at structural sites of the reciprocal lattice and nonzero in the rest of reciprocal space. In accordance with [21], the summand $I_1(\boldsymbol{q})$, which describes the intensity distribution over the whole reciprocal space, can be written as the quantity $n(\boldsymbol{r})$, which stands for the probability that an atom of a given species occupies a site $\boldsymbol{r}$:

$$I_1(\boldsymbol{q}) = |F(\boldsymbol{q})|^2 \left|\sum_{\boldsymbol{r}} [n(\boldsymbol{r}) - c] \exp(-i\boldsymbol{kr})\right|^2, \tag{8.62}$$

where $F(\boldsymbol{q})$ is the effective atomic factor and c is the relative fraction of atoms of a given species in the solid solution. In a disordered solid solution, $n(\boldsymbol{r}) = c$ and the intensity $I_1(\boldsymbol{q})$ is zero in accordance with (8.62). In an ordered solution, where $n(\boldsymbol{r})$ depends on the coordinates of sites $\boldsymbol{r}$, the intensity $I_1(\boldsymbol{q})$ is nonzero at superstructure sites of the reciprocal lattice. This is possible if the function $n(\boldsymbol{r})$ represents a superposition of standing static plane concentration waves $\Delta_{\boldsymbol{k}}(\boldsymbol{r})$ whose wave vectors are superstructure vectors of the reciprocal lattice, which belong to the first Brillouin zone of a disordered crystal:

$$n(\boldsymbol{r}) = c + \sum_{\boldsymbol{k}} \Delta_{\boldsymbol{k}}(\boldsymbol{r})\,, \tag{8.63}$$

where

$$\Delta_{\boldsymbol{k}}(\boldsymbol{r}) = \tfrac{1}{2}\, \eta(\boldsymbol{k})[\gamma(\boldsymbol{k})\exp(i\boldsymbol{kr}) + \gamma^*(\boldsymbol{k})\exp(-i\boldsymbol{kr})]\,. \tag{8.64}$$

In (8.64), $\exp(i\boldsymbol{kr})$ is a plane static wave with amplitude $\eta(\boldsymbol{k})\gamma(\boldsymbol{k})$, where $\eta(\boldsymbol{k})$ is the long-range order parameter and $\gamma(\boldsymbol{k})$ is a coefficient chosen such that the long-range order parameter is unity in a fully ordered state. The summation in (8.63) is taken over all superstructure wave vectors in the first Brillouin zone of the disordered crystal.

Thus, (8.63) and (8.64) show that the homogeneous atomic distribution over lattice sites experiences a spatial-periodic modulation upon order–disorder phase transformation. In other words, the probability $n(\boldsymbol{r})$ is other than in the case of the statistic distribution. Let it be noted that unlike (5.19), expression (8.63) is written for a single star only, i.e. for the simplest case.

In accordance with (8.64), an order–disorder phase transformation results from the loss of stability by a homogeneous disordered solid solution with respect to the formation of static concentration waves ($\Delta_{\boldsymbol{k}}(\boldsymbol{r}) = 0$ for a disordered solution). It follows from (8.64) that the amplitude $\eta(\boldsymbol{k})\gamma(\boldsymbol{k})$ determines the atomic distribution inhomogeneity over lattice sites. If static concentration waves with finite amplitudes appear at the phase transition point, the phase transition is of the first kind. The appearance of concentration waves with infinitely small amplitudes at the transition point suggests a phase transition of the second kind. Thus, the difference between the internal parameters of the disordered and ordered phases at the point of the phase transition of the second kind is infinitely small, and the crystal symmetry changes thanks to an infinitely small variation of the crystal structure. It also follows from relation (8.64) that all amplitudes are zero in a homogeneous disordered solid solution.

The change in the free energy ΔF, which is connected with the appearance of atomic distribution inhomogeneity over the crystal lattice and which is referred to as the free energy of the homogeneous disordered solid solution, can be presented as amplitudes $\eta(\boldsymbol{k})\gamma(\boldsymbol{k})$ in the form

$$\Delta F = \tfrac{1}{2}\sum_{\boldsymbol{k}} b(\boldsymbol{k})\,|\,\eta(\boldsymbol{k})\gamma(\boldsymbol{k})\,|^2\,, \tag{8.65}$$

where $b(\boldsymbol{k})$ is a function that depends on external thermodynamic parameters (temperature, pressure, composition). It represents the coefficient in expansion of the free energy into long-range order parameters. The summation in (8.65) is taken over all vectors $\boldsymbol{k}$ of the first Brillouin zone of a disordered crystal. The function $b(\boldsymbol{k})$ has the disordered-phase symmetry and assumes the smallest $\boldsymbol{k}$ values not for a single vector $\boldsymbol{k}_0$ but for several vectors $\boldsymbol{k}_{01}, \boldsymbol{k}_{02}, ..., \boldsymbol{k}_{0j}$ simultaneously. Each of these vectors is derived from the first vector $\boldsymbol{k}_{01}$, for which purpose all disordered-phase symmetry transformations are applied to this vector. This set of nonequivalent vectors is called the star $\{\boldsymbol{k}_s\}$ of the wave vector $\boldsymbol{k}_s$ (see Sect. 2.2).

The free energy F of a stable homogeneous solid solution should be a minimum and $\Delta F = 0$ in a homogeneous state when all amplitudes are zero. In this case any distribution inhomogeneity, i.e. any set of nonzero amplitudes, will cause an increase in the free energy, and therefore $\Delta F > 0$. Hence, the stability condition for a homogeneous solid solution can be written as the inequality

$$\Delta F = \tfrac{1}{2}\sum_{\boldsymbol{k}} b(\boldsymbol{k})\,|\,\eta(\boldsymbol{k})\gamma(\boldsymbol{k})\,|^2 \geq 0\,. \tag{8.66}$$

Relation (8.66) holds good if all coefficients $b(\boldsymbol{k}) > 0$. Let the coefficients satisfy $b(\boldsymbol{k}_0) \leq 0$ at least for one wave vector $\boldsymbol{k} = \boldsymbol{k}_0$. In this case we have an inhomogeneous atomic distribution when the amplitude $\eta(\boldsymbol{k}_0)\gamma(\mathbf{k}_0) \neq 0$ and all other amplitudes $\eta(\boldsymbol{k})\gamma(\boldsymbol{k}) \equiv 0$ at $\boldsymbol{k} \neq \boldsymbol{k}_0$. The quantity $\Delta F \leq 0$, i.e. any increase in the amplitude $\eta(\boldsymbol{k}_0)\gamma(\boldsymbol{k}_0)$ reduces the free energy, because $|\eta(\boldsymbol{k}_0)\gamma(\boldsymbol{k}_0)|^2 > 0$ at $b(\boldsymbol{k}_0) \neq 0$, and enhances the system stability. Thus, in the case under consideration, a disordered state turns out to be unstable relative to infinitely small concentration inhomogeneities. The stability loss condition actually reduces to the presence of a minimum of the function $b(\boldsymbol{k})$, i.e.

$$\left(\frac{\partial b(\boldsymbol{k})}{\partial \boldsymbol{k}}\right)_{\boldsymbol{k}=\boldsymbol{k}_0} = 0\,. \tag{8.67}$$

Since $b(\boldsymbol{k})$ is a function of the external thermodynamic parameters, a stable solution can be rendered absolutely unstable if some of those parameters (for example, temperature) are changed. For a phase transition of the second kind, the temperature of absolute stability loss $T_0 = T_c$ is the phase transition temperature. Each phase can exist only on one side of the transition point. In the case of a phase transition of the first kind, the equilibrium transition temperature $T = T_c$ (upon cooling) proves to be higher than the absolute stability loss temperature T_0. Consequently, two phases coexist in the temperature interval $T_0 < T < T_c$: an undercooled disordered phase and an ordered phase. The last phase can appear in this temperature interval only through fluctuation-induced formation of its nucleation centers having a critical size. Similar considerations can be given to a phase transition of the first kind, which takes place during heating when $T_c < T_0'$.

Thus, a significant indication of a phase transition of the first kind is the existence of a two-phase region in the T–c equilibrium diagram. The interval of the two-phase region is limited by points T_0 and T_0' of the absolute stability loss. A two-phase region cannot exist in the case of a phase transition of the second kind under equilibrium conditions, because the transition and stability loss temperatures coincide: $T_c = T_0$. However, in the last case a two-phase region can be formed due to purely kinetic factors, namely a slow rate of diffusion processes responsible for the transformation. Hence it follows that a two-phase region, which may be observed upon a phase transition of the second kind, is always nonequilibrium.

According to Landau's phenomenological phase transition theory dealt with in Sect. 8.2, a necessary condition for a phase transition of the second kind is an identical equality to zero of the coefficient of the cubic term in the free energy expansion in terms of the long-range order parameter. If the phase transition of the second kind is associated with a star of nonzero wave vectors, i.e. is accompanied by changes in translational symmetry, the Landau group theoretical criterion is equivalent to the following necessary condition for a phase transition of the second kind [21]: A phase transition of the second kind is possible only if three (not necessarily different) vectors, whose sum is zero or equal to a vector of the reciprocal lattice of the disordered phase, cannot be picked out of the star vectors associated with the phase transition. This criterion is applicable if the space group of one phase is a symmetry subgroup of the other phase.

Generally, a phase transition, which is or is not accompanied by changes in translational symmetry, can be realized as a transition of the second kind if the star associated with the phase transformation contains no three vectors whose sum is zero or equal to a structural vector of the reciprocal lattice of a disordered crystal. If there is one or more sets of three vectors whose sum is zero then, in order to determine the kind of transition, it will be necessary to compute the nature χ of the irreducible representation associated with the transition and find the sum

$$\sigma = \frac{1}{N} \sum_{h \in G_0} \chi(g), \tag{8.68}$$

where $g = \{h|t_h + t\} \in G$ is a symmetry element of the space group G; and $h \in G_0$ denotes rotational symmetry elements of the point group G_0 ($G_0 \subset G$) having order N. If this sum is zero, a phase transition of the second kind is possible. If, however, the sum (8.68) is positive, a phase transition of the second or lower kind is impossible and, consequently, a phase transition of the first kind occurs [22].

The free energy of a binary substitutional solid solution, which is expressed [22] in terms of the distribution function $n(\boldsymbol{r})$, has the form

$$F = \tfrac{1}{2}\sum_{i,j} \tilde{V}(\boldsymbol{r}_i - \boldsymbol{r}_j) n(\boldsymbol{r}_i) n(\boldsymbol{r}_j)$$
$$+ k_{\mathrm{B}} T \sum_{\boldsymbol{r}} \{n(\boldsymbol{r}) \ln n(\boldsymbol{r}) + [1 - n(\boldsymbol{r})] \ln[1 - n(\boldsymbol{r})]\} - \mu \sum_{\boldsymbol{r}} n(\boldsymbol{r}), \tag{8.69}$$

where $\tilde{V}(\boldsymbol{r}_i - \boldsymbol{r}_j)$ is the atomic interaction potential (mixing energy) of two atoms located at lattice sites $\boldsymbol{r}_i$ and $\boldsymbol{r}_j$; μ is the Lagrange indeterminate factor acting as the chemical potential. The first summand in (8.69) is the internal energy, and the second summand

$$S = -k_{\mathrm{B}} \sum_{\boldsymbol{r}} \{n(\boldsymbol{r}) \ln n(\boldsymbol{r}) + [1 - n(\boldsymbol{r})] \ln[1 - n(\boldsymbol{r})]\} \tag{8.70}$$

denotes the entropy. The summation in (8.70) is taken with respect to all lattice sites. In the mean field approximation, the probabilities $n(\boldsymbol{r})$ are given by the equation

$$n(\boldsymbol{r}) = \left\{ \exp\left[-\frac{\mu}{k_{\mathrm{B}} T} + \sum_{\boldsymbol{r}} \frac{\tilde{V}(\boldsymbol{r}_i - \boldsymbol{r}_j)}{k_{\mathrm{B}} T} n(\boldsymbol{r}_j) \right] + 1 \right\}^{-1} \tag{8.71}$$

having several solutions. Each of these solutions provides the free energy extremum and describes the ordered or disordered distribution of atoms.

In terms of the static concentration wave method, the solution to (8.71) is sought as a superposition of several static concentration waves $\Delta_k(\boldsymbol{r})$.

In the mean field approximation, the total potential of all the atoms acting on an atom at a site $\boldsymbol{r}$ can be presented, in view of (8.63) and (8.64), as

$$\sum_{\boldsymbol{r}} \tilde{V}(\boldsymbol{r} - \boldsymbol{r}') n(\boldsymbol{r}') = V(0) c + \sum_{s} V(\boldsymbol{k}_s) \Delta_k(\boldsymbol{r}), \tag{8.72}$$

where $V(\boldsymbol{k}_s) = \sum_{\boldsymbol{r}} V(\boldsymbol{r}) \exp(-i\boldsymbol{k}_s \boldsymbol{r})$ is the Fourier component of the potential $V(\boldsymbol{r})$. The summation in (8.72) is with respect to all stars of wave vectors associated with ordering.

The mean field equation (8.71) for the probabilities $n(\boldsymbol{r})$ can be written differently by substituting the total potential (8.72):

$$c + \sum_{k} \Delta_k(\boldsymbol{r}) = \left\{ \exp\left[-\frac{\mu}{k_{\mathrm{B}} T} + \frac{V(0) c}{k_{\mathrm{B}} T} + \sum_{s} \frac{V(\boldsymbol{k}_s)}{k_{\mathrm{B}} T} \Delta_k(\boldsymbol{r}) \right] + 1 \right\}^{-1} . \tag{8.73}$$

If the coordinates of the sites $\boldsymbol{r}$ are given specific nimerical values, (8.73) can be replaced by a system of transcendental equations with respect to the long-range order parameters $\eta(\boldsymbol{k})$. The number of transcendental equations is equal to that of the different values assumed by the distribution function (8.63) over the whole set of lattice sites.

The set of equations derived from (8.73) using the static concentration wave method is fundamentally simpler than the initial nonlinear finite-difference equation of integral type (8.71).

The theoretical determination of the structure of ordered phases by the static concentration wave method is associated with the analysis for the free energy minimum (8.69). The internal energy decreases and the entropy term increases upon ordering of a solid solution. The competition of these components of the free energy (8.69) brings about an order–disorder phase transition and determines the transition temperature and the structure of the ordered phase. Phases corresponding to a minimum internal energy will appear first. According to [21], the internal energy can be written as

$$U = \tfrac{1}{2} N[V(0)c^2 + \sum_s V(\boldsymbol{k}_s)\eta_s^2] \,, \tag{8.74}$$

where s denotes the wave vector star. It follows from (8.74) that the internal energy is a minimum if long-range order parameters, which correspond to the minimum Fourier component of the mixing energy $V(\boldsymbol{k})$, are nonzero. In their turn, the vectors $\boldsymbol{k}_s$ ensuring minimum values of $V(\boldsymbol{k})$ form stars associated with ordering. Expression (8.74) is written in the approximation of paired atomic interactions. If multiparticle interactions are taken into account, quadratic terms of the order parameters $V(\boldsymbol{k}_s)\eta_s^2$ will be replaced by terms exhibiting a more complex dependence on the parameters η_s in the expression for the internal energy.

Thus, in the static concentration wave method, the theoretical determination of the structure of ordered phases in a solid solution is reduced to the following steps: determination of the mixing energy $V(\boldsymbol{k})$, ascertaining the minima of the mixing energy, and construction of distribution functions of the form (8.63) using wave-vector stars that ensure a minimum of $V(\boldsymbol{k})$. It should be noted that the problem of determining the potentials of paired or multiparticle interactions and their Fourier components cannot be solved in the static concentration wave method. For this reason, a theoretical method has not yet been realized for superstructures of real systems. However, the static concentration wave method is effective for the construction of distribution functions when superstructure wave vectors or stars associated with ordering are known from theoretical or experimental data. This method is applicable not only to substitutional solid solutions but also to interstitial solid solutions.

In general, the concept of the distribution function $n(\boldsymbol{r})$ is the key point of the static concentration wave method, and is very helpful in the detailed analysis of crystal structures.

8.4 Cluster Methods in Ordering Theory

A feature in common for all cluster approximations in the microscopic theory of ordering is that a group of particles (a cluster) is treated such that the interaction of particles in the cluster is described exactly, while their interaction with the environment is approximated by some effective field determined from conditions of self-consistency of the free energy minimum. The cluster approximation proves to be most efficient for the description of order–disorder structural phase transitions in compounds where particular emphasis is placed on short-range correlations associated with competing interparticle interactions or prohibitions for certain configurations of neighboring particles. A variety of cluster approximations is used to account roughly for correlations in structural phase transitions: the quasichemical approximation [23], the Bethe method [24], and the cluster variation method due to Kikuchi [25].

Results of the mean field approximation are improved in cluster methods, because the same reasoning is applied to a cluster as to a single lattice site occupied by a certain atomic species. As a result, local interactions and correlations are accounted for through consideration of a compact group of atoms located in a medium whose properties correspond to the mean macroscopic state of a substance. Although the cluster approximation often proves to be reasonably effective, a cluster itself presents (since it cannot be chosen unambiguously) a fairly artificial construction as opposed to a site occupied by some atom.

The Bethe-Guggenheim Approximation. Bethe proposed [24] the simplest approach, which is equivalent to Guggenheim's quasichemical approximation [23] and constitutes a generalized concept of the mean field. The Bethe method deals with the central particle and its nearest neighborhood in the first coordination sphere. The interaction with the nearest neighbors is taken into account exactly, while the effect of more distant atoms is allowed for in terms of the mean field theory using some unknown mean field H. In other words, a cluster behaves as if it corresponds to the following Hamiltonian:

$$\mathcal{H}_{\text{cluster}} = -\tfrac{1}{2}\sum_{i=1}^{z} Vc(\boldsymbol{r}_0)c(\boldsymbol{r}_i) - H\sum_{i=1}^{z} c(\boldsymbol{r}_i) - H_0 c(\boldsymbol{r}_0)\,. \tag{8.75}$$

The summation in (8.75) is taken over all z nearest neighbors of the atom at the site $\boldsymbol{r}_0$. From (8.75) it is clear that the external field H_0 acts directly only on the central atom, while the effect of this field H_0 on the nearest neighbors is included in the mean field H.

The total number of system configurations needs be found to estimate the statistical sum. Consider a binary solid solution A-B of an arbitrary composition where the lattice has an equal number of sites of the first and second types. Assume that sites of one type are surrounded only by sites of the other type. In the nearest-neighbor approximation, the energy of such a solid solution can be given, in accordance with (8.42), as

$$E_\kappa = -N_{AA}\varepsilon_{AA} - N_{BB}\varepsilon_{BB} - (N_{AB}^{(12)} + N_{BA}^{(12)})\varepsilon_{AB} \,, \tag{8.76}$$

where $N_{AB}^{(12)}$ is the number of pairs where atom A is at a site of the first type and atom B is at a neighboring site of the second type. The total number of different configurations of the system $W(\eta, N_{AB})$ will be proportional to the number of methods by which the total number of atoms $zN/2$ can be broken down into groups of pairs AA, BB, AB and BA, i.e.

$$W(\eta, N_{AB}) = c(\eta)\frac{(zN/2)!}{[(zN_A - N_{AB})/2]![(zN_B - N_{AB})/2]!N_{AB}^{(12)}!N_{BA}^{(12)}!} \,. \tag{8.77}$$

In (8.77) the factor $c(\eta)$ depends on the long-range order parameter, i.e. on the distribution of atoms A and B over sites of two nonequivalent sublattices. This factor can be found if one looks at the atomic distribution in a completely disordered crystal with $\eta = 0$. In this case the number of different configurations can be determined from (8.54). Equating W found from (8.54) and (8.77), we obtain an expression for the factor $c(\eta)$. Equilibrium values of the long-range order parameter η and N_{AB} can be found, in turn, from the condition of a maximum of the statistical sum $Z(\eta, N_{AB})$ or the free energy minimum equal to

$$F = E_\kappa - k_B T \ln W(\eta, N_{AB}) \,. \tag{8.78}$$

Minimization of the free energy leads to the expression

$$N_{AB}^{(12)} N_{BA}^{(12)} = N_{AA} N_{BB} \exp(V / k_B T) \,, \tag{8.79}$$

where $V = 2\varepsilon_{AB} - \varepsilon_{AA} - \varepsilon_{BB}$. The number of pairs N_{AA}, N_{BB}, $N_{AB}^{(12)}$, and $N_{BA}^{(12)}$ can be found using the probabilities P that sites of one or the other type are occupied and correlations σ ($\sigma_{AB}^{(12)} = \sigma_{BA}^{(12)} = \sigma$) as follows:

$$\left.\begin{aligned} N_{AB}^{(12)} &= zN(P_A^{(1)}P_B^{(2)} + \sigma)/2, & N_{BA}^{(12)} &= zN(P_B^{(1)}P_A^{(2)} + \sigma)/2, \\ N_{AA} &= zN(P_A^{(1)}P_A^{(2)} - \sigma)/2, & N_{BB} &= zN(P_B^{(1)}P_B^{(2)} - \sigma)/2, \end{aligned}\right\} \tag{8.80}$$

where the probabilities P are determined from (8.52). Considering (8.80), equation (8.79) yields the correlation parameter as a function of the long-range order parameter η, the solid solution composition $c = c_A$, and temperature T:

$$\sigma = \frac{1 - 2c(1-c)[1 - \exp(-V/k_B T)] - \eta^2[1 - \exp(-V/k_B T)]/2 - \sqrt{D}}{2[1 - \exp(-V/k_B T)]} \,, \tag{8.81}$$

where $D = (1 - 2c)^2 + 4c(1 - c)\exp(-V/k_B T) - \eta^2[1 - \exp(-V/k_B T)]\exp(-V/k_B T)$. In the particular case of a disordered solid solution AB having a stoichiometric composition ($c = 0.5$), the correlation parameter above the order–disorder transition temperature is

$$\sigma = [\tanh(V/4k_B T)]/4 \,. \tag{8.82}$$

Consideration of the interaction in the first coordination sphere is known as the first Bethe approximation. The second Bethe approximation deals with a large cluster, which accounts exactly for the interaction in two coordination spheres relative to the central atom. The second Bethe approximation however has not been in wide use, since it does not offer any considerable advantages over the first approximation.

Obviously, the Bethe method can be generalized by constructing larger clusters taking into account exactly the interaction in several coordination spheres. The mean field in each sphere is determined by self-consistency conditions.

The Cluster Variation Method. A more systematic method for the introduction of larger clusters, which takes into account implicitly multiparticle correlations, consists in successive refinement of the combinatorial factor in the notation of the free energy. Such a refinement is provided by various versions of the cluster variation (CV) method [25, 26] which gives the best self-consistent approximation for combinatorial factors describing the statistical distribution in the lattice. Researchers [27] believe the Kikuchi approximation can provide the most accurate results for transition temperatures, since it successively allows for clusters of different sizes and bonds between these clusters, and also takes into account multiparticle correlations in particle arrangement at cluster sites. In accordance with [28], the CV method is a generalization of the whole quasichemical line in studies of atomic ordering.

The CV method was pioneered [25] just as a technique for constructing the combinatorial factor of the Ising lattice with due account of multiparticle correlations of atoms at cluster sites.

In the case of a three-dimensional lattice, the method proposed for the determination of the combinatorial factor turned out to be complicated and cumbersome. In this connection, the authors [26] came up with another version of the Kikuchi method referred to as the pseudoensembles method. The concepts developed in [26] provide a better understanding of the causes responsible for the roughness of the pseudoensembles method and permit the classification of various quasichemical methods as to how they work to refine the exact expression for the statistical sum in the Ising model.

Hijmans and de Boer [29] further advanced and generalized the CV method. Replacing the crystal lattice by an ensemble of certain figures such that the number of sites in the lattice and in the ensemble is equal, they proposed a general formalism of the CV method eliminating combinatorial reasoning.

The CV method describes a crystal as a set of figures of different types ranging from a basis cluster figure to overlapping figures. The latter are understood as various combinations of crystal sites. Overlapping figures ensure that each lattice site is taken once only. Since clusters overlap one another, some lattice sites may be shared by several clusters. When only a basis cluster figure is used to model a crystal, those sites may be taken several times each. Overlapping figures help overcome this situation. A group of figures shared by adjacent clusters forms the first generation of overlapping figures. Those figures, in turn, mutually overlap and form the next generation. If this procedure continues to involve sites, it is possible to obtain a genetic sequence of overlapping figures for a basis cluster figure chosen. Each generation of figures includes so-called special figures that do not appear in the next generation. It is special overlapping figures, which are arranged in accordance with the number of sites they include, in combination with a basis cluster figure that form a sequence of special figures $\{s\}$ needed to describe a crystal.

Sites of any figure of type s in a given sequence $\{s\}$ may be occupied by different atomic species in different ways. Therefore it is necessary to consider alternatives of mutual spatial arrangement of particles, i.e. configurations i. Each configuration i of a figure s is characterized by the probability $P_i^{(s)}$, which is equal to the ratio between the number of figures s with the configuration i and the total number of type s figures in the crystal. Configurations, which can be obtained from a given configuration i by applying symmetry operations such as rotation and reflection, are equivalent. The number of equivalent configurations i of a figure s (configuration multiplicity) will be labeled as $\lambda_i^{(s)}$. In general the multiplicity $\lambda_i^{(s)}$ is equal, in accordance with [30, 31], to the index of the point symmetry group $G_i^{(s)}$ of a figure s with an i-configuration relative to the point group $G_0^{(s)}$ of a figure s where all sites are occupied by atoms of the same species, i.e. $\lambda_i^{(s)} = n(G_0^{(s)})/n(G_i^{(s)})$. Here $n(G)$ is the order of the group G. The definitions of the probability $P_i^{(s)}$ and the multiplicity $\lambda_i^{(s)}$ suggest the probability normalization condition:

$$\sum_{i\in s} \lambda_i^{(s)} P_i^{(s)} = 1 . \tag{8.83}$$

Special figures of the sequence $\{s\}$ are related by the matrix $|u^{(m)(s)}|$ whose elements show how many m-type figures are contained in a figure of type s. The overestimated number of microstates due to overlapping of figures is taken into account by the coefficients $y^{(s)}$, which can be found by solving a linear system comprising equations of the form

$$\sum_{s=a}^{q} u^{(m)(s)} y^{(s)} = x^{(m)} \quad (m = a, b, c, ..., q), \tag{8.84}$$

where $x^{(m)}$ is a coefficient showing how many times the number of m-type figures is larger than the number of sites N in the Ising lattice. The coefficient $x^{(m)}$ equals the number of m-type figures, which share any randomly chosen site of the lattice in question, divided by the number of sites in the m-type figure.

The probabilities $P_i^{(s)}$ obey so-called constant relationships that connect the probabilities of figures of different generations, because the probability of a geometrically simpler figure in a given sequence is always expressible in terms of the probabilities of more complex (with more sites) figures in the same sequence:

$$\lambda_g^{(n)} P_g^{(n)} = \sum_{i\in s} k_{gi}^{(n)(s)} \lambda_i^{(s)} P_i^{(s)} , \tag{8.85}$$

where $k_{gi}^{(n)(s)}$ is the relative share of g-configurations of a figure n among all configurations of that figure included in the i-configuration of a figure s.

Thus, the CV method places major emphasis on the choice of the basis cluster figure, because it determines the sequence of special figures, which is typical of a given lattice. A far more significant point is that the probabilities of different configurations of the basis cluster figure are variables that determine the equilibrium state of a crystal. Necessary conditions for the selection of a basis cluster postulated in [26] were refined elsewhere [29]: A geometrical figure to be chosen as a cluster should be "closed", i.e. should be free of separately bound sites. A cluster site is considered separately bound if it belongs to no more than one overlapping figure. In addition, the basis cluster figure should fit the crystal lattice geometry, allow for the character of interparticle interactions in the crystal, and be large enough to take into account most significant correlations in the lattice.

To calculate the energy of a crystal whose lattice is described by a set of figures having the probabilities $P_i^{(s)}$, each i-configuration of some figure s will be assumed to be characterized by the energy $\varepsilon_i^{(s)}$. If all $x^{(n)}N$ clusters of particles n in the lattice can be isolated from one another, the energy of such an n-figure ensemble of $x^{(n)}N$ clusters is

$$E = x^{(n)}N\sum_i \lambda_i^{(n)} P_i^{(n)}\varepsilon_i^{(n)} . \tag{8.86}$$

Expression (8.86) estimates the lattice energy incorrectly, because many figures, which are smaller than n-figures, are counted several times due to overlapping of those smaller figures. The energy (8.86) may be brought closer to the lattice energy if the coefficients $y^{(s)}$ calculated from (8.84) are introduced to avoid the overestimate. To this end, a correction to the energy of the largest figure in a sequence $\{s\}$ describing the crystal is found. Subsequently a correction is found to the energy of the second-largest figure and so on down to the smallest figure, which is a lattice site. The lattice energy can be eventually presented as the sum

$$E = N[x^{(n)}\sum_i \lambda_i^{(n)} P_i^{(n)}\varepsilon_i^{(n)} + y^{(m)}\sum_i \lambda_i^{(m)} P_i^{(m)}\varepsilon_i^{(m)} + \ldots + y^{(a)}\sum_i \lambda_i^{(a)} P_i^{(a)}\varepsilon_i^{(a)}] . \tag{8.87}$$

In conformity with (8.84), $y^{(n)} = x^{(n)}$ for a basis n-figure. Considering this, the final expression for the energy is written as

$$E(P_i^{(n)}) = N\sum_{s=a}^{n} y^{(s)} \sum_{i\in s} \lambda_i^{(s)} P_i^{(s)}\varepsilon_i^{(s)} . \tag{8.88}$$

To compute the entropy of a lattice described by a set of probabilities $P_i^{(s)}$, it is necessary to determine the number of different microstates corresponding to a given macrostate, i.e. count the number of different configurations of a crystal with energy $E(P_i^{(s)})$. Consider an n-figure ensemble containing $x^{(n)}N$ figures of particles n with probabilities $P_i^{(n)}$. The configuration factor for this ensemble is found using known rules of combinatorial analysis:

$$g^{(n)}(P_i^{(n)}) = \frac{(x^{(n)}N)!}{\prod_i [(P_i^{(n)}x^{(n)}N)!]^{\lambda_i^{(n)}}} . \tag{8.89}$$

Considering the Stirling formula in the form $(xN)! = x^{xN}(N!)^x$ and the probability normalization condition (8.83), expression (8.89) can be rearranged into

$$g^{(n)}(P_i^{(n)}) = \frac{(N!)^{x(n)}}{\prod_i [(P_i^{(n)}N)!]^{\lambda_i^{(n)}x(n)}} . \tag{8.90}$$

The configuration factor (8.90) was calculated disregarding figure overlap. As a result, the number of possible configurations is greatly overestimated. If an n-figure is the basis figure of a sequence $\{s\}$ and the next figures in this sequence are figures m, l, etc. down to a figure a (lattice site), then, with due regard to the overestimated number of microstates resulting from figure overlap, the lattice configuration factor is

$$g(P_i^{(n)}) = \prod_{s=a}^{n} Q(P_i^{(s)}) , \tag{8.91}$$

where

$$Q(P_i^{(s)}) = \frac{(N!)^{y(s)}}{\prod_i [(P_i^{(s)}N)!]^{\lambda_i^{(s)}y(s)}} . \tag{8.92}$$

The configuration entropy of a macrostate is

$$S(P_i^{(n)}) = -k_B \ln g(P_i^{(n)}) = k_B \sum_{s=a}^{n} \ln Q(P_i^{(s)}) . \tag{8.93}$$

Hence, considering (8.92), the probability normalization condition (8.83), and the Stirling formula $\ln(N!) \approx N \ln N - N$, we obtain

$$S(P_i^{(n)}) = -k_B N \sum_{s=a}^{n} y^{(s)} \sum_{i=s} \lambda_i^{(s)} P_i^{(s)} \ln P_i^{(s)} . \tag{8.94}$$

Taking into account the crystal lattice energy (8.88) and the configuration entropy (8.94), the expression for the free energy of a crystal having N atoms can be written as

$$F = N \sum_{s=a}^{n} y^{(s)} \sum_{i\in s} \lambda_i^{(s)} P_i^{(s)} [\varepsilon_i^{(s)} + k_B T \ln P_i^{(s)}] . \tag{8.95}$$

From the notation of the free energy (8.95) it is clear that the first summand (lattice energy E) is the internal energy of a crystal. Ordering in a solid body represents an isobaric process ($\Delta p = 0$). Therefore it should be remembered that when the pressure p ($p = 0$ for solids) is constant, the internal energy or its variation upon ordering coincides with the enthalpy H or its variation ΔH respectively. Consequently, as will be shown in the next chapter, the cluster energy $\varepsilon_i^{(s)}$ can be found from experimental thermodynamic quantities.

The correction $y^{(s)}$ and multiplicity $\lambda_i^{(s)}$ coefficients in the expression for the free energy depend on the choice of the basis cluster and overlapping figures, while the figure energies $\varepsilon_i^{(s)}$ in the CV method are considered as model parameters and are thought to be known. Therefore the free energy of a crystal depends only on the probabilities $P_i^{(s)}$, which are variables. Minimization of the free energy by these probabilities determines the equilibrium state of a crystal. The probabilities of multiparticle figures corresponding to the free energy minimum provide information about multiparticle correlations in the arrangement of particles at lattice sites.

It should be noted however that not all probabilities $P_i^{(s)}$ of the basis cluster figure with different configurations are independent variables because of the normalization condition (8.83). At least in the absence of additional boundary conditions the number of independent variables $P_i^{(s)}$ is one less than the number of nonequivalent configurations of the basis cluster figure.

So, the ordering problem in the CV method formally reduces to solving a system of nonlinear equations of the form $\partial F / \partial P_i^{(s)} = 0$. Clearly, the number of such equations grows with the size of the basis cluster figure to entail snowballing mathematical difficulties and scope of computations. In this connection, whenever specific problems are solved by the CV method, it is critical to choose an optimal variant providing the best approximation of the free energy with the use of a basis cluster figure including the least possible number of particles.

In the single-cluster approximation, results obtained for a lattice with an R-particle cluster can be easily improved by taking a larger cluster and, thus, allowing for correlations in the arrangement of a greater number of particles.

One more method [32] of improving the approximation is to use a multicluster approximation when an R'-particle cluster is added to an R-particle cluster. In this case improvement is achieved thanks to consideration of correlations which are disregarded by the R-particle cluster. In a multi-cluster approximation, the lattice free energy is presented, as in a single-cluster approximation, as an expansion in terms of ensembles of lattice figures comprising a certain sequence. Figures, which do not mutually include one another, are chosen as clusters in a multicluster approximation.

The CV method describing atomic ordering can be refined if changes in the atomic interaction upon an order–disorder transition are taken into account [33]. In terms of the Ising model, atoms are rigidly fixed at sites of an ideal lattice, while statistical ordering theory suggests that lattice parameters remain unchanged upon an order–disorder transition. Actually, atomic rearrangement changes the lattice parameters and symmetry upon ordering. Hence it follows that the crystal energy should change as a result of both atomic rearrangement and different interparticle interaction upon an order–disorder transition. Using the CV method and a paired approximation to write the energies $\varepsilon_i^{(s)}$ of various figures, researchers [33] showed that consideration of the energy variation upon ordering tends to refine the crystal description if the ordered phase exists within a narrow concentration interval. It was also intimated that the assumption of an abrupt change in the ordering energy upon an order–disorder transition essentially implies its dependence on the degree of order.

Let us dwell on the method [33], which takes into account changes in atomic interactions upon an order–disorder transition in a binary solid solution A–B. Analysis of atomic ordering can be reasonably focused only on the configuration part of the energy, which depends on the mutual arrangement of atoms. This part of the energy can be presented in a paired approximation as

$$E = -\tfrac{1}{2}\sum_j N_{\mathrm{AB}}(\rho_j)V_j\,, \tag{8.96}$$

where $N_{\mathrm{AB}}(\rho)$ is the number of A–B pairs in the j-th coordination sphere with radius ρ_j and $V_j = 2\varepsilon_{\mathrm{AB}}(\rho_j) - \varepsilon_{\mathrm{AA}}(\rho_j) - \varepsilon_{\mathrm{BB}}(\rho_j)$ is the mixing energy in the j-th sphere expressed as energies of paired interactions. In order to define the configuration energy E of a crystal as (8.96), the energies $\varepsilon_i^{(s)}$ of different figure configurations, which describe the crystal, should also be presented using a paired approximation:

$$\varepsilon_i^{(s)} = -\sum_j n_i^{(s)}(\rho_j)V_j\,, \tag{8.97}$$

where $n_i^{(s)}(\rho_j)$ is the number of A–B pairs in the j-th coordination sphere of the i-configuration of a figure s.

Considering (8.95) and (8.97), the crystal free energy can be written in the form:

$$F = N\sum_s y^{(s)}\sum_{i\in s}\lambda_i^{(s)}P_i^{(s)}[-n_i^{(s)}(\rho_1)V_1 - n_i^{(s)}(\rho_2)V_2 + k_{\mathrm{B}}T\ln P_i^{(s)}]\,. \tag{8.98}$$

The free energy is a function of temperature T and the probabilities $P_i^{(s)}$. If the temperature and the solid solution composition are preset, the independent variables will be probabilities $P_i^{(s)}$ only. The equilibrium values of the probabilities can be found from the condition of the minimum free energy of ordered F^{ord} and disordered F^{disord} phases, i.e.

$$\frac{\partial F^{\mathrm{ord}}}{\partial P_i^{(s)\,\mathrm{ord}}} = 0\,; \qquad \frac{\partial F^{\mathrm{disord}}}{\partial P_i^{(s)\,\mathrm{disord}}} = 0\,. \tag{8.99}$$

Expressions (8.99) are necessary but insufficient conditions for a free energy minimum. To determine the temperature-concentration interval, in which the free energy of each phase has an absolute minimum, the phase equilibrium equations

$$\left.\begin{aligned} \mu_{\mathrm{A}}^{\mathrm{ord}} &= \mu_{\mathrm{A}}^{\mathrm{disord}} \\ \mu_{\mathrm{B}}^{\mathrm{ord}} &= \mu_{\mathrm{B}}^{\mathrm{disord}} \end{aligned}\right\} \tag{8.100}$$

should be solved simultaneously with (8.99). The chemical potentials μ_{A} and μ_{B} of solid solution components are defined in terms of the free energy:

$$\left.\begin{aligned} \mu_{\mathrm{A}} &= F + (1-c)\partial F/\partial c \\ \mu_{\mathrm{B}} &= F - c\partial F/\partial c \end{aligned}\right\}. \tag{8.101}$$

Solving the system of nonlinear equations (8.100) and (8.101), it is possible to construct a phase diagram taking into account the equilibrium between the disordered and ordered phases.

Application of the model was demonstrated [33] with atomic ordering in a hexagonal closely packed lattice where a superstructure of WC type was formed. Calculations were made [33] for different values of the $V_1^{\rm ord}/V_1^{\rm disord}$ ratio characterizing changes in the atomic interaction energy in the first coordination sphere upon an order–disorder transition. In accordance with those calculations, the order–disorder transformation is a phase transition of the second kind when $V_1^{\rm ord}/V_1^{\rm disord} = 1$ and becomes a phase transition of the first kind when $V_1^{\rm ord}/V_1^{\rm disord} > 1$. The concentration interval of the ordered phase narrows when a two-phase region appears in the equilibrium phase diagram.

Although the CV method was first proposed [25] more than 50 years ago, it enjoys a sustained interest. Different variants of this method were used, for example, to compute the critical temperature of a ferromagnetic with an fcc lattice [34] and to describe a two-dimensional model of the lattice gas with multiparticle interactions [35].

Widespread use of the CV method is due to the fact that this method offers a compact presentation of model properties and provides reasonably accurate results. The analysis of the CV method shows that the statistical sum can be approximated more accurately by this method if the number of figures included in the sequence of overlapping figures is reduced [36]. This can be achieved thanks to one-dimensional enlargement of the basis cluster or selection of a cluster allowing for bonds in more than one coordination sphere. However, enlargement of the basis cluster figure entails snowballing computational difficulties. For this reason, basis cluster figures in the form of tetrahedra or octahedra are used most frequently to describe ordering in bcc and fcc lattices.

In general, Kikuchi's cluster variation method represents the most accurate description of the three-dimensional Ising model, has a simple mathematical formulation, and can be used to analyze ordering in a variety of alloys and compounds. The mathematical formalism of the CV method allows for multiparticle correlations in the arrangement of particles over lattice sites, and permits calculating directly the location of ordered-phase intervals in phase diagrams. However, the energy of various configurations of the basis cluster and overlapping figures need be known to perform specific calculations for real ordered systems by the CV method. The CV method has not been used so far to directly describe ordering in nonstoichiometric compounds. Researchers [37–40] have proposed using the CV method in combination with the static concentration wave method to describe atomic ordering in highly nonstoichiometric interstitial compounds. This combination method is known as the order parameter functional method (OPF method) [37, 40–42].

References

1 P. Ehrenfest: Phasenumwandlungen im üblichen und erweiterten Sinn, klassifiziert nach den entsprechenden Singularitäten des thermodynamischen Potenziales, Commun. Kamerlingh Onnes Lab. Univ. Leiden. Suppl. B **20**, 628–632 (1933); Proc. Amsterdam Acad. **36**, 153–157 (1933)
2 A. B. Pippard: *The Elements of Classical Thermodynamics* (Cambridge University Press, Cambridge 1964) 264 pp.
3 L. D. Landau: Theory of phase transitions, in *Collected Papers* **1** (Nauka, Moscow 1969) pp. 234–261 (in Russian)
4 E. M. Lifschitz: Theory of second kind phase transitions, ZhETF **11**, 255–268 (1941) (in Russian).
5 L. D. Landau, E. M. Lifschitz: *Statistical Physics. Course of Theoretical Physics* **5** (Pergamon, Oxford 1980) 544 pp.
6 A. P. Levanyuk: Theory of light scattering near phase transition points, ZhETF **36**, 810–817 (1959) (in Russian)
7 V. L. Ginzburg: Some notes on phase transitions of second kind and on microscopic theory of ferroelectric, Fiz. Tverd. Tela **2**, 2031–2043 (1960) (in Russian)
8 I. E. Dzyaloshinskii: Thermodynamic theory of weak ferromagnetism of antiferromagnets, ZhETF **32**, 1547–1562 (1957) (in Russian)
9 I. E. Dzyaloshinskii, V. I. Man'ko: Non-linear effects in antiferromagnets, ZhETF **46**, 1352–1359 (1964) (in Russian)
10 I. E. Dzyaloshinskii: Theory of helicoidal structures of antiferromagnets, ZhETF **46**, 1420–1438 (1964) (in Russian)
11 V. L. Indenbom: Thermodynamic theory of ferroelectricity, Izv. AN SSSR. Seriya Fizicheskaya **24**, 1180–1183 (1960) (in Russian)
12 V. L. Indenbom: Phase transitions without change of number of atoms in unit cell of a crystal, Kristallografiya **5**, 115–125 (1960) (in Russian)
13 I. L. Birman: Simplified theory of symmetry change in second-order phase transitions applications to V_3Si, Phys. Rev. Lett. **17**, 1216–1219 (1966)
14 Yu. M. Gufan: *Structural Phase Transitions* (Nauka, Moscow 1982) 304 pp. (in Russian)
15 Yu. A. Izyumov, V. N. Syromyatnikov: *Phase Transitions and Symmetry of Crystals* (Nauka, Moscow 1984) 248 pp. (in Russian)
16 E. Ising: Beitrag zur Theorie des Ferromagnetismus, Z. Phys. **31**, 253–258 (1925)
17 W. Gorsky: Röntgenographische Üntersuchung von Umwandlungen in der Legierung CuAu, Z. Phys. **50**, 64–81 (1928)
18 W. L. Bragg, E. J. Williams: The effect of thermal agitation on atomic arrangement in alloys, Proc. Roy. Soc. London A **145**, 699–730 (1934)
19 E. J. Williams: The effect of thermal agitation on atomic arrangement in alloys-III, Proc. Roy. Soc. London A **152**, 231–252 (1935)
20 M. A. Krivoglaz, A. A. Smirnov: *The Theory of Order-Disorder in Alloys* (American Elsevier, New York 1964) 427 pp.
21 A. G. Khachaturian: *Theory of Structural Transformations in Solids* (John Wiley & Sons, New York 1983) 574 pp.
22 W. Ludwig, C. Falter: *Symmetries in Physics* (Springer-Verlag, Berlin – Heidelberg 1988) 461 pp.

23 R. H. Fowler, E. A. Guggenheim: *Statistical Thermodynamics* (Cambridge University Press, Cam-bridge 1939) 480 pp.
24 H. A. Bethe: Statistical theory of superlattices, Proc. Roy. Soc. London A **150**, 552–575 (1935)
25 R. Kikuchi: Theory of cooperative phenomena, Phys. Rev. **81**, 988–1003 (1951)
26 M. Kurata, R. Kikuchi, T. Watari: Theory of cooperative phenomena. Detailed discussions of the cluster variation method, J. Chem. Phys. **21**, 434–448 (1953)
27 V. G. Vaks, V. I. Zinenko, V. E. Shneider: Microscopic theories of the order-disorder structural phase transitions in crystals, Uspekhi Fiz. Nauk **141**, 629–673 (1983) (in Russian)
28 N. S. Golosov: Cluster variation method in the theory of atomic ordering, Izv. Vyssh. Ucheb. Zaved. Fizika No 8, 64–82 (1976) (in Russian)
29 J. Hijmans, J. de Boer: Approximation method for order-disorder problem, Physica **21**, 471–516 (1955)
30 A. A. Rempel, A. I. Gusev: Relation between short-range and long-range order in solid solutions with b.c.c. and f.c.c. structures, Phys. Stat. Sol. (b) **130**, 413–420 (1985)
31 A. A. Rempel, A. I. Gusev: The relationship between short-range and long-range order in ordered alloys, Fiz. Metall. Metalloved. **60**, 847–854 (1985) (in Russian). (Engl. transl.: Phys. Met. Metallogr. **60**, 11–17 (1985))
32 N. S. Golosov, V. N. Udodov: Many-cluster approximation in CV method, Izv. Vyssh. Ucheb. Zaved. Fizika No 12, 93–97 (1975) (in Russian)
33 N. S. Golosov, A. V. Ushakov: Model of alloy in statistical theory of ordering, Fiz. Tverd. Tela **18**, 1262–1268 (1976) (in Russian)
34 J. M. Sanchez, D. de Fontaine: The f.c.c. Ising model in the cluster variation approximation, Phys. Rev. B **17**, 2926–2936 (1978)
35 A. Surda, I. Karasova: Cluster variation method for lattice gas model, Czechosl. J. Phys. B **33**, 1289–1302 (1983)
36 R. Kikuchi, S. G. Brush: Improvement of the cluster-variation method, J. Chem. Phys. **47**, 195–203 (1967)
37 A. I. Gusev, A. A. Rempel: Order parameter functional method in the theory of atomic ordering, Phys. Stat. Sol. (b) **131**, 43–51 (1985)
38 A. I. Gusev, A. A. Rempel: Thermodynamic model of atomic ordering. 1. Basic equations, Zh. Fiz. Khimii **60**, 1349–1352 (1986) (in Russian). (Engl. transl.: Russ. J. Phys. Chem. **60**, 808–810 (1986))
39 A. A. Rempel, A. I. Gusev: Thermodynamic model of atomic ordering. 2. The order-disorder structural phase transition in nonstoichiometric niobium carbide, Zh. Fiz. Khimii **60**, 1353–1357 (1986) (in Russian). (Engl. transl.: Russ. J. Phys. Chem. **60**, 810–813 (1986))
40 A. I. Gusev, A. A. Rempel: Calculating the energy parameters for CV and OPF methods, Phys. Stat. Sol. (b) **140**, 335–346 (1987)
41 A. I. Gusev, A. A. Rempel: *Structural Phase Transitions in Nonstoichiometric Compounds* (Nauka, Moscow 1988) 308 pp. (in Russian)
42 A. I. Gusev: Atomic ordering and the order parameter functional method, Philosoph. Mag. B **60**, 307–324 (1989)

9. Order-Parameter Functional Method

It was shown in the previous chapter that the static concentration wave method [1], which is a version of the mean field approximation [2], is effective for describing such complex phenomenon as atomic ordering. Correlations arising as a result of ordering can be taken into account by the cluster variation method [3–9]. However, these methods cannot be used directly for the description of the equilibrium distribution of interstitial atoms and structural vacancies in strongly nonstoichiometric compounds for the following reasons. In the static concentration wave method, energy parameters expressing the free energy are Fourier components of interparticle interaction potentials, which cannot be determined in the framework of this method. In this connection, in order to determine interatomic interaction potentials, it was proposed [1] to use the pseudopotential method, which has been insufficiently developed with respect to transition-metal compounds. As for the cluster variation (CV) method, the correct qualitative and quantitative description of structural order–disorder phase transitions calls for the choice of a basis cluster figure whose minimum size is limited by the unit cell. In complex lattices with unit cells containing a few tens of atoms, the description of order–disorder transformations by such a large cluster encounters insurmountable computational difficulties. Moreover, for quantitative applications of the CV method it is necessary to have a large number of energy parameters.

The order-parameter functional (OPF) method was proposed [10–19] to describe structural order–disorder phase transformations in substitutional solid solutions A_yB_{1-y} and strongly nonstoichiometric interstitial compounds $MX_y\square_{1-y}$, where the substitutional solution is formed by nonmetallic interstitial atoms X and structural vacancies □ (unoccupied sites of the nonmetallic sublattice). This method provides not only a quantitative, but also a qualitative description of atomic ordering and combines features of both the cluster variation method, which accurately takes into account the interaction of particles inside the cluster and their multiparticle correlations, and the static concentration wave method permitting detailed allowance for the lattice symmetry.

9.1 Main Approaches of the Method

Like the cluster variation method [5], the order-parameter functional method preserves the concept of a crystal as a combination of figures s with various configurations i each having corresponding values of the probability $P_i^{(s)}$ and the energy $\varepsilon_i^{(s)}$. Therefore the lattice configuration energy of an ordered crystal (internal energy of a crystal) is considered as an algebraic sum (8.88) of values $\lambda_i^{(s)} P_i^{(s)} \varepsilon_i^{(s)}$ proportional to the probabilities of clusters and overlapping figures. It should be pointed out that when ordering is described as an isobaric process (for solids the pressure p is assumed to be 0), the internal energy, i.e. the lattice configuration energy E, coincides with the crystal enthalpy H.

Problems, which are associated in the cluster variation method with the choice of a basis cluster figure involving a large number of particles, are eliminated in the OPF method by representing the probabilities of any figures by long-range order parameters with the help of the atomic distribution function $n(\boldsymbol{r})$ in the crystal lattice. Indeed, the CV method may include many independent variables (for example, cluster probabilities) whose quantity depends on the size of the basis cluster figure. Since cluster probabilities are represented by the distribution function, the number of independent variables in the OPF method used to minimize the free energy is not great and coincides with the number of long-range order parameters. In addition, along with accurate allowance for the interaction of particles in a cluster of any size, the OPF method permits taking into account the lattice symmetry of the ordered phase and, consequently, long-range correlations in the crystal.

A significant feature of the order-parameter functional method, which distinguishes it from other microscopic techniques of atomic ordering theory, is that the OPF method allows for the temperature versus enthalpy dependence of a real crystal. This dependence can be taken into account by considering the cluster energy $\varepsilon_i^{(s)}(T)$ as a function of temperature. From general thermodynamic considerations it is clear that if the temperature dependence of the ordered crystal enthalpy is taken into account, the entropy in the OPF method is an explicit function not only of the order parameters, but also of temperature. In this case the temperature dependent part of the entropy should be given by cluster probabilities in the form analogous to that for the crystal enthalpy (configuration energy). To put it differently, if the temperature dependence of the cluster energy is taken into consideration, the entropy of the ordered crystal consists of two parts: the residual (configurational) entropy S_c and the entropy $S(T)$ which obviously depends on temperature and is determined as the algebraic sum of the values $\lambda_i^{(s)} P_i^{(s)} S_i^{(s)}(T)$, where $S_i^{(s)}(T) = \int_0^T d\varepsilon_i^{(s)}(T)/T = S_{i\,298}^{(s)} + \int_{298}^T d\varepsilon_i^{(s)}(T)/T$ is the cluster entropy. If the cluster energy is viewed as independent of temperature, the crystal entropy is also a function of the composition and order parameters only, but not of temperature.

Let us consider the basic concepts of the order parameter functional method.

Suppose we have a crystal consisting of N atoms of different species $\nu = \alpha, \beta, ..., \tau$ located at exactly N sites of the Ising lattice.

To describe atomic ordering in a crystal, it is necessary to calculate the temperature dependences of the long-range order parameters $\{\eta\}$. In the general case, if a system is in thermodynamic equilibrium, the values of the long-range order parameters observed at a given temperature T are determined by averaging over the canonical assembly

$$\eta = Z^{-1} \sum_{\kappa=1}^{N_{st}} \eta_\kappa \exp(-E_\kappa / k_B T), \tag{9.1}$$

where $N_{st} = N! / \prod_{\nu=\alpha}^{\tau} (N - N_\nu)!$ is the total number of all crystal states (possible optional locations of atoms of species α, β, ..., τ in a lattice of N sites); $Z = \sum_{\kappa=1}^{N_{st}} \exp(-E_\kappa / k_B T)$ is the statistical sum; and E_κ is the configurational energy of the crystal in some state κ.

The configurational energy E_κ is determined not only by energy characteristics, but also by the whole set of values used to assign the atomic arrangement in a specific model describing the crystal. These values may include long- and short-range order parameters in different coordination spheres, and various multiparticle correlations in the atomic arrangement. In the order parameter functional method, the crystal configurational energy E_κ in the state κ corresponding to the long-range order parameter η_κ is given, like in the CV method (see Sect. 8.4), through the energies $\varepsilon_i^{(s)}$ of finite groups of particles

$$E_\kappa = N \sum_s y^{(s)} \sum_{i \in s} \varepsilon_i^{(s)} \lambda_i^{(s)} P_{i\kappa}^{(s)}, \tag{9.2}$$

where $P_{i\kappa}^{(s)}$ is the probability of a figure s with the configuration i in a crystal being in the state κ.

Approximation (9.2) representing the configurational energy of a crystal is preferable to the analogous expression in the method [1], since it easily allows for both two- and multiparticle interactions. In addition, this approximation accurately takes into account the energy and probability of overlapping figures.

To determine the probabilities $P_{i\kappa}^{(s)}$ in a crystal in the state κ, averaging over the lattice should be carried out. Assume that a crystal is described by a certain sequence of special figures $\{s\}$, which was determined according to [5]. If $x^{(s)}$ is the value showing how by many times the number of s-type figures is greater than the number of sites in the Ising lattice, the total number of s-type figures in the lattice with N sites equals $Nx^{(s)}$. Therefore, the probability $P_{i\kappa}^{(s)}$ can be given as the mean probability over the whole lattice

$$P_{i\kappa}^{(s)} = \frac{1}{Nx^{(s)}} \sum_{j=1}^{Nx^{(s)}} P_{ij\kappa}^{(s)}, \tag{9.3}$$

where $P_{ij\kappa}^{(s)}$ is the probability that the i-th configuration of the figure s is formed instead of the j-th figure s. The summation in (9.3) is taken over all s-type figures in the crystal.

In the order-parameter functional method [10–18], the probabilities of figures and, consequently, the energy E_κ (9.2) depend in fact only on long-range order parameters. That is why the energies of crystal states having similar values of long-range order parameters are equal. Let us label the number of such states (degree of degeneration) as $g\{\eta\}$. If we switch from summation over all κ states of the crystal to summation over states ξ with different $\{\eta\}$, equation (9.1) can be written as

$$\eta = Z^{-1}\sum_{\xi}\eta_\xi g_\xi\{\eta\}\exp(-E_\xi\{\eta\}/k_{\mathrm{B}}T). \tag{9.4}$$

If the entropy of the state ξ is equal to

$$S_\xi\{\eta\} = -k_{\mathrm{B}}\ln(g_\xi\{\eta\}), \tag{9.5}$$

and the degree of degeneration is expressed through the entropy and substituted into (9.4), we have

$$\eta = Z^{-1}\sum_{\xi}\eta_\xi \exp(-F_\xi\{\eta\}/k_{\mathrm{B}}T), \tag{9.6}$$

where $F_\xi\{\eta\}$ is the nonequilibrium free energy of the crystal in the state ξ characterized by the whole set of long-range order parameters $\{\eta\}$.

From statistical physics it is known that for a system comprising N particles, the difference between the mean energy of the crystal and the energy corresponding to the maximum term in (9.6) is proportional to $1/\sqrt{N}$ [20]. In other words, at large N the function $\exp(-F_\xi\{\eta\}/k_{\mathrm{B}}T)$ exhibits a sharp peak at some value of $\{\eta\}_{\min}$. Consequently, if energy fluctuations of the crystal in the equilibrium state are neglected, the statistical sum and the sum (9.6) can be substituted for the maximum terms, which correspond to a minimum free energy. From the above it is apparent that the observed values of the long-range order parameters $\{\eta\}$ correspond to the set of long-range order parameters describing the crystal state with a minimum free energy, i.e. $\{\eta\} = \{\eta\}_{\min}$. So, calculations of the sum (9.6), i.e. the exact solution to the Ising problem, can be replaced by minimization of the free energy F over parameters determining the distribution of atoms in the crystal. In the order parameter functional method discussed here, minimization of F over the long-range order parameters $\{\eta\}$ can be substituted for calculations of this sum. Since any κ (or ξ) state of the crystal is unambiguously characterized by a set of long-range order parameters $\{\eta\}$, the index κ (or ξ) at $P_{i\kappa}^{(s)}$, E_κ and other values will be omitted hereafter for simplicity.

Let us consider in greater detail the representation of cluster probabilities in terms of the values of the distribution function $n(\boldsymbol{r})$ expressed directly as long-range order parameters in accordance with (8.63) and (8.64).

Think of the ordered Ising lattice as a set of figures (clusters) s, which allow for all N sites. Each figure s has one of the possible configurations $i \in s$ and each configuration i occupies one of the possible locations $\lambda_i^{(s)}$. The total number of figures s in the lattice equals $Nx^{(s)}$.

Align the origin of coordinates with any site of some figure s in the crystal and let the pseudosite $\boldsymbol{r}_{\varphi}^{(s)}$ correspond to this figure (Fig. 9.1). A certain set of radius-vectors $\{\boldsymbol{r}_{\varphi l}^{(s)}\}$ referred to the pseudosite $\boldsymbol{r}_{\varphi}^{(s)}$ will correspond to sites l of this figure. Translation of all sites s to one and the same vector of the Ising lattice gives another figure s_j. Correlate the last figure to a pseudosite with a radius-vector $\boldsymbol{r}_{\varphi j}^{(s)}$ such that vectors of s_j figure sites referred to this radius-vector constitute the same set of radius-vectors $\{\boldsymbol{r}_{\varphi l}^{(s)}\}$. In this case the lattice sites have coordinates $\{\boldsymbol{r}_{\varphi j}^{(s)}+\boldsymbol{r}_{\varphi l}^{(s)}\}$. Since different crystal sites are connected with one another by translation vectors, pseudosite coordinates can be written as

$$\boldsymbol{r}_{\varphi j}^{(s)}=\boldsymbol{r}_{\varphi}^{(s)}+\sum_{\mu=1}^{3} n_{j\mu}\boldsymbol{a}_{\mu}+\sum_{\mu=1}^{\omega-1} h_{j\mu}\boldsymbol{u}_{\mu}, \tag{9.7}$$

where $n_{j\mu}$ and $h_{j\mu}$ are integral numbers determining translations to primitive $\boldsymbol{a}_{\mu}$ and basis $\boldsymbol{u}_{\mu}$ vectors of the crystal; and ω is the number of atoms in a unit cell of the Ising lattice. As is seen from (9.7), the pseudolattice made up of pseudosites $\boldsymbol{r}_{\varphi j}^{(s)}$ is equivalent to the Ising lattice, is displaced relative to the last lattice by the vector $\boldsymbol{r}_{\varphi}^{(s)}$, and contains N pseudosites (Fig. 9.1). Generally the lattice has $Nx^{(s)}$ s-type figures and therefore $x^{(s)}$ pseudolattices will correspond to all s figures. Each pseudolattice contains N pseudosites and is displaced relative to the Ising lattice by the vector $\boldsymbol{r}_{\varphi}^{(s)}$, where φ ranges from unity to $x^{(s)}$ (Fig. 9.2). Recall that $x^{(s)}$ is equal to the number of figures s, which share a site and do not transform into each other during translation to vectors of the Ising lattice.

The probability $P_i^{(s)}$ of an i-th configuration figure s can be represented as a successive summation and averaging over pseudosites of all $x^{(s)}$ pseudolattices with allowance for averaging over $\lambda_i^{(s)}$ equivalent configurations, i.e.

$$P_i^{(s)}=\frac{1}{x^{(s)}}\sum_{\varphi=1}^{x^{(s)}}\frac{1}{N}\sum_{j=1}^{N}\frac{1}{\lambda_i^{(s)}}\sum_{k=1}^{R^{(s)}} c_{ik}(\boldsymbol{r}_{\varphi j}^{(s)}), \tag{9.8}$$

where $c_{ik}(\boldsymbol{r}_{\varphi j}^{(s)})$ is the eigenvalue of the generalized projection operator $\hat{P}_{ik}(\boldsymbol{r}_{\varphi j}^{(s)})$, which equals unity, if the i-th configuration figure s located in a position k out of $\lambda_i^{(s)}$ possible positions corresponds to the pseudosite $\boldsymbol{r}_{\varphi j}^{(s)}$; in all other cases $c_{ik}(\boldsymbol{r}_{\varphi j}^{(s)})=0$. The eigenvalues of the generalized projection operator can be expressed as the product of occupancy numbers (eigenvalues $\sigma_{\nu}(\boldsymbol{r})$ of the projection operator $\hat{P}_{\nu}(\boldsymbol{r})$) of the i-configuration sites of an s figure:

$$c_{ik}(\boldsymbol{r}_{\varphi j}^{(s)})=\prod_{l=1}^{R^{(s)}}\sigma_{\nu_{ikl}}(\boldsymbol{r}_{\varphi j}^{(s)}+\boldsymbol{r}_{\varphi l}^{(s)}), \tag{9.9}$$

where $R^{(s)}$ is the number of sites comprising the s-type figure, ν_{ikl} is the atom species at a site l of the s-type figure with the i-configuration occupying position k, $\boldsymbol{r}_{\varphi j}^{(s)}+\boldsymbol{r}_{\varphi l}^{(s)}$ are the coordinates of a site l of the j-th figure s corresponding to pseudolattice pseudosite φ, and $\sigma_{\nu}(\boldsymbol{r})=1$ if the site with coordinates $\boldsymbol{r}$ contains an atom of species ν, otherwise $\sigma_{\nu}(\boldsymbol{r})=0$.

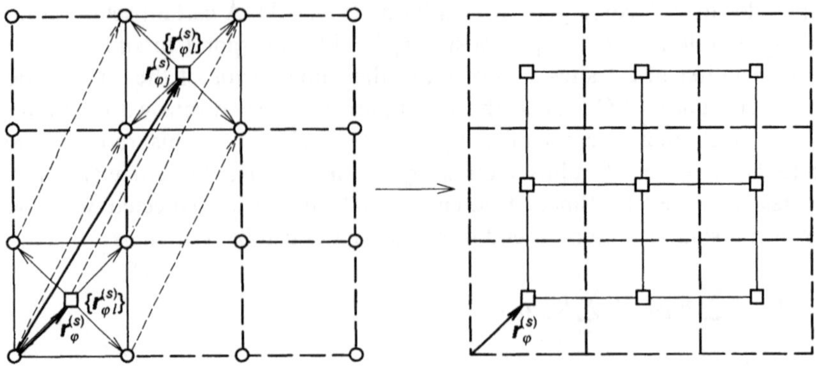

Fig. 9.1. Disordered plane lattice and the corresponding pseudolattice for a cluster in the form of a square ($x^{(s)} = 1$) with four sites: (○) disordered lattice site, (□) pseudolattice site

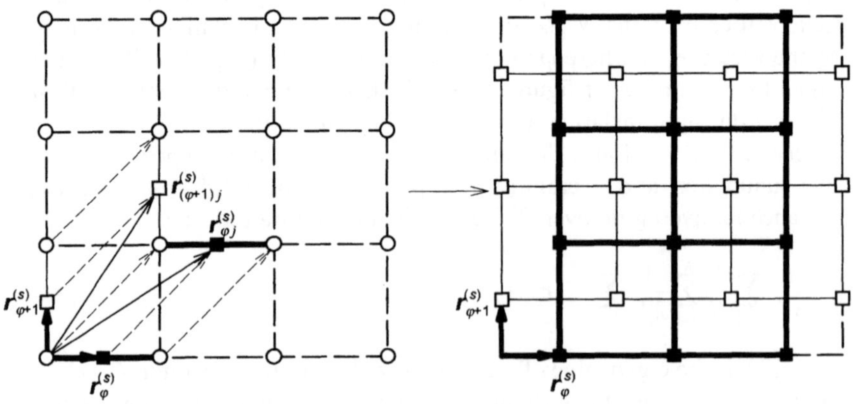

Fig. 9.2. Representation of a disordered plane lattice as a set of one-dimensional figures ($x^{(s)} = 2$) containing two sites: (○) disordered lattice site, (□) and (■) pseudosites of two pseudolattices corresponding to one-dimensional figures which do not transform into one another upon translation to the Ising lattice vectors

The distribution of atoms in the lattice is conveniently described using the distribution function $n_\nu(\boldsymbol{r})$. In the process of ordering the Ising lattice is divided into u nonequivalent sublattices which are characterized by different probabilities $n_\nu(\boldsymbol{r}_u)$ of their $\boldsymbol{r}_u$ sites being filled with ν-species atoms and contain N_u sites each. The value of the distribution function at sublattice sites u is determined as the average of the occupancy numbers, i.e. $n_\nu(\boldsymbol{r}_u) = \langle\sigma_\nu(\boldsymbol{r}_{ju})\rangle$.

Any sublattice of an ordered crystal can be presented as a set of primitive sublattices whose translation vectors are equal primitive translation vectors of an ordered crystal. The number of primitive sublattices for a given sublattice u of the ordered crystal coincides with the multiplicity g_u of positions occupied by atoms of this sublattice. Obviously, the total number of primitive sublattices in the ordered crystal is equal to the sum of multiplicities $\sum_u g_u$ of crystallographically non-equivalent positions, i.e. Φ_s is the number of ordered lattice sites constituting a unit cell of the ordered crystal. Since all atoms of one and the same sublattice are in crystallographically equivalent positions, the primitive sublattices that the given sublattice u is broken into are equivalent. This means that the average occupancy numbers of sites of a primitive sublattice p coincide with the value of the distribution function at sites of the u sublattice, i.e.

$$(1/N_p)\sum_{j_p=1}^{N_p}\sigma_\nu(\boldsymbol{r}_{j_p}) = n_\nu(\boldsymbol{r}_u)\,, \tag{9.10}$$

where $N_p = N/\Phi_s$ is the number of atoms in the primitive sublattice p.

In the process of ordering the pseudolattice is broken into $f = \Phi_s$ translational nonequivalent pseudosublattices. Each pseudosublattice contains the same number of pseudosites $N_f = N_p$, possesses the ordered crystal translational symmetry, is identical to the primitive sublattices p of the ordered crystal, and is displaced relative to these primitive sublattices by a vector $\boldsymbol{r}_\varphi^{(s)}$ (Fig. 9.3).

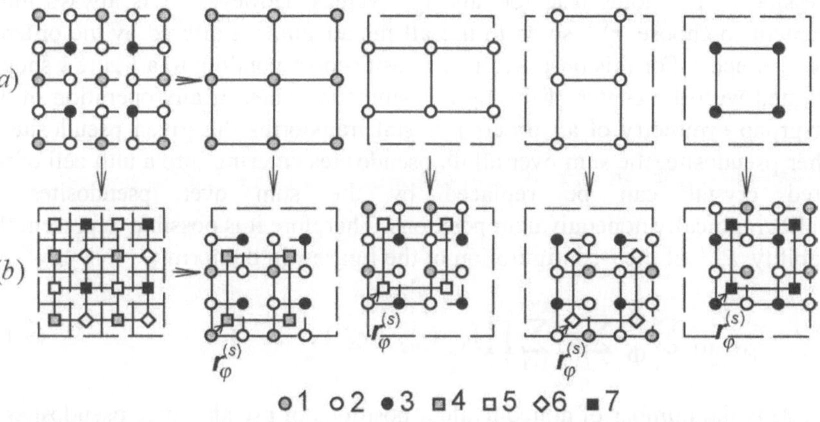

Fig. 9.3. Breakdown of the lattice and of the corresponding pseudolattice during ordering: (*a*) the ordered plane lattice and its primitive sublattices (1, 2 and 3 are the first, second and third sublattice sites with multiplicities $g_1 = 1$, $g_2 = 2$ and $g_3 = 1$; the second sublattice is composed of two primitive sublattices displaced relative to one another by the vector of the disordered lattice); (*b*) the ordered plane pseudolattice and its pseudosublattices displaced relative to primitive sublattices of the ordered plane lattice by the vector $\boldsymbol{r}_\varphi^{(s)}$ (4 to 7 are pseudosublattice sites)

Since the pseudolattice can be broken down into pseudosublattices, in (9.8) it is possible to replace the summation over pseudosites j of the pseudolattice φ by summation over pseudosites j_f of pseudosublattices f. Considering this fact and (9.9), the probability $P_i^{(s)}$ can be written as

$$P_i^{(s)} = \frac{1}{x^{(s)}\lambda_i^{(s)}} \sum_{\varphi=1}^{x^{(s)}} \sum_{k=1}^{\lambda_i^{(s)}} \frac{1}{\Phi_s} \sum_{f=1}^{\Phi_s} \frac{1}{N_f} \sum_{j_f=1}^{N_f} \prod_{l=1}^{R^{(s)}} \sigma_{\nu_{ikl}} (\boldsymbol{r}_{\varphi j_f}^{(s)} + \boldsymbol{r}_{\varphi l}^{(s)}) . \tag{9.11}$$

Since pseudosites $\boldsymbol{r}_{\varphi j_f}^{(s)}$ have the translational symmetry of the ordered crystal and a certain figure s corresponds to each pseudosite, all sites of this figure possess the same translational symmetry. Hence it follows that at fixed φ and l values, the vectors $\boldsymbol{r}_{\varphi j_f}^{(s)} + \boldsymbol{r}_{\varphi l}^{(s)} = \boldsymbol{r}_{\varphi j_p}^{(s)}$ of a given pseudosublattice f correspond to sites of one and the same primitive sublattice p. Consequently, in (9.11) it is possible to change from summation over pseudosites j_f of the pseudosublattice f to summation over sites j_p of the primitive sublattice p and, using the mean field approximation, replace the mean product of occupancy numbers p in (9.11) by the product of mean values. As a result, considering (9.11), we have an expression determining the probability $P_i^{(s)}$ through the distribution function

$$P_i^{(s)} = \frac{1}{x^{(s)}\lambda_i^{(s)}} \sum_{\varphi=1}^{x^{(s)}} \sum_{k=1}^{\lambda_i^{(s)}} \frac{1}{\Phi_s} \sum_{f=1}^{\Phi_s} \prod_{l=1}^{R^{(s)}} n_{\nu_{ikl}} (\boldsymbol{r}_{\varphi f}^{(s)} + \boldsymbol{r}_{\varphi l}^{(s)}) . \tag{9.12}$$

Expression (9.12) holds true for any $\boldsymbol{r}_{\varphi}^{(s)}$ values. However, it is always more convenient to choose $\boldsymbol{r}_{\varphi}^{(s)}$ so as to use all the advantages offered by the ordered phase symmetry. For this purpose, a pseudosite corresponding to a figure s should be aligned with the center of the figure symmetry. Then, if any operation of the point group symmetry of an ordered crystal transforms the given pseudosite to another pseudosite, the sum over all Φ_s pseudosites entering into a unit cell of the ordered crystal can be replaced by the sum over pseudosites in crystallographically nonequivalent positions. Therefore it is possible to present the probability $P_i^{(s)}$ of the i-configuration of the figure s in the form

$$P_i^{(s)} = \frac{1}{x^{(s)}\lambda_i^{(s)}} \sum_{\varphi=1}^{x^{(s)}} \frac{1}{\Phi_s} \sum_{f=1}^{M} g_f \sum_{k=1}^{\lambda_i^{(s)}} \prod_{l=1}^{R^{(s)}} n_{\nu_{ikl}} (\boldsymbol{r}_{\varphi f}^{(s)} + \boldsymbol{r}_{\varphi l}^{(s)}) , \tag{9.13}$$

where M is the number of nonequivalent positions of pseudolattice pseudosites φ located at centers of symmetry of the corresponding figures; g_f is the multiplicity of nonequivalent positions ($\sum_{f=1}^{M} g_f = \Phi_s$). The product over sites l of the figure s in (9.13) can be broken down into products over sites of the figure s occupied by atoms of one species. Then

$$P_i^{(s)} = \frac{1}{x^{(s)}\lambda_i^{(s)}} \sum_{\varphi=1}^{x^{(s)}} \frac{1}{\Phi_s} \sum_{f=1}^{M} g_f \sum_{k=1}^{\lambda_i^{(s)}} \prod_{\nu=\alpha}^{\tau} \prod_{l_\nu=0}^{L_\nu^{(s)}} n_\nu (\boldsymbol{r}_{\varphi f}^{(s)} + \boldsymbol{r}_{\varphi i k l_\nu}^{(s)}) , \tag{9.14}$$

where $\nu = \alpha, \beta, ..., \tau$ is the atomic species; l_ν and $L_\nu^{(s)}$ denote the designation and the number of sites in the s-type figure occupied by ν-species atoms ($\sum_{\nu=\alpha}^{\tau} L_\nu^{(s)} = R^{(s)}$); and $(\boldsymbol{r}_{\varphi f}^{(s)} + \boldsymbol{r}_{\varphi i k l_\nu}^{(s)})$ stands for the coordinates of a site l_ν belonging to the i-configuration of the figure s, which occupies the position k and corresponds to nonequivalent pseudosites with radius-vector $\boldsymbol{r}_{\varphi f}^{(s)}$. It is easy to see that in the absence of long-range order, when $\eta = 0$ and $n_\nu(\boldsymbol{r}) = c_\nu$, equations (9.12), (9.13) and (9.14) describe the probabilities $P_i^{(s)}$ as the binomial distribution.

In accordance with (8.63) and (8.64), the distribution function $n(\mathbf{r})$ directly depends on the order parameters η. Therefore the probability $P_i^{(s)}$ (9.14) expressed as the distribution function and dependent on all values the function $n(\boldsymbol{r})$ assumes at unit cell sites of an ordered crystal is a functional of the long-range order parameters.

When the temperature dependence of the crystal enthalpy is taken into account, the lattice energy is determined by (9.2) where the cluster energies $\varepsilon_i^{(s)}(T)$ are functions of temperature. Representing the probabilities $P_i^{(s)}$ of nonequivalent configurations of various figures with the aid of equation (9.14) permits the configurational energy to be expressed in terms of the distribution function

$$E = \frac{N}{\Phi_s} \sum_s \frac{y^{(s)}}{x^{(s)}} \sum_{i \in s} \varepsilon_i^{(s)}(T) \sum_{\varphi=1}^{x^{(s)}} \sum_{f=1}^{M} g_f \sum_{k=1}^{\lambda_i^{(s)}} \prod_{\nu=\alpha}^{\tau} \prod_{l_\nu=0}^{L_\nu^{(s)}} n_\nu(\boldsymbol{r}_{\varphi f}^{(s)} + \boldsymbol{r}_{\varphi i k l_\nu}^{(s)}) . \tag{9.15}$$

Long-range correlations are taken into account correctly if the combinatorial entropy (the residual part of the configurational energy) is also expressed as the distribution function

$$S_c = -k_B \sum_{\boldsymbol{r}} \sum_{\nu=\alpha}^{\tau} n_\nu(\boldsymbol{r}) \ln n_\nu(\boldsymbol{r}) , \tag{9.16}$$

where the summation is performed over all sites $\boldsymbol{r}$ of the ordered Ising lattice. When the long-range order is a maximum, i.e. when $\eta = 1$, the combinatorial entropy S_c calculated from (9.16) becomes to zero. Thus, the function (9.16) correctly describes the dependence of the entropy S_c on long-range order parameters since the boundary conditions are fulfilled: at $\eta = 0$ the entropy is $S_c = S_c(y,0)$ and at $\eta = 1$ the entropy is $S_c = 0$. Equation (9.16) is equivalent to (8.94) in the cluster variation method. It is worth noting, however, that (8.94) yields correct results if the probabilities $P_i^{(s)}$ are calculated from (9.8) and (9.11)–(9.14) without averaging.

In the case of an isobaric process the configurational energy of the lattice coincides with the enthalpy of the crystal. Therefore the temperature-dependent part of the entropy equals $S(T) = \int_0^T dH/T = \int_0^T dE/T$. Using (9.15), $S(T)$ can be expressed as

$$S(T) = \frac{N}{\Phi_s} \sum_s \frac{y^{(s)}}{x^{(s)}} \sum_{i \in s} \left[\int_0^T \frac{d\varepsilon_i^{(s)}(T)}{T} \right] \sum_{\varphi=1}^{x^{(s)}} \sum_{f=1}^{M} g_f \sum_{k=1}^{\lambda_i^{(s)}} \prod_{\nu=\alpha}^{\tau} \prod_{l_\nu=0}^{L_\nu^{(s)}} n_\nu(\boldsymbol{r}_{\varphi f}^{(s)} + \boldsymbol{r}_{\varphi i k l_\nu}^{(s)}) . \tag{9.17}$$

With allowance for (9.15)–(9.17), the free energy of the ordered crystal can be written in the general form

$$F = N\left\{\frac{1}{\Phi_{\mathrm{s}}}\sum_{s}\frac{y^{(s)}}{x^{(s)}}\sum_{i\in s}\left[\varepsilon_i^{(s)}(T) - T\int_0^T \frac{d\varepsilon_i^{(s)}(T)}{T}\right]\sum_{\varphi=1}^{x^{(s)}}\sum_{f=1}^{M} g_f \sum_{k=1}^{\lambda_i^{(s)}}\prod_{\nu=\alpha}^{\tau}\prod_{l_\nu=0}^{L_\nu^{(s)}} n_\nu(\boldsymbol{r}_{\varphi f}^{(s)} + \boldsymbol{r}_{\varphi i k l_\nu}^{(s)}) + k_{\mathrm{B}}T\sum_{\boldsymbol{r}}\sum_{\nu=\alpha}^{\tau} n_\nu(\boldsymbol{r})\ln n_\nu(\boldsymbol{r})\right\}. \tag{9.18}$$

Equation (9.18) for the free energy of a crystal with any degree of ordering is the fundamental relationship in the order-parameter functional method. The key feature of this technique is that the probabilities of different figures (clusters) are given as a distribution function. Consequently, long-range correlations in the crystal can be taken into account indirectly. Considering that values of the distribution function, which are expressed as long-range order parameters, are taken into account explicitly in (9.18) and the free energy F of a crystal is minimized over these parameters, it is possible to determine the main characteristics of the order–disorder phase transition: the transition temperature and variations of the free energy, entropy and long-rang order parameters upon transition and, ultimately, the kind of phase transition. From the aforesaid it is clear that a general method for formulation of the free energy of crystals with an arbitrary structure can be consistently developed within the order-parameter functional method. In this method it is very efficient to consider the pseudolattice of cluster figures, making it possible to use a simple formalism for probabilities of various atomic configurations.

Generally, from the above discussion it is evident that the physical basis of the order-parameter functional method is the mean field approximation and, as for its formalism, it refers to cluster methods but favorably differs from them since the OPF method allows in detail for the symmetry of a crystal with any degree of long-range order.

A practical implementation of the OPF method is possible if one knows the cluster energies $\varepsilon_i^{(s)}$, which can be assigned as model parameters or calculated from some considerations. Let us examine in greater detail the method proposed [13] for calculating the energies $\varepsilon_i^{(s)}$ from experimental thermodynamical data characterizing a disordered phase.

9.2 Energies of Clusters

While in the cluster variation method the free energy is minimized over variables represented by probabilities, in the OPF method the probabilities $P_i^{(s)}$ are expressed through the distribution function $n(\boldsymbol{r})$ in accordance with (9.13) and (9.14). The distribution function directly depends on the long-range order parameters η, and therefore it is precisely these parameters that are variables in the

OPF method. Obviously, the number of variables in the OPF method and, consequently, the volume of calculations associated with minimization of the free energy is essentially smaller than in the CV method.

Let us examine the method of calculating the energy of different R-particle interactions in a crystal, i.e. the energy of different configurations of a figure containing R sites of an ordering lattice. The basis for this method is the concept of the configurational energy E being the sum of energies of individual configurations. This concept is valid for both the cluster variation method and the order-parameter functional method.

According to (9.18), the free energy of a crystal, whose lattice is broken down into u sublattices during ordering, is the sum of the configurational energy and the entropy term. From (9.18) it is evident that in the OPF method the free energy of a crystal with any degree of ordering is determined by its composition, temperature and long-range order parameters. Under isobaric conditions the configurational energy (9.15) of a crystal has the meaning of enthalpy. The enthalpy of compounds at 298 K is equal to the enthalpy of their formation from elements, i.e. $H^{\circ}_{298} \equiv \Delta H^{\circ}_{f298}$, and, with allowance for the temperature versus enthalpy dependence, $H(T) = H^{\circ}_{298} + (H^{\circ}_{T} - H^{\circ}_{298}) = \Delta H^{\circ}_{f298} + (H^{\circ}_{T} - H^{\circ}_{298})$. Hence, it is possible to determine the energy $\varepsilon_i^{(s)}(T)$ if for some crystal state we know all the probabilities $P_i^{(s)}$, the enthalpy of crystal formation in this state, and the temperature dependence of the enthalpy or heat capacity. As was shown in [13–15], the best choice here is a disordered state when long-range order parameters η are zero and, as a result, the probabilities $n_v(\boldsymbol{r})$ are the same for all sites $\boldsymbol{r}$ and are equal to c_v. Taking this into account, it is easy to show that the free energy of a disordered crystal is written similarly both in the order-parameter functional method and in the cluster variation method. Since a disordered state was chosen in the model for calculating $\varepsilon_i^{(s)}(T)$, the formula for the cluster energy derived in [13, 15] can be equally applied in both procedures. Let us consider the deduction of the formula for determining the cluster energy following the lines of reasoning presented in [13, 15].

For a disordered binary substitutional solid solution A_yB_{1-y} or a disordered nonstoichiometric compound MX_y where the substitutional solution is formed by interstitial atoms and structural vacancies, $n_v(\boldsymbol{r}) = y$ (v denoting A-species atoms in the solid solution or interstitial atoms X in the nonstoichiometric compound). As a result, summation over equivalent configurations k, crystallographically nonequivalent positions of pseudosites of pseudosublattices f, and pseudo-sublattices φ is omitted from (9.14). Therefore the probabilities of different i-configurations in a disordered crystal are determined by the binomial distribution

$$P_i^{(s)}(y,0) = P_i^{(s)\text{bin}} = y^{(R^{(s)}-m)}(1-y)^m \,, \tag{9.19}$$

where m is the number of B-species atoms (or structural vacancies in the nonmetallic sublattice of the nonstoichiometric compound MX_y) in the i-configuration of the s-type figure.

According to (9.19), the probabilities $P_i^{(s)}(y,0)$ depend on the crystal composition only. Therefore the temperature dependence of the enthalpy $H(y,0,T)$ of a disordered crystal is indicative of the temperature dependence of energies $\varepsilon_i^{(s)}$ of different figure configurations. Hence, the equation relating the energies $\varepsilon_i^{(s)}$ and the enthalpy of a disordered crystal takes the form

$$N\sum_s y^{(s)}\sum_{i\in s}\varepsilon_i^{(s)}(T)\lambda_i^{(s)}P_i^{(s)}(y,0)=H(y,0,T). \tag{9.20}$$

However, the system of linear equations of the form (9.20) written for different compositions y cannot be solved with respect to $\varepsilon_i^{(s)}(T)$ directly. The point here is that in the OPF method (like in the CV method) a crystal is described by a sequence of special figures including the basis cluster figure and overlapping figures. According to (8.85), the probabilities of any subfigures of the basis figure (overlapping figure probabilities inclusive) present a linear combination of probabilities of different configurations of the basis figure. As a consequence, the system of equations (9.20) is indeterminate. It can be solved only if relationships between the energies of different-type figures are known additionally. This makes it possible to divide the configurational energy of a crystal into several terms, each accounting for the energy of configurations of figures of one type only and having the form

$$N\sum_{i\in s}\varepsilon_i^{(s)}(T)\lambda_i^{(s)}P_i^{(s)}(y,0)=\kappa^{(s)}H(y,0,T), \tag{9.21}$$

where $\kappa^{(s)}$ is a coefficient relating the enthalpy and energies of all figures of a given type s with different configurations. The coefficients $\kappa^{(s)}$ and $y^{(s)}$ are related to one another by the normalization condition $\sum_s \kappa^{(s)}y^{(s)}=1$.

In the general case the system of linear equations (9.21) can be solved numerically, but with certain assumptions it is possible to solve it analytically.

Let us expand the enthalpy $H(y,0,T)$ in a power series with respect to the composition y, i.e.

$$H(y,0,T)=N\sum_{n=0}^{R^{(s)}}y^n H_n(T). \tag{9.22}$$

Assume that nonequivalent configurations with one and the same number m of atoms, which are similar but located differently, have equal energies $\varepsilon_{i\in m}^{(s)}=\varepsilon_m^{(s)}$. From (9.19) it is obvious that in the disordered state the probabilities of configurations $i\in m$ with one and the same number m of the like-species atoms are equal, i.e. $P_{i\in m}^{(s)}=P_m^{(s)}$. In conformity with known combinatorial relationships, the multiplicities $\lambda_m^{(s)}=\sum_{i\in m}\lambda_i^{(s)}$ of the m-configuration of a figure s are equal to binomial distribution coefficients $C_{R^{(s)}}^m$.

Transforming (9.21) to allow for the above approximations and substituting (9.19) and (9.22), we have

$$\sum_{m=0}^{R^{(s)}} \varepsilon_m^{(s)}(T) C_{R^{(s)}}^m y^{(R^{(s)}-m)} (1-y)^m = \kappa^{(s)} \sum_{n=0}^{R^{(s)}} y^n H_n(T) . \tag{9.23}$$

Expand $(1-y)^m$ in the series

$$(1-y)^m = \sum_{k=0}^{m} (-1)^k C_m^k y^k . \tag{9.24}$$

Considering this expansion, (9.23) takes the form

$$\sum_{m}^{R^{(s)}} \sum_{k=0}^{m} (-1)^k C_{R^{(s)}}^m C_m^k y^{(R^{(s)}-m+k)} \varepsilon_m^{(s)}(T) = \kappa^{(s)} \sum_{n=0}^{R^{(s)}} y^n H_n(T) . \tag{9.25}$$

The summation in the left part of (9.25) is performed over the lines of a triangular matrix $\boldsymbol{T} = [\boldsymbol{t}_{mk}]$ with elements $\boldsymbol{t}_{mk} = (-1)^k C_{R^{(s)}}^m C_m^k y^{(R^{(s)}-m+k)} \varepsilon_m^{(s)}(T)$. Let us group the terms with the same degrees y in the left part of (9.25). This grouping corresponds to a transition from summation over lines of the matrix $\boldsymbol{T}$ to summation over the main diagonal and lines parallel to it. The sum $(R^{(s)} - m + k) = n$ is a constant for each of the lines. The said grouping yields the following formula

$$\sum_{n=0}^{R^{(s)}} y^n \sum_{m=0}^{R^{(s)}} (-1)^{(n-R^{(s)}+m)} C_{R^{(s)}}^m C_m^{R^{(s)}-n} \varepsilon_m^{(s)}(T) = \kappa^{(s)} \sum_{n=0}^{R^{(s)}} y^n H_n(T) . \tag{9.26}$$

Equating coefficients with similar degrees y in the left and right parts of (9.26), we shall pass to a system comprising $(R^{(s)} + 1)$ linear equations written for all values n from zero to $R^{(s)}$ and having the form

$$\sum_{m=0}^{R^{(s)}} (-1)^{(n-R^{(s)}+m)} C_{R^{(s)}}^m C_m^{R^{(s)}-n} \varepsilon_m^{(s)}(T) = \kappa^{(s)} H_n(T) . \tag{9.27}$$

Since $C_{R^{(s)}}^m C_m^{R^{(s)}-n} = C_{R^{(s)}}^n C_n^{R^{(s)}-m}$,

$$\sum_{m=0}^{R^{(s)}} (-1)^{(n-R^{(s)}+m)} C_n^{R^{(s)}-m} \varepsilon_m^{(s)}(T) = \kappa^{(s)} H_n(T) / C_{R^{(s)}}^n , \tag{9.28}$$

where $(-1)^{(n-R^{(s)}+m)} C_n^{R^{(s)}-m} = \boldsymbol{a}_{nm}$ is an element of the triangular matrix $\boldsymbol{A} = [\boldsymbol{a}_{nm}]$. Then the system of equations (9.28) written for all n values can be represented in the matrix form

$$\boldsymbol{A} \varepsilon_m^{(s)}(T) = \kappa^{(s)} H_n(T) / C_{R^{(s)}}^n , \tag{9.29}$$

whence it follows that

$$\varepsilon_m^{(s)}(T) = \kappa^{(s)} \boldsymbol{A}^{-1} H_n(T) / C_{R^{(s)}}^n \equiv \kappa^{(s)} \sum_n \boldsymbol{a}_{kn}^{-1} H_n(T) / C_{R^{(s)}}^n . \tag{9.30}$$

To determine the element $\boldsymbol{a}_{kn}^{-1}$ of the reciprocal lattice A^{-1}, let us prove the lemma according to which $\sum_l (-1)^{(l-m)} C_l^m C_k^l = \delta_{mk}$, where C_l^m and C_k^l are binomial coefficients and

$$\delta_{mk} = \begin{cases} 1, \text{ if } k = m, \\ 0, \text{ if } k \neq m. \end{cases}$$

The product of the binomial coefficients is transformed in the following way:

$$C_l^m C_k^l = C_k^m C_{k-m}^{l-m}, \tag{9.31}$$

hence

$$\sum_l (-1)^{(l-m)} C_l^m C_k^l = C_k^m \sum_l (-1)^{(l-m)} C_{k-m}^{l-m}. \tag{9.32}$$

In compliance with known combinatorial relationships,

$$\sum_l (-1)^{(l-m)} C_{k-m}^{l-m} = \begin{cases} 1, \text{ if } k = m, \\ 0, \text{ if } k \neq m. \end{cases} \tag{9.33}$$

Since at $k = m$ the coefficient $C_k^m = C_m^m = 1$,

$$C_k^m \sum_l (-1)^{(l-m)} C_{k-m}^{l-m} = \begin{cases} 1, \text{ if } k = m, \\ 0, \text{ if } k \neq m, \end{cases} \tag{9.34}$$

or

$$C_k^m \sum_l (-1)^{(l-m)} C_{k-m}^{l-m} = \delta_{mk}. \tag{9.35}$$

Allowance for (9.32) yields

$$\sum_l (-1)^{(l-m)} C_l^m C_k^l = \delta_{mk}, \tag{9.36}$$

as we set out to prove.

In the particular case when $k = m$,

$$\sum_l (-1)^{(l-m)} C_l^m C_m^l = \delta_{mm}. \tag{9.37}$$

The element $\boldsymbol{a}_{kn}^{-1}$ in the reciprocal matrix A^{-1} included in (9.30) can be found from (9.37), which should be written in this case as

$$\sum_n (-1)^{(n-R^{(s)}+m)} C_n^{R^{(s)}-m} C_{R^{(s)}-m}^n = \delta_{mm},$$

and the condition

$$AA^{-1} = \sum_n (-1)^{(n-R^{(s)}+m)} C_n^{R^{(s)}-m} \boldsymbol{a}_{kn}^{-1} = \delta_{km}.$$

It follows that at $k = m$ the element $\boldsymbol{a}_{kn}^{-1} = C_{R^{(s)}-m}^n$. Substitution of $\boldsymbol{a}_{kn}^{-1}$ into (9.30) gives

$$\varepsilon_m^{(s)}(T) = \kappa^{(s)} \sum_n H_n(T) C^n_{R^{(s)}-m} / C^n_{R^{(s)}} \,. \tag{9.38}$$

In view of the transformation $C^n_{R^{(s)}-m} / C^n_{R^{(s)}} = C^m_{R^{(s)}-n} / C^m_{R^{(s)}}$ the final expression for the energy $\varepsilon_m^{(s)}$ is

$$\varepsilon_m^{(s)}(T) = (\kappa^{(s)} / C^m_{R^{(s)}}) \sum_n C^m_{R^{(s)}-n} H_n(T) \,. \tag{9.39}$$

From (9.39) it is clear that a particular form of the cluster energy functions depends on the number of sites $R^{(s)}$ of the chosen basis cluster, the degree n of the polynomial (9.22) describing the y dependence of the formation enthalpy $H(y,0,T)$, and the relationship (which is determined by the coefficients $\kappa^{(s)}$) between the energies of the sequence of figures $\{s\}$.

Let us consider, by way of example, the calculation of the cluster energy for the nonstoichiometric compound MX_y. In Sects. 9.3 and 9.5 below it will be shown that an octahedron comprising six sites ($R^{(s)} = 6$) of the nonmetallic sublattice with a metal atom at the center is used as the basis cluster for the description of nonstoichiometric compounds MX_y. In this case the sequence of special figures $\{s\}$ includes a basis octahedral cluster (type c figure) and overlap figures: a bond (type b figure comprising two adjacent sites of the octahedron) and a site (type a figure). The coefficients are $\kappa^{(a)} = \kappa^{(b)} = 0$ and $\kappa^{(c)} = 1$ for nonstoichiometric compounds MX_y. Let the polynomial (9.22) be of the sixth degree, i.e. $n = 6$. Considering $R^{(s)}$, $\kappa^{(s)}$ and n, and using (9.39), find the energies of octahedron clusters with vacant sites from $m = 0$ to $m = 6$:

$$\left.\begin{aligned}
\varepsilon_0^{(c)} &= H_0(T) + H_1(T) + H_2(T) + H_3(T) + H_4(T) + H_5(T) + H_6(T) \\
\varepsilon_1^{(c)} &= H_0(T) + \tfrac{5}{6} H_1(T) + \tfrac{4}{6} H_2(T) + \tfrac{3}{6} H_3(T) + \tfrac{2}{6} H_4(T) + \tfrac{1}{6} H_5(T) \\
\varepsilon_2^{(c)} &= H_0(T) + + \tfrac{10}{15} H_1(T) + \tfrac{6}{15} H_2(T) + \tfrac{3}{15} H_3(T) + \tfrac{1}{15} H_4(T) \\
\varepsilon_3^{(c)} &= H_0(T) + \tfrac{10}{20} H_1(T) + \tfrac{4}{20} H_2(T) + \tfrac{1}{20} H_3(T) \\
\varepsilon_4^{(c)} &= H_0(T) + \tfrac{5}{15} H_1(T) + \tfrac{1}{15} H_2(T) \\
\varepsilon_5^{(c)} &= H_0(T) + \tfrac{1}{6} H_1(T) \\
\varepsilon_6^{(c)} &= H_0(T)
\end{aligned}\right\} . \tag{9.39a}$$

It is seen that the cluster energy can be rather simply expressed in the binomial approximation as the coefficients of the polynomial (9.22) describing the dependence of the formation enthalpy $H(y,0,T)$ on the relative concentration y of interstitial atoms.

In the OPF method, the entropy of an ordered crystal consists of two parts: the combinatorial energy S_c and the entropy $S(T)$, which explicitly depends on temperature. Since the enthalpy $H(T)$ is determined as the standard value of the enthalpy H_{298} at 298 K and the variation of the enthalpy with temperature ($H_T - H_{298}$), the entropy $S(T)$ should be represented by its standard value S_{298} and its variation with temperature to preserve the universal character of the formula:

$$S(T) = S_{298} + \int_{298}^{T} d(H_T - H_{298})/T\ . \tag{9.40}$$

It is easy to see that for a disordered compound (9.17) takes the following form:

$$N\sum_{s} y^{(s)} \sum_{i\in s} \lambda_i^{(s)} P_i^{(s)}(y,0) S_i^{(s)}(T) = S(y,0,T)\ . \tag{9.41}$$

If we expand the entropy of a disordered compound into a power series analogous to (9.22):

$$S(y,0,T) = N\sum_{n}^{R(s)} y^n S_n(T)\ , \tag{9.42}$$

where $S_n(T) = S_{n298} + \int_{298}^{T} dH_n(T)/T$, and make the same manipulations as with the deduction of the expression for the cluster energy, we shall have:

$$S_m^{(s)}(T) = (\kappa^{(s)} / C_{R(s)}^m) \sum_{n} C_{R(s)-n}^m S_n(T)\ , \tag{9.43}$$

i.e.

$$S_m^{(s)}(T) = S_{m298}^{(s)} + \int_{298}^{T} d\varepsilon_m^{(s)}(T)/T\ . \tag{9.44}$$

Thus, we have derived an analytical formula for calculating the cluster energies through the coefficients of a polynomial, which approximates the concentration dependence of the disordered crystal enthalpy, and also an analogous formula for the cluster entropy. As opposed to existing laborious and approximate techniques, the above method provides an easy way for calculating absolute values of the cluster energy taking account for its temperature dependence. The use of the energy values obtained by the proposed method permits a highly accurate determination of absolute values of order–disorder transformation temperatures. Phase diagrams calculated by this method need not be fitted to experimental phase diagrams with the use of energy parameters.

9.3 The Choice of the Basis Cluster Figure

In order to describe the atomic ordering in strongly nonstoichiometric interstitial compounds by the OPF method, it is necessary to choose the basis cluster figure, find overlapping figures and determine the sequence of special figures $\{s\}$.

A specific feature of strongly nonstoichiometric compounds with the $B1$ structure is that they contain metal and nonmetal fcc sublattices displaced relative to one another. In nonstoichiometric compounds with such a structure, the first coordination sphere of any metal atom is formed by six sites of the nonmetal sublattice. Each site may be vacant or occupied by an interstitial atom. Disregarding atomic displacements, the six sites of the nonmetal sublattice make up a regular octahedron with a metal atom at the center. The choice of such an octahedral cluster as the basis figure enables one to take into account metal–nonmetal interatomic interactions and multiparticle correlations in the first coordination sphere of the metal atom, which are most significant for nonstoichiometric compounds. Moreover, the choice of an octahedron as the basis cluster figure makes it possible to allow for all conceivable sites of the crystal lattice, since each site of the lattice belongs to at least one octahedron.

The OPF method imposes the same requirements upon the basis cluster figure as the cluster variation method. In particular, the basis cluster figure should be "closed", i.e. it should not contain separately bound sites. To verify if this condition can be met, it is necessary to find the sequence of overlapping figures. In a lattice with the $B1$ structure the first generation of overlapping figures is octahedron edges coupling two nearest sites of the nonmetal sublattice, i.e. paired bonds. Nonmetal sublattice sites serve, in their turn, as bonding overlapping figures. Thus, in the case under consideration, the sequence of special figures $\{s\}$ includes three figures: a nonmetal sublattice site (a-type figure); a bond between adjoining nonmetal sublattice sites located at a distance equal to the octahedron edge (b-type figure); and an octahedron constituted by six nonmetal sublattice sites with the metal atom at its center (c-type figure). The octahedron satisfies the definition of a "closed" figure since any of its six sites is at the top and belongs simultaneously to four edge-bonds, i.e. to more than one special figure b.

To avoid overestimation of lattice sites as a result of figure overlapping (for example, each nonmetal sublattice site belongs to six adjoining octahedra and twelve bonds simultaneously), correction coefficients $y^{(s)}$ should be found. For this purpose it is necessary to solve the system of linear equations (8.84) written for a chosen sequence of figures $\{s\}$:

$$\left.\begin{aligned} u^{(a)(a)}y^{(a)}+u^{(a)(b)}y^{(b)}+u^{(a)(c)}y^{(c)}&=x^{(a)}\\ u^{(b)(a)}y^{(a)}+u^{(b)(b)}y^{(b)}+u^{(b)(c)}y^{(c)}&=x^{(b)}\\ u^{(c)(a)}y^{(a)}+u^{(c)(b)}y^{(b)}+u^{(c)(c)}y^{(c)}&=x^{(c)} \end{aligned}\right\}. \qquad (9.45)$$

Necessary matrix elements $|u^{(m)(s)}|$ establishing the relationship between special figures of the sequence $\{s\}$ and coefficients $x^{(m)}$, as well as coefficients $y^{(s)}$ calculated on their basis for the $B1$-type lattice, are given in Table 9.1.

Table 9.1. The matrix elements $|u^{(m)(s)}|$, the values of $x^{(m)}$ and the overestimate coefficients $y^{(s)}$ for a $B1$-type crystal lattice in the case of an octahedral basis cluster

Figure of m-type	Figure of s-type a	b	c	$X^{(m)}$
A	1	2	6	1
B	0	1	12	6
C	0	0	1	1
$y^{(s)}$	7	−6	1	

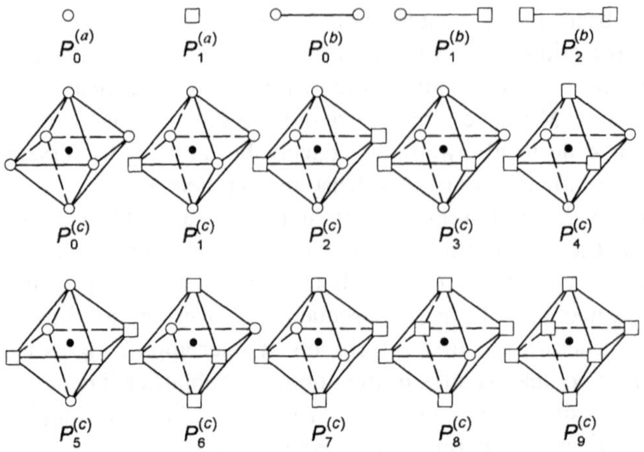

Fig. 9.4. Nonequivalent configurations of figures *a* (site), *b* (bond) and *c* (octahedron) of the sequence $\{s\}$ necessary to describe strongly nonstoichiometric interstitial compounds MX_y with the basic structure *B*1: (○) interstitial atom X, (□) vacancy, (●) metal atom M

Sites of one and the same figure can be vacant or occupied by atoms, therefore several alternative mutual locations of vacant and occupied sites, i.e. several different *i*-configurations with probabilities $P_i^{(s)}$ correspond to each figure of the chosen sequence $\{s\}$. For the figure *a* (site), two nonequivalent configurations are possible – a site occupied by an atom X (configuration probability $P_0^{(a)}$) and a vacant site □ (configuration probability $P_1^{(a)}$). The probabilities $P_0^{(b)}$, $P_1^{(b)}$, and $P_2^{(b)}$ correspond to the figure *b* (bond) with the following configurations: X–X, X–□, and □–□. For the basis cluster figure in the form of an octahedron, ten nonequivalent configurations are possible – a complete cluster (all six sites are occupied by atoms), a cluster with one vacancy, etc., up to a completely defective cluster (all six sites are vacant) with the corresponding probabilities $P_0^{(c)}, \ldots, P_9^{(c)}$. All nonequivalent configurations of the chosen sequence $\{s\} = \{a, b, c\}$ are shown in Fig. 9.4. Nonequivalent configurations exist in addition to equivalent ones. Equivalent configurations are those that coincide with one another upon operations of point group symmetry corresponding to a disordered lattice – operations of $m\bar{3}m$ point group symmetry in the case at hand. For example, the configurations X–□ and □–X are equivalent. The multiplicities $\lambda_i^{(s)}$ are calculated by the following equation

$$\lambda_i^{(s)} = n(G_0^{(s)}) / n(G_i^{(s)}), \tag{9.46}$$

where $n(G)$ is the order of the group *G*. The multiplicities $\lambda_i^{(s)}$ of all configurations of the chosen sequence of figures are presented in Table 9.2.

Table 9.2. The multiplicities $\lambda_i^{(s)}$ of i-configurations of s-type figures of the sequence of special figures $\{s\}$ which are used for describing a lattice with the $B1$ structure

Type of figure s	I 0	1	2	3	4	5	6	7	8	9
a	1	1	–	–	–	–	–	–	–	–
b	1	2	1	–	–	–	–	–	–	–
c	1	6	3	12	8	12	12	3	6	1

9.4 Probabilities of Clusters and Subfigures

A necessary condition for the application of the OPF method is the correct representation of the probabilities of different configurations of figures in the sequence $\{s\}$ as the distribution function $n(\boldsymbol{r})$ or, to be more exact, as values the function $n(\boldsymbol{r})$ assumes at sites of a figure with the given configuration.

In the general case the distribution function assumes several values. For a binary substitutional solid solution A_yB_{1-y} or a nonstoichiometric compound MX_y, these values can be presented as

$$n_j = y + \sum_s \mu_{js}\eta_s \,, \tag{9.47}$$

where s is the number of the order parameter coinciding with the number of the star $\{\boldsymbol{k}_s\}$; and μ_{js} is a numerical coefficient preceding η_s in the formulas for the distribution function values (see Tables 5.1–5.11, 5.13, 5.14, and 5.16–5.18).

Each structure can be characterized by a set of values of the corresponding distribution function and K_j values, which stand for the relative share of ordered lattice sites at which the distribution function acquires the value n_j. For any distribution function taking u values, K_j values obey the usual normalization condition $\sum_{j=1}^{u} K_j = 1$ and the composition normalization condition:

$$\sum_{j=1}^{u} K_j n_j = y \,. \tag{9.48}$$

In the order parameter functional method it is shown [15] that the probabilities $P_i^{(s)}$ are determined as the mean value over the probabilities $P_{if}^{(s)}$ of crystallographically nonequivalent figures f corresponding to crystallographically nonequivalent pseudosites:

$$P_i^{(s)} = \frac{1}{\Phi_s}\sum_{f=1}^{M} g_f P_{if}^{(s)} \,. \tag{9.49}$$

Let K_{jf} be the share of sites f at which the distribution function acquires the value n_j. The definition of K_{jf} suggests the normalization condition $\sum_{j=1}^{u} K_{jf} = 1$ and the equation relating K_{jf} and K_j:

$$K_j = \frac{1}{\Phi_s} \sum_{f=1}^{M} g_f K_{jf} \,. \tag{9.50}$$

Taking into account (9.48) and (9.50), it is easy to deduce another useful relationship

$$\sum_{j=1}^{u} \mu_{js} K_j = 0 \,, \tag{9.51}$$

which readily shows that the obtained values of the distribution function are correct.

Let us consider the practical determination of the probability $P_i^{(s)}$ for some configuration of the basis cluster taking superstructures of strongly stoichiometric carbides under discussion as an example.

In the process of ordering, redistribution of carbon atoms and vacancies takes place. For figures in the form of an octahedron the coefficient $x^{(c)} = 1$ (see Table 9.1). Therefore one pseudolattice corresponds to c-type figures and in formulas (9.12)–(9.14) summation over pseudolattices φ is omitted. If pseudosites corresponding to octahedral configurations are made coincident with metal atoms located at symmetry centers of octahedra, summation over all Φ_s pseudosites of the ordered crystal unit cell can be replaced by summation over pseudosites which coincide with metal atoms occupying crystallographically nonequivalent positions. The types of nonequivalent positions of metal atoms and their multiplicity g_f for different superstructures of solid solutions (alloys) and nonstoichiometric compounds are given in Tables 5.1–5.11, 5.13, 5.14, and 5.16–5.18. The distribution of sites in ordered crystal sublattices is found using the corresponding distribution function. The distribution of sites in the sublattices and the values of the distribution function at these sites are also listed in the above tables.

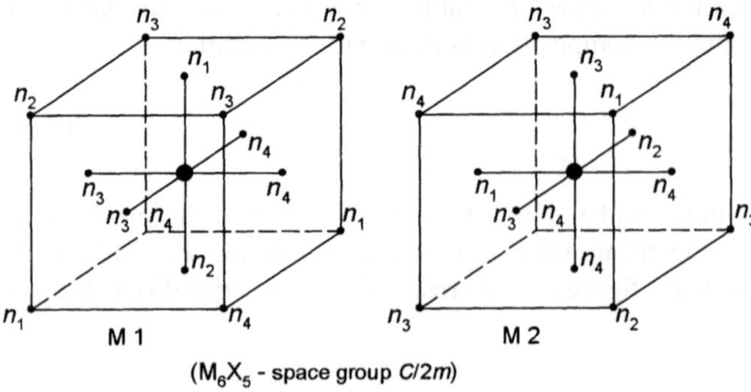

Fig. 9.5. Values of the distribution function (5.72) at the nonmetal sublattice sites forming the first and third coordination spheres of metal atoms occupying crystallographically nonequivalent positions in the monoclinic (space group $C2/m$) superstructure M_6X_5 (for numbering of metal atoms see Table 5.11)

For instance, the monoclinic ordered structure M_6X_5 (space group *C*2/*m*) has two nonequivalent positions of metal atoms with multiplicities 4 and 8 (see Table 5.11). The distributions of nonmetal sublattice sites, which correspond to these positions and form the first and third coordination spheres of metal atoms (the second coordination sphere is made up of metal atoms), are shown in Fig. 9.5. The octahedral cluster contains six sites of the first coordination sphere, whereas a unit cell of the superstructure M_6X_5 (space group *C*2/*m*) has twelve pseudosites corresponding to *c*-type figures (octahedra). Let us determine, as an example, the probability $P_2^{(c)}$ of a cluster with two vacancies (see Fig. 9.4) in this superstructure. The superstructure M_6X_5 (space group *C*2/*m*) is described in the most general case by three long-range order parameters η_9, η_4 and η_3. Given different values of the parameters η, the distribution function $n(\mathbf{r})$ (5.72) takes four values (Table 5.11) which directly depend on η_9, η_4 and η_3. The distribution of vacancies is described by the function $n_\square(\mathbf{r}) = 1 - n(\mathbf{r})$. With allowance made for the values of $n(\mathbf{r})$ (Table 5.11) and $\lambda_2^{(c)} = 3$ (Table 9.2), the probability $P_2^{(c)}$ will be equal, in accordance with (9.14), to

$$P_2^{(c)} = \tfrac{1}{3}\times\tfrac{1}{12}\times\{4[(1-n_1)(1-n_2)n_3^2n_4^2 + 2(1-n_3)(1-n_4)n_1n_2n_3n_4]$$

$$+8[(1-n_1)(1-n_4)n_2n_3^2n_4 + (1-n_2)(1-n_3)n_1n_3n_4^2 + (1-n_3)(1-n_4)n_1n_2n_3n_4]\}$$

$$= \tfrac{1}{9}\times[(1-n_1)(1-n_2)n_3^2n_4^2 + 2(1-n_1)(1-n_4)n_2n_3^2n_4 + 2(1-n_2)(1-n_3)n_1n_3n_4^2$$

$$+4(1-n_3)(1-n_4)n_1n_2n_3n_4]. \qquad (9.52)$$

The probability $P_i^{(c)}$ of any other octahedral configuration of ordered phases in strongly nonstoichiometric compounds can be determined analogously.

It is often necessary to determine the configuration probability of a figure smaller than the basis cluster. This problem was solved [13] employing a combinatorial method.

Consider an ordered nonstoichiometric compound MX_y. To describe the crystal, let us choose an *s*-type figure containing $R^{(s)}$ sites of the ordered lattice. Any figure formed by *n* sites of the figure *s* ($n < R^{(s)}$) is its *n*-subfigure. The least possible subfigure is a site ($n = 1$); the next subfigure is a pair of sites ($n = 2$), etc. Note that sites of the figure *s* can be located in different coordination spheres relative to one another. For this reason, the *n*-subfigure may have a different geometric form or (in the same geometric form) different dimensions. The total number of all configurations of a given *n*-subfigure (taking account for all possible geometric forms) in the figure *s* equals the number of combinations of $R^{(s)}$ sites including *n* sites, i.e. $N^{(n)(s)} = C^n_{R(s)}$.

Since each configuration of the *n*-subfigure enters one or several configurations of the figure *s*, the probability $P_g^{(n)}$ of some *g*-configuration of the *n*-subfigure can be expressed in terms of the probabilities $P_i^{(s)}$ of different *i*-configurations of the figure *s* with the help of the constant relationships (8.85).

In view of (8.85), for the n-subfigure the sum of the probabilities of all configurations $g \in h$ with the same number of vacancies h is determined by the expression

$$\sum_{g\in h}\lambda_g^{(n)}P_g^{(n)(s)}=\sum_{g\in h}\sum_{i\in s}k_{gi}^{(n)(s)}\lambda_i^{(s)}P_i^{(s)}\,. \tag{9.53}$$

The left part of (9.53) can be transformed by changing the order of summation

$$\sum_{g\in h}\sum_{i\in s}k_{gi}^{(n)(s)}\lambda_i^{(s)}P_i^{(s)}=\sum_{i\in s}\lambda_i^{(s)}P_i^{(s)}\sum_{g\in h}k_{gi}^{(n)(s)}=\sum_{i\in s}\lambda_i^{(s)}P_i^{(s)}k_{hi}^{(n)(s)}\,. \tag{9.54}$$

The coefficient $k_{hi}^{(n)(s)}=N_{hi}^{(n)(s)}/N^{(n)(s)}$ is the relative share of h-configurations of the n-subfigure among all configurations entering the i-configuration of the figure s. All configurations of the n-subfigure which contain h vacancies belong to h-configurations of this subfigure irrespective of the mutual location of occupied and vacant sites. The total number $N_{hi}^{(n)(s)}$ of such h-configurations in the i-configuration of the figure s for all $i \in m$ depends on m only (m is the number of vacancies in the i-configuration of the figure s). Hence it follows that the equations $N_{hi}^{(n)(s)}=N_{hm}^{(n)(s)}$ and $k_{hi}^{(n)(s)}=k_{hm}^{(n)(s)}$ are valid for all configurations $i \in m$ of the figure s. Taking into account the latter equation, break down the sum taken over $i \in s$ in (9.54) into the sums taken over configurations $i \in m$:

$$\sum_{i\in s}\lambda_i^{(s)}P_i^{(s)}k_{hi}^{(n)(s)}=\sum_{m=0}^{R^{(s)}}\sum_{i\in m}\lambda_i^{(s)}P_i^{(s)}k_{hm}^{(n)(s)}=\sum_{m=0}^{R^{(s)}}k_{hm}^{(n)(s)}\sum_{i\in m}\lambda_i^{(s)}P_i^{(s)}\,. \tag{9.55}$$

The value $N_{hm}^{(n)(s)}$ is the product of the number of combinations of $(R^{(s)}-m)$ occupied sites of the i-configuration of the figure s including $(n-h)$ occupied sites by the number of combinations of m vacancies of the i-configuration of the figure s including h, i.e. $N_{hm}^{(n)(s)}=C_{R^{(s)}-m}^{n-h}C_m^h$. If all transformations are taken into consideration, (9.53) can be written as

$$\sum_{g\in h}\lambda_g^{(n)}P_g^{(n)(s)}=\sum_{m=0}^{R^{(s)}}(C_{R^{(s)}-m}^{n-h}C_m^h/C_{R^{(s)}}^n)\sum_{i\in m}\lambda_i^{(s)}P_i^{(s)}\,. \tag{9.56}$$

Since $\sum_{g\in h}\lambda_g^{(n)}P_g^{(n)(s)}=C_n^hP_h^{(n)(s)}$, the probability of any subfigure can be found from the formula

$$C_n^hP_h^{(n)(s)}=\sum_{m=0}^{R^{(s)}}(C_{R^{(s)}-m}^{n-h}C_m^h/C_{R^{(s)}}^n)\sum_{i\in m}\lambda_i^{(s)}P_i^{(s)}\,. \tag{9.57}$$

In the particular case when $h = 0$, the formula for the probability of a complete (vacancy-free) n-subfigure of the figure s, with allowance for $C_{R^{(s)}-m}^n/C_{R^{(s)}}^n=C_{R^{(s)}-n}^m/C_{R^{(s)}}^m$, is

$$P_0^{(n)(s)}=\sum_{m=0}^{R^{(s)}}(C_{R^{(s)}-n}^m/C_{R^{(s)}}^m)\sum_{i\in m}\lambda_i^{(s)}P_i^{(s)}\,. \tag{9.58}$$

Note that at $n=0$ (9.58) reduces to the usual probability normalization condition:

$$P_0^{(0)} = P_0^{(0)(s)} = \sum_{i\in s} \lambda_i^{(s)} P_i^{(s)} \equiv 1 . \tag{9.59}$$

Taking $n = 1$, determine the probability of a site being occupied by atoms, i.e. derive the second normalization condition relating probabilities $P_i^{(s)}$ to the crystal composition. For a nonstoichiometric compound MX_y the probability is $P_0^{(1)} = y$. Taking into account that $C^m_{R^{(s)}-1} = C^m_{R^{(s)}}[(R^{(s)} - m)/R^{(s)}]$, where $(R^{(s)} - m)/R^{(s)}$ is the relative share of sites of the configuration $i \in m$ of the figure s, which are occupied by interstitial atoms, we shall have

$$P_0^{(1)} = P_0^{(1)(s)} = \sum_{i\in s} [(R^{(s)} - m)/R^{(s)}]\lambda_i^{(s)} P_i^{(s)} \equiv y . \tag{9.60}$$

9.5 Free Energy Minimum Conditions

Free Energy Transformation. In the case of ordered structures, the probabilities $P_i^{(s)}$ of different configurations of the basis cluster figure are described by simple but cumbersome and not always convenient expressions. Using some approximations, it is possible to bring the free energy of a crystal with any degree of order to such a form when there is no need to determine the probabilities $P_i^{(s)}$ of numerous clusters.

Considering the temperature dependence of energies $\varepsilon_i^{(s)}(T)$, the free energy of a crystal with any degree of order expressed as probabilities $\varepsilon_i^{(s)}(T)$ has the form

$$F = N\left\{\sum_s y^{(s)} \sum_{i\in s} [\varepsilon_i^{(s)}(T) - TS_i^{(s)}(T)]\lambda_i^{(s)} P_i^{(s)} - TS_c\right\}. \tag{9.61}$$

Break down the sum taken over $i \in s$ in (9.61) into the sum over $i \in m$. Then, in the approximation of equality of energies $\varepsilon_{i\in m}^{(s)} = \varepsilon_m^{(s)}$ and entropies $S_{i\in m}^{(s)} = S_m^{(s)}$, we shall have:

$$F = N\left\{\sum_s y^{(s)} \sum_{m=0}^{R^{(s)}} [\varepsilon_m^{(s)}(T) - TS_m^{(s)}(T)] \sum_{i\in m} \lambda_i^{(s)} P_i^{(s)} - TS_c\right\}. \tag{9.62}$$

In Sect. 9.2 it was shown that when the enthalpy of a disordered crystal is expanded in a power series (9.22), the cluster energy $\varepsilon_m^{(s)}$ can be found in the analytical form (9.39). An analogous expression (9.43) was obtained for the entropy of clusters. Substitution of (9.39) and (9.43) into (9.62) and replacement of $F_n(T) = H_n(T) - TS_n(T)$ gives

$$F = N\left[\sum_s \kappa^{(s)} y^{(s)} \sum_{n=0}^{R(s)} F_n(T) \sum_{m=0}^{R(s)} (C^m_{R(s)-n} / C^m_{R(s)}) \sum_{i\in m} \lambda_i^{(s)} P_i^{(s)} - TS_c\right]. \tag{9.63}$$

In view of (9.58), the free energy (9.63) can be written as

$$F = N\left[\sum_s \kappa^{(s)} y^{(s)} \sum_{n=0}^{R(s)} F_n(T) P_0^{(n)(s)} - TS_c\right]. \tag{9.64}$$

According to (9.59) and (9.60), the probabilities $P_0^{(0)} = 1$ and $P_0^{(1)} = y$, therefore the free energy can be expressed as

$$F = N\left\{\sum_s \kappa^{(s)} y^{(s)} [F_0(T) + yF_1(T) + \sum_{n\geq 2} F_n(T) P_0^{(n)(s)}] - TS_c\right\}. \tag{9.65}$$

At η = 0 the probability $P_0^{(n)(s)} \equiv y^n$. Considering this inference and the normalization condition $\sum_s \kappa^{(s)} y^{(s)} = 1$ (see Sect. 9.2), the free energy $F(y,0,T)$ of a disordered crystal ($\eta = 0$) has the form

$$F(y,0,T) = N[\sum_n y^n F_n(T) - TS_c(y,0)]. \tag{9.66}$$

The terms $F_0(T)$ and $yF_1(T)$ do not depend on the degree of order in the crystal, therefore the difference of free energies of ordered and disordered phases $\Delta F = F(y, \eta, T) - F(y,0,T)$ is equal to

$$\Delta F = N\{\sum_s \kappa^{(s)} y^{(s)} \sum_{n\geq 2} [P_0^{(n)(s)}(y,\eta) - P_0^{(n)(s)}(y,0)] F_n(T)$$

$$-T[S_c(y,\eta) - S_c(y,0)]\}, \tag{9.67}$$

where $S_c(y, \eta)$ and $S_c(y, 0)$ denote the combinatorial entropy (9.16) of ordered and disordered crystals respectively; $P_0^{(n)(s)}(y,0) \equiv y^n$.

It follows from (9.67) that the difference ΔF does not depend on the coefficients $H_0(T)$, $H_1(T)$, $S_0(T)$, and $S_1(T)$ when the expansion members (9.22) and (9.42) are linear. In the general case $[S_c(y,\eta) - S_c(y,0)] < 0$ and therefore when the enthalpy and entropy of a disordered crystal exhibit a linear dependence on the composition, $H(y,0,T) = N[H_0(T) + yH_1(T)]$ and $S(y,0,T) = N[S_0(T) + yS_1(T)]$, the difference $\Delta F = F(y,\eta,T) - F(y,0,T) > 0$ and the ordered phase is not formed.

Thus, in the framework of the aforementioned approximations, a necessary condition for ordering of a nonstoichiometric compound is the nonlinear dependence of enthalpy and entropy on the MX_y composition. Similarly, a necessary condition for ordering of a substitutional solid solution is the nonlinear dependence of the displacement heat on the A_yB_{1-y} composition, i.e. nonideal character of the solid solution.

To calculate the free energy of a crystal using (9.61), the probabilities $P_i^{(s)}$ of all configurations of the basis figure and overlapping figures should be determined in terms of the distribution function. Transformation of the free energy to the form (9.65) reduces the volume of calculations since in this case only the probabilities $P_0^{(n)}$ of several complete subfigures of the basis figure should be expressed through the distribution function. The number of complete subfigures taken into consideration is determined by the order of polynomials (9.22) and (9.42) approximating the composition dependence of the enthalpy and entropy of a crystal. The majority of available experimental data on the enthalpy and entropy for nonstoichiometric compounds and on the heat of mixing for solid solutions (alloys) can be described rather accurately by low-power polynomials. Therefore (9.65) permits minimizing the volume of numerical calculations necessary to determine the equilibrium thermodynamic parameters of structural order–disorder and order–order phase transitions and to construct phase diagrams of ordered systems.

Free Energy Minimum Conditions. As was shown in Sect. 9.3, the sequence of figures $\{s\}$ including the basis cluster in the form of octahedra and overlapping figures (bond and site) is employed in the OPF method to describe strongly nonstoichiometric interstitial compounds. The use of the basis cluster in the form of a regular octahedron with a metal atom located at its center makes it possible to allow for all (metal and nonmetal) sites of the crystal lattice. In addition, the basis cluster energy $\varepsilon_i^{(c)}$ implicitly includes the energies of M–M and M–X interactions, which are the main interactions in nonstoichiometric compounds.

Results of numerous studies generalized in [15, 21, 22] indicate that the energies $\varepsilon_i^{(a)}$ and $\varepsilon_i^{(b)}$ are very small compared to $\varepsilon_{\text{M-M}}$ and $\varepsilon_{\text{M-X}}$. Hence, $\varepsilon_i^{(a)}$ and $\varepsilon_i^{(b)}$ are negligible in comparison to $\varepsilon_i^{(c)}$ and therefore the corresponding coefficients $\kappa^{(a)} = \kappa^{(b)} = 0$. The coefficient $\kappa^{(c)}$ can be found from the condition $\sum_s \kappa^{(s)} y^{(s)} = 1$. Since $y^{(c)} = 1$ (see Table 9.1), at the coefficients $\kappa^{(a)} = \kappa^{(b)} = 0$ the coefficient $\kappa^{(c)} = 1$.

The analysis [15, 17] of experimental data [22–26] on the formation enthalpy and on temperature dependencies of the enthalpy and heat capacity of disordered nonstoichiometric compounds MX_y showed that the expansion of $H(y,0,T)$ into a power series (9.22) for these compounds is a quadratic function

$$H(y,0,T) = N[H_0(T) + yH_1(T) + y^2H_2(T)]. \tag{9.68}$$

Considering (9.68) and the coefficients $\kappa^{(a)} = \kappa^{(b)} = 0$, $\kappa^{(c)} = 1$ and $y^{(c)} = 1$, the free energy (9.65) calculated per mole of the nonstoichiometric compound MX_y with any degree of order has the form

$$F\{n_j\} = F_0(T) + yF_1(T) + F_2(T)P_0^{(2)}\{n_j\} - TS_c\{n_j\}, \tag{9.69}$$

where $\{n_j\}$ is the multitude of values the distribution function $n(\mathbf{r})$ assumes at the crystal lattice sites. In (9.69) and in the further description of nonstoichiometric compounds the probability of a complete pair ($n = 2$) of the basis cluster (c-type figure) will be written for simplicity as $P_0^{(2)}$ ($P_0^{(2)} \equiv P_0^{(2)(c)}$).

It was noted in Sect. 9.2 that the enthalpy of compounds at 298 K is equal to the enthalpy of their formation from constituent elements, i.e. $H_{298}^{\circ} \equiv \Delta H_{f\,298}^{\circ}$. Taking into account the temperature dependence of the enthalpy $(H_T^{\circ} - H_{298}^{\circ})$, the enthalpy is $H(T) = H^\circ_{298} + (H^\circ_T - H^\circ_{298}) \equiv \Delta H^\circ_{f\,298} + (H^\circ_T - H^\circ_{298})$. In the case of nonstoichiometric disordered ($\eta = 0$) compounds MX_y the enthalpy term of the free energy $F(y,0,T)$ is equal to $H(y,0,T) = H^\circ_{298}(y,0) + (H^\circ_T - H^\circ_{298})_{MX_y} \equiv \Delta H^\circ_{f\,298}(y,0) + (H^\circ_T - H^\circ_{298})_{MX_y}$. From this it is seen that a sufficient condition for the expansion of the function $H(y,0,T)$ into the power series of the second or higher order is a nonlinear dependence of the formation heat $\Delta H^\circ_{f\,298}(y)$ on the relative concentration of nonmetal atoms in compounds MX_y. The authors [15, 17, 27] have shown that the dependencies $\Delta H^\circ_{f\,298}(y)$ of nonstoichiometric carbides MC_y are quadratic functions of y and determined their numerical form (functions $\Delta H^\circ_{f\,298}(y)$ of nonstoichiometric carbides and nitrides MX_y are given in numerical form in Table 10.4 of Chap. 10). The conclusion about a nonlinear dependence of the heat of formation, $\Delta H^\circ_{f\,298}(y)$, on the composition of titanium carbide TiC_y was also drawn by Seifert *et al* [28] from the analysis of experimental data [29–34] concerning $\Delta H^\circ_{f\,298}(y)$ of disordered titanium carbide having different carbon contents.

For solid solutions (alloys) the energy of paired bonds makes the main contribution to the total energy, therefore $\kappa^{(b)} \neq 0$. The analysis of experimental data [35] on the formation (mixing) enthalpy $H(y,0,T)$ of disordered solid solutions (alloys) A_yB_{1-y} shows that expansion of $H(y,0,T)$ in a power series with respect to composition y is also described for these solutions by the quadratic function (9.68). Therefore the free energy (9.65) calculated per mole of the solid solution A_yB_{1-y} with any degree of order can be presented as

$$F\{n_j\} = \sum_s \kappa^{(s)} y^{(s)} [F_0(T) + yF_1(T) + F_2(T) P_0^{(2)(s)}\{n_j\}] - TS_c\{n_j\}\,. \tag{9.70}$$

In the disordered state when all long-range order parameters are equal to 0 and all values of the distribution function equal y, $P_0^{(n)(s)} = y^n$ and $P_0^{(2)(s)} = y^2$. Taking into account these formulas and the normalization condition $\sum_s \kappa^{(s)} y^{(s)} = 1$ (see Sect. 9.2), the free energy of the disordered nonstoichiometric compound MX_y or solid solution A_yB_{1-y} has the form

$$F(y,0,T) = F_0(T) + yF_1(T) + y^2F_2(T) - TS_c(y,0). \tag{9.71}$$

Let us consider the dependence of the free energy $F(y)$ (disregarding the configurational term) on the composition of the disordered compound MX_y. The very existence of the nonstoichiometric compound MX_y indicates that its free energy over the whole homogeneity interval is less than the additive sum of free energies of components comprising this compound, i.e. $F_{MX_y}(T) < F_M(T) + yF_X(T)$. Moreover, the composition dependence of the free energy of the nonstoichiometric compound should be described by a concave curve (Fig. 9.6). Indeed, only in this case will the free energy of the compound MX_y (or $M_{1-z}X_z$, where $z = y/(1+y)$) be smaller than the free energy of any two-phase system irrespective of the composition of its constituent phases (M and X; M and $M_{1-z_4}X_{z_4}$; $M_{1-z_1}X_{z_1}$ and X; $M_{1-z_2}X_{z_2}$ and $M_{1-z_3}X_{z_3}$) (see Fig. 9.6).

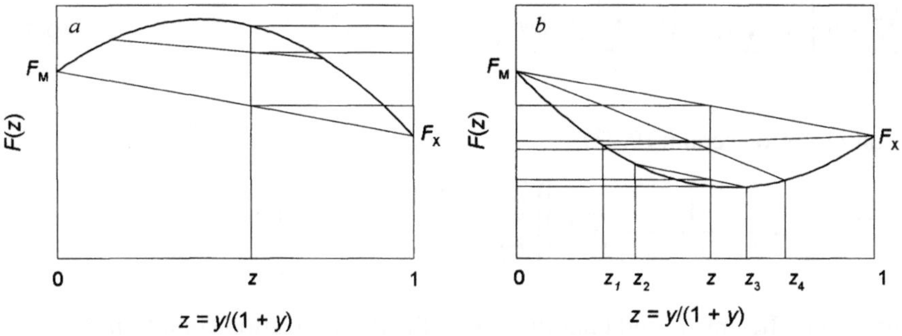

Fig. 9.6. Dependence of the free energy $F(z) \equiv F(y)$ on the composition of the nonstoichiometric compound $MX_y \equiv M_{1-z}X_z$: (*a*) nonstoichiometric compound whose free energy is described by a convex curve cannot exist since it will precipitate to a heterophase mixture with a lower free energy; (*b*) nonstoichiometric compound $MX_y \equiv M_{1-z}X_z$ whose free energy is described by a concave curve always has a smaller free energy than a mixture of any two phases (M and X; M and $M_{1-z4}X_{z4}$; $M_{1-z1}X_{z1}$ and X; $M_{1-z2}X_{z2}$ and $M_{1-z3}X_{z3}$)

Thus, the dependence of the free energy (disregarding the configurational term) on the composition of the disordered compound or solid solution is a concave curve. Otherwise, a compound possessing a homogeneity interval or a solid solution will be thermodynamically unstable and should decompose spontaneously [17, 36, 37]. Considering (9.71), the curve $F(y)$ may be concave only if $F_2(T) > 0$. This suggests that for disordered nonstoichiometric compounds and solid solutions under discussion whose free energy is described by (9.71), the energy parameter $F_2(T)$ is always positive, i.e. $F_2(T) > 0$.

We estimate the probability $P_0^{(2)(s)}\{n_j\}$ of a complete pair, which is a subfigure of the figure s (for example, of the basis cluster) with $R^{(s)}$ sites of the ordering lattice. Let the crystal lattice of an ordered compound MX_y have several configurations of the basis cluster differing by the number of sites at which the distribution function $n(\mathbf{r})$ takes particular values. For instance, the cubic (space group $Pm\bar{3}m$) superstructure M_4X_3 includes octahedral clusters at all six sites of which the distribution function (5.68) assumes the value n_2 (such clusters amount to one-quarter of all clusters in the superstructure); for the remaining octahedral clusters (they are 3/4 of the total number) the distribution function (5.68) takes the value n_1 at two sites and the value n_2 at four sites (see Fig. 5.20). Label the share of clusters with different configurations in the crystal lattice of a certain superstructure as l_q. It is easily understood that in the figure s the probability $P_0^{(2)(s)}\{n_j\}$ is equal to

$$P_0^{(2)(s)}\{n_j\} = \frac{1}{R^{(s)}(R^{(s)}-1)} \sum_q l_q \sum_{\substack{i,j=1 \\ i \neq j}}^{R^{(s)}} n_i^{(q)} n_j^{(q)} . \qquad (9.72)$$

Since

$$\left[\sum_{j=1}^{R^{(s)}} n_j^{(q)}\right]^2 = \sum_{j=1}^{R^{(s)}} (n_j^{(q)})^2 + \sum_{\substack{i,j=1 \\ i\neq j}}^{R^{(s)}} n_i^{(q)} n_j^{(q)} , \tag{9.73}$$

then, taking into account (9.73),

$$P_0^{(2)(s)}\{n_j\} = \frac{1}{R^{(s)}(R^{(s)}-1)} \sum_q l_q \left\{\left[\sum_{j=1}^{R^{(s)}} n_j^{(q)}\right]^2 - \sum_{j=1}^{R^{(s)}} (n_j^{(q)})^2\right\}. \tag{9.74}$$

For the coefficients l_q and quantities $n_j^{(q)}$ the normalization condition is

$$\sum_q l_q \sum_{j=1}^{R^{(s)}} n_j^{(q)} = yR^{(s)} . \tag{9.75}$$

Therefore

$$\left[\sum_q l_q \sum_{j=1}^{R^{(s)}} n_j^{(q)}\right]^n = (yR^{(s)})^n . \tag{9.76}$$

From (9.76) it follows that

$$\sum_q l_q \left[\sum_{j=1}^{R^{(s)}} n_j^{(q)}\right]^n \leq (yR^{(s)})^n . \tag{9.77}$$

The right and left parts of the function (9.77) are equal when the superstructure is described by a sole configuration of the basis cluster, i.e. when $q = 1$ and $l_q = 1$. In the general case superstructures are described by several configurations of the basis cluster and for such superstructures from formula (9.77) there follows an equation, which is valid for any superstructure:

$$\sum_q l_q \left[\sum_{j=1}^{R^{(s)}} n_j^{(q)}\right]^n = (yR^{(s)})^n + \sum_q l_q \Delta_q \sum_{j=1}^{R^{(s)}} (n_j^{(q)})^n , \tag{9.78}$$

where $0 \leq \Delta_q \leq 1$. Substitution of (9.78) written for $n = 2$ into (9.74) gives

$$P_0^{(2)(s)}\{n_j\} = \frac{y^2 R^{(s)}}{R^{(s)}(R^{(s)}-1)} - \frac{1}{R^{(s)}(R^{(s)}-1)} \sum_q l_q (1-\Delta_q) \sum_{j=1}^{R^{(s)}} (n_j^{(q)})^2 . \tag{9.79}$$

If u is the maximum number of different values of the distribution function at all sites of the ordered lattice, the combinatorial entropy $S_c\{n_j\}$ equals

$$S_c\{n_j\} = -k_B \sum_{j=1}^{u} K_j [n_j \ln n_j + (1-n_j)\ln(1-n_j)] . \tag{9.80}$$

Note that if the composition of the ordered compound MX_y or solid solution A_yB_{1-y} is fixed, the restriction (9.48) is imposed on the distribution function. In this case the search for the free energy minimum reduces to finding the arbitrary minimum of the function $F\{n_j\}$ and can be carried out using the Lagrangian indefinite multiplier method. Considering the normalization condition (9.48), the Lagrange function takes the form

$$\Phi_{\text{L}} = F\{n_j\} + \lambda_{\text{L}}(y - \sum_{j=1}^{u} K_j n_j), \tag{9.81}$$

where λ_{L} is the Lagrangian multiplier.

The necessary condition for the minimum of the free energy F is determined by solving the system of equations $\partial\Phi_{\text{L}}/\partial n_j = 0$. Consider the free energy F written in a more general form (9.70), of which the dependence (9.69) is a particular case. Taking into account (9.70) and (9.80),

$$\partial\Phi_{\text{L}} / \partial n_j = \partial F / \partial n_j - \lambda_{\text{L}} K_j = K_j[X(n_j) - \lambda_{\text{L}}], \tag{9.82}$$

where

$$X(n_j) = \frac{1}{K_j}\sum_{s \neq a} \kappa^{(s)} y^{(s)} F_2(T) \frac{\partial P_0^{(2)(s)}\{n_j\}}{\partial n_j} + k_{\text{B}}T \ln[n_j/(1-n_j)], \tag{9.83}$$

with

$$\frac{\partial P_0^{(2)(s)}\{n_j\}}{\partial n_j} = -\frac{2R^{(s)}}{R^{(s)}(R^{(s)}-1)}\sum_q l_q(1-\Delta_q)n_j^{(q)} \equiv -\frac{R^{(s)}}{C^2_{R^{(s)}}}\sum_q l_q(1-\Delta_q)n_j^{(q)}. \tag{9.84}$$

Thus, the necessary condition for the minimum of the free energy F is determined by a system of equations of the type

$$X(n_j) - \lambda_{\text{L}} = 0. \tag{9.85}$$

The system of equations (9.84) can be written differently:

$$X(n_j) - X(n_1) = 0. \tag{9.86}$$

Consider the form of the function $X(n_j)$ (9.83). It can be seen from (9.84) that the derivatives $\partial P_0^{(2)(s)}\{n_j\}/\partial n_j < 0$ and represent linear functions of n_j. The energy parameter $F_2(T) > 0$ and the coefficient $K_j > 0$. If in the sequence $\{s\}$ of special figures each figure, which has a greater number of sites than the previous one, contributes to a decrease of the total energy of the crystal, all $\kappa^{(s)}$ values are positive and do not exceed unity, i.e. $0 \leq \kappa^{(s)} \leq 1$. Since K_j, $F_2(T)$ and $\kappa^{(s)}$ are positive and the derivative $\partial P_0^{(2)(s)}\{n_j\}/\partial n_j$ is negative, the first term in (9.83) is a negative linear function of n_j. Therefore, the function $X(n_j)$ (9.83) can be presented as

$$X(n_j) = -bn_j + k_{\text{B}}T \ln\frac{n_j}{1-n_j}, \tag{9.87}$$

where $b > 0$ is a proportionality coefficient.

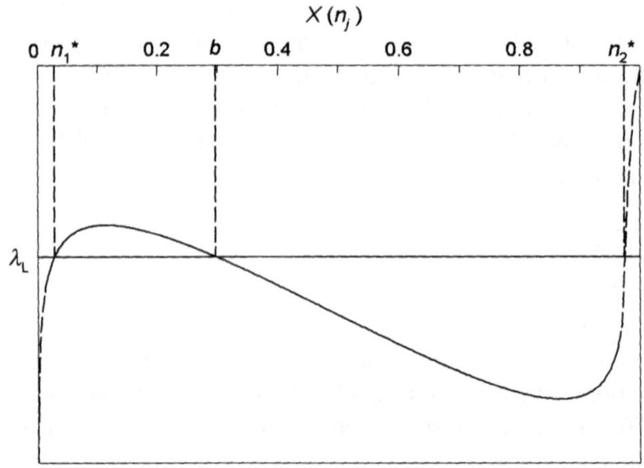

Fig. 9.7. The function $X(n_j)$ and possible solutions of the system of equations of the type (9.85)

The form of the function $X(n_j)$ (9.83) (or (9.87)) is shown in Fig. 9.7. From this figure it is apparent that below the transition temperature to the ordered state any equation of the type (9.85) or (9.86) has no more than three solutions that are values of the distribution function.

To fulfil the sufficient conditions of the local minimum of the free energy F, the second differential of the Lagrange function

$$d^2\Phi_{\mathrm{L}} = \sum_j \sum_m \frac{\partial^2 \Phi_{\mathrm{L}}}{\partial n_j \partial n_m} dn_j dn_m \tag{9.88}$$

should have a positively determined quadratic form. The second derivative is

$$\frac{\partial^2 \Phi_{\mathrm{L}}}{\partial n_j \partial n_m} = \delta_{jm} K_j \partial X(n_j) / \partial n_j \, , \tag{9.89}$$

where

$$\delta_{jm} = \begin{cases} 1, \text{ if } & j = m \\ 0, \text{ if } & j \neq m \end{cases}$$

is the Kronecker symbol. Taking into account (9.89), bring the quadratic form $d^2\Phi_{\mathrm{L}}$ to the canonical form

$$d^2\Phi_{\mathrm{L}} = \sum_j K_j \frac{\partial X(n_j)}{\partial n_j} (dn_j)^2 \, . \tag{9.90}$$

It is clear that the quadratic form (9.90) is positively determined only if $\partial X(n_j)/\partial n_j > 0$. Consequently, the necessary and sufficient condition for the existence of the local minimum of the free energy F at points satisfying the system of equations

$$\left.\begin{aligned} &X(n_j)-X(n_1)=0, \\ &y-\sum_{j=1}^{u}K_j n_j=0, \end{aligned}\right\} \tag{9.91}$$

is positive values of the function $\partial X(n_j)/\partial n_j$. To put it differently,

$$\partial X(n_j)/\partial n_j > 0 \tag{9.92}$$

for the n_j values assumed by the distribution function $n(\boldsymbol{r})$.

In view of the condition (9.92), it is obvious from Fig. 9.7 that in the equilibrium state the distribution function has no more than two different values (n_1* and n_2*), because for the third solution of (9.85) (point $n_j = b$ in Fig. 9.7) the derivative $\partial X(n_j)/\partial n_j < 0$.

Thus, the necessary and sufficient conditions for the local minimum of the free energy F place very rigid restrictions on values of the distribution function and, hence, on long-range order parameters. The above analysis showed that for a crystal in the equilibrium state all possible values of the distribution function degenerate to two values n_1* and n_2*. This is always the case for ordered structures described by one long-range order parameter. This situation is possible for superstructures described by more than one long-range order parameter if all order parameters except one are equal to zero. However, the distribution function is most likely to degenerate when all long-range order parameters are equal. Thus, the necessary and sufficient condition for the local minimum of the free energy is fulfilled when long-range order parameters are equal. For this reason the further analysis of the free energy of ordered phases described by several long-range order parameters will include only one parameter $\eta = \eta_1 = \eta_2 = ... = \eta_m$ instead of a set of order parameters. This theoretical conclusion about the equality of long-range order parameters under equilibrium conditions is confirmed by computer-aided numerical calculations. Indeed, long-range order parameters always become equal as a result of minimization of the free energy over several long-range order parameters.

For superstructures described by one long-range order parameter, the minimum of the free energy is unambiguously determined by the equilibrium value η_{equil} of this parameter.

Analysis of Disorder–Order Transformation Conditions. When long-range order parameters are equal, any distribution function $n(\boldsymbol{r})$ degenerates and takes two values

$$\left.\begin{aligned} n_1^{(\mathrm{d})} &= y-\eta(2t-1)/2t \\ n_2^{(\mathrm{d})} &= y+\eta/2t \end{aligned}\right\} \tag{9.93}$$

at all sites of the ordered lattice of superstructures type $M_{2t}X_{2t-1}$ or $A_{2t-1}B$ formed in the compound MX_y or solid solution A_yB_{1-y}. The $n_1^{(\mathrm{d})}$ value is the probability that an interstitial atom occupies a site in the vacancy sublattice (or an A atom fills a site in the sublattice of B atoms) and $n_2^{(\mathrm{d})}$ is the probability of finding an interstitial atom at a site in the sublattice of interstitial atoms (or an A atom at a site in the sublattice of A atoms) of the ordered structure $M_{2t}X_{2t-1}$ (or $A_{2t-1}B$).

Considering that the minimum of the free energy is reached when long-range order parameters are equal, the probability $P_0^{(2)(s)}$ depends only on two parameters (9.93) of the distribution function. The probability $P_0^{(2)(s)}$ of a complete pair can be determined from the formula (9.12). Taking into account $n_1^{(\mathrm{d})}$ and $n_2^{(\mathrm{d})}$, we have

$$P_0^{(2)} \equiv P_0^{(2((s)} = y^2 - a^{(s)}\eta^2/4t^2, \tag{9.94}$$

where the coefficient $a^{(s)}$ depends on the type and the structure of the ordered phase, as well as on the size and the form of the basis cluster [38, 39]. Formula (9.94) for the probability of a complete pair being a subfigure of the basis cluster can be derived using the concept of the superstructural short-range order (see Sect. 6.2) introduced by the authors of [40, 41].

The superstructural short-range order is part of the short-range order, which directly depends on the presence of long-range order. The superstructural short-range order parameter in the j-th coordination sphere $\alpha_j(y,\eta)$ of the superstructure $M_{2t}X_{2t-1}$ ($A_{2t-1}B$) is described by (6.21) or (6.22). When correlations in the atomic arrangement are absent, the probability of finding an interstitial atom X (or A atom) in the j-th coordination sphere constructed around an X atom of the compound MX_y (or an A atom of the solid solution A_yB_{1-y}) ordered after the $M_{2t}X_{2t-1}$ (or $A_{2t-1}B$) pattern is determined by the superstructural short-range order only:

$$P_{\mathrm{X(X)}} = P_{\mathrm{A(A)}} = y + \alpha_j(y,\eta)(1-y). \tag{9.95}$$

Sites of the ordering lattice and atoms occupying these sites are spaced differently from one another (in different coordination spheres) in the cluster s. Let a site q in the basis cluster have the environment $\{z_{jq}^{(s)}\}$, where $z_{jq}^{(s)}$ is the number of sites of the basis cluster s included in the j-th coordination sphere of the site q. The probability $P_{\mathrm{X(X)}}^{\mathrm{aver}}$ averaged over the cluster s can be obtained by successive averaging over coordination spheres j and sites q:

$$P_{\mathrm{X(X)}}^{\mathrm{aver}} = \frac{\sum_q R_q^{(s)} \sum_j z_{jq}^{(s)} P_{\mathrm{X(X)}}}{\sum_q R_q^{(s)} \sum_j z_{jq}^{(s)}}. \tag{9.96}$$

For the compound MX_y, the probability that an X interstitial atom occupies a site, which is the center of the j-th coordination sphere, is equal to y. Therefore the probability of a complete pair $P_0^{(2)(s)} = yP_{\mathrm{X(X)}}^{\mathrm{aver}}$ or, considering (9.95) and (9.96),

$$P_0^{(2)(s)} = y^2 + y(1-y)\frac{\sum_q R_q^{(s)} \sum_j z_{jq}^{(s)} \alpha_j(y,\eta)}{\sum_q R_q^{(s)} \sum_j z_{jq}^{(s)}}. \tag{9.97}$$

Summation over j and q in (9.97) is performed within the limits of cluster s, and

$\sum_q R_q^{(s)} = R^{(s)}$ and $\sum_j z_{jq}^{(s)} = (R^{(s)} - 1)$. Taking this into account and considering (6.22) for $\alpha_j(y,\eta)$, the probability of a complete pair can be given as

$$P_0^{(2)(s)} = y^2 - \frac{\eta^2}{4t^2}\left\{1 - 2t + \left[\frac{2t}{R^{(s)}(R^{(s)}-1)}\right]\sum_q R_q^{(s)} \sum_j z_{jq}^{(s)} m_{21}^{(j)}\right\}. \quad (9.98)$$

Comparison of (9.94) and (9.98) shows that the coefficient $a^{(s)}$ in (9.94) is described by the expression derived in [39]:

$$a \equiv a^{(s)} = 1 - 2t + \left[\frac{2t}{R^{(s)}(R^{(s)}-1)}\right]\sum_q R_q^{(s)} \sum_j z_{jq}^{(s)} m_{21}^{(j)}. \quad (9.99)$$

The coefficients a for all considered superstructures can be calculated from (9.99). The necessary values of $m_{21}^{(j)}$ characterizing the short-range order in the superstructures can be obtained from the relationship $m_{21}^{(j)} = y_{st}[1 - \alpha_j(y_{st}, \eta_{max})]$ using numerical values of $\alpha_j(y_{st}, \eta_{max})$ listed in Tables 6.4 and 6.5.

Now that we have determined the probability $P_0^{(2)}$ for equal long-range order parameters, we may analyze the conditions of disorder–order transition MX_y–$M_{2t}X_{2t-1}$ (or A_yB_{1-y}–$A_{2t-1}B$).

So, under equilibrium conditions (i.e. when long-range order parameters are equal) the free energy of the nonstoichiometric compound MX_y with any degree of order is described by the expression

$$F(y,\eta,T) = F_0(T) + yF_1(T) + P_0^{(2)} F_2(T) - TS_c(y,\eta), \quad (9.100)$$

where the configurational entropy $S_c(y,\eta)$ depends only on two values of the distribution function

$$S_c(y,\eta) = -(k_B/2t)\{n_1^{(d)} \ln n_1^{(d)} + (1 - n_1^{(d)})\ln(1 - n_1^{(d)}) + (2t-1)[n_2^{(d)} \ln n_2^{(d)} + (1 - n_2^{(d)})\ln(1 - n_2^{(d)})]\}. \quad (9.101)$$

It is easy to see that at $\eta = 0$ entropy (9.101) may be rearranged to the form (9.71) describing only the disordered state of a nonstoichiometric compound. Clearly, $F_0(T)$, $F_1(T)$ and $F_2(T)$, which represent the description parameters in the OPF method, can be found by expanding the free energy of a disordered compound (disregarding the configurational entropy) in a power series in y.

When the disorder–order transformation temperature T_{trans} is reached, the free energies of disordered and ordered phases become equal, therefore $\Delta F = F(y,\eta,T) - F(y,0,T) = 0$. On the other hand, when η reaches the equilibrium value, the free energy $F(y,\eta,T)$ is at a minimum. Consequently, $\partial F(y,\eta,T)/\partial\eta = 0$. Thus, the equilibrium conditions of the disorder–order transformation can be determined, with allowance for (9.93) and $P_0^{(2)}$ in the form (9.94), by

$$\Delta F(\eta) = -a^{(s)}\eta^2 F_2(T)/4t^2 - T[S_c(y,\eta) - S_c(y,0)] = 0, \quad (9.102)$$

$$\partial F(y,\eta,T)/\partial\eta = -(a^{(s)}/2t^2)\eta F_2(T) - T\partial S_c(y,\eta)/\partial\eta = 0. \quad (9.103)$$

The equilibrium value of the long-range order parameter at the disorder–order transition point η_{trans} and the disorder–order transition temperature T_{trans} are the solution of the system of equations (9.102) and (9.103). The solution can be tangibly simplified if we find T from these equations and equate their right parts. Then we shall obtain an equation for η_{trans}:

$$\frac{\eta_{\text{trans}}}{2}\left(\frac{\partial S_{\text{c}}(y,\eta)}{\partial \eta}\right)_{\eta=\eta_{\text{trans}}} - S_{\text{c}}(y,\eta_{\text{trans}}) + S_{\text{c}}(y,0) = 0 . \tag{9.104}$$

Equilibrium conditions of the order–order transition have the form

$$\Delta F(y,\eta_\ell,\eta_{\text{h}},T) = \left[\frac{a_{\text{h}}^{(s)}\eta_{\text{h}}^2}{4t_{\text{h}}^2} - \frac{a_\ell^{(s)}\eta_\ell^2}{4t_\ell^2}\right]F_2(T) - T[S_{\text{c},\ell}(y,\eta_\ell) - S_{\text{c,h}}(y,\eta_{\text{h}})] = 0 , \tag{9.105}$$

$$\frac{\partial F(y,\eta_\ell,T)}{\partial \eta_\ell} = -\frac{a_\ell^{(s)}}{2t_\ell^2}\eta_\ell F_2(T) - T\frac{\partial S_{\text{c},\ell}(y,\eta_\ell)}{\partial \eta_\ell} = 0 , \tag{9.106}$$

$$\frac{\partial F(y,\eta_{\text{h}},T)}{\partial \eta_{\text{h}}} = -\frac{a_{\text{h}}^{(s)}}{2t_{\text{h}}^2}\eta_{\text{h}} F_2(T) - T\frac{\partial S_{\text{c,h}}(y,\eta_{\text{h}})}{\partial \eta_{\text{h}}} = 0 , \tag{9.107}$$

where the indices ℓ and h correspond to low- and high-temperature phases. The transition temperature T_{trans} and equilibrium values of long-range order parameters at the transition point η_{trans} for each of the phases are the solution of the system of equations (9.105)–(9.107). In the case of the order–order transition, values of long-range order parameters at transition points, i.e. values of $\eta_{\text{trans},\ell}$ and $\eta_{\text{trans,h}}$ corresponding to low- and high-temperature phases, can be found from the self-consistent solution of the following system of equations:

$$\left.\begin{aligned} &\left\{\frac{\eta_\ell}{2}\frac{\partial S_{\text{c},\ell}(y,\eta_\ell)}{\partial \eta_\ell} - S_{\text{c},\ell}(y,\eta_\ell)\right\}_{\eta_\ell=\eta_{\text{trans},\ell}} = \\ &\qquad = \left\{\frac{\eta_{\text{h}}}{2}\frac{\partial S_{\text{c,h}}(y,\eta_{\text{h}})}{\partial \eta_{\text{h}}} - S_{\text{c,h}}(y,\eta_{\text{h}})\right\}_{\eta_{\text{h}}=\eta_{\text{trans,h}}} \\ &\left\{\frac{t_\ell^2}{a_\ell^{(s)}\eta_\ell}\frac{\partial S_{\text{c},\ell}(y,\eta_\ell)}{\partial \eta_\ell}\right\}_{\eta_\ell=\eta_{\text{trans},\ell}} = \left\{\frac{t_{\text{h}}^2}{a_{\text{h}}^{(s)}\eta_{\text{h}}}\frac{\partial S_{\text{c,h}}(y,\eta_{\text{h}})}{\partial \eta_{\text{h}}}\right\}_{\eta_{\text{h}}=\eta_{\text{trans,h}}} \end{aligned}\right\} . \tag{9.108}$$

The fact that $F_2(T)$ characterizing the thermodynamic properties of a crystal is not included in (9.104) and (9.108) indicates that η_{trans} depends only on the composition of the ordered compound and the type of the superstructure formed.

Having determined η_{trans} from (9.104) and substituted it in (9.103), we shall find the disorder–order transition temperature T_{trans}. To obtain the order–order transition temperature, determine $\eta_{\text{trans},\ell}$ and $\eta_{\text{trans,h}}$ from (9.108) and insert them into (9.106) and (9.107), respectively. If disorder–order or order–order transformations are transitions of the first kind, the transition temperature is equal to

$$T_{\text{trans}} = -\frac{a^{(s)}}{2t^2} F_2(T_{\text{trans}})\eta_{\text{trans}} \left[\frac{\partial\, S_{\text{c}}(\eta)}{\partial \eta}\right]^{-1}_{\eta=\eta_{\text{trans}}}$$

$$= \frac{2a^{(s)}}{(2t-1)} \frac{F_2(T_{\text{trans}})}{k_{\text{B}}} \eta_{\text{ttans}} \left\{ \ln\left[\frac{n_2^{(\text{d})}(1-n_1^{(\text{d})})}{n_1^{(\text{d})}(1-n_2^{(\text{d})})}\right]_{\eta=\eta_{\text{trans}}} \right\}^{-1} . \qquad (9.109)$$

Note that the derivative $\partial S_{\text{c}}(y,\eta)/\partial\eta$ is always less than zero and the expression in (9.109) in curly brackets is always positive. Therefore, (9.103) and (9.109) suggest that at $F_2(T) > 0$ ordering is possible only if the coefficient $a^{(s)} > 0$. If the coefficient $a^{(s)}$ is negative, ordering is impossible, because no nonstoichiometric compounds with $F_2(T) < 0$ exist (see Fig. 9.6).

Two ordered phases are in equilibrium upon order–order transitions. The long-range order parameter $\eta_{\text{equil}} = \eta_{\text{trans}}^{\text{ord-ord}}$ acquires its own equilibrium value at the transition point in each ordered phase. Inserting $\eta_{\text{trans}}^{\text{ord-ord}}$, as well as $a^{(s)}$ and t for any of the two equilibrium ordered phases into (9.109), we shall find the order–order transition temperature.

The entropy jump ΔS_{trans} upon order–order or order–disorder transitions of the first kind can be determined from the formula

$$\Delta S_{\text{trans}} = \left[\frac{a_\ell^{(s)}\eta^2_{\text{trans},\ell}}{4t_\ell^2} - \frac{a_{\text{h}}^{(s)}\eta^2_{\text{trans,h}}}{4t_{\text{h}}^2}\right] S_2(T_{\text{trans}}) + S_{\text{c}}(y,\eta_{\text{trans,h}}) - S(y,\eta_{\text{trans},\ell}), \qquad (9.110)$$

where $\eta_{\text{trans},\ell}$ and $\eta_{\text{trans,h}}$ are equilibrium values of long-range order parameters at the transition point corresponding to low- and high-temperature phases; $S_2(T) \equiv -\partial F_2(T)/\partial T$. The high-temperature phase is disordered and has $\eta_{\text{trans,h}} = 0$ upon order–disorder transitions.

In the case of disorder–order phase transitions of the second kind, when the long-range order parameter smoothly changes from $\eta = 0$ to η_{equil}, (9.109) is rearranged at $T \leq T_{\text{trans}}$ to the form

$$T_{\text{trans}} = \frac{2a^{(s)}}{(2t-1)} y(1-y) \frac{F_2(T_{\text{trans}})}{k_{\text{B}}}, \qquad (9.111)$$

which is analogous to the formula for the transition temperature in the mean-field approximation. It follows from (9.111) that for phase transitions of the second kind the dependence of the ordering temperature on the composition of the compound MX_y or solid solution A_yB_{1-y} is a symmetrical parabola with a maximum at $y = 1/2$. From (9.111) it is obvious that ordering by the mechanism of phase transitions of the second kind is possible only when $a^{(s)} > 0$. If $a^{(s)} < 0$, $F_2(T)$ should be negative but, as was already stated above, this is impossible.

The inference that superstructures with $a^{(s)} < 0$ cannot be formed refers only to the case at hand when the formation enthalpy of the compound MX_y is a quadratic function of y and the free energy F of a disordered compound is written as (9.71). If the expansion of the formation enthalpy in a power series in y includes members higher than those of the second order, equilibrium conditions of the disorder–order transformation will be different and restrictions imposed on the formation of ordered phases with $a^{(s)} < 0$ may be less rigid.

The analysis of (9.103) shows that in the general case it may have one, two or three solutions, one of them always being $\eta = 0$. If (9.103) has one solution $\eta = 0$, the equilibrium state of a crystal at a preset temperature and an assigned composition is a disordered one. If (9.103) has two or three roots which are equal to zero, the equilibrium state corresponds to the η value at which $\Delta F(\eta)$ is a minimum.

References

1 A. G. Khachaturian: *Theory of Structural Transformations in Solid* (John Wiley & Sons, New York 1983) 574 pp.

2 M. A. Krivoglaz, A. A. Smirnov: *The Theory of Order–Disorder in Alloys* (American Elsevier, New York 1964) 427 pp.

3 R. Kikuchi: Theory of cooperative phenomena, Phys. Rev. **81**, 988–1003 (1951)

4 M. Kurata, R. Kikuchi, T. Watari: Theory of cooperative phenomena. Detailed discussions of the cluster variation method, J. Chem. Phys. **21**, 434–448 (1953)

5 J. Hijmans, J. de Boer: Approximation method for order–disorder problem, Physica **21**, 471–516 (1955)

6 R. Kikuchi, S. G. Brush: Improvement of the cluster-variation method, J. Chem. Phys. **47**, 195–203 (1967)

7 J. M. Sanchez, D. de Fontaine: The f.c.c. Ising model in the cluster variation approximation, Phys. Rev. B **17**, 2926–2936 (1978)

8 V. G. Vaks, V. I. Zinenko, V. E. Shneider: Microscopic theories of the order–disorder structural phase transitions in crystals, Uspekhi Fiz. Nauk **141**, 629–673 (1983) (in Russian)

9 N. S. Golosov: Cluster variation method in the theory of atomic ordering, Izv. Vyssh. Ucheb. Zaved. Fizika No 8, 64–82 (1976) (in Russian)

10 A. I. Gusev, A. A. Rempel: Order parameter functional method in the theory of atomic ordering, Phys. Stat. Sol. (b) **131**, 43–51 (1985)

11 A. I. Gusev, A. A. Rempel: Thermodynamic model of atomic ordering. 1. Basic equations, Zh. Fiz. Khimii **60**, 1349–1352 (1986) (in Russian). (Engl. transl.: Russ. J. Phys. Chem. **60**, 808–810 (1986))

12 A. A. Rempel, A. I. Gusev: Thermodynamic model of atomic ordering. 2. The order-disorder structural phase transition in nonstoichiometric niobium carbide, Zh. Fiz. Khimii **60**, 1353–1357 (1986) (in Russian). (Engl. transl.: Russ. J. Phys. Chem. **60**, 810–813 (1986))

13 A. I. Gusev, A. A. Rempel: Calculating the energy parameters for CV and OPF methods, Phys. Stat. Sol. (b) **140**, 335–346 (1987)
14 A. I. Gusev, A. A. Rempel, V. N. Lipatnikov: Thermodynamic model of atomic ordering. 3. Calculation of cluster energies, Zh. Fiz. Khimii **61**, 916–921 (1987) (in Russian). (Engl. transl.: Russ. J. Phys. Chem. **61**, 476–479 (1987))
15 A. I. Gusev, A. A. Rempel: *Structural Phase Transitions in Nonstoichiometric Compounds* (Nauka, Moscow 1988) 308 pp. (in Russian)
16 A. I. Gusev: Atomic ordering and the order parameter functional method, Philosoph. Mag. B **60**, 307–324 (1989)
17 A. I. Gusev: *Physical Chemistry of Nonstoichiometric Refractory Compounds* (Nauka, Moscow 1991) 286 pp. (in Russian)
18 A. A. Rempel: *Effects of Ordering in Nonstoichiometric Interstitial Compounds* (Nauka, Ekaterinburg 1992) 232 pp. (in Russian)
19 A. I. Gusev, A. A. Rempel: Phase diagrams of metal–carbon and metal–nitrogen systems and ordering in strongly nonstoichiometric carbides and nitrides, Phys. Stat. Sol. (a) **163**, 273–304 (1997)
20 L. D. Landau, E. M. Lifschitz: *Statistical Physics. Course of Theoretical Physics* **5** (Pergamon, Oxford 1980) 544 pp.
21 H. J. Goldschmidt: *Interstitial Alloys* (Plenum Press, New York 1967) 820 pp.
22 L. Toth: *Transition Metal Carbides and Nitrides* (Academic Press, New York and London 1971) 280 pp.
23 G. P. Shveikin, S. I. Alyamovskii, Yu. G. Zainulin, A. I. Gusev, V. A. Gubanov, E. Z. Kurmaev: *Compounds of Variable Composition and Their Solid Solutions* (Ural Scientific Centre, Sverdlovsk 1984) 292 pp. (in Russian)
24 A. I. Gusev, A. A. Rempel: *Thermodynamics of the Structural Vacancies in Nonstoichiometric Interstitial Phases* (Ural Scientific Centre, Sverdlovsk 1987) 114 pp. (in Russian)
25 A. S. Bolgar, A. G. Turchanin, V. V. Fesenko: *Thermodynamic Properties of Carbides* (Naukova Dumka, Kiev 1973) 272 pp. (in Russian)
26 A. S. Bolgar, V. F. Litvinenko: *Thermodynamic Properties of Nitrides* (Naukova Dumka, Kiev 1980) 282 pp. (in Russian)
27 V. N. Lipatnikov, A. I. Gusev: *Ordering of Titanium and Vanadium Carbides* (Ural Division of the RAS, Ekaterinburg 2000) 265 pp. (in Russian)
28 H. J. Seifert, H. L. Lukas, G. Petzow: Thermodynamic optimization of the Ti–C system, J. Phase Equilibria **17**, 24–35 (1996)
29 G. L. Humphrey: The heat of combustion and formation of titanium nitride TiN and titanium carbide TiC, J. Amer. Ceram. Soc. **73**, 2261–2263 (1951)
30 S. Fujishiro, N. A. Goksen: Thermodynamic properties of TiC at high temperatures, J. Phys. Chem. **65**, 161–163 (1961)
31 C. E. Lowell, W. S. Williams: High temperature calorimeter for the determination of heats of formation of refractory compounds, Rev. Sci. Instrum. **32**, 1120–1123 (1961)
32 M. P. Morozova, M. K. Khripun, S. M. Ariya: Enthalpy of formation of titanium carbides and oxycarbides, Zh. Obshchey Khimii **32**, 2072–2076 (1962) (in Russian)
33 V. M. Maslov, A. S. Neganov, I. P. Borovinskaya, A. G. Merganov: Self-propagating high-temperature synthesis as a method for determination of the heat of formation of refractory compounds, Fizika Goreniya i Vzryva **14**, 73–82 (1978) (in Russian)

34 R. Berkane: Thermodynamic study of chromium, titanium, zirconium and hafnium carbides by high temperature calorimetry numerical modeling of the phase diagrams, Ph.D. Thesis, Université de Nancy I, France (1989)
35 A. E. Vol, I. K. Kagan: *Structure and Properties of Binary Metallic Systems* **3** (Nauka, Moscow 1976) 816 pp. (in Russian)
36 A. I. Gusev: Thermodynamic conditions for the congruent vaporization of unordered nonstoichiometric compounds, Doklady Akad. Nauk SSSR **311**, 631–635 (1990) (in Russian). (Engl. transl.: Doklady Phys. Chem. **311**, 264–268 (1990))
37 A. I. Gusev: Thermodynamics of the congruent vaporization of nonstoichiometric refractory compounds, Teplofiz. Vysokikh Temp. **29**, 695--701 (1991) (in Russian). (Engl. transl.: High temperature **29**, 541--546 (1991))
38 A. I. Gusev: Phase diagrams for ordering systems in order-parameter functional method, Fiz. Tverd. Tela **32**, 2752–2760 (1990) (in Russian). (Engl. transl.: Sov. Physics – Solid State **32**, 1595–1599 (1990))
39 A. I. Gusev, A. A. Rempel: Thermodynamic model of atomic ordering. Phase diagrams of ordered systems, Zh. Fiz. Khimii **65**, 625–633 (1991) (in Russian). (Engl. transl.: Russ. J. Phys. Chem. **65**, 330–334 (1991))
40 A. A. Rempel, A. I. Gusev: Short-range order in ordered alloys and interstitial phases, Fiz. Tverd. Tela **32**, 16–24 (1990) (in Russian). (Engl. transl.: Sov. Physics – Solid State **32**, 8–13 (1990))
41 A. A. Rempel, A. I. Gusev: Short-range order in superstructures, Phys. Stat. Sol. (b) **160**, 389–402 (1990)

10. Equilibrium Phase Diagrams

Calculations of order–disorder phase transformations reduce to determining and comparing the free energies of disordered and various ordered phases. Consequently, it is possible to find temperature and concentration intervals within which one or another phase is in a thermodynamic equilibrium state. In the case of order–order transformations, the free energies of different ordered phases are compared and the temperature and concentration intervals of their existence are determined. Thus, such calculations result in the construction of a phase diagram. Calculation of phase diagrams of ordering systems is a highly complicated task. It is only since 1985 that this task was first tackled for nonstoichiometric interstitial compounds.

The order-parameters functional method (OPF method) [1–5], also known as the thermodynamic model of atomic ordering [6–8], was proposed for the description of structural disorder–order phase transitions in substitutional solid solutions A_yB_{1-y} and nonstoichiometric interstitial compounds $MX_y\square_{1-y}$. This method is comprehensively covered in the previous chapter. In physical terms, the OPF method is based on the mean field approximation. Because of its formalism, this method belongs to the group of cluster methods, but favourably differs from these by providing a detailed account of the symmetry of a crystal with any degree of long-range order. The principal feature of the OPF method is that it represents probabilities of different figures (clusters) in terms of values of the distribution function $n_v(\boldsymbol{r})$ which directly depends on the long-range order parameters η. The distribution function $n_v(\boldsymbol{r})$ itself is the probability of finding atoms of a given species v at a site $\boldsymbol{r}$ of an ordering lattice. Using the OPF method, it was possible to describe quantitatively and qualitatively first- and second-kind disorder–order phase transitions in some nonstoichiometric interstitial compounds MX_y and solid solutions A_yB_{1-y}, to determine thermodynamic equilibrium types of superstructures in these substances [1, 2, 5, 9], and to construct phase diagrams of M–X (with X = C, N) systems taking into account the ordering of nonstoichiometric carbides MC_y and nitrides MN_y [10–17].

In recent decades the molecular dynamics and Monte Carlo methods have been extensively used for mathematical modeling in the theory of phase transitions. Specifically, the Monte Carlo method was used to analyze ordering in nonstoichiometric titanium carbide and nitride [18–20]. The Monte Carlo method is a universal numerical tool for solving mathematical problems by modeling random variables [21]. The large volume of computational operations prevented the widespread use of the Monte Carlo method before the advent of computers. At present this method is employed to analyze any cooperative phenomena where probabilistic factors are significant.

In the Monte Carlo method we deal with part of a crystal containing a large number of atoms (at least 1000). This part is considered as a representative system for the whole infinite crystal. Any equilibrium characteristic of this system consisting of N particles can be found by averaging over the canonical configurational ensemble $g(1, ..., N) \sim \exp[-\beta U(1, ..., N)]$, where U is the internal energy of the system. Let some initial configuration have atoms at points $\boldsymbol{R}_1$, $\boldsymbol{R}_2$, ..., $\boldsymbol{R}_N$. Using a random number generator, the i-th atom is displaced by the vector $\boldsymbol{R}_i' - \boldsymbol{R}_i = \delta$. The step length δ is selected arbitrarily so that the whole system can be covered in a finite number of steps. The change of the system energy ΔU is computed after each step. If $\Delta U \leq 0$, the new configuration is suitable for the next step. Systems formed due to random motion of particles make up a canonical ensemble. Generally, the energy of a crystal is represented by paired (two-particle) interactions. As a result of the successive procedure, at preset parameters of the interparticle interactions, the system achieves some equilibrium distribution of atoms over the lattice sites, which corresponds to a minimum energy. The foregoing suggests two principal features of the Monte Carlo method: first, the relatively simple structure of the computational algorithm and, second, the calculation error is proportional to $1/\sqrt{N}$.

A reverse Monte Carlo method was used in [18–20] devoted to ordering in nonstoichiometric titanium and niobium carbides and titanium nitride. The energy of paired interactions in the nonmetal sublattice was computed from neutron diffraction data on short-range order. A cell of $18 \times 17 \times 17 = 5202$ fcc unit cells comprising 20 808 nonmetal sublattice sites was used for modelling.

This chapter will discuss the description of phase equilibrium by the OPF method, summarize data on phase equilibrium in M–C and M–N binary systems with their phase diagrams computed taking account of ordering by the OPF and the Monte Carlo methods, and also present phase diagrams of pseudobinary carbide systems computed in the approximation of subregular solutions. Construction of phase diagrams has both theoretical and practical significance, because it is possible to determine temperature and composition intervals where disordered and ordered phases exist whose structures and properties differ markedly [1, 2, 9, 22].

10.1 Computation of Disorder–Order Phase Transformations in Nonstoichiometric Compounds

As shown in Sect. 8.1, the equilibrium condition (8.10) for multiphase systems is met if chemical potentials μ_i of a substance are equal in all phases, i.e. $\mu_1 = \mu_2 = \ldots = \mu_n$. The equilibrium condition (8.10) holds for each component of multicomponent systems. In other words, chemical potentials of each given component must be equal in all phases under equilibrium conditions of a multiphase multicomponent system.

Consider ordering of a nonstoichiometric compound $MX_y\square_{1-y}$, i.e. ordering in a two-component system X–□. In this case the set of equations

$$\left.\begin{aligned} \mu_1^{(X)} &= \mu_2^{(X)} \\ \mu_1^{(\square)} &= \mu_2^{(\square)} \end{aligned}\right\} \tag{10.1}$$

should be solved to analyze the two-phase equilibrium. Here chemical potentials $\mu^{(X)}$ and $\mu^{(\square)}$ are expressed in terms of the free energy of the corresponding phase:

$$\left.\begin{aligned} \mu^{(X)} &= F(y,T)+(1-y)\frac{\partial F(y,T)}{\partial y} \\ \mu^{(\square)} &= F(y,T)-y\frac{\partial F(y,T)}{\partial y} \end{aligned}\right\}. \tag{10.2}$$

As shown in the previous section, in terms of the OPF method, the free energy of a nonstoichiometric compound MX_y ($MX_y\square_{1-y}$) with any degree of order is given by the (9.70). When $\eta = 0$, (9.70) is rearranged to the form (9.71) describing a disordered state only. Taking this and (10.2) into account, consider equilibrium conditions for the disordered phase MX_y and the ordered phase $M_{2t}X_{2t-1}$, which have free energies $F(y,0,T)$ and $F(y,\eta,T)$ respectively.

Equilibrium between disordered and ordered phases is described by the set of equations

$$\frac{\partial F(y_1,0,T)}{\partial y_1} = \frac{\partial F(y_2,\eta,T)}{\partial y_2}, \tag{10.3}$$

$$F(y_1,0,T) - y_1\frac{\partial F(y_1,0,T)}{\partial y_1} = F(y_2,\eta,T) - y_2\frac{\partial F(y_2,\eta,T)}{\partial y_2}, \tag{10.4}$$

where y_1 and y_2 are the relative contents of the component X in the disordered and ordered phases respectively.

Let a crystal be in the equilibrium state such that all possible values of the distribution function degenerate to two values $n_1^{(d)}$ and $n_2^{(d)}$ (9.93) corresponding to the equality of long-range order parameters $\eta_1 = \eta_2 = \ldots = \eta_m = \eta$. In accordance with [10], substituting (9.65), (9.66) and (9.101) into the set of equations (10.3), (10.4) and rearranging them taking account of (9.93), we obtain equilibrium conditions of the disordered phase A_yB_{1-y} (or MX_y) and the ordered phase $A_{2t-1}B$ (or $M_{2t}X_{2t-1}$):

$$\begin{aligned} &\sum_{n=0}^{R(s)} y_1^{(n-1)}[n+(1-n)y_1]F_n(T) + k_B T\ln y_1 \\ &= \sum_s \kappa^{(s)} y^{(s)} \Bigg\{ \sum_{n=0}^{R(s)} \left[P_0^{(n)(s)} + (1-y_2)\frac{\partial P_0^{(n)(s)}}{\partial y_2} \right] F_n(T) \\ &\quad - \eta \sum_{n=0}^{R(s)} \frac{\partial P_0^{(n)(s)}}{\partial \eta} F_n(T) + \frac{k_B T}{2t}[\ln(1-n_1^{(d)}) + (2t-1)\ln(1-n_2^{(d)})] \Bigg\}_{\eta=\eta_{\mathrm{equil}}}, \end{aligned} \tag{10.5a}$$

$$\sum_{n=0}^{R^{(s)}}(1-n)y_1^n F_n(T)+k_B T\ln(1-y_1)$$

$$=\sum_s \kappa^{(s)}y^{(s)}\left\{\sum_{n=0}^{R^{(s)}}\left[P_{0i}^{(n)(s)}-y_i\frac{\partial P_{0i}^{(n)(s)}}{\partial y_i}\right]F_n(T)\right.$$

$$\left.-\eta\sum_{n=0}^{R^{(s)}}\frac{\partial P_0^{(n)(s)}}{\partial\eta}F_n(T)+\frac{k_B T}{2t}[\ln(1-n_1^{(d)})+(2t-1)\ln(1-n_2^{(d)})]\right\}_{\eta=\eta_{\rm equil}}. \quad (10.5b)$$

Equilibrium conditions for two ordered phases of the type $A_{2t_i-1}B$ or $M_{2t_i}X_{2t_i-1}$, where i = 1 or 2, can be written similarly:

$$\sum_{i=1}^{2}(-1)^i\sum_s \kappa^{(s)}y^{(s)}\left\{\sum_{n=0}^{R^{(s)}}\left[P_{0i}^{(n)(s)}+(1-y_i)\frac{\partial P_{0i}^{(n)(s)}}{\partial y_i}\right]F_n(T)\right.$$

$$\left.-\eta_i\sum_{n=0}^{R^{(s)}}\frac{\partial P_{0i}^{(n)(s)}}{\partial\eta_i}F_n(T)+\frac{k_B T}{2t_i}[\ln n_{1,i}^{(d)}+(2t_i-1)\ln n_{2,i}^{(d)}]\right\}_{\eta_i=\eta_{i,\rm equil}}=0, \quad (10.6a)$$

$$\sum_{i=1}^{2}(-1)^i\sum_s \kappa^{(s)}y^{(s)}\left\{\sum_{n=0}^{R^{(s)}}\left[P_{0i}^{(n)(s)}-y_i\frac{\partial P_{0i}^{(n)(s)}}{\partial y_i}\right]F_n(T)-\eta_i\sum_{n=0}^{R^{(s)}}\frac{\partial P_{0i}^{(n)(s)}}{\partial\eta_i}F_n(T)\right.$$

$$\left.+\frac{k_B T}{2t_i}[\ln(1-n_{1,i}^{(d)})+(2t_i-1)\ln(1-n_{2,i}^{(d)})]\right\}_{\eta_i=\eta_{i,\rm equil}}=0. \quad (10.6b)$$

In (10.6) y_i, $P_{0i}^{(n)(s)}$, η_i, t_i, $n_{1,i}^{(d)}$ and $n_{2,i}^{(d)}$ denote corresponding quantities for the ordered phase $A_{2t_i-1}B$ (or $M_{2t_i}X_{2t_i-1}$).

The equilibrium value of the long-range order parameter $\eta_{\rm equil}$ for the ordered phase $A_{2t-1}B$ (or $M_{2t}X_{2t-1}$) can be found from the condition of the minimum of the free energy of that phase, for which a thermodynamic equilibrium is reached, i.e. from the condition $\partial F(y,\eta,T)/\partial\eta = 0$. Considering (9.65) and (9.101), it has the form

$$\sum_s \kappa^{(s)}y^{(s)}\left\{\sum_{n=0}^{R^{(s)}}\frac{\partial P_0^{(n)(s)}}{\partial\eta}F_n(T)\right.$$

$$\left.-k_B T\frac{2t-1}{4t^2}\ln\left[\frac{n_2^{(d)}(1-n_1^{(d)})}{n_1^{(d)}(1-n_2^{(d)})}\right]\right\}_{\eta=\eta_{\rm equil}}=0. \quad (10.7)$$

Equations (10.5)–(10.7) allow the computation of equilibrium phase diagrams of binary systems in which ordering is possible. The calculation parameters are the coefficients $F_0(T)$, $F_1(T)$, ..., $F_n(T)$ of the expansion (9.66) of the free energy of a disordered crystal. So, in terms of the OPF method, it suffices to know the free energy of the disordered state of a crystal to compute disorder–order and order–order transitions, and equilibrium structural states. This energy can be preset in model terms or determined from experimental thermodynamic data.

As was shown in Sect. 9.5, expansions of the formation enthalpy (9.22) and the free energy (9.66) of disordered nonstoichiometric compounds MX_y are limited by the second-order terms in y and have the form (9.68) and (9.71). Because $F_{n>2}(T) \equiv 0$, $\kappa^{(a)} = \kappa^{(b)} = 0$ and $\kappa^{(c)} = 1$ for real nonstoichiometric compounds, the set of equations (10.5) describing equilibrium of the disordered phase MX_y and the ordered phase $M_{2t}X_{2t-1}$ assumes the form

$$y_1(2-y_1)+\frac{k_B T}{F_2(T)}\ln y_1$$

$$= y_2(2-y_2)+\frac{a^{(s)}}{4t^2}\eta^2_{\rm equil}+\frac{1}{2t}\frac{k_B T}{F_2(T)}\left[\ln n_1^{(\rm d)}+(2t-1)\ln n_2^{(\rm d)}\right]_{\eta=\eta_{\rm equil}}, \quad (10.8a)$$

$$y_1^2-\frac{k_B T}{F_2(T)}\ln(1-y_1)$$

$$= y_2^2-\frac{a^{(s)}}{4t^2}\eta^2_{\rm equil}-\frac{1}{2t}\frac{k_B T}{F_2(T)}\left[\ln(1-n_1^{(\rm d)})+(2t-1)\ln(1-n_2^{(\rm d)})\right]_{\eta=\eta_{\rm equil}}. \quad (10.8b)$$

Equilibrium conditions for two ordered phases $M_{2t_i}X_{2t_i-1}$ (where $i = 1$ or 2) have the form

$$\sum_{i=1}^{2}(-1)^i\left\{y_i(2-y_i)+\frac{a_i^{(s)}}{4t_i^2}\eta^2_{i,\rm equil}\right.$$

$$\left.+\frac{1}{2t_i}\frac{k_B T}{F_2(T)}\left[\ln n_{1,i}^{(\rm d)}+(2t_i-1)\ln n_{2,i}^{(\rm d)}\right]_{\eta_i=\eta_{i,\rm equil}}\right\}=0, \quad (10.9a)$$

$$\sum_{i=1}^{2}(-1)^i\left\{y_i^2-\frac{a_i^{(s)}}{4t_i}\eta^2_{i,\rm equil}\right.$$

$$\left.-\frac{1}{2t_i}\frac{k_B T}{F_2(T)}\left[\ln(1-n_{1,i}^{(\rm d)})+(2t_i-1)\ln(1-n_{2,i}^{(\rm d)})\right]_{\eta_i=\eta_{i,\rm equil}}\right\}=0. \quad (10.9b)$$

Phase boundaries can be found and the equilibrium phase diagram can be constructed by solving the sets of equations (10.8) and (10.9) for all binary equilibria possible in an ordering compound MX_y.

Considering (9.94), (9.100) and the values of $\kappa^{(s)}$ and $y^{(s)}$, expression (10.7) for the equilibrium long-range order parameter can be written as

$$\frac{1}{\eta_{\rm equil}}\ln\left[\frac{n_2^{(\rm d)}(1-n_1^{(\rm d)})}{n_1^{(\rm d)}(1-n_2^{(\rm d)})}\right]_{\eta=\eta_{\rm equil}} = \frac{2a^{(s)}}{(2t-1)}\frac{F_2(T)}{k_B T}. \quad (10.10)$$

From (10.10) it follows that the equilibrium long-range order parameter $\eta_{\rm equil}$ is a function not only of y and T, but also of the energy parameter $F_2(T)$, i.e. $\eta_{\rm equil}$ depends on the nature of the ordering nonstoichiometric compound.

Generally, (9.71), (9.93), (9.94), (9.100), (9.101), (9.104), (9.109)–(9.111), and (10.8)–(10.10) allow not only the construction of phase diagrams of ordering systems, but also the computation of the free energy, entropy and the long-range order parameter as a function of the temperature and the MX_y composition, the determination of abrupt changes in entropy and the order parameter at the point of a phase transition of the first kind, or the calculation of the enthalpy (heat) of the phase transformation.

10.2 Ordering of a MX_y Compound in a Generalized Phase Diagram of the M–X System

In accordance with [1, 2, 9, 22–25] and results of structural studies summarized in Chap. 3, superstructures of the types M_2X, M_2X_3, M_4X_3, M_6X_5 or M_8X_7 can be formed, from the crystallographic point of view, in nonstoichiometric compounds MX_y having the basic structure $B1$. A detailed crystallographic description of these superstructures $M_{2t}X_{2t-1}$ and their interstitial atomic distribution functions is given in Chap. 5. Table 10.1 presents the coefficients t and $a^{(s)}$ for all the superstructures. It was shown in Sect. 9.5 that the energy parameter $F_2(T)$ is positive for disordered nonstoichiometric compounds MX_y whose free energy has the form (9.71). Since T_{trans} is always positive, formulas (9.109) and (9.111) suggest that the coefficient $a^{(s)}$ can be positive only. Thus, the assumption that a trigonal (space group $P\bar{3}m1$) superstructure of the type M_3X_2 for which $a^{(s)} < 0$ (Table 10.1) is formed in a compound MX_y with the basic structure $B1$ can be dropped *a priori*. Therefore only rhombic (space groups *Immm* and $C222_1$) and monoclinic (space group $P2$) superstructures of this type will be meant when the ordered phase M_3X_2 is mentioned further in the text.

Table 10.1. Some structural characteristics of possible ordered $M_{2t}X_{2t-1}$ phases of strongly nonstoichiometric compounds MX_y with a basic $B1$ (NaCl)-type crystal structure

Type of ordered phase	Space group	$m_{21}^{(1)}$	$m_{21}^{(2)}$	t	$a^{(s)}$	$2a^{(s)}/(2t-1)$
M_2X	$R\bar{3}m$, $Fd\bar{3}m$	1/2	1	1	0.2	0.4
M_2X	$I4_1/amd$	2/3	1/3	1	0.2	0.4
M_2X	$P4/mmm$	2/3	0	1	0.067	0.133
M_3X_2	$Immm$, $P2$	5/6	2/3	1.5	0.4	0.4
M_3X_2	$C222_1$	3/4	1	1.5	0.4	0.4
M_3X_2	$P\bar{3}m1$	1/2	1	1.5	−0.2	−0.2
M_4X_3	$Pm\bar{3}m$	1	0	2	0.2	0.133
M_4X_3	$I4/mmm$	1	1/3	2	0.467	0.311
M_6X_5	$C2/m$, $P3_1$, $C2$	1	1	3	1.0	0.4
M_8X_7	$Fm\bar{3}m$, $P4_332$	1	1	4	1.0	0.286

If $a^{(s)}$ and t are known, formulas (9.104) and (9.108) can be used to calculate η_{trans} for superstructures of the types M_3X_2, M_4X_3, M_6X_5 or M_8X_7 formed in a nonstoichiometric compound MX_y (Table 10.2), and for order–order transitions M_2X–M_3X_2 and M_3X_2–M_6X_5 in ordered phases (Table 10.3).

The form of a generalized equilibrium phase diagram of the M–X system with an ordering nonstoichiometric compound MX_y, in which all nonmetal interstitial atoms X can be replaced by structural vacancies □, i.e. y changes from 0 to 1, was found in [11].

Table 10.2. Equilibrium values of long-range order parameter at the transition point, η_{trans}, in the case of formation superstructures of type $M_{2t}X_{2t-1}$ in a nonstoichiometric compound MX_y

y	η_{trans}				y	η_{trans}			
	M_3X_2	M_4X_3	M_6X_5	M_8X_7		M_3X_2	M_4X_3	M_6X_5	M_8X_7
	$t = 1.5$	$t = 2$	$t = 3$	$t = 4$		$t = 1.5$	$t = 2$	$t = 3$	$t = 4$
0.01	0.000	0.002	0.002	0.001	0.26	0.003	0.025	0.025	0.025
0.02	0.000	0.002	0.003	0.003	0.27	0.003	0.025	0.025	0.025
0.03	0.001	0.004	0.003	0.004	0.28	0.003	0.025	0.025	0.025
0.04	0.001	0.005	0.004	0.004	0.29	0.003	0.025	0.025	0.025
0.05	0.001	0.006	0.005	0.005	0.30	0.003	0.025	0.025	0.025
0.06	0.001	0.007	0.006	0.006	0.31	0.003	0.025	0.025	0.025
0.07	0.002	0.008	0.007	0.007	0.32	0.003	0.025	0.025	0.025
0.08	0.002	0.009	0.008	0.008	0.33	0.004	0.025	0.025	0.025
0.09	0.002	0.010	0.009	0.009	0.34	0.004	0.026	0.025	0.025
0.10	0.002	0.011	0.009	0.010	0.35	0.004	0.026	0.025	0.026
0.11	0.002	0.011	0.010	0.011	0.36	0.004	0.026	0.026	0.032
0.12	0.002	0.012	0.010	0.013	0.37	0.005	0.027	0.035	0.036
0.13	0.002	0.012	0.011	0.013	0.38	0.005	0.039	0.039	0.038
0.14	0.002	0.013	0.012	0.014	0.39	0.005	0.040	0.040	0.039
0.15	0.002	0.014	0.013	0.015	0.40	0.006	0.040	0.040	0.039
0.16	0.002	0.015	0.014	0.016	0.41	0.006	0.041	0.040	0.040
0.17	0.002	0.016	0.016	0.017	0.42	0.007	0.042	0.041	0.041
0.18	0.003	0.017	0.018	0.018	0.43	0.007	0.043	0.042	0.042
0.19	0.003	0.018	0.018	0.019	0.44	0.008	0.044	0.043	0.043
0.20	0.003	0.019	0.019	0.020	0.45	0.009	0.046	0.044	0.044
0.21	0.003	0.020	0.020	0.020	0.46	0.010	0.047	0.046	0.046
0.22	0.003	0.021	0.021	0.021	0.47	0.011	0.050	0.046	0.048
0.23	0.003	0.022	0.022	0.022	0.48	0.012	0.052	0.051	0.051
0.24	0.003	0.023	0.023	0.023	0.49	0.013	0.056	0.055	0.054
0.25	0.003	0.025	0.025	0.024	0.50	0.015	0.061	0.059	0.059

(*Continued*)

Table 10.2 (continued)

y	η_{trans} M_3X_2 $t = 1.5$	M_4X_3 $t = 2$	M_6X_5 $t = 3$	M_8X_7 $t = 4$	y	η_{trans} M_3X_2 $t = 1.5$	M_4X_3 $t = 2$	M_6X_5 $t = 3$	M_8X_7 $t = 4$
0.51	0.027	0.062	0.066	0.065	0.76	0.299	0.467	0.566	0.571
0.52	0.039	0.071	0.075	0.074	0.77	0.293	0.467	0.583	0.591
0.53	0.056	0.084	0.087	0.085	0.78	0.287	0.465	0.598	0.611
0.54	0.074	0.100	0.102	0.100	0.79	0.279	0.459	0.612	0.631
0.55	0.096	0.118	0.119	0.117	0.80	0.270	0.451	0.623	0.650
0.56	0.114	0.139	0.139	0.137	0.81	0.260	0.440	0.632	0.668
0.57	0.134	0.160	0.160	0.158	0.82	0.250	0.427	0.636	0.684
0.58	0.153	0.181	0.182	0.179	0.83	0.239	0.411	0.635	0.699
0.59	0.172	0.203	0.205	0.201	0.84	0.227	0.394	0.629	0.712
0.60	0.190	0.225	0.228	0.223	0.85	0.215	0.375	0.616	0.721
0.61	0.205	0.246	0.250	0.245	0.86	0.202	0.354	0.596	0.725
0.62	0.219	0.267	0.273	0.267	0.87	0.190	0.332	0.570	0.720
0.63	0.233	0.288	0.295	0.289	0.88	0.176	0.309	0.538	0.705
0.64	0.247	0.308	0.317	0.312	0.89	0.162	0.286	0.502	0.678
0.65	0.260	0.328	0.340	0.334	0.90	0.148	0.261	0.462	0.638
0.66	0.271	0.347	0.362	0.356	0.91	0.134	0.236	0.420	0.588
0.67	0.281	0.365	0.383	0.378	0.92	0.120	0.211	0.377	0.531
0.68	0.289	0.382	0.405	0.400	0.93	0.106	0.185	0.332	0.470
0.69	0.296	0.398	0.427	0.421	0.94	0.091	0.159	0.285	0.406
0.70	0.301	0.414	0.448	0.443	0.95	0.075	0.133	0.239	0.340
0.71	0.304	0.427	0.469	0.465	0.96	0.060	0.106	0.191	0.273
0.72	0.307	0.439	0.489	0.486	0.97	0.045	0.080	0.144	0.205
0.73	0.306	0.450	0.509	0.508	0.98	0.030	0.053	0.096	0.137
0.74	0.305	0.458	0.529	0.529	0.99	0.015	0.027	0.048	0.069
0.75	0.303	0.464	0.548	0.550	1.00	0	0	0	0

In accordance with thermodynamic calculations [1, 10, 11, 15, 16] and a crystallographic analysis [25] (see also Sect. 5.5), the formation of superstructures of the types M_2X_3, M_4X_3, M_6X_5 and M_8X_7 represents a phase transition of the first kind, while a superstructure type M_2X can appear through a phase transition of the second kind. Transformation of one ordered phase into another is a phase transition of the first kind.

It follows from (9.109) and (9.111) that T_{trans} depends on the coefficient $2a^{(s)}/(2t-1)$. This coefficient is largest and equals 0.4 for superstructures of types M_2X (space groups $R\bar{3}m$, $Fd\bar{3}m$, $I4_1/amd$), M_3X_2 (space groups *Immm*, $C222_1$, $P2$) and M_6X_5 (space groups $C2/m$, $C2$, $P3_1$), while it assumes its smallest positive value for tetragonal (space group $P4/mmm$) M_2X and cubic (space group $Pm\bar{3}m$) M_4X_3 superstructures (Table 10.1). From the foregoing it is clear that order–disorder transition temperatures of superstructures of types M_2X (space groups $R\bar{3}m$, $Fd\bar{3}m$, $I4_1/amd$), M_3X_2 (space groups *Immm*, $C222_1$, $P2$) and M_6X_5 (space groups $C2/m$, $C2$, $P3_1$) are comparable. These superstructures are stable within similar temperature intervals but different concentration intervals of the equilibrium phase diagram. Tetragonal (space group $I4/mmm$) M_4X_3, cubic (space groups $Fm\bar{3}m$, $P4_332$) M_8X_7 and, particularly, tetragonal (space group $P4/mmm$) M_2X and cubic (space group $Pm\bar{3}m$) M_4X_3 superstructures can be stable in the low-temperature range only.

Table 10.3. Equilibrium values of long-range order parameter at the transition point, η_{trans} , in the case of M_2X–M_3X_2 and M_3X_2–M_6X_5 order–order transitions

y	η_{trans} transition M_2X–M_3X_2		y	η_{trans} transition M_3X_2–M_6X_5	
	M_2X	M_3X_2		M_3X_2	M_6X_5
0.52	0.259	0.240	0.52	0.075	0.078
0.53	0.380	0.363	0.53	0.078	0.094
0.54	0.499	0.480	0.54	0.120	0.136
0.55	0.591	0.582	0.55	0.161	0.178
0.56	0.682	0.6825	0.56	0.203	0.219
0.57	0.777	0.7917	0.57	0.245	0.261
0.58	0.835	0.8667	0.58	0.283	0.295
0.583	0.8338	0.8744	0.59	0.319	0.335
			0.60	0.357	0.375
			0.61	0.395	0.415
			0.62	0.433	0.454
			0.63	0.471	0.494
			0.64	0.509	0.535
			0.65	0.548	0.576
			0.66	0.586	0.616
			0.67	0.624	0.655
			0.68	0.659	0.694
			0.69	0.690	0.732
			0.70	0.718	0.767
			0.71	0.744	0.801
			0.72	0.763	0.833
			0.73	0.775	0.862
			0.74	0.770	0.8847
			0.75	0.7496	0.9000
			0.752	0.7439	0.9024

A generalized phase diagram of an ordering stoichiometric compound MX_y ($0 < y < 1$) was calculated and constructed for the temperature interval from 300 K to the transition point T_{trans} [11]. At these temperatures the quantity $k_B T/F_2(T)$ is a monotonically increasing function of T for nonstoichiometric carbides and nitrides. Therefore it is possible to construct the phase diagram using a dimensionless reduced temperature $k_B T/F_2(T)$. (The reduced temperature

$k_BT/F_2(T) = 0.02$ corresponds to about 300 K for these compounds.) It follows from the calculated phase diagram (Fig. 10.1) that ordered phases types M_2X, M_3X_2 and M_6X_5 appear in a nonstoichiometric compound MX_y upon ordering. This phase diagram also includes four two-phase regions: $M_2X + M_3X_2$, $M_3X_2 + M_6X_5$, $M_6X_5 + MX_y$, and $M_3X_2 + MX_y$. The M_6X_5 phase can be formed by the peritectoid reaction $M_3X_2 + MX_y \rightarrow M_6X_5$ in a narrow concentration interval $0.5236 \le y \le 0.5255$ ($y \equiv X/M$) at a reduced temperature $k_BT/F_2(T) = 0.09981$ (inset in Fig. 10.1). Concentration dependences of free energies of phases M_3X_2, MX_y and M_6X_5 have a common tangent at the point (2) of the peritectoid transformation (inset in Fig. 10.1). The disordered phase MX_y has a minimum of the free energy at a larger y than the ordered phases M_3X_2 and M_6X_5 .

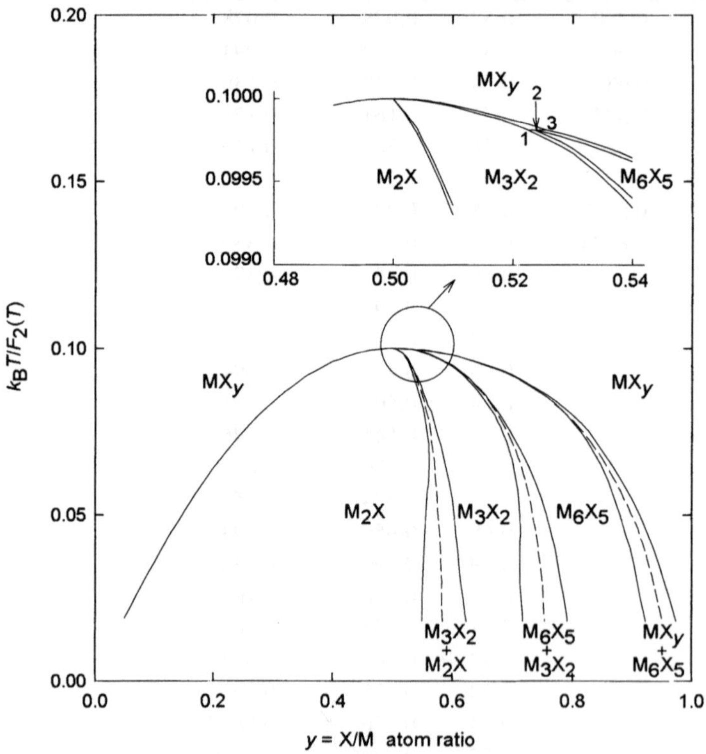

Fig. 10.1. Generalized equilibrium phase diagram of the M–X system constructed [11] taking account of ordering of interstitial atoms X and structural vacancies □ in nonstoichiometric interstitial compound MX_y ($MX_y□_{1-y}$). It is assumed that all nonmetallic interstitial atoms X are replaced by vacancies □. Solid lines are the phase boundaries, dotted lines are the lines of order–order and order–disorder phase transitions of the first kind; the phase boundary and the phase transition lines coincide for a $MX_y \leftrightarrow M_2X$ transition of the second kind. The inset magnifies the ordering region where the peritectoid transformation $M_3X_2 + MX_y \rightarrow M_6X_5$ takes place at the reduced temperature $k_BT/F_2(T) = 0.09981$

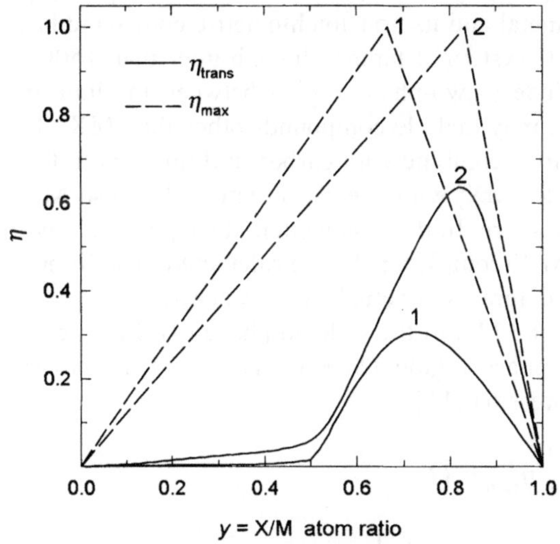

Fig. 10.2. Equilibrium long-range order parameter at the transition point η_{trans} and the maximum value of the long-range order parameter η_{max} vs. the MX_y composition upon formation of ordered phases of the M_3X_2 (1) and M_6X_5 (2) types

The parameter η_{equil} of ordered phases M_2X (space group *P4/mmm*), M_4X_3 type and M_8X_7 type is small even in the low-temperature range ($k_BT/F_2(T)$ approaching 0.02), while their free energy is higher than the free energy of other superstructures. For this reason, ordered phases M_2X (space group *P4/mmm*), M_4X_3 type and M_8X_7 type cannot appear under equilibrium conditions in nonstoichiometric compounds MX_y with the basic structure *B*1.

Thus, ordered phases of the M_2X, M_3X_2 and M_6X_5 types are formed in a nonstoichiometric compound MX_y upon ordering. Figure 10.2 shows the parameters η_{trans} and η_{max} versus composition of a nonstoichiometric compound MX_y, when M_3X_2 and M_6X_5 superstructures are formed through phase transitions of the first kind.

Calculations of ordering of actual nonstoichiometric compounds should allow additionally for boundary equilibrium conditions. For example, $VC_{0.875}$ corresponds to the upper boundary of the homogeneity interval of disordered cubic vanadium carbide VC_y in the V–C system. Note that for all other carbides and nitrides with the *B*1 structure the upper homogeneity boundary coincides with the composition $MX_{1.00}$. It is for the boundary effect that the ordered cubic phase V_8C_7 (space group $P4_132$ or $P4_332$) is formed only in vanadium carbide. An equilibrium ordered cubic phase of the M_8X_7 type does not appear in other carbides and nitrides.

Vacancies cannot completely replace nonmetal atoms in actual nonstoichiometric compounds MX_y, because the crystal lattice symmetry of the metal M differs from the metal sublattice symmetry of a nonstoichiometric compound [1, 2]. Crystal lattices of a transition metal and its nonstoichiometric compound have similar symmetry only in the Th–C system at temperatures higher than 1500 K. This system indeed does not include a two-phase region between thorium and thorium carbide. The M–X system may include compounds other than MX_y. For example, transition metals of group V combine with carbon and nitrogen to form not only cubic carbides and nitrides MX_y with the $B1$ structure, but also M_2X_y compounds having a hexagonal structure. In this connection, the equilibrium with the metal M or M_2X_y compound (M_2X(comp)) need to be calculated in addition to ordering when constructing phase diagrams of actual M–X systems.

The equilibrium between the metal M and the ordered phase $M_{2t}X_{2t-1}$, i.e. the phase boundary between the two-phase region ($M + M_{2t}X_{2t-1}$) and the ordered phase $M_{2t}X_{2t-1}$, is given by the expression [11]

$$F_{\mathrm{M}}(T) = F_0(T) - y^2 F_2(T) + \frac{a^{(s)}}{4t^2}\eta^2_{\mathrm{equil}} F_2(T) + \frac{k_{\mathrm{B}}T}{2t}\left[\ln(1-n_1^{(\mathrm{d})}) + (2t-1)\ln(1-n_2^{(\mathrm{d})})\right]_{\eta=\eta_{\mathrm{equil}}}, \qquad (10.11)$$

where $F_{\mathrm{M}}(T)$ is the free energy of the metal M.

The equilibrium between the ordered phase $M_{2t}X_{2t-1}$ and the MX_{y_b} compound (y_b being the maximum relative concentration of nonmetal atoms X at the upper boundary of the homogeneity range) is [15]

$$F_{\mathrm{MX}_{y_b}(\mathrm{comp})}(T) = F_0(T) + y_b F_1(T) + y(2y_b - y)F_2(T) + \frac{a^{(s)}}{4t^2}\eta^2_{\mathrm{equil}} F_2(T) + \frac{k_{\mathrm{B}}T}{2t}\left\{\ln\left[(n_1^{(\mathrm{d})})^{y_b}(1-n_1^{(\mathrm{d})})^{(1-y_b)}\right] + (2t-1)\ln\left[(n_2^{(\mathrm{d})})^{y_b}(1-n_2^{(\mathrm{d})})^{(1-y_b)}\right]\right\}_{\eta=\eta_{\mathrm{equil}}}. \qquad (10.12)$$

In accordance with (10.12), the position of the lower homogeneity boundary of the ordered phase $M_{2t}X_{2t-1}$, which is in equilibrium with the compound M_2X(comp) (or $MX_{0.5}$(comp), i.e. $y_b = 0.5$), can be calculated from the equation

$$0.5F_{\mathrm{M_2X(comp)}}(T) = F_0(T) + 0.5F_1(T) + y(1-y)F_2(T) + \frac{a^{(s)}}{4t^2}\eta^2_{\mathrm{equil}} F_2(T) + \frac{k_{\mathrm{B}}T}{4t}\left\{\ln\left[n_1^{(\mathrm{d})}(1-n_1^{(\mathrm{d})})\right] + (2t-1)\ln\left[n_2^{(\mathrm{d})}(1-n_2^{(\mathrm{d})})\right]\right\}_{\eta=\eta_{\mathrm{equil}}}, \qquad (10.13)$$

where $F_{\mathrm{M_2X(comp)}}$ is the free energy of a lower hexagonal carbide or nitride. For the disordered state, when $\eta = 0$, (10.11) and (10.13) describe the position of the lower boundary of the homogeneity interval for a disordered nonstoichiometric compound MX_y. Thus, these equations allow the computation of the equilibrium with a nonstoichiometric compound MX_y that may have any degree of long-range order from $\eta_{\max}$ to 0.

Let us consider in more detail the possible methods for determining parameters $F_0(T)$, $F_1(T)$ and $F_2(T)$ of the free energy of a disordered nonstoichiometric compound, which are necessary to calculate phase equilibrium.

Describing the thermodynamics of a solid body, pressure is assumed to be a constant value (usually $p = 0$ is taken). Under isobaric conditions, the internal energy, i.e. the crystal configuration energy E, coincides with the enthalpy H. The enthalpy of a compound at 298 K is equal to the enthalpy of formation of this compound from its elements, i.e. $H^\circ_{298} \equiv \Delta H^\circ_{f\,298}$. Hence, considering the temperature dependence of the enthalpy of a compound MX_y, we have

$$E(y,T) = H(y,T) = H^\circ_{298}(y) + (H^\circ_T - H^\circ_{298})_{MX_y}$$

$$\equiv \Delta H^\circ_{f\,298}(y) + (H^\circ_T - H^\circ_{298})_{MX_y}\,. \qquad (10.14)$$

As was shown in Sect. 9.5, the free energy of a disordered nonstoichiometric compound MX_y is described by the expression (9.71):

$$F(y,0,T) = E(y,T) - T[S(y,0,T) + S_c(y,0)]$$

$$\equiv F_0(T) + yF_1(T) + y^2F_2(T) - TS_c(y,0) \qquad (10.15)$$

and includes the configuration energy and the entropy term. The quantities $F_0(T)$, $F_1(T)$ and $F_2(T)$ in (10.15) represent energy parameters of the free energy. The crystal entropy comprises the configuration entropy $S_c(y,0)$, which is independent of temperature, and the nonconfiguration entropy $S(y,0,T)$, which explicitly depends on temperature. The temperature dependence of the entropy can be presented as its standard value and variation with temperature in the form $S(T) = S^\circ_{298} = \int_{298}^{T} [d(H^\circ_T - H^\circ_{298})/T] \equiv S^\circ_{298} + \int_{298}^{T} (C_p/T)dT$. Therefore

$$S(y,0,T) = S^\circ_{298}(y) + \int_{298}^{T} \frac{d(H^\circ_T - H^\circ_{298})_{MX_y}}{T} \equiv S^\circ_{298}(y) + \int_{298}^{T} \frac{C_p(T)dT}{T}\,. \qquad (10.16)$$

As is seen from (10.15), part of the free energy of a disordered compound MX_y, which does not contain the term allowing for the contribution from the configuration entropy $S_c(y,0)$, is equal to

$$F_0(T) + yF_1(T) + y^2F_2(T) \equiv E(y,T) - TS(y,0,T) \qquad (10.17)$$

or, taking into account (10.14) and (10.16),

$$F_0(T) + yF_1(T) + y^2F_2(T)$$

$$= \Delta H^\circ_{f\,298}(y) + (H^\circ_T - H^\circ_{298})_{MX_y} - T\left\{S^\circ_{298}(y) + \int_{298}^{T} \frac{d(H^\circ_T - H^\circ_{298})_{MX_y}}{T}\right\}$$

$$\equiv \Delta H^\circ_{f\,298}(y) + \int_{298}^{T} C_p(y)dT - T\left\{S^\circ_{298}(y) + \int_{298}^{T} \frac{C_p(y)dT}{T}\right\}. \qquad (10.18)$$

Table 10.4. Dependencies of the formation enthalpy $\Delta H^\circ_{f\,298}(y)$ and nonconfiguration entropy $S^\circ_{298}(y)$ on the composition of disordered nonstoichiometric carbides and nitrides MX_y (M = Ti, Zr, Hf, V, Nb, Ta), formation enthalpy $\Delta H^\circ_{f\,298}$ and entropy S°_{298} of lower hexagonal carbides M_2C

Compound	Formation enthalpy $\Delta H^\circ_{f\,298}(y)$ or $\Delta H^\circ_{f\,298}$ (kJ mol^{-1})	Entropy $S^\circ_{298}(y)$ or S°_{298} (J mol^{-1} K^{-1})
TiC_y	$16.75 - 304.49y + 89.04y^2 \pm 10.0$	$20.62 + 3.62y$
ZrC_y	$27.97 - 347.40y + 112.80y^2 \pm 10.0$	$29.43 + 4.08y$
HfC_y	$42.52 - 378.00y + 101.41y^2 \pm 10.0$	$35.22 + 4.36y$
VC_y	$104.78 - 437.73y + 224.02y^2 \pm 10.0$	$40.21 - 32.77y + 17.22y^2$
β-V_2C	-147.4 ± 21.0	$58.53 \cdot 10^{-3}$
NbC_y	$27.63 - 297.05y + 128.74y^2 \pm 10.0$	$32.73 - 9.05y + 11.29y^2$
β-Nb_2C	-195.1 ± 14.0	67.0 ± 4.0
TaC_y	$56.39 - 343.40y + 144.40y^2 \pm 6.0$	$42.14 - 14.90y + 16.18y^2$
α-Ta_2C	-197.6 ± 14.0	81.6 ± 4.0
TiN_y	$3.75 - 417.90y + 92.81y^2 \pm 10.0$	$6.6 + y + 13.7y^2$

Using explicitly experimental or model dependencies of the formation enthalpy $\Delta H^\circ_{f\,298}(y)$, enthalpy $(H^\circ_T - H^\circ_{298})_{MX_y}$ or heat capacity $C_p(y)$, and nonconfiguration entropy $S^\circ_{298}(y)$ on the MX_y composition, and considering (10.18), it is possible to derive an expression describing numerically part of the free energy $[F_0(T) + yF_1(T) + y^2F_2(T)]$ of a disordered nonstoichiometric compound MX_y.

Dependencies of the formation enthalpy $\Delta H^\circ_{f\,298}(y)$, and the nonconfiguration entropy $S^\circ_{298}(y)$ and the enthalpy $(H^\circ_T - H^\circ_{298})_{MX_y}$ on the composition of disordered nonstoichiometric carbides and nitrides MX_y are given in Tables 10.4 and 10.5. The formation enthalpy $\Delta H^\circ_{f\,298}$, entropy S°_{298} and the enthalpy $(H^\circ_T - H^\circ_{298})_{M_2C}$ of lower hexagonal carbides M_2C are given in the same tables. All these dependencies are determined using experimental data reported in the literature.

Numerical values of the energy parameters $F_0(T)$, $F_1(T)$ and $F_2(T)$ (Table 10.6) required to calculate phase diagrams were obtained [1, 10, 15] by expanding the free energy (the contribution from the configuration entropy being neglected) of disordered compounds MC_y and MN_y in a second-order power series of y. Each energy parameter $F(T)$ was represented by the polynomial

$$F(T) = f_0 + f_1T + f_2T^2 + f_3T^{-1} + f_4T\ln T \ . \qquad (10.19)$$

Free energies of disordered carbides TiC_y, ZrC_y, HfC_y, VC_y, NbC_y, TaC_y and nitride TiN_y were found [1, 10, 15, 26, 27] from experimental thermodynamic data [28–37]. Temperature dependencies of free energies of the metals α-Ti, β-Ti, α-Zr, β-Zr, α-Hf and lower carbides β-Nb_2C, β-V_2C and α-Ta_2C, which are in equilibrium with nonstoichiometric compounds MX_y, were also represented by polynomials (10.19) determined from the data in [34, 38–40]. The literature does not provide experimental thermodynamic data characterizing tetragonal nitride ε-Ti_2N, which does not have a homogeneity interval. Therefore the temperature dependence of its free energy was found as a model parameter proceeding from the data [41, 42] concerning the position of the phase boundary between TiN_y and the two-phase region (ε-Ti_2N + TiN_y).

Table 10.5. Dependence of enthalpy $(H_T^\circ - H_{298}^\circ)_{MX_y}$ on the composition of disordered nonstoichiometric carbides and nitrides MX_y (M = Ti, Zr, Hf, V, Nb, Ta), and enthalpy $(H_T^\circ - H_{298}^\circ)_{M_2C}$ of lower hexagonal carbides M_2C

Compound	Enthalpy $(H_T^\circ - H_{298}^\circ)_{MX_y}$ or $(H_T^\circ - H_{298}^\circ)_{M_2C}$ (kJ mol^{-1})
TiC_y (298–1500 K)	$(15.87 + 1.20\cdot10^{-2}T + 9.67\cdot10^{-6}T^2 - 6052T^{-1})$ $+ (-22.80 + 0.62\cdot10^{-2}T - 3.09\cdot10^{-6}T^2 + 6328T^{-1})y$ $+ (-5.68 + 1.91\cdot10^{-2}T)y^2$
ZrC_y (298–1500 K)	$(59.97 - 4.46\cdot10^{-2}T + 30.38\cdot10^{-6}T^2 - 14714T^{-1})$ $+ (-74.32 + 9.01\cdot10^{-2}T - 40.84\cdot10^{-6}T^2 + 15228T^{-1})y$ $+ (-1.20 + 13.50\cdot10^{-6}T^2)y^2$
HfC_y (298–1500 K)	$(96.51 - 11.17\cdot10^{-2}T + 119.49\cdot10^{-6}T^2 - 22001T^{-1})$ $+ (-109.59 + 16.42\cdot10^{-2}T - 243.25\cdot10^{-6}T^2 + 24513T^{-1})y$ $+ (-11.24 + 126.54\cdot10^{-6}T^2)y^2$
VC_y (298–2000 K)	$(-7.69 + 2.70\cdot10^{-2}T + 11.84\cdot10^{-6}T^2 - 416T^{-1})$ $+ (0.24 - 1.83\cdot10^{-2}T - 11.75\cdot10^{-6}T^2 + 1864T^{-1})y$ $+ (-12.66 + 4.25\cdot10^{-2}T)y^2$
β-V_2C (298–1600 K)	$-33.58 + 8.31\cdot10^{-2}T + 1.70\cdot10^{-6}T^2 + 2580T^{-1}$
NbC_y (298–1800 K)	$(-13.65 + 1.21\cdot10^{-2}T + 8.00\cdot10^{-6}T^2 + 2781T^{-1})$ $+ (12.06 + 3.26\cdot10^{-2}T - 4.40\cdot10^{-6}T^2 - 6373T^{-1})y$ $+ 4493T^{-1}y^2$
β-Nb_2C (298–2500 K)	$-23.25 + 6.65\cdot10^{-2}T + 6.28\cdot10^{-6}T^2 + 858T^{-1}$
TaC_y (298–1800 K)	$(-15.05 + 3.13\cdot10^{-2}T - 3.04\cdot10^{-6}T^2 + 1785T^{-1})$ $+ (12.62 + 0.69\cdot10^{-2}T + 10.48\cdot10^{-6}T^2 - 4650T^{-1})y$ $+ (-1.80\cdot10^{-6}T^2 + 3000T^{-1})y^2$
α-Ta_2C (298–2500 K)	$-23.31 + 6.65\cdot10^{-2}T + 6.90\cdot10^{-6}T^2 + 858T^{-1}$
TiN_y (298–2200 K)	$(-9.75 + 3.02\cdot10^{-2}T + 8.40\cdot10^{-6}T^2)$ $+ (-3.52 + 0.94\cdot10^{-2}T + 8.20\cdot10^{-6}T^2)y$ $+ (1.67 - 18.8\cdot10^{-6}T^2)y^2$

Table 10.6 lists values of the coefficients f_0, f_1, f_2, f_3 and f_4 for all phases taken into account in calculating phase diagrams that allow for ordering of nonstoichiometric compounds.

Table 10.6. Coefficients f_0 (kJ mol^{-1}), f_1 (kJ mol^{-1} K^{-1}), f_2 (kJ mol^{-1} K^{-2}), f_3 (kJ K mol^{-1}), and f_4 (kJ mol^{-1} K^{-1}) of the polynomials $F(T) = f_0 + f_1T + f_2T^2 + f_3T^{-1} + f_4T\ln T$ (10.19) describing the parameters $F_0(T)$, $F_1(T)$ and $F_2(T)$ of the free energy $F(y,0,T)$ of the disordered nonstoichiometric compounds MX_y and free energies $F(T)$ of the metals M and lower hexagonal carbides and nitrides M_2X

Phase	Free energy parameters	Coefficients				
		f_0	$f_1 \times 10^2$	$f_2 \times 10^6$	f_3	$f_4 \times 10^2$
TiC_y	F_0(298–1500 K)	32.61	3.16	–9.67	–3026	–1.20
	F_1(298–1500 K)	–327.29	7.15	3.09	3164	–0.62
	F_2(298–1500 K)	83.36	12.77	–	–	–1.91
α-Ti	F(298–1150)	–7.02	12.00	–5.28	–	2.20
β-Ti	F(1150–2000 K)	–6.91	18.05	–	–	–3.14
ZrC_y	F_0(298–1500 K)	87.93	–39.28	–30.38	–7357	4.46
	F_1(298–1500 K)	–421.72	66.07	40.84	7614	–9.01
	F_2(298–1500 K)	111.60	0.80	–13.50	–	–
α-Zr	F(298–1140 K)	–9.93	15.60	–2.34	182	–2.86
β-Zr	F(1140–2200 K)	–4.90	16.19	–	–	–3.04
HfC_y	F_0(298–1500 K)	139.03	–83.59	–119.49	–11001	11.17
	F_1(298–1500 K)	–487.59	108.80	243.25	12257	–16.42
	F_2(298–1500 K)	90.18	7.54	–126.54	–	–
α-Hf	F(298–2500 K)	–7.58	11.37	–1.09	–	–2.56
VC_y	F_0(298–2000 K)	97.09	14.50	–11.84	–208	–2.70
	F_1(298–2000 K)	–437.49	–8.62	11.75	932	1.83
	F_2(298–2000 K)	211.36	26.72	0	0	–4.25
β-V_2C	F(298–1600 K)	–180.96	51.37	–1.70	1290	–8.31
V_8C_7	F(298–1400 K)	–123.89	27.41	–1.56	608	–4.35
NbC_y	F_0(298–1800 K)	13.98	6.87	–8.00	1390	–1.21
	F_1(298–1800 K)	–284.99	18.89	4.40	–3186	–3.26
	F_2(298–1800 K)	128.74	1.40	–	2246	–
β-Nb_2C	F(298–2500 K)	–218.35	38.69	–6.28	429	–6.65
TaC_y	F_0(298–1800 K)	41.34	17.57	3.04	893	–3.13
	F_1(298–1800 K)	–330.78	4.12	–10.48	–2325	–0.69
	F_2(298–1800 K)	144.40	–0.036	1.80	1500	–
α-Ta_2C	F(298–2500 K)	–220.93	37.26	–6.97	429	–6.65
TiN_y	F_0(298–2200 K)	–6.00	20.09	–8.40	–	–3.02
	F_1(298–2200 K)	–421.42	5.76	–8.20	–	–0.94
	F_2(298–2200 K)	94.48	–2.49	18.80	–	–
ε-Ti_2N	F(500–1350 K)	–1067.24	707.07	306.89	127931	–96.45

10.3 Phase Diagrams of M–X Systems (M = Ti, Zr, Hf, V, Nb, Ta; X = C, N)

Ti–C System. The Ti–C system has only one compound, namely cubic titanium carbide TiC_y, which is capable of ordering. Titanium carbide is in equilibrium with metallic α-Ti (at $T < 1150$ K) or β-Ti (at $T > 1150$ K) at the lower boundary of the homogeneity interval. In accordance with [43], the carbide $TiC_{0.32}$–$TiC_{0.37}$ corresponds to the lower boundary of the homogeneity interval of the disordered phase TiC_y at 1900 K > T > 1000 K. The position of the lower boundary was not determined exactly in experiments. According to different estimates the lower boundary corresponds to $TiC_{0.48}$ [28] or $TiC_{0.47}$ [44, 45] at 1900 K. Titanium carbide was calculated to have the composition $TiC_{0.52}$ [46] or $TiC_{0.45}$ [47] at the lower boundary of the homogeneity interval at 1900 K. In accordance with [28, 44, 46], when the temperature decreases, the position of the lower boundary of the homogeneity interval of disordered carbide TiC_y shifts toward larger y, i.e. toward carbide with a higher carbon content, a little faster than the calculated value [43].

Figure 10.3 shows free energies of disordered and crystallographically possible ordered phases as a function of the carbon concentration y of titanium carbide at 600 and 1200 K. Ordered phases Ti_2C, Ti_3C_2 and Ti_6C_5, and also disordered carbide TiC_y (at $y > 0.94$) have the lowest free energy in different concentration intervals at 600 (curves 2, 3, 4 and 1 respectively in Fig. 10.3).

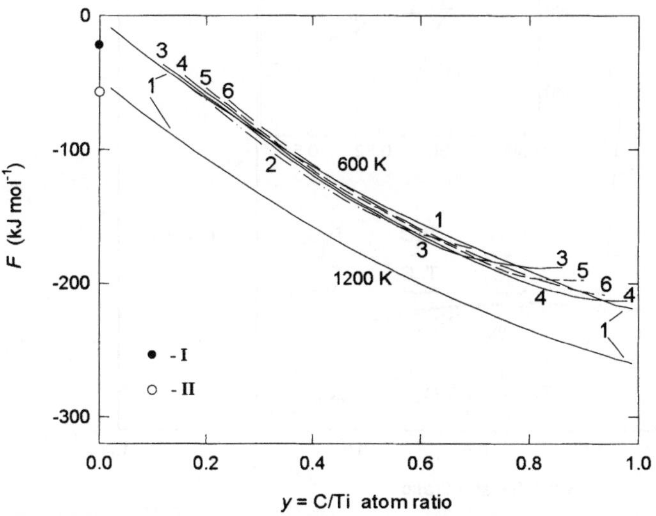

Fig. 10.3. Comparison of free energies of the disordered phase and possible ordered phase of titanium carbide at 600 and 1200 K: (1) disordered titanium carbide TiC_y, (2) ordered phase of the Ti_2C type, (3) ordered phase of the Ti_3C_2 type, (4) ordered phase of the Ti_6C_5 type, (5) ordered phase of the Ti_4C_3 type, (6) ordered phase of the Ti_8C_7 type. Points I and II correspond to free energies of α-Ti and β-Ti at 600 and 1200 K respectively

The ordered tetragonal phase Ti_4C_3 (space group *I4/mmm*) and cubic phases Ti_8C_7 possess higher free energies (curves 5 and 6 in Fig. 10.3) and cannot appear under equilibrium conditions. Cubic (space group $Pm\bar{3}m$) phase Ti_4C_3 and tetragonal (space group *P4/mmm*) phase Ti_2C cannot be formed at all at 600 K, because their transition temperature T_{trans} is lower than 300 K. For this reason, free energy curves corresponding to those phases are not given in Fig. 10.3. Only the disordered phase TiC_y can exist over the whole homogeneity range of titanium carbide at a temperature of 1000 K or higher.

Points I and II marked in Fig. 10.3 at $y = 0$ correspond to the free energy of α-Ti at 600 K and β-Ti at 1200 K. A tangent to the free energy $F(y, 600\text{ K})$ passing through point I determines the free energy and the width of the two-phase region that borders on metallic titanium. As can be seen from Fig. 10.3, a two-phase mixture (α-Ti + Ti_2C) dominates at the boundary with metallic titanium at 600 K.

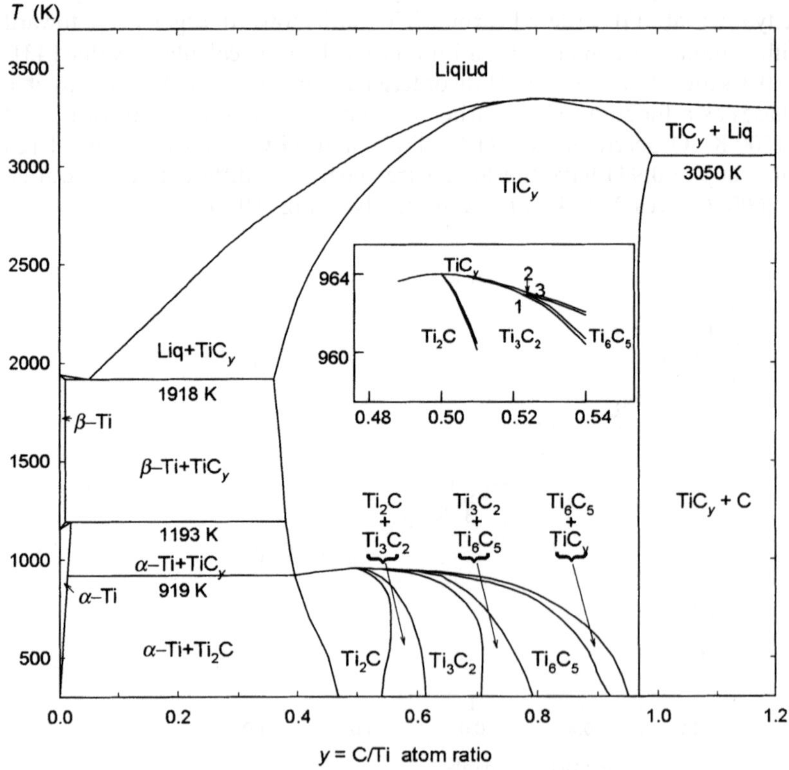

Fig. 10.4. Equilibrium phase diagram of the Ti–C system taking account of atomic ordering of nonstoichiometric cubic titanium carbide TiC_y. The inset magnifies the ordering region where the peritectoid transformation $Ti_3C_2 + TiC_y \rightarrow Ti_6C_5$ takes place at 963 K. The coordinates y = C/Ti of points 1, 2 and 3 are 0.5236, 0.524 and 0.5255 respectively [15]

The low-temperature part of the phase diagram of the Ti–C system (Fig. 10.4) was calculated by the OPF method [43]. The position of phase boundaries outside the ordering range is shown in accordance with [28, 44].

Phases Ti_2C, Ti_3C_2 and Ti_6C_5 (Fig. 10.4) are formed upon ordering under thermodynamic equilibrium conditions thanks to the wide homogeneity interval of the carbide TiC_y. Calculations show that the ordered phase Ti_2C appears by transition of the second kind at T_{trans} = 920–950 K. The homogeneity interval of the phase Ti_2C is fairly wide ($TiC_{0.42}$–$TiC_{0.56}$ at 700 K and $TiC_{0.40}$–$TiC_{0.54}$ at 800 K) and is bounded by two-phase regions (α-Ti + Ti_2C) and (Ti_2C + Ti_3C_2) on the left and on the right. Ordered phases Ti_3C_2 and Ti_6C_5 have homogeneity intervals comparable in width ($TiC_{0.59}$–$TiC_{0.71}$ for Ti_3C_2 and $TiC_{0.74}$–$TiC_{0.87}$ for Ti_6C_5 at 700 K). They are formed as a phase transition of the first kind. All order–order transformations in titanium carbide are also phase transitions of the first kind.

Let us compare the phase diagram of the Ti–C system (Fig. 10.4) with experimental findings. Different studies show that the ordered phase Ti_2C with cubic (space group $Fd\bar{3}m$) [48–52] or trigonal (space group $R\bar{3}m$) [20, 53] symmetry is formed in carbide TiC_y in the interval $0.5 \leq y \leq 0.65$ at $T < 1100$ K. Reports of the trigonal phase Ti_2C (Ti_6C_{3+x}) with space group $P3_121$ result from a delusion [54] and successive uncritical reproduction of this mistake in other work. We will not cite those reports. Indeed, such a trigonal (space group $P3_121$) phase of M_2C type with unit cell constants $a = b = a_{B1}/\sqrt{2}$ ($\mathbf{a} = \{1/2\ 1/2\ 0\}_{B1}$) and $c = 3\sqrt{2} \times a_{B1}$ ($\mathbf{c} = \{2\ 2\ 2\}_{B1}$) is fundamentally implausible for the basic structure $B1$, because sites of the metallic and nonmetallic sublattices partially overlap. The analysis of structural experiments [20, 48–57] suggests that the cubic (space group $Fd3m$) phase Ti_2C mostly occurs in annealed TiC_y samples with $y < 0.55$–0.56, while trigonal ordering is common in annealed carbide TiC_y with $0.58 \leq y \leq 0.65$. It is worth noting that powder diffraction patterns of ordered cubic (space group $Fd\bar{3}m$) and trigonal (space group $R\bar{3}m$) phases Ti_2C exhibit similar sets of superstructure reflections [25]. They can be distinguished only if trigonal distortions are present in the phase with space group $R\bar{3}m$ and directions of static atomic displacements are taken into account. This was probably the reason why in the pioneering papers [49–52], which were concerned with the cubic ordering model only [48], the observed superstructure reflections were attributed to the cubic phase Ti_2C even in annealed carbide TiC_y ($y \geq 0.59$). Later research [20, 55–57] showed that the trigonal phase Ti_2C is the main ordered phase in TiC_y ($y \leq 0.6$).

It is assumed that an ordered phase Ti_3C_2 exists in the $TiC_{0.60}$–$TiC_{0.70}$ interval. Some experimental evidence confirms its existence:
- superstructure reflections (2/3 2/3 0) observed by researchers [55] who examined an annealed single crystal of $TiC_{0.61}$ by the method of elastic neutron scattering;
- weak superstructure lines with the diffraction vector $|\mathbf{q}| \approx 2.03$ typical of the rhombic (space group $C222_1$) phase Ti_3C_2 in X-ray patterns of annealed carbide $TiC_{0.70}$ [58];
- diffusion peaks of neutron scattering due to short-range order in $TiC_{0.76}$ whose position corresponds to reflections (2/3 2/3 0) [59];

- evaluation of short-range order parameters [20] for a $TiC_{0.64}$ single crystal from neutron diffuse scattering data showed that the best agreement between theory and experiment is achieved when annealed carbide $TiC_{0.64}$ contains two ordered phases Ti_2C and Ti_3C_2.

The existence of the trigonal ordered phase Ti_2C and the rhombic ordered phase Ti_3C_2 also follows from calculations of the phase diagram for the Ti–C system performed by the Monte Carlo method for the interval $TiC_{0.57}$–$TiC_{0.70}$ (Fig. 10.5).

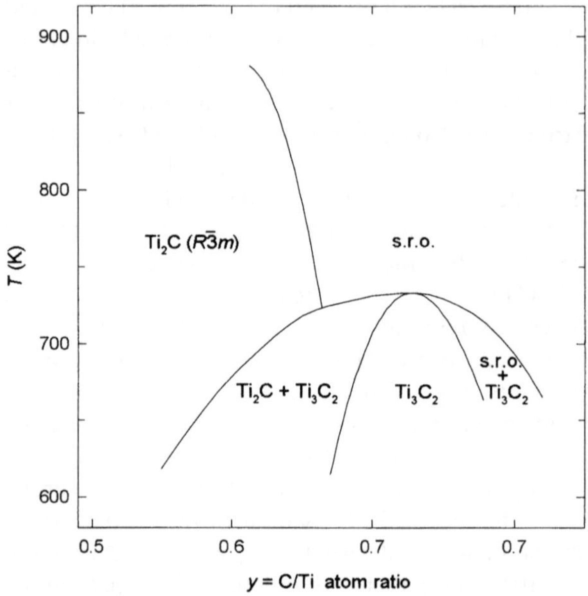

Fig. 10.5. Part of the phase diagram of the Ti–C system calculated by the Monte Carlo method allowing for ordering of nonstoichiometric titanium carbide TiC_y in the interval $TiC_{0.57}$–$TiC_{0.70}$ [18, 20]

It was shown [60] that short-range order and order–disorder transitions in nonstoichiometric carbides and nitrides can be described using the Ising Hamiltonian on an fcc sublattice. In this case the Hamiltonian is defined as the sum of effective energies of paired interactions V_n between sites of the nonmetal sublattice, which are n-th neighbors. In [18, 20, 61] energies of paired interactions V_n in the nonmetal sublattice of carbides $TiC_{0.76}$ and $TiC_{0.65}$ were computed not only for the first-nearest (V_1) and second-nearest (V_2) neighbors but also for more distant sites located in the third (V_3) and fourth (V_4) coordination spheres relative to one another. Energies of paired interactions were computed from experimental high-temperature data [61] on neutron diffusion scattering by three methods: the mean field approximation [62, 63], the reverse Monte Carlo method [64], and the cluster variation method [65–68]. The phase diagram calculated by the Monte

Carlo method fitted the experiment best. In accordance with [20, 61], formation of the ordered rhombic phase Ti_3C_2 is a phase transition of the first kind. This coincides with the calculation [43] of ordering in TiC_y performed by the OPF method.

Some difference in mutual positions of the ordered phases Ti_2C and Ti_3C_2 in the phase diagrams of the Ti–C system calculated by the OPF method [43] (Fig. 10.4) and the Monte Carlo method [20, 64] (Fig. 10.5) can be explained as follows. When calculated by the OPF method, the free energy of an ordering compound is a function of its composition and long-range order parameters, whereas in the Monte Carlo method the energies of interatomic interactions were obtained from short-range order parameters. Clearly, the observed small variance in the calculation results can be decreased provided each method takes into account both long-range and short-range order parameters. However, the problem of taking simultaneous account for long- and short-range orders remains unsolved in the theory of phase transformations of the atomic ordering type.

Generally, experimental and theoretical studies [18, 20, 48–59, 61] support calculations of the Ti–C phase diagram [15, 16] (Fig. 10.4), according to which ordered phases Ti_2C and Ti_3C_2 can be formed in nonstoichiometric titanium carbide at temperatures below 1000 K. The existence of the ordered phase Ti_2C above 2150 K as shown in the Ti–C phase diagram [45] is more than doubtful. In accordance with [15], carbide TiC_y ($0.78 < y < 0.88$) can also have the ordered phase Ti_6C_5, which refers to the type of ordered phases M_6X_5 common to all nonstoichiometric interstitial compounds. A hexagonal ordered phase with the presumed composition Ti_6C_5 has been found recently [69] in thin films of cubic titanium carbide. Discontinuites of $C_p(T)$ dependencies at ~1065 and 1034 K observed when measuring heat capacity on heating and cooling respectively [70] can be considered as indirect evidence of ordering of nonstoichiometric titanium carbide $TiC_{0.82}$ too. Thus, all ordered phases of nonstoichiometric titanium carbide as predicted by the researchers [43] have been spotted by now. Disorder–order transition temperatures calculated in [15, 43] for the phases Ti_3C_2 and Ti_6C_5 do not exceed 950 K.

Note that the calculated phase diagram of the Ti–C system includes a wide two-phase region at $y < 0.5$ (α-Ti + Ti_2C at $T < 919$ K and α-Ti + TiC_y at $T > 919$ K) (Fig. 10.4), which is absent in the generalized phase diagram (Fig. 10.1). This two-phase region appeared because equilibrium boundary conditions were taken into account. Carbon atoms cannot be replaced completely by structural vacancies in nonstoichiometric titanium carbide TiC_y, because titanium and titanium carbide have different lattice symmetries. Metallic α-Ti has an hcp structure, while nonstoichiometric titanium carbide TiC_y possesses a cubic structure of the $B1$ type. As a result, the lower homogeneity boundary of titanium carbide corresponds (depending on temperature) to $TiC_{0.4\text{-}0.5}$. The two-phase "titanium–titanium carbide" mixture is in equilibrium when the carbon concentration is small. The generalized phase diagram in Fig. 10.1 is constructed for the ideal case of complete substitution of interstitial atoms and structural vacancies, which is possible if the metal and the nonstoichiometric compound have similar symmetry. The two-phase region is absent in this case.

Table 10.7 lists the calculated values of T_{trans}, ΔS_{trans} and phase transition heat (enthalpy) ΔH_{trans} for all order–order and order–disorder transitions of the first kind possible in titanium carbide. Generally, ordering in titanium carbide TiC_y takes place at $T < 1000$ K.

Table 10.7. Temperatures T_{trans} (K), entropies ΔS_{trans} (J mol^{-1} K^{-1}) and heats ΔH_{trans} (kJ mol^{-1}) of the order-order and order-disorder phase transitions for nonstoichiometric titanium carbide TiC_y

y	Ti_2C–Ti_3C_2			Ti_3C_2–Ti_6C_5			Ti_6C_5–TiC_y		
	T_{trans}	ΔS_{trans}	ΔH_{trans}	T_{trans}	ΔS_{trans}	ΔH_{trans}	T_{trans}	ΔS_{trans}	ΔH_{trans}
0.52	948	0.09	0.08	–	–	–	–	–	–
0.53	926	0.15	0.14	961	0.003	0.003	962	0.02	0.02
0.54	893	0.25	0.22	959	0.013	0.013	961	0.03	0.03
0.55	853	0.28	0.24	956	0.03	0.03	960	0.04	0.04
0.56	795	0.32	0.25	951	0.05	0.05	959	0.06	0.05
0.58	460	0.29	0.13	940	0.12	0.12	955	0.10	0.09
0.60				924	0.19	0.18	949	0.15	0.14
0.62				902	0.29	0.26	942	0.22	0.21
0.63				891	0.35	0.31	938	0.26	0.24
0.64				877	0.41	0.36	934	0.30	0.28
0.65				860	0.48	0.42	929	0.35	0.32
0.66				841	0.56	0.47	924	0.40	0.37
0.68				794	0.74	0.59	912	0.50	0.46
0.70				733	0.91	0.66	898	0.62	0.56
0.72				639	0.99	0.63	882	0.75	0.67
0.73				566	1.00	0.57	874	0.82	0.72
0.74				472	0.90	0.42	864	0.90	0.78
0.75				320	0.69	0.22	854	0.97	0.83
0.76							843	1.05	0.88
0.78							818	1.20	0.98
0.80							789	1.34	1.06
0.82							755	1.46	1.10
0.83							728	1.49	1.08
0.84							713	1.49	1.07
0.85							689	1.48	1.02
0.86							662	1.43	0.95
0.88							661	1.27	0.76
0.90							528	1.05	0.55
0.92							444	0.82	0.36
0.94							348	0.58	0.20
0.95							294	0.46	0.14

A recent study of ordering in carbide TiC_y [71] is of interest. Using data on the structure and electrical resistivity of annealed and quenched TiC_y samples, it is shown [71] that the cubic (space group $Fd\bar{3}m$) ordered phase Ti_2C has a homogeneity interval between $TiC_{0.49-0.51}$ and $TiC_{0.54-0.55}$, while the trigonal (space group $R\bar{3}m$) superstructure Ti_2C is formed between $TiC_{0.55}$ and $TiC_{0.59}$. The $TiC_{0.59}$–$TiC_{0.63}$ interval corresponds to a two-phase region (Ti_2C (space group $R\bar{3}m$) + Ti_3C_2 (space group $C222_1$)). The range of the rhombic (space group $C222_1$) ordered phase Ti_3C_2 is relatively narrow and does not go beyond $TiC_{0.63}$–$TiC_{0.67}$. The lowest annealing temperature used in [71] was 770 K. Because ordered phase Ti_6C_5 could not be detected even after such low-temperature annealing of $TiC_{0.83}$ and $TiC_{0.85}$ carbides, the researchers [71] supposed that the transition temperature of this phase is below 770 K. Figure 10.6 shows a section of the equilibrium phase diagram of the Ti–C system where nonstoichiometric titanium carbide TiC_y is ordered.

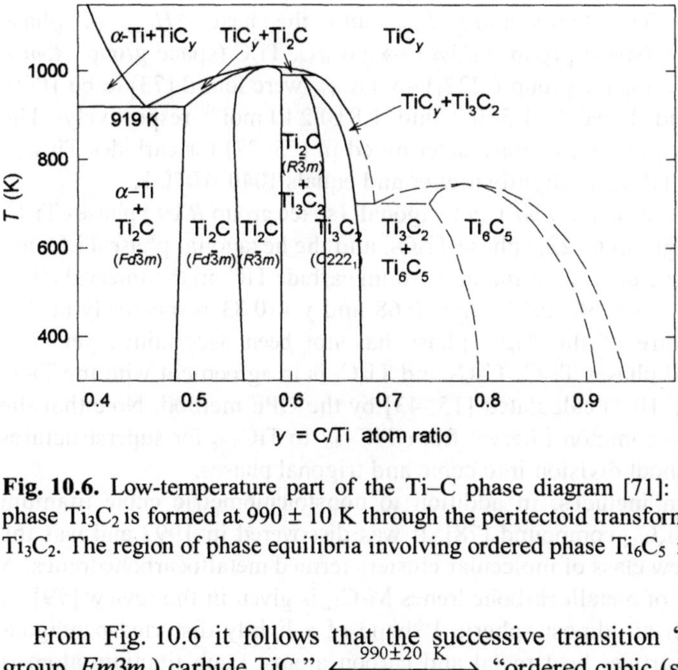

Fig. 10.6. Low-temperature part of the Ti–C phase diagram [71]: The rhombic ordered phase Ti_3C_2 is formed at 990 ± 10 K through the peritectoid transformation $Ti_2C + TiC_y \rightarrow Ti_3C_2$. The region of phase equilibria involving ordered phase Ti_6C_5 is shown tentatively

From Fig. 10.6 it follows that the successive transition "disordered (space group $Fm\bar{3}m$) carbide TiC_y" $\xleftrightarrow{990\pm20\ \text{K}}$ "ordered cubic (space group $Fd\bar{3}m$) phase Ti_2C" $\xleftrightarrow{960\pm20\ \text{K}}$ "ordered trigonal (space group $R\bar{3}m$) phase Ti_2C" is possible at $0.54 \le y \le 0.57$. The ordered rhombic phase Ti_3C_2 is formed most likely by peritectoid transformation $Ti_2C + TiC_y \rightarrow Ti_3C_2$ at a temperature 990±10 K in the interval $0.61 \le y < 0.63$. It follows from the constructed part of the Ti–C phase diagram (Fig. 10.6) that the ordered phase Ti_2C with cubic or trigonal symmetry can be observed over a broad composition interval of nonstoichiometric titanium carbide between $TiC_{0.40}$ and $TiC_{0.63}$. But only the $TiC_{0.49-0.50}$–$TiC_{0.58-0.59}$ interval is the one-phase region of the ordered phase Ti_2C. The region, where an ordered phase type Ti_6C_5 can appear, is shown conditionally, because no experimental evidence of its existence is available in [71]. Hysteresis in the temperature dependencies of electrical resistivity suggests that transformations $TiC_y \leftrightarrow Ti_2C$ and $TiC_y \leftrightarrow Ti_3C_2$ are phase transitions of the first kind with $T_{trans} = 980–1000$ K [71]. This is in good agreement with $T_{trans} \approx 1000$ K determined from measurements of electrical resistivity of carbide $TiC_{0.55}$ [72].

Measurements of the heat capacity of carbides $TiC_{0.54}$, $TiC_{0.58}$ and $TiC_{0.62}$ [73, 74] showed that the heat capacity changed abruptly at the transition temperature T_{trans} as a result of an equilibrium order–disorder transformation. The heat capacity peak near T_{trans} has a fairly symmetric shape more typical of a transition of the first kind. Moreover, a weak hysteresis of heat capacity is observed on cooling, a fact which also attests to a transition of the first kind. In accordance with [73, 74], the reversible order–disorder transformation Ti_2C (space group $Fd\bar{3}m$) $\leftrightarrow TiC_{0.54}$ can be viewed as a weak phase transition of the first kind with a small latent transformation heat, while the $Ti_3C_2 \leftrightarrow TiC_y$ transformation is a normal transition of the first kind. The temperature T_{trans} and the heat ΔH_{trans} of phase transformations Ti_2C (space group $Fd\bar{3}m$) $\leftrightarrow TiC_{0.54}$, Ti_2C (space group $R\bar{3}m$) $\leftrightarrow TiC_{0.58}$, and Ti_3C_2 (space group $C222_1$) $\leftrightarrow TiC_{0.62}$ were found [73] to be 1000, 980 and 970 K, and 1.5±0.2, 1.5±0.2 and 1.8±0.2 kJ mol^{-1} respectively. The disorder–order transition temperature determined in [75–77] for carbides $TiC_{0.49}$, $TiC_{0.55}$, $TiC_{0.60}$ and $TiC_{0.63}$ is slightly higher and equals 1040–1070 K.

Thus, cubic (space group $Fd\bar{3}m$) and trigonal (space group $R\bar{3}m$) phases Ti_2C, the rhombic (space group $C222_1$) phase Ti_3C_2, and the hexagonal phase Ti_6C_5 are formed in experiments upon ordering of titanium carbide TiC_y in the intervals $0.52 \le y \le 0.55$, $0.56 \le y \le 0.58$, $0.62 \le y < 0.68$ and $y \approx 0.83$ respectively at $T <$ 1000 K. The structure of the Ti_6C_5 phase has not been ascertained yet. The formation of ordered phases Ti_2C, Ti_3C_2 and Ti_6C_5 is in agreement with the Ti–C phase diagram (Fig. 10.4) calculated [15, 43] by the OPF method. Note that the OPF method gives a common interval from $TiC_{0.46}$ to $TiC_{0.58}$ for superstructures of the Ti_2C type without division into cubic and trigonal phases.

The Ti–C system includes, in addition to nonstoichiometric cubic titanium carbide TiC_y, the Ti_8C_{12} compound [78]. It was discovered in 1992 and was the first member of a new class of molecular clusters termed metallocarbohedrenes. A detailed description of metallocarbohedrenes M_8C_{12} is given in the review [79]. A M_8C_{12} molecule has an almost spherical shape of a lightly distorted pentagon-dodecahedron. Its vertices host metal and carbon atoms. Each titanium atom is bound with three carbon atoms, while each carbon atom is bound with the nearest carbon atom and two titanium atoms. The length of Ti–C and C–C bonds in Ti_8C_{12} differs by about 40% and is equal to 1.99 and 1.39 nm respectively. However, carbon and titanium atoms are nearly equidistant from the cluster center. The structure of metallocarbohedrenes M_8C_{12} is highly symmetric: its point symmetry group includes 24 symmetry elements of the cubic group. Metallocarbohedrene molecules are likely to be very stable thanks to the high symmetry.

The Ti_8C_{12} clusters were obtained by plasmochemical gas phase synthesis involving laser heating of the metal and hydrocarbon (methane, ethylene, propylene, benzene etc.) plasma. Molecular clusters M_8C_{12} (M = Ti, Zr, Hf, V, Nb, Ta) are formed using strong laser heating of the plasma with a high concentration of hydrocarbons [78–83], whereas conventional plasmochemical synthesis provides particles of cubic carbide MC_y only. Atoms form strong bonds in metallocarbohedrenes M_8C_{12}: the binding energy per one atom of a Ti_8C_{12} molecule is 6.1–6.7 eV atom^{-1} [84–86]. For comparison, the binding energy equals 7.4–7.6 eV atom^{-1} in a molecule of fullerene C_{60} [87, 88] and 7.2 eV atom^{-1} for carbide TiC with the *B*1 structure [84]. The Ti_8C_{12} compound is not shown in the Ti–C phase diagram in Fig. 10.4, because it is outside the ordering interval.

Zr–C System. The Zr–C system contains only one compound whose nonmetal sublattice can undergo ordering. It is nonstoichiometric cubic zirconium carbide ZrC_y with a broad homogeneity interval. The phase diagram of the Zr–C system (Fig. 10.7) is very similar to that of the Ti–C system.

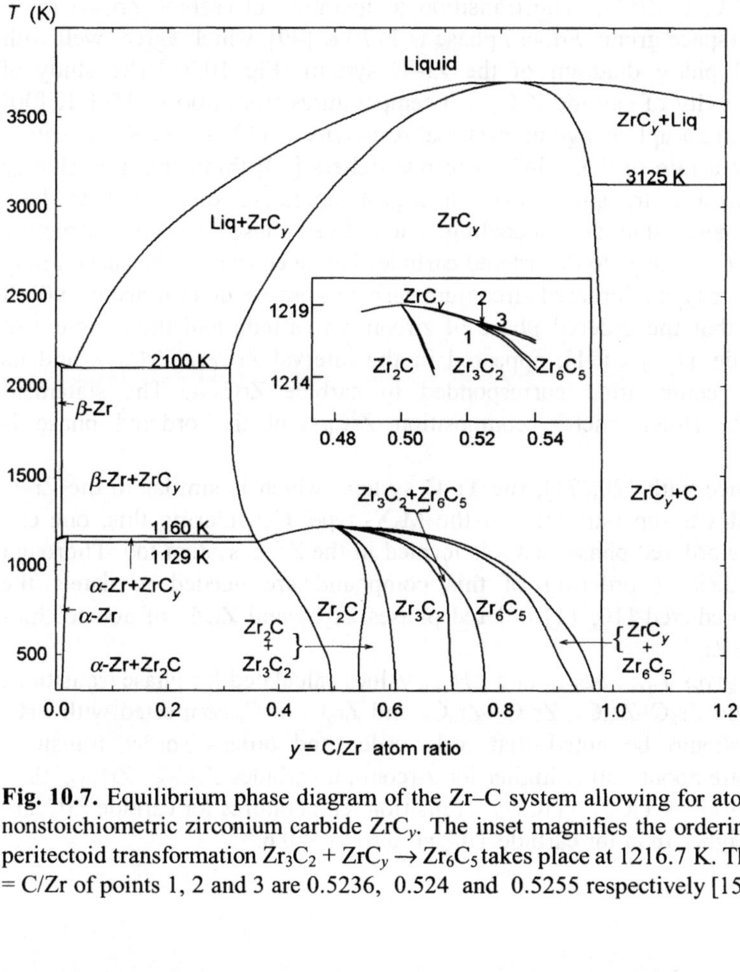

Fig. 10.7. Equilibrium phase diagram of the Zr–C system allowing for atomic ordering of nonstoichiometric zirconium carbide ZrC_y. The inset magnifies the ordering region where peritectoid transformation $Zr_3C_2 + ZrC_y \rightarrow Zr_6C_5$ takes place at 1216.7 K. The coordinates *y* = C/Zr of points 1, 2 and 3 are 0.5236, 0.524 and 0.5255 respectively [15, 16]

Calculations [1, 10–12, 14] show that ordered phases Zr_2C, Zr_3C_2 and Zr_6C_5 appear in this system under thermodynamic equilibrium conditions at $T < 1200$ K in different concentration intervals. The disorder–order transformation ZrC_y–Zr_2C can proceed as a phase transition of the second kind, whereas all other transformations due to ordering of nonstoichiometric zirconium carbide ($Zr_2C \rightarrow Zr_3C_2$, $Zr_3C_2 \rightarrow Zr_6C_5$ and $Zr_6C_5 \rightarrow ZrC_y$) are transitions of the first kind. Direct disorder–order transformation $ZrC_y \rightarrow Zr_3C_2$ can take place in a narrow interval of compositions $ZrC_{0.505}$–$ZrC_{0.522}$ at $T = 1217$ K (Fig. 10.7). Given other concentrations of carbon in carbide ZrC_y, the superstructure Zr_3C_2 can be formed only as a result of order–order transitions $Zr_2C \rightarrow Zr_3C_2$ or $Zr_6C_5 \rightarrow Zr_3C_2$. Note also that peritectoid transformation $Zr_3C_2 + ZrC_y \rightarrow Zr_6C_5$ occurs in the ordering interval of zirconium carbide at $T = 1216.7$ K and $y = 0.524$ (see inset in Fig. 10.7).

Ordered cubic (space group $Fd\bar{3}m$) [48, 49] and trigonal (space group $R\bar{3}m$) [89] phases, which may be identified with the superstructure Zr_2C, were observed experimentally in zirconium carbide near the lower homogeneity boundary within the interval $ZrC_{0.63}$–$ZrC_{.74}$. The transition temperature of carbide $ZrC_{0.63}$ to an ordered cubic (space group $Fd\bar{3}m$) phase is 1170 K [49], which agrees well with the calculated phase diagram of the Zr–C system (Fig. 10.7). The study of electrical resistivity of carbide $ZrC_{0.70}$ at temperatures from 300 to 1500 K [90] revealed a small abrupt change in electrical resistivity at 1330–1350 K on heating and cooling at a rate of 2 K min^{-1}. The researchers [90] think that this change corresponds to the transition from a high-temperature disordered to low-temperature ordered state during cooling. The ordered phase has a cubic structure with a double (compared to disordered carbide) lattice constant. The space group and the structure type of ordered zirconium carbide were not determined in [90]. It was assumed that the ordered phase of zirconium carbide had the structure of thorium carbide $ThC_{0.76}$ [91], appeared in the interval $ZrC_{0.69}$–$ZrC_{0.77}$, and its stoichiometric composition corresponded to carbide $ZrC_{0.75}$. The statement concerning the stoichiometric composition $ZrC_{0.75}$ of the ordered phase is questionable.

In accordance with [20, 71], the Ti–C system, which is similar to the Zr–C system, includes a superstructure of the M_3X_2 type. Considering this, one can expect that the ordered phase Zr_3C_2 is formed in the Zr–C system too. Thorough structural studies of ordering of this compound are needed to detect the theoretically predicted [10, 11] ordered phases Zr_3C_2 and Zr_6C_5 of nonstoichiometric carbide ZrC_y.

Table 10.8 gives T_{trans}, ΔS_{trans} and ΔH_{trans} values calculated for phase transitions of the first kind Zr_2C–Zr_3C_2, Zr_3C_2–Zr_6C_5 and Zr_6C_5–ZrC_y connected with ZrC_y ordering. It should be noted that order-order and order-disorder transition temperatures are about 250 K higher for zirconium carbides $ZrC_{0.52}$–$ZrC_{0.60}$ than for titanium carbides $TiC_{0.52}$–$TiC_{0.60}$. Transition temperatures for carbide ZrC_y are about 100 K higher than for carbide TiC_y at $0.6 < y < 0.6$.

Table 10.8. Temperatures T_{trans} (K), entropies ΔS_{trans} (J mol^{-1} K^{-1}) and heats ΔH_{trans} (kJ mol^{-1}) of the order-order and order-disorder phase transitions for nonstoichiometric zirconium carbide ZrC_y

y	Zr_2C–Zr_3C_2			Zr_3C_2–Zr_6C_5			Zr_6C_5–ZrC_y		
	T_{trans}	ΔS_{trans}	ΔH_{trans}	T_{trans}	ΔS_{trans}	ΔH_{trans}	T_{trans}	ΔS_{trans}	ΔH_{trans}
0.52	1198	0.09	0.10	–	–	–	–	–	–
0.53	1171	0.15	0.18	1215	0.003	0.003	1216	0.02	0.03
0.54	1130	0.25	0.28	1213	0.014	0.017	1215	0.03	0.04
0.55	1080	0.28	0.31	1208	0.03	0.04	1214	0.04	0.05
0.56	1008	0.32	0.32	1203	0.05	0.07	1212	0.06	0.07
0.58	582	0.29	0.17	1188	0.13	0.15	1207	0.10	0.12
0.60				1169	0.20	0.23	1200	0.16	0.19
0.62				1143	0.30	0.34	1191	0.23	0.27
0.63				1128	0.35	0.40	1186	0.27	0.32
0.64				1110	0.41	0.46	1181	0.31	0.36
0.65				1089	0.49	0.53	1175	0.36	0.42
0.66				1065	0.57	0.60	1168	0.41	0.47
0.68				1005	0.74	0.75	1154	0.51	0.59
0.70				928	0.91	0.84	1137	0.63	0.72
0.72				810	0.99	0.80	1117	0.76	0.85
0.73				717	1.00	0.71	1106	0.83	0.92
0.74				597	0.89	0.53	1094	0.91	0.99
0.75				405	0.69	0.28	1081	0.98	1.06
0.76							1067	1.06	1.13
0.78							1036	1.21	1.25
0.80							1000	1.35	1.35
0.82							956	1.46	1.39
0.83							922	1.49	1.47
0.84							903	1.49	1.35
0.85							872	1.47	1.28
0.86							838	1.43	1.19
0.88							761	1.26	0.96
0.90							668	1.04	0.70
0.92							561	0.82	0.46
0.94							439	0.59	0.26
0.95							373	0.48	0.18

The compound Zr_8C_{12} was obtained in the Zr–C system in 1992 [80, 81]. It has no homogeneity interval, enters the metallocarbohedrenes group M_8C_{12}, and has the same structure as the molecular cluster Ti_8C_{12} described in the section devoted to the Ti–C system.

Hf–C System. The Hf–C system has only one compound, which serves as the basis for ordering. It is nonstoichiometric carbide HfC_y having a relatively broad homogeneity interval and the *B*1 structure. In accordance with [13, 14], Hf_3C_2 and Hf_6C_5 phases appear in hafnium carbide upon ordering under thermodynamic equilibrium conditions (Fig. 10.8). An ordered phase of M_2C type is not formed in HfC_y, because its carbon content is smaller than those in disordered carbide HfC_y which corresponds to the lower boundary of the homogeneity interval. In accordance with [1, 2, 23, 24, 92, 93], hafnium carbide has the composition $HfC_{0.56}$ at the lower boundary, while calculations [13] show that $HfC_{0.54}$ corresponds to the lower boundary of the homogeneity interval of disordered cubic hafnium carbide at 1000 K. The Hf–C phase diagram (Fig. 10.8) also includes a newly discovered [80, 81] compound Hf_8C_{12} belonging to the metallocarbohedrene group M_8C_{12}.

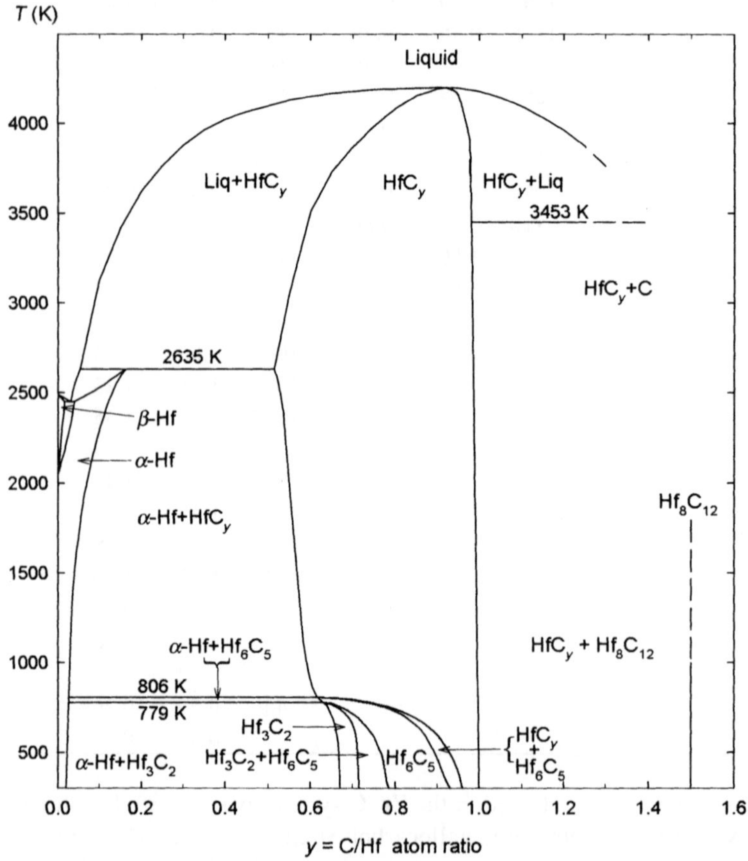

Fig. 10.8. Equilibrium phase diagram of the Hf–C system allowing for atomic ordering of nonstoichiometric cubic hafnium carbide HfC_y. Compound Hf_8C_{12} has no homogeneity interval and is not an ordered phase of carbide HfC_y. The position of phase boundaries outside the ordering region at $y < 1.2$ is shown in accordance with [92]

Calculations of the phase diagram for the Hf–C system showed that the main ordered phase of nonstoichiometric hafnium carbide is the Hf_6C_5 phase having a wide homogeneity interval. The Hf_3C_2 phase has a narrower homogeneity interval $HfC_{0.65}$–$HfC_{0.70}$, is a low-temperature one relative to the Hf_6C_5 phase, and can appear only as a result of successive transformations $HfC_y \rightarrow Hf_6C_5 \rightarrow Hf_3C_2$ (Fig. 10.8). The formation of the Hf_3C_2 and Hf_6C_5 phases proceeds as a transition of the first kind and is accompanied by a thermal effect ΔH_{trans}. Values of T_{trans}, ΔS_{trans} and ΔH_{trans} for phase transformations connected with ordering of nonstoichiometric cubic hafnium carbide HfC_y are given in Table 10.9.

Table 10.9. Temperatures T_{trans} (K), entropies ΔS_{trans} (J mol^{-1} K^{-1}) and heats ΔH_{trans} (kJ mol^{-1}) of the order–order and order–disorder phase transitions for nonstoichiometric hafnium carbide HfC_y

y	Hf_3C_2–Hf_6C_5			Hf_6C_5–HfC_y		
	T_{trans}	ΔS_{trans}	ΔH_{trans}	T_{trans}	ΔS_{trans}	ΔH_{trans}
0.62	–	–	–	805	0.44	0.36
0.63	–	–	–	803	0.52	0.42
0.64	775	0.77	0.59	801	0.60	0.48
0.65	767	0.89	0.68	799	0.69	0.55
0.66	758	1.02	0.77	797	0.78	0.62
0.68	734	1.28	0.94	792	0.98	0.77
0.70	700	1.46	1.02	785	1.19	0.94
0.72	641	1.47	0.94	778	1.42	1.11
0.73	589	1.37	0.81	774	1.54	1.19
0.74	511	1.11	0.57	769	1.67	1.28
0.75	363	0.73	0.26	764	1.79	1.37
0.76				759	1.91	1.45
0.78				747	2.14	1.60
0.80				731	2.33	1.70
0.82				712	2.44	1.73
0.83				697	2.42	1.69
0.84				688	2.40	1.65
0.85				673	2.31	1.55
0.86				656	2.18	1.43
0.88				614	1.81	1.11
0.90				559	1.38	0.77
0.92				486	0.98	0.47
0.94				392	0.63	0.25
0.95				335	0.49	0.17

As can be seen from Fig. 10.8, ordered phases of hafnium carbide are formed at temperatures below 800 K. The diffusion rate is low at $T < 800$ K and therefore it is very difficult to reach an ordered equilibrium state in the Hf–C system by long-time annealing in experiments. The difficulties involved in producing a disordered equilibrium state and the impracticability of studying the structure of ordered phases by X-ray or neutron diffraction methods might be the reason why experimental evidence of any kind concerning ordering of hafnium carbide has been unavailable in the literature for quite a long time. Indeed, the relative intensity of possible superstructure reflections in an X-ray experiment is very small because hafnium atom has more larger scattering amplitude than carbon atom has. Strong absorption of neutrons by massive hafnium nuclei in a neutron diffraction experiment considerably decreases the integral intensity of the diffraction spectrum and, thus, makes superstructure reflections virtually undetectable. The NMR method is inapplicable to the analysis of atomic distribution in the HfC_y lattice, because the most commonly encountered hafnium isotopes ^{180}Hf, ^{176}Hf and ^{174}Hf, and carbon ^{12}C have zero-spin nuclei, which do not possess a magnetic moment.

A pioneering experimental study of ordering of hafnium carbide HfC_y by the magnetic susceptibility method was reported recently [94–96]. The findings [95] suggest that a phase of Hf_3C_2 type appears in a broader interval of compositions than the one calculated in [13, 14], namely from HfC_y ($0.62 < y < 0.71$) to $HfC_{0.78}$. The range of the Hf_6C_5 phase is not wider than $0.80 < y < 0.87$. Whether the Hf_3C_2 phase is formed through the intermediate phase Hf_6C_5 or directly from HfC_y cannot be inferred from the data reported in [95], since the temperature interval between two successive measurements of χ is too large and the measurement accuracy is too low to allow detection of the intermediate phase having the temperature interval of 30–50 K or lower (~780–805 K (Fig. 10.8)). Temperatures measured in [94–96] for the equilibrium disorder–order transition HfC_y–Hf_3C_2 in carbides $HfC_{0.71}$ and $HfC_{0.78}$ are 870–890 and 970–1000 K, which is about 100–110 and 220–230 K higher than calculated [13–16] disorder–order transition temperatures for $HfC_{0.71}$ and $HfC_{0.78}$. The analysis of magnetic susceptibility of nonstoichiometric carbide HfC_y and calculations of short-range order parameters [94, 96] confirm the formation of ordered phases of the Hf_3C_2 and Hf_6C_5 types. From the findings [94–96] it follows that the main ordered phase of hafnium carbide is Hf_3C_2, which is formed through a phase transition of the first kind.

V–C System. Although ordering of nonstoichiometric vanadium carbide VC_y has been studied long enough, a reliable phase diagram of the V–C system, which would allow for formation of ordered phases, is still lacking in the literature. The first attempt to construct a section of the V–C phase diagram with coexistent ordered phases V_6C_5 and V_8C_7 was made in [97]. Later researchers [98] tried to account for phase equilibrium in VC_y in a narrow interval of compositions $0.84 \le y \le 0.93$ and temperatures $1273 \text{ K} < T < 1473$ K. The ordering interval in the V–C phase diagram [99] is shown hypothetically and actually replicates the data of [97]. Ordering is not discussed at all in the review [100] and is not reflected in the V–C phase diagram. The main difficulty encountered in constructing the V–C phase diagram is that the upper homogeneity boundary of disordered cubic vanadium carbide VC_y is the carbide $VC_{0.875}$ rather than the stoichiometric $MC_{1.0}$ compound as in carbides of other transition metals. The unusual position of the upper boundary of the homogeneity interval of carbide VC_y is due to the edge effect that shows up as the formation of an ordered phase of M_8C_7 type, which is absent in other carbides.

Experimental results have been reported recently on temperatures and heats of phase transformations $V_6C_5 \leftrightarrow VC_y$ and $V_8C_7 \leftrightarrow VC_y$ [26, 37], which allowed [101, 102] taking into account the ordering of carbide VC_y in the V–C phase diagram. Phase equilibria in the ordering interval of nonstoichiometric cubic carbide VC_y were calculated by the order-parameters functional method [1].

In the first version of calculations it is assumed that the ordered phase V_8C_7 did not have a homogeneity interval. Indeed, one-phase samples with cubic superstructure V_8C_7 are formed only for carbide $VC_{0.87-0.875}$, while samples with the ordered phase V_8C_7 always contain the V_6C_5 phase having a wide homogeneity interval at a low carbon concentration ($0.83 < y < 0.875$) [26, 37, 103, 104]. Moreover, according to [35, 105, 106], the $V_8C_7 \leftrightarrow VC_y$ transition temperature depends little on the VC_y composition within measurement error (Fig. 10.9). To allow for specific features of the ordered phase V_8C_7 having no homogeneity interval, its free energy was presented not in the standard form (9.71) but as

$$F_{V_8C_7(VC_{0.875})}(T) = F(T) - TS_c + (\Delta H_{trans} / T_{trans})(T - T_{trans}), \qquad (10.20)$$

where $F(T) = F_0(T) + xF_1(T) + x^2F_2(T)$ at $x = 0.875$. The parameters $F_0(T)$, $F_1(T)$ and $F_2(T)$ are the same as those for the disordered carbide VC_y; $S_c = 0.00313$ kJ mol^{-1} is the configuration entropy of carbide $VC_{0.875}$; $\Delta H_{trans} = 3$ kJ mol^{-1} and $T_{trans} = 1380$ K are the heat and temperature of the $V_8C_7 \rightarrow VC_{0.875}$ transition estimated from experimental data [26, 35, 37, 104].

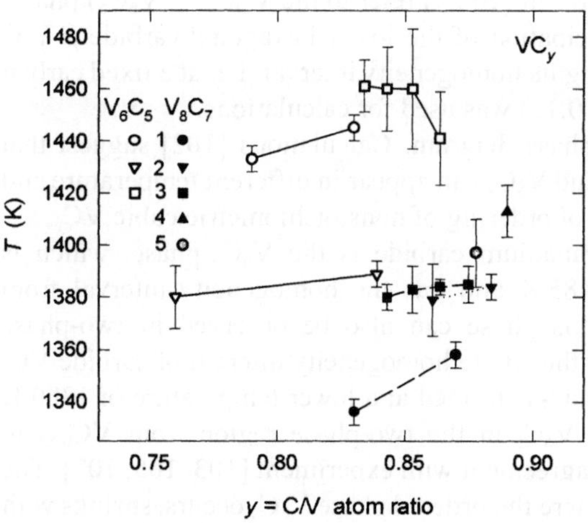

Fig. 10.9. Temperatures T_{trans} of phase transitions $V_6C_5 \leftrightarrow VC_y$ and $V_8C_7 \leftrightarrow VC_y$ for vanadium carbide VC_y with different carbon content: (1) [26], (2) [35], (3) [105], (4) [98], (5) [106]

The phase boundary separating the homogeneity interval of disordered carbide VC_y and the two-phase region $(VC_y+V_8C_7)$ was found from the equation

$$F_{V_8C_7\,(VC_{0.875})}(T) = F(y,0,T) - (y - y_b)\frac{\partial F(y,0,T)}{\partial y}\,, \tag{10.21}$$

where $y_b = 0.875$ and

$$F(y,0,T) - (y - y_b)\frac{\partial F(y,0,T)}{\partial y}$$

$$= F_0(T) + y_bF_1(T) + y(2y_b - y)F_2(T) + k_BT\{\ln(1-y) + y_b\ln[y/(1-y)]\}. \tag{10.22}$$

Analogously, the position of the phase boundary between the homogeneity interval of the ordered phase $V_{2t}C_{2t-1}$ (V_6C_5) and the two-phase region $(V_8C_7 + V_{2t}C_{2t-1})$ is described by the equation

$$F_{V_8C_7\,(VC_{0.875})}(T) = F(y,\eta,T) - (y - y_b)\frac{\partial F(y,\eta,T)}{\partial y}\,, \tag{10.23}$$

where $y_b = 0.875$ and

$$F(y,\eta,T) - (y - y_b)\frac{\partial F(y,\eta,T)}{\partial y}$$

$$= F_0(T) + y_bF_1(T) + y(2y_b - y)F_2(T) + \frac{a^{(s)}}{4t^2}\eta^2_{\text{equil}}F_2(T)$$

$$+\frac{k_BT}{2t}\{\ln[n_1^{y_b}(1-n_1)^{(1-y_b)}] + (2t-1)\ln[n_2^{y_b}(1-n_2)^{(1-y_b)}]\}\,. \tag{10.24}$$

The quantities n_1, n_2, η_{equil}, t, and $a^{(s)}$ in (10.24) refer to the $V_{2t}C_{2t-1}$ (V_6C_5) phase.

Phase equilibria with the participation of the lower hexagonal carbide β-V_2C (V_2C_y) were calculated disregarding its homogeneity interval, i.e. at a fixed carbon concentration $y = 0.5$. Equation (10.13) was used for calculations.

Figure 10.10 shows the V–C phase diagram. Calculations [102] suggest that three ordered phases V_3C_2, V_6C_5 and V_8C_7 can appear in different temperature and concentration intervals as a result of ordering of nonstoichiometric cubic VC_y.

The main ordered phase of vanadium carbide is the V_6C_5 phase, which is formed at temperatures below 1485 K and has the homogeneity interval from $VC_{0.745}$ to $VC_{0.831}$ at 1100 K. This phase can also be observed in two-phase sections and, hence, it exists over the whole homogeneity interval of carbide VC_y (Fig. 10.10). The ordered phase V_8C_7 is formed at a lower temperature of 1380 K and can be observed along with V_6C_5 in the two-phase region from $VC_{0.83}$ to $VC_{0.875}$ at 1330 K. This is in good agreement with experiment [103, 105, 107]. The two-phase region $V_6C_5 + V_8C_7$, where the ordered phase V_8C_7 occurs, shrinks with decreasing temperature. Ordered phases V_6C_5 and V_8C_7 form a eutectic with T_e = 1331 K and y_e = 0.852. This is very close to the estimate [98] of $T_e \approx 1300$ K and $y_e \approx 0.852$. It also follows from the calculations that the ordered phase V_3C_2 with a narrow homogeneity interval ($VC_{0.667}$ to $VC_{0.710}$ at 800 K) can appear in vanadium carbide by the peritectoid reaction $V_2C_y + V_6C_5 \rightarrow V_3C_2$ under equilibrium conditions at 1155 K or lower. This phase has not been observed experimentally to date. In accordance with calculations, the range of the ordered phase V_3C_2 approximates the concentration interval where the rhombic phase ζ-V_4C_{3-x} is actually observed. Ordering of carbide VC_y was calculated disregarding the existence of ζ-V_4C_{3-x}, since its thermodynamic properties were unknown. The V_3C_2 phase cannot exist if the free energy of the ζ-phase is lower than that of the ordered phase V_3C_2 at $T < 1150$ K. Considering this, the range of the ordered phase V_3C_2 is shown hypothetically.

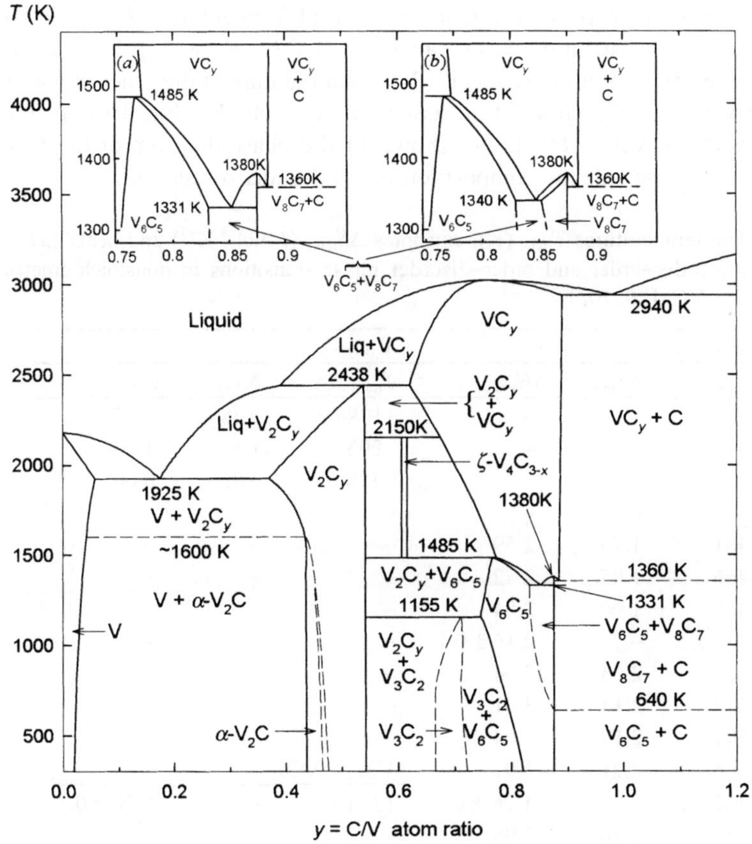

Fig. 10.10. Equilibrium phase diagram of the V–C system allowing for atomic ordering of nonstoichiometric cubic vanadium carbide VC_y. Phase boundaries outside the ordering region are shown after [99, 110]; the position of the ζ-V_4C_{3-x} phase is given according to [108]. V_8C_7, V_6C_5 and V_3C_2 are ordered phases of cubic carbide VC_y; α-V_2C is an ordered phase of lower hexagonal carbide V_2C_y (β-V_2C). Inset (*a*) magnifies part of the ordering region near the upper homogeneity boundary of carbide VC_y (it is assumed that the V_8C_7 phase has no homogeneity interval); inset (*b*) shows part of the ordering region of carbide VC_y calculated on the assumption that the V_8C_7 phase has a narrow homogeneity interval

Position of the ζ-V_4C_{3-x} phase is given according to the data [108]. This phase contains 37.9 at.% carbon and has a very narrow (around 0.4 at.% C) homogeneity interval $VC_{0.605}$–$VC_{0.615}$. The minimum temperature, at which it was detected after 78-hour annealing, is 1540 K [108]. This is close to the calculated maximum temperature of the disorder–order transition VC_y–V_6C_5, which equals 1485 K and is taken as the lower temperature boundary of the existence of the ζ-V_4C_{3-x} phase. The peritectoid decomposition temperature of the ζ-phase on heating is 2150 K.

An ordered rhombic phase α-V_2C [109] appears in the homogeneity interval of the lower carbide V_2C_y (β-V_2C). The exact ordering temperature is unknown, but, according to [99, 110], it is below 1870 K. When measuring magnetic susceptibility of ordered carbide α-V_2C in the interval from 300 to 1300 K, we are unable to detect the structural phase transition α-$V_2C \rightarrow V_2C_y$ (β-V_2C) up to 1300 K. Considering this and the data [110], the temperature of the order–disorder transition α-$V_2C \rightarrow V_2C_y$ (β-V_2C) is assumed at ~1600 K (Fig. 10.10). The metallocarbohedrene V_8C_{12} [80] is not shown in the phase diagram of the V–C system, because it is outside the composition interval under discussion.

Table 10.10. The temperatures T_{trans} (K), entropies ΔS_{trans} (J mol^{-1}·K^{-1}) and heats ΔH_{trans} (kJ mol^{-1}) of the order–order and order–disorder phase transitions in nonstoichiometric vanadium carbide VC_y [26, 102]

Y	V_6C_5–VC_y			V_3C_2–V_6C_5		
	T_{trans}	ΔS_{trans}	ΔH_{trans}	T_{trans}	ΔS_{trans}	ΔH_{trans}
0.73	–	–	–	1138	1.38	1.57
0.74	–	–	–	993	1.17	1.16
0.75	–	–	–	728	0.82	0.60
0.76	–	–	–	–	–	–
0.77	1481	1.75	2.59	–	–	–
0.78	1466	1.85	2.72	–	–	–
0.79	1450	1.96	2.84	–	–	–
[1] 0.79	1430 ± 5	–	2.16 ± 0.1	–	–	–
0.80	1433	2.05	2.94	–	–	–
0.81	1413	2.13	3.01	–	–	–
0.82	1391	2.19	3.04	–	–	–
0.83	1366	2.21	3.02	V_8C_7 – VC_y		
[1] 0.83	1442 ± 5	–	[2] 1.26 ± 0.1	1334 ± 5	–	[2] 0.70 ± 0.1
0.84	1338	2.20	2.95	–	–	–
0.85	1307	2.15	2.81	–	–	–
0.86	1272	2.06	2.62	–	–	–
[1] 0.87	–	–	–	1355 ± 5	–	2.30 ± 0.1
0.875	–	–	–	1380	2.17	3.00

[1] Experimental data [26, 37].

[2] Authors [37] reported total thermal effect ΔH_{trans} = 2.11 kJ mol^{-1} for transformation of carbide $VC_{0.83}$ at cooling; this value of ΔH_{trans} includes the heats of transformations $VC_{0.83} \rightarrow V_6C_5$ ($\Delta H_{trans} \approx$ 1.37 kJ mol^{-1}) and $VC_{0.83} \rightarrow V_8C_7$ ($\Delta H_{trans} \approx$ 0.75 kJ mol^{-1}).

As is apparent from the V–C phase diagram (Fig. 10.10), disorder–order transformations in vanadium carbide VC_y are all phase transitions of the first kind. Researchers [26, 37], who examined the variation of the heat capacity of carbides $VC_{0.79}$, $VC_{0.83}$ and $VC_{0.87}$ on heating and cooling, came to the same conclusion. Abrupt changes in the $C_p(T)$ curves of carbides $VC_{0.79}$, $VC_{0.83}$ and $VC_{0.87}$ in the interval of reversible equilibrium disorder–order transitions and an almost symmetric peak confirm that transformations $V_6C_5 \leftrightarrow VC_y$ and $V_8C_7 \leftrightarrow VC_{0.875}$ are phase transitions of the first kind. Moreover, a weak hysteresis of the heat capacity is observed on cooling, a fact which points to transition of the first kind.

Table 10.10 lists calculated values of T_{trans}, ΔS_{trans} and ΔH_{trans} for phase transformations connected with ordering of nonstoichiometric vanadium carbide VC_y. For comparison, Table 10.10 gives experimentally measured values of T_{trans} and ΔH_{trans} [26, 37]. The calculated temperature T_{trans} and heat ΔH_{trans} of the $V_6C_5 \rightarrow VC_{0.79}$ transition in carbide $VC_{0.79}$ are 1450 K and 2.84 kJ mol^{-1}. The experimentally measured value of ΔH_{trans} (Table 10.10) is a little smaller, because, as follows from calorimetric measurements [37], a small quantity of ζ-V_4C_{3-x} is formed in carbide $VC_{0.79}$.

Additional model calculations were performed to determine the form of the V–C phase diagram when the V_8C_7 phase has a narrow homogeneity interval. In those calculations the free energy of the ordered phase V_8C_7 was presented as

$$F_{V_8C_7\,(VC_{0.875})}(T) = F(T) - TS_c + \frac{\Delta H_{trans}(y)}{T_{trans}(y)}\left[T - T_{trans}^{max}(a + by + cy^2)\right] \qquad (10.25)$$

where $F(T) = F_0(T) + xF_1(T) + x^2F_2(T)$ at $x = 0.875$ (the parameters $F_0(T)$, $F_1(T)$ and $F_2(T)$ are the same as those for the disordered carbide VC_y); S_c = 0.00313 kJ mol^{-1} is the configuration entropy of carbide $VC_{0.875}$; $[(\Delta H_{trans}(y)/T_{trans}(y)] \equiv \Delta S_{trans} = 2.174$ J mol^{-1}K^{-1}); $T_{trans}^{max} = 1380$ K is the maximum temperature of the $V_8C_7 \rightarrow VC_y$ transition corresponding to carbide $VC_{0.875}$; a = 40.7455, $b = -93.367$ and $c = 54.793$.

Phase boundaries of the two-phase region ($VC_y + V_8C_7$) with the V_8C_7 phase having a homogeneity interval were calculated from the equations

$$y_1^2F_2(T) - k_BT\ln(1 - y_1) = y_2^2F_2(T) - k_BT\ln(1 - y_2) - \Delta S_{trans}[T - (a - cy_2^2)T_{trans}^{max}]\,, \qquad (10.26a)$$

$$y_1(2 - y_1)F_2(T) + k_BT\ln y_1 = y_2(2 - y_2)F_2(T) + k_BT\ln y_2 + \Delta S_{trans}[T - (a + b + 2cy_2 - cy_2^2)T_{trans}^{max}]. \qquad (10.26b)$$

Phase boundaries of the two-phase region ($V_6C_5 + V_8C_7$) with the V_8C_7 phase having a homogeneity interval were determined analogously from the equations

$$y_1^2 F_2(T) - \frac{a^{(s)}}{4t^2}\eta_{\text{equil}}^2 F_2(T) - \frac{k_B T}{2t}\left[\ln(1-n_1^{(d)}) + (2t-1)\ln(1-n_2^{(d)})\right]_{\eta=\eta_{\text{equil}}}$$

$$= y_2^2 F_2(T) - k_B T \ln(1-y_2) - \Delta S_{\text{trans}}[T - (a - cy_2^2)T_{\text{trans}}^{\text{max}}], \quad (10.27a)$$

$$y_1(2-y_1)F_2(T) + \frac{a^{(s)}}{4t^2}\eta_{\text{equil}}^2 F_2(T) + \frac{k_B T}{2t}\left[\ln n_1^{(d)} + (2t-1)\ln n_2^{(d)}\right]_{\eta=\eta_{\text{equil}}}$$

$$= y_2(2-y_2)F_2(T) + k_B T \ln y_2 + \Delta S_{\text{trans}}[T - (a + b + 2cy_2 - cy_2^2)T_{\text{trans}}^{\text{max}}]. \quad (10.27b)$$

The parameters n_1, n_2, t, η_{equil}, and $a^{(s)}$ in (10.27) refer to ordered phase V_6C_5.

A section of the V–C phase diagram calculated considering the homogeneity interval of the V_8C_7 phase is shown in Fig. 10.10 (inset (*b*)). In accordance with calculations, the homogeneity interval of V_8C_7 is very narrow: its maximum width is $VC_{0.852}$ to $VC_{0.875}$ at 1340 K. The homogeneity interval of the V_8C_7 phase shrinks as the temperature drops below 1340 K. The phases V_6C_5 and V_8C_7 form a eutectic with $y_e \approx 0.8476$ at 1340 K. The two-phase region ($V_6C_5 + V_8C_7$) at $T <$ 1340 K is nearly twice as narrow as the region calculated by the first method of calculations. Otherwise the V–C phase diagram remained unchanged. The problem of whether or not the V_8C_7 phase has a homogeneity interval calls for extensive experimental studies of the structure of vanadium carbide within a narrow interval of compositions $VC_{0.85}$–$VC_{0.875}$.

Generally, the calculations showed that the ordering of vanadium carbide is followed by formation of three phases: an ordered monoclinic or trigonal phase V_6C_5 having the homogeneity interval $VC_{0.75}$–$VC_{0.83}$ (at 1100 K); an ordered cubic phase V_8C_7 that may be observed along with V_6C_5 in the two-phase region; and an ordered phase V_3C_2, which appears at $T <$ 1150 K and has a very narrow homogeneity interval. The temperatures of the $V_6C_5 \leftrightarrow VC_y$ and $V_8C_7 \leftrightarrow VC_{0.875}$ transformations are 1400–1450 and 1360–1380 K respectively.

Nb–C System. The first version of the Nb–C phase diagram, which showed hypothetically an ordered phase Nb_6C_5, is proposed by the authors [111]. A more precise and correct Nb–C phase diagram is constructed [15, 16] using the calculations from [10, 11]. Thermodynamic evaluation of the Nb–C phase diagram disregarding ordering of carbide NbC_y is performed by Huang [112].

In addition to cubic carbide NbC_y with a wide homogeneity interval, the Nb–C system includes the lower hexagonal carbide β-Nb_2C with a narrow homogeneity interval, which was neglected in calculations [10–12, 14]. Figure 10.11 shows the calculated Nb–C phase diagram where ordering of nonstoichiometric cubic carbide NbC_y is observed. The position of phase boundaries outside the ordering region is shown in compliance with [92, 113].

As is seen from Fig. 10.11, an ordered phase Nb_3C_2 with a very narrow homogeneity interval ($0.67 < y < 0.71$ at 400 K) exists at a temperature below 900 K under thermodynamic equilibrium conditions. No ordered phase of the M_3C_2 type has been observed experimentally in niobium carbide so far. It is very difficult to detect this low-temperature ordered phase, because diffusion mobility of atoms is low at $T < 900$ K and NbC_y samples need to be annealed for a long time to reach a thermodynamically equilibrium structural state. Moreover, the composition of nonstoichiometric carbide NbC_y should correspond exactly to the concentration interval of the superstructure Nb_3C_2.

The main ordered phase of niobium carbide is Nb_6C_5, which appears as a result of the disorder–order transition at $T < 1594$ K and has a relatively broad range: the ordered phase Nb_6C_5 possesses the homogeneity interval $0.80 \le y \le 0.88$ at $T = 800$ K. Calculations are in good agreement with experiment for the Nb_6C_5 phase [114–119]. The experimental temperature T_{trans} for carbide $NbC_{0.83}$ is 1304 K [116–118]. The calculated range of the superstructure Nb_6C_5 at $T > 800$ K is a little wider than the range observed in experiments [2, 115–118]. The ordered phase Nb_6C_5 was also observed in [120–124].

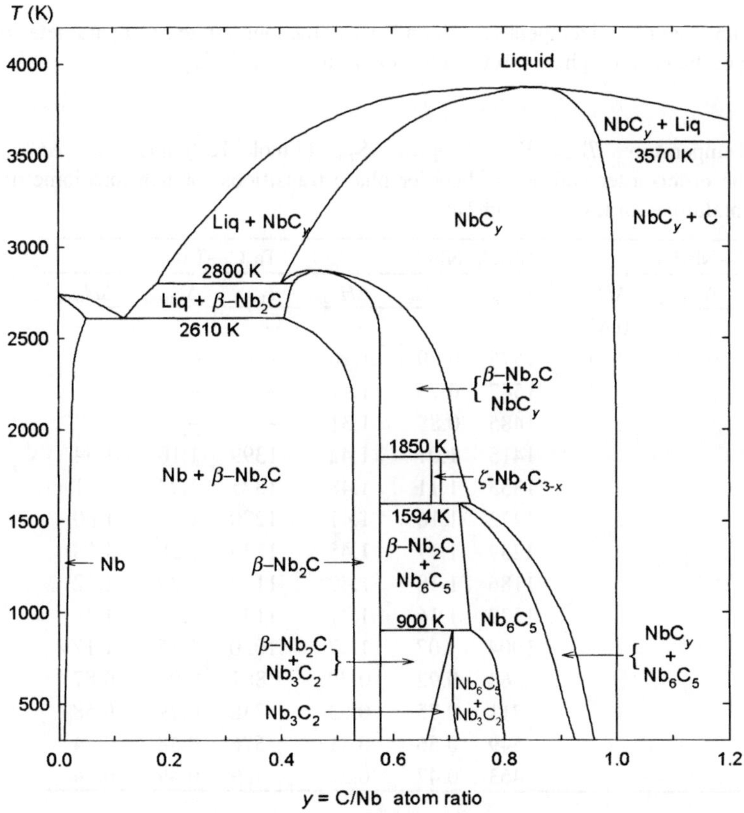

Fig. 10.11. Equilibrium phase diagram of the Nb–C system allowing for atomic ordering of nonstoichiometric cubic niobium carbide NbC_y: Nb_3C_2 and Nb_6C_5 are ordered phases of carbide NbC_y. The position of the rhombohedral phase ζ-Nb_4C_{3-x}, which is not an ordered phase of carbide NbC_y, is shown after [108]

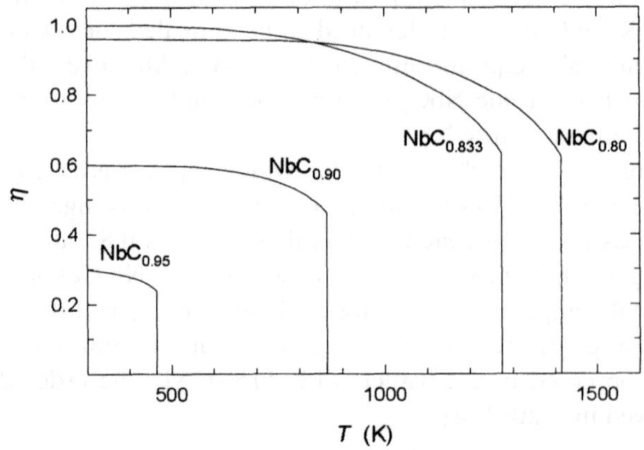

Fig. 10.12. Abrupt change and temperature dependence of the long-range order parameter η upon formation of the ordered phase Nb_6C_5 in niobium carbide NbC_y [2]

Table 10.11. Temperatures T_{trans} (K), entropies ΔS_{trans} (J mol^{-1} K^{-1}) and heats ΔH_{trans} (kJ mol^{-1}) of the order-order and order–disorder phase transitions for nonstoichiometric niobium and tantalum carbides NbC_y and TaC_y

Y	Nb_3C_2–Nb_6C_5			Nb_6C_5–NbC_y			Ta_6C_5–TaC_y		
	T_{trans}	ΔS_{trans}	ΔH_{trans}	T_{trans}	ΔS_{trans}	ΔH_{trans}	T_{trans}	ΔS_{trans}	ΔH_{trans}
0.74	761	0.81	0.62	–	–	–	–	–	–
0.75	505	0.67	0.34	1575	0.70	1.10	–	–	–
0.76				1547	0.76	1.17	–	–	–
0.78				1485	0.88	1.31	–	–	–
0.80				1415	1.01	1.42	1399	1.10	1.54
0.82				1333	1.11	1.48	1325	1.22	1.61
0.833				1272	1.16	1.47	1270	1.26	1.60
0.84				1239	1.17	1.45	1239	1.28	1.58
0.85				1186	1.18	1.40	1190	1.28	1.52
0.86				1129	1.16	1.31	1137	1.26	1.43
0.88				1004	1.07	1.07	1020	1.15	1.17
0.90				864	0.92	0.79	887	0.98	0.87
0.92				711	0.75	0.53	738	0.79	0.58
0.94				549	0.56	0.31	576	0.58	0.34
0.95				463	0.47	0.22	489	0.49	0.24

Transformation of the disordered phase NbC_y to the ordered phase Nb_6C_5 at the transition temperature T_{trans} is accompanied by a discontinuous increase in the long-range order parameter η from 0 to η_{trans}. This fact points to a disorder–order transformation of the first kind. As the temperature decreases from T_{trans} to 0 K, the degree of long-range order in the carbide smoothly increases from η_{trans} and nearly reaches $\eta_{max}(y)$ at 300 K (Fig. 10.12). At the Nb_3C_2–Nb_6C_5 and Nb_6C_5–NbC_y transition points, the temperature dependence of the free energy $F(T)$ exhibits a break, while the temperature dependencies of the enthalpy and entropy have clearly pronounced jumps characteristic of phase transitions of the first kind. Table 10.11 lists calculated values of T_{trans}, ΔS_{trans} and ΔH_{trans} for phase transformations connected with ordering of nonstoichiometric cubic niobium carbide NbC_y.

Calculated transition temperatures are in good agreement with T_{trans} determined experimentally using the differential thermal analysis method [116, 121]. Calculated temperatures of the Nb_6C_5–NbC_y transition for carbides $NbC_{0.81}$ and $NbC_{0.83}$ are 1375 and 1288 K, while experimental T_{trans} values are 1258 ± 10 and 1304 ± 10 K. Experimental T_{trans} values determined in [119, 125] for the Nb_6C_5–$NbC_{0.83}$ order–disorder transition are 1298 ± 25 and 1303 ± 50 K respectively. A neutron diffraction examination [126] showed that the order–disorder and disorder–order transition temperatures T_{trans} for carbide $NbC_{0.83}$ are 1298 ± 5 and 1273 ± 10 K. Niobium carbide NbC_y was the first nonstoichiometric compound in which disorder–order transformations were detected [127] by measuring magnetic susceptibility.

The position of the ζ-Nb_4C_{3-x} phase, which exists at a temperature from 1600 to 1850 K, is shown in the Nb–C phase diagram (Fig. 10.11) using the data of [108]. The homogeneity interval of ζ-Nb_4C_{3-x} is between 40.1 and 40.7 at.% carbon. Rhombohedral carbide ζ-Nb_4C_3 [128] is not an ordered phase of nonstoichiometric niobium carbide NbC_y. It should be noted that the calculated concentration range of the ordered phase Nb_3C_2 is close to the concentration interval where the ζ-Nb_4C_{3-x} phase is actually observed. Calculations of ordering of carbide NbC_y disregarded the formation of the ζ-Nb_4C_{3-x} phase, because its thermodynamic properties are unknown.

A new compound Nb_8C_{12} was synthesized in the Nb–C system (as in Ti–C, Zr–C, and Hf–C systems) by the gas phase synthesis method combined with laser heating [83]. This compound is a molecular cluster and enters the metallo-carbohedrene group M_8C_{12}. It is stable at a temperatures below 1500 K. The Nb_8C_{12} compound is not shown in the Nb–C phase diagram (Fig. 10.11), because it appears at $y = C/Nb > 1.2$.

Ta–C System. Nonstoichiometric tantalum carbide TaC_y is a challenging material for experimental study of ordering. The theoretical phase diagram constructed for the Ta–C system taking into account ordering of carbide TaC_y [10–12, 14] is therefore quite remarkable. In addition to cubic carbide TaC_y having a broad homogeneity interval, this system includes a lower hexagonal carbide α-Ta_2C_y with a narrow homogeneity interval. The literature provides thermodynamic characteristics only of the stoichiometric composition of lower tantalum carbide, i.e. α-Ta_2C. Therefore (10.13) was used to calculate phase equilibria with participation of this compound.

As was noted in the foregoing, ordered phases of the types M_2X, M_3X_2, M_4X_3, M_6X_5 and M_8X_7 can appear generally (in crystallographic terms) in a nonstoichiometric compound MX_y. Free energies of disordered carbide TaC_y and possible ordered phases Ta_2C, Ta_3C_2, Ta_4C_3, Ta_6C_5 and Ta_8C_7 were calculated at temperatures of 700 and 1600 K in order to determine which ordered phases can actually appear in tantalum carbide TaC_y (Fig. 10.13). Equations (9.93), (9.94), (9.100), (9.101) and (10.10) were used for the calculations. The ordered phase Ta_4C_3 does not exist at 700 K, because its transition temperature $T_{trans} < 600$ K when $0.5 \le y \le 1.0$. Ordered phases Ta_2C, Ta_3C_2 and Ta_6C_5, and disordered carbide TaC_y ($y > 0.96$) have the lowest free energy in different concentration intervals at $T = 700$ K. The free energy of the possible ordered phase Ta_8C_7 is higher than the free energy of other ordered phases in the whole interval $0.5 \le y \le 1.0$. Only a disordered phase can exist in the whole homogeneity interval of tantalum carbide at a temperature above 1430 K. Points I and II corresponding to the free energy $0.5F_{\alpha\text{-}Ta_2C}$ at 700 and 1600 K are marked on the ordinate axis in Fig. 10.13 at $y = 0.5$. The tangent to the free energy $F(y, 700\text{ K})$ at 700 K passing through point I will define the width of the possible two-phase region (α-Ta_2C + $Ta_{2t}C_{2t-1}$ or α-Ta_2C + TaC_y) and the free energy of the two-phase mixture. It is readily seen that the two-phase mixture (α-Ta_2C + Ta_6C_5) has the lowest free energy at 700 K. Therefore ordered phases Ta_2C and Ta_3C_2 cannot exist.

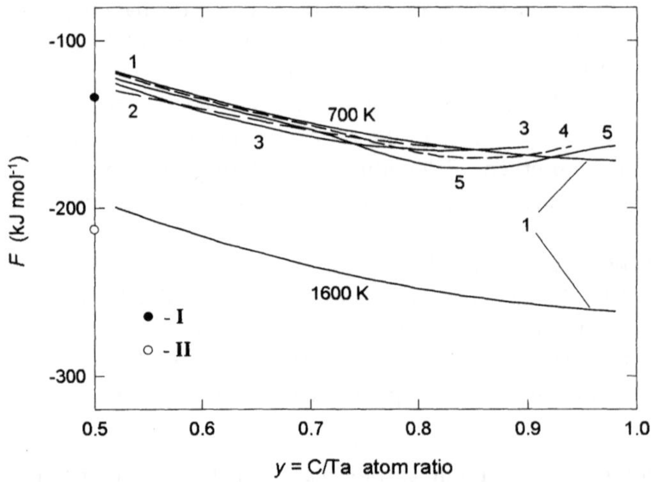

Fig. 10.13. Comparison of free energies of disordered phase and possible ordered phase of tantalum carbide at 700 and 1600 K: (1) disordered tantalum carbide TaC_y; (2) ordered phase of Ta_2C type; (3) ordered phase of Ta_3C_2 type; (4) ordered phase of Ta_8C_7 type; (5) ordered phase of Ta_6C_5 type. Points I and II correspond to the free energy $0.5F_{\alpha\text{-}Ta_2C}$ at 700 and 1600 K respectively

Figure 10.14 shows a Ta–C equilibrium phase diagram calculated taking account of the atomic ordering in nonstoichiometric cubic tantalum carbide TaC_y.

Phase boundaries outside the ordering region are given in line with [28]. The position of the rhombohedral phase ζ-Ta_4C_{3-x} [128] in Fig. 10.14 is given in accordance with [108]. This ζ-phase is not an ordered phase of TaC_y carbide, has a narrow homogeneity interval (carbon concentration between 38.2 and 39.0 at.%), and appears at a temperature below 2450 K [108]. The free energy of ζ-Ta_4C_{3-x} is unknown and therefore calculations of the Ta–C phase diagram disregarded the formation of this phase. A new compound Ta_8C_{12} was synthesized recently [80, 81]. This compound is a molecular cluster and belongs to the metallocarbohedrene group M_8C_{12} (M = Ti, Zr, Hf, V, Nb, Ta, Cr). The relative carbon content of the compound Ta_8C_{12} ($y = 1.5$) is outside the diagram section considered and therefore this compound is not shown in the Ta–C phase diagram (Fig. 10.14).

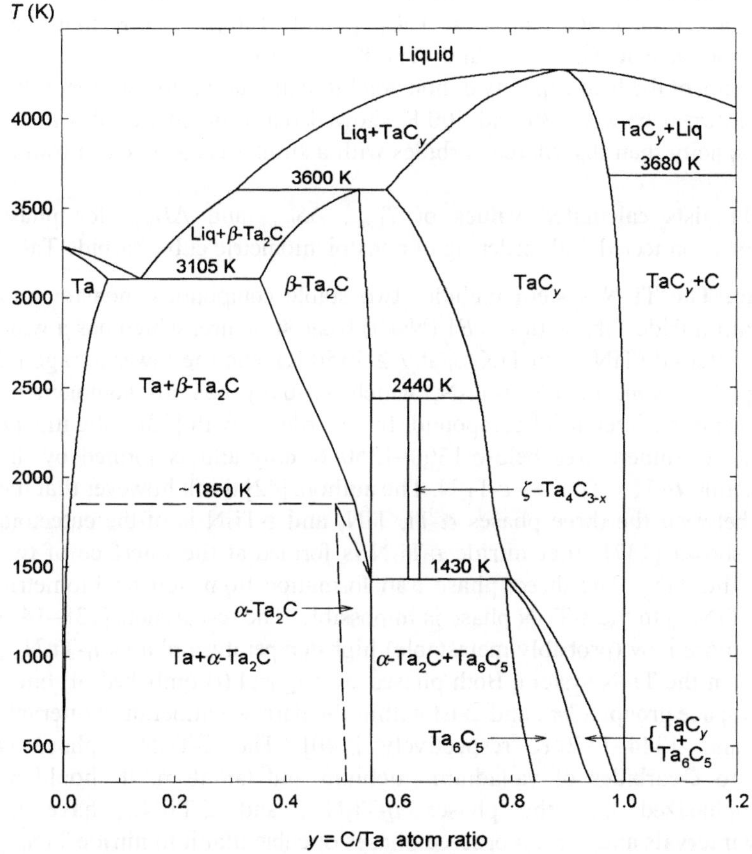

Fig. 10.14. Equilibrium phase diagram of the Ta–C system allowing for atomic ordering of nonstoichiometric cubic tantalum carbide TaC_y [131]. The position of the rhombohedral phase ζ-Ta_4C_{3-x} is given in accordance with [108]. This phase is not an ordered phase of carbide TaC_y and is formed below 2450 K

As is seen from Fig. 10.14, the Ta_6C_5 phase is the only ordered phase of nonstoichiometric tantalum carbide. It has a narrower homogeneity interval than the analogous ordered phase Nb_6C_5. An electron diffraction study of carbide $TaC_{0.83}$ [119] revealed a diffuse band whose geometry corresponds to ordering of M_6C_5 type with a very small degree of order. The structure of ordered tantalum carbide was examined most comprehensively and, essentially, for the first time by the neutron diffraction method in [129–131]. Those studies showed that an incommensurable ordered phase, which closely approximates known superstructures of M_6C_5 type, is formed in nonstoichiometric tantalum carbide TaC_y after long-time slow annealing at a temperature of 1600–750 K. Magnetic susceptibility studies [132, 133] also confirmed appearance of an ordered phase of M_6C_5 type in tantalum carbide. The Ta_6C_5–TaC_y transition temperature T_{trans}, which was determined for carbides $TaC_{0.83}$ and $TaC_{0.85}$ using magnetic susceptibility measurements, is 1110 and 1130 K respectively.

Measurements of the heat capacity of nonstoichiometric tantalum carbide in the temperature interval between 80 and 300 K showed that ordered carbides had a larger heat capacity than disordered carbides with a similar carbon concentration [134, 135].

Table 10.11 lists calculated values of T_{trans}, ΔS_{trans} and ΔH_{trans} for phase transformations connected with ordering of nonstoichiometric cubic carbide TaC_y.

Ti–N System. The Ti–N system includes two stable compounds: nonstoichiometric titanium nitride TiN_y with the $B1$ (NaCl) basic structure, which has a wide homogeneity interval ($TiN_{0.38}$ to $TiN_{1.00}$ at $T \geq 1350$ K), and the lower tetragonal (space group $P4_2/mnm$) nitride ε-Ti_2N, which virtually has no homogeneity interval and is not an interstitial compound. In accordance with [136], the nitride ε-Ti_2N exists at temperatures below 1300–1350 K only and is formed by the peritectic reaction α-Ti + $TiN_y \rightarrow \varepsilon$-$Ti_2N$. The authors [42] think however that the equilibrium between the three phases α-Ti, TiN_y and ε-Ti_2N is of the eutectoid type. It was shown [137] that nitride ε-Ti_2N is formed at the interface of two phases α-Ti and TiN_y. The direct phase transformation from nonstoichiometric cubic nitride $TiN_{0.50}$ to the ε-Ti_2N phase is impossible. The researchers [138–141] detected two more new (probably metastable) high-temperature phases η-Ti_3N_{2-x} and ζ-Ti_4N_{3-x} in the Ti–N system. Both phases are trigonal (rhombohedral) ones, belong to the space group $R\bar{3}m$, and exist within the narrow temperature intervals 1335–1374 and 1344–1570 K respectively [140]. The ζ-Ti_4N_{3-x} phase is isostructural to ζ-carbides of vanadium, niobium and tantalum. It should be specially emphasized that the phases η-Ti_3N_{2-x} and ζ-Ti_4N_{3-x} have no homogeneity intervals and are not ordered phases of cubic titanium nitride TiN_y.

Calculations [13] showed that phases Ti_2N, Ti_3N_2 and Ti_6N_5 appear in different temperature and concentration intervals upon ordering of cubic nitride TiN_y (Fig. 10.15). The position of phase boundaries outside the ordering region is shown in accordance with [136]. The position of phases η-Ti_3N_{2-x} and ζ-Ti_4N_{3-x} is given in accordance with [137, 138, 140]. The ordered phase Ti_2N has a narrow homogeneity interval ($TiN_{0.51}$–$TiN_{0.53}$ at 1000 K and $TiN_{0.52}$–$TiN_{0.57}$ at 800 K).

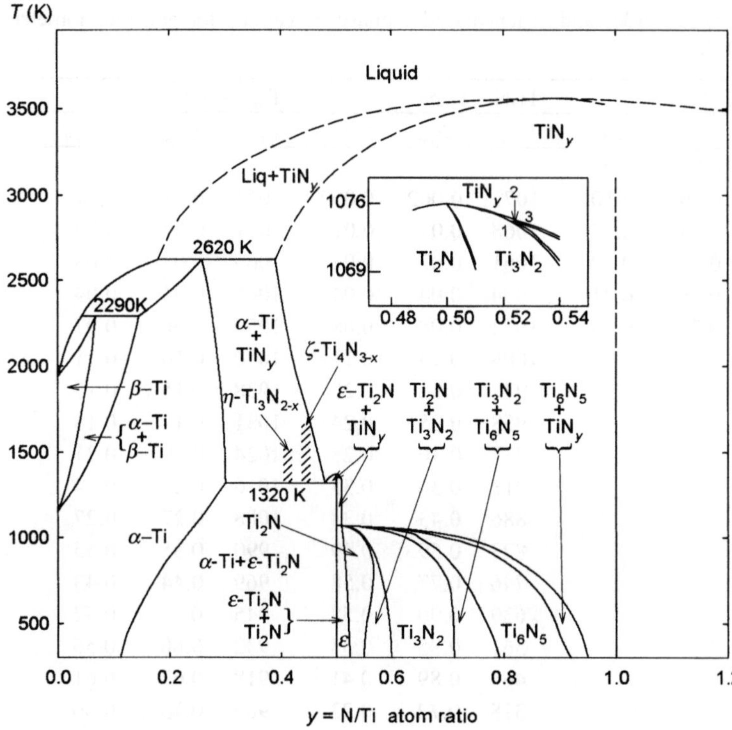

Fig. 10.15. Equilibrium phase diagram of the Ti–N system allowing for atomic ordering of nonstoichiometric cubic titanium nitride TiN_y [13]. The inset magnifies part of the ordering region where the peritectoid transformation $Ti_3N_2 + TiN_y \rightarrow Ti_6N_5$ takes place at 1073.4 K. The coordinates y = N/Ti of points 1, 2 and 3 are 0.5236, 0.524 and 0.5255 respectively. The high-temperature trigonal phases η-Ti_3N_{2-x} and ζ-Ti_4N_{3-x} [108] are not ordered phases of nonstoichiometric titanium nitride TiN_y and are stable at $T > 1300$ K

The two-phase regions (ε-$Ti_2N + Ti_2N$) and ($Ti_2N + Ti_3N_2$) are located on the left and on the right of the Ti_2N phase respectively. The direct disorder–order transformation TiN_y–Ti_2N is possible only for nitride $TiN_{0.51}$. The $TiN_{0.51}$–Ti_2N transition temperature is ~1068 K. The ordered phase Ti_2N with a high nitrogen concentration $y > 0.51$ is formed with decreasing temperature as a result of successive transformations $TiN_{0.511-0.524} \rightarrow Ti_3N_2 \rightarrow Ti_2N$ and $TiN_{0.525-0.570} \rightarrow Ti_6N_5 \rightarrow Ti_3N_2 \rightarrow Ti_2N$. In accordance with calculations, the Ti_6N_5 phase is the main ordered phase of cubic titanium nitride TiN_y. This phase is formed directly through the TiN_y–Ti_6N_5 disorder-order transition of the first kind. The peritectoid transformation $Ti_3N_2 + TiN_y \rightarrow Ti_6N_5$ takes place in the ordering region of TiN_y at T = 1073.4 K and y = 0.524 (see inset in Fig. 10.15). The transformation TiN_y–Ti_2N represents a phase transition of the second kind. All other disorder–order and order–order transitions are phase transitions of the first kind.

Table 10.12. Temperatures T_{trans} (K), entropies ΔS_{trans} (J mol^{-1} K^{-1}) and heats ΔH_{trans} (kJ mol^{-1}) of the order–order and order–disorder phase transitions for nonstoichiometric titanium nitride TiN_y

y	$Ti_2N-Ti_3N_2$			$Ti_3N_2-Ti_6N_5$			$Ti_6N_5-TiN_y$		
	T_{trans}	ΔS_{trans}	ΔH_{trans}	T_{trans}	ΔS_{trans}	ΔH_{trans}	T_{trans}	ΔS_{trans}	ΔH_{trans}
0.52	1048	0.06	0.06	–	–	–	–	–	–
0.53	1012	0.10	0.10	1072	0.002	0.002	1073	0.01	0.01
0.54	961	0.18	0.17	1068	0.01	0.01	1071	0.02	0.02
0.55	902	0.21	0.19	1061	0.02	0.02	1069	0.03	0.03
0.56	825	0.26	0.21	1054	0.04	0.04	1066	0.04	0.04
0.58	453	0.29	0.13	1034	0.08	0.08	1059	0.06	0.07
0.60				1009	0.13	0.13	1050	0.10	0.11
0.62				977	0.21	0.20	1038	0.15	0.15
0.63				958	0.25	0.24	1031	0.17	0.18
0.64				937	0.30	0.28	1024	0.20	0.21
0.65				912	0.36	0.33	1016	0.24	0.24
0.66				886	0.43	0.38	1008	0.27	0.27
0.68				823	0.59	0.49	990	0.35	0.35
0.70				746	0.77	0.58	969	0.44	0.43
0.72				639	0.90	0.58	945	0.55	0.52
0.73				561	0.95	0.53	932	0.60	0.56
0.74				465	0.89	0.41	918	0.67	0.61
0.75				318	0.72	0.23	904	0.73	0.66
0.76							888	0.80	0.71
0.78							855	0.94	0.80
0.80							817	1.08	0.88
0.82							773	1.21	0.94
0.833							741	1.27	0.94
0.84							723	1.29	0.93
0.85							695	1.30	0.90
0.86							664	1.28	0.85
0.88							597	1.18	0.71
0.90							522	1.02	0.53
0.92							437	0.82	0.36
0.94							344	0.60	0.21
0.95							294	0.50	0.15

Table 10.12 lists T_{trans}, ΔS_{trans} and ΔH_{trans} values for disorder–order and order–order phase transitions of the first kind.

The formation of an ordered tetragonal (space group $I4_1/amd$) phase Ti_2N (δ'-$TiN_{0.50}$) in nitride TiN_y was experimentally established in [142] and confirmed in later works [143–145]. In accordance with [145], the temperature of the Ti_2N–$TiN_{0.50}$ transition is 1070–1140 K and this transition represents a phase transformation of the first kind. According to [146], the ordered phase Ti_2N has no homogeneity interval and its nitrogen content is the same as in $TiN_{0.51}$. The transition from disordered cubic nitride TiN_y to the ordered phase Ti_2N takes place in the interval $TiN_{0.5}$–$TiN_{0.6}$ at 1070 ± 10 K. Below that temperature (at ~1000–1070 K), two-phase regions (ε-Ti_2N + Ti_2N) and (Ti_2N + TiN_y) are located on the left and on the right of the ordered phase having the composition $TiN_{0.51}$. The data on thermal decomposition [147, 148] suggest that the ordered phase Ti_2N exists at a temperature not higher than 1140–1180 K.

Calculations [13] of the Ti–N phase diagram show that the ordered phase Ti_2N has a small homogeneity interval. The two-phase region (Ti_2N + TiN_y) at 1000 K is possible only if the equilibrium state with two more ordered phases of titanium nitride Ti_3N_2 and Ti_6N_5 could not be reached. The ordered phase Ti_6N_5 has not been observed in experiments so far. Nevertheless, discontinuites on heat capacity versus temperature dependencies $C_p(T)$ of nonstoichiometric titanium nitride $TiN_{0.82}$ [70] are indirect evidence of the formation of an ordered phase Ti_6N_5. Discontinuity temperatures are 973 K on heating and 939 K on cooling. The presence of a temperature hysteresis of the heat capacity indicates that the phase transformation of $TiN_{0.82}$ is of the first kind. This coincides with theoretical conclusions [13] according to which the transformation $TiN_y \leftrightarrow Ti_6N_5$ is a phase transitions of the first kind. The formation of the superstructure Ti_6N_5 also follows from calculations [18, 19] performed by the Monte Carlo and cluster variation methods (Fig. 10.16).

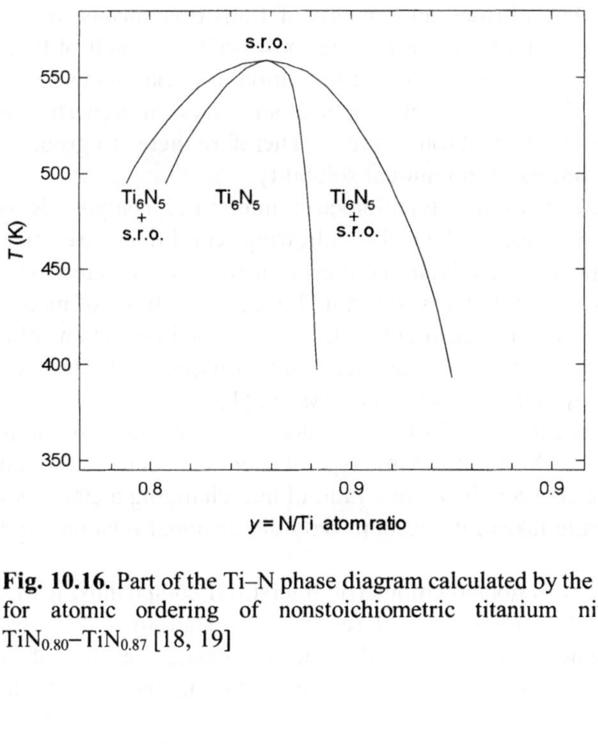

Fig. 10.16. Part of the Ti–N phase diagram calculated by the Monte Carlo method allowing for atomic ordering of nonstoichiometric titanium nitride TiN_y in the interval $TiN_{0.80}$–$TiN_{0.87}$ [18, 19]

In accordance with [18], the Ti_6N_5–$TiN_{0.83}$ transition temperature is ~560 K, which is likely to be an underestimation. Interestingly, calculations of the Ti–N phase diagram [18, 19] revealed the structure of the ordered phase Ti_6N_5 where atomic arrangement was fully identical to that of the monoclinic (space group *C2/m*) superstructure Nb_6C_5 [116–118, 121, 122].

10.4 Phase Diagrams of Pseudobinary Systems

A significant feature of highly nonstoichiometric compounds is that they form solid solutions by mutual substitution of atoms of either a metal or a nonmetal, or else both at a time. The mutual solubility of nonstoichiometric compounds changes widely and depends on their crystal and electronic structure as well as external conditions. The distribution of mutually substituting components in the solid solution lattice can be either statistical or ordered.

Experimental studies of equilibrium phase diagrams of pseudobinary systems formed by nonstoichiometric compounds are very tedious. It is next to impossible to locate the liquidus in a system formed by compounds with melting points at 3000 K or higher, or else find a latent solid-phase precipitation interval in that system. Attempted use of a variety of empirical and semi-empirical relationships for estimating mutual solubility of nonstoichiometric compounds is of interest in this connection.

Nonstoichiometric carbides, nitrides and oxides of transition metals are so similar in their structures and atomic sizes that the most probable result of their interaction is unlimited mutual solubility, i.e. the formation of a continuous series of solid solutions. However the similarity of structures and physical properties of these compounds do not render them absolutely akin. Therefore there is a group of pseudobinary systems with limited or no mutual solubility.

Unlimited mutual solubility of nonstoichiometric interstitial compounds is generally believed to be the case when the following conditions are met: isomorphism of the compounds involved; similar dimensions and electronic structures of mutually substitutional atoms; unlimited mutual solubility of metals making up the compound. Structural vacancies must have a certain effect on solid solution formation, since they add new interstices and, consequently, facilitate mutual solubility of components in a pseudobinary system [1].

According to the “15 percent” rule [149], continuous series of solid solutions between metal-like compounds having the same type of chemical bond are formed when the relative difference between the atomic radii of interchanging metals does not exceed 15%. But this rule takes into account only dimensional relations and does not hold for all cases.

To assess mutual solubility of nonstoichiometric interstitial compounds, it was proposed [150] using the relative difference of electronic densities $\Delta d_e = (d_e^{(2)} - d_e^{(1)})/d_e^{(2)}$, where $d_e^{(2)} \geq d_e^{(1)}$. By the electronic density of a compound MX_y is meant the quantity $d_e = cz/v$, where z is the number of formula units in a unit cell, v is the unit cell volume, $c = c_M + yc_X$ is the number of valence electrons per formula unit of the compound MX_y, and c_M and c_X denote the number of valence electrons of atom M and atom X respectively. The electronic density d_e in the generalized form simultaneously allows for the electronic structure and size of a unit cell and, consequently, can be used to evaluate the mutual solubility of nonstoichiometric compounds. The electronic density d_e changes widely for the compounds MX_y in question whose composition corresponds to the upper boundary of the homogeneity interval of a cubic phase (Table 10.13).

Table 10.13. Electronic density, d_e, for cubic carbides, nitrides and oxides of transition metals of IV and V group [1]

Compound	$d_e \times 10^{29}$ (m^{-3})	Compound	$d_e \times 10^{29}$ (m^{-3})	Compound	$d_e \times 10^{29}$ (m^{-3})
TiC	3.93	NbC	4.02	ThN	2.49
ZrC	3.08	TaC	3.92	VN	5.58
HfC	3.19	TiN	4.55	NbN	4.56
ThC	2.06	ZrN	3.72	TiO	4.64
$VC_{0.88}$	4.75	HfN	3.88	VO	5.53

Comparison of the Δd_e value with experimental data on mutual solubility in pseudobinary systems of carbides, nitrides, oxides and borides showed that unlimited solid solutions are formed if $\Delta d_e \leq 32.5\%$ [150]. Table 10.14 gives Δd_e values for 105 pseudobinary systems comprising near-stoichiometric carbides, nitrides and oxides with the $B1$-type structure. Carbide $VC_{0.88}$ is a component in systems with vanadium carbide. Values Δd_e suggest that unlimited mutual solubility is characteristic of 72 systems, while limited mutual solubility should be observed for 33 systems (Table 10.14).

The elastic deformation energy is another quantity directly related to immiscibility (decomposition) in binary solid-phase systems. The study of solid solutions of transition metals [151] showed that the deformation energy is a maximum for a solid solution whose composition corresponds to the miscibility gap vertex in a solid state. The temperature of the miscibility gap vertex is proportional to the elastic deformation energy. Thus, the larger the elastic deformation energy, the higher the temperature at which the miscibility gap occurs. The dependence of the elastic deformation energy on the composition of solid solution $A_{1-x}B_x$ is given by the formula

$$E^{(d)} = 2x(1-x)(V_B - V_A)^2[(1-x)G_{sh\text{-}A}/V_A + xG_{sh\text{-}B}/V_B]/3\,, \qquad (10.28)$$

where $G_{sh\text{-}A}$ and $G_{sh\text{-}B}$ are the shear moduli of components A and B of solid solution $A_{1-x}B_x$; V_A and V_B are the molar volumes of components A and B; and x is the relative concentration of component B in solid solution.

Table 10.14. Relative difference of electronic densities, Δd_e, and mutual solubility of compounds for pseudobinary systems [1, 150]

System	Δd_e (%)	System	Δd_e (%)	System	Δd_e (%)
Unlimited mutual solubility					
TiC–ZrC	21.6	TiC–ZrN	5.4	NbC–NbN	11.8
TiC–HfC	18.8	TiC–HfN	1.3	TaC–TiN	13.8
TiC–$VC_{0.88}$	17.2	TiC–VN	29.6	TaC–ZrN	5.1
TiC–NbC	12.2	TiC–NbN	13.8	TaC–HfN	1.0
TiC–TaC	0.3	ZrC–TiN	32.3	TaC–VN	29.6
ZrC–HfC	3.4	ZrC–ZrN	20.6	TaC–NbN	14.0
ZrC–NbC	23.4	ZrC–HfN	17.2	TiC–TiO	15.3
ZrC–TaC	21.4	ZrC–ThN	19.1	TiC–VO	29.0
HfC–NbC	20.6	ZrC–NbN	32.4	HfC–TiO	31.3
HfC–TaC	18.1	HfC–TiN	29.9	$VC_{0.88}$–TiO	2.3
$VC_{0.88}$–NbC	15.4	HfC–ZrN	14.2	$VC_{0.88}$–VO	14.1
$VC_{0.88}$–TaC	17.5	HfC–HfN	17.8	NbC–TiO	13.3
NbC–TaC	2.5	HfC–ThN	21.9	NbC–VO	27.4
TiN–ZrN	18.2	HfC–NbN	30.0	TaC–TiO	15.5
TiN–HfN	14.7	ThC–ThN	17.3	TaC–VO	29.1
TiN–VN	18.5	$VC_{0.88}$–TiN	4.2	TiN–TiO	1.9
TiN–NbN	0.2	$VC_{0.88}$–ZrN	21.7	TiN–VO	17.7
ZrN–HfN	4.1	$VC_{0.88}$–HfN	18.3	ZrN–TiO	19.8
ZrN–NbN	18.4	$VC_{0.88}$–VN	14.9	HfN–TiO	16.4
HfN–VN	30.4	$VC_{0.88}$–NbN	4.0	HfN–VO	29.8
NfN–NbN	14.9	NbC–TiN	7.5	VN–TiO	16.8
VN–NbN	18.3	NbC–ZrN	11.6	VN–VO	1.0
TiO–VO	16.1	NbC–HfN	3.5	NbN–TiO	1.7
TiC–TiN	13.6	NbC– VN	28.0	NbN–VO	17.5
Finite mutual solubility					
TiC–ThC	47.6	HfN–ThN	35.8	TiC–ThN	36.6
ZrC–ThC	33.1	ThN–VN	55.4	ZrC–VN	44.8
ZrC–$VC_{0.88}$	35.1	ThN–NbN	45.4	HfC–VN	42.7
HfC–ThC	35.4	ZrC–TiO	33.6	ThC–TiN	54.7
HfC–$VC_{0.88}$	32.8	ZrC–VO	44.3	ThC–ZrN	44.7
ThC–$VC_{0.88}$	56.5	HfC–VO	42.3	ThC–HfN	46.9
ThC–NbC	48.7	ThC–TiO	55.6	ThC–VN	63.0
ThC–TaC	47.0	ThC–VO	62.7	ThC–NbN	54.8
TiN–ThN	45.3	ZrN–VO	32.7	$VC_{0.88}$–ThN	47.5
ZrN–ThN	33.1	ThN–TiO	46.4	NbC–ThN	38.0
ZrN–VN	33.3	ThN–VO	55.0	TaC–ThN	36.5

The relative concentration x of component B in solid solution, i.e. the solid solution composition corresponding to the maximum elastic deformation energy E_{max} and the miscibility gap, can be found from the condition $dE^{(d)}/dx = 0$ by the equation

$$3\left(\frac{G_{\text{sh-A}}}{V_A}-\frac{G_{\text{sh-B}}}{V_B}\right)x^2-2\left(\frac{G_{\text{sh-A}}}{V_A}-\frac{G_{\text{sh-B}}}{V_B}\right)x+\frac{G_{\text{sh-B}}}{V_B}=0\,. \tag{10.29}$$

Substituting the x value into (10.28), we obtain the maximum elastic deformation energy E_{max}. Thus, analyzing the elastic deformation energy, it is possible to determine the miscibility gap from the concentration x and estimate the maximum temperature T_{max} at which the miscibility gap appears.

An analogous interrelation between the elastic deformation energy on the one hand and the miscibility gap on the other hand is realized in solid solutions of nonstoichiometric compounds. This is demonstrated [152] taking six pseudobinary carbide systems as an example (Table 10.15). Calculations [152] showed that the maximum elastic deformation energy actually correlates with the temperature at which the miscibility gap appears in a solid state. The maximum deformation energy corresponds to a eutectic solid solution in systems with limited solubility (Table 10.15). Thus, calculated values of the maximum elastic deformation energy are in good agreement with experiment [153]. Analogous calculations were made for 36 carbide, nitride and carbonitride pseudobinary systems [1, 152]. The comparison of mutual solubility of components in a pseudobinary system with calculated values of the maximum elastic deformation energy E_{max} showed that unlimited solid solutions are formed if $E_{max} < 37.5$ kJ mol^{-1} (Table 10.16).

Table 10.15. Maximum elastic deformation energy, E_{max} , of carbide solid solutions

System A - B	Composition corresponding to E_{max} (mol. % A)		Temperature of the miscibility gap or eutectic temperature (K) [153]	E_{max} (kJ mol^{-1}) [1, 152]
	[153]	[1, 152]		
Continuous (unlimited) solid solutions				
TiC – ZrC	59.0	54.3	2270	27.3
TiC – HfC	58.0	53.6	2130	20.1
$VC_{0.88}$ – NbC	57.0	53.9	1740	19.1
$VC_{0.88}$ – TaC	62.0	51.6	1610	18.7
Finite solid solutions				
$VC_{0.88}$ – ZrC	34.8	25.7	2820	49.0
$VC_{0.88}$ – HfC	26.5	26.3	2920	39.9

The effect of the structural vacancy concentration in nonstoichiometric compounds MX_y on the mutual solubility of these compounds was analyzed in [154]. One hundred and five pseudobinary systems comprising cubic carbides, nitrides and oxides of titanium, zirconium, hafnium, thorium, vanadium, niobium and tantalum were analyzed. The electronic density $d_e(y)$ was calculated as a function of the relative concentration of nonmetal atoms for each nonstoichiometric compound MX_y. In accordance with calculations, unlimited mutual solubility of compounds $M^{(1)}X^{(1)}_{y_1}$ and $M^{(2)}X^{(2)}_{y_2}$ is observed if the electronic density ratio is $1.481 \geq d_e^{(1)}(y_1)/d_e^{(2)}(y_2) \geq 0.675$.

Table 10.16. Maximum elastic deformation energies, E_{max}, for carbide, carbonitride and nitride solid solutions and solid solution compositions corresponding to E_{max}

System A – B	Composition (mol. % A)	E_{max} (kJ mol^{-1})	System A – B	Composition (mol. % A)	E_{max} (kJ mol^{-1})
TiC–ZrC	54.3	27.3	HfC–TiN	48.3	27.4
TiC–HfC	53.6	20.1	HfC–ZrN	52.4	0.8
TiC–$VC_{0.88}$	47.8	5.3	HfC–NbN	50.2	8.4
TiC–NbC	52.0	4.3	$VC_{0.88}$–NbC	53.9	19.1
TiC–TaC	49.0	3.8	$VC_{0.88}$–TaC	51.6	18.7
TiC–TiN	51.9	1.0	$VC_{0.88}$–TiN	53.8	1.3
TiC–ZrN	55.9	11.3	$VC_{0.88}$–ZrN	57.3	31.9
TiC–NbN	53.8	1.7	$VC_{0.88}$–NbN	58.7	12.5
ZrC–HfC	49.4	0.5	NbC–TaC	47.6	0.1
ZrC–$VC_{0.88}$	25.7	49.0	NbC–TiN	49.3	8.9
ZrC–NbC	47.7	9.4	NbC–ZrN	54.0	1.8
ZrC–TaC	45.5	12.0	NbC–NbN	51.9	0.4
ZrC–TiN	47.6	35.2	TaC–TiN	51.6	8.5
ZrC–ZrN	51.8	2.4	TaC–ZrN	56.1	2.7
ZrC–NbN	49.6	12.8	TaC–NbN	54.2	0.2
HfC–$VC_{0.88}$	26.3	39.9	TiN–ZrN	54.1	17.0
HfC–NbC	48.3	5.5	TiN–NbN	52.2	4.9
HfC–TaC	45.8	7.2	ZrN–NbN	47.2	3.6

The form of the phase diagram of a pseudobinary system can be determined in most detail by calculations, provided the composition and temperature dependencies of parameters describing the interaction in the system are known. The interchange (mixing) energy B is often used in the description of substitutional solid solutions as a quantity characterizing the interaction between interchanging components. Physically, this energy represents a quantity, which is well known in the paired interaction model, and is equal to the excess of the mutual potential energy of unlike atoms over the mean energy of like atoms (see Sect. 8.3).

Systems based on nonstoichiometric compounds do not fit models of ideal or regular solutions. A subregular solution model was proposed for calculating their phase diagrams. This model allows for composition and temperature dependencies of the interchange energy in different phases [155–158].

Atomic ordering in the lattice of a solid represents an isobaric process. Therefore it is common practice to use the free energy F for thermodynamic analysis of ordering. Indeed, the pressure $p = 0$ for a solid and the internal (configuration) energy E of the lattice coincides with the crystal enthalpy H, because $H = E + pV = E$ ($pV = 0$ at $p = 0$), while the isobaric potential (Gibbs free energy G) equals the free energy F, because $G = F + pV = F$. The Gibbs free energy is used directly in thermodynamic analysis of equilibria allowing for possible change of the aggregation state (e.g. melting or crystallization).

Let us consider the subregular solution model in more detail.

If the pressure is constant, the Gibbs free energy of the j-th phase in a multicomponent system can be written as

$$\Delta G_j(\{x_i\},T) = \sum_i x_i \Delta G_j(x_i = 1,T) + RT\sum_i x_i \ln x_i + G_j^{e}(\{x_i\},T), \qquad (10.30)$$

where x_i is the i-component concentration of the j-th phase; $\Delta G_j(x_i = 1,T)$ denotes the free energies of the system components; $G_j^{e}(\{x_i\},T)$ is the excess free energy of mixing of the j-th phase. Phase equilibrium conditions remain invariant if any linear function of the composition is added to the free energy of mixing $\Delta G_j^{\mathrm{mix}} = G_j^{e}(\{x_i\},T) + RT\sum_i x_i \ln x_i$. Consequently, phase equilibria can be described using the relationship between free energies of mixing of equilibrium phases.

The degree to which a system deviates from ideality is defined by the excess free energy of mixing $G_j^{e} = B_j(\{x_i\},T)\prod x_i^{(j)}$, where $B_j(\{x_i\},T)$ is the interchange energy in the j-th phase, i.e. an energy parameter characterizing the interaction of components in the j-th phase and depending on the composition and temperature. The interchange energy in a binary system A–B is taken to mean the energy difference of paired interactions of unlike (ε_{AB}) and like (ε_{AA} and ε_{BB}) atoms: $B = N_A[2\varepsilon_{AB} - (\varepsilon_{AA} + \varepsilon_{BB})]$. The interchange energy $B_j = 0$ in the ideal solution model and, consequently, the excess free energy of mixing of an ideal solution, G_j^{e}, is zero. The interchange energy B_j = const in the regular solution model, i.e. is nonzero and is a constant immune to temperature and the composition. Therefore, in terms of the regular solution model, the excess free energy of mixing of a binary solid solution is $G_j^{e} = x_A x_B B_j$. Finally, in the more complicated subregular solution model the interchange energy B_j depends on the solid solution temperature and composition. Therefore the excess free energy of mixing for a binary solution has the form $G_j^{e} = x_A x_B B_j(x_A, x_B, T)$.

Interchange energies of all phases in equilibrium should be known to calculate phase boundaries using the subregular solution model. In accordance with [159], the interchange energy of the liquid phase $B_\ell = e_0 + e_p$ is the sum of the electronic interaction parameter e_0 and the internal pressure parameter e_p. The interchange energy of the solid phase includes, in addition to e_0 and e_p, the electronic interaction parameter of the solid phase e_1 and the elastic distortion parameter of the lattice e_2, i.e. $B_s = e_0 + e_p + e_1 + e_2$.

Interaction parameters of pseudobinary carbide systems were calculated in terms of the subregular solution model [157] using the atomization energy (enthalpy) E^{at}, i.e. the energy needed to break a compound into atoms and separate them to an infinitely long distance (in other words, the energy consumed for the transfer of a compound from its standard state to a monatomic gas), as a quantity characterizing the strength of interatomic bonds. The nonstoichiometric compound MX_y has the atomization energy $E^{at}(y) = \Delta H^{o}_{s,M} + y\Delta H^{o}_{s,X} - \Delta H^{o}_{f\,298}(y)$, where ΔH^{o}_{s} stands for evaporation heats of components and $\Delta H^{o}_{f\,298}(y)$ is the formation enthalpy of the compound MX_y. In accordance with [160–162], the internal pressure parameter e_p can be presented as

$$e_p = [(1-x)V_1(T) + xV_2(T)]\left\{[E_1^{at}/V_1(T)]^{1/2} - [E_2^{at}/V_2(T)]^{1/2}\right\}^2 \qquad (10.31)$$

if the change in the system volume with the composition and temperature is taken into account. The internal pressure parameter of the liquid phase, e_p, is related to the difference between atomic binding energies of components in the liquid solution and to the difference between molar volumes of these components (atomic volume if elements are concerned). The parameter e_p is always positive and characterizes the immiscibility of solution components in the liquid state.

The electronic interaction parameter of the liquid phase e_0 is defined as

$$e_0 = [(1-x)V_1(T) + xV_2(T)]\left\{\left[\frac{E_1^{at}}{V_1(T)}\right]^{1/2} - \left[\frac{E_1^{at}+E_2^{at}}{V_1(T)+V_2(T)}\right]^{1/2}\right\}\times$$
$$\times\left\{\left[\frac{E_2^{at}}{V_2(T)}\right]^{1/2} - \left[\frac{E_1^{at}+E_2^{at}}{V_1(T)+V_2(T)}\right]^{1/2}\right\}. \qquad (10.32)$$

It can be seen from (10.32) that the parameter e_0 is always negative for two-component (binary or pseudobinary) systems.

The interchange energy of the solid phase includes, in addition to parameters e_0 and e_p, the electronic interaction parameter e_1 and the elastic distortion parameter e_2. Since elastic distortions are absent in a liquid, the quantity directly related to the solid–liquid transition should only be the parameter e_1. The parameter e_1 is as negative as that the parameter e_0. The electronic interaction parameters e_0 and e_1 depend on the charge redistribution between unlike atoms and their Coulomb interaction. The parameter e_1 is computed from the formula

$$e_1 = -[(1-x)V_1(T) + xV_2(T)]\left|\frac{\Delta H_1}{V_1(T)} - \frac{\Delta H_2}{V_2(T)}\right|. \qquad (10.33)$$

The lattice elastic distortion parameter e_2 characterizes the deformation of the solid solution lattice and stresses arising in the lattice due to the difference in molar volumes and elasticity moduli of the solid solution components. The parameter e_2 generally makes the largest positive contribution to the total energy of the system and is therefore responsible for precipitation in the solid state. The parameter e_2 is

$$e_2 = \tfrac{1}{3}[V_1(T) - V_2(T)]^2\left[(1-x)\frac{G_{sh-1}(T)}{V_1(T)} + x\frac{G_{sh-2}(T)}{V_2(T)}\right]. \qquad (10.34)$$

In (10.31)–(10.34) x is the concentration of the second component (in mole fractions), $V_{1,2}(T) = V_{1,2}(300)[1 + 3\alpha_{1,2}(T - 300)]$ are the molar volumes of the components 1 and 2 at temperature T (or atomic volumes for elements), $V_{1,2}(300)$ and $\alpha_{1,2}$ denote the molar volumes at 300 K and the linear thermal expansion coefficients of the components, $E^{\mathrm{at}}_{1,2}$ are the atomization energies of the components (atomization energy coincides with evaporation heat for elements), $\Delta H_{1,2}$ are the melting heats of the components, $G_{\mathrm{sh}\text{-}1,2}(T) = G_{\mathrm{sh}\text{-}1,2}(300)[1 - \beta_{1,2}(T - 300)]$ and $G_{\mathrm{sh}\text{-}1,2}(300)$ stand for the shear modulus of the components at temperatures T and 300 K, $\beta_{1,2}$ are the thermal coefficients of the shear modulus of the components.

The electronic interaction parameters e_0 and e_1 are negative in sign and describe the binding part of the interchange energy, while the positive parameters e_p and e_2 describe the antibinding part of the interchange energy. The interchange energy can be positive, zero (in the case of ideal solutions) or negative depending on the relationship between the binding and antibinding parameters. In turn, interchange energies unlike in sign and different in absolute value determine diverse forms of phase diagrams: from a phase diagram with complete mutual solubility of components in liquid or solid states through phase diagrams with regions of precipitation in liquid or solid states to diagrams with a eutectic.

Equilibrium conditions of phases j_1 and j_2 are defined by a conventional set of equations of the form

$$\frac{\partial \Delta G_{j_1}(x_1,T)}{\partial x_1} = \frac{\partial \Delta G_{j_2}(x_2,T)}{\partial x_2} \tag{10.35a}$$

$$\Delta G_{j_1}(x_1,T) - x_1 \frac{\partial \Delta G_{j_1}(x_1,T)}{\partial x_1} = \Delta G_{j_2}(x_2,T) - x_2 \frac{\partial \Delta G_{j_2}(x_2,T)}{\partial x_2} \tag{10.35b}$$

where x_1 and x_2 are relative contents of the second component in these phases. Let us consider these conditions as applied to a two-component system.

In terms of the subregular solution model, the free energy of the j-phase (10.30) for a two-component system assumes the form

$$\begin{aligned}\Delta G_j(x,T) &= (1-x_j)\Delta G_j(0,T) + x_j \Delta G_j(1,T) \\ &\quad + RT[x_j \ln x_j + (1-x_j)\ln(1-x_j)] + G^{\mathrm{e}}_j(x,T),\end{aligned} \tag{10.36}$$

where x_j is the concentration of the second component in the j-th phase; $\Delta G_j(0,T)$ and $\Delta G_j(1,T)$ are the free energies of the initial components of the system; and $G^{\mathrm{e}}_j(x,T) = x_j(1 - x_j)B_j(x,T)$ is the excess energy of mixing of the j-th phase. In the particular case of equilibrium between solid and liquid phases having free energies (10.36), the equilibrium conditions (10.35) can, after some rearrangement, be written as equations for calculating the position of liquidus and solidus:

$$\begin{aligned}RT\ln(1-x_\ell) + x_\ell^2 B_\ell(x_\ell,T) - x_\ell^2(1-x_\ell)\frac{\partial B_\ell(x_\ell,T)}{\partial x_\ell} &= RT\ln(1-x_s) \\ + x_s^2 B_s(x_s,T) - x_s^2(1-x_s)\frac{\partial B_s(x_s,T)}{\partial x_s} &+ \frac{\Delta H_1(T-T_1)}{T_1},\end{aligned} \tag{10.37a}$$

$$RT \ln x_\ell + (1-x_\ell)^2 B_\ell(x_\ell,T) + x_\ell(1-x_\ell)^2 \frac{\partial B_\ell(x_\ell,T)}{\partial x_\ell} = RT \ln x_s$$

$$+(1-x_s)^2 B_s(x_s,T) + x_s(1-x_s)^2 \frac{\partial B_s(x_s,T)}{\partial x_s} + \frac{\Delta H_2(T-T_2)}{T_2} \quad (10.37b)$$

where x_ℓ and x_s denote the concentration of the second component in the liquid and solid phases; T_1 and T_2 are the melting temperatures of the first and second components; and B_ℓ and B_s are interchange energies in the liquid and solid phases.

When precipitation of solid solutions is described as a subregular approximation, it is assumed that transition from one phase to another during diffusion precipitation proceeds as a phase transformation of the second kind without an abrupt change in the entropy. Also, precipitated phases with x_1 and x_2 of the second component are assumed to have equal interchange energies $B_1(x_1,T) = B_2(x_2,T) = B_s(x,T)$. In this case the set of equations (10.37) rearranges to

$$RT \ln(1-x_1) + x_1^2 B_s - x_1^2(1-x_1)\frac{\partial B_s}{\partial x_1}$$

$$= RT \ln(1-x_2) + x_2^2 B_s - x_2^2(1-x_2)\frac{\partial B_s}{\partial x_2},$$

$$RT \ln x_1 + (1-x_1)^2 B_s + x_1(1-x_1)^2 \frac{\partial B_s}{\partial x_1}$$

$$= RT \ln x_2 + (1-x_2)^2 B_s + x_2(1-x_2)^2 \frac{\partial B_s}{\partial x_2}. \quad (10.38)$$

The authors [163, 164] analyzed specific features of the subregular solution model as applied to the description of precipitation in binary solid solutions. Consider a system with concentration x of the second component in the solid phase. In accordance with [1, 156], the interchange energy is $B(x_i) = B_0 + xB_1$ in systems based on nonstoichiometric compounds. Therefore the excess free energy of mixing in the solid phase has the form $G_s^e = x(1-x)(B_0 + xB_1)$ in the subregular solution model. The free energy of mixing $G(x)$ in the solid phase is

$$G(x) = x(1-x)(B_0 + xB_1) + RT[x\ln x + (1-x)\ln(1-x)]. \quad (10.39)$$

Theoretical analysis of the dependence of the solid-phase free energy on the parameters B_0 and B_1 is facilitated if (10.39) is rearranged to the form

$$G^*(x) = x(1-x)(b_0 + xb_1) + T_r\,[x\ln x + (1-x)\ln(1-x)], \quad (10.40)$$

where $G^*(x) = G(x)/RT^*$, $b_0 = B_0/RT^*$, $b_1 = B_1/RT^*$ and $T_r = T/T^*$ are dimensionless quantities.

Figure 10.17 shows the concentration dependence of the solid-phase free energy of mixing $G^*(x)$ as a function of temperature at fixed values of the parameters b_0 and b_1. When the temperature is high ($T_r > 0.6$), the free energy curve has one minimum, which corresponds to a single-phase state of the pseudobinary system, i.e. to unlimited mutual solubility of the components. One more minimum in the $G^*(x)$ curve at $T_r < 0.6$ means that the homogeneous solid solution precipitates in this temperature interval to two phases having different compositions. As can be seen from Fig. 10.17, the $G^*(x)$ dependencies are asymmetric in the subregular solution model ($b_1 \neq 0$) and are symmetric with respect to $x = 0.5$ in the regular solution approximation ($b_1 \equiv 0$). Note that symmetric dependencies of the free energy of mixing $G^*(x)$ also determine the symmetry of the precipitation range boundaries in the phase diagram. This is far from reality and suggests limited applicability of the regular approximation.

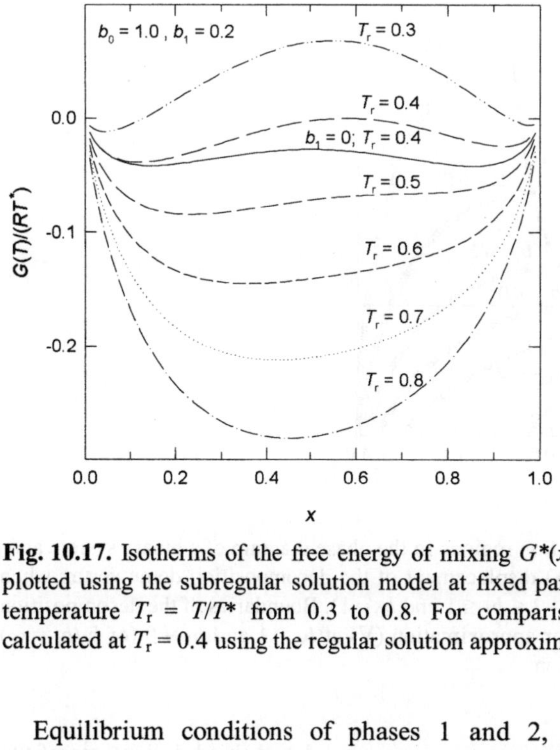

Fig. 10.17. Isotherms of the free energy of mixing $G^*(x) = G(x)/RT^*$ of a binary system plotted using the subregular solution model at fixed parameters $b_0 = 1$ and $b_1 = 0.2$ and temperature $T_r = T/T^*$ from 0.3 to 0.8. For comparison, the solid line shows $G^*(x)$ calculated at $T_r = 0.4$ using the regular solution approximation ($b_1 \equiv 0$) [164]

Equilibrium conditions of phases 1 and 2, to which the solid solution precipitates, are defined by two equations $\partial G_1(x)/\partial x = 0$ and $\partial G_2(x)/\partial x = 0$. Considering $B(x_i) = B_0 + xB_1$ and performing standard manipulations with these equations or equations (10.38), the equilibrium conditions can be written as

$$RT\ln(1-x_1)+x_1^2(B_0+x_1B_1)-x_1^2(1-x_1)B_1$$

$$=RT\ln(1-x_2)+x_2^2(B_0+x_2B_1)-x_2^2(1-x_2)B_1, \tag{10.41a}$$

$$RT\ln x_1+(1-x_1)^2(B_0+x_1B_1)+x_1(1-x_1)^2B_1$$

$$=RT\ln x_2+(1-x_2)^2(B_0+x_2B_1)+x_2(1-x_2)^2B_1, \tag{10.41b}$$

where x_1 and x_2 denote the concentration of the second component in phases 1 and 2, to which the solid solution precipitates. The solution of the set of equations (10.41) gives the position of the decomposition range boundaries. The calculations of parameters b_0 and b_1 show that the decomposition range has asymmetric boundaries. The decomposition dome vertex is displaced to either side depending on the ratio between the parameters b_0 and b_1 (Fig. 10.18): $x_{max} < 0.5$ if $b_0/b_1 < 0$, and $x_{max} > 0.5$ if $b_0/b_1 > 0$ (x_{max} being the solid solution composition corresponding to the decomposition dome vertex.)

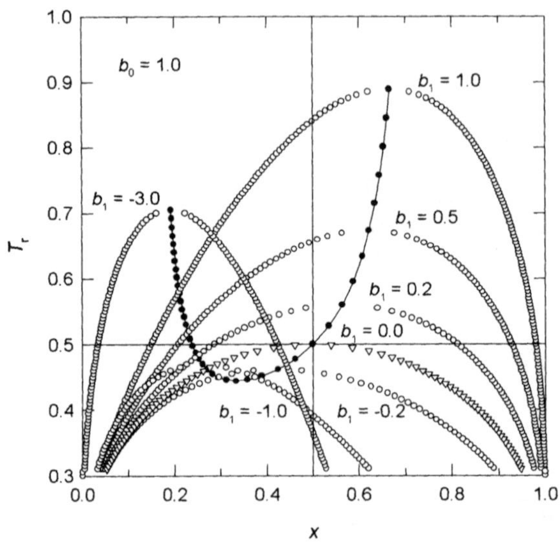

Fig. 10.18. Position of decomposition regions in the phase diagram of a binary system at $b_0 = 1$ and different values of b_1; (●) displacement of the decomposition dome vertex when b_0/b_1 changes from −1/3 to +1 (or $-3 \le b_1 \le 1$ and $b_0 = 1$). Boundaries of the decomposition region calculated using the regular approximation (∇) at $b_0 = 1$ and $b_1 \equiv 0$ (or $b_0/b_1 = \infty$) [163,164] are shown for comparison

The critical miscibility conditions are zero second and third derivatives of the free energy of mixing with respect to composition:

$$\left.\begin{aligned} \partial G^2(x)/\partial x^2 &= 0 \\ \partial G^3(x)/\partial x^3 &= 0 \end{aligned}\right\}. \tag{10.42}$$

These expressions yield the maximum decomposition temperature of the solid solution, i.e. the immiscibility temperature T_{decomp}^{max} above which components form unlimited solid solutions and below which a miscibility gap appears. From these equations it is also possible to determine the solid solution composition x_{max} corresponding to the decomposition dome vertex. Solving the set of equations (10.42) taking account of (10.39), we have

$$T_{\text{decomp}}^{\max} = \frac{6B_1 x_{\max}^2 (1-x_{\max})^2}{(2x_{\max}-1)R}, \tag{10.43}$$

$$x_{\max} = \frac{1}{9}\left\{4 - \frac{B_0}{B_1} \pm \left[7 + \frac{B_0}{B_1} + \left(\frac{B_0}{B_1}\right)^2\right]^{1/2}\right\}. \tag{10.44}$$

Only those solutions to (10.43) and (10.44) are physically meaningful for which $T_{\text{decomp}}^{\max} \geq 0$ and $1 \geq x_{\max} \geq 0$. From the analysis of equations (10.43) and (10.44) it follows that $T_{\text{decomp}}^{\max} \geq 0$ if

$$\begin{cases} 0 \leq x_{\max} < 0.5, B_1 < 0, & x_{\max} = \dfrac{1}{9}\left\{4 - \dfrac{B_0}{B_1} - \left[7 + \dfrac{B_0}{B_1} + \left(\dfrac{B_0}{B_1}\right)^2\right]^{1/2}\right\} \\ 0.5 < x_{\max} \leq 1.0, B_1 > 0, & x_{\max} = \dfrac{1}{9}\left\{4 - \dfrac{B_0}{B_1} + \left[7 + \dfrac{B_0}{B_1} + \left(\dfrac{B_0}{B_1}\right)^2\right]^{1/2}\right\} \end{cases} \tag{10.45}$$

The analysis of (10.44) shows that limiting values of $x_{\max}$ equal to 1 and 0 are reached when $B_0/B_1 = -2$ and $B_0/B_1 = 1$ respectively. Note that $B_0/B_1 \equiv b_0/b_1$. The limiting values $x_{\max} = 0$ (at $B_1 < 0$) and $x_{\max} = 1$ (at $B_1 > 0$) correspond to $T_{\text{decomp}}^{\max} \equiv 0$, i.e. degeneration of the decomposition range. The quantity $x_{\max}$ asymptotically tends to 0.5 when $B_0/B_1 \to \pm\infty$ (Fig. 10.19). It is readily seen that $B_0/B_1 \to \pm\infty$ is realized if $B_1 \to 0$. It is equivalent to the regular approximation when $x_{\max} \equiv 0.5$. Thus, the regular solution approximation is a particular case of the subregular solution model when $B_1 = 0$.

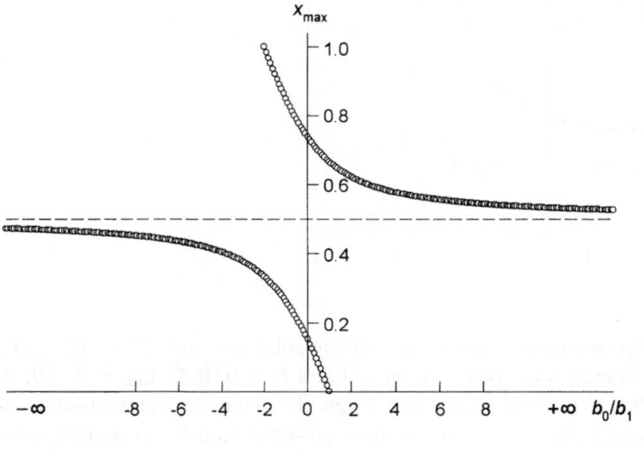

Fig. 10.19. The $x_{\max}$ (solid solution composition corresponding to the decomposition dome vertex) vs. the ratio $b_0/b_1 \equiv B_0/B_1$: the upper branch corresponds to a positive and the lower branch to a negative values of b_1 [164]

Figure 10.20 shows the x_{max} dependence of the maximum decomposition temperature of the solid solution $T_r^{max} = T_{decomp}^{max}/T^*$.

The left-hand region of Fig. 10.20 corresponds to negative values of the parameters $b_0 < 0$, $b_1 < 0$ and $b_0/b_1 > 0$. The left-hand branch in Fig. 10.20 was plotted for $b_0 = -1$ and b_1 from -7 to -1. When $b_0 < 0$, $b_1 < 0$ and $b_0/b_1 > 0$, the quantity x_{max} changes between 0 and $(4 - \sqrt{7})/9 \approx 0.150$. The value of $x_{max} = 0.150$ is reached asymptotically when $b_0/b_1 \to 0$, i. e. when $b_0 = 0$.

The middle branch of the x_{max} versus T_r^{max} plot corresponds to positive values of the parameter $b_0 > 0$, whereas the parameter b_1 may be either positive or negative. This branch was plotted for $b_0 = 1$ and b_1 from -3 to $+3$ in Fig. 10.20. The analysis shows that when $b_0 = \text{const} > 0$, the lowest value of $T_{decomp}^{max} = -8B_1/9R \equiv 4B_0/9R$ is reached when $B_1 < 0$ and $B_0/B_1 \equiv b_0/b_1 = -2$, which correspond to $x_{max} = 1/3$. In the case of the middle branch, $(4 - \sqrt{7})/9 < x_{max} < (4 + \sqrt{7})/9$. Extreme values of x_{max} equal to 0.150 and $(4 + \sqrt{7})/9 \approx 0.738$ are reached asymptotically when $b_0 = 0$ and $b_1 < 0$ or when $b_0 = 0$ and $b_1 > 0$.

The right-hand branch of the x_{max} versus T_r^{max} plot corresponds to $b_0 < 0$ and $b_1 > 0$, i.e. $b_0/b_1 < 0$. The right-hand branch in Fig. 10.20 was plotted for $b_0 = -1$ and b_1 from 4.5 to 0.5.

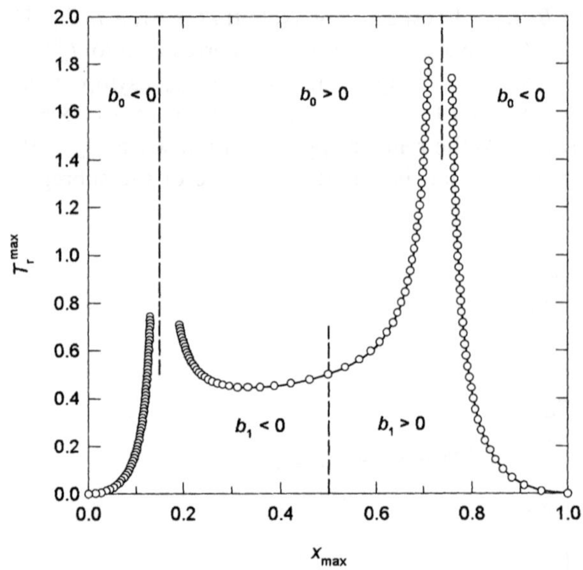

Fig. 10.20. Maximum decomposition temperature of the solid solution $T_r^{max} = T_{decomp}^{max}/T^*$ vs. x_{max} : the left-hand branch corresponds to $b_0 < 0$ and $b_1 < 0$, $0 \le x_{max} < 0.150$; the middle branch corresponds to positive values $b_0 > 0$, and the parameter b_1 may assume any positive or negative values, $0.150 < x_{max} < 0.738$; the right-hand branch corresponds to $b_0 < 0$ and $b_1 > 0$, $0.738 < x_{max} \le 1$ [163, 164]

Elements or compounds forming a particular phase may have different crystal structures. This difference should be taken into account when computing all interaction parameters by the method of volume reduction. To this end, the crystal structure of one of the components involved in the phase equilibrium is selected as the basic structure, and parameters of unit cells of the other components are reduced to this structure. The reduction is accomplished such that surface areas of reduced and unreduced unit cells are equal, while angles and lattice constant ratios of a reduced cell are the same as their counterparts in the unit cell of the basic structure. With this in mind, one can calculate lattice constants and the volume of a reduced unit cell $V_i^{\text{red.c.}}$, and also the reduced molar volume $V_i^{\text{red}} = N_A V_i^{\text{red.c.}}/z$ necessary for calculating the energy parameters e_p, e_0, e_1 and e_2 (z being the number of formula units in the unit cell). If the unit cell of the basic structure has angles α, β, and γ, and lattice constant ratios are $a/c = m$ and $b/c = n$, while the surface area of the unreduced cell is S, then

$$a_{\text{red}} = \left(\frac{mnS}{\sin\alpha + m\sin\beta + n\sin\gamma}\right)^{1/2} \tag{10.46}$$

and

$$V_i^{\text{red}} = \frac{N_A a_{\text{red}}^3 \sin\alpha \sin\beta \sin\gamma}{mnz}. \tag{10.47}$$

The crystal structure is chosen arbitrarily, since reduction of the molar volumes of the components to one and the same crystal structure ensures an equal ratio between reduced volumes V_i^{red}, i.e.

$$V_1 : V_2^{\text{red}(1)} : \ldots : V_n^{\text{red}(1)} = V_1^{\text{red}(2)} : V_2 : \ldots : V_n^{\text{red}(2)} = \cdots = V_1^{\text{red}(i)} : V_2^{\text{red}(i)} : \ldots : V_i : \ldots V_n^{\text{red}(i)}.$$

Clearly, absolute values of the energy interaction parameters and the interchange energy are independent of the crystal structure to which molar volumes of the phase components are reduced. Indeed, from (10.31)–(10.34) it follows that the k-fold increase in the molar volume of one component or the corresponding decrease in the molar volume of another component will give one and the same value of the energy parameter e.

The above method was used [155–158, 165] to determine the interaction parameters and interchange energies in 15 pseudobinary carbide systems formed by titanium, zirconium, hafnium, vanadium, niobium and tantalum carbides. Values of lattice constants, atomization energy, melting heat and temperature, thermal expansion coefficient and shear modulus of these compounds, B_4C and TiB_2, which were used in calculations, are given in Table 10.17.

Calculated interaction parameters $e(x,T)$ and the interchange energy $B(x,T)$ of the pseudobinary systems under discussion are presented by polynomials $B_0(T) + xB_1(T) = [a_0 + a_1T + a_2T^2 + x(c_0 + c_1T + c_2T^2)]$ (Table 10.18).

Table 10.17. Lattice constant a_{B1}, molar volume V, melting temperature T_m, heat of melting ΔH, atomization energy E^{at}, shear modulus G_{sh}, mean linear thermal expansion coefficient α and mean shear modulus temperature coefficient β of some carbides and titanium boride

Compound	a_{B1} (nm)	$V \times 10^6$ (m^3)	T_m (K)	ΔH (kJ mol^{-1})	E^{at} (kJ mol^{-1})	$G_{sh}(300)$ (GPa)	$\alpha \times 10^6$ (K^{-1})	$\beta \times 10^5$ (K^{-1})
TiC	0.4328	12.207	3340	80.2	1370	198	9.0	8.0
ZrC	0.4698	15.609	3690	87.5	1530	178	7.0	10.0
HfC	0.4640	15.044	4220	98.7	1510	193	7.0	7.0
$VC_{0.88}$	0.4166	10.887	2970	75.7	1350	197	8.2	12.0
NbC	0.4470	13.452	3870	91.7	1570	188	7.8	10.0
TaC	0.4456	13.318	4270	93.5	1635	220	7.4	8.0
[1] B_4C	$a = 0.563$ $c = 1.214$	25.025	2720	104.6	3020	190	7.5	6.0
[2] TiB_2	$a = 0.3025$ $b = 0.3233$	15.424	3500	92.2	1860	230	5.0	6.0

[1] Space group $R\bar{3}m$, $z = 9$, reduced molar volume is given for B_4C.
[2] Space group $P6/mmm$, $z = 1$.

Table 10.18. Coefficients of polynomials $[(a_0 + a_1T + a_2T^2) + x(c_0 + c_1T + c_2T^2)]$ approximating the energy parameters of interaction, e, and the mutual exchange energies, B, for pseudobinary carbide and carboboride systems [1]

	$a_0 \times 10^{-3}$ (J mol^{-1})	a_1 (J $mol^{-1}K^{-1}$)	$a_2 \times 10^3$ (J $mol^{-1}K^{-2}$)	$c_0 \times 10^{-3}$ (J mol^{-1})	c_1 (J $mol^{-1}K^{-1}$)	$c_2 \times 10^3$ (J $mol^{-1}K^{-2}$)
1	2	3	4	5	6	7
TiC–ZrC						
E_0	–1.489	0.126	–0.005	–0.418	0.047	–0.003
e_p	6.022	–0.514	0.020	1.692	–0.191	0.010
B_ℓ	4.533	–0.388	0.015	1.274	–0.144	0.007
E_1	–11.895	0.415	–0.008	–3.342	0.209	–0.007
E_2	64.580	–6.856	0.173	–18.994	1.426	–0.063
B_s	57.218	–6.828	0.180	–21.062	1.491	–0.063
TiC–HfC						
E_0	–1.034	0.107	–0.005	–0.242	0.033	–0.002
e_p	4.169	–0.433	0.019	0.978	–0.132	0.008
B_ℓ	3.135	–0.326	0.014	0.736	–0.099	0.006
E_1	–0.258	0.173	–0.163	–0.061	0.039	–0.035
E_2	45.055	–5.165	0.158	–9.584	1.643	–0.066
B_s	47.932	–5.318	0.009	–8.909	1.583	–0.095

(Continued)

Table 10.18 (continued)

1	2	3	4	5	6	7
$VC_{0.88}$–TiC						
E_0	−0.786	−0.038	0	−0.094	−0.007	0
e_p	3.149	0.152	−0.002	0.379	0.027	0
B_ℓ	2.363	0.114	−0.002	0.285	0.020	0
E_1	−4.120	−0.174	0.004	−0.496	−0.032	0
E_2	10.663	−0.485	−0.084	−1.209	0.387	0.037
B_s	8.906	−0.545	−0.082	−1.420	0.375	0.037
TiC–NbC						
E_0	−0.126	−0.025	−0.001	−0.013	−0.002	0
e_p	0.506	0.099	0.002	0.052	0.008	0
B_ℓ	0.380	0.074	0.001	0.039	0.006	0
E_1	−2.923	−0.303	0.007	−0.301	−0.019	0.001
E_2	8.714	−1.121	0.043	−1.170	0.034	−0.003
B_s	6.171	−1.350	0.051	−1.432	0.021	−0.002
TiC–TaC						
E_0	−0.695	−0.079	0	−0.064	−0.003	0
e_p	2.790	0.316	0.001	0.258	0.014	−0.001
B_ℓ	2.095	0.237	0.001	0.194	0.011	−0.001
e_1	−5.376	−0.416	0.008	−0.498	−0.010	0.002
e_2	7.016	−1.157	0.062	0.119	0.014	−0.004
B_s	3.735	−1.336	0.071	−0.185	0.015	−0.003
ZrC–HfC						
e_0	−0.054	0	0	0.002	0	0
e_p	0.217	0	0	−0.008	0	0
B_ℓ	0.163	0	0	0.006	0	0
e_1	−14.907	0	0	0.540	0	0
e_2	1.242	−0.094	−0.003	0.143	0.029	0.001
B_s	−13.502	−0.094	−0.003	0.677	0.029	0.001
$VC_{0.88}$–ZrC						
e_0	−4.094	0.117	−0.003	−1.782	0.072	−0.002
e_p	16.752	−0.486	0.013	7.292	−0.299	0.009
B_ℓ	12.658	−0.369	0.010	5.510	−0.227	0.007
e_1	−14.737	0.222	−0.004	−6.415	0.174	−0.004
e_2	139.300	−16.027	−0.002	−52.115	7.945	−0.032
B_s	137.221	−16.174	0.004	−53.020	7.892	−0.029

(*Continued*)

Table 10.18 (continued)

1	2	3	4	5	6	7
ZrC–NbC						
E_0	−3.199	0.092	−0.002	0.444	−0.019	0.001
e_p	12.832	−0.371	0.010	−1.781	0.078	−0.003
B_ℓ	9.632	−0.277	0.007	−1.337	0.058	−0.002
E_1	−18.981	0.259	−0.006	2.635	−0.075	0.002
E_2	18.365	−1.937	0.018	4.136	−0.493	−0.011
B_s	9.325	−1.997	0.020	8.834	−0.876	0.014
ZrC–TaC						
E_0	−5.433	0.062	−0.001	0.799	−0.015	0
e_p	21.789	−0.248	0.005	−3.205	0.059	−0.001
B_ℓ	16.356	−0.186	0.004	−2.406	0.044	−0.001
E_1	−22.124	0.133	−0.003	3.254	−0.043	0.001
E_2	20.506	−1.847	−0.013	9.037	0.290	−0.001
B_s	14.738	−1.900	−0.012	9.885	0.291	−0.001
$VC_{0.88}$–HfC						
E_0	−3.367	0.108	−0.003	−1.291	0.058	−0.002
e_p	13.716	−0.455	0.012	5.258	−0.239	0.008
B_ℓ	10.349	−0.337	0.009	3.967	−0.181	0.006
E_1	−4.350	0.260	−0.005	−1.668	0.121	−0.004
E_2	108.032	−12.674	0.029	−32.630	7.664	−0.034
B_s	114.031	−12.751	0.033	−30.331	7.604	−0.032
NbC–HfC						
E_0	−2.087	0.064	−0.002	−0.249	0.013	−0.001
e_p	8.360	−0.257	0.007	0.996	−0.053	0.002
B_ℓ	6.273	−0.193	0.005	0.747	−0.040	0.001
E_1	−3.509	0.214	−0.004	−0.418	0.035	−0.001
E_2	12.242	−1.457	0.030	−1.110	0.476	−0.012
B_s	15.006	−1.436	0.031	−0.781	0.471	−0.012
HfC–TaC						
E_0	−4.247	0.054	−0.001	0.488	−0.011	0
e_p	17.012	−0.216	0.005	−1.957	0.043	−0.001
B_ℓ	12.765	−0.162	0.004	−1.469	0.032	−0.001
E_1	−6.956	0.128	−0.003	0.800	−0.022	0.001
E_2	12.997	−0.859	−0.001	3.794	−0.432	0.001
B_s	18.806	−0.893	0	3.125	−0.422	0.001

(*Continued*)

Table 10.18 (continued)

1	2	3	4	5	6	7
$VC_{0.88}$–NbC						
E_0	−0.301	0.012	0	−0.071	0.003	0
e_p	1.216	−0.048	0.001	0.287	−0.013	0
B_ℓ	0.915	−0.036	0.001	0.216	−0.010	0
E_1	−1.512	0.090	−0.002	−0.357	0.024	0
E_2	40.964	−4.250	−0.057	−9.250	1.615	0.016
B_s	40.367	−4.196	−0.058	−9.661	1.629	0.016
$VC_{0.88}$–TaC						
E_0	−0.010	0.004	0	−0.002	0.001	0
e_p	0.039	−0.017	0.002	0.008	−0.004	0
B_ℓ	0.029	−0.013	0.002	0.006	−0.003	0
E_1	−0.678	−0.186	0.004	−0.146	−0.038	0.002
E_2	33.061	−3.958	0.005	−3.339	1.618	−0.008
B_s	33.312	−4.157	0.011	−3.479	1.577	−0.006
NbC–TaC						
E_0	−0.254	−0.012	0	0.002	0.001	0
e_p	1.015	0.050	−0.001	−0.009	−0.002	0
B_ℓ	0.761	0.038	−0.001	−0.007	−0.001	0
E_1	−2.706	−0.115	0.003	0.026	0.004	0
E_2	0.080	0.014	−0.001	0.014	0.004	0
B_s	−1.865	−0.063	0.001	0.033	0.007	0
TiC–TiB_2						
E_0	−0.010	0.020	−0.011	−0.004	0.007	−0.004
e_p	0.040	−0.082	0.044	0.015	−0.028	0.015
B_ℓ	0.030	−0.062	0.033	0.011	−0.021	0.011
E_1	−11.816	0.838	−0.020	−4.401	0.506	−0.023
E_2	104.886	−14.642	0.650	−16.630	5.041	−0.305
B_s	93.100	−13.866	0.644	−21.020	5.526	−0.317
B_4C–TiB_2						
E_0	0	0	0	0	0	0
e_p	0	0.001	0.002	0	0	0
B_ℓ	0	0.001	0.002	0	0	0
E_1	−45.060	0.226	−0.003	17.300	−0.129	0.002
E_2	236.868	−11.900	−0.120	228.565	−12.187	−0.072
B_s	191.809	−11.673	−0.123	245.865	−12.316	−0.070

By way of example, Fig. 10.21 shows the temperature dependence of parameters $B_{0\ell}$, $B_{1\ell}$, B_{0s} and B_{1s} of the interchange energies B_ℓ and B_s in the liquid and solid phases of the ZrC–NbC system.

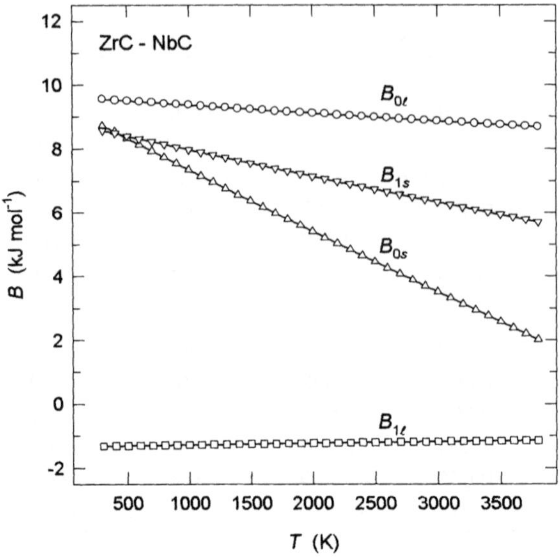

Fig. 10.21. Temperature dependence of the parameters $B_{0\ell}$, $B_{1\ell}$, B_{0s}, and B_{1s} of the interchange energies B_ℓ and B_s in the liquid and solid phases of the pseudobinary system ZrC–NbC

Calculated phase diagrams of pseudobinary carbide systems [155–158, 163–165] are shown in Figs. 10.22, 10.23 and 10.24. Titanium, zirconium, hafnium, vanadium, niobium and tantalum monocarbides have a similar crystal structure $B1$. Therefore the volume reduction operation was skipped in calculations of the energy parameters. Calculations of phase diagrams confirmed the limited solubility in systems $VC_{0.88}$–ZrC and $VC_{0.88}$–HfC (Fig. 10.22) and formation of a series of continuous solid solutions in systems TiC–HfC, TiC–ZrC, $VC_{0.88}$–TaC, $VC_{0.88}$–NbC (Fig. 10.22), TiC–TaC, ZrC–HfC, NbC–TaC (Fig. 10.23), ZrC–TaC, ZrC–NbC, HfC–TaC, NbC–HfC, $VC_{0.88}$–TiC, TiC–NbC (Fig. 10.24). Calculated positions of phase boundaries agree well with experimental data [153, 166] (see Fig. 10.22). Phase diagrams of nine pseudobinary carbide systems (Figs. 10.23 and 10.24) were calculated [155–158, 165] for the first time ever.

The form of calculated equilibrium phase diagrams (Fig. 10.22) agrees with the concept of a smooth transition from continuous solid solutions to systems with limited solubility in the solid state and a eutectic. Mutual solubility is largely affected by the dimensional factor: The system $VC_{0.88}$–TaC is nearly ideal, because interchanging metals have almost equal atomic radus, whereas the system $VC_{0.88}$–HfC has very low limited solubility, because atomic radius of vanadium and hafnium are largely different.

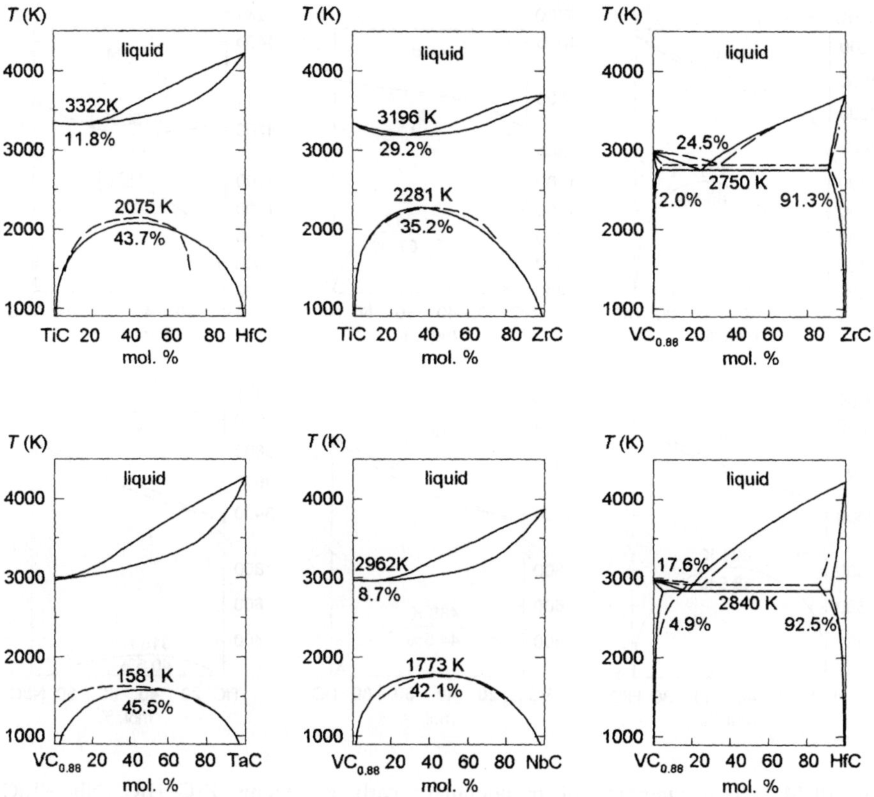

Fig. 10.22. Phase diagrams of pseudobinary carbide systems TiC–HfC, TiC–ZrC, $VC_{0.88}$–ZrC, $VC_{0.88}$–TaC, $VC_{0.88}$–NbC, $VC_{0.88}$–HfC. The solid line is the result of calculations [155, 158, 165]; the dotted line is experimental data of [153, 166]

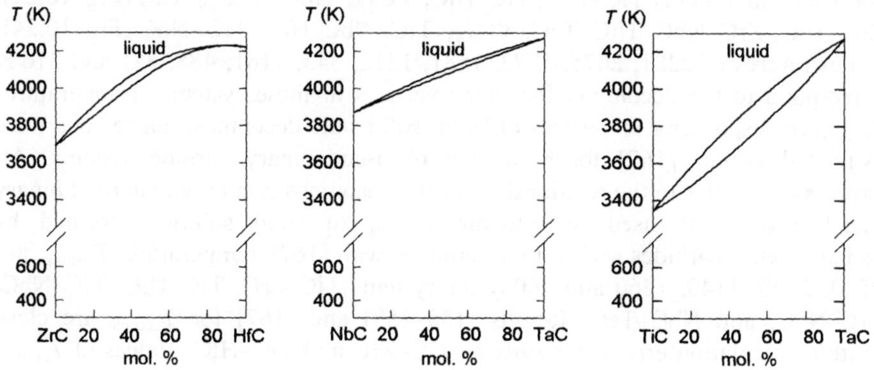

Fig. 10.23. Phase diagrams of pseudobinary carbide systems ZrC–TaC, ZrC–NbC, HfC–TaC, NbC–HfC, $VC_{0.88}$–TiC, TiC–NbC (calculations [155, 157, 158])

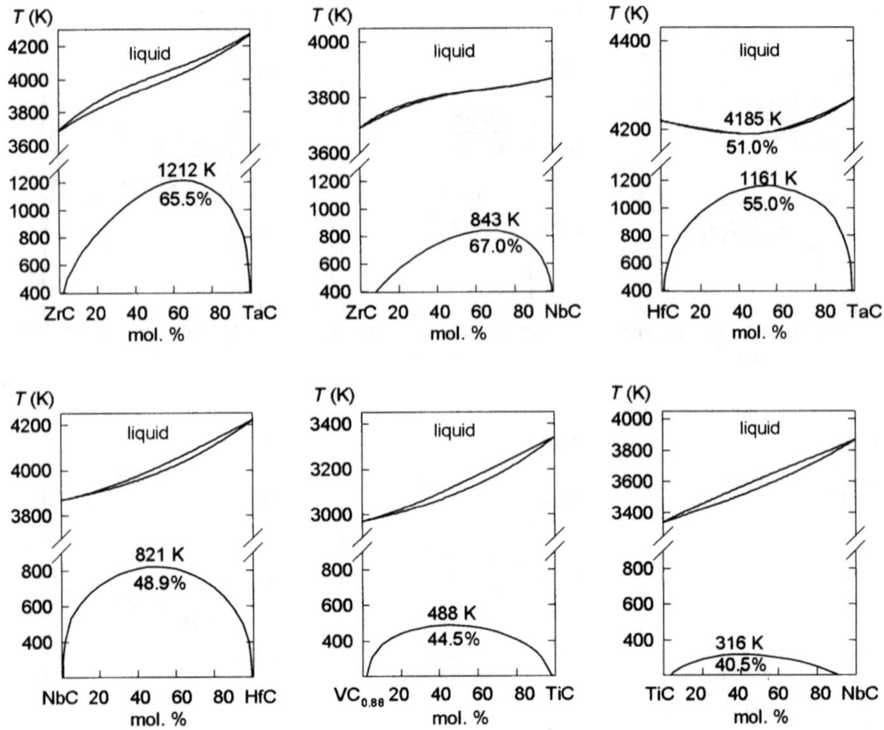

Fig. 10.24. Phase diagrams of pseudobinary carbide systems ZrC–HfC, NbC–TaC, TiC–TaC with nearly ideal solubility (calculations [155, 156, 158])

The analysis of phase equilibria revealed latent decomposition ranges in the solid state in systems TiC–ZrC, TiC–HfC, $VC_{0.88}$–NbC, $VC_{0.88}$–TaC (Fig. 10.22), ZrC–TaC, ZrC–NbC, HfC–TaC, $VC_{0.88}$–TiC, NbC–HfC, TiC–NbC (Fig. 10.24). Temperatures of 2281, 2075, 1773, 1581, 1212, 843, 1161, 488, 821 and 316 K correspond to the decomposition dome vertices in these systems. Temperatures T_{decomp} corresponding to vertices of latent solid-state decomposition regions were estimated earlier [167] for a number of pseudobinary carbide systems. An expression similar to the relationship for the decomposition temperature of binary metal alloys was used to estimate T_{decomp} of solid solutions formed by stoichiometric carbides $MC_{1.0}$. In accordance with [167], temperatures T_{decomp} are 2530, 2150, 1140, 1260 and 760 K for systems TiC–ZrC, TiC–HfC, TiC–NbC, ZrC–NbC, and NbC–TaC. Results [155–158] and [167] for T_{decomp} are close (within calculation errors) for systems TiC–ZrC and TiC–HfC. Values of T_{decomp} for systems TiC–NbC, ZrC–NbC and NbC–TaC [167] are probably overestimated. Probably, the reason is that only differences of carbide formation heats were taken into account in [167], whereas calculations [155–158] also allowed for carbide molar volumes and their change with temperature, as well as a number of other characteristics.

Decomposition regions are absent and solid solutions are stable from 300 K up to melting points in systems ZrC–HfC, NbC–TaC and TiC–TaC (Fig. 10.23).

Experimental data [168–170] on short-range order in solid solutions of systems TiC–NbC, TiC–TaC, and $VC_{0.88}$–NbC obtained by the X-ray diffuse scattering method agree with calculated phase diagrams of these systems. In accordance with [168–170], solid solutions $Ti_{0.5}Nb_{0.5}C$ and $Ti_{0.5}Ta_{0.5}C$ quenched from 2200 K have a negative short-range order parameter in the first coordination sphere of the metal sublattice ($\alpha_1 = -0.06$). Put another way, solid solutions do not tend to decomposition in this temperature interval. On the other hand, the parameter α_1 is positive ($\alpha_1 = 0.01$) in solid solutions $V_{0.5}Nb_{0.5}C$ quenched from 2100 K. This is an indication that these solid solutions are prone to decomposition. Indeed, the calculated phase diagram of the $VC_{0.88}$–NbC system suggests that decomposition in this system already begins at 1773 K (Fig. 10.22). Experimental studies of phase equilibria in ZrC–TaC and VC–NbC systems [171, 172] showed that calculations [156, 157, 165] performed in terms of the subregular solution model accurately predict the form of phase diagrams of these carbide systems.

The subregular solution model was applied to calculate phase boundaries not only in pseudobinary carbide systems but also in systems MO–Na_3AlF_6 (MO = Al_2O_3, BeO, SnO_2) [173] and in the ternary system Ti–B–C [162, 174].

Figure 10.25 shows the general form of the calculated phase diagram of the Ti–B–C system in the temperature interval from 300 to 3500 K and liquidus surfaces for pseudobinary equilibria TiC_y–TiB_2 and B_4C_y–TiB_2 of this system.

The phase diagram of the binary system Ti–C taking account of ordering of nonstoichiometric titanium carbide TiC_y was calculated by the order-parameters functional method. Pseudobinary equilibria (including TiC_y–TiB_2 and B_4C_y–TiB_2) were calculated using the subregular solution model. The operation of molar volume reduction was used [160–162] in calculating the equilibria of compounds having different crystal structures.

The composition and temperature of the eutectic of the pseudobinary system TiC_y–TiB_2 depend on the carbon concentration of carbide TiC_y. The eutectic temperature T_e increases from 2910 to 2961 K as the carbide composition changes from $TiC_{0.6}$ to $TiC_{0.8}$. If the carbon concentration of TiC_y increases further, T_e drops to 2936 K in the $TiC_{1.0}$–TiB_2 cross-section (Fig. 10.26). The eutectic contains 40.1, 40.5 and 40.2 mol.% TiB_2 in cross-sections $TiC_{0.6}$–TiB_2, $TiC_{0.8}$–TiB_2 and $TiC_{1.0}$–TiB_2 respectively. Note that only carbide $TiC_{0.8}$ melts with its composition unchanged (congruently). Titanium carbides with other carbon concentrations melt incongruently, i.e. with a change in their composition. As a result, the composition of titanium carbide in equilibrium with liquid changes from $TiC_{0.6}$ to $TiC_{0.7}$ in the $TiC_{0.6}$–TiB_2 cross-section with increasing temperature (Fig. 10.26*a*). The composition of carbide in equilibrium with liquid changes from $TiC_{1.0}$ to $TiC_{0.905}$ in the $TiC_{1.0}$–TiB_2 cross-section as temperature rises from 2936 to 3287 K (Fig. 10.26*c*).

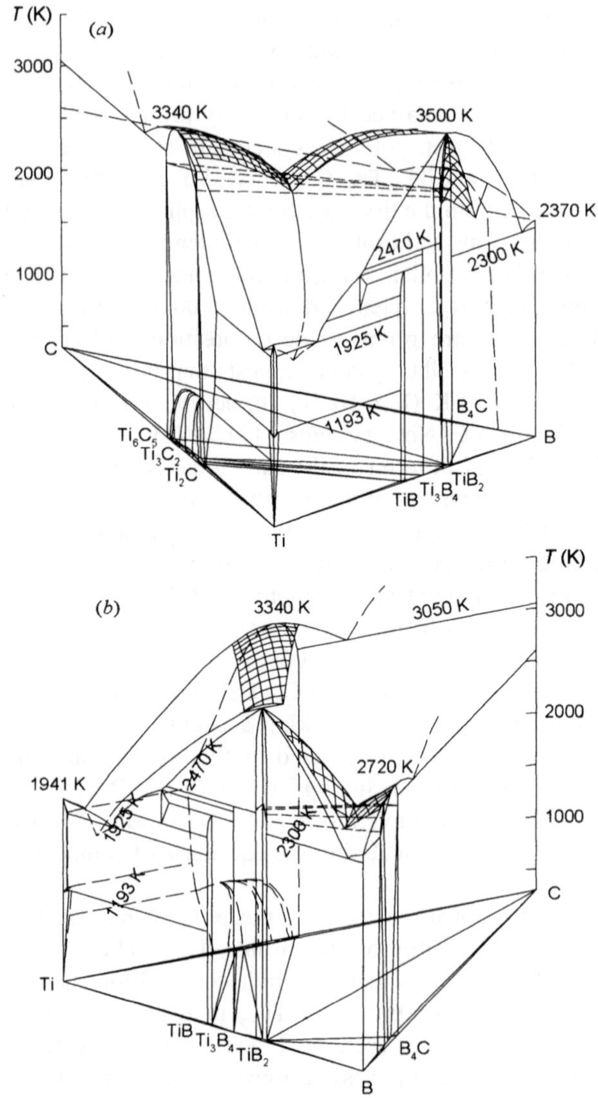

Fig. 10.25. Three-dimensional phase diagram of the ternary system Ti–B–C in the temperature interval from 300 to 3500 K [160-162, 174]: (*a*) view of the liquidus surface in the region of pseudobinary equilibrium TiC_y–TiB_2 (Ti_2C, Ti_3C_2, Ti_6C_5 are ordered phases of nonstoichiometric titanium carbide TiC_y), (*b*) view of the liquidus surface in the region of pseudobinary equilibrium B_4C_y–TiB_2

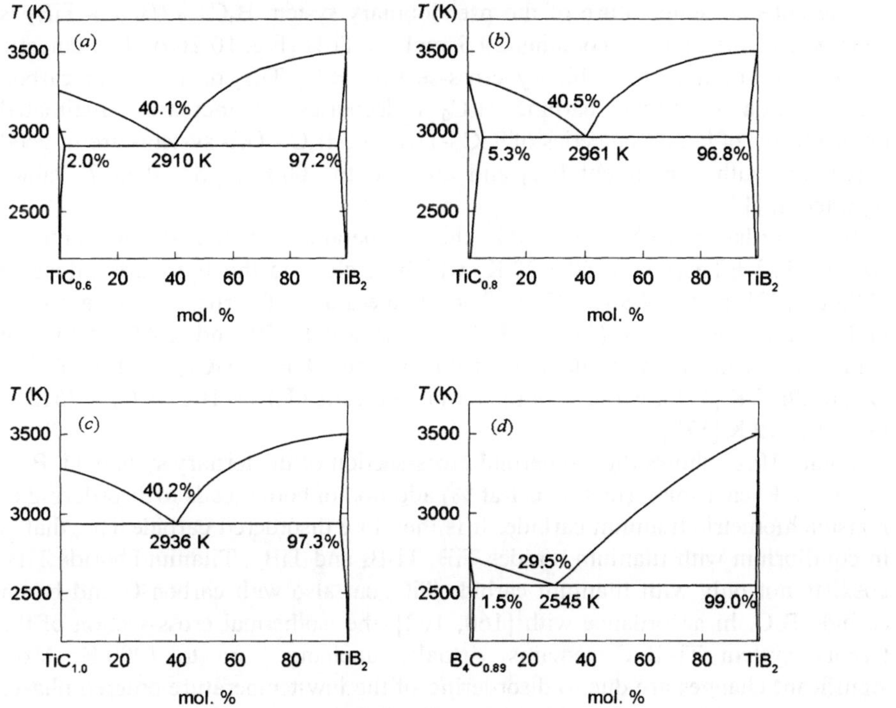

Fig. 10.26. Polythermal pseudobinary cross-sections of the ternary system Ti–B–C [161, 162]: (*a*) $TiC_{0.6}$–TiB_2, (*b*) $TiC_{0.8}$–TiB_2, (*c*) $TiC_{1.0}$–TiB_2, (*d*) $B_4C_{0.89}$ ($B_{4.5}C$)–TiB_2

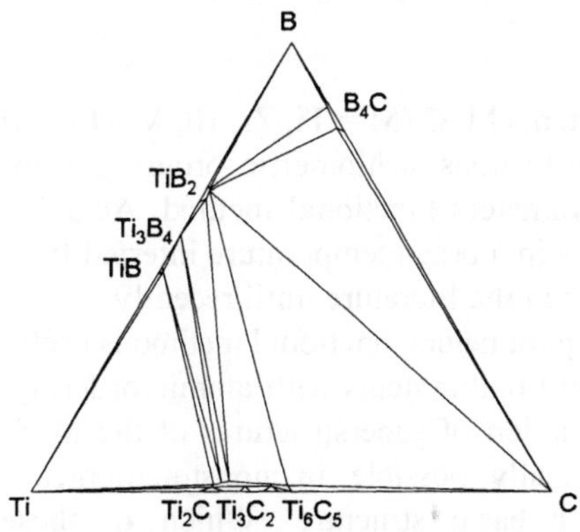

Fig. 10.27. Isothermal cross-section of the ternary system Ti–B–C at 300 K [161, 162]: ordered phases Ti_2C, Ti_3C_2 and Ti_6C_5 of nonstoichiometric carbide TiC_y survive up to 0.8 at.% boron dissolved in these phases. More boron leads to disordering

The eutectic temperature of the pseudobinary system $B_4C_{0.89}$ ($B_{4.5}C$)–TiB_2 is 2545 K and the eutectic contains 29.5 mol. % TiB_2 (Fig. 10.26*d*). The eutectic temperature of the pseudobinary cross-section B_4C_y–TiB_2 drops if the carbon concentration of boron carbide $B_4C_{0.89}$ decreases or increases. Estimated temperatures of binary eutectics of TiC_y–TiB_2 and B_4C_y–TiB_2 systems are in good agreement with experiment [92] and are 100 to 200 K higher than T_e values reported in [175].

In accordance with [159–161], the temperature of the ternary eutectic (β–Ti + TiB + TiC_y) is 1720 ± 20 K, which is close to the literature values of 1783 K [92] and 1808 K [176]. The temperature of ternary field eutectics (TiB_2 + $TiC_{1.0}$ + C) and (TiB_2 + B_4C_y + C) is 2650 ± 50 and 2380 ± 50 K. In compliance with literature data, T_e of the eutectic (TiB_2 + $TiC_{1.0}$ + C) is 2673 K [92] or 2667 K [176], while T_e of the ternary eutectic (TiB_2 + B_4C_y + C) is 2513 K [92] or 2519 K [176].

Figure 10.27 shows the isothermal cross-section of the ternary system Ti–B–C at 300 K. Even a small (less than 1 at.%) addition of boron leads to disordering of nonstoichiometric titanium carbide. It is therefore disordered carbide TiC_y that is in equilibrium with titanium borides TiB, Ti_3B_4 and TiB_2. Titanium boride TiB_2 coexists not only with titanium carbide TiC_y but also with carbon C and boron carbide B_4C. In accordance with [160, 162], the isothermal cross-section of the ternary system Ti–B–C remains virtually unchanged up to 1900 K. Most significant changes are due to disordering of the low-temperature ordered phases Ti_2C, Ti_3C_2 and Ti_6C_5 of carbide TiC_y at $T > 950$ K.

Calculated [160–162] phase equilibria in the ternary system Ti–B–C fit well experimental data summarized in [176–178] and confirm the applicability of the subregular solution model for calculation and construction of phase diagrams of binary, pseudobinary and ternary systems comprising both stoichiometric and nonstoichiometric compounds.

Calculated phase diagrams of binary systems M–C (M = Ti, Zr, Hf, V, Nb, Ta) and Ti–N [10–16,43,102,131] with strongly nonstoichiometric ordering compounds is the main result of the order parameters functional method. Actually, equilibrium phase diagrams of these systems in a broad temperature interval from 300 K to the melting point were unavailable in the literature until recently.

The obtained results show that the order-parameters functional method is useful for calculating phase diagrams in model and real systems with atomic ordering. The theoretical analysis suggests that formation of superstructures of the M_2X, M_3X_2 and M_6X_5 types is thermodynamically possible in nonstoichiometric carbides and nitrides MX_y with the *B*1 basic structure. Which of these superstructures will actually form upon ordering of a particular nonstoichiometric compound depends on the width of the homogeneity interval of the compound MX_y and on phases in equilibrium with the compound MX_y. Nonstoichiometric cubic carbides or nitrides MX_y cannot have superstructures of the M_4X_3 or M_8X_7 type. An exception is carbide VC_y where the ordered phase V_8C_7 appears due to the edge effect: In disordered vanadium carbide the upper homogeneity boundary corresponds to carbide MC_y with $y = 0.875$ rather than $y = 1.00$ as in other nonstoichiometric interstitial compounds MX_y. Formation of the V_8C_7 phase was used recently [180] to produce vanadium carbide in a nanocrystalline state. Vanadium carbide grains were refined thanks to a stepwise increase in the unit cell volume upon the disorder–order transition VC_y–V_8C_7.

Nonstoichiometric hafnium, titanium and zirconium carbides HfC_y, TiC_y and ZrC_y have the lowest temperatures of transition to the ordered state. This means that ordered carbides $M^{(IV)}C_y$ of group IV transition metals are more difficult to produce than vanadium, niobium and tantalum carbides, because annealing temperatures sufficient for ordering of carbides $M^{(IV)}C_y$ are 300 to 700 K lower than for carbides $M^{(V)}C_y$ with the same carbon concentration y. Consequently, diffusion mobility of carbon atoms in carbides $M^{(IV)}C_y$ will be much lower than in carbides $M^{(V)}C_y$. The atomic weight of metal atoms is a critical factor affecting diffusion mobility of interstitial atoms, since the diffusion rate of nonmetal atoms decreases with increasing atomic weight of the metal. The mutual account of transition temperatures and atomic weight of the metal makes one think that hafnium carbide HfC_y is the most challenging material for ordering among nonstoichiometric carbides: its annealing time at a temperature of 750 K or lower must run into thousands of hours. Also, ordering of hafnium carbide is very difficult to track instrumentally. Great effort is needed to reach ordering in tantalum and zirconium carbides TaC_y and ZrC_y.

Generally, fundamentally correct estimates of the main thermodynamic characteristics of order–disorder and order–order phase transitions in the nonstoichiometric compounds under discussion confirm the soundness of the basic concepts of the order-parameters functional method and its approximations. The theoretical analysis and calculations suggest the applicability of this method to a qualitative and quantitative description of ordering in substitutional and interstitial solid solutions and nonstoichiometric compounds.

However, the following should be observed. Researchers [1, 2, 10, 11, 15, 16, 179] used a modification of the OPF method, which takes into account only long-range order parameters and correlations due to these parameters (or superstructural short-range order [181, 182]). But ordered phases have, in addition to long-range correlations, short-range correlations that do not vanish at the order–disorder transition temperature but survive in the disordered phase within a certain temperature interval above T_{trans} and decrease smoothly with increasing temperature. Clearly, phase boundaries in calculated phase diagrams can be located more accurately if short-range order in ordered and disordered phases is fully taken into account. Thus, the order-parameters functional method can be improved if it allows for short-range order effects.

Phase diagrams of pseudobinary carbide systems, which were calculated using the subregular solution model, are in good agreement with available experimental data and suggest the appropriateness of this model for calculating phase diagrams of other pseudobinary systems with complete or partial atomic substitution.

References

1 A. I. Gusev: *Physical Chemistry of Nonstoichiometric Refractory Compounds* (Nauka, Moscow 1991) 286 pp. (in Russian)

2 A. I. Gusev, A. A. Rempel: *Structural Phase Transitions in Nonstoichiometric Compounds* (Nauka, Moscow 1988) 308 pp. (in Russian)

3 A. I. Gusev, A. A. Rempel: Order parameter functional method in the theory of atomic ordering, Phys. Stat. Sol. (b) **131**, 43–51 (1985)

4 A. I. Gusev, A. A. Rempel: Calculating the energy parameters for CV and OPF methods, Phys. Stat. Sol. (b) **140**, 335–346 (1987)

5 A. I. Gusev: Atomic ordering and the order parameter functional method, Philosoph. Mag. B **60**, 307–324 (1989)

6 A. I. Gusev, A. A. Rempel: Thermodynamic model of atomic ordering. 1. Basic equations, Zh. Fiz. Khimii **60**, 1349–1352 (1986) (in Russian). (Engl. transl.: Russ. J. Phys. Chem. **60**, 808–810 (1986))

7 A. A. Rempel, A. I. Gusev: Thermodynamic model of atomic ordering. 2. The order-disorder structural phase transition in nonstoichiometric niobium carbide, Zh. Fiz. Khimii **60**, 1353–1357 (1986) (in Russian). (Engl. transl.: Russ. J. Phys. Chem. **60**, 810–813 (1986))

8 A. I. Gusev, A. A. Rempel, V. N. Lipatnikov: Thermodynamic model of atomic ordering. 3. Calculation of cluster energies, Zh. Fiz. Khimii **61**, 916–921 (1987) (in Russian). (Engl. transl.: Russ. J. Phys. Chem. **61**, 476–479 (1987))

9 A. A. Rempel: *Effects of Ordering in Nonstoichiometric Interstitial Compounds* (Nauka, Ekaterinburg 1992) 232 pp. (in Russian)

10 A. I. Gusev: Phase equilibria and phase diagrams of systems with atomic ordering, Doklady Akad. Nauk SSSR **313**, 887–893 (1990) (in Russian). (Engl. transl.: Doklady Phys. Chem. **313**, 615–620 (1990))

11 A. I. Gusev: Phase diagrams for ordering systems in order-parameter functional method, Fiz. Tverd. Tela **32**, 2752–2760 (1990) (in Russian). (Engl. transl.: Sov. Physics – Solid State **32**, 1595–1599 (1990))

12 A. I. Gusev, A. A. Rempel: Thermodynamic model of atomic ordering. Phase diagrams of ordered systems, Zh. Fiz. Khimii **65**, 625–633 (1991) (in Russian). (Engl. transl.: Russ. J. Phys. Chem. **65**, 330–334 (1991))

13 A. I. Gusev: Phase diagrams of ordered nonstoichiometric hafnium carbide and titanium nitride, Doklady Akademii Nauk **322**, 918–923 (1992) (in Russian). (Engl. transl.: Doklady Phys. Chem. **322**, 84–88 (1992))

14 A. I. Gusev, A. A. Rempel: Calculation of phase diagrams of interstitial compounds, J. Phys. Chem. Solids **55**, 299–304 (1994)

15 A. I. Gusev, A. A. Rempel: Phase diagrams of metal–carbon and metal–nitrogen systems and ordering in strongly nonstoichiometric carbides and nitrides, Phys. Stat. Sol. (a) **163**, 273–304 (1997)

16 A. I. Gusev, A. A. Rempel: Atomic ordering and phase equilibria in strongly nonstoichiometric carbides and nitrides, in *Materials Science of Carbides, Nitrides and Borides*, Y. G. Gogotsi, R. A. Andrievski (Eds.) (Kluwer Academic Publishers, Dordrecht 1999) pp. 47–64

17 A. I. Gusev, A. A. Rempel: OPF-method for calculation phase diagrams of hard materials, in *Proc. Intern. Conf. on Advances in Hard Materials Production*, Bonn,

Germany, May 4-6, 1992 (MPR Publ. Serv. Ltd, Shrewsbury, England 1992) pp. 39-1–39-11

18 T. Priem: *Etude de l'ordre à courte distance dans les carbures et nitrures non-stoechiometriques de metaux de transition par diffusion diffuse de neutrons*, Rapport CEA-R-5499 (Commissariat à l'Energie Atomique, Centre d'Etudes Nucléaires de Saclay, Gif-sur-Yvette (France) 1989) 162 pp.

19 T. Priem, B. Beuneu, C. H. de Novion, A. Finel, F. Livet: Short-range order, atomic displacements and effective interatomic ordering energies in $TiN_{0.82}$, J. Physique (France). **50**, 2217–2242 (1989)

20 C. H. de Novion, B. Beuneu, T. Priem, N. Lorenzelli, A. Finel: Defect structures and order-disorder transformations in transition metal carbides and nitrides, in *The Physics and Chemistry of Carbides, Nitrides and Borides*, R. Freer (Ed.), (Kluwer Acad. Publ., Netherlands 1990) pp. 329–355

21 D. Raabe: *Computational Materials Science: The Simulation of Materials, Microstructures and Properties* (Wiley – VCH, Weinheim 1998) 379 pp.

22 A. A. Rempel: Atomic and vacancy ordering in nonstoichiometric carbides, Uspekhi Fiz. Nauk **166**, 33-62 (1996) (in Russian). (Engl. transl.: Physics – Uspekhi **39**, 31–56 (1996))

23 A. I. Gusev: Short-range order in nonstoichiometric transition metal carbides, nitrides and oxides, Phys. Stat. Sol. (b) **156**, 11–40 (1989)

24 A. I. Gusev: Disorder and long-range order in nonstoichiometric interstitial compounds: transition metal carbides, nitrides and oxides, Phys. Stat. Sol. (b) **163**, 17–54 (1991)

25 A. I. Gusev, A. A. Rempel: Superstructures of non-stoichiometric interstitial compounds and the distribution functions of interstitial atoms, Phys. Stat. Sol. (a) **135**, 15–58 (1993)

26 V. N. Lipatnikov, A. I. Gusev, P. Ettmayer, W. Lengauer: Phase transformation in non-stoichiometric vanadium carbide, J. Phys.: Condens. Matter **11**, 163–184 (1999)

27 V. N. Lipatnikov, A. I. Gusev, P. Ettmaier, W. Lengauer: Order-disorder phase transformations and specific heat of nonstoichiometric vanadium carbide, Fiz. Tverd. Tela **41**, 529–536 (1999) (in Russian). (Engl. transl.: Physics of the Solid State **41**, 474–480 (1999))

28 E. Storms: *The Refractory Carbides* (Academic Press, New York and London 1967) 300 pp.

29 L. Toth: *Transition Metal Carbides and Nitrides* (Academic Press, New York and London 1971) 280 pp.

30 A. S. Bolgar, A. G. Turchanin, V. V. Fesenko: *Thermodynamic Properties of Carbides* (Naukova Dumka, Kiev 1973) 272 pp. (in Russian)

31 A. S. Bolgar, V. F. Litvinenko: *Thermodynamic Properties of Nitrides* (Naukova Dumka, Kiev 1980) 282 pp. (in Russian)

32 A. N. Kornilov, N. V. Chelovskaya, V. I. Zhelankin, G. P. Shveikin: Enthalpies of formation of hafnium carbides, J. Chem. Thermodyn. **9**, 629–642 (1977)

33 S. Z. Nazarova, A. I. Gusev: *Stoichiometric Solid Solutions of Niobium and Tantalum Carbides and Their Properties* (Ural Scientific Centre, Sverdlovsk 1987) 63 pp. (in Russian)

34 I. Barin: *Thermochemical Data of Pure Substance*, 3rd ed. (VCH, Weinheim 1995) 1880 pp.

35 N. M. Volkova, P. V. Geld: Temperature dependencies of enthalpy of vanadium carbide, Trudy Inst. Khimii Ural Fil. Akad. Nauk SSSR **14**, 41–46 (1967) (in Russian)
36 A. G. Turchanin, E. A. Guseva, V. V. Morozov, A. S. Bolgar, V.V. Fesenko: Enthalpy and heat capacity of nonstoichiometric vanadium monocarbides at high temperatures, Izv. AN SSSR. Neorgan. Materialy **10**, 1619–1622 (1974) (in Russian)
37 V. N. Lipatnikov, W. Lengauer, P. Ettmayer, E. Keil, G. Groboth, E. Kny: Effects of vacancy ordering on structure and properties of vanadium carbide, J. Alloys Comp. **261**, 192–197 (1997)
38 C. E. Wicks, F. E. Block: *Thermodynamic Properties of 65 Elements, Their Oxides, Halides, Carbides and Nitrides* (US Bureau of Mines, US Government Printing Office, Washington 1963) 240 pp.
39 R. F. Voitovich: *Refractory Compounds: Thermodynamic Characteristics* / Handbook (Naukova Dumka, Kiev 1971) 220 pp. (in Russian)
40 V. P. Glushko (Ed.): *Thermodynamic Properties of Individual Substances* / Handbook in 4 volumes (Nauka, Moscow) **II**, book 2 (1979) 340 pp.; **IV**, book 2 (1982) 560 pp. (in Russian)
41 H. A. Wriedt, J. L. Murray: The nitrogen–titanium system, Bull. Alloy Phase Diagrams **8**, 378–388 (1987)
42 H. Othani, M. Hillert: Thermodynamic assesment of the Ti–N system, CALPHAD: Comp. Coupl. Phase Diagrams and Thermochem. **14**, 289–306 (1990)
43 Gusev A.I., Rempel A.A. Phase diagrams for Ti–C and Ti–N systems and atomic ordering of nonstoichiometric titanium carbide and nitride, Doklady Akademii Nauk **332**, 717–721 (1993) (in Russian). (Engl. transl.: Doklady Chemistry **332**, 456–461 (1993))
44 J. L. Murray: The C–Ti system, in *Phase diagrams of Binary Titanium Alloys*, J. L. Murray (Ed.) (ASM Intern. Publ., Metals Park, Ohio 1987) pp. 47–51
45 M. Enomoto: The C–Ti–V system (carbon-titanium-vanadium), J. Phase Equilibria **17**, 237–247 (1996)
46 S. Jonsson: Phase relations in quaternary hard materials, Ph.D. Thesis, Royal Institute of Technology, Stockholm (Sweden) (1993) 18 pp.
47 H. J. Seifert, H. L. Lukas, G. Petzow: Thermodynamic optimization of the Ti–C system, J. Phase Equilibria **17**, 24–35 (1996)
48 H. Goretzki: Neutron diffraction studies of titanium-carbon and zirconium-carbon alloys, Phys. Stat. Sol. **20**, K141–K143 (1967)
49 I. Karimov, V. T. Em, I. Khidirov, I. S. Latergaus: Neutron diffraction investigation of ordering in titanium and zirconium carbides, Izv. AN Uzb. SSR. Seriya Fiz. Mat. Nauk No 4, 81–83 (1979) (in Russian)
50 M. P. Arbuzov, B. V. Khaenko, E. T. Kachkovskaya, S. Ya. Golub: X-ray study of titanium carbide in homogeneity region, Ukrain. Fiz. Zh. **19**, 497–501 (1984) (in Russian)
51 V.T. Em, I. A. Karimov, V. F. Petrunin, I. Khidirov, I. S. Latergaus, A. G. Merzhanov, I. P. Borovinskaya, V. K. Prokudina: Neutron diffraction study of ordering in titanium carbides, Kristallografiya **20**, 320–323 (1975) (in Russian)
52 A. Sh. Remeev, I. Karimov: Neutron diffraction study of kinetics of structural phase transition, Izv. AN Uzb. SSR. Seriya Fiz. Mat. Nauk No 2, 87–88 (1986) (in Russian)
53 N. Lorenzelli, R. Caudron, J. P. Landesman, C. H. de Novion: Influence of the ordering of carbon vacancies on the electronic properties of $TiC_{0.625}$, Solid State Commun. **59**, 765–769 (1986)

54 B. V. Khaenko, V. V. Kukol': Real structure of ordering of titanium carbide, Kristallografiya **34**, 1513–1517 (1989) (in Russian)
55 V. Moisy-Maurice: *Structure atomique des carbures non-stoechiometriques de metaux de transition*, Rapport CEA-R-5127 (Commissariat à l'Energie Atomique, Gif-sur-Yvette (France) 1981) 184 pp.
56 B. V. Khaenko, V. V. Kukol', O. A. Gnatetskii, S. V. Sirichenko: Structures of ordering of monocarbides of IV group transition metals, Poroshkovaya Metallurgiya No 6, 63–68 (1990) (in Russian)
57 V. Moisy-Maurice, N. Lorenzelli, C. H. de Novion, P. Convert: High temperature neutron diffraction study of the order-disorder transition in TiC_{1-x}, Acta Metall. **30**, 1769–1779 (1982)
58 V. N. Lipatnikov, A. A. Rempel, A. I. Gusev: Atomic ordering and hardness of nonstoichiometric titanium carbide, Int. J. Refract. Met. Hard Mater. **15**, 61–64 (1997)
59 V. Moisy-Maurice, C. H. de Novion, A. N. Christensen, W. Just: Elastic diffuse neutron scattering study of the defect structure of $TiC_{0.76}$ and $NbC_{0.73}$, Solid State Commun. **39**, 661–665 (1981)
60 J. P. Landesman, G. Treglia, P. Turchi, F. Ducastelle: Electronic structure and pairwise interactions in substoichiometric transition metal carbides and nitrides, J. Physique (France) **46**, 1001–1015 (1985)
61 T. Priem, B. Beuneu, C. H. de Novion, J. Chevrier, F. Livet, A. Finel, S. Lefevbre: Short-range order and static displacements in non-stoichiometric transition metal carbides and nitrides, Physica B **156–157**, 47–49 (1989)
62 P. C. Clapp, S. C. Moss: Correlation functions of disordered binary alloys, Phys. Rev. **142**, 418–427 (1966)
63 R. Fletcher, M. J. D. Powell: A rapidly convergent descent method for minimixation, The Computer J. **6**, 163–168 (1963)
64 F. Livet: Linearized inverse Monte Carlo method for applied to Ni_3Fe, Acta Metall. **35**, 2915–2919 (1987)
65 M. Kurata, R. Kikuchi, T. Watari: Theory of cooperative phenomena. Detailed discussions of the cluster variation method, J. Chem. Phys. **21**, 434–448 (1953)
66 J. Hijmans, J. de Boer: Approximation method for order-disorder problem, Physica **21**, 471-516 (1955)
67 R. Kikuchi: The cluster-variation method, J. Physique Colloq. (France) **38**, C7-307–C7-313 (1977)
68 A. Finel: Contribution à l'étude des effets d'ordre dans le cadre du modèle d'Ising: états de base et diagrammes de phase, Ph.D. Thesis, Université Pierre et Marie Curie, Paris (France) (1987)
69 N. V. Dzhalabadze, B. G. Eristavi, N. I. Maisuradze, E. R. Kuteliya: Structure of crystalline phases in Ti–C thin films, Fiz. Metall. Metalloved. **86**, 85–92 (1998) (in Russian). (Engl. transl.: Phys. Met. Metallogr. **86**, 59–64 (1998))
70 W. Lengauer, S. Binder, K. Aigner, P. Ettmayer, A. Guillou, J. Debuigne: Solid state properties of group IVb carbonitrides, J. Alloys Comp. **217**, 137–147 (1995)
71 V. N. Lipatnikov, A. Kottar, L. V. Zueva, A. I. Gusev: Disorder–order phase transformations and electrical resistivity of a nonstoichiometric titanium carbide, Fiz. Tverd. Tela **40**, 1332–1340 (1998) (in Russian). (Engl. transl.: Physics of the Solid State **40**, 1211–1218 (1998))
72 A. N. Emel'yanov: Features of order–disorder phase transition in nonstoichiometric carbides of transition metals, Fiz. Tverd. Tela **38**, 3678–3682 (1996) (in Russian)

73 V. N. Lipatnikov, A. I. Gusev: Effect of ordering on the structure and specific heat of nonstoichiometric titanium carbide, Pis'ma v ZhETF **69**, 631–637 (1999) (in Russian). (Engl. transl.: JETP Letters **69**, 669–675 (1999))

74 V. N. Lipatnikov, A. I. Gusev: High-temperature heat capacity and order–disorder phase transformations in nonstoichiometric titanium carbide, Phys. Stat. Sol. (b) **212**, R11–R12 (1999)

75 A. N. Emel'yanov: Temperature conductivity of nonstoichiometric titanium carbide in the region of order–disorder phase transition, Teplofizika Vysok. Temp. **28**, 269–278 (1990) (in Russian)

76 A. V. Karpov, V.P. Kobyakov, E.A. Chernomorskaya: Dilatation of nonstoichiometric titanium carbide in the region of order-disorder phase transition, Neorgan. Materialy **31**, 655–659 (1995) (in Russian)

77 A. V. Karpov, V. P. Kobyakov: Order–disorder phase transition in $TiC_{0.55}$, Teplofizika Vysok. Temp. **34**, 965–968 (1996) (in Russian). (Engl. transl.: High Temp. **34**, 950–953 (1996))

78 B. C. Guo, K. P. Kerns, A. W. Castleman: Ti_8C_{12}-metallo-carbohedrenes: a new class of molecular clusters?, Science **255**, 1411–1413 (1992)

79 A. V. Eletskii, B. M. Smirnov: Fullerenes and carbon structures, Uspekhi Fiz. Nauk **165**, 977–1009 (1995) (in Russian)

80 B. C. Guo, S. Wei, J. Purnell, S. Buzza, A. W. Castleman: Metallo-carbohedrenes [M_8C_{12} (M = V, Zr, Hf, and Ti)]: a class of stable molecular cluster ions, Science **256**, 515–516 (1992)

81 S. Wei, B. C. Guo, J. Purnell, S. Buzza, A. W. Castleman: Metallocarbohedrenes as a class of stable neutral clusters: formation mechanism of M_8C_{12} (M = Ti and V), J. Phys. Chem. **96**, 4166–4168 (1992)

82 J. S. Pilgrim, M.A.Duncan: Metallo-carbohedrenes: chromium, iron, and molybdenum analogues, J. Amer. Chem. Soc. **115**, 6958–6961 (1993)

83 S. Wei, B. C. Guo, H. T. Deng, K. Kerns, J. Purnell, S. Buzza, A. W. Castleman: Formation of metcars and face-centered cubic structures: thermodynamically or kinetically controlled?, J. Amer. Chem. Soc. **116**, 4475–4476 (1994)

84 M. Methfessel, M. van Schilfgaarde, S. Scheffler: Electronic structure and bonding in the metallocarbohedrene Ti_8C_{12}, Phys. Rev. Lett. **70**, 29–32 (1993)

85 B. V. Reddy, S. N. Khanna, P. Jena: Electronic, magnetic, and geometric structure of metallo-carbohedrenes, Science **258**, 1640–1643 (1992)

86 S. N. Khanna, B. V. Reddy: Geometry, stability and properties of metallo-carbohedrenes, Comput. Mater. Sci. **2**, 638–642 (1994)

87 J. H. Weaver, J. L. Martins, T. Komeda, Y. Chen, T. R. Ono, G. H. Kroll, N. Troullier, R. E. Haufler, R. E. Smalley: Electronic structure of solid sixty-atom carbon (C_{60}): experiment and theory, Phys. Rev. Lett. **66**, 1741–1744 (1991)

88 S. Saito, A. Oshiyama: Cohesive mechanism and energy bands of a solid of carbon sixty-atom molecules, Phys. Rev. Lett. **66**, 2637–2640 (1991)

89 B. V. Khaenho: Order in cubic carbides and nitrides of transition metals of groups IV and V, Izv. AN SSSR. Neorgan. Materialy **15**, 1952–1960 (1979) (in Russian). (Engl. transl.: Inorganic Materials **15**, 1535–1543 (1979))

90 N. Obata, N. Nakasawa: Superlattice formation in zirconium-carbon system, J. Nucl. Mater. **60**, 39–42 (1976)

91 R. Lorenzelli, I. de Dieuleveult: Existence d'une surstructure dans le carbure de thorium sous-stoechiometrique: $ThC_{0.76}$, J. Nucl. Mater. **29**, 349–353 (1969)

92 E. Rudy: *Compendium of Phase Diagram Data*, Final Tech. Report AFML TR-65-2. Part V (Metals and Ceramics Division, Air Force Materials Laboratory, Wright-Patterson Air Force Base (Ohio) 1969) 735 pp.

93 A. I. Gusev: Structural stability boundaries for nonstoichiometric compounds, Phys. Stat. Sol. (a) **111**, 443–450 (1989)

94 A. N. Zyryanova, S. Z. Nazarova, A. I. Gusev: Magnetic susceptibility anomaly as evidence for ordering of nonstoichiometric hafnium carbide HfC_y, Doklady Akad. Nauk **359**, 348–353 (1998) (in Russian). (Engl. transl.: Doklady Phys. Chem. **359**, 91–96 (1998))

95 A. N. Zyryanova, A. I. Gusev: Magnetic susceptibility and ordering in nonstoichiometric hafnium carbide, Zh. Fiz. Khimii **72**, 2234–2242 (1998) (in Russian). (Engl. transl.: Russ. J. Phys. Chem. **72**, 3034–2041 (1998))

96 A. I. Gusev, A. N. Zyryanova: Atomic-vacancy ordering and magnetic susceptibility of nonstoichiometric hafnium carbide, Pis'ma v ZhETF **69**, 2096–3001 (1999) (in Russian). (Engl. transl.: JETP Letters **69**, 324–329 (1999))

97 J. Billingham, P. S. Bell, M. H. Lewis: Superlattice with monoclinic symmetry based on compounds V_6C_5, Philosoph. Mag. **25**, 661–671 (1972)

98 T. Athanassiadis, N. Lorenzelli, C. H. de Novion: Diffraction studies of the order-disorder transformation in V_8C_7, Ann. Chim. France **12**, 129–142 (1987)

99 O. N. Carlson, A. N. Ghaneya, J. F. Smith: The vanadium–carbon system, Bull. Alloy Phase Diagrams **6**, 115–124 (1985)

100 W. Huang: An assessment of the V–C system, Z. Metallkunde **82**, 174–181 (1991)

101 A. I. Gusev: Phase diagram of V–C mixtures, Doklady Akademii Nauk **365**, 644–648 (1999) (in Russian). (Engl. transl.: Doklady Phys. Chem. **365**, 107–111 (1999))

102 A. I. Gusev: A phase diagram of the vanadium–carbon system taking into account ordering in nonstoichiometric vanadium carbide, Zh. Fiz. Khimii **74**, 600–606 (2000) (in Russian). (Engl. transl.: Russ. J. Phys. Chem. **74**, 510–516 (2000))

103 M. P. Arbuzov, B. V. Khaenko, V. G. Fak, Yu. F. Nosachev: Study of ordering process in alloys on a base of vanadium monocarbide, Ukrain. Fiz. Zh. **22**, 291–297 (1977) (in Russian)

104 N. M. Volkova, P. V. Geld, S. I. Alyamovskii: Phase transformation of highest vanadium carbide, Zh. Neorgan. Khimii **10**, 1758–1759 (1965) (in Russian). (Engl. transl.: Russ. J. Inorg. Chem. **10**, 960–961 (1965))

105 G. H. Emmons, W. S. Williams: Thermodynamics of order-disorder transformations in vanadium carbide, J. Mater. Science **18**, 2589–2602 (1983)

106 L. W. Shacklette, W. S. Williams: Influence of order-disorder transformations on the electrical resistivity of vanadium carbide, Phys. Rev. B **7**, 5041–5053 (1973)

107 K. Hiraga: Vacancy ordering in vanadium carbides based on V_6C_5, Philosoph. Mag. **27**, 1301–1311 (1973)

108 H. Wiesenberger, W. Lengauer, P. Ettmayer: Reactive diffusion and phase equilibria in the V–C, Nb–C, Ta–C and Ta–N systems, Acta Mater. **46**, 651–666 (1998)

109 K. Yvon, W. Rieger, H. Nowotny: Die Kristallstrukture V_2C, Monatsh. Chem. **97**, 689–694 (1966)

110 J. F. Smith: The C–V system, in *Phase Diagrams of Binary Vanadium Alloys*, J. F. Smith (Ed.) (ASM Intern. Publication, Materials Park (Ohio) 1989) pp. 33–42

111 J. F. Smith, O. N. Carlson, R. R. de Avillez: The niobium–carbon system, J. Nucl. Mater. **148**, 1–16 (1987)

112 W. Huang: Thermodynamic evaluation of Nb–C system, Mater. Sci. Technol. **6**, 687–694 (1990)
113 E. K. Storms, B. Calcen, A. Yencha: The vaporization behavior of the defect carbides. Part I. The Nb–C system, High Temp. Science **1**, 430–455 (1969)
114 J. Billingham, P. S. Bell, M. H. Lewis: Vacancy short-range order in substoichiometric transition metal carbides and nitrides with the NaCl structure. Electron diffraction studies of short-range ordered compounds, Acta Crystallogr. A **28**, 602–606 (1972)
115 J. P. Landesman, A. N. Christensen, C. H. de Novion, N. Lorenzelli, P. Convert: Order-disorder transition and structure of the ordered vacancy compound Nb_6C_5: powder neutron diffraction studies, J. Phys. C: Solid State Phys. **18**, 809–823 (1985)
116 A. I. Gusev, A. A. Rempel: Ordering in the carbon sublattice of nonstoichiometric niobium carbide, Fiz. Tverd. Tela **26**, 3622–3627 (1984) (in Russian). (Engl. transl.: Sov. Physics – Solid State **26**, 2178–2181 (1984))
117 A. A. Rempel, A. I. Gusev: Order-disorder phase transition in nonstoichiometric niobium carbide, Kristallografiya **30**, 1112–1115 (1985) (in Russian). (Engl. transl.: Sov. Physics – Crystallography **30**, 648–650 (1985))
118 A. I. Gusev, A. A. Rempel: Order-disorder phase transition channel in niobium carbide, Phys. Stat. Sol. (a) **93**, 71–80 (1986)
119 J. D. Venables, M. H. Meyerhoff: Ordering effect in NbC and TaC, in *Solid State Chemistry*, Proc. of 5th Intern. Mater. Res. Symp. (NBS Special Publ. 364). (NBS Publ., Berkeley (California) 1972) pp. 583–590
120 M. H. Lewis, J. Billingham, P. S. Bell: Non-stoichiometry in ceramic compounds, in *Solid State Chemistry*, Proc. of 5th Intern. Mater. Res. Symp. (NBS Special Publ. 364). (NBS Publ., Berkeley (California) 1972) pp. 1084–1114
121 A. A. Rempel, A. I. Gusev: *Ordering in Nonstoichiometric Niobium Monocarbide* (Urals Scientific Centre, Sverdlovsk 1983) 68 pp. (in Russian)
122 A. A. Rempel, A. I. Gusev, V. G. Zubkov, G. P. Shveikin: Structure of ordered niobium carbide Nb_6C_5, Doklady AN SSSR **275**, 883–887 (1984) (in Russian). (Engl. transl.: Sov. Physics Doklady **29**, 257–259 (1984))
123 A. N. Christensen: Vacancy order in Nb_6C_5, Acta Chem. Scand. A **39**, 803–804 (1985)
124 B. V. Khaenko, O. P. Sivak: Structure of order of niobium monocarbide, Kristallografiya **35**, 1110–1115 (1990) (in Russian)
125 L. C. Dy, W. S. Williams: Resistivity, superconductivity and order-disorder transformations in transition metal carbides and hydrogen-doped carbides, J. Appl. Phys. **53**, 8915–8927 (1982)
126 J.-P. Landesman: *Etude de la stabilité de phases ordonnées dans les carbures et nitrures sous-stoechiometriques de metaux de transition*, Rapport CEA-R-5342 (Commissariat à l'Energie Atomique, Centre d'Etudes Nucléaires de Fontenay-aux-Roses, Gif-sur-Yvette, 1986) 176 pp.
127 A. I. Gusev, A. A. Rempel: A study of the atomic ordering in the niobium carbide using the magnetic susceptibility method, Phys. Stat. Sol. (a) **84**, 527–534 (1984)
128 K. Yvon, E. Parthe: Crystal structure of the ζ-V, Nb and Ta carbides, Acta Crystallogr. B **26**, 149–152 (1970)
129 A. A. Rempel, V. N. Lipatnikov, A. I. Gusev: The superstructure in nonstoichiometric tantalum carbide, Doklady AN SSSR **310**, 878–882 (1990) (in Russian). (Engl. Transl.: Sov. Physics Doklady **35**, 103–106 (1990))

130 A. I. Gusev, A. A. Rempel, V. N. Lipatnikov: Incommensurate superlattice and superconductivity in tantalum carbide, Fiz. Tverd. Tela **33**, 2298–2305 (1991) (in Russian). (Engl. transl.: Sov. Physics – Solid State **33**, 1295–1299 (1991))

131 A. I. Gusev, A. A. Rempel, V. N. Lipatnikov: Incommensurate ordered phase in nonstoichiometric tantalum carbide, J. Phys.: Condens. Matter **8**, 8277–8293 (1996)

132 V. N. Lipatnikov, A. A. Rempel, A. I. Gusev: The influence of ordering of the vacancies on the magnetic susceptibility of tantalum carbide, Zh. Neorgan. Khimii **33**, 1860–1863 (1988) (in Russian). (Engl. transl.: Russ. J. Inorg. Chem. **33**, 1058–1059 (1988))

133 A. I. Gusev, A. A. Rempel, V. N. Lipatnikov: Magnetic susceptibility and atomic ordering in tantalum carbide, Phys. Stat. Sol. (a) **106**, 459–466 (1988)

134 A. I. Gusev, A. A. Rempel, V. N. Lipatnikov: Thermodynamic model of atomic ordering. 5. Specific heat of non-stoichiometric compounds in different structural states, Zh. Fiz. Khimii **64**, 2343–2348 (1990) (in Russian). (Engl. transl.: Russ. J. Phys. Chem. **64**, 1265–1267 (1990))

135 A. I. Gusev, A. A. Rempel, V. N. Lipatnikov: Heat capacity of niobium and tantalum carbides NbC_y and TaC_y in disordered and ordered states below 300 K, Phys. Stat. Sol. (b) **194**, 467–482 (1996)

136 H. A. Wriedt, J. L. Murray: The N–Ti (nitrogen–titanium) system, Bull. Alloy Phase Diagrams **8**, 378–388 (1987)

137 W. Lengauer, P. Ettmayer: Some aspects of the formation of ε-Ti_2N, Revue de Chimie Miner. **24**, 707–713 (1987)

138 W. Lengauer, P. Ettmayer: Investigations of phase equilibria in the Ti – N and Ti–Mo–N systems, Mater. Sci. Engineer. A **105–106**, 257–263 (1988)

139 W. Lengauer: The temperature gradient diffusion couple technique: an application of solid-solid phase reactions for phase diagram imaging, J. Solid State Chem. **91**, 279–285 (1991)

140 W. Lengauer: The titanium–nitrogen system: a study of phase reactions in the sub-nitride region by means of diffusion couples, Acta Metall. Mater. **39**, 2985–2995 (1991)

141 W. Lengauer, P. Ettmayer: Phase diagram imaging by means of temperature-gradient diffusion couples, J. Phase Equilibria **14**, 162–166 (1993)

142 C. Lobier, J. P. Marcon: Etude et structure d'une nouvelle phase du sous-nitrure de titane Ti_2N, Compt. Rend. Acad. Sci. Paris. Ser. C **268**, 1132–1135 (1969)

143 S. Nagakura, T. Kusunoki: Structure of titanium nitride TiN_x studied by electron diffraction and microscopy, J. Appl. Crystallogr. **10**, 52–56 (1977)

144 A. N. Christensen, A. Alamo, J. P. Landesman: Structure of vacancy-ordered titanium heminitride δ'-Ti_2N by powder neutron diffraction, Acta Crystallogr. C **41**, 1009–1011 (1985)

145 A. Alamo, C. H. de Novion: Ordered titanium heminitride Ti_2N, in *Proc. 7th Intern. Conf. Solid Compounds of the Transition Elements*, Grenoble, France, 1982, Paper II A1

146 E. Etchessahar, Sohn Yong-Un, M. Harmelin, J. Debuigne: The Ti–N system: kinetic, calorimetric, structural and metallurgical investigations of the δ'-$TiN_{0.51}$ phase, J. Less-Common Met. **167**, 261–281 (1991)

147 W. Lengauer, P. Ettmayer: Thermal decomposition of ε-Ti_2N and δ'-$TiN_{0.50}$ investigated by high-temperature x-ray diffraction, High Temp. – High Pressure **19**, 673–676 (1987)

148 W. Lengauer, P. Ettmayer: Recent advances in the field of transition-metal refractory nitrides, High Temp. – High Pressure **22**, 13-24 (1990)
149 W. Hume-Rothery, G. V. Raynor: *The Structure of Metals and Alloys* (Inst. of Metals, London 1955) 460 pp.
150 A. I. Gusev, G. P. Shveikin: Formation of solid solutions by isomorphic carbides, nitrides, oxides, and borides of transition metals of group IV and V, Izv. AN SSSR. Neorgan. Materialy **10**, 2144–2147 (1974) (in Russian). (Engl. transl.: Inorganic Materials **10**, 1840–1843 (1974))
151 A.G. Knapton: Niobium and tantalum alloys, J. Less-Common Met. **2**, 113–124 (1960)
152 A. I. Gusev, G. P. Shveikin: Energy of elastic lattice deformation in the formation of solid solutions of transition metal carbides and nitrides, Izv. AN SSSR. Neorgan. Materialy **12**, 1565–1568 (1976) (in Russian). (Engl transl.: Inorganic Materials **12**, 1283–1286 (1976))
153 R. Kieffer, H. Nowotny, A. Neckel, P. Ettmayer, L. Usner: Zur Entmischung von kubischen Mehrstoffkarbiden, Monatsh. Chem. **99**, 1020–1027 (1968)
154 A. I. Gusev: Influence of structural defects on the character of the mutual solubility of high-melting compounds of variable composition, Izv. AN SSSR. Neorgan. Materialy **19**, 1319–1324 (1983) (in Russian). (Engl. transl.: Inorganic Materials **19**, 1183–1188 (1983))
155 A. I. Gusev: Calculation of phase diagrams of pseudobinary systems based on high-melting carbides of titanium, zirconium, hafnium, and vanadium, Izv. AN SSSR. Neorgan. Materialy **20**, 1132–1137 (1984) (in Russian).
156 A. I. Gusev: Phase diagrams of the pseudobinary TiC–NbC, TiC–TaC, ZrC–NbC, ZrC–TaC, and HfC–TaC carbide systems, Zh. Fiz. Khimii **59**, 579–584 (1985) (in Russian). (Engl. transl.: Russ. J. Phys. Chem. **59**, 336–340 (1985))
157 A. I. Gusev: Prediction and calculation of phase diagrams of pseudobinary systems on a base of refractory compounds of transition metals, in *Calculations and Experimental Methods of Constructing of Phase Diagrams* (Nauka, Moscow 1985) pp. 42–47 (in Russian)
158 A. I. Gusev: Formation of solid solutions by transition-metal carbides and nitrides and calculation of their phase diagrams, in *Vortrage* **3**, VIII Intern. Pulvermet. Tagung. (Zentralinstitut Festkörperphysik und Werkstofforschung, Dresden (GDR) 1985) pp. 169–179
159 L. Kaufman, H. Bernstein: *Computer Calculation of Phase Diagrams* (Academic Press, New York and London 1970) 334 pp.
160 A. I. Gusev: Phase diagram of the Ti–B–C system in the temperature range 300–3500 K, Doklady Akademii Nauk **350**, 209–212 (1996) (in Russian). (Engl. transl.: Doklady Phys. Chem. **350**, 230–233 (1996))
161 A. I. Gusev: Phase equilibria in the ternary Ti–B–C system, Zh. Fiz. Khimii **71**, 1177–1181 (1997) (in Russian). (Engl. transl.: Russ. J. Phys. Chem. **71**, 1049–1053 (1997))
162 A. I. Gusev: Phase equilibria in the ternary system titanium–boron–carbon: the sections TiC_y–TiB_2 and B_4C_y–TiB_2 , J. Solid State Chem. **133**, 205–210 (1997)
163 S. V. Rempel, A. I. Gusev: Decomposition of solid solutions in ZrC–NbC system, Doklady Akademii Nauk **367**, 85–89 (1999) (in Russian). (Engl. transl.: Doklady Phys. Chem. **367**, 187–191 (1999))
164 S. V. Rempel, A. A. Rempel, A. I. Gusev: Latent decomposition regions in the model of subregular solutions: the ZrC–NbC system, Zh. Fiz. Khimii **74**, 412–417 (2000) (in Russian). (Engl. transl.: Russ. J. Phys. Chem. **74**, 341–346 (2000))

165 A. I. Gusev, G. P. Shveikin: Thermodynamic calculation of the phase diagrams of the $VC_{0.88}$–NbC, $VC_{0.88}$–TaC, $VC_{0.88}$–HfC, NbC–TaC, and NbC–HfC systems, Zh. Fiz. Khimii **58**, 2163–2167 (1984) (in Russian) (Engl. transl.: Russ. J. Phys. Chem. **58**, 1322–1325 (1984))
166 R. Kieffer: Preparation and properties of interstitial compounds, J. Inst. Metals **97**, 164–171 (1969)
167 E. V. Fedorov, R. A. Andrievski: Thermodynamic stability of carbide solid solutions, Izv. AN SSSR. Neorgan. Materialy **15**, 454–457 (1979) (in Russian)
168 Ya. S. Umanski, A. S. Muller: Study of short-range order in high-melting compounds. Report I, Izv. Vyssh. Ucheb. Zaved. Chernaya Metallurgiya No 7, 130–133 (1969) (in Russian)
169 Ya. S. Umanski, A. S. Muller: Study of short-range order in high-melting compounds. Report II, Izv. Vyssh. Ucheb. Zaved. Chernaya Metallurgiya No 1, 128–131 (1970) (in Russian)
170 V. I. Iveronova, A. A. Katsnelson: *Short-Range Order in Solid Solutions* (Nauka, Moscow 1977) 256 pp. (in Russian)
171 A. K. Shurin, N. A. Razumova, G. P. Dmitrieva, E. L. Handros: Phase diagram of the Ni–ZrC–TaC system, Poroshkovaya Metallurgia No 1, 56–61 (1987) (in Russian)
172 A. K. Shurin, G. P. Dmitrieva, N. A. Razumova, E. L. Handros: Phase diagram of the Ni–VC–NbC system, Poroshkovaya Metallurgia No 8, 67–69 (1987) (in Russian)
173 G. D. Milova, A. I. Gusev: A method of calculating the solubility in cryolite, Rasplavy No 5, 79–83 (1991) (in Russian). (Engl. transl.: Melts **5**, 398–401 (1991))
174 A. I. Gusev: The ternary system titanium-boron-carbon: TiC_y–TiB_2 and B_4C–TiB_2 cross-sections, in *12th Intern. Symp. on Boron, Borides and Related Compounds*, P. Rogl (Ed.), Proc. of the ISBB'96, 25-30 August 1996, Baden, Austria (Universität Wien, Wien (Austria) 1996) p.23
175 S. S. Ordan'yan, V. I. Unrod, A. I. Avgustinnik: Reaction in the system TiC_x–TiB_2 , Poroshkovaya Metallurgia No 9, 40–43 (1975) (in Russian)
176 H. Duschanek, P. Rogl, H. L. Lukas: A critical assessment and thermodynamic calculation of the boron-carbon-titanium (B–C–Ti) ternary system, J. Phase Equilibria **16**, 46–60 (1995)
177 M. Schouler, M. Dicarroir, C. Bernard: Mise au point sur la topographie et les propriétés des systèmes metal-carbone-azote (M–C–N) et metal-carbone-bore (M–C–B), Rev. Intern. Hautes Temp. Refr. **20**, 261–311 (1983)
178 P. Villars, A. Prince, H. Okamoto: *Handbook of Ternary Alloy Phase Diagrams* **5** (ASM Publication, Metals Park (Ohio) 1995) pp. 5373–5379
179 A. I. Gusev: Order–disorder transformations and phase equilibrium in strongly nonstoichiometric compounds, Uspekhi Fiz. Nauk **170**, 3–40 (2000) (in Russian). (Engl. transl.: Physics – Uspekhi **43**, 1–37 (2000))
180 A. A. Rempel, A. I. Gusev: Nanostructure and atomic ordering in vanadium carbide, Pis'ma v ZhETF **69**, 436–442 (1999) (in Russian). (Engl. transl.: JETP Letters **69**, 472–478 (1999))
181 A. A. Rempel, A. I. Gusev: Short-range order in ordered alloys and interstitial phases, Fiz. Tverd. Tela **32**, 16–24 (1990) (in Russian). (Engl. transl.: Sov. Physics – Solid State **32**, 8–13 (1990))
182 A. A. Rempel, A. I. Gusev: Short-range order in superstructures, Phys. Stat. Sol. (b) **160**, 389–402 (1990)

11. Effects of Ordering on the Properties of Strongly Nonstoichiometric Compounds

Atomic ordering of strongly nonstoichiometric interstitial compounds is a widely occurring but still understudied phenomenon (as compared to its counterpart in metallic alloys). This statement especially applies to the effect of ordering on properties of strongly nonstoichiometric compounds. Indeed, most work dealing with ordering of nonstoichiometric compounds is dedicated to the structure of ordered phases and their range in phase diagrams. Studies concerned with properties of nonstoichiometric compounds in both ordered and disordered states or properties vs. degree of ordering are dozens of times less in number. This is probably due to the fact that the effects of ordering and those associated with variation of the nonstoichiometric compound composition are very difficult to resolve. Experimental studies should be performed with the utmost care and accuracy. However, the effect of ordering on properties of nonstoichiometric compounds is comparable with the variation of properties in the homogeneity interval of a disordered compound where an ordered phase is formed. In other words, ordering and degree of nonstoichiometry have a comparable effect on properties of strongly nonstoichiometric compounds. Knowledge of the dependence of properties of strongly nonstoichiometric compounds MX_y on both the compound composition and the degree of order will open up new opportunities for producing materials with preset properties. By varying the structural state and the degree of order, it will be possible to realize fine adjustment of properties of strongly nonstoichiometric interstitial compounds without any impact on the chemical composition.

The effect of ordering on properties of strongly nonstoichiometric interstitial compounds was considered earlier in [1–4]. In the period from 1990 to 2000 the interest in properties of ordered nonstoichiometric compounds quickened. Therefore this monograph presents a large body of new experimental data and the review is considerably extended with a detailed discussion of post-ordering changes in various properties of strongly nonstoichiometric compounds.

11.1 The Basic Lattice Constant

Many researchers analyzed the effect of nonstoichiometry on the basic lattice constant of nonstoichiometric compounds MX_y with the *B*1 (NaCl) structure. Investigations show that the lattice constant a_{B1} of these compounds decreases nonlinearly with decreasing concentration of interstitial atoms or, which is the same, with increasing concentration of structural vacancies. Variation of the lattice constant a_{B1} in the homogeneity interval of nonstoichiometric carbides of titanium, zirconium, hafnium, vanadium, niobium and tantalum is shown in Fig. 3.12. Let it be noted that a weak maximum is present in the dependencies $a_{B1}(y)$ of disordered carbides TiC_y, ZrC_y and HfC_y with a small concentration of structural vacancies ($z = 1 - y < 0.1$). Most reliable data on the lattice constant of disordered carbides MC_y (M = Ti, Zr, Hf, V, Nb, Ta) were analyzed statistically [5–11] to show that within the accuracy of the experimental findings, the concentration dependencies of the lattice constant, $a_{B1}(y)$, represent quadratic dependencies of the content of carbon atoms y or structural vacancies $z = 1 - y$. Coefficients of the dependencies $a_{B1}(y) = a_0 + a_1 y + a_2 y^2$ for disordered carbides are given in Table 3.6.

Probably, the fact that thermal treatment conditions affect the lattice constant of nonstoichiometric carbide was established for the first time in [12]. It was found that vanadium carbide $VC_{0.74}$, which was quenched from 1700 to 300 K, had the lattice constant $a_{B1} = 0.4125 \pm 0.0001$ nm. The same carbide, which was annealed at a slow temperature rate from 1620 to 300 K, had the lattice constant $a_{B1} = 0.4131 \pm 0.0001$ nm. Researchers [12] could not account for that effect.

An ordered phase appeared in thorium carbide $ThC_{0.76}$ when it was annealed at a temperature of 1000 K for 150 h [13]. The phase was thought to have a cubic structure with lattice constant $a = 2a_{B1}$ (a_{B1} being the lattice constant of disordered carbide with a *B*1-type structure). Measurements of the lattice constant of the basic structure of ordered thorium carbide, which was heated from 300 to 1300 K, revealed an abrupt decrease in a_{B1} at a temperature of ~1020 K (Fig. 11.1).

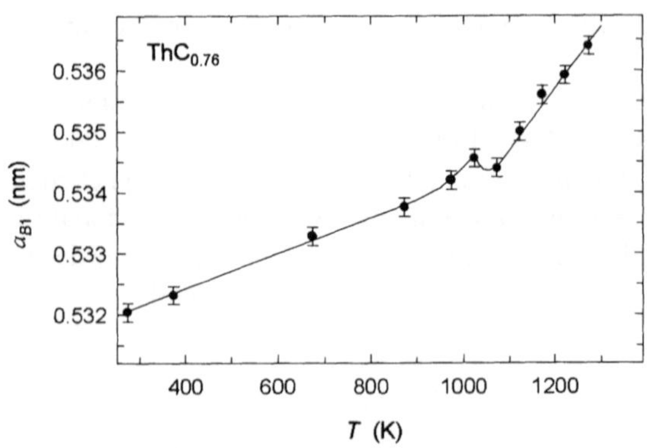

Fig. 11.1. Variation of the basic lattice constant a_{B1} of ordered thorium carbide $ThC_{0.76}$ during heating and order–disorder transition at a temperature $T_{trans} = 1020$ K [13]

Lorenzelli and de Dieuleveult [13] related this sudden change to the transition of thorium carbide from a disordered to an ordered state at the transition temperature T_{trans} = 1020–1070 K. In accordance with [14], T_{trans} = 1100 K. When $T < T_{trans}$, the thermal expansion coefficient α of ordered thorium carbide is 5.8×10^{-6} K^{-1}. When $T > T_{trans}$, it is much larger and equals 18.4×10^{-6} K^{-1}. In accordance with [14], the thermal expansion coefficient of disordered thorium carbide $ThC_{0.76}$ is 6.6×10^{-6} K^{-1} at $T > T_{trans}$.

Authors [15–18] were the first to detect the effect of ordering on the basic lattice constant over the whole interval of composition of nonstoichiometric carbide forming an ordered phase. They analyzed the structural disorder–order phase transition in nonstoichiometric niobium carbide NbC_y. It turned out that within the homogeneity interval of the ordered phase Nb_6C_5 at a temperature of 300 K the basic lattice constant a_{B1} of ordered niobium carbide is larger than a_{B1} of disordered carbide NbC_y with a similar carbon content (Fig. 11.2). The same effect of ordering is detected on the basic lattice constant of nonstoichiometric tantalum carbide TaC_y. The lattice constant difference $\Delta a_{B1} = a_{B1}(y,\eta) - a_{B1}(y,0)$ of ordered and disordered niobium and tantalum carbides changed with the carbon content.

The increase in the basic lattice constant during the $VC_{0.89} \rightarrow V_8C_7$ transformation was detected later by Athanassiadis *et al* [19] who analyzed ordering of nonstoichiometric vanadium carbide. They found that at room temperature the basic lattice constant of quenched disordered carbide $VC_{0.89}$ was 0.41662 nm while its counterpart for ordered carbide of a similar composition was equal to 0.41667 nm (or 0.83334 nm for the cubic superstructure V_8C_7 having a double lattice constant).

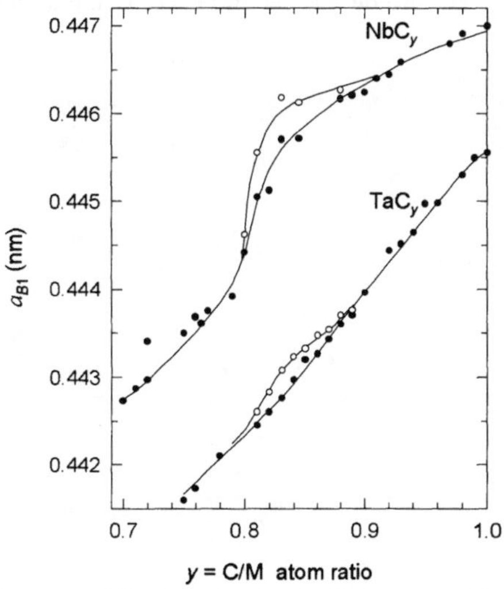

Fig. 11.2. Dependence of the basic cubic lattice constant a_{B1} on the composition of niobium NbC_y and tantalum TaC_y carbides in disordered (•) and ordered (○) states at $T = 300$ K [8]

These values agree well with the sudden change in the basic lattice constant of vanadium carbide from 0.4202 to 0.4200 nm at the order–disorder transition temperature T_{trans} = 1413 ± 10 K (Fig. 11.3), which was revealed during high-temperature X-ray examination [19].

In accordance with [10, 20, 21], at a temperature of 300 K the basic lattice constant of ordered carbide V_8C_7 is larger than the lattice constant of quenched disordered carbide $VC_{0.87}$. It was shown [10] that at room temperature the basic lattice constant of ordered carbide V_6C_5 was larger in the whole homogeneity interval of the ordered phase than a_{B1} of the disordered carbide VC_y (Fig. 11.4).

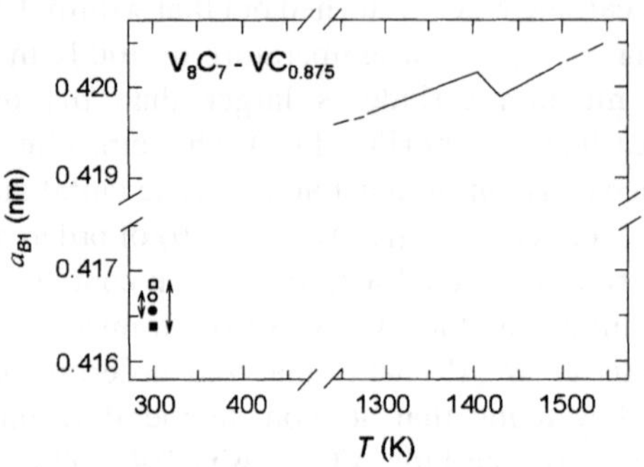

Fig. 11.3. Variation of the basic lattice constant a_{B1} of ordered cubic (space group $P4_332$) phase V_8C_7 during heating and $V_8C_7 \rightarrow VC_{0.875}$ order–disorder transition at a temperature T_{trans} = 1413 ± 10 K (plotted considering data in [19]): (■) and (●) denote the lattice constant a_{B1} of disordered carbide $VC_{0.875}$ at 300 K according to the data [10] and [19], (□) and (○) stand for the lattice constant a_{B1} of ordered carbide V_8C_7 at 300 K according to the data [10] and [19] respectively

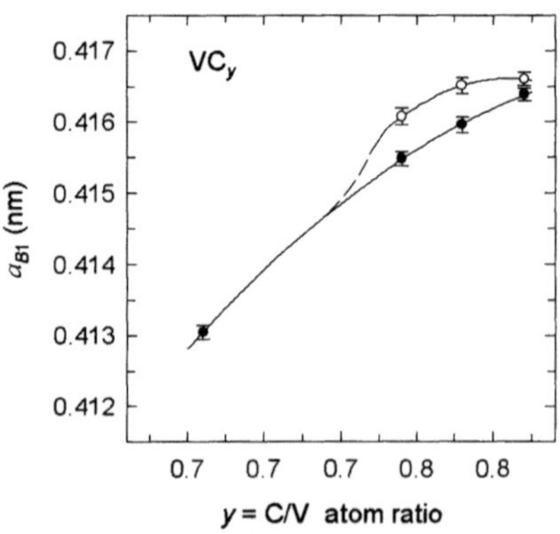

Fig. 11.4. Dependence of the basic cubic lattice constant a_{B1} on the composition of vanadium monocarbide VC_y in disordered (●) and ordered (○) states at a temperature of 300 K [10]

An anomalous increase in the lattice constant of nonstoichiometric tantalum carbide in the interval $TaC_{0.81}$–$TaC_{0.89}$ was reported in [22]. Considering the preparation conditions of TaC_y samples [22], it may be thought that this anomalous increase was due to partial ordering of tantalum carbide.

A larger lattice constant a_{B1} in ordered carbides TiC_y as compared to disordered carbides of a similar composition was observed at a temperature of 300 K [23–27]. For example, a_{B1} increased after ordering annealing of carbides $TiC_{0.53}$, $TiC_{0.58}$, $TiC_{0.63}$ and $TiC_{0.67}$ [23, 24]. Lipatnikov *et al* [25] reported the increase in the lattice constant of carbide TiC_y ($0.6 \leq y \leq 0.9$) annealed at 1000 and 900 K for 15 h at each temperature. A similar effect was noted [26] for carbides $TiC_{0.49}$ and $TiC_{0.55}$ annealed at 773 K. An electron microscopic examination of long-range order in carbide $TiC_{0.58}$ [27] showed that 140-hour annealing of disordered carbide at a temperature of 970 K led to the appearance of the trigonal (space group $R\bar{3}m$) ordered phase Ti_2C. The basic lattice constant a_{B1} of carbide $TiC_{0.59}$ increased upon ordering from 0.43189 to 0.43242 nm. Zueva and Gusev [11] studied in detail the effect of nonstoichiometry and ordering on the basic lattice constant of nonstoichiometric titanium carbide in the interval $TiC_{0.50}$–$TiC_{1.00}$. They showed that at 300 K the lattice constant a_{B1} of ordered titanium carbides was larger than the lattice constant of disordered carbides (Fig. 11.5).

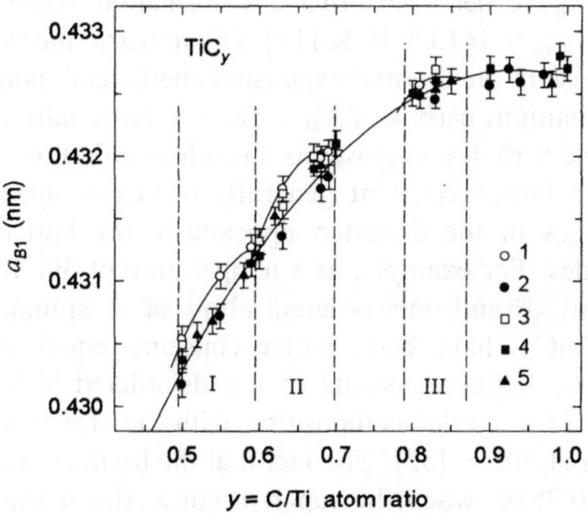

Fig. 11.5. Dependence of the basic lattice constant a_{B1} on the composition of titanium monocarbide TiC_y in quenched disordered (black symbols) and annealed ordered (white symbols) states [11]: (1, 2) samples produced by sintering of Ti and $TiC_{0.94}$, (3, 4) samples produced by sintering of Ti and C, (5) samples produced by hot pressing of Ti and $TiC_{0.98}$; (I, II, III) the regions of existence of ordered phases Ti_2C, Ti_3C_2 and Ti_6C_5 respectively

On the strength of the aforementioned experimental results it may be asserted that ordering of nonstoichiometric carbides MC_y not only leads to the lowering of the crystal lattice symmetry, but also causes changes in the basic cubic lattice constant. In the majority of works the constant a_{B1} of the basic cubic lattice of ordered and disordered nonstoichiometric carbides was measured at 300 K. In this case the positive difference of the lattice constant $\Delta a_{B1} = a_{B1}(y,\eta) - a_{B1}(y,0) > 0$, which is observed at $T = 300$ K for ordered and disordered carbides with one and the same carbon concentration, can be explained in three ways: 1) the observed effect results from the increase in the lattice constant a_{B1} immediately at the temperature of transition from the disordered to ordered state; 2) the lattice constant a_{B1} remains unchanged upon the equilibrium reversible disorder ↔ order transformation and the observed effect is due to the fact that the thermal expansion coefficient α of disordered carbide is larger than α of ordered carbide, i.e. $\alpha(y,0) > \alpha(y,\eta)$; 3) the increase in the lattice constant a_{B1} is followed by the decrease in the thermal expansion coefficient α upon transition from the disordered to ordered state, i.e. $a_{B1}(y,\eta,T_{trans}) > a_{B1}(y,0,T_{trans})$ and $\alpha(y,\eta,T_{trans}) < \alpha(y,0,T_{trans})$.

The third explanation is most probable. It is confirmed by investigations when the lattice constant of the basic phase was measured in the temperature interval from 300 K to $T > T_{trans}$. Indeed, in accordance with [13], at $T_{trans} \approx 1020$ K disordered carbide $ThC_{0.76}$ has the lattice constant $a_{B1} \approx 0.5340$ nm and the coefficient $\alpha = 18.4 \times 10^{-6}$ K^{-1}, while ordered carbide of a similar composition has $a_{B1} \approx 0.5345$ and $\alpha = 5.8 \times 10^{-6}$ K^{-1}. The lattice constant a_{B1} decreases from 0.4202 to 0.4200 nm and the thermal expansion coefficient α increases from 6.2×10^{-6} to 15.2×10^{-6} K^{-1} upon the $V_8C_7 \rightarrow VC_{0.89}$ order → disorder transformation, which takes place in vanadium carbide at $T_{trans} = 1413 \pm 10$ K [19]. Dilatometric measurements [28–30] revealed the increase in the thermal expansion coefficient upon the order → disorder transition in titanium carbide $TiC_{0.55}$. So, the basic lattice constant of strongly nonstoichiometric carbides increases as a result of ordering.

It is worth noting that the basic lattice constant generally decreases upon ordering of metal alloys, i.e. changes in the direction opposite to the lattice constant of nonstoichiometric carbides. For example, at a temperature of 300 K the disordered fcc alloy $Au_{0.248}Cu_{0.752}$ and the ordered alloy of a similar composition having the structure $AuCu_3$ have basic lattice constants equal to 0.37525 and 0.37465 nm [31]. Basic lattice constants of the disordered alloy $Au_{0.744}Cu_{0.256}$ and the ordered alloy of a similar composition with the structure Au_3Cu are 0.39843 and 0.39821 nm at 300 K [32]. The fact that the basic lattice constant decreased from 0.3844 to 0.3810, when the superstructure Au_3Cu was formed in the alloy $Au_{0.751}Cu_{0.249}$, was noted in the reference book [33]. The opposite change in the basic lattice constant of ordered metal alloys on the one hand and strongly nonstoichiometric carbides on the other hand is due to the more complicated structure of nonstoichiometric interstitial compounds.

Before considering the causes responsible for the increase in the basic lattice constant upon ordering of nonstoichiometric carbides, we have to understand why the lattice constant of strongly nonstoichiometric compounds diminishes with growing concentration of structural vacancies.

The lattice constant of the metallic sublattice of the nonstoichiometric compound MX_y is measured in diffraction experiments. The measured lattice constant represents the statistical average of the sizes of a multitude of unit cells or, which is the same, a multitude of interatomic distances between metal atoms M–X–M and M–□–M. Clearly, the lattice constant value depends on the direction and magnitude of displacements of metal atoms relative to a structural vacancy. In Sect. 3.2 it was shown that the lattice constant a_{B1} may decrease with growing concentration of structural vacancies in two cases. If metal atoms shift toward a vacancy and this vacancy shrinks, the lattice constant a_{B1} diminishes with increasing concentration of vacancies. The lattice constant will decrease if static displacements of metal atoms decrease monotonically and tend asymptotically to zero with distance to a vacancy. However, in this case it is impossible to account for a weak maximum in the dependence $a_{B1}(y)$ of the lattice constant on the composition of titanium, zirconium and hafnium carbides. If metal atoms nearest to a vacancy shift away from this vacancy, metal atoms comprising the next coordination sphere should shift toward the vacancy for the lattice constant to decrease. In this case damping of perturbations with distance from a vacancy represents Friedel oscillations. If the vacancy concentration is small, vacancy-induced regions of perturbations do not overlap and the lattice constant a_{B1} increases with growing concentration of vacancies. When the vacancy concentration is sufficiently large, perturbation regions overlap and compensate static atomic displacements caused by adjacent vacancies. Consequently, the lattice constant diminishes. Thus, a maximum appears in the dependence $a_{B1}(y)$ of disordered nonstoichiometric compounds MX_y in the interval of small vacancy concentrations. The location of this maximum depends on the effective radius of the perturbation around a vacancy.

A maximum in the dependence $a_{B1}(y)$ could not be observed in experiments concerned with carbides $M^{(V)}C_y$ of group V transition metals. Consequently, vacancy-induced perturbations are efficient within a very long range and overlap at a small vacancy concentration $c_{\square} = 1 - y < 0.01$. In the interval $0.9 < y < 1.0$ the derivative da_{B1}/dy is larger for nonstoichiometric carbide TaC_y than for niobium carbide NbC_y (see Fig. 11.2). This means that even a small concentration of vacancies causes a considerable distortion of the metallic sublattice in tantalum carbide.

The effective radius of perturbations in carbides $M^{(IV)}C_y$ is smaller than in $M^{(V)}C_y$ and a maximum in $a_{B1}(y)$ is observed at a vacancy concentration $c_{\square} = 1 - y \approx 0.05$–$0.07$ (see Fig. 11.5). Experimental findings [23, 34–38] confirm that metal atoms of the first coordination sphere shift from a vacancy in disordered nonstoichiometric carbides and atomic displacements oscillate in direction and magnitude (see Sect. 4.3 and Fig. 4.14).

Nonstoichiometric carbides MC_y are ordered at a relatively high concentration of structural vacancies $(1 - y) > 0.12$. Given this concentration of vacancies, vacancy-induced perturbation regions overlap in disordered carbide. As a result, perturbations cover the whole volume of the crystal and cause a rapid decrease in the lattice constant a_{B1}. Upon ordering, vacancies and carbon atoms rearrange themselves over sites of the nonmetallic sublattice such that the number of nearest-neighbor vacancies proves to be smaller in ordered carbide than in disordered carbide of a similar composition. Consequently, the overlap of perturbation regions decreases and the basic lattice constant a_{B1} of ordered carbide turns out to be larger than that of disordered carbide having a similar composition.

Let us show that the relative number of neighboring vacancies is indeed less in ordered carbides than in disordered ones.

When the vacancy content is small ($(1-y) << 1$), vacancies are spaced from one another by a distance larger than the radius r_{110} of the first coordination sphere and are isolated. In the case of an fcc nonmetallic sublattice with the coordination number 12, the probability that a given vacancy is isolated equals $P = y^{12}$ and is close to unity at $(1-y) << 1$. On the other hand, the probability that an arbitrary vacancy is not isolated equals $P = 1 - y^{12}$ and approaches zero at $(1-y) << 1$.

As the vacancy concentration increases ($0 < (1-y) < 1$), the probability of one vacancy being isolated from another vacancy diminishes quickly and vacancy clusters of different sizes are formed in the sublattice. When some critical concentration of vacancies $(1-y) = p_c$ is reached, a single vacancy cluster appears in the system and this cluster goes throughout the lattice. It is referred to as an infinite cluster. It was shown [39, 40] that at $p \geq p_c$ one infinite cluster exists in a disordered system with probability equal to 1. The critical fraction of free sites in an fcc lattice is 0.20 [41]. Clearly, for the disordered fcc nonmetallic sublattice at hand, the critical concentration of structural vacancies needed for percolation is $(1-y) = 0.20$.

Thus, from percolation theory [6, 41, 42] it follows that vacancies form an infinite vacancy cluster with probability equal to 1 in disordered carbides MC_y containing $y \leq 0.80$ carbon. Upon ordering, an infinite vacancy cluster is "broken up" and the lattice includes isolated or bound vacancy groups of different sizes. The probability of those groups can be easily determined if the structure of ordered phases is known.

The fcc nonmetallic sublattice of nonstoichiometric carbides can be visualized as a set of clusters in the form of regular tetrahedrons including four sites, which can be occupied with interstitial atoms or be vacant. All these sites are located in the first nonmetallic coordination sphere (110) relative to one another and are spaced $(a_{B1}\sqrt{2})/2$. Thus, the least possible distance between neighboring vacancies is $(a_{B1}\sqrt{2})/2$ and corresponds to the radius r_{110} of the first nonmetallic coordination sphere (110) relative to the vacancy located at the center.

Let us consider the occurrence probability of configurations, which are formed by vacancies located relative to one another in the first nonmetallic coordination sphere (110) with radius $r_{110} = (a_{B1}\sqrt{2})/2$. Let us label the probabilities of these configurations formed by two (tetrahedron edge), three (tetrahedron face) and four (tetrahedron) vacancies as $P_{2\square}$, $P_{3\square}$ and $P_{4\square}$. Table 11.1 gives numerical values of the probabilities $P_{2\square}$, $P_{3\square}$ and $P_{4\square}$ for ordered phases M_2C (space groups $R\bar{3}m$ and $Fd3m$), M_3C_2 (space group $C222_1$) and M_6C_5 (space groups $C2/m$, $P3_1$, and $C2$), and also for disordered carbides $MC_{0.50}$, $MC_{0.667}$ and $MC_{0.833}$ with a similar carbon concentration as in the ordered phases.

Table 11.1. Probabilities of vacancy configurations in disordered and ordered carbides

Carbide	$P_{2\square}$	$P_{3\square}$	$P_{4\square}$
$MC_{0.50}$	0.250	0.125	0.063
M_2C (space group $R\bar{3}m$)	0.250	0.125	0
M_2C (space group $Fd\bar{3}m$)	0.250	0.125	0.125
$MC_{0.667}$	0.111	0.037	0.012
M_3C_2 (space group $C222_1$)	0.125	0	0
$MC_{0.833}$	0.028	0.005	0.001
M_6C_5 (space groups $C2/m$, $P3_1$, $C2$)	0	0	0

Comparison shows that probabilities of configurations containing neighboring vacancies diminish upon ordering (see Table 11.1). A different situation arises with cubic (space group $Fd\bar{3}m$) superstructure M_2C, for which probabilities $P_{2\square}$ and $P_{3\square}$ are similar to those for disordered carbide $MC_{0.50}$, while the probability $P_{4\square}$ is even larger than its counterpart for disordered carbide. However, this contradiction is just imaginary: if one takes into account both the first and second nonmetallic coordination spheres in cubic (space group $Fd\bar{3}m$) superstructure M_2C, the probability is $P_{4\square} = 0.043$, i.e. is less than $P_{4\square} = 0.063$ for disordered carbide $MC_{0.50}$.

Thus, ordering decreases the probability of configurations containing neighboring vacancies and, hence, reduces the overlap of vacancy-induced perturbation regions. As a result, the basic lattice constant a_{B1} of ordered carbide exceeds the lattice constant of disordered carbide with a similar carbon content.

A phenomenological model was proposed by Lipatnikov, Rempel *et al.* [8, 9] for quantitative analysis of changes in the lattice constant a_{B1} upon formation of an ordered phase type M_6C_5 in carbides NbC_y and TaC_y. Later [11] the said model was extended to superstructures of type $M_{2t}C_{2t-1}$. Let us consider application of this model to various ordered phases $M_{2t}C_{2t-1}$ (t = 1, 1.5, 2, 3 and 4) that can be formed in strongly nonstoichiometric carbides.

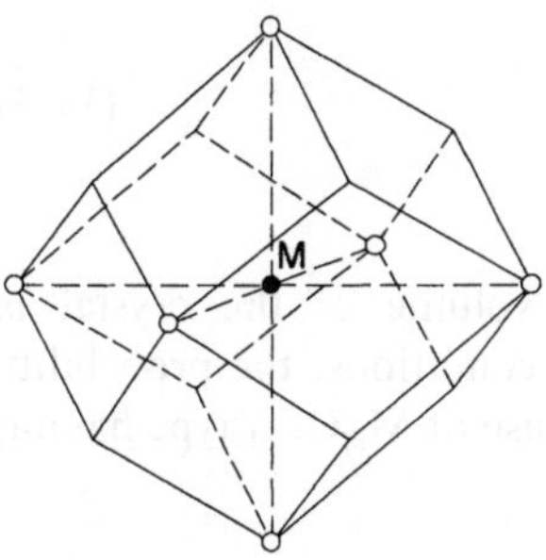

Fig. 11.6. Cluster figure in the form of a rhombododecahedron (distortions are neglected) used for describing the volume of the crystal lattice of nonstoichiometric compounds MX_y with the basic structure of the $B1$ type: (○) nonmetallic sublattice site, (●) atom of metal M

In nonstoichiometric carbides having the $B1$ basic structure, each metal is surrounded by six sites of the nonmetallic sublattice, which can be occupied with carbon atoms or can be vacant. Then a crystal with the $B1$ structure can be visualized as a set of clusters in the form of Dirichlet–Voronoi polyhedra, a distorted Seitz cell (the Seitz cell is a dodecahedron in the $B1$ structure). Each cluster includes a metal atom located at the center and six sites of the nonmetallic sublattice (Fig. 11.6). The nonmetallic sublattice sites may be occupied with carbon atoms or may be vacant. The magnitude of distortions depends on the number and mutual location of vacant sites in a cluster and also on the arrangement of vacancies in the nearest neighborhood of this cluster. Such clusters in the form of Dirichlet–Voronoi polyhedra fill the whole volume of the crystal, provide the continuum, and take into account all sites of the crystal lattice.

Assume as a first approximation that the cluster volume V_m depends only on the number of vacancies m in the cluster and is independent of the mutual location of the vacancies. In this case the crystal volume V can be written as

$$V = N\sum_{m=0}^{6} \lambda_m P_m(y,\eta)V_m \ , \tag{11.1}$$

where $P_m(y,\eta)$ is the probability that a cluster with m vacancies is formed in the crystal; η is the long-range order parameter; $\lambda_m = C_6^m$ denotes the multiplicity of the m-configuration of the cluster; and N is the number of metal sublattice sites. On the other hand, the volume of disordered carbide MC_y can be expressed through the basic structure constant $a_{B1}(y) = a_0 + a_1 y + a_2 y^2$ as $V = (N/4)a_{B1}^3(y)$. Considering this, (11.1) for disordered carbide takes the form

$$\sum_{m=0}^{6} \lambda_m P_m(y,0)V_m = \frac{(a_0 + a_1 y + a_2 y^2)^3}{4} \ , \tag{11.2}$$

where $P_m(y,0) = (1-y)^m y^{(6-m)}$ is the probability that a cluster with m vacancies is formed in disordered carbide. The solution of (11.2) yields an expression for the cluster volume:

$$V_m = \tfrac{1}{4}\sum_{k=m}^{6} A_{6-k}\frac{k!(6-m)!}{6!(k-m)!} \ , \tag{11.3}$$

where A_{6-k} at y^k on the right of equation (11.2).

The contribution of each cluster to the total volume of the crystal is proportional to its probability P_m. Under equilibrium conditions, the probability $P_m(y,\eta)$ of a cluster with m vacancies in an ordered phase of $M_{2t}C_{2t-1}$ type having any degree of long-range order can be presented [2] as

$$P_{m,t}(y,\eta) = \frac{1}{\Phi_s}\sum_f \frac{g_f}{C_6^{\nu(t,f)}} \sum_{\nu=0}^{\nu(t,f)} C_{6-m}^{\nu(t,f)-\nu} C_m^{\nu} n_1^{[\nu(t,f)-\nu]} n_2^{[6-m-\nu(t,f)+\nu]}$$
$$\times (1-n_1)^{\nu}(1-n_2)^{(m-\nu)} \ , \tag{11.4}$$

where $n_1 = y - (2t-1)\eta/2t$ and $n_2 = y + \eta/2t$ are the probabilities that a carbon atom is present at a site in the vacancy or carbon sublattice respectively, when a superstructure of $M_{2t}C_{2t-1}$ type is formed; g_f is the multiplicity of nonequivalent positions of metal atoms located at the center of each cluster ($\sum_f g_f = \Phi_s$); and $v(t,f)$ is the number of sites in the vacancy sublattice, which belong to a cluster with the multiplicity g_f in a superstructure of $M_{2t}C_{2t-1}$ type.

Using (11.1), (11.3) and (11.4), it is possible to find the crystal volume and the basic lattice constant a_{B1} of nonstoichiometric carbide with any degree of order η. The η value of ordered carbide can change between η_{trans} (the order parameter at the transition point) and η_{max}. Therefore the lattice constant a_{B1} of ordered carbide can be calculated for these two limiting cases.

The first option assumes that nonstoichiometric carbide MC_y has a maximum degree of long-range order η_{max}. In accordance with [1, 2], η_{max} depends on the composition of carbide MC_y in the presence of an ordered phase type $M_{2t}C_{2t-1}$ as

$$\eta_{max}(y) = \begin{cases} 2t(1-y), & \text{if } y \geq \dfrac{2t-1}{2t} \\ \dfrac{2ty}{2t-1}, & \text{if } y < \dfrac{2t-1}{2t}. \end{cases} \quad (11.5)$$

The second option assumes that the long-range order parameter of ordered carbide has the same value as at the disorder–order phase transition temperature T_{trans}, i.e. $\eta = \eta_{trans}$.

Table 11.2 contains experimental values of the lattice constant difference Δa_{B1} for vanadium [10], niobium [8, 15, 16, 18] and tantalum [8, 9] carbides in ordered and disordered states at a temperature of 300 K. The same Table 11.2 also gives Δa_{B1} values calculated in two different approximations. Values of Δa_{B1} calculated in the approximation $\eta = \eta_{trans}$ are closest to the experimental values of the difference Δa_{B1}. This means that annealed nonstoichiometric carbides of vanadium, niobium and tantalum acquire a long-range order corresponding to a thermodynamically equivalent distribution of carbon atoms and vacancies in the lattice. The largest and lowest experimental values of Δa_{B1} are observed for vanadium carbide VC_y and tantalum carbide TaC_y respectively. The annealing temperature and time were nearly equal for carbides VC_y, NbC_y and TaC_y [8–10, 15, 16, 18]. Therefore one may think that nonstoichiometric tantalum carbide has a smaller degree of long-range order under similar annealing conditions.

The variation of the lattice constant of vanadium, niobium and tantalum carbides during ordering is comparable with the variation of the lattice constants in the homogeneity interval where ordered phases are formed. For example, the basic lattice constant of disordered vanadium carbide changes by 0.00048 nm in the region $VC_{0.79}$–$VC_{0.83}$, whereas at 300 K the lattice constant changes by 0.00059 nm after ordering. The basic lattice constant of disordered niobium carbide varies from 0.4445 to 0.4466 nm in the region $NbC_{0.81}$–$NbC_{0.88}$, i.e. by 0.0021 nm, whereas ordering causes the lattice constant to change by 0.0005 nm.

The lattice constant of disordered tantalum carbide TaC_y changes by 0.0014 nm in the region $TaC_{0.81}$–$TaC_{0.89}$, and by 0.0003 nm after ordering.

Table 11.2. Change of basis lattice constant, $\Delta a_{B1} = a_{B1}(y,\eta) - a_{B1}(y,0)$, as a result of ordering of nonstoichiometric vanadium, niobium and tantalum carbides VC_y, NbC_y and TaC_y

Carbide MC_y	Phase transformation	$\Delta a_{B1} = a_{B1}(y,\eta) - a_{B1}(y,0)$ (nm) experimental ± 0.00005 nm	calculation $\eta = \eta_{trans}$	calculation $\eta = \eta_{max}$	References
$VC_{0.66}$	–	0	–	–	10
$VC_{0.79}$	$VC_y \to V_6C_5$	0.00059	0.00031	0.00079	10
$VC_{0.83}$	$VC_y \to V_6C_5$	0.00055	0.00033	0.00086	10
$VC_{0.87}$	$VC_y \to V_8C_7$	0.00022	0.00027	0.00047	10
$NbC_{0.79}$	–	0	–	–	8, 15, 16, 18
$NbC_{0.80}$	$NbC_y \to Nb_6C_5$	0.00020	0.00034	0.00082	8, 15, 16, 18
$NbC_{0.81}$	$NbC_y \to Nb_6C_5$	0.00050	0.00035	0.00084	8, 15, 16, 18
$NbC_{0.82}$	$NbC_y \to Nb_6C_5$	–	0.00036	0.00086	8, 15, 16, 18
$NbC_{0.83}$	$NbC_y \to Nb_6C_5$	0.00048	0.00036	0.00089	8, 15, 16, 18
$NbC_{0.84}$	$NbC_y \to Nb_6C_5$	–	0.00035	0.00083	8, 15, 16, 18
$NbC_{0.845}$	$NbC_y \to Nb_6C_5$	0.00041	0.00034	0.00079	8, 15, 16, 18
$NbC_{0.86}$	$NbC_y \to Nb_6C_5$	–	0.00032	0.00063	8, 15, 16, 18
$NbC_{0.87}$	$NbC_y \to Nb_6C_5$	–	0.00029	0.00054	8, 15, 16, 18
$NbC_{0.88}$	$NbC_y \to Nb_6C_5$	0.00010	0.00026	0.00046	8, 15, 16, 18
$NbC_{0.89}$	$NbC_y \to Nb_6C_5$	–	0.00022	0.00039	8, 15, 16, 18
$TaC_{0.78}$	–	0	–	–	8, 9
$TaC_{0.79}$	$TaC_y \to Ta_6C_5$	0.00005	0.00010	0.00023	8, 9
$TaC_{0.81}$	$TaC_y \to Ta_6C_5$	0.00015	0.00010	0.00025	8, 9
$TaC_{0.82}$	$TaC_y \to Ta_6C_5$	0.00023	0.00011	0.00025	8, 9
$TaC_{0.83}$	$TaC_y \to Ta_6C_5$	0.00032	0.00011	0.00026	8, 9
$TaC_{0.84}$	$TaC_y \to Ta_6C_5$	0.00026	0.00010	0.00024	8, 9
$TaC_{0.85}$	$TaC_y \to Ta_6C_5$	0.00022	0.00010	0.00021	8, 9
$TaC_{0.87}$	$TaC_y \to Ta_6C_5$	0.00011	0.00008	0.00016	8, 9
$TaC_{0.88}$	$TaC_y \to Ta_6C_5$	0.00010	0.00008	0.00014	8, 9
$TaC_{0.89}$	$TaC_y \to Ta_6C_5$	0.00006	0.00007	0.00012	8, 9

Calculations of $\Delta a_{B1} = a_{B1}(y,\eta) - a_{B1}(y,0)$ for titanium carbide TiC_y [11] in the approximations $\eta = \eta_{max}$ and $\eta = \eta_{trans}$ (Table 11.3) showed that experimental values of Δa_{B1} correspond to an intermediate value of the long-range order parameter $\eta_{trans} < \eta < \eta_{max}$. Table 11.3 gives calculated values of the parameter $\eta = \eta_{equilibr}$, at which the constant Δa_{B1} changes as in experiments, and the temperature T, at which the long-range parameter acquires an equilibrium value. Values of T_{trans}, η_{trans} and $\eta_{equilibr}$ were calculated by the order-parameters functional method [1, 2, 43, 44] using formulas given in Chaps. 9 and 10.

Table 11.3. Change of basis $B1$-type crystal lattice constant, $\Delta a_{B1} = a_{B1}(y,\eta) - a_{B1}(y,0)$, as a result of ordering of nonstoichiometric titanium carbide TiC_y [11]

Phase transformation	$\Delta a_{B1} = a_{B1}(y,\eta) - a_{B1}(y,0)$ (nm)							
	Experi mental	$\eta = \eta_{equilibr}$			$\eta = \eta_{trans}$		$\eta = \eta_{max}$	
		η	T (K)	Δa_{B1}	η	Δa_{B1}	η	Δa_{B1}
$TiC_{0.50} \rightarrow Ti_2C$	0.00010	0.64	940	0.00010	–	–	1.00	0.00062
$TiC_{0.50} \rightarrow Ti_2C$	0.00035	0.74	760	0.00035	–	–	1.00	0.00062
$TiC_{0.55} \rightarrow Ti_2C$	0.00032	0.71	730	0.00032	–	–	0.90	0.00051
$TiC_{0.59} \rightarrow Ti_2C$	0.00013	0.51	1020	0.00023	–	–	0.82	0.00042
$TiC_{0.60} \rightarrow Ti_3C_2$	0.00010	0.34	800	0.00010	0.19	0.00002	0.90	0.00040
$TiC_{0.63} \rightarrow Ti_3C_2$	0.00014	0.40	760	0.00014	0.23	0.00003	0.95	0.00050
$TiC_{0.67} \rightarrow Ti_3C_2$	0.00010	0.37	760	0.00010	0.28	0.00004	0.99	0.00055
$TiC_{0.68} \rightarrow Ti_3C_2$	0.00024	0.56	600	0.00024	0.29	0.00005	0.96	0.00052
$TiC_{0.69} \rightarrow Ti_3C_2$	0.00020	0.52	610	0.00020	0.30	0.00005	0.93	0.00049
$TiC_{0.81} \rightarrow Ti_6C_5$	0.00008	<0.6	<600	0.00008	0.63	0.00014	0.97	0.00033
$TiC_{0.83} \rightarrow Ti_6C_5$	0.00020	0.75	610	0.00020	0.64	0.00014	≤1.0	0.00035

Variations of the lattice constant of the carbide TiC_y during ordering [11] are close to Δa_{B1} values (0.00052 nm for $TiC_{0.60}$, 0.00028 nm for $TiC_{0.70}$, and 0.00012 nm for $TiC_{0.80}$) measured in [25]. In accordance with [24], the difference Δa_{B1} for carbides $TiC_{0.53}$ and $TiC_{0.58}$ after long-term ordering annealing was 0.00168 and 0.00112 nm. However, 11.3 at.% and 3.3 at.% metallic titanium precipitated during annealing of those carbides and the actual composition of the samples was $TiC_{0.58}$ and $TiC_{0.60}$. Considering this, the difference Δa_{B1} caused by ordering of carbides $TiC_{0.58}$, $TiC_{0.60}$, $TiC_{0.63}$ and $TiC_{0.67}$ was 0.00088, 0.00082, 0.00032, and 0.00008 nm [24].

The variation of the lattice constant of titanium carbide TiC_y during ordering is comparable in value with the variation of the lattice constant a_{B1} in the homogeneity interval of TiC_y where ordered phases are formed. For example, the lattice constant a_{B1} increases by 0.00089 nm when the composition of disordered carbide changes from $TiC_{0.50}$ to $TiC_{0.59}$. The lattice constant increases by 0.00035 nm after formation of the ordered phase Ti_2C. In the region from $TiC_{0.63}$ to $TiC_{0.68}$, when a superstructure of the Ti_3C_2 type is formed, the lattice constant of disordered carbide changes by 0.00036 nm. The difference Δa_{B1} = 0.00024 nm during ordering. Clearly, the effects of nonstoichiometry and ordering on the basic lattice constant of carbide TiC_y are comparable in value.

So, the experimental data concerning the effect of ordering on the basic lattice constant of nonstoichiometric carbides and the analysis of changes in the volume of the crystal lattice of carbides upon formation of the ordered phases $M_{2t}C_{2t-1}$ show that transition to the ordered state is followed by an increase in the basic lattice constant. Variation of the crystal volume during ordering is an indication that disorder–order transformations $TiC_y \leftrightarrow Ti_2C$, $TiC_y \leftrightarrow Ti_3C_2$, $VC_y \leftrightarrow V_6C_5$, $VC_y \leftrightarrow V_8C_7$, $NbC_y \leftrightarrow Nb_6C_5$ and $TaC_y \leftrightarrow Ta_6C_5$ most probably represent transitions of the first kind. This conclusion agrees with results of symmetry analysis (see Sect. 5.5), calculations of phase diagrams (see Chap. 10), and the experimental data discussed below.

Variation of the crystal volume upon formation of the ordered phase V_8C_7 in vanadium carbide $VC_{0.875}$ was used by Rempel and Gusev [45] to refine grains and produce a nanocrystalline structure in nonstoichiometric VC_y carbide. This carbide was the first to have a nanocrystalline structure. Earlier the ordering effect was not used for the formation of nanosized structures. Vanadium carbide was chosen among all nonstoichiometric carbides because ordering is most illustrative in this carbide.

Initial disordered vanadium carbide $VC_{0.875}$ in the form of a powder with particles 1 to 2 μm in size was obtained by carbothermal reduction of oxide V_2O_5 and then was aged for a long time at 300 K in a closed vessel preventing ingress of water vapor from air. The aged powder of vanadium carbide proved to be very hygroscopic: immediately after the powder was removed from the vessel, it contained not more than 0.2 mass.% physically adsorbed water. After the powder was left in contact with air for several months, the water content reached 2.0 mass.% and came to saturation. The normal coarse-grain powder of vanadium carbide does not possess such hygroscopicity.

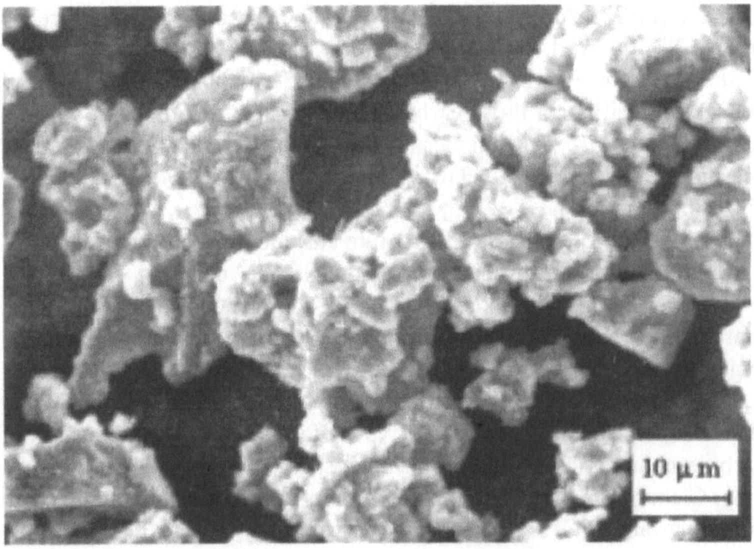

Fig. 11.7. Scanning electron micrograph of aged vanadium carbide $VC_{0.875}$ powder. Large agglomerates with irregular shape have size from 5 to 50 μm and consist of particles with a size of about 1 μm

Quantitative gas chromatographic analysis showed that the aged powder of $VC_{0.875}$ contained 3.1 ± 0.1 mass.% chemisorbed oxygen in addition to water. A negligibly small quantity of oxygen (about 0.1 mass %) was dissolved in the vanadium carbide lattice and oxygen atoms were located at sites of the nonmetallic sublattice. A very small quantity of oxygen formed an oxide film having thickness of 2 or 3 atomic monolayers. This film contained oxide V_2O_5 and homologous oxides V_nO_{2n-1} located between V_2O_3 and VO_2.

According to the chemical analysis, the aged powder of vanadium carbide had the composition $VC_{0.875}$ and corresponded to the upper boundary of the homogeneity interval of the cubic phase $B1$. The carbon sublattice contained ~13 at.% structural vacancies.

As seen in a microscope at ×100 magnification, the vanadium carbide powder was made up of separate irregularly shaped agglomerates 5–50 μm in size comprising particles of about 1 μm in size (Fig. 11.7).

However, at a larger magnification it was obvious that those particles had a complex structure and actually represented an assembly of a great number of tiny particles of nanometer size (hereinafter referred to as "nanoparticles" or "nanocrystals"). The SEM micrograph in Fig. 11.8 was taken at ×10000 magnification in a high-resolution scanning electron microscope type DSM 982 Gemini.

Fig. 11.8. Microstructure of the powder of vanadium carbide $VC_{0.875}$ aged for a long time at ambient temperature in an ambient atmosphere (×10000) [45]. The scale bar corresponds to 2 μm

Fig. 11.9. Morphology of particles of aged powder of carbide $VC_{0.875}$ (×50000): particles about 1 μm in size are nanocrystallites in the form of curved petal-disks 400–600 nm in diameter and 15–20 nm thick [45]. The scale bar corresponds to 500 nm

It is readily seen that each object about 1 μm in size looks like an open rose or a loose head of cabbage and consists of nanocrystallites (see Fig. 11.8). At ×50000 magnification it is seen that nanocrystallites are shaped as bent petals or leafs, which can be modeled in a first approximation by a disk 400–600 nm in diameter and ~15–20 nm thick (Fig. 11.9). Despite the fact that the volume of this disk corresponds to a relatively large spherical particle 150–220 nm in diameter, the ratio between the surface area S and the volume V is $S/V = 0.107–0.143\ nm^{-1}$ thanks to the small thickness of the disk. This value corresponds to the specific surface of the powder 19–26 $m^2\ g^{-1}$. Electron microscopic examination also shows a small grain size distribution of nanocrystallites in the powder.

The X-ray pattern of vanadium carbide $VC_{0.875}$ contains, along with structural reflections, clearly pronounced superstructure reflections of the cubic ordered phase V_8C_7 having the space group $P4_332$. The intensity ratio of structural and superstructure reflections suggests that the ordered phase occupies the whole volume of the material, i.e. the powder contains one phase. The lattice constant of the ordered phase is 0.8337 ± 0.0001 nm. The lattice constant of an ideal cubic superstructure of M_8C_7 type (space group $P4_332$) is twice as large as the lattice constant of the disordered basic phase $B1$. Consequently, the lattice constant of the basic phase in vanadium carbide [45] is $a_{B1} = 0.41685$ nm. This lattice constant is larger (by 0.00047 nm) than the lattice constant of disordered carbide $VC_{0.875}$.

These largely different lattice constants of ordered and disordered carbides $VC_{0.875}$ may be observed only at maximum or nearly maximum degree of ordering. From the intensity ratio of structure and superstructure reflections it follows that long-range order is nearly at a maximum in vanadium carbide and oxygen is virtually absent in the carbide lattice.

Measurements of the positron lifetime in a nanocrystalline powder of carbide $VC_{0.875}$ [45] showed that the positron lifetime spectrum had two components. The short component of 157 ± 2 ps corresponds to annihilation of positrons at a structural vacancy in the carbon sublattice. The long component of 500 ps having the relative intensity $I_2 = 7\%$ is due to annihilation of positrons in large defects on the surface of particles. Trapping of positrons in structural vacancies means the absence of the positron diffusion to long distances. In this case, the intensities of the components are proportional to the volume fractions of phases containing different types of defects. Thus, the relative intensity of the long component I_2 coincides with the volume fraction of the surface $\Delta V_{surf} = \Delta D \times S/V$ in the nanopowder of vanadium carbide. In accordance with estimates, the surface layer has thickness $\Delta D = 0.5$–0.7 nm, which corresponds to 3–4 atomic monolayers.

Generally, the data on the positron lifetime suggest that the bulk of the nanocrystallites contains nonmetal structural vacancies only, while the surface layer of the nanocrystallites includes defects like vacancy agglomerates.

Rempel and Gusev [45] proposed a model describing the structure of vanadium carbide nanocrystallites. The nanocrystallites represent strongly bent plate-disks 400–600 nm in diameter and 15–20 nm thick. The bulk of a nanocrystallite contains ordered carbide V_8C_7 with a high degree of long-range order and a negligibly small quantity of oxygen. The surface layer of the nanocrystallites includes a large quantity of chemisorbed oxygen and a great number of vacancy agglomerates testifying to a loose structure of the layer. The depth of the surface phase does not exceed 0.7 nm or four atomic layers. The observed morphology of the nanocrystalline powder of nonstoichiometric vanadium carbide may be due to cracking of grains at the interface between disordered and ordered phases. Indeed, high-temperature X-ray measurements [19] showed that the fcc sublattice constant increased abruptly by 0.0004 nm after the $VC_{0.875} \rightarrow V_8C_7$ disorder–order phase transition at a temperature of 1413 ± 20 K. The domains of the ordered phase are ~20 nm in size. Different volumes of the disordered and ordered phases give rise to stresses, which cause cracking at the interface between the phases. One more mechanism is possible: cracking along domain boundaries of the ordered phase thanks to the mismatch of the atomic structures of antiphase domains and build-up of stresses. Special research needs be performed to determine the mechanism responsible for refinement of grains during ordering.

Results of investigations [45] give us grounds to believe that disorder–order phase transitions, which are accompanied by changes in the volume, can be used for producing the nanostructured state not only of vanadium carbide, but also of other strongly nonstoichiometric compounds.

11.2 X-Ray Photoelectron and Emission Spectra

The possibility of calculating electron energy spectra of strongly nonstoichiometric compounds with an arbitrary concentration of structural vacancies and a random function of their spatial distribution in the lattice has been limited so far. In this connection, experimental studies dealing with specific features of the electronic structure present special interest. The electron spectroscopic method is widely used to analyze the electronic structure of disordered nonstoichiometric compounds. Just a few papers have been dedicated to ordered nonstoichiometric compounds.

Researchers [46, 47] analyzed $L\beta_{2,15}$ and $L\gamma_1$ X-ray spectra of niobium in carbides $NbC_{0.81}$, $NbC_{0.83}$ and $NbC_{0.88}$ in disordered and ordered states, and also in stoichiometric niobium carbide $NbC_{1.00}$ and disordered carbide $NbC_{0.71}$. The spectra were measured using fluorescence excitation, which excluded heating of samples and alteration of their structural state during experiments.

The low-energy C2*p*-like band *A*, high-energy *d*-like band *B*, and band *C* in the middle of the emission spectrum can be separated in the spectra of disordered carbides NbC_y. X-ray emission Nb$L\beta_{2,15}$-spectra of ordered carbides NbC_y are simpler: band *C* in the middle of the spectrum is absent (Fig. 11.10).

The largest intensity of the high-energy *d*-like band *B* relative to the C2*p*-like band *A* is observed in carbide $NbC_{0.71}$ [47]. The increase in the intensity of the *d*-like band *B* with decreasing concentration of carbon in carbide NbC_y points to the growing occupancy of this band. The analogous conclusion that the increase in the concentration of structural vacancies in the carbon sublattice of carbides is followed by filling of the *d*-like band was drawn earlier in [48]. In the series from $NbC_{1.00}$ to $NbC_{0.71}$, i.e. as the concentration of structural vacancies rises, the energy separation of bands *A* and *B* decreases from 3.3 to 1.9 eV in $L\beta_{2,15}$-spectra and from 3.1 to 1.9 eV in $L\gamma_1$-spectra.

Ordering of niobium carbides leads to vanishing of the fine structure in the middle of the $L\beta_{2,15}$- and $L\gamma_1$-spectra (Figs. 11.10 and 11.11). The effect of ordering is more pronounced in $L\gamma_1$-spectra, which reflect the distribution of $4d_{3/2}$-states of niobium in the valence band. For example, the middle part *C* of the emission band in $L\gamma_1$-spectra decreases considerably and the energy separation of subbands *A* and *B* shrinks from 3.5 to 2.4 eV in the series from disordered carbide $NbC_{0.83}$ to ordered carbide Nb_6C_5 ($NbC_{0.83}$) (Fig. 11.11).

What are the causes of changes in X-ray emission spectra during ordering of carbon atoms and vacancies in nonstoichiometric niobium carbide?

In accordance with cluster calculations [49, 50], the appearance of vacancies in carbide NbC_y leads to an increase in the Fermi level by ~0.19 eV. Perturbations of the electronic distribution, which are due to single vacancies, are localized to a great extent. Analysis of X-ray photoemission spectra of carbides $NbC_{0.89}$ and $NbC_{0.85}$ [51] showed that a new peak corresponding to vacancy states appeared in the region below the Fermi level upon deviation from stoichiometry.

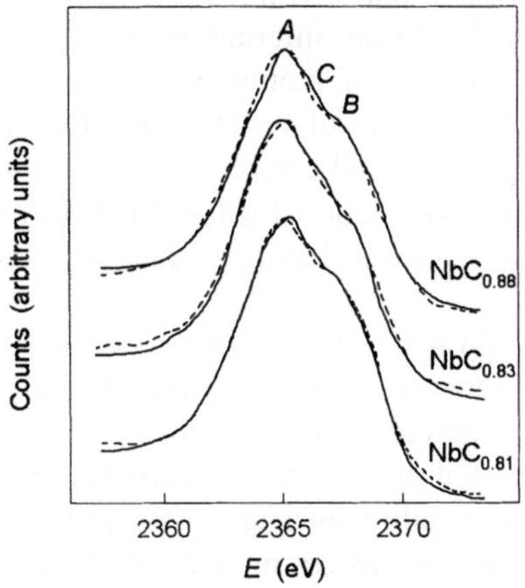

Fig. 11.10. X-ray emission Nb$L\beta_{2,15}$ spectra of carbides $NbC_{0.81}$, $NbC_{0.83}$ and $NbC_{0.88}$ in disordered (solid line) and ordered (dashed line) states [46]

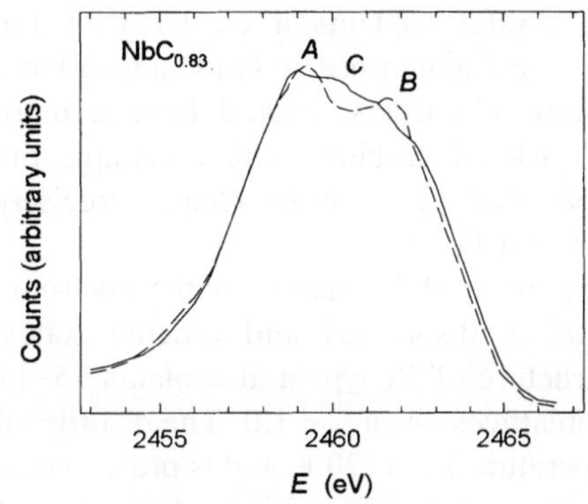

Fig. 11.11. X-ray emission Nb$L\gamma_1$ spectra of carbide $NbC_{0.83}$ in disordered (solid line) and ordered (dashed line) states [46]

The maximum *B*, which is clearly seen in the spectrum of ordered carbide $NbC_{0.83}$ (Fig. 11.11), resembles very closely the peak of vacancy states detected in [51]. Formation of a local peak of the density of states in the presence of vacancies is also confirmed by the data [52] on superconductivity of niobium carbide.

However, the appearance of a vacancy in the nonmetallic sublattice may result not only in the formation of vacancy states themselves, but also in the splitting of some *d*-levels of metal atoms nearest to the vacancy. The growth of the number of vacancies in the neighborhood of a metal atom causes changes in the splitting of *d*-levels. This means that a certain emission spectrum should correspond to each type of environment of a metal atom. Indeed, in accordance with [53], emission *L*-spectra of transition metals in compounds are due to the interaction between *d* wave functions of metal atoms and wave functions of atoms in the nearest environment. Therefore each type of nearest environment of a metal atom has its own spectrum with a certain degree of splitting of the *d*-like subband *B* from the C2*p*-like subband *A*. This fact provides an easy explanation for the different *L*-spectra of disordered and ordered niobium carbides having equal carbon concentrations.

Each niobium atom in niobium carbide has an octahedral environment of six sites of the nonmetallic sublattice. Some of those sites may be vacant. Fully disordered carbide NbC_y contains all types of octahedral clusters: from complete NbC_6 to fully defective $Nb\square_6$. However, the probability of their occurrence in a crystal is different. NMR and magnetic susceptibility measurements [54–58] suggest that ordered carbide most probably contains clusters of one type $NbC_5\square$ (one vacancy and five carbon atoms in the nearest environment of the niobium atom).

If it is assumed that X-ray emission *L*-spectra depend on the type of nearest environment of the metal atom and are determined by superposition of contributions from different types of clusters, each contribution should be proportional to the probability that the crystal contains a cluster with the corresponding configuration. In this case, *L*-spectra of disordered niobium carbide, which receive contributions from all types of clusters, should have a more complicated shape than the spectra of ordered carbide, which contain the contribution mostly from clusters of one type $NbC_5\square$. *L*-spectra change precisely in this way during ordering (see Figs. 11.10 and 11.11).

Titanium monoxide TiO_y is an interesting material for analysis of the electronic structure of nonstoichiometric compounds in disordered and ordered states. Disordered monoxide TiO_y has a cubic structure of *B*1 type and contains 15–17 at. % vacancies in titanium and oxygen sublattices when $y \approx 1.0$. The disordered state is thermodynamically stable at a temperature $T > 1520$ K and is preserved as metastable at room temperature after quenching from 1500 to 300 K. Annealing of TiO_y $(0.9 < y < 1.3)$ at a temperature below 1200 K leads to formation of the monoclinic (space group *C*2/*m*) ordered phase Ti_5O_5.

Barman and Sarma [59] examined oxide TiO_y $(0.8 < y < 1.3)$ by the method of X-ray photoelectron spectroscopy. Spectra were measured on TiO_y samples in ordered and disordered states. They contained a Ti3*d*-like band close to the Fermi level and the main O 2*p*-like band with the intensity maximum corresponding to the binding energy of 7.3 ± 0.3 eV referred to the Fermi level. The intensity of the Ti 3*d*-like band decreased and the intensity of the O 2*p*-band increased in the spectra of disordered oxide TiO_y with *y* varying between 0.81 and 1.30. The spectra of ordered oxide TiO and disordered oxide $TiO_{1.03}$ differ little and the difference shows up as a small narrowing of the *d*-band in ordered oxide. Also, the intensity ratio of 3*d*- and 2*p*-bands in ordered oxide TiO is higher than in disordered oxide.

Electron structures of perfect (with defect free lattice) monoxide $TiO_{1.0}$ and ordered monoclinic monoxide Ti_5O_5 were analyzed using methods of X-ray emission and photoelectron spectroscopy [60]. Defectless titanium monoxide was obtained from oxide $Ti_{0.85}O_{0.85}$ treated under a pressure of 80 kbar at a temperature of 1950 K. The authors [60] did not find any considerable differences in the spectra of defectless cubic and ordered monoclinic oxides of titanium.

The effect of vacancies on the electron structure of TiO was studied repeatedly using quantum mechanical calculations. The method of linearized augmented plane waves (LAPW) was used [61] to calculate the structure of hypothetical defectless cubic monoxide TiO and oxide $Ti_{0.85}O_{0.85}$. The calculated structure of oxide $Ti_{0.85}O_{0.85}$ correlated well with the shape of the experimental X-ray emission Ti $K\beta_5$-spectrum, but the bands were separated wider in the experimental spectrum. However, comparison with experimental emission Ti L_3-spectra and photoelectron spectra showed that the model of defect free cubic titanium monoxide TiO described the energy location of peaks better. It is worth noting that the width of the spectrum calculated [61] for defectless TiO correlated well with the width of experimental spectra [60] of monoclinic ordered and defectless cubic oxides of titanium.

The main finding of the experiments [56, 60] was the absence of vacancy states in spectra of disordered monoxide TiO_y and little difference between the spectra of disordered $TiO_{1.03}$ and ordered TiO or ordered monoclinic and defectless cubic oxides. Probably, experimental studies into the effects of nonstoichiometry and ordering in cubic titanium monoxide TiO_y should be continued.

Experimental and theoretical studies of the electron energy spectrum of nonstoichiometric compounds of transition 5*d*-metals are much more difficult than those of compounds with 3*d*- and 4*d*-metals. Indeed, X-ray emission spectroscopy is virtually inapplicable owing to the absence of narrow lines of internal levels (for example, the width of internal levels of tantalum is about 10 eV [62]), while theoretical quantum mechanical calculations should account comprehensively for relativistic effects of *f*-electrons [63]. For these reasons, the method of X-ray photoelectron spectroscopy was used to examine the energy spectrum of electrons in disordered tantalum carbide TaC_y [64–67].

Experimental photoelectron spectra of tantalum carbide [65, 66] contain a band with the intensity maximum corresponding to the binding energy $E_b \approx 12$ eV referred to the Fermi level. When the carbon concentration of carbide TaC_y decreases, this band shifts slightly towards the Fermi level and its intensity diminishes [66]. It was shown [68–70] that the band with $E_b \approx 12$ eV in tantalum carbide and analogous bands in other carbides and nitrides of transition metals correspond to two valence 2*s*-electrons of the nonmetal. The ratio between impurity states of metal valence electrons and 2*s*-states of the nonmetal is negligibly small. Therefore it may be assumed that 2*s*-electrons of carbon are not involved in any component of the chemical bond, i.e. are not valence electrons in carbides. Considering this, researchers [67] introduced the notion of the concentration of bonding electrons $z = z_M + 2y$, which determines the changes in the properties of carbides MC_y in their homogeneity interval (z_M = 4 or 5 for transition metals of groups IV and V respectively; the factor 2 at y is equal to the number of carbon atom electrons involved in the chemical bond in carbides).

In [67] samples of tantalum carbide TaC_y with different carbon concentrations were prepared by the method of solid-state vacuum sintering under conditions shown in Fig. 3.1. All the synthesized samples were quenched in vacuum from 2200 K to 300 K at a rate of 2000 K min^{-1}. These thermal treatment conditions ensured a disordered distribution of structural vacancies and carbon atoms in the crystal lattice of tantalum carbide.

X-ray photoelectron spectra of valence bands of Ta and carbides TaC_y have four singularities *A*, *B*, *C* and *D* (Fig. 11.12). In the series of carbides from $TaC_{1.00}$ to $TaC_{0.79}$ band *A* shifts towards the Fermi level for 0.2 eV. The shift of band *A* towards the Fermi level for 0.6 eV was detected earlier [66] in the series from $TaC_{0.97}$ to $TaC_{0.52}$. The shift of band *A* towards the Fermi level for 0.35 eV in the series of carbides from $TaC_{1.00}$ to $TaC_{0.72}$ also follows from self-inconsistent relativistic calculations [63] using Green's functions and the KKR formalism. Band *B* is spaced nearly 0.7 eV from the Fermi level. The position of band *B* remains unchanged with the composition of TaC_y, but the relative intensity of this band is enhanced with the concentration of structural vacancies [63, 66, 67].

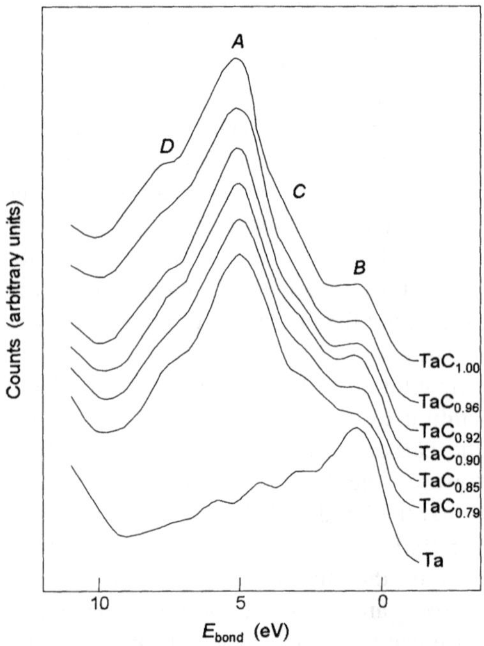

Fig. 11.12. X-ray photoelectron spectra of valence bands of disordered tantalum carbides TaC_y ($1.00 \geq y \geq 0.79$) and tantalum Ta normalized to the integral intensity of the Ta4*f* spin doublet [67]. The energy scale zero for each sample was its corresponding Fermi level

Band C is absent in spectra of near-stoichiometric carbides $TaC_{1.00}$ and $TaC_{0.96}$ (Fig. 11.12). This band appears in the spectrum of carbide $TaC_{0.92}$ and its location changes from 2.7 eV for $TaC_{0.92}$ to 2.4 eV for $TaC_{0.79}$. The intensity of band C is enhanced with decreasing concentration of carbon. Kink D is seen in all X-ray photoelectron spectra of tantalum carbide [65, 67].

The analysis of experimental data [64–67] and theoretical calculations [63, 68–70] of the electron spectrum of tantalum carbide shows that band A reflects the state of C 2*p*- and Ta 5*d*-electrons of M–C covalent bonds, while band B reflects the state of Ta 5*d*-electrons participating in M–M bonds. Kink D probably results from the superposition of narrow band A on a wide conduction band. Band C is directly connected with the formation of vacancies in the carbon sublattice of carbide and corresponds to vacancy states of electrons with *p*-symmetry [63]. The absence of band C in the spectra of carbides $TaC_{1.00}$ and $TaC_{0.96}$ confirms this statement.

Information about the redistribution of the electron charge between carbon and metal atoms was derived [67] from the spectra of internal levels of tantalum Ta $4f_{7/2,\,5/2}$ and carbon C 1*s*. Indeed, the study [67] showed that spin-doublets Ta 4*f* occupy largely different positions in metallic tantalum Ta, stoichiometric carbide $TaC_{1.00}$, and oxide Ta_2O_5. Doublets of stoichiometric carbide are located between doublets of tantalum and tantalum oxide on the energy scale. Doublets of different samples of nonstoichiometric tantalum carbide TaC_y are also located between doublets of tantalum and tantalum oxide and occupy different positions on the energy scale (Table 11.4).

Figure 11.13 presents the variation of the binding energy of Ta $4f_{7/2}$ in the homogeneity interval of disordered carbide TaC_y [64, 66, 67, 71]. All the experimental data (except [64]) unambiguously testify to the increase in the shift ΔE_b with growing concentration of carbon in carbide. Researchers [66, 67] obtained most accurate and close results, which were approximated by linear functions $\Delta E_b = 22.31 + 1.21y$ and $\Delta E_b = 21.96 + 1.65y$ [eV] respectively.

Table 11.4. Position of the inner levels of Ta and C in disordered tantalum carbides TaC_y and of inner levels of Ta in Ta_2O_5 oxide and metal tantalum Ta [67]

Composition	Binding energy E_b (eV)			Shift ΔE_b of the Ta $4f_{7/2}$ level relative to metal tantalum Ta (eV)	Width of the Ta $4f_{7/2,5/2}$ spin doublet on half of height (eV)
	Ta $4f_{5/2}$	Ta $4f_{7/2}$	C 1*s*		
Ta_2O_5	28.5	26.6	–	5.0	3.30
$TaC_{1.00}$	25.53	23.49	283.05	1.87	3.40
$TaC_{0.96}$	25.42	23.52	283.05	1.90	3.43
$TaC_{0.92}$	25.30	23.43	283.05	1.81	3.36
$TaC_{0.90}$	25.25	23.38	283.05	1.76	3.34
$TaC_{0.85}$	25.17	23.35	283.05	1.73	3.34
$TaC_{0.79}$	25.00	23.27	283.05	1.65	3.32
Ta	23.49	21.62	–	0	3.28
error	±0.03	±0.03	±0.05	±0.04	±0.05

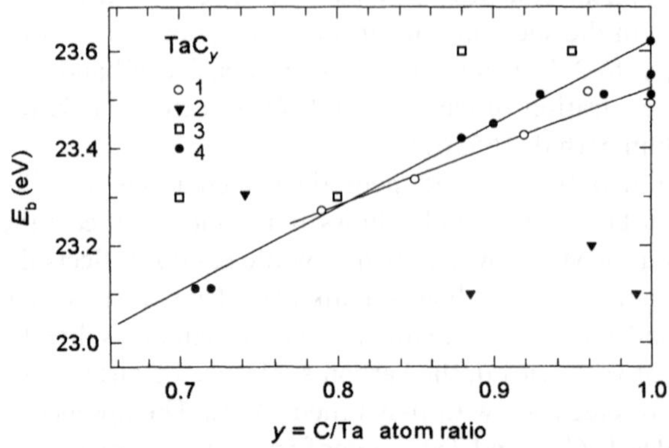

Fig. 11.13. Binding energy E_b of the internal level Ta $4f_{7/2}$ for disordered tantalum carbide TaC_y: (1) data from [67], determination error of E_b is ±0.03 eV; (2) data from [64], error is ±0.2 eV; (3) data from [71], error ±0.1 eV; (4) data from [66], error is unknown. Values [66, 67] were approximated by linear functions

All metal atoms have a similar environment in stoichiometric carbide. Since carbon atoms and vacancies are distributed randomly in nonstoichiometric carbides, metal atoms occupy nonequivalent positions and, consequently, can possess different charges. In this case, internal levels may broaden in nonstoichiometric carbides. However, widening of the doublets Ta $4f_{7/2,\ 5/2}$ is not observed experimentally (see Table 11.4), which is an indication that all metal atoms in carbide TaC_y of the given composition have equal charges. Taking into account this circumstance, Rempel and Sinelnichenko [67] estimated the effective charge of tantalum atoms in carbide TaC_y.

Let us assume that tantalum atoms in metallic tantalum have a zero charge, while the charge of tantalum atoms in tantalum oxide Ta_2O_5 is positive and is nearly equal in absolute value to the charge of five electrons $5e$. If the difference of Madelung energies in tantalum, tantalum oxide and tantalum carbides is neglected, the dependence between the atomic charge q and the shift of the internal level ΔE_b can be considered as linear [72]. Calculations of the effective charge of a tantalum atom $q_{Ta}(y)$ in TaC_y, which were made [67] in the approximation of the linear dependence $\Delta E_b = kq$, yielded the following results: $q_{Ta}(y) = 0.34 + 1.65y$ [e] [66] and $q_{Ta}(y) = 0.69 + 1.21y$ [e] [67]. Hence, the charge of tantalum atoms in stoichiometric tantalum carbide $TaC_{1.00}$ is about $1.9e$. The charge q_{Ta} decreases with growing concentration of structural vacancies in TaC_y.

As can be seen from Table 11.4, the binding energy of the internal level of carbon C 1s is independent of the carbide composition and, consequently, the charge of carbon atoms in the carbide is independent of the concentration of structural vacancies as a first approximation. The charges of tantalum and carbon atoms in stoichiometric carbide $TaC_{1.00}$ are equal in absolute value and are opposite in sign. Therefore $q_C = -1.9e$. The vacancy charge $q_\square$, which was determined [67] from the crystal electroneutrality condition $q_{Ta}(y) + yq_C + (1-y)q_\square = 0$ using data on $q_{Ta}(y)$ and q_C, is independent of the carbide composition, is negative, and is equal to $-0.34e$ [66] or $-0.69e$ [67]. The charge of a structural vacancy is much smaller in absolute value than the charges of tantalum and carbon atoms. This means that metal and carbon atoms rather than vacancies make the main contribution to the ionic component of the chemical bond in tantalum carbide.

Results of an X-ray emission spectroscopic examination of tantalum carbide exposed to synchrotron radiation [73] are remarkable. Synchrotron radiation opens up fundamentally new perspectives for spectroscopic analysis of elements since it provides a required wavelength chosen from a broad spectral range. X-ray synchrotron radiation with an energy of 420 eV was used to excite Ta $N_3(4p_{3/2})$-spectra in tantalum and disordered tantalum carbides $TaC_{0.75}$, $TaC_{0.79}$ and $TaC_{0.85}$.

The analysis of the spectra (Fig. 11.14) showed that they comprise three main bands with intensity maxima corresponding to ~392, ~394 and 400 eV. A wide smeared band is present in the energy region from 388 to 389 eV. Let it be noted that four bands were also observed in X-ray photoelectron spectra (see Fig. 11.12) of tantalum carbide [67]. However, electrons of carbon atoms contribute, along with electrons of tantalum atoms, to photoelectron spectra. Consequently, bands in emission and photoelectron spectra of tantalum carbide are not fully identical.

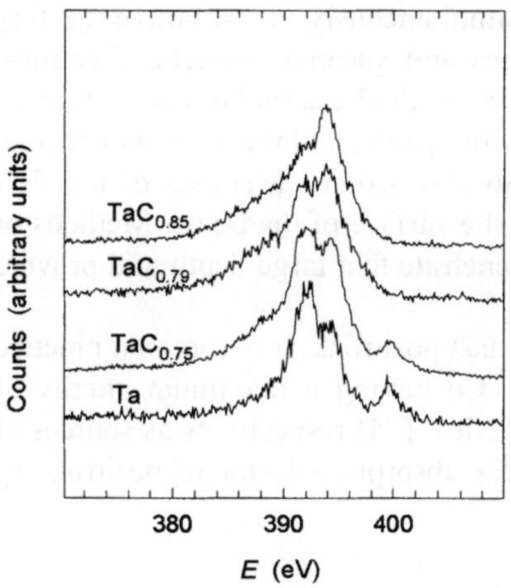

Fig. 11.14. X-ray emission Ta $N_3(4p_{3/2})$ spectra of tantalum Ta and nonstoichiometric tantalum carbides $TaC_{0.75}$, $TaC_{0.79}$ and $TaC_{0.85}$ [73]

Emission spectra of tantalum Ta and carbides TaC_y differ mainly by the intensity of the bands. In going from tantalum to carbide $TaC_{0.75}$ and with growing concentration of carbon in carbides TaC_y, the intensity of the low-energy (392 eV) band drops considerably, the intensity of the 394 eV band rises, and the intensity of the high-energy (400 eV) band decreases quickly. The last band is absent in the spectra of carbides $TaC_{0.79}$ and $TaC_{0.85}$. The intensity and the width of the smeared band near 388 eV increase simultaneously. The bands shift to the region of lower energies for 0.3–0.5 eV in the series Ta $\rightarrow TaC_{0.75} \rightarrow TaC_{0.85}$.

The observed modification of the valence spectrum with the chemical composition of tantalum carbide is related to the redistribution of valence electrons of tantalum atoms between metal–metal and metal–carbon bonds. One may think that the most intense bands with energies of ~392 and ~394 eV reflect the distribution of Ta 5*d*-electrons in the *d*-like band of carbide, while the low-energy band (388 eV) reflects the distribution of Ta 5*d*-electrons in the binding C 2*p*-Ta 5*d* band. Relativistic quantum mechanical calculations need be performed for detailed quantitative analysis of emission spectra of carbide TaC_y.

11.3 Electron–Positron Annihilation

Electron–positron annihilation is one of the most sophisticated methods of analyzing defects in nonstoichiometric compounds. However, this method is little known to a wide circle of researchers. Let us discuss briefly the specific features, capabilities and applicability of the method.

Annihilation of positrons allows the determination of the characteristics of the electron system of perfect crystals and, simultaneously, is sensitive to tiny imperfections in solids, specifically vacancies and vacancy clusters. Electron–positron annihilation (EPA) has three main methods: angular correlation of annihilation radiation, Doppler broadening of the γ-line, and the positron lifetime. These three methods can be subdivided into two groups. Methods of the first group use slow positrons for examination of the surface of materials. Methods of the second group use fast positrons, which penetrate to a large depth and provide information about volume properties of solids.

Let us consider annihilation methods with fast positrons. It is common practice to use radioactive isotopes ^{22}Na, ^{44}Ti and ^{58}Co having a maximum energy of emitted positrons equal to 0.54, 1.5 and 0.47 MeV [74] respectively as sources of positrons. In accordance with [75], the linear absorption factor of positrons in solids is

$$\alpha = \frac{(16 \pm 1)\rho}{E_{max}^{1.43}} \ (\mathrm{cm}^{-1}) . \tag{11.6}$$

Substituting the maximum energy of positrons E_{max} (MeV) and the density ρ (g cm^{-3}) into formula (11.6), it is possible to calculate the quantity $x = \alpha^{-1}$, i.e. the penetration depth of a positron in a solid at a given intensity of the beam. The most popular isotope ^{22}Na emits positrons with E_{max} = 0.54 MeV and the carbide density ρ is from 5 to 15 g cm^{-3}. From (11.6) it follows that a positron penetrates in carbides to a maximum depth of 100–300 μm (attenuation of e^6 times), while the average penetration depth of a positron in carbides is about 15–45 μm (attenuation of e times).

A positron released from a radioactive source (emitter) is thermalized upon penetration in a solid, i.e. quickly loses its speed and energy, which becomes equal to the crystal temperature. The thermalization time is about 1–5 ps and is negligibly small compared to the positron lifetime in a solid. If the mean dimension of grains in carbides is several microns, thermalized positrons are uniformly distributed over the volume of many grains occurring at a small depth from the crystal surface. Thanks to the large size of grains, the fraction of positrons thermalized near the surface of grains is negligibly small compared to the fraction of positrons thermalized in the volume of grains.

A thermalized positron starts diffusing in a crystalline grain in the so-called free or delocalized state and annihilates from this state in the characteristic time τ^{free}. The free-state lifetime τ^{free} of a positron in carbides is estimated at 89–114 ps [76, 77]. A positron can move in carbides to a distance of about 100 nm during the time τ^{free}. This estimate follows from experimental data concerning the diffusion coefficient of positrons $D_0^+ = 1{-}2$ cm^2 s^{-1} in such metals as Al, Cu, Mo, Ag, and silicon carbide SiC [78, 79] where positrons move to a distance of 100–180 nm and 70 ± 10 nm respectively during their free-state lifetime. In most cases, the size of grains in carbides is several dozen times larger than the diffusion length. Therefore a negligible number of positrons thermalized in grains will reach the grain surface [80]. They will be still fewer if grains contain defects that capture positrons.

From the fundamental research of positron annihilation in metals [81], semiconductors [82], nanocrystals [83, 84], amorphous materials [85, 86] and ceramics [87–89] it is known that positrons are captured by vacancies in metallic and nonmetallic sublattices, vacancy clusters, grain boundaries, dislocations, etc. Interstitial atoms, which also represent defects in a solid, normally do not capture positrons because of the large positive charge of the atomic nucleus. When captured by a defect, a positron annihilates from the localized state in a time exceeding τ^{free}. The electron density of a defect is smaller than the electron density in the interatomic space of a defectless crystal. As a result, the positron lifetime in a defect is longer than the free lifetime.

Vacancies in nonmetallic and metallic sublattices serve as the main defects in transition-metal compounds. Positrons may be captured by these defects if the defect-to-defect distance is smaller than the positron diffusion length equal to ~100 nm. This distance corresponds to about 10^{-6} defects per atom in a compound. Carbon vacancies are the most probable traps of positrons in carbides, because the carbon vacancy concentration is always $> 10^{-6}$ even in near-stoichiometric carbides. The equilibrium concentration of metal vacancies in carbides is assumed to be negligibly small, because the formation enthalpy of those vacancies is very high. However, trapping of positrons by metal vacancies cannot be disclaimed without special investigations. The trapping rate of positrons by metal vacancies may be higher than the trapping rate of positrons by nonmetal vacancies if one considers the high charge of the metal atom nucleus compared to the carbon charge.

Until recently, nonstoichiometric carbides of transition metals were studied using two methods of positron annihilation: the positron lifetime [88–90] and the angular correlation of annihilation radiation [91–93]. First, we shall consider results concerned with the positron lifetime showing which defects capture positrons and, correspondingly, the distribution of electronic momenta provided by the angular correlation of annihilation radiation.

A fast–slow correlation γ-spectrometer was used to measure the positron lifetime. The source of positrons was radioactive salt ^{22}NaCl deposited on an aluminum foil 0.8 μm thick. The source component about 350 ps long accounted for nearly 5% in the positron lifetime spectra. It was considered during multi-component analysis of the spectra. The positron lifetime spectra were accumulated up to the integral intensity of $(2–7) \times 10^6$ coincidence counts. The value of one channel was 12.3 ps. The time resolution of the spectrometer was determined using annealed Al. An aluminum polycrystal was annealed in a 10^{-3} Pa vacuum at a temperature of 900 K for 2 h and was then cooled slowly to room temperature. The measured lifetime of positrons in annealed Al was 163 ps, which was in good agreement with literature data [81]. The full width at half-maximum (FWHW) of the spectrometer resolution function was 260 ps. The spectra were processed numerically using the POSFIT and RESOLUTION software packages [94].

The measured lifetime of positrons in near-stoichiometric carbides is given in Table 11.5. The analysis of experimental data (Table 11.5) shows that the lifetime of positrons in carbides is much longer than the positron lifetime in corresponding metals, although the mean density of electrons is by far larger in carbides than in metals. From this it may be inferred that the measured lifetime of positrons in carbides corresponds to localization of positrons in defects rather than to the free delocalized state. This conclusion is also confirmed by quantum mechanical calculations [77, 95].

The lifetime spectra of positrons in carbides also depend on the composition of nonstoichiometric carbide. This is readily seen on the example of titanium carbide (Fig. 11.15). The temperature dependence of the lifetime of positrons in nonstoichiometric niobium carbide is illustrated in Fig. 11.16. The dependence of the positron lifetime on the composition of cubic (with *B*1 structure) carbides of titanium and niobium is shown in Fig. 11.17.

Table 11.5. Lattice constant, a_{B1}, and positron lifetime, τ_{MC_y}, for groups IV and V transition-metal carbides with *B*1 type structure and free positron lifetime, τ_M^{free}, for transition metals

Property	group IV			group V		
	$TiC_{1.00}$	$ZrC_{0.98}$	$HfC_{1.00}$	$VC_{0.87}$	$NbC_{1.00}$	$TaC_{0.99}$
a_{B1} (nm)	0.4328	0.4698	0.4639	0.4163	0.4470	0.4456
$\tau_{MC_y} \pm 1$ (ps)	160	176	173	157	161	155
$\tau_M^{free} \pm 1$ (ps)	147	165	174	130	119	116

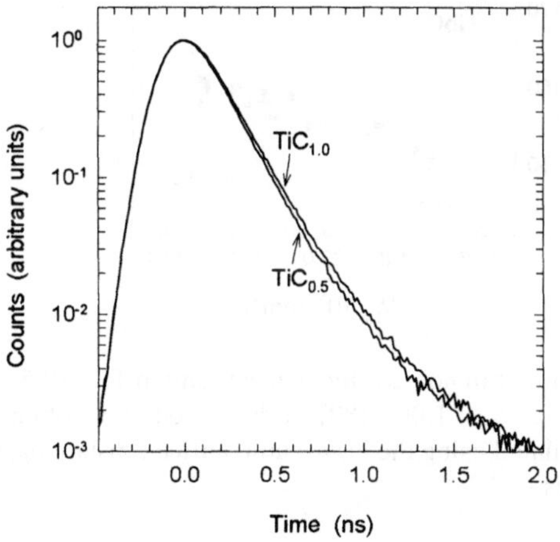

Fig. 11.15. Spectra of the positron lifetime in titanium carbides with largely different compositions $TiC_{1.0}$ and $TiC_{0.5}$. The positron lifetime in nonstoichiometric carbide $TiC_{0.5}$ is shorter than in near-stoichiometric carbide $TiC_{1.0}$

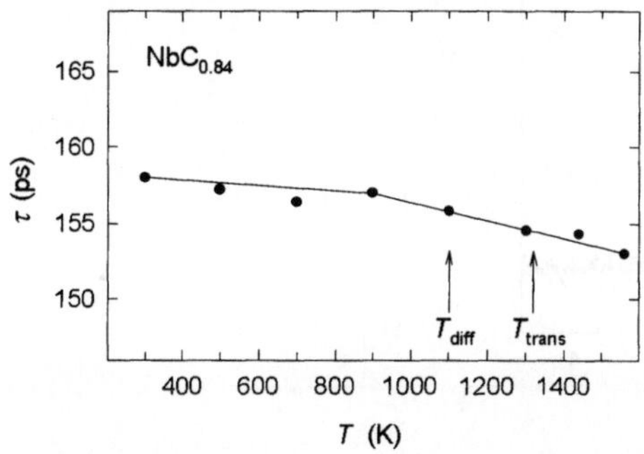

Fig. 11.16. Decrease in the positron lifetime τ in ordered niobium carbide $NbC_{0.84}$ after heating under vacuum [76]: T_{dif} is the diffusion start temperature of carbon atoms in the nonmetallic sublattice during the disorder–order phase transition, T_{trans} is the $Nb_6C_5 \rightarrow NbC_{0.84}$ order–disorder transition temperature. The lattice volume decreases during heating of carbide as a result of disordering

One more argument in favor of the localized state of positrons in carbides is adduced by the study of nanocrystalline vanadium carbide $VC_{0.875}$ [45, 96]. Figure 11.18 presents lifetime spectra of positrons in a coarse-grain powder of carbide $VC_{0.875}$ and a nanocrystalline (the mean size of particles being about 150–200 nm) powder having the same composition.

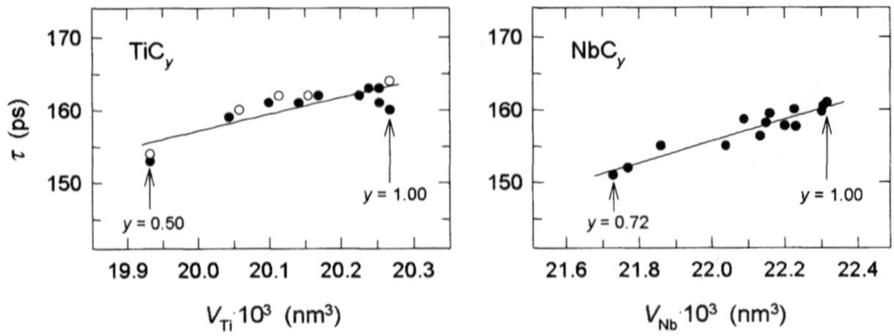

Fig. 11.17. Dependence of the positron lifetime τ in nonstoichiometric titanium TiC_y ($0.5 \le y \le 1.0$) [90] and niobium NbC_y ($0.72 \le y \le 1.00$) [89] carbides on the volume $V_M \equiv V_{MC_y} = a^3_{B1}/4$ per metal atom as calculated using the basic cubic lattice constant a_{B1}: (○) annealed samples, (●) quenched samples

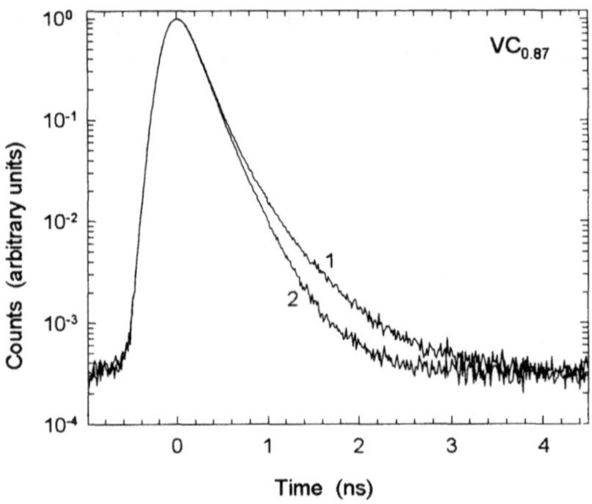

Fig. 11.18. Positron lifetime spectra of nanocrystalline (1) and coarse-crystalline (2) vanadium carbide $VC_{0.875}$ [45]: in addition to the short component, the spectrum of nanocrystalline carbide contains a long component responsible for annihilation of positrons on the surface of nanoparticles

In addition to the short (157 ps) component, which is characteristic of the coarse-grain sample, the spectrum of nanocrystalline carbide (Fig. 11.18, spectrum 1) contains a long (500 ps) component, which is typical of the surface state of positrons. The intensity of the short component exceeds 90%, while the intensity of the long component is about 7%. The low intensity of the long component suggests that the free path of positrons in nanocrystalline vanadium carbide is smaller than the free path of positrons in the free state. Consequently, a small number of positrons, which were thermalized in a nanocrystalline grain of vanadium carbide, can reach the grain surface. This means that vanadium carbide contains positron trapping centers. These centers may be presented by carbon vacancies whose concentration amounts to 13% and by metal vacancies if they are present in the sample.

So, experimental and theoretical data suggest that the positron lifetime in carbides corresponds to localization of positrons in defects. Today this opinion has been universally accepted, but the question of whether nonmetal or metal vacancies capture positrons is open [97].

Let us consider arguments in favor of the capture of positrons by carbon vacancies in carbides since these arguments are most convincing. The study of the positron lifetime τ_{MC_y} in near-stoichiometric carbides $TiC_{1.00}$, $ZrC_{0.98}$, $HfC_{1.00}$, $NbC_{1.00}$ and $TaC_{0.99}$ [88, 89] showed that positrons annihilate mainly with valence electrons of metal atoms and, to a lesser extent, with electrons of carbon atoms. This means that positrons are captured by structural vacancies in the carbon sublattice. Carbon vacancies are surrounded by six metal atoms and the total electron density of these vacancies is determined by electrons of metal atoms.

Approximation of the experimental spectra showed that they are best described by one exponential component. When the spectra were described by two components with different values of the positron lifetime, the convergence was impaired. Thus, positrons in carbides annihilate only on one type of defect.

From the analysis of the data [88, 89] it follows that the normalized rate of positron annihilation [98]

$$\lambda^* = (\lambda - \lambda_\infty) / \lambda_\infty \tag{11.7}$$

($\lambda_\infty = 2.004 \times 10^9$ s^{-1} and $\lambda = 1/\tau$) does not correlate with the total density of electrons in metals and their carbides. This means that not all electrons of the material but only part of them participate in the annihilation of positrons. Positrons annihilate mostly with electrons whose density is relatively high at a distance from positively charged nuclei, i.e. valence electrons.

Figure 11.19 [89] depicts the positron annihilation rate λ^* versus the density of valence electrons $\eta^{\rm val} = zV_{\rm B}/V$, where $z = z_{\rm M}^{\rm val} + 2y$ is the number of valence electrons participating in annihilation and corresponding to one formula unit of the material occupying a volume $V = (a_{B1})^3/4$ and $V_{\rm B} = 6.208 \times 10^{-31}$ m^3 is the volume of a sphere with the Bohr radius. The points near straight line 1 correspond to the dependence of the positron annihilation rate λ^* in carbides $TiC_{1.00}$, $ZrC_{0.98}$, $HfC_{1.00}$, $NbC_{1.00}$ and $TaC_{0.99}$ on the total density of valence bonding electrons $\eta_{\rm tot}^{\rm val}$ in these carbides. The points near straight line 2 correspond to the dependence of the annihilation rate λ^* in transition metals Ti, Zr, Hf, Nb, Ta and their carbides $TiC_{1.00}$, $ZrC_{0.98}$, $HfC_{1.00}$, $NbC_{1.00}$ and $TaC_{0.99}$ on the density of valence electrons of metal atoms $\eta_{\rm M}^{\rm val} = z_{\rm M}^{\rm val} V_{\rm B} / V$.

Experimental values of λ^* as a function of $\eta_{\rm tot}^{\rm val}$ for carbides [88, 89] fit well straight line 1 (Fig. 11.19), a fact which attests to annihilation of positrons with valence electrons. However, those values lie below the analogous dependence known for metals [98]. Moreover, straight line 1 does not pass through the origin of coordinates. Hence, not all valence electrons are involved in annihilation.

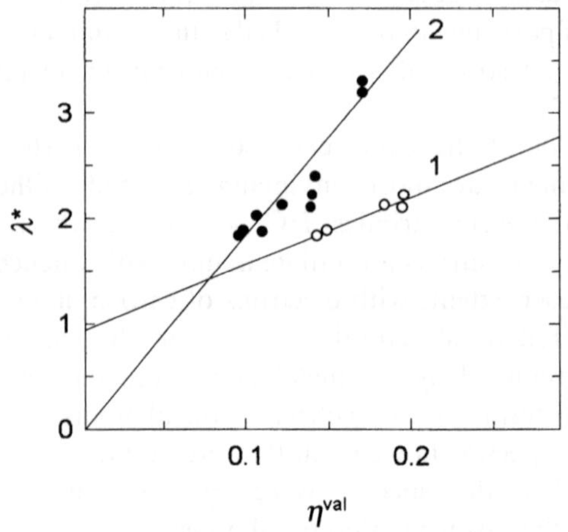

Fig. 11.19. Normalized positron annihilation rate λ^* [88, 89]: (1) dependence of λ^* on the total concentration of valence electrons η^{val}_{tot} in near-stoichiometric carbides $TiC_{1.00}$, $ZrC_{0.98}$, $HfC_{1.00}$, $NbC_{1.00}$ and $TaC_{0.99}$; (2) dependence of λ^* on the concentration of valence electrons of metal atoms η^{val}_{M} in metals Ti, Zr, Hf, Nb, Ta and their near-stoichiometric carbides

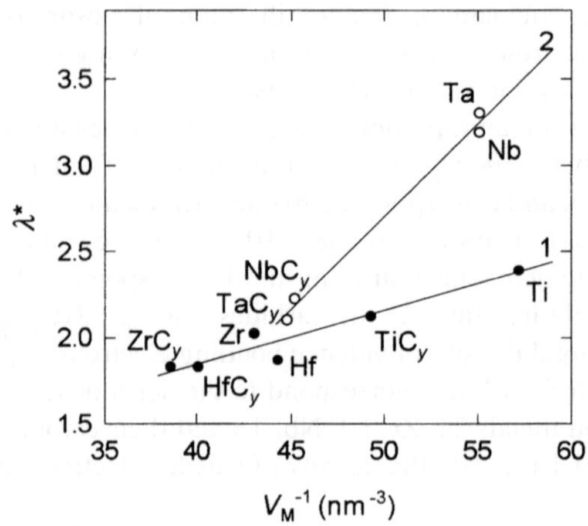

Fig. 11.20. Dependence of the normalized positron annihilation rate λ^* on the inverse volume V_M^{-1} per one metal atom [88]: (1) IV group metals and their near-stoichiometric carbides; (2) V group metals and their near-stoichiometric carbides

Straight line 2 passing through the origin of the coordinates is drawn through the experimental points on the assumption that positrons in carbides annihilate only with valence electrons of metal atoms. This experimental line is described by the function

$$\lambda^* = 17.9\eta_{\mathrm{M}}^{\mathrm{val}}, \quad (11.8)$$

which is similar to the function known for metals [98].

The conclusion about annihilation of positrons with valence electrons of metal atoms in carbides and the absence of annihilation with electrons of carbon atoms was confirmed [88] by the linear behavior of the λ^* vs. V_{M}^{-1} plot (V_M being the volume per metal atom) constructed for near-stoichiometric carbides of IV- and V-group transition metals (Fig. 11.20). From Fig. 11.20 it is seen that the slope of the two straight lines describing the normalized annihilation rate of positrons in materials is the steeper, the larger the number of valence electrons of the metal atoms.

The concentration of valence electrons of metal atoms is the main factor affecting the dependence of the positron lifetime on the composition of strongly nonstoichiometric titanium [90] and niobium [76, 88, 89] carbides. In those works the lifetime was measured using quenched and annealed samples of titanium carbide TiC_y with $0.50 \le y \le 1.00$ and annealed samples of niobium carbide NbC_y with $0.72 \le y \le 1.00$ (see Fig. 11.17). Annealed carbides $TiC_{0.50}$ and $TiC_{0.60}$ contained the ordered phase Ti_2C, while samples of annealed carbide $TiC_{0.63}$ had the phase Ti_3C_2 in addition to the phase Ti_2C. Diffraction patterns of annealed carbide $TiC_{0.67}$ contained superstructure reflections of the ordered phase Ti_3C_2 only. Samples of annealed carbides TiC_y with $y \ge 0.70$ were disordered.

The analysis of experimental spectra of the positron lifetime (see, e.g., Fig. 11.15) showed that these spectra are described best by one exponential component irrespective of the carbon concentration. Hence, positrons in strongly nonstoichiometric carbides annihilate on defects of the same type, namely, structural vacancies of the carbon sublattice. Since positrons are captured by negative defects only, it may be stated that carbon vacancies in titanium and niobium carbides bear a negative charge, i.e. possess some electron density.

When the carbon concentration rises within the homogeneity interval of titanium and niobium carbides, the positron lifetime τ increases from 153 to 164 ps in TiC_y and from 151 to 161 ps in NbC_y. The increase in the carbon concentration is followed by the growth of the lattice constant a_{B1} in carbides. As a result, the volume of a structural vacancy changes. The relative increase in the volume of a structural vacancy with variation of the carbide composition is proportional to the increase in the volume $V_{\mathrm{M}} \equiv V_{\mathrm{MC}_y}/4 = a_{B1}^3/4$ corresponding to one metal atom. For this reason, the positron lifetime is shown as a function of V_{M} in Fig. 11.17. From Fig. 11.17 it is seen that the positron lifetime in titanium and niobium carbides increases linearly with the vacancy volume. This behavior of the positron lifetime can be expected if positrons annihilate only with electrons of metal atoms. In this case, given a constant vacancy charge, the increase in the volume V_M should be followed by a decrease in the volume electron density of a structural vacancy. The positron lifetime should increase correspondingly. If electrons of carbon atoms participated in annihilation of electrons, i.e. if annihilation occurred in metal vacancies, the positron lifetime would decrease, because the electron density in a metal vacancy should grow with the carbon concentration y thanks to the large number of carbon atoms surrounding the vacancy.

The effect of ordering of structural vacancies on the positron lifetime in titanium and niobium carbide was studied by Rempel *et al.* [76, 90]. Annealing of titanium carbide samples and the formation of ordered phases Ti_2C and Ti_3C_2 in those samples led to a small increase (by 1 ps) in the positron lifetime [90] (see Fig. 11.17). Ordering in the titanium carbide samples was accompanied by a small increase (by 0.0001 nm) in the basic lattice constant a_{B1} of titanium carbide. Therefore the observed increase in τ can be viewed as an immediate consequence of the increase in the lattice constant a_{B1} and the volume of the basic crystal lattice of titanium carbide during ordering. An analogous result was obtained in the work [76] dealing with the effect of ordering on the positron lifetime *in situ* in nonstoichiometric niobium carbide $NbC_{0.84}$. The positron lifetime at a high temperature was measured in a Pelletron accelerator (Stuttgart) using a beam of high-energy positrons (4 MeV). The $\beta^+\gamma$ coincidence scheme was used for measurements [99]. The measurements showed that the positron lifetime was nearly constant in the temperature interval from 300–900 K (Fig. 11.16). The positron lifetime decreased upon reaching the temperature of active diffusion of carbon atoms and vacancies and the temperature of the order-disorder phase transition. The observed decrease in the positron lifetime by ~10 ps may be related to the decrease in the crystal volume and, correspondingly, structural vacancies during disordering of niobium carbide. The positron lifetime in niobium carbide differs little at high and low temperatures, but is largely different from the free lifetime. This means that positrons do not detach from structural vacancies and, consequently, the binding energy of a positron and a structural vacancy is high, i.e. a structural vacancy presents a deep trap for a positron.

One more argument in favor of trapping of positrons at structural vacancies is adduced by experiments concerned with exposure of nonstoichiometric and stoichiometric carbides to high-energy electrons [75, 90, 100, 101]. The lifetime of positrons in irradiated carbides is given in Table 11.6. It is seen that the positron lifetime increased by 15 ps in vanadium carbide $VC_{0.87}$ exposed to an especially high dose of electrons with energy 2.5 MeV. This fact suggests that carbide $VC_{0.87}$ contains a new type of irradiation-induced point defect with the positron lifetime longer than in the initial vanadium carbide. The positron lifetime increased little in other nonstoichiometric carbides both after irradiation with high- (2.5 to 2.8 MeV) and low-energy (1 MeV) electrons. Tungsten and silicon carbides virtually do not have homogeneity intervals. As distinct from nonstoichiometric carbides, the positron lifetime in WC and SiC changes considerably both at high and low energies of irradiation particles.

Table 11.6. Change of positron lifetime, $\Delta\tau_{irr} = \tau_{irr} - \tau_{unirr}$, after irradiation by electrons with energy E and dose Φ at a temperature below 200 K [76, 90, 100, 101] (τ_{unirr} and τ_{irr} are the positron lifetime before and after irradiation)

Carbide	E (MeV)	$\Phi \times 10^{-23}$ (m^{-2})	$\tau_{irr} \pm 2$ (ps)	$\Delta\tau_{irr}$ (ps)
$TiC_{0.99}$	2.8	1.6	163	3
$VC_{0.87}$	1.0	1.8	157	0
$VC_{0.87}$	2.5	3.0	172	15
$NbC_{0.98}$	1	0.5	160	0
$NbC_{0.98}$	2.8	1.6	160	0
$NbC_{0.84}$	1	0.5	159	1
$NbC_{0.84}$	2.8	1.6	155	−3
$TaC_{0.99}$	2.8	1.6	155	0
[1] WC	1.0	1.8	136	14
[1] WC	2.5	1.2	137	15
[1] SiC	0.5	1.9	159	13
[1] SiC	2.5	1.1	210	64

[1] Changes of the positron lifetime in irradiated WC and SiC carbides are given relative to the lifetime of positrons in the free state in these compounds

The energy E_t^{max}, which is transferred by incident electrons to atoms of a material, can be estimated by a formula which describes elastic collisions of particles taking account of relativistic effects [102]: $E_t^{max} = 2(E/Mc^2)(E+2mc^2)$, where m is the electron mass, M is the mass of an irradiated atom, and c is the light velocity. Calculations show that Ti, V and Nb atoms receive an energy equal to 90, 84 and 46 eV, while C atoms receive an energy of 360 eV even if the irradiation electrons have an energy E = 1 MeV. These energy values are much higher than the displacement threshold T_d estimated [103] for titanium and carbon atoms at 4–16 and 20–44 eV respectively. The displacement threshold of vanadium is small and is close to T_d of titanium. Consequently, metal vacancies and additional carbon vacancies should appear in vanadium carbides exposed to electrons with energy 1 MeV. However, the positron lifetime remained unchanged. Thanks to stoichiometry and the absence of defects, a different situation is observed for tungsten and silicon carbides irradiated with 1-MeV electrons. Irradiation-induced vacancies, which capture positrons, are formed in the carbon sublattice of these defectless carbides [100, 101]. This conclusion follows unambiguously for tungsten carbide. An energy of 23 eV transferred to W atoms is smaller than the displacement threshold of a Ta atom equal to 42 eV [104] in tantalum carbide. Displacement thresholds of a tungsten atom in tungsten carbide and a tantalum atom in tantalum carbide should be nearly equal thanks to the similarity of physical properties of those compounds. Therefore it may be inferred that only carbon atoms are displaced in tungsten carbide exposed to 1-MeV electrons and no vacancies appear in the metallic sublattice. The positron lifetime is much longer after irradiation and this new lifetime of positrons can be attributed to carbon vacancies only.

When carbides are irradiated with electrons having an energy of 2.5–2.8 MeV, both types of atoms are displaced in all the carbides studied and vacancies appear in both sublattices thanks to the large transfer energy E_t^{max}. The concentration of radiation vacancies can be found from the formula $c_v = \sigma_d \Phi_{irr}$, where σ_d is the displacement cross-section and Φ_{irr} is the irradiation dose. In accordance with [103], in titanium carbide $TiC_{0.97}$ the displacement cross-section σ_d is 32–90 barns for titanium atoms at the displacement threshold from 44 to 20 eV, and 60–320 barns for carbon atoms at the displacement threshold from 16 to 4 eV. Considering the similarity of titanium and vanadium atoms, it may be conjectured that electron irradiation at a dose $\Phi_{irr} = 3.0 \times 10^{23}$ m^{-2} induces up to 0.1–0.3 at.% radiation vacancies in the vanadium sublattice and up to 0.2–1.0 at.% additional vacancies in the carbon sublattice. The positron lifetime in irradiated carbide $VC_{0.87}$ increased by 15 ps. Defects (13% carbon vacancies) were present in vanadium carbide before irradiation and, consequently, the new lifetime value can be related only to radiation-induced metal vacancies. Metal vacancies can be observed against the background of a large concentration of carbon vacancies only if the trapping rate of positrons at metal vacancies is much larger than its counterpart at carbon vacancies. The positron lifetime did not increase in other irradiated nonstoichiometric carbides. This means that positrons were not captured at irradiation-induced metal vacancies.

It is interesting to compare experimental and calculated [77, 95] values of the positron lifetime in cubic carbides. The LMTO-PAS method was used in calculations. Hexagonal tungsten carbide was modeled by carbide with a *B*1 cubic lattice [95]. Vacancies in transition-metal carbides were modeled as a cluster in the form of a cubic unit cell containing four sites of the metallic and nonmetallic sublattices each. The central site of the cell was vacant. This cluster allowed modeling carbide $MC_{0.75}$ (M_4C_3) or $M_{0.75}C$ (M_3C_4), which contained 25 at.% vacancies in the carbon or metal sublattice respectively. Calculated values of the positron lifetime at metal and carbon vacancies in ideal defect-free carbides are given in Table 11.7.

The comparison of experimental and calculated values of the positron lifetime shows that the calculated values correlate well (±5 ps) with the free lifetime of positrons in stoichiometric carbides. The positron lifetime calculated for carbon and metal vacancies is underestimated for 20 ps on the average. An exception is the positron lifetime at metal vacancies in tungsten carbide and at carbon vacancies in silicon carbide. In this case, the calculated and experimental values are in satisfactory agreement. Calculations yielded a correct qualitative relation between the positron lifetime in the free state, at carbon vacancies, and metal vacancies: $\tau^{free} < \tau_{\square C} < \tau_{\square M}$. The experimental and calculated values of the positron lifetime at vacancies disagree qualitatively probably because the relaxation of atoms near vacancies was neglected and, also, the change of the lattice parameter upon formation of vacancies was disregarded in calculations [77, 95].

Table 11.7. Calculated positron lifetime in transition metal carbides and silicon carbide [77, 79, 95]: τ^{free} free positron lifetime; $\tau_{\square C}$ positron lifetime in carbon vacancy; $\tau_{\square M}$ positron lifetime in metal vacancy

Carbide	τ^{free} (ps)	$\tau_{\square C}$ (ps)	$\tau_{\square M}$ (ps)
TiC	98	124	161
ZrC	114	148	196
HfC	105	137	185
VC	89	117	149
NbC	105	138	181
TaC	99	130	177
[1] WC	95	116	161
SiC	141	153	[2] 192

[1] Tungsten carbide WC is simulated as a nonequilibrium cubic phase with a *B*1-type structure in the calculation [79].
[2] Positron lifetime for Si-vacancy in silicon carbide SiC.

One more reason for the disagreement between experimental and theoretical values of the positron lifetime is that experimental results are probably impaired due to the presence of impurities in carbides. Oxygen is the most frequent impurity in carbides. The oxygen concentration is as high as 0.1 mass.% even in the purest carbides of group V transition metals and 0.2–0.3 mass.% in carbides of group IV transition metals. From the literature [1, 2] it is known that impurity oxygen occupies sites in the carbon sublattice. If the oxygen concentration grows high, one may assume, by analogy with cubic titanium monoxide TiO_y [105, 106], the appearance of equilibrium metal vacancies in carbides, which are capable of trapping a positron. The method of positron annihilation in carbides provides a maximum sensitivity of about 1 ppm, i.e. 10^{-6} defects per atom. In nonstoichiometric compounds with the *B*1 structure each site of the metallic sublattice has an octahedral environment of six sites of the nonmetallic sublattice. The probability that an octahedron with n sites occupied by carbon atoms is formed in the disordered compound MC_yO_{1-y} is $C_6^n y^n (1-y)^{(6-n)}$. If oxygen atoms are arranged randomly in the carbon sublattice of a $MC_{0.99}O_{0.01}$ sample, then a vacancy should appear in the octahedron with three oxygen atoms so that metal vacancies achieve the concentration of 10^{-6}. However, the oxygen impurity can hardly exert such a strong effect on the metallic sublattice. Most probably, a metal vacancy may be formed in an octahedral cluster with five oxygen atoms, as is the case with titanium monoxide. Then the concentration of metal vacancies in carbides containing 0.1–0.3 mass.% impurity oxygen should be much smaller than 1 ppm. Such a low concentration of metal vacancies is outside the sensitivity range of the method of positron annihilation and cannot be detected by this method. This means that impurities present in carbides cannot be the cause of the disagreement between measured and calculated values of the positron lifetime.

Experimental data obtained for nonstoichiometric carbides of transition metals suggest unambiguously that positrons are captured at vacancies of the carbon sublattice and annihilate at these vacancies with the average lifetime τ = 153–176 ps. The method of positron lifetime is applicable to the study of electron properties of structural vacancies. The method of angular correlation of annihilation radiation (ACAR) provides more detailed information about the electron structure of vacancies. The ACAR method was first used to analyze electron properties of nonstoichiometric carbides in [91–93]. The goal of those studies was to determine the spectra of the kinetic energy of valence electrons in disordered cubic carbides MC_y of group IV and V transition metals. An effort was also made to compare electron states of carbon atoms in graphite and those of metal atoms near a carbon vacancy in carbides. Quantitative differences of the spectra of various carbides were large enough to necessitate the study of their electron structure depending on the species of transition metals.

The angular correlation of annihilation radiation was measured [91–93] using a spectrometer with a long-gap geometry. The source of positrons was the ^{22}Na isotope having an activity of 400 MBq. ACAR spectra were measured at an angle of −8.0 to 25.0 mrad (θ being the anticollinearity deviation angle of annihilation gamma-quanta). The statistical error at the maximum of the ACAR curve did not exceed 1%. The random coincidence background was 0.4–0.8%. Measurements were performed at ambient temperature. The angular resolution function of the spectrometer, $R(\theta)$, was determined in a special experiment on annihilation of positrons in teflon. The narrow component of the ACAR spectrum, which corresponded to annihilation of parapositronium formed in teflon [107], was approximated by the Gaussian

$R(\theta) = \frac{1}{\theta_R \sqrt{2\pi}} \exp\left(-\frac{\theta^2}{2\theta_R^2}\right)$ normalized under the condition $\int_{-\infty}^{+\infty} R(\theta) d\theta = 1$.

Separating the narrow component of the ACAR spectrum of teflon and taking into account the natural FWHM of this line (0.5 mrad), the FWHM of the angular resolution function of the spectrometer, θ_R, was found to be 1.57 mrad. Researchers [93] measured ACAR spectra of titanium, zirconium, vanadium, niobium and tantalum, which were free of defects and thermal vacancies. The measured spectra are in good agreement with the spectra of annealed tantalum Ta [108] and unirradiated vanadium V and niobium Nb [109].

The comparison of the spectra of niobium carbide $NbC_{0.83}$ with the spectra of niobium and carbon (graphite) (Fig. 11.21), whose atoms form niobium carbide, shows that these spectra differ considerably, especially in the region $\theta < 10$ mrad. However, the similarity of the spectra of $NbC_{0.83}$ and Nb in the high-energy region (10–13 mrad) is an indication that positrons in carbide annihilate at a carbon vacancy surrounded by niobium atoms.

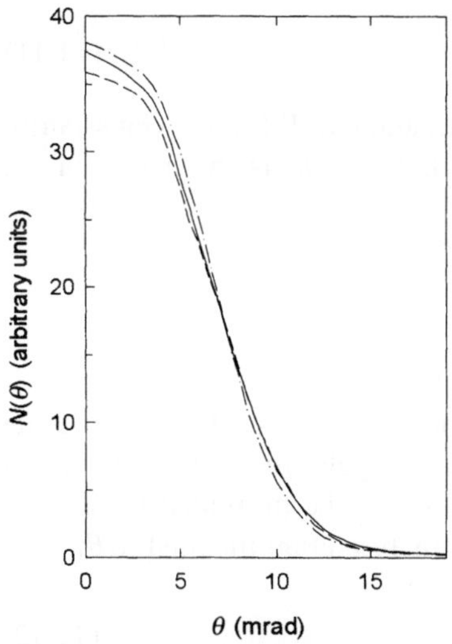

Fig. 11.21. Angular correlation spectra of annihilation radiation of nonstoichiometric niobium carbide $NbC_{0.83}$ (solid line), niobium Nb (dashed line), and graphite C (dash-dot line) [92]. These spectra are normalized to an equal area of 1000

In the literature, ACAR spectra are most frequently approximated by the superposition of a truncated inverted parabola and a Gaussian [110, 111]. In the case of nontransition metals, this description has an easy explanation making it possible to analyze theoretical data in theoretical terms. The parabola corresponds to the contribution of positron annihilation with free electrons, while the Gaussian denotes the contribution of positron annihilation with core electrons. In the case of transition metals and their compounds, the use of this model is complicated by hybridization of *z*, *p* and *d* bands. Consequently, only part of the valence electrons can be described by the truncated inverted parabola. The other electrons involved in covalent metal–metal bonds can be described by a Gaussian. Considering the Coulomb repulsion of positrons from atomic nuclei, the contribution of core electrons to angular correlation spectra normally does not exceed a few percent and, therefore, this component of the spectrum can be neglected against the background of the other two intense components.

Let us consider in more detail the model describing ACAR spectra by the superposition of two contributions

$$I(\theta) = I_p(\theta) + I_g(\theta). \tag{11.9}$$

The first term $I_p(\theta)$ is the truncated inverted parabola

$$I_p(\theta) = h_p\left[1 - (\theta/\theta_p)^2\right] f(|\theta| - \theta_p) \tag{11.10}$$

and the second term $I_g(\theta)$ is the Gaussian

$$I_g(\theta) = h_g \exp\left(-\theta^2 / 2\theta_g^2\right). \tag{11.11}$$

In (11.10) and (11.11) h_g and θ_g denote the height and FWHM of the Gaussian; θ_p is the angle of the parabola truncation, at which $I_p = 0$; h_p is the height of the parabola; and

$$f(|\theta| - \theta_p) = \begin{cases} 1, & \text{if } |\theta| \le \theta_p \\ 0, & \text{if } |\theta| > \theta_p \end{cases}$$

is the Heavyside function, respectively.

The experimental ACAR function $N(\theta)$ widens as a result of the finite resolution of the detectors. Taking into account the angular resolution function of the spectrometer, $R(\theta)$, the experimental function can be presented [111] as a convolution of the true ACAR function $I(\theta)$ and the resolution function $R(\theta)$:

$$N(\theta) = \int_{-\infty}^{+\infty} R(\theta' - \theta) I(\theta) d\theta'. \tag{11.12}$$

Expression (11.12) is a Fredholm integral equation of the first kind with regard to the true ACAR function $I(\theta)$.

Integrating expression (11.12) and using functions (11.9)–(11.11) [91], we obtain the following dependence:

$$\begin{aligned} N(\theta) = {} & \frac{h_p(\theta_p^2 - \theta_R^2)}{2\theta_p^2}\left[1 - \frac{(\theta-\theta_0)^2}{(\theta_p^2 - \theta_R^2)}\right]\left\{\mathrm{erf}\left[\frac{\theta_p - (\theta-\theta_0)}{\theta_R\sqrt{2}}\right] + \mathrm{erf}\left[\frac{\theta_p + (\theta-\theta_0)}{\theta_R\sqrt{2}}\right]\right\} \\ & + \frac{h_p\theta_R}{\theta_R^2\sqrt{2\pi}}\left\{[\theta_p + (\theta-\theta_0)]\exp\left[-\frac{(\theta_p - \theta + \theta_0)^2}{2\theta_R^2}\right]\right. \\ & \left. + [\theta_p - (\theta-\theta_0)]\exp\left[-\frac{(\theta_p + \theta - \theta_0)^2}{2\theta_R^2}\right]\right\} \\ & + \frac{h_g\theta_g}{\sqrt{\theta_R^2 + \theta_g^2}}\exp\left[-\frac{(\theta-\theta_0)^2}{2(\theta_R^2+\theta_g^2)}\right] + U, \end{aligned} \tag{11.13}$$

where θ_0 is the angle corresponding to the maximum in the ACAR spectrum, U is the random coincidence background and $\mathrm{erf}(x) = (2/\sqrt{\pi})\int_0^x \exp(-t^2)dt$ is the error function integral. To use this model, it is necessary to determine its parameters by the method of weighted least squares, i.e. perform fitting of the function

$$\chi^2 = \sum_i \frac{[N(\theta_i) - N_{\rm exper}(\theta_i)]^2}{N_{\rm exper}(\theta_i)}, \tag{11.14}$$

where i is the number of the experimental point. Values of the function (11.14), which were obtained [92] by the fitting procedure using the criterion χ^2/v [112] (v being the number of degrees of freedom, which is equal to the difference between the number of experimental points and the number of model parameters), showed that the function (11.13) provided an adequate quantitative description of the ACAR spectra. Calculated values of $\theta_{\rm p}$, $\theta_{\rm g}$, and the ratio of the areas under the parabola and the Gaussian $S_{\rm p}/S_{\rm g}$ ($S_{\rm p} = \frac{2}{3} h_{\rm p}\theta_{\rm p}$, $S_{\rm g} = h_{\rm g}\theta_{\rm g}\sqrt{(\pi/2)}$) for the transition metal carbides studied are given in Table 11.8.

Table 11.8. Parameters of the spectra of angular correlation of annihilation radiation and of kinetic spectra of valence electrons in nonstoichiometric transition-metal carbides [92, 93]: Gaussian halfwidth $\theta_{\rm g}$, inverted parabola cutt-off angle $\theta_{\rm p}$, ratio of areas under the truncated parabola and the Gaussian curve $S_{\rm p}/S_{\rm g}$, Fermi energy $E_{\rm F}$ and covalent band width $\Delta E_{\rm g}$

Carbide	$\theta_{\rm g} \pm 0.05$ (mrad)	$\theta_{\rm p} \pm 0.10$ (mrad)	$S_{\rm p}/S_{\rm g} \pm 0.01$	$E_{\rm F} \pm 0.05$ (eV)	$\Delta E_{\rm g} \pm 1$ (eV)
$TiC_{0.94}$	4.6	6.6	0.35	11	19
$ZrC_{0.88}$	4.2	6.4	0.36	10	16
$HfC_{0.90}$	4.5	6.3	0.51	10	19
$VC_{0.85}$	4.8	6.7	0.25	11	21
$NbC_{0.93}$	4.5	7.1	0.23	13	18
$TaC_{1.00}$	4.5	7.0	0.16	12	19
$TaC_{0.92}$	4.6	6.9	0.30	12	19
Ordered $TaC_{0.85}$	4.7	7.1	0.38	13	20
Disordered $TaC_{0.85}$	4.8	7.6	0.48	15	21

Approximation of experimental ACAR spectra by the aforementioned model actually means reconstruction of the true angular correlation function. Figure 11.22 illustrates the reconstruction of the angular correlation function and its resolution into a truncated parabola and a Gaussian for tantalum carbide $TaC_{1.00}$.

If one needs a more comprehensive analysis of experimental data, it is possible to pass from angular spectra to momentum and energy spectra of electrons, in which ACAR parameters acquire an illustrative physical sense. To this end, authors [92] used the equations derived in [113]. In accordance with [92], the momentum density distribution $\rho(p)$ has the form

$$\rho(p) = 4\pi(m_0c)^2 A^{-1} p^2 \left[\frac{2h_{\rm p}}{p_{\rm p}^2} f(|\boldsymbol{p}| - p_{\rm p}) + \frac{h_{\rm g}}{p_{\rm g}^2} \exp\left(-\frac{p^2}{2p_{\rm g}^2} \right) \right], \tag{11.15}$$

where $p_{\rm p} = m_0 c\theta_{\rm p}$ and $p_{\rm g} = m_0 c\theta_{\rm g}$.

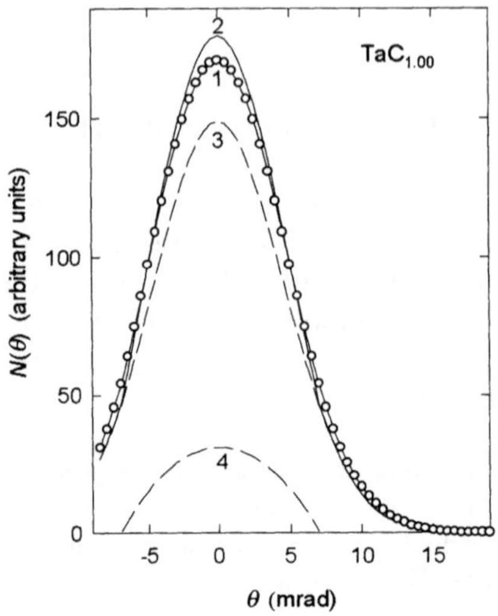

Fig. 11.22. Components of the angular correlation spectra of annihilation radiation of tantalum carbide $TaC_{1.00}$ [91]: (1) experimental spectrum; (2) deconvoluted spectrum; (3) Gaussian component; (4) parabolic component

It is possible to change from the distribution (11.15) to the kinetic energy distribution $\rho(E_k)$ of electrons. The kinetic energy of electrons expressed as a momentum is $E_k = p^2/(2m_0)$. Considering this and using the identity $\rho(E_k)dE_k = \rho(p)dp$, we obtain the known dependence of the density of states of conduction electrons:

$$\rho_p(E_k) = 2\pi(m_0c)^2\sqrt{2m_0}\,A^{-1}\sqrt{E_k}\,\frac{2h_p}{E_p}f(E_k - E_p) \tag{11.16}$$

and the dependence of the density of electron states on the kinetic energy for electrons in covalent bonds:

$$\rho_g(E_k) = 2\pi(m_0c)^2\sqrt{2m_0}\,A^{-1}\sqrt{E_k}\,\frac{h_g}{E_g}\exp\left(-\frac{E_k}{2E_g}\right), \tag{11.17}$$

where $E_p = (m_0c\theta_p^2)/2$, $E_g = (m_0c\theta_g^2)/2$, m_0 is the mass of a free electron and c is the light velocity. If θ_p and θ_g values are substituted as milliradians, we have $E_p = 0.256\theta_p^2$ and $E_g = 0.256\theta_g^2$ eV. It is worth noting that direct summation of expressions (11.16) and (11.17) is devoid of the physical sense, because this summation disregards different values of potential energies in different electron subsystems.

From expression (11.16) it is clear that the energy E_p actually represents the Fermi energy $E_F = E_p$ and, simultaneously, the width of the conduction band. The analysis of (11.17) shows that the energy band of covalent electrons has width $\Delta E_g \approx 3.6 E_g$. Values of E_F and ΔE_g, which were calculated from these relationships using parameters θ_p and θ_g found for carbides, are given in Table 11.8.

From the comparison of S_p/S_g and E_F it follows that carbides of groups IV and V transition metals differ considerably (see Table 11.8). For example, the Fermi energy of group V metal carbides is on the average 1–2 eV higher than the Fermi energy of group IV metal carbides. This result agrees qualitatively with results of quantum calculations due to Korringa–Kohn–Rostoker [114, 115] and is in quantitative agreement with X-ray emission spectroscopic data for zirconium and niobium carbides [116]. According to calculations [114, 115], the Fermi energy increases in transition from carbides of group IV to group V metals because an additional *d*-electron occupies the metal–metal band. However, S_p/S_g for carbides of group V metals is 1.5 to 2 times smaller and the fraction of positrons annihilating with conduction electrons diminishes. This means that an additional *d*-electron in carbides $M^{(V)}C$ also causes an increase in the concentration of the covalent component of metal–metal bonds in a structural vacancy.

From Table 11.8 it is seen that the kinetic spectra of valence electrons are considerably different despite the fact that carbides of metals of one group are isoelectronic. The S_p/S_g value increases in the series TiC_y–ZrC_y–HfC_y and decreases in the series VC_y–NbC_y–TaC_y. Other characteristics of the spectra of the kinetic energy of electrons exhibit a nonmonotonic behavior in going from 3*d*-metal carbide to 5*d*-metal carbide. Probably, these changes can be explained only after annihilation of positrons in nonstoichiometric carbides is studied comprehensively as a function of the carbon concentration.

The dependence of ACAR parameters on the carbon concentration was analyzed in detail taking nonstoichiometric tantalum carbide as an example [91, 92]. Results [91, 92] show that the increase in the concentration of structural vacancies does not cause a considerable narrowing of angular correlation spectra, as was observed for other types of defects [117]. It should be noted also that when the carbon concentration of tantalum carbide decreases from $TaC_{1.00}$ to $TaC_{0.79}$, ACAR parameters of the carbide spectra approximate in value to those of the spectrum of pure Ta. These two facts agree with conclusions about the annihilation of positrons at structural vacancies, which were drawn from the study of the positron lifetime in nonstoichiometric carbides (see also [118]).

Remarkably, angular correlation spectra depend on the structural state of the carbon sublattice in tantalum carbide [92]. Indeed, the Fermi energy of ordered tantalum carbide was found [92] to be 2 eV higher than the Fermi energy of disordered carbide having a similar composition (see Table 11.8). This difference between the Fermi energies of ordered and disordered states testifies to a redistribution of electrons of metal atoms during atomic ordering, namely, points to the increase in the electron density of metal–metal bonds and the decrease in the electron density of covalent bonds. If one considers that the number of electrons in a compound remains unchanged during ordering, the growth of the fraction of metal bonds can cause an increase in the lattice parameter of tantalum carbide during ordering.

The following conclusions may be drawn from the study of carbides by the positron annihilation method: a structural carbon vacancy in nonstoichiometric carbides of transition metals bears a negative effective charge, which does not change as a first approximation with the vacancy concentration. A considerable contribution of the Gaussian to ACAR spectra suggests that the density of electrons involved in the covalent component of metal–metal bonds is higher than the density of conduction electrons in a vacancy. As a result, electron properties of a structural vacancy differ from properties of thermal vacancies in pure metals.

11.4 Electrokinetic Properties

A combined covalent-metallic-ionic type of chemical bonds is realized in strongly nonstoichiometric interstitial compounds MX_y and their solid solutions. The covalent and metallic components are comparable in value, while the ionic component is relatively small (see Sect. 3.2). Electrical conductivity and other kinetic properties of strongly nonstoichiometric compounds are governed mainly by collectivized *d*-electrons of the metal participating in M–M bonds.

Indeed, most carbides and nitrides of transition metals possess electronic conductivity. At a temperature of 300 K the electrical resistivity of these compounds near the upper boundary of their homogeneity interval is 10–100 μΩ cm. Conductivity of carbides and nitrides is usually considered as movement of electrons in the conduction band taking into account their scattering on thermal vibrations of the lattice. This mechanism is responsible for part of the resistivity, which depends primarily on temperature.

Given a constant temperature T, the resistivity ρ of nonstoichiometric carbides MC_y rises quickly with decreasing concentration of carbon. The derivative $\partial\rho/\partial y$ is relatively high. For example, $\partial\rho/\partial y$ for titanium carbide in the interval $TiC_{0.95}$–$TiC_{1.00}$ is (according to different authors) ~16 [119], ~24 [103, 120] or ~14 [121] μΩ cm per at.% vacancies. The derivative $\partial\rho/\partial y$ for near-stoichiometric carbides of niobium and tantalum is 6–8 μΩ cm [22, 120, 122-126]. If the change of electrical resistivity in the homogeneity interval of carbide is assumed to be due only to scattering of electrons on vacancies, the scattering ability of a carbon vacancy in titanium carbide is nearly ten times larger than in copper [119]. Considering this, Williams [119] supposed that the rapid growth of electrical resistivity with the concentration of structural vacancies in titanium carbide is caused by a small number of conduction electrons and a large negative effective charge of vacancies ($q_{\square} \approx 1e$). Recent measurements of the positron lifetime in nonstoichiometric titanium carbide TiC_y [90] showed that carbon vacancies do have a negative charge, but this charge is small. Measurements of residual electrical resistivity of nonstoichiometric niobium NbC_y [120, 127] and titanium TiC_y [120, 128] carbides revealed that ρ of crystals having structural vacancies was much higher than the electrical resistivity of stoichiometric carbides. The decrease in the carbon concentration of vanadium carbide VC_y is also followed by an increase in electrical resistivity [20, 21, 122]. These facts confirm that structural vacancies are quite reasonably considered as defects leading to additional scattering of electrons.

Let it be noted that almost all studies concerned with electrical resistivity of nonstoichiometric carbides as a function of the concentration of carbon or structural vacancies were performed disregarding the structural (disordered or ordered) state of those carbides. In some cases, nonstoichiometric carbides were ordered as suggested by their synthesis conditions. The analysis of most reliable literature data shows that the increase in the vacancy concentration of disordered nonstoichiometric compounds MX_y is accompanied by the growth of electrical resistivity ρ (Fig. 11.23), which is a maximum at $0.2 < 1 - y \leq 0.3$.

Theoretical work discussing the relation between electrical resistivity and atomic ordering is scarce. One is dedicated to the influence of partial long- and short-range order on electrical resistivity of substitutional alloys [129]. Christoph *et al.* [129] established the analytical dependence of electrical resistivity on short- and long-range order parameters of fcc alloys. One of the main results [129] is the conclusion that long and short-range orders may have an equal effect on electrical resistivity. It should be remembered however that those results were obtained for alloys and not for nonstoichiometric compounds.

Experimental studies concerned with the effect of ordering on electrical resistivity of nonstoichiometric compounds are few. However, results of those studies are highly accurate and reliable.

All metal atoms, which are nearest neighbors in nonstoichiometric compounds with the $B1$ structure, are located in planes $(111)_{B1}$ or equivalent planes $(\bar{1}11)_{B1}$, $(1\bar{1}1)_{B1}$ and $(11\bar{1})_{B1}$, and provide hexagonal close packing in these planes. One may think therefore that in such compounds conduction electrons are distributed nonuniformly in the crystal volume and have maxima of the density of spatial distribution near closely packed metal planes [1]. Since metal and nonmetal atomic planes alternate in the direction $[111]_{B1}$ and equivalent directions in the compounds at hand, conduction electrons propagate within some spatial network formed by closely packed metal planes. Depending on the MX_y composition, nonmetal planes may be complete (free of structural vacancies) or defective (structural vacancies, i.e. vacant sites, are present along with nonmetal atoms). Results of research into electrical resistivity and other electrokinetic properties, which attest to a strong scattering of electrons on vacancies, suggest that the free path length of electrons along metal planes depends on the type of nonmetal planes (complete or defective) adjacent to metal atomic planes. Gusev and Rempel [1] assumed that the free path length of conduction electrons is larger when a metal plane borders on complete nonmetal planes and is smaller otherwise. Consequently, the effect of ordering on electrokinetic properties of nonstoichiometric compounds can be considered using the model of the two-dimensional distribution of conduction electrons proposed in [1].

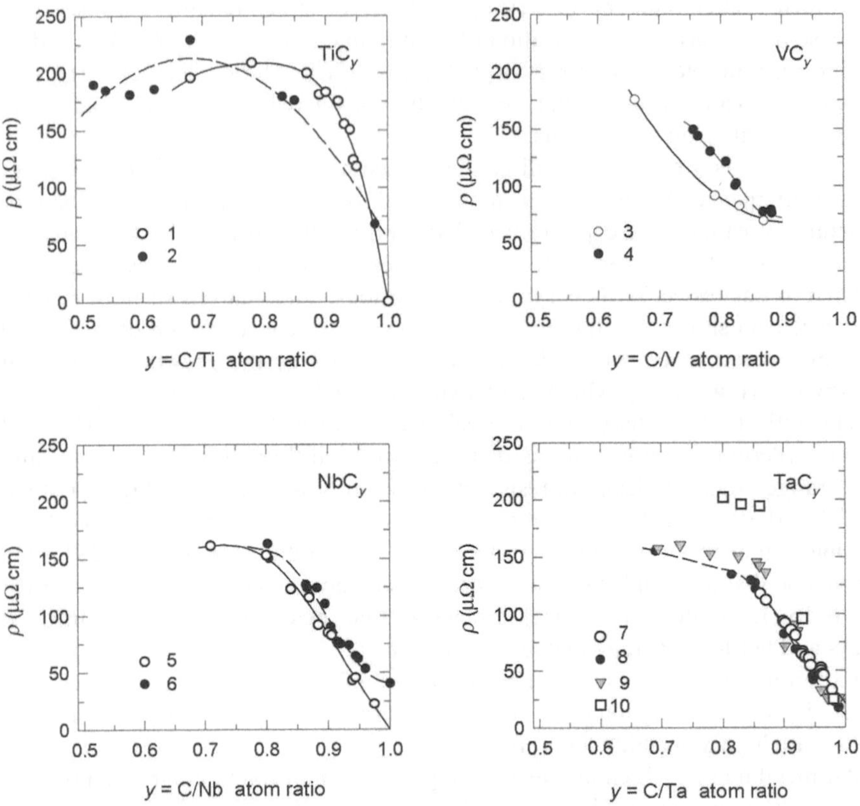

Fig. 11.23. Isotherms of electrical resistivity $\rho(y,T)$ in the homogeneity interval of disordered cubic carbides MC_y at a temperature of 0 and 300 K. TiC_y: (1) $\rho(y, 0)$ [120, 128], (2) $\rho(y, 300)$ [121]. VC_y: (3) $\rho(y, 300)$ [20, 21], (4) $\rho(y, 300)$ [122]. NbC_y: (5) $\rho(y, 0)$ [120], (6) $\rho(y, 300)$ [122]. TaC_y: (7) $\rho(y, 300)$ [123], (8) $\rho(y, 300)$ [122], (9) $\rho(y, 300)$ [124], (10) $\rho(y, 300)$ [126]. Measurements are made on single crystals of TiC_y and NbC_y [120, 128], and on sintered samples of TiC_y and VC_y with porosity less than 1% [20, 21, 121]. In [122] poreless samples of VC_y, NbC_y and TaC_y were produced by carburization of metallic fibers in methane CH_4. Overestimated values [126] of resistivity of carbides TaC_y with $y > 0.85$ are probably due to the large (up to 28%) porosity of the samples

In Sect. 5.4 it is shown that ordering of nonstoichiometric compounds causes a redistribution of nonmetal atoms and structural vacancies such that complete and ordered defective nonmetal atomic planes alternate in a special way in the lattice. All nonmetal atomic planes are defective in disordered nonstoichiometric compounds. Considering the differences of ordered and disordered nonstoichiometric compounds, from the model of two-dimensional distribution of electrons it follows that the electrical resistivity ρ of the ordered compound should be lower than ρ of the disordered compound. This is confirmed by experimental data.

The effect of ordering on electrical resistivity was studied in most detail for nonstoichiometric titanium carbide TiC_y.

Lipatnikov *et al* [121, 130] analyzed the electrical resistivity of carbide TiC_y ($0.5 \le y \le 0.98$) as a function of temperature, composition and distribution of carbon vacancies. The samples were prepared by hot pressing of a powder mixture of carbide $TiC_{0.98}$ and metallic titanium. The porosity of the samples was less than 1%. All the synthesized samples were homogeneous and contained a disordered cubic phase TiC_y only. To switch to the ordered state, they were annealed for 340 h while the temperature was decreased smoothly from 1070 to 770 K at a rate of 1 K h^{-1}. The temperature was reduced from 770 to 300 K at a rate of 1 K min^{-1}. The structural state of the titanium carbide samples before and after annealing, and also after electrical resistivity measurements, was determined by the X-ray diffraction method with step-by-step scanning. Electrical resistivity was measured at a temperature from 300 to 1200 K with a step 1 K on TiC_y samples in disordered and ordered states. The mean heating and cooling rates during measurements were the same as those during annealing, i.e. 1 K min^{-1}.

Temperature dependencies of electrical resistivity $\rho(T)$ of some samples of nonstoichiometric titanium carbide TiC_y are shown in Figs. 11.24 and 11.25.

Measurements of the electrical resistivity of disordered carbide $TiC_{0.52}$ showed that ρ increased trivially as a result of scattering of charge carriers on phonons as the temperature was increased up to ~800 K. The electrical resistivity ρ decreased anomalously at $T \approx 815$ K. When $T > 960$ K, the resistivity ρ increased first quickly (up to ~1030 K) and then slowly (Fig. 11.24, curve 1). The resistivity ρ decreased during cooling of the samples and it dropped abruptly in the region from 900–1020 K (Fig. 11.24, curve 2). If ordered carbide $TiC_{0.52}$ was heated subsequently, ρ followed curve 3, which is similar to curve 2.

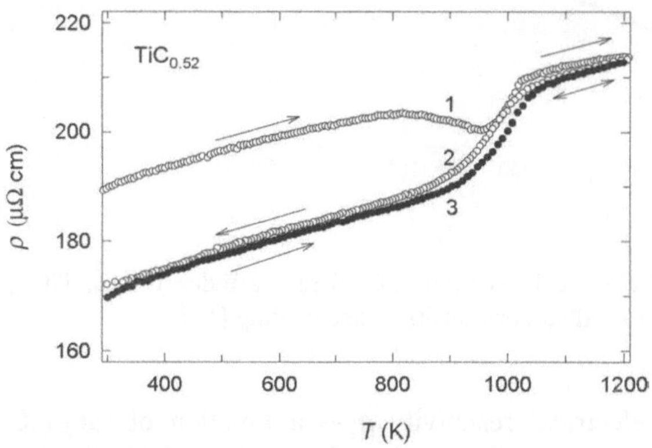

Fig. 11.24. Effect of ordering on the electrical resistivity ρ of $TiC_{0.52}$: (1) dependence $\rho(T)$ for disordered carbide $TiC_{0.52}$ upon heating and the nonequilibrium disorder → order transition; (2) change of ρ upon cooling and the equilibrium disorder ↔ order transition; (3) change of ρ upon heating of ordered carbide $TiC_{0.52}$ [121, 130]

Curve 1 is characteristic of the irreversible transition from the disordered nonequilibrium state to the ordered equilibrium state and of subsequent disordering at $T > 960$ K. Measured dependencies $\rho(T)$ (Fig. 11.24, curves 2 and 3) are typical of the equilibrium reversible disorder–order transition. Electrical resistivity hysteresis is observed in the transition region. Formation of an ordered phase Ti_2C in the $TiC_{0.52}$ sample during resistivity measurements is confirmed by the fact that the diffraction pattern of this sample contains the same set of superstructure lines as the pattern measured after long-term annealing of $TiC_{0.52}$.

Figure 11.25 presents temperature dependencies of electrical resistivity of carbides $TiC_{0.54}$, $TiC_{0.58}$ and $TiC_{0.62}$, which were ordered by preliminary annealing. At $940 < T < 1060$ K the dependencies $\rho(T)$ exhibit an abrupt change and hysteresis of resistivity which are due to the reversible order–disorder transition. The presence of hysteresis is an indication that $TiC_y \leftrightarrow Ti_2C$ and $TiC_y \leftrightarrow Ti_3C_2$ transitions are close to a transition of the first kind. The conclusion that $TiC_y \leftrightarrow Ti_2C$ represents a transition of the first kind was also drawn in [131], although, for symmetry considerations [132, 133], this transformation satisfies the Landau criterion of phase transitions of the second kind. Taking into account also the data [132] concerning the variation of the heat capacity near the transition temperature T_{trans}, the reversible order–disorder transformation $TiC_y \leftrightarrow Ti_2C$ may be viewed as a weak phase transition of the first kind with a small latent heat of transformation.

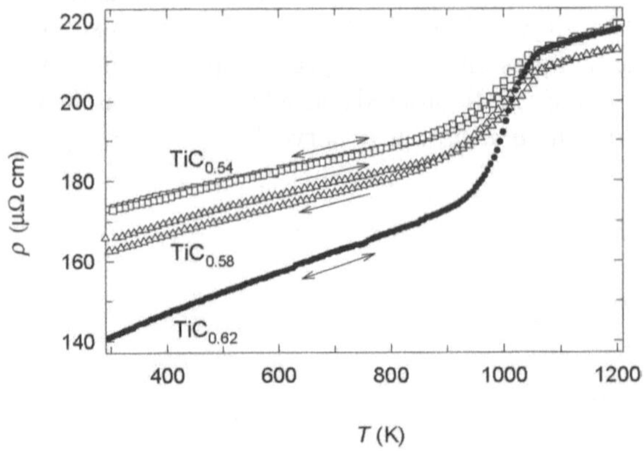

Fig. 11.25. Reversible change of electrical resistivity of ordered carbides $TiC_{0.54}$, $TiC_{0.58}$ and $TiC_{0.62}$ upon heating, the order ↔ disorder transition, and cooling [121]

Figure 11.26 presents the electrical resistivity ρ as a function of the TiC_y composition at 300 K. As the carbon content y rises from 0.52 to 0.98, the ρ of disordered carbide TiC_y first grows, then passes a maximum at $y = 0.68$, and finally diminishes. Ordered carbides possess a smaller resistivity ρ than disordered carbides of the same composition. Two regions corresponding to the homogeneity intervals for ordered phases Ti_2C and Ti_3C_2 can be distinguished in the dependence $\rho(y,300)$ of ordered carbide TiC_y. When y varies, the resistivity $\rho(y,300)$ in each of the regions tends to some minimum value corresponding to a stoichiometric composition of the ordered phase. The electrical resistivity $\rho(y,300)$ decreases in the homogeneity interval of the phase Ti_2C when $y \to 0.5$. It diminishes in the homogeneity interval of the phase Ti_3C_2 when y varies between the value corresponding to the upper boundary of the homogeneity interval ($y \approx 0.58$) and $y = 2/3$ (Fig. 11.26).

From the dependence $\rho(y,300)$ of the ordered phases of titanium carbide it follows that $y \approx 0.58$–0.59 corresponds to the boundary between the homogeneity interval of the phase Ti_2C and the two-phase region ($Ti_2C + Ti_3C_2$).

The dependence $\rho(y,300)$ established for disordered carbide TiC_y agrees well with the literature data. For comparison, Fig. 11.26 displays the dependencies $\rho(y,300)$ measured by the authors [119, 120, 128] for disordered single crystals of titanium carbide TiC_y.

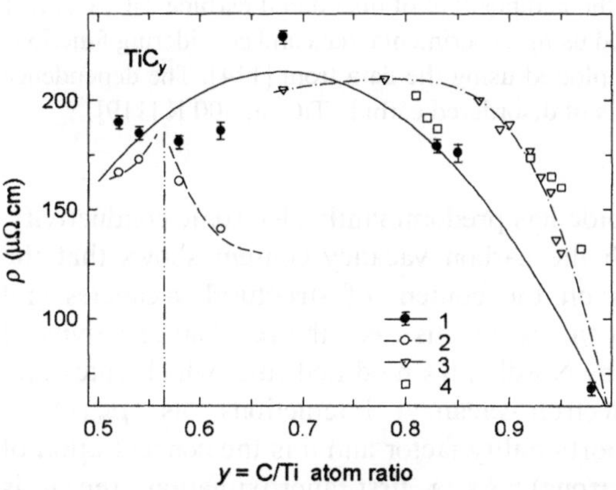

Fig. 11.26. Dependence of electrical resistivity ρ on the composition of titanium carbide TiC_y at 300 K in disordered (1) and ordered (2) states [121]; (3) and (4) are the electrical resistivity of disordered carbide TiC_y at 300 K according to [120] and [119] respectively. The dash-dot line shows the boundary between ordered phases Ti_2C and Ti_3C_2

The temperature coefficient of electrical resistivity $\partial\rho/\partial T$ is small and decreases insignificantly from 0.05 to 0.03 μΩ cm K^{-1} in the series from $TiC_{0.52}$ to $TiC_{0.62}$. It increases up to 0.06 μΩ cm K^{-1} for $TiC_{0.98}$. Considering $\partial\rho/\partial T$, the concentration dependence of the residual (at $T = 4.2$ K) resistivity $\rho_{res}(y)$ of disordered carbide TiC_y (Fig. 11.27, curve 1) is similar to the dependence $\rho(y,300)$.

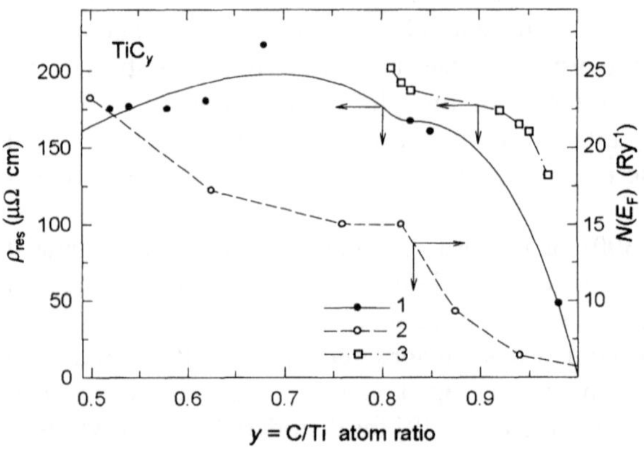

Fig. 11.27. Residual electrical resistivity $\rho_{res}(y)$ (1) and the density of electron states at the Fermi level $N(E_F)$ (2) as a function of the composition of disordered carbide TiC_y. Curve 1 is calculated by the least squares method using experimental data and considering functions (11.18) and (11.19) [121]. Curve 2 is plotted using the data from [114]. The dependence $\rho_{res}(y)$ (3) is measured on single crystals of disordered carbide TiC_y at 300 K [119]

Nonstoichiometric titanium carbide has predominantly electronic conductivity. Variation of the conductivity with the carbon vacancy content shows that the concentration of carriers depends on the content of structural vacancies and scattering of electrons on these vacancies. In this case, the residual resistivity of disordered TiC_y may be described by Nordheim's modified rule, which takes into account atom–vacancy and electron–vacancy interactions as $\rho_{res}(y) = Ay(1-y)/n(y)$, where A is the proportionality factor and n is the concentration of current carriers (conduction electrons). As a first approximation, the n is proportional to $[N(E_F)]^3$, where $N(E_F)$ is the density of electron states at the Fermi level. The dependence of residual electrical resistivity on the composition of disordered carbide TiC_y has the form

$$\rho_{res}(y) \propto \frac{Ay(1-y)}{[N(E_F)(y)]^3} . \qquad (11.18)$$

Calculations of the electron structure of TiC_y, which were performed by the Korringa–Kohn–Rostoker method in the coherent potential approximation (KKR-CPA) [114], showed that when the vacancy concentration $(1-y)$ increased from 0 to 0.5, the Fermi energy E_F dropped from 0.69 to 0.66 Ry (9.4 to 9.0 eV), while the density of electron states at the Fermi level $N(E_F)$ increased from 5.7 to 23.2 Ry^{-1} in terms of one unit cell (or from 0.105 to 0.426 eV^{-1} in terms of one formula unit). The plot of $N(E_F)$ versus the carbon concentration y of carbide TiC_y exhibits a singularity at $y = 0.82$, which brings about an almost linear section (Fig. 11.27, curve 2). This section with $N(E_F) \approx 15$–17 Ry^{-1} corresponds to the interval $TiC_{0.625}$–$TiC_{0.82}$. The growth of the vacancy content is accompanied by smoothing of all sharp peaks of the density of states (especially below E_F), a decrease in the occupancy of the low-energy subband of Ti–C interactions, and an increase in the occupancy of the high-energy subband of conductivity. Considering the singularity at $y = 0.82$, the dependence of $N(E_F)$ on the carbon concentration y in carbide TiC_y can be described by the simple expression

$$N(E_F) = c + \alpha y + \beta(y - 0.82)^{2/3}, \tag{11.19}$$

where c, α and β are fitting parameters. Taking into account the relation (11.19) and $N(E_F) = 0.105$ eV^{-1} formula^{-1}, fitting of residual resistivity $\rho_{res}(y)$ by the function (11.18) gives the following values for stoichiometric carbide $TiC_{1.0}$: $A = 3.006$, $c = 0.248$, $\alpha = -0.132$, and $\beta = -0.033$. Curve 1 in Fig. 11.27 shows the dependence of the residual resistivity of disordered carbide TiC_y on carbon content y. It is plotted using experimental data by the method of least squares. When $y = 0.82$, the dependence $\rho_{res}(y)$ exhibits a break, which is related to an analogous singularity of the density of electron states. For comparison, curve 3 depicts ρ of single crystals of disordered carbide TiC_y at 300 K [119]. The pronounced break in the dependence $\rho(y,300)$ at $y \approx 0.82$-0.83 was not explained in [119], because the shape of the electron spectrum of carbide TiC_y and the changes of density of electron states $N(E_F)$ depending on the carbon vacancy content were unknown.

The study of the structure and electrical resistivity of titanium carbide TiC_y ($0.5 < y \le 0.98$) [121, 130] showed that ordering of TiC_y, which is followed by the formation of cubic (space group $Fd\bar{3}m$) and trigonal (space group $R\bar{3}m$) superstructures Ti_2C and the rhombic (space group $C222_1$) superstructure Ti_3C_2, occurred at a temperature of 970–1000 K. Resistivity dropped during ordering. These results correlate with the phase diagram of the Ti–C system [44], which takes into account ordering of nonstoichiometric titanium carbide TiC_y.

The authors [131] measured the electrical resistivity of titanium carbides $TiC_{0.51}$–$TiC_{0.65}$ with a high (up to 1 mass.%) content of impurity oxygen. Measurements were made at a temperature of 300–1350 K. Readings were taken at each temperature after the sample reached the equilibrium state. The mean heating rate was 2 K min^{-1}. When quenched samples TiC_y were heated to a temperature of 820–870 K, resistivity dropped thanks to arising long-range order, which is an equilibrium at the given temperature. Further heating led to disordering and to increasing ρ. Variation of ρ during ordering of carbide $TiC_{0.55}$ was also analyzed in [134]. Carbide $TiC_{0.55}$ was produced by the method of self-propagating high-temperature synthesis and subsequent hot pressing. The total concentration of impurities did not exceed 0.05 mass.%. Electrical resistivity was measured in the temperature interval from 300 to 1500 K. In accordance with [134], the order–disorder transition temperature of carbide $TiC_{0.55}$ is ~1000 K, which is in agreement with the data reported in [121, 130, 131] and results of theoretical calculations [44]. A common drawback of the work [131, 134] is that the structure of the samples of nonstoichiometric carbide TiC_y was not certified.

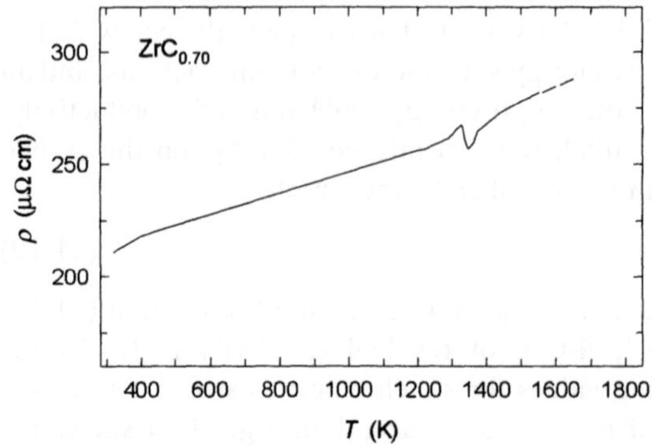

Fig. 11.28. Temperature dependence of electrical resistivity of nonstoichiometric zirconium carbide $ZrC_{0.70}$ [136]

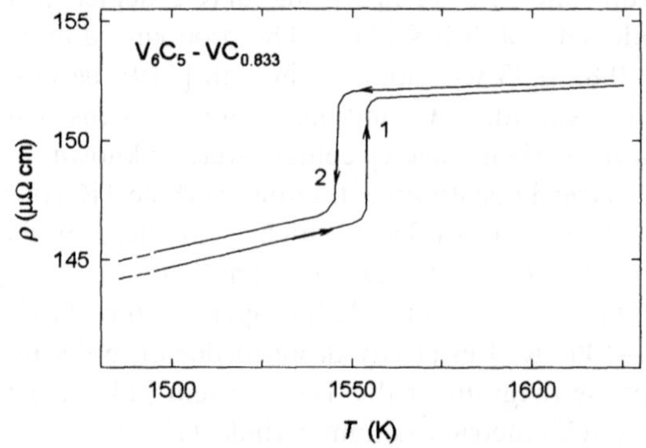

Fig. 11.29. Change of electrical resistivity ρ of deeply annealed carbide V_6C_5 during heating (1) and cooling (2) at a rate of 0.5 K min^{-1} [137]

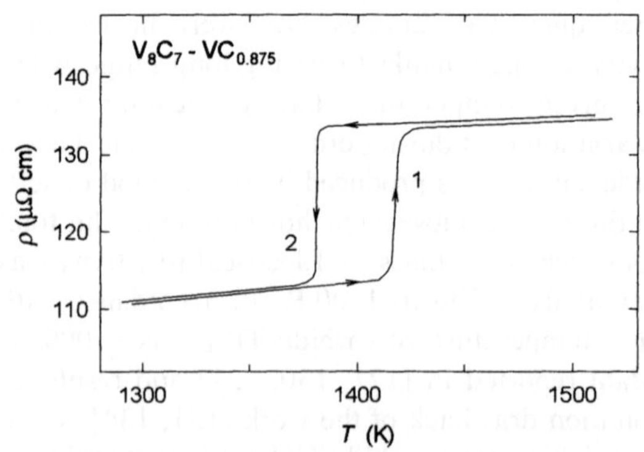

Fig. 11.30. Temperature dependence of electrical resistivity ρ of ordered vanadium carbide V_8C_7 during slow heating (1) and cooling (2) in the $VC_{0.875} \leftrightarrow V_8C_7$ transition region [137]

Lorenzelli *et al.* [135] measured the resistivity ρ of disordered and ordered carbides $TiC_{0.625}$ in the temperature interval from 1.6 to 300 K. Disordered carbide $TiC_{0.625}$ was produced by annealing of a synthesized sample at 1270 K and subsequent quenching of the sample to 300 K. To prepare ordered carbide, a sample was annealed at 1070 K for four days, cooled to 1020 K for 10 days, and annealed at 1020 K for four days. The crystal structure of disordered and ordered carbides was carefully examined in [135]. All thermal treatment procedures and measurements were realized in a good vacuum of 10^{-4} Pa. Therefore results of that study may be viewed as reliable. The decrease in resistivity $\Delta\rho(y,300)$ during ordering of carbide $TiC_{0.625}$ was ~20 μΩ cm (~10%). The quantity $\Delta\rho(300)$ for $TiC_{0.62}$ was ~40 μΩ cm (~24%) [121]. This fact suggests a lower degree of ordering achieved in [135]. The temperature coefficient of resistivity $\partial\rho/\partial T$ in the interval from 100 to 300 K was 1.8 times larger for ordered carbide than for disordered one.

The study of the dependence $\rho(T)$ for zirconium carbide $ZrC_{0.70}$ during slow heating and cooling (heating and cooling rates did not exceed 2 K min^{-1}) in the temperature interval from 1333 to 1353 K revealed an abrupt change of ρ [136] (Fig. 11.28). In accordance with [136], this abrupt change of ρ corresponds to the transition from the high-temperature disordered state to the low-temperature ordered state, which takes place during cooling. The ordered phase Zr_4C_3 has a cubic structure with a double lattice constant $a = 2a_{B1}$ (as compared to disordered carbide $ZrC_{0.70}$ having the *B*1 structure). The conclusion [136] about the type and the symmetry of the superstructure in ordered zirconium carbide is questionable. Remarkably, researchers [136] did not observe any singularities in the temperature dependence $\rho(T)$ of carbide $ZrC_{0.51}$, which would attest to ordering. However, structural studies (see Sect. 3.4) provided evidence of the formation of the ordered phase $Zr_2C \equiv ZrC_{0.50}$ in zirconium carbide ZrC_y with $y \approx 0.55$–0.65.

It was found [137] that electrical resistivity decreased by 5-10% during the $VC_{0.833}$–V_6C_5 disorder–order transition and ρ increased during disordering of vanadium carbide (Fig. 11.29). The presence of hysteresis in the dependence $\rho(T)$ suggests that ordering $VC_{0.833} \rightarrow V_6C_5$ presents a phase transition of the first kind. The behavior of the electrical resistivity ρ near the temperature of the $VC_{0.875} \leftrightarrow V_8C_7$ transition (Fig. 11.30) is similar to the behavior during the $VC_{0.833} \leftrightarrow V_6C_5$ transition. However, the hysteresis is about 25 K wide, which is nearly three times larger than the hysteresis width for the $VC_{0.833} \leftrightarrow V_6C_5$ transition.

From the analysis of kinetic properties of carbide VC_y with different content of structural vacancies it was inferred [137] that the ordered phase V_8C_7 was formed even when the composition of vanadium carbide was close to $VC_{0.833}$. This means that the free energy of the phase V_8C_7 is smaller than that of the ordered phase V_6C_5. This inference is confirmed by measurements of the enthalpy (heat) of the phase transitions [138]: the heat $\Delta H_{trans} = -1.45 \pm 0.7$ kJ mol^{-1} is released during the $VC_{0.833} \rightarrow V_6C_5$ transition and a larger heat $\Delta H_{trans} = -1.54 \pm 0.6$ kJ mol^{-1} is released during the $VC_{0.875} \rightarrow V_8C_7$ transition. A similar relation between latent heats ΔH_{trans} was found [10, 20, 21, 139] for vanadium carbide: ΔH_{trans} was −1.37 and −2.35 kJ mol^{-1} for $VC_{0.833} \rightarrow V_6C_5$ and $VC_{0.875} \rightarrow V_8C_7$ transitions.

Interesting results are obtained [128] for the electrical resistivity of annealed single crystals of vanadium carbide VC_y ($0.75 \le y \le 0.88$). Measurements are made in the temperature interval from 4.2 to 300 K. The plot of resistivity $\rho(T = 4.2\ K)$ versus the VC_y composition (Fig. 11.31) has clear-cut minima of ρ, which correspond to stoichiometric compositions of ordered phases $V_6C_5 \equiv VC_{0.833}$ and $V_8C_7 \equiv VC_{0.875}$. For comparison, Fig. 11.31 depicts the electrical resistivity of carbides $VC_{0.79}$, $VC_{0.83}$ and $VC_{0.87}$ in disordered and ordered states at a temperature of 300 K [20, 21]. One can readily see that the electrical resistivity of carbides decreases markedly as a result of annealing and subsequent ordering. The smaller value of resistivity obtained [20, 21] for disordered vanadium carbides is probably due to the presence of some long-range order in these carbides. Indeed, it is difficult to produce fully disordered vanadium carbide by quenching from a high temperature. This is especially true of the phase V_8C_7: even when carbide $VC_{0.875}$ is quenched from a temperature of 1400 K at a high cooling rate, weak superstructure reflections are still preserved in the diffraction pattern.

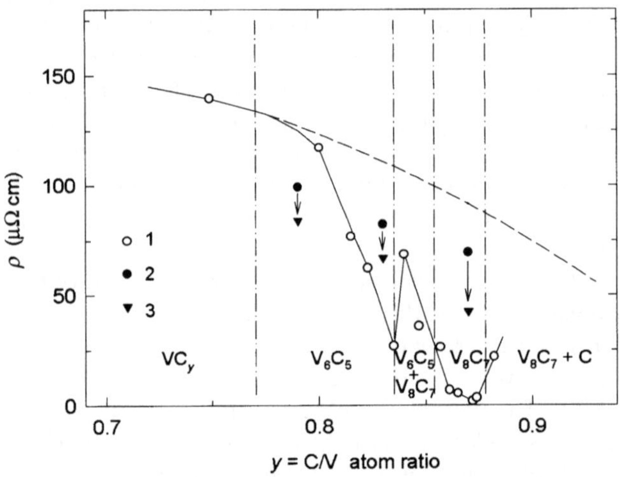

Fig. 11.31. Effect of ordering on the electrical resistivity ρ of vanadium carbide VC_y: (1) annealed carbide at a temperature of 4.2 K [128]; (2) disordered carbide at a temperature of 300 K [20, 21]; (3) annealed ordered carbide at 300 K [20, 21]. The dotted line shows the dependence of ρ on the composition of a disordered carbide VC_y [20, 21, 128]. Regions of ordered phases V_6C_5 and V_8C_7 are marked in the homogeneity interval of cubic carbide VC_y. The arrows show the change of ρ after annealing. Measurements [128] are made on single crystals of VC_y. Samples of VC_y with porosity less than 1% are used in [20, 21]

The study of resistivity as a function of the composition of niobium carbide NbC_y [140] at a temperature of 300 K revealed an anomalous drop of ρ in the interval $NbC_{0.83}$–$NbC_{0.87}$. The observed effect might be due to ordering in the carbon sublattice, but the authors [140] did not examine the structure of the carbide.

Figure 11.32 presents polythermal curves of the electrical resistivity of quenched disordered niobium carbide $NbC_{0.83}$ and partially ordered carbide Nb_6C_5 [127]. Partially ordered niobium carbide has a large temperature coefficient of electrical resistivity $\partial\rho/\partial T$. The resistivity ρ of ordered carbide is smaller than ρ of disordered carbide $NbC_{0.83}$ below the order–disorder transition temperature. The dependence $\rho(T)$ observed [127] for ordered carbide Nb_6C_5 is characteristic of compounds having a superconductivity transition temperature of about 10 K.

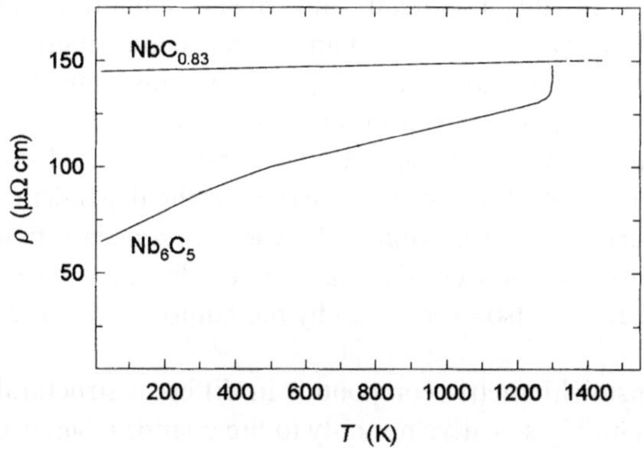

Fig. 11.32. Temperature dependencies of electrical resistivity ρ of disordered carbide $NbC_{0.83}$ and partially ordered carbide Nb_6C_5 [127]

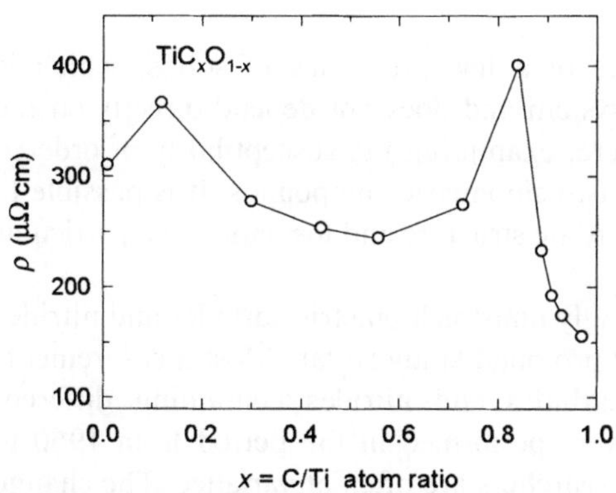

Fig. 11.33. Electrical resistivity of solid solutions of the $TiC_{1.0}$–$TiO_{1.0}$ system at 300 K [141, 142]. The variation of the vacancy concentration in the metal and nonmetal sublattices as a function of the composition of oxycarbides TiC_xO_{1-x} was not taken into account in [141, 142]

Ordering in solid solutions of nonstoichiometric compounds is also accompanied by a decrease in electrical resistivity. For example, the dependence of resistivity on the composition of titanium oxycarbides $TiC_xO_{1-x} \equiv (TiC)_x(TiO)_{1-x}$ [141, 142] exhibits a pronounced minimum ρ (Fig. 11.33) which is typical of ordering binary metal systems. Neutron diffraction examination of $TiC_{0.44}O_{0.57}$, whose composition was close to equimolar oxycarbide $(TiC)_{0.5}(TiO)_{0.5}$, indeed revealed superstructure reflections pointing to the formation of an ordered cubic phase [142]. However, no information is reported in [141, 142] concerning the content of structural vacancies in the nonmetal and metal sublattices of oxycarbides TiC_xO_y of different compositions. The point is that cubic titanium monoxide $TiO_{1.0}$ actually has the composition $Ti_{0.85}O_{0.85}$ and contains structural vacancies in both sublattices. Therefore the concentration of vacancies in the solid solution $(TiC)_x(TiO)_{1-x}$ should change as a function of x. Structural vacancies have a great effect on electrical resistivity and the features observed in the dependence $\rho(x)$ can be explained (at least partially) by variation of the vacancy concentration with the solid solution composition. In addition, titanium monoxide $TiO_{1.0}$ forms the ordered phase Ti_5O_5. This fact was also overlooked by the authors [141, 142] in interpretation of their results.

Investigations of strongly nonstoichiometric compounds in different structural states show that the resistivity is highly sensitive not only to the quantity, but also to the distribution of interstitial atoms and structural vacancies in a crystal lattice.

11.5 Magnetic Susceptibility

The magnetic susceptibility χ is one of a few properties of solids, which is determined only by the electron subsystem and does not depend directly on the energy spectrum of phonons. Therefore, examining the susceptibility of ordered and disordered phases of strongly nonstoichiometric compounds, it is possible to establish the relation between the electronic structure and formation of a particular superstructure.

The magnetic susceptibility of strongly nonstoichiometric carbides and nitrides MX_y has not received systematic and profound study so far. Most measurements of the magnetic susceptibility of carbides and nitrides containing different concentrations of interstitial atoms were performed in the period from 1960 to 1975. Results obtained by different researchers are often at variance. The change of the magnetic susceptibility χ in the homogeneity intervals of disordered cubic carbides TiC_y, ZrC_y, HfC_y, VC_y, NbC_y, and TaC_y [143–150] is shown in Fig. 11.34.

It is the usual practice to interpret susceptibility data disregarding possible ordering of interstitial atoms and structural vacancies. In other words, the crystal structure $B1$ is assumed. Consequently, the actual structural state (degree of order) of a compound is neglected. However, any ordering-induced change in the mutual arrangement of atoms in the crystal lattice may lead to alterations of the electronic structure. Such alterations may influence, specifically, the magnetic susceptibility.

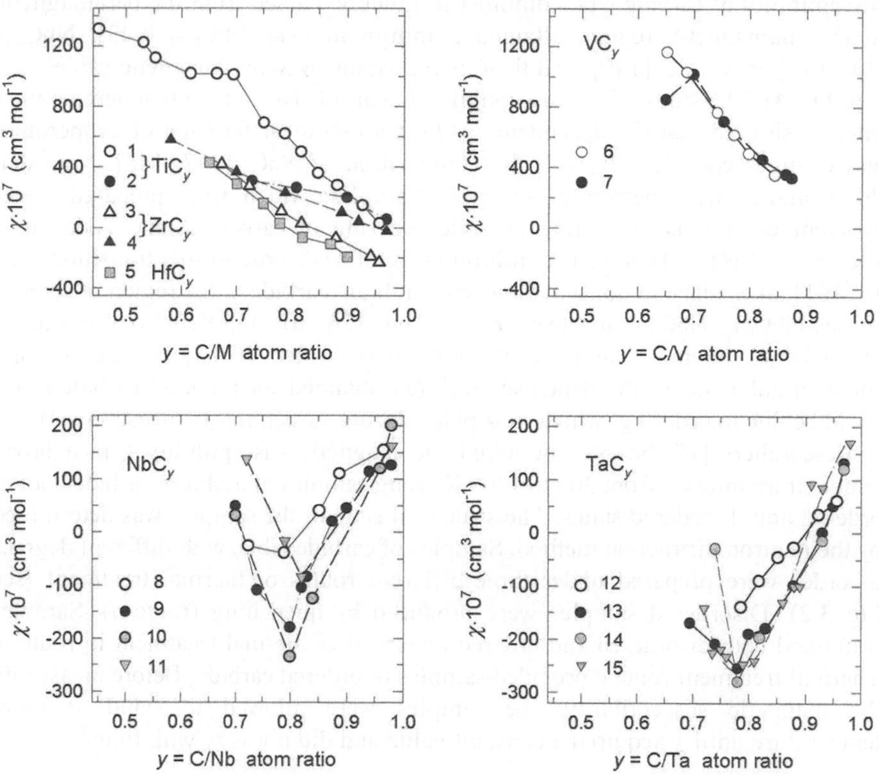

Fig. 11.34. Isotherms of magnetic susceptibility $\chi(y,T)$ in homogeneity intervals of cubic carbides MC_y. TiC_y: (1) $\chi(y,300)$ [143], (2) $\chi(y,300)$ [144]. ZrC_y: (3) $\chi(y,300)$ [143], (4) $\chi(y,300)$ [145]. HfC_y: (5) $\chi(y,300)$ [143]. VC_y: (6) $\chi(y,300)$ [143], (7) $\chi(y, 300)$ [146]. NbC_y: (8) $\chi(y,300)$ [143], (9) $\chi(y,300)$ [147], (10) $\chi(y,100)$ [148], (11) $\chi(y,300)$ [149]. TaC_y: (12) $\chi(y,300)$ [143], (13) $\chi(y,300)$ [150], (14) $\chi(y,100)$ [148], (15) $\chi(y,300)$ [22]

The fact that ordering affects the magnetic susceptibility of nonstoichiometric compounds was first detected and explained by researchers [15, 56–58] who studied niobium carbide NbC_y. Later, the subject of study was variation of the magnetic susceptibility χ during disorder–order phase transitions in nonstoichiometric carbides of titanium TiC_y [151, 152], hafnium HfC_y [153–156], and tantalum TaC_y [157–160]. The observed differences in the susceptibility of disordered and ordered carbides TiC_y, HfC_y, NbC_y and TaC_y were explained [151–160] using a model proposed for the susceptibility of niobium carbide NbC_y. Therefore we shall look at the results of [15, 56–58] more closely.

The magnetic susceptibility χ of niobium carbide NbC_y in the homogeneity interval of the cubic phase ($NbC_{0.70}$ to $NbC_{1.00}$) was examined in [143, 147–149]. It was found [143, 147–149] that when the carbon concentration decreased, the susceptibility of carbide NbC_y diminished quickly, passed from the paramagnetic to the diamagnetic region, attained a minimum near $NbC_{0.79}$ [143], $NbC_{0.80}$ [147,148] or $NbC_{0.82}$ [149], and then increased again to paramagnetic values (see Fig. 11.34). The study of the susceptibility of carbide NbC_y in a broad temperature interval showed that the dependence $\chi(T)$ is a nonlinear function of temperature and changes considerably with the composition of NbC_y [147, 148]. Note that those studies were performed on NbC_y samples, which were prepared under different conditions neglecting possible ordering of carbon atoms. Today it is known (see Sect. 3.4) that an equilibrium reversible disorder–order transition [15–18, 161] takes place in nonstoichiometric niobium carbide in the region of compositions $NbC_{0.81}$–$NbC_{0.88}$ at a temperature from 1260 to 1355 K. Also, the ordered phase Nb_6C_5 having a homogeneity interval is formed. It may be assumed that some singularities in the dependencies $\chi(T)$ obtained for niobium carbide [147] could be due to ordering, which took place during susceptibility measurements.

Researchers [15, 56–58] measured the magnetic susceptibility χ in a broad temperature interval from 300 to 1300 K using samples of niobium carbide both in ordered and disordered states. The structural state of the samples was determined by the neutron diffraction method. Samples of carbide NbC_y with different degrees of order were prepared under three different routes of thermal treatment (see Fig. 3.2). Disordered samples were produced by quenching (route *a*). Samples contained both disordered and ordered phases after thermal treatment in route *b*. Thermal treatment route *c* provided samples of ordered carbide. Before measuring the magnetic susceptibility, the samples were allowed to stand at room temperature until χ acquired a constant value and did not vary with time.

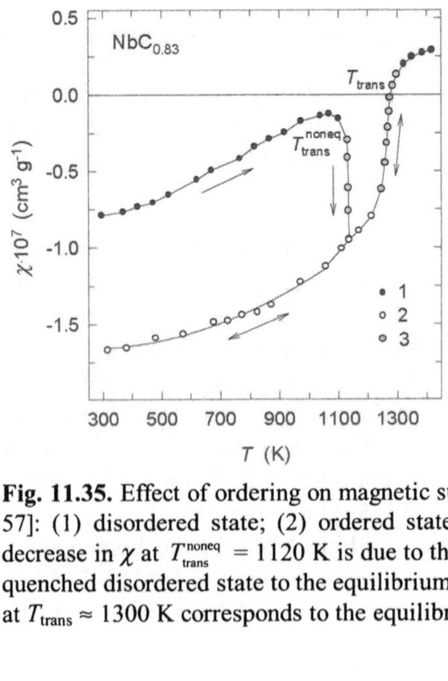

Fig. 11.35. Effect of ordering on magnetic susceptibility χ of niobium carbide $NbC_{0.83}$ [56, 57]: (1) disordered state; (2) ordered state; (3) transition state. The sharp irreversible decrease in χ at $T_{trans}^{noneq} = 1120$ K is due to the irreversible transition from a nonequilibrium quenched disordered state to the equilibrium ordered one. An abrupt reversible change of χ at $T_{trans} \approx 1300$ K corresponds to the equilibrium reversible order–disorder transition

A typical dependence $\chi(T)$ for ordering niobium carbide is shown in Fig. 11.35. Susceptibility measurements on a disordered sample of $NbC_{0.83}$, which was quenched from a temperature much higher than T_{trans}, showed that the susceptibility χ decreased abruptly to the susceptibility of the ordered sample having a similar composition. Neutron diffraction patterns, which were taken for NbC_y samples ($0.81 \le y \le 0.88$) after measuring the temperature dependence of the susceptibility, contained superstructure reflections attesting to ordering that tooks place during measurements. It is worth noting that the observed temperature of the irreversible transition from the quenched nonequilibrium disordered state to the equilibrium ordered state, T_{trans}^{noneq}, should be distinguished from the temperature of the reversible equilibrium order–disorder transition, T_{trans} (Fig. 11.35).

Temperature dependencies of the susceptibility of other samples of niobium carbide (Fig. 11.36) are similar to $\chi(T)$ of carbide $NbC_{0.83}$.

Measurements showed that minima in dependencies $\chi(y)$ of carbides NbC_y in disordered and ordered states corresponded to carbides $NbC_{0.83}$ and $NbC_{0.81}$ (Fig. 11.37). Therefore the discrepancy in experimental data [147–149] can be explained by the fact that the susceptibility was measured on samples of niobium carbide with different degrees of order.

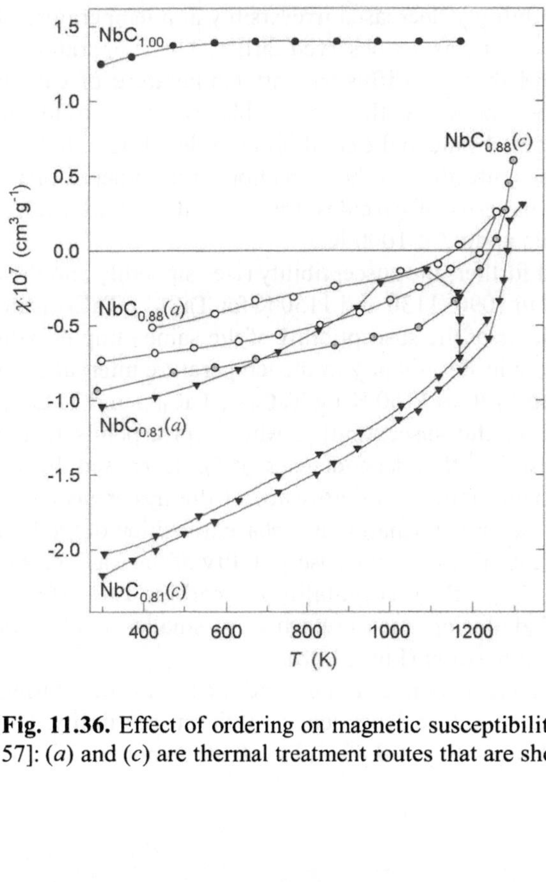

Fig. 11.36. Effect of ordering on magnetic susceptibility χ of niobium carbides NbC_y [56, 57]: (*a*) and (*c*) are thermal treatment routes that are shown in Fig. 3.2

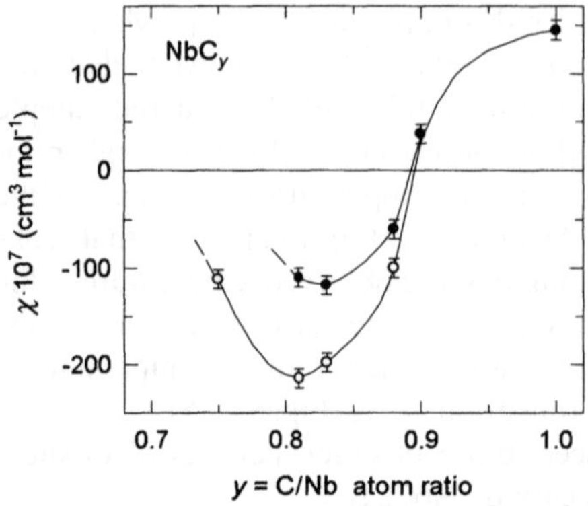

Fig. 11.37. Variation of magnetic susceptibility χ with the composition of disordered (•) and ordered (○) niobium carbide NbC_y at a temperature of 300 K [56]

The effect of ordering on the magnetic susceptibility of tantalum carbide TaC_y was analyzed in [157–160]. When samples of carbides $TaC_{0.82}$, $TaC_{0.83}$ and $TaC_{0.85}$ were heated slowly, the susceptibility χ decreased irreversibly at a temperature of 960–1000 K (Fig. 11.38). This effect was not observed earlier. The temperature of the abrupt decrease in χ is probably the diffusion start temperature of carbon atoms in carbide TaC_y and corresponds to the irreversible transition from the quenched nonequilibrium disordered state to the equilibrium ordered state. Indeed, a set of superstructure reflections attesting to the formation of an ordered phase appears in neutron diffraction patterns of carbides $TaC_{0.82}$, $TaC_{0.83}$ and $TaC_{0.85}$ after susceptibility measurements up to $T < 1000$ K.

If the temperature is elevated further, the susceptibility rises smoothly and then increases abruptly at T_{trans} equal to 1090, 1130 and 1150 K for $TaC_{0.82}$, $TaC_{0.83}$ and $TaC_{0.85}$ respectively. Measurements of the susceptibility of the same samples with decreasing temperature revealed a hysteresis of χ in the temperature interval from 1070 to 1090, 1100 to 1130, and 1120 to 1150 K for $TaC_{0.82}$, $TaC_{0.83}$ and $TaC_{0.85}$. The reversible abrupt increase in the susceptibility, which corresponds to the region of temperature hysteresis in the dependencies $\chi(T)$, is caused by an equilibrium order–disorder transformation. The presence of the hysteresis and a two-phase region suggests that this transformation is a phase transition of the first kind. If the temperature is decreased further, the susceptibility of the TaC_y drops. In the temperature interval $T < T_{trans}^{noneq}$ the susceptibility χ of carbide TaC_y ($0.82 \le y \le 0.85$), which was annealed during measurements, is smaller than χ of quenched carbide of the same composition (Fig. 11.38).

Samples of tantalum carbide annealed under route c and those annealed during measurements have coincident temperature dependencies of the susceptibility. The dependencies $\chi(T)$ of carbides TaC_y with $y < 0.80$ and $y > 0.90$ do not exhibit any singularities, because these compositions are outside the region $TaC_{0.79}$–$TaC_{0.89}$ where an incommensurate ordered phase close to the M_6C_5 type is formed.

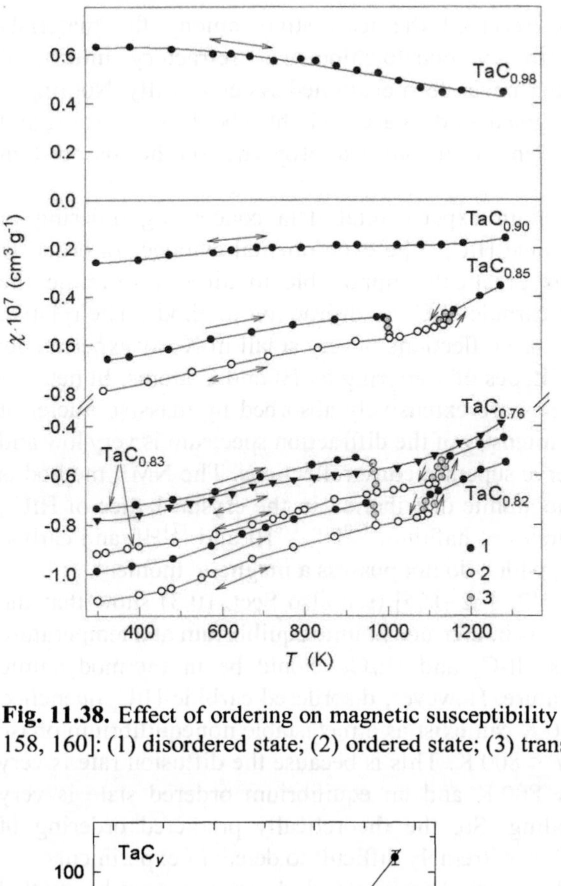

Fig. 11.38. Effect of ordering on magnetic susceptibility χ of tantalum carbide TaC_y [157, 158, 160]: (1) disordered state; (2) ordered state; (3) transition state

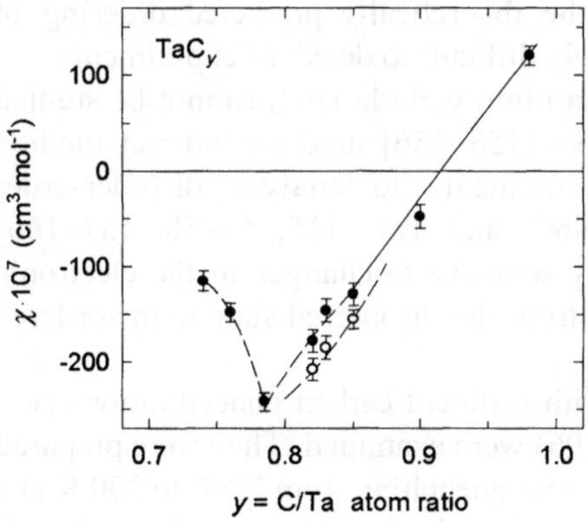

Fig. 11.39. Variation of magnetic susceptibility χ with the composition of disordered (•) and ordered (○) tantalum carbide TaC_y at a temperature of 300 K [160]

Measurements showed that ordering is accompanied by a decrease in the magnetic susceptibility of tantalum carbide (Fig. 11.39). The largest difference of susceptibilities of carbides in disordered and ordered states, $\Delta\chi = \chi_{\rm disord} - \chi_{\rm ord}$, is observed for carbide $TaC_{0.83}$.

Hafnium carbide HfC_y has received the least study among the materials comprising the group of strongly nonstoichiometric refractory interstitial compounds. Hafnium carbide has never been examined systematically. Nothing is known or 30- to 40-year old fragmented data are available about its electronic and crystal structure, and the dependence of its physical properties on the composition and temperature.

The literature does not contain experimental data concerning ordering of nonstoichiometric hafnium carbide HfC_y. The experimental evidence of ordering of HfC_y is lacking since it is practically impossible to directly examine the structure of ordered phases of carbide HfC_y by diffraction methods. The relative intensity of possible superstructure reflections is very small in X-ray experiments owing to largely different amplitudes of scattering by Hf and C atoms. In neutron diffraction experiments, neutrons are extensively absorbed by massive nuclei of hafnium. As a result, the total intensity of the diffraction spectrum is very low and it is next to impossible to observe superstructure reflections. The NMR method is inapplicable to the study of the atomic distribution in the crystal lattice of HfC_y, because the most frequent isotopes of hafnium ^{180}Hf, ^{176}Hf and ^{174}Hf, and carbon ^{12}C have nuclei with zero spin, which do not possess a magnetic moment.

Theoretical calculations [2, 43, 162–165] (see also Sect. 10.3) show that the disordered state of carbide HfC_y is in thermodynamic equilibrium at a temperature above 800 K. Ordered phases Hf_3C_2 and Hf_6C_5 should be in thermodynamic equilibrium at a lower temperature. However, disordered carbide HfC_y quenched from a temperature above 1000 K can exist as a metastable nonequilibrium phase for an infinitely long time at $T < 800$ K. This is because the diffusion rate is very small at a temperature below 800 K and an equilibrium ordered state is very difficult to achieve by annealing. So, the theoretically predicted ordering of nonstoichiometric carbide HfC_y is extremely difficult to detect in experiments.

Since ordering of nonstoichiometric hafnium carbide HfC_y cannot be studied directly by structural methods, researchers [153–156] used an indirect method involving magnetic susceptibility measurements to analyze disorder-order transitions. Investigations of carbides NbC_y and TaC_y [15, 56–58, 157–160] showed that the susceptibility χ is highly sensitive to changes in the electronic structure, which occur during transitions from the disordered state to the ordered one and vice versa.

Samples of disordered carbide HfC_y with different carbon concentrations (y = 0.62, 0.71, 0.78, 0.80, 0.83, 0.87, 0.90, 0.96) were examined. They were prepared by solid-state vacuum synthesis at 2200 K and quenching from 2200 to 300 K at a rate of 2000 K min^{-1} (the temperature route of synthesis is shown in Fig. 3.1). All the samples were homogeneous and contained one phase having the $B1$ structure. The susceptibility was measured using a high-sensitivity magnetic balance intended for examination of weak para- and diamagnetics. Measurements were made in fields of 7.2, 8.05 and 8.8 kOe at a temperature from 300 to 1250 K under a vacuum of 0.001 Pa or better. The sensitivity of the magnetic balance was 2×10^{-10} cm^3 g^{-1}. The susceptibility of the hafnium carbide samples did not depend on the magnetic field intensity, a fact which attested to the absence of ferromagnetic impurities. The relative measurement error did not exceed 3%. Prior to measurements, the samples were held for 1.5–2.0 h at each temperature until χ was constant and did not vary with time.

The analysis of the dependencies $\chi(T)$ showed that the susceptibility of carbides $HfC_{0.62}$, $HfC_{0.80}$, $HfC_{0.87}$, $HfC_{0.90}$ and $HfC_{0.96}$ increased nonlinearly with elevating temperature. No effects testifying to phase transitions were observed. The dependencies $\chi(T)$ of carbides $HfC_{0.71}$, $HfC_{0.78}$ and $HfC_{0.83}$ exhibited an anomalous decrease in χ at 835–860 K. The susceptibility χ increased as the temperature is elevated further (Fig. 11.40). In accordance with calculations [162, 165], hafnium carbide forms ordered phases Hf_3C_2 and Hf_6C_5 whose transition temperature T_{trans} is 600–800 K depending on the composition of disordered carbide. One may think that the observed anomalous drop and the subsequent rise of the susceptibility of HfC_y are due to ordering and disordering during measurements.

Considering hysteresis present in the region of the equilibrium order–disorder transformation, ordering in HfC_y may be classified as a phase transition of the first kind. We cannot but mention such an interesting experimental finding as the zero susceptibility of carbide $HfC_{0.78}$ in the interval from 910 to 980 K (Fig. 11.40).

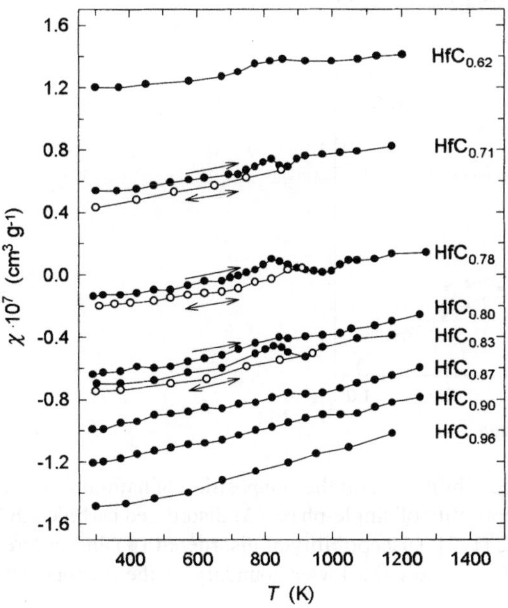

Fig. 11.40. Effect of ordering on magnetic susceptibility χ of hafnium carbides HfC_y [153–156]: (•) disordered state; (○) ordered state

The absence of any singularities in the dependence $\chi(T)$ of carbide $HfC_{0.80}$ probably means that the composition of this carbide is at the boundary of ordered phases Hf_3C_2 and Hf_6C_5 and, consequently, is not ordered. To produce short- or long-range order, the powder of carbide $HfC_{0.80}$ was annealed additionally at 750 K for 20 h and was examined by the method of transmission electron microscopy using a JEM-200C microscope [155, 156]. Electron diffraction patterns of some grains of carbide $HfC_{0.80}$ had only systems of point reflections from planes of the reciprocal lattice of the cubic phase HfC_y. Thus, the electron diffraction analysis confirmed the absence of short- or long-range order in the distribution of vacancies in carbide $HfC_{0.80}$.

The dependence of the magnetic susceptibility χ on the relative concentration of carbon in hafnium carbide HfC_y at 300 K is shown in Fig. 11.41. The magnetic susceptibility of HfC_y in its homogeneity interval at room temperature was studied earlier in [143]. In accordance with [143], near-stoichiometric hafnium carbide is a diamagnetic. The susceptibility is enhanced upon deviation from stoichiometry and changes from diamagnetic to paramagnetic values in the region of $HfC_{0.82}$. The susceptibility of carbide HfC_y at $0.6 < y < 0.9$ [153, 154, 156] is somewhat less than the susceptibility measured in [143]. The susceptibility data of [143] and [153, 154, 156] are nearly coincident at $y = 0.9$ (Fig. 11.41). The slightly overestimated values of the magnetic susceptibility of hafnium carbide [143] may be due to the presence of a small (~0.01 at. %) ferromagnetic impurity.

Ordering of carbides $HfC_{0.71}$, $HfC_{0.78}$ and $HfC_{0.83}$ causes an insignificant decrease in the magnetic susceptibility (Fig. 11.41).

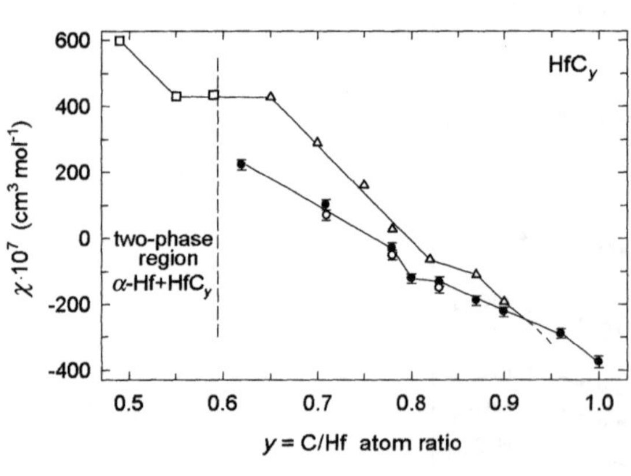

Fig. 11.41. Variation of magnetic susceptibility χ with the composition of hafnium carbide HfC_y at a temperature of 300 K: susceptibility of single-phase (Δ) disordered carbide HfC_y and two-phase (□) samples Hf + HfC_y [143], susceptibility of disordered (●) and ordered (○) carbide HfC_y [156]. The dotted line shows the lower boundary of the homogeneity interval of hafnium carbide HfC_y

Titanium carbide TiC_y is one more compound whose magnetic susceptibility was measured to analyze ordering. Carbide TiC_y has the widest homogeneity interval $0.48 \le y \le 1.00$ among all cubic carbides (see Sect. 3.2). The experimental data (see Sect. 3.4) and theoretical calculations (see Sect. 10.3) show that this carbide forms ordered phases Ti_2C, Ti_3C_2 and Ti_6C_5.

Carbide $TiC_{0.96}$ is a paramagnetic and its susceptibility is nearly zero [143]. The magnetic susceptibility χ of titanium carbide increases with deviation from stoichiometry. In the interval $TiC_{0.62}$–$TiC_{0.70}$ the susceptibility remains almost unchanged (see Fig. 11.34).

The study of the temperature dependencies $\chi(T)$ of titanium carbide samples with different carbon contents [151, 152] showed that the susceptibility increased nonlinearly with growing temperature (Fig. 11.42). The increasing χ of carbides $TiC_{0.83}$, $TiC_{0.90}$, $TiC_{0.925}$, $TiC_{0.94}$ and $TiC_{1.00}$ is not accompanied by any effects attesting to the occurrence of phase transformations. The dependencies $\chi(T)$ of carbides TiC_{050}, $TiC_{0.55}$, $TiC_{0.59}$, $TiC_{0.63}$, $TiC_{0.68}$, $TiC_{0.69}$ and $TiC_{0.81}$ exhibit an anomalous decrease in χ at 750–950 K. The susceptibility is enhanced again when the temperature is elevated further (Fig. 11.42). The diffraction analysis showed that the observed anomalous drop and the growth of the magnetic susceptibility of titanium carbide are due to ordering and disordering during heating. The small hysteresis present in the transformation region (Fig. 11.42) suggests that ordering of carbide TiC_y is a phase transition that is close to a transition of the first kind.

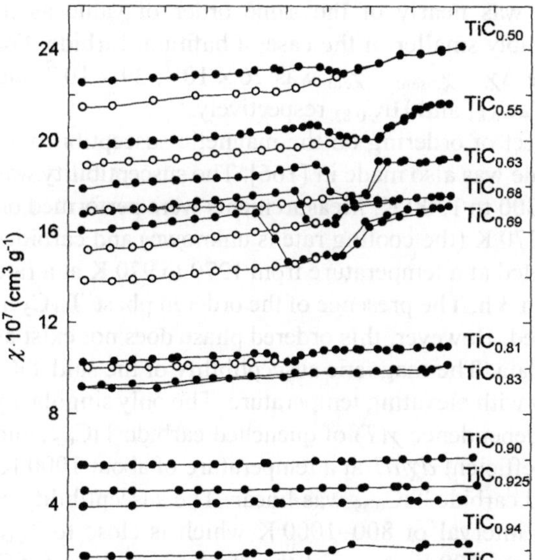

Fig. 11.42. Temperature dependencies of magnetic susceptibility χ of nonstoichiometric titanium carbide TiC_y in disordered (•) and ordered (○) states [151, 152]

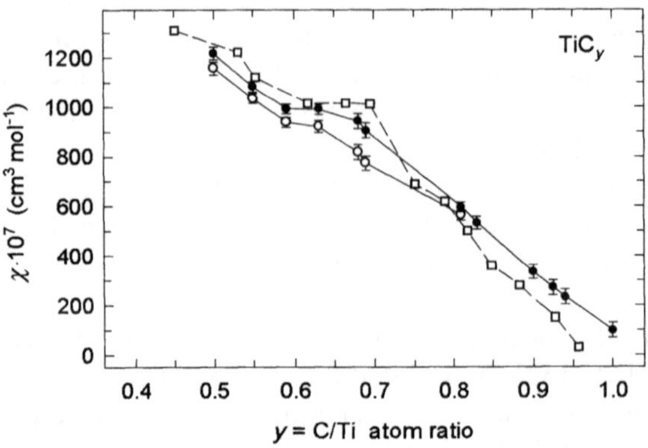

Fig. 11.43. Variation of magnetic susceptibility χ at 300 K within the homogeneity interval of cubic titanium carbide TiC_y: susceptibility of disordered (•) and ordered (○) carbide TiC_y according to [151, 152], (□) data from [143] on susceptibility of disordered titanium carbide

Ordering of titanium carbide TiC_y was followed by the decrease in the magnetic susceptibility by (40–120) × 10^{-7} cm^3 mol^{-1} depending on the carbide composition (Fig. 11.43). An analogous effect was observed during ordering of niobium, tantalum and hafnium carbides. The effect of ordering on the susceptibility of niobium and tantalum carbides was nearly of the same order of value as in titanium carbide. It was considerably smaller in the case of hafnium carbide. For example, at 300 K the difference $\Delta\chi = \chi_{disord} - \chi_{ord}$ was 76 × 10^{-7}, 39 × 10^{-7} and 9 × 10^{-7} cm^3 mol^{-1} for $NbC_{0.83}$, $TaC_{0.83}$ and $HfC_{0.83}$ respectively.

An attempt to analyze the effect of ordering on the magnetic susceptibility of nonstoichiometric titanium carbide was also made in [166]. The susceptibility was measured at a temperature from 300 to 1300 K. Measurements were performed on carbide $TiC_{0.59}$ quenched from 1270 K (the cooling rate is unknown) and carbides $TiC_{0.59}$, $TiC_{0.65}$ and $TiC_{0.74}$ annealed at a temperature from 1270 to 970 K at a rate of 50 K h^{-1} and cooled to 300 K in 3 h. The presence of the ordered phase Ti_8C_5 in the annealed samples was assumed. However, this ordered phase does not exist as follows from modern structural data. The magnetic susceptibility of the said TiC_y samples increased insignificantly with elevating temperature. The only singularity was presented by a break in the dependence $\chi(T)$ of quenched carbide $TiC_{0.59}$ and the growth of the temperature coefficient $d\chi/dT$ at a temperature of about 1000 K. The dependence $\chi(T)$ of annealed carbide $TiC_{0.59}$ was linear. The susceptibility of quenched carbide $TiC_{0.59}$ in the interval of 800–1000 K which is close to T_{trans} [130, 151, 152], was measured at a 100 K step and therefore researchers [166] could not detect a clear-cut effect of ordering. The absence of any effects of ordering in the dependence $\chi(T)$ of carbide $TiC_{0.74}$ is due to the fact that the composition of this carbide is outside the homogeneity intervals of ordered phases. Indeed, the main ordered phase of TiC_y carbide is Ti_2C, which exists at $T < 1000$ K in the region $TiC_{0.48}$–$TiC_{0.58-0.59}$. The ordered phase Ti_3C_2 may appear in the region $TiC_{0.63}$–$TiC_{0.67}$ [130]. The change of the susceptibility might probably be observed in carbide $TiC_{0.65}$ during transition from the quenched disordered state to the equilibrium ordered state, but authors [166] did not measure the susceptibility of disordered carbide $TiC_{0.65}$. So, the erroneous experimental technique did not allow [166] elucidating the effect of ordering on the susceptibility of TiC_y carbide.

The changes of the magnetic susceptibility χ observed in [15, 56–58, 151–160] are explained by variation of the composition of nonstoichiometric carbides MC_y on the one hand and the effect of ordering on the other. Let us consider possible causes of this change.

The Effect of Nonstoichiometry. The measured magnetic susceptibility of carbides is the sum of several contributions [2]:

$$\chi = \chi_p^+ + \chi_d^+ + \chi_p^- + \chi_d^- + \chi_{orb} , \qquad (11.20)$$

where χ_p^+ and χ_d^+ denote para- and diamagnetism of atomic cores ($\chi_p^+ \equiv 0$; χ_d^+ for Ti^{4+}, Hf^{4+}, Nb^{5+}, Ta^{5+} and C^{4+} is equal to -40×10^{-7}, -160×10^{-7}, -90×10^{-7}, -140×10^{-7} and -1.5×10^{-7} $cm^3\, mol^{-1}$ [167] respectively); χ_p^- is the Pauli paramagnetism of conduction electrons; $\chi_d^- = -(m_0/m^*)^2 \chi_p^-(0)/3$ is the Landau diamagnetism of conduction electrons (m_0 being the mass of a free electron and m^* the effective mass); and χ_{orb} is the Van Flek orbital paramagnetism. Among all the terms of (11.20), only the Pauli paramagnetism depends on temperature [168]:

$$\chi_p^-(T) = 2N_A \mu_B^2 N(E_F) + \frac{\pi^2 k_B^2}{3} N_A \mu_B^2 \left\{ N''(E_F) - \frac{[N'(E_F)]^2}{N(E_F)} \right\} T^2 , \qquad (11.21)$$

where μ_B is the Bohr magneton. The Landau diamagnetism χ_d^- is proportional to the Pauli paramagnetism $\chi_p^-(0)$ at 0 K. If the measured susceptibility is extrapolated to 0 K and corrections for paramagnetism and diamagnetism of atomic cores are introduced, $\chi^*(0)$ will be the sum of two paramagnetic contributions and one diamagnetic contribution: $\chi^*(0) = \chi_p^-(0) + \chi_d^-(0) + \chi_{orb}$. From (11.21) it is clear that at 0 K the Pauli paramagnetism depends directly only on the density of electron states at the Fermi level $N(E_F)$.

When the carbon content decreases, the susceptibility of carbides NbC_y and TaC_y is reduced and achieves a minimum for $NbC_{0.83}$ and $TaC_{0.78}$ (see Figs. 11.37 and 11.39). The quantity χ_{orb} exhibits a similar dependence on the composition of these carbides. The orbital paramagnetism χ_{orb} is positive and can increase only thanks to the distorted symmetry of the nearest environment of metal atoms when the composition of disordered carbides deviates from stoichiometry. Thus, the observed decreasing χ of carbides NbC_y and TaC_y and the sign reversal of the susceptibility, which takes place with decreasing carbon content, can be explained only by the change in the contributions χ_p^- and χ_d^-.

The electron energy spectrum of carbides of group V transition metals is formed by two overlapping bands corresponding to M–C and M–M interactions. The high-energy M–M band is less than a half filled even in stoichiometric carbide. When the carbide composition deviates from stoichiometry, the degree of filling of the M–M band changes and the density of electron states at the Fermi level $N(E_F)$ decreases [169]. As a result, χ_p^- decreases. The M–M band becomes almost empty as the lower boundary of the homogeneity interval is approached. In accordance with [168], the effective mass of collectivized electrons decreases quickly with decreasing degree of occupancy of the energy band, which is filled less than a half. Quite reasonably, diamagnetism is enhanced. The deviation of the composition of carbides NbC_y and TaC_y from stoichiometry is also accompanied by the decrease in the effective mass of conduction electrons and by a rapid rise of diamagnetism of the electron gas. When unoccupied electron levels appear in the overlap region of M–C and M–M bands, the effective mass of electrons starts increasing and the diamagnetic contribution decreases. This variation of the density of electron states $N(E_F)$ and the effective mass m^* results in a minimum of the susceptibility corresponding to the overlap region of M–C and M–M bands. The quantity $\chi^*(0) = \chi_p^- + \chi_d^-$ is less than zero for disordered carbides NbC_y ($0.79 \le y \le 0.87$) and TaC_y ($0.76 < y < 0.83$). This is possible if and only if the Landau diamagnetism χ_d^- of conduction electrons is larger in absolute value than the Pauli paramagnetism χ_p^-. Since $\chi_d^- = -(m_0/m^*)^2 \chi_p^-(0)/3$, then $|\chi_d^-| > \chi_p^-$ if $m_0/m^* > \sqrt{3}$. Thus, large diamagnetism of disordered nonstoichiometric niobium and tantalum carbides is due to a small effective mass of conduction electrons. Generally, the susceptibility of disordered NbC_y and TaC_y carbides changes as a result of variation of para- and diamagnetism of conduction electrons.

Let us consider the effect of nonstoichiometry on the magnetic susceptibility of disordered nonstoichiometric carbides of hafnium HfC_y and titanium TiC_y.

When the carbon concentration y decreases, the susceptibility of carbides HfC_y and TiC_y is enhanced (see Figs. 11.41 and 11.43). The quantities $\chi^*(0)$ change analogously as a function of the carbide composition. The orbital paramagnetism χ_{orb} is positive and increases thanks to the distorted symmetry of the nearest environment of metal atoms when the carbon content decreases. When the composition of carbides TiC_y and HfC_y deviates from stoichiometry, the degree of occupancy of the M–C band decreases and electrons of metal atoms, which are released from M–C bonds, fill the band of M–M interactions. As a result, the density of states $N(E_F)$ rises [114, 170–172] and $\chi_p^-(0)$ increases. The M–M conduction band is less than a half filled even at the lower boundary of the homogeneity interval of carbides of group IV transition metals. Therefore filling of this band is followed by the increase in the effective mass of collectivized electrons and the decrease (in absolute value) of the Landau diamagnetism. Thus, when the composition of carbides HfC_y and TiC_y deviates from stoichiometry towards the lower boundary of the homogeneity interval, the positive contributions χ_{orb} and $\chi_p^-(0)$ increase, while the negative contribution χ_d^- decreases in absolute value. The result is a general tendency to an increase in the magnetic susceptibility of disordered carbides HfC_y and TiC_y with decreasing y.

The quantity $\chi^*(0) = \chi_p^-(0) + \chi_d^- + \chi_{orb}$ is negative for disordered carbide HfC_y with $y \ge 0.76$ (Fig. 11.41). This is possible only if the Landau diamagnetism χ_d^- is larger in absolute value than the Pauli paramagnetism χ_p^-, i.e. if

$m_0/m^* > \sqrt{3}$. Thus, diamagnetism of disordered carbide HfC_y with $y \geq 0.76$ is due to the small effective mass of conduction electrons.

The dependence $\chi(y, 300)$ of disordered carbide TiC_y has a section where the susceptibility changes insignificantly with the composition ($0.59 \leq y \leq 0.63$, Fig. 11.43). This fact suggests that the density of states at the Fermi level $N(E_F)$ depends little on the carbide composition. This is in agreement with theoretical calculations of the electronic structure of TiC_y [114], which are performed in the coherent potential approximation by the Korringa–Kohn–Rostoker method.

In accordance with [114], the Fermi level of the stoichiometric carbide $TiC_{1.00}$ is located at the minimum of the overlap region of Ti–C and Ti–Ti subbands. When the vacancy concentration $(1 - y)$ rises from 0 to 0.5, the Fermi energy E_F decreases from 0.69 to 0.66 Ry (9.4 to 9.0 eV), while the density of electron states at the Fermi level $N(E_F)$ increases in terms of one unit cell from 5 to 23 Ry^{-1} (0.05 to 0.282 eV^{-1} $atom^{-1}$) (Fig. 11.44). The growth of the vacancy concentration is accompanied by smoothing of all sharp peaks of the density of states (especially below E_F), the increase of occupancy of the high-energy subband of conductivity, and the decrease of occupancy of the low-energy subband of Ti–C interactions. The shape of the high-energy subband changes slightly upon implantation of vacancies. However, an additional vacancy peak appears near the minimum of the density of states. This peak splits into two smeared peaks when the vacancy content increases. A nearly linear section is observed in the calculated dependence [114] of the density of states $N(E_F)$ on the relative content of carbon y in carbide TiC_y (Fig. 11.44). This section with $N(E_F) \approx 15$–17 Ry^{-1} (0.152–0.192 eV^{-1} $atom^{-1}$) corresponds to the compositions $TiC_{0.82}$–$TiC_{0.625}$, i.e. nearly to the same region, in which an almost linear section is observed in the concentration dependence of the susceptibility $\chi(y, 300)$ of disordered titanium carbide (Fig. 11.43).

Indeed, in the interval from $TiC_{0.6}$ to $TiC_{0.8}$ the density of states $N(E_F)$ is about 15–16 Ry^{-1} and changes insignificantly with the carbide composition (Fig 11.44).

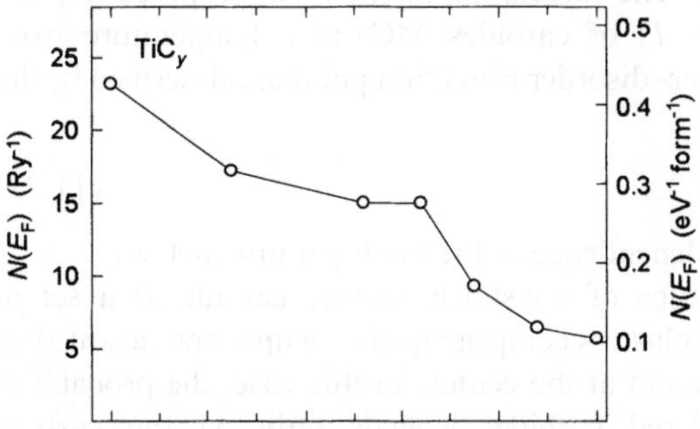

Fig. 11.44. Dependence of the density of electron states at the Fermi level, $N(E_F)$, on the composition of disordered nonstoichiometric titanium carbide TiC_y [114]

The Effect of Ordering. What are the causes of the observed decrease in the magnetic susceptibility of nonstoichiometric carbides NbC_y, TaC_y, HfC_y and TiC_y during ordering?

The susceptibility χ may decrease during ordering if the effective mass of a collectivized electron in ordered carbide is smaller than in disordered carbide, i.e. $m^*_{ord} < m^*_{disord}$. If the effective mass of conduction electrons changes little during ordering, the decrease in the susceptibility may be related to the change in the density of electron states at the Fermi level $N(E_F)$. However, $N(E_F)$ changes insignificantly during ordering of niobium carbide [52, 173, 174] and there are no grounds to think that the situation will be different for tantalum, titanium and hafnium carbides. So, the decrease in the density of electron states at the Fermi level can be disregarded as a cause of the decrease in the susceptibility χ of carbides during disorder–order transitions.

The most probable cause of the decrease in the susceptibility during ordering may be the decrease in the contribution of the orbital paramagnetism, which is connected with the appearance of short-range order [56, 160]. Short-range order necessarily arises upon formation of long-range order, i.e. during ordering. This short-range order was called [175, 176] superstructural short-range order.

The magnetic susceptibility of nonstoichiometric compounds MX_y having the *B*1 basic structure is determined mainly by the electronic subsystem of metal atoms. Short-range order is absent in disordered nonstoichiometric carbide and the nearest environment of a metal atom may have different symmetries. Several types (frequently only one type) of the nearest environment of metal atoms are possible in ordered carbide (see Sect. 5.4). Since the nearest environment of metal atoms in ordered carbide has a higher symmetry than in disordered carbide, the contribution of the Van Vleck orbital paramagnetism to the susceptibility of ordered carbide is less in value than the analogous contribution to χ of disordered carbide. The fact that the nearest environment symmetry affects the magnetic susceptibility is proved by calculations of the short-range order parameters using experimental data on the magnetic susceptibility. The calculation method was proposed in [56].

Experimental dependencies $\chi(y,T)$ of carbides MC_y at a temperature from 300 K to the disorder–order or order–disorder transition point are described by the function

$$\chi(y,T) = a(y) + b(y)T^2 \tag{11.22}$$

corresponding to the temperature dependence of the Pauli paramagnetism.

Let us consider the crystal lattice of nonstoichiometric carbide as a set of noninteracting octahedron-shaped clusters comprising six occupied or vacant sites of the carbon lattice and a metal atom at the center. In this case, the probability $P_i(y)$ of each cluster in disordered carbide depends only on the carbide composition and is determined by a binomial distribution. If the susceptibility (11.22) of disordered carbide is expressed through the susceptibility $\chi_i = a_i + b_iT^2$ of separate clusters as

$$\chi(y,T) = \Sigma\chi_i(T)\lambda_iP_i(y), \tag{11.23}$$

then, knowing the probabilities $P_i(y)$ and experimental coefficients $a(y)$ and $b(y)$ for several disordered carbides with different carbon contents, it is possible to calculate the coefficients a_i and b_i for each cluster (λ_i being the number of equivalent configurations of a cluster having the i-configuration, in which n sites out of six sites are vacant; $\sum_{i \in n} \lambda_i = C_6^n$). Indeed, considering (11.22), equation (11.23) can be written as

$$\chi(y,T) = \Sigma(a_i + b_i T^2)\lambda_i P_i(y), \tag{11.24}$$

whence

$$a(y) = \Sigma a_i \lambda_i P_i(y)\,, \quad b(y) = \Sigma b_i \lambda_i P_i(y). \tag{11.25}$$

Interstitial atoms and vacancies are distributed statistically in a disordered compound. Therefore the probability of octahedral clusters containing n vacancies and $(6-n)$ interstitial atoms is described by the binomial distribution $P_i(y) = P_i^{\text{bin}}(y) = y^{(6-n)}(1-y)^n$ of the form (9.19). Using calculated values of $P_i^{\text{bin}}(y)$ and substituting them into (11.25) written for disordered compound MX_y, we determine the coefficients a_i and b_i. Now, using experimental values of the coefficients $a(y)$ and $b(y)$ for ordered carbide, calculated values of a_i and b_i, and the normalization conditions $\Sigma\lambda_i P_i(y) = 1$ and $\Sigma\ell_i\lambda_i P_i(y) = y$ (ℓ_i being the number of sites in an i-configuration cluster occupied by carbon atoms), it is possible to calculate the cluster probability $P_i^{\text{ord}}(y)$ for ordered carbide.

Short-range order parameters α_1 and α_2 for the first and second coordination spheres of the carbon sublattice can be found from the formula

$$\alpha_j = 1 - \frac{P_{\text{C}-\square}^{\text{ord}\,(j)}}{P_{\text{C}-\square}^{\text{bin}}}\,, \tag{11.26}$$

where $P_{\text{C}-\square}^{\text{bin}} = \lambda_{\text{C}-\square}\, y(1-y)$ is the binomial probability that carbon–vacancy pairs C–□ ($\lambda_{\text{C}-\square}$ = 2) are formed. This binomial probability corresponds to the disordered state of carbide MC_y and depends only on the carbide composition. The probability that C–□ pairs are formed in any j-th coordination sphere of the nonmetallic sublattice of an ordered compound is given by the relation

$$P_{\text{C}-\square}^{\text{ord}\,(j)} = \sum_i n_i^{(j)} \lambda_i P_i^{\text{ord}}(y)\,, \tag{11.27}$$

where $n_i^{(j)}$ is the relative fraction of "carbon atom–vacancy" pairs in the j-th coordination sphere of a cluster with i-configuration. Values of the coefficients λ_i, ℓ_i and $n_i^{(j)}$ are given in Table 6.2.

In accordance with [1–4], the correct description of carbides MC_y with $0.7 \le y \le 1.0$ requires taking into account four configurations of an octahedral cluster: a complete (free of vacancies) cluster with probability P_0, a cluster with one vacancy having probability P_1, and clusters with two inadjacent and two adjacent vacancies having probabilities P_2 and P_3. The homogeneity interval of titanium carbide TiC_y is wider than the homogeneity intervals of niobium, tantalum and hafnium carbides. Therefore the correct description for the region $0.5 \le y \le 1.0$ requires taking into account eight configurations of an octahedral cluster: the aforementioned four configurations having probabilities P_0, P_1, P_2 and P_3, clusters with three vacancies each having probabilities P_4 and P_5 and also clusters with four vacancies each having probabilities P_6 and P_7. All these configurations of an octahedral basis cluster are shown in Fig. 9.4.

Using experimental dependencies $\chi(y,T)$, the coefficients a and b were found from (11.22) for disordered and ordered carbides NbC_y, TaC_y, HfC_y and TiC_y (Tables 11.9 and 11.11). Calculated probabilities of clusters $P_i(y)$ and short-range order parameters α_1 and α_2 for ordered carbides NbC_y, TaC_y, HfC_y and TiC_y are given in Tables 11.10 and 11.12.

From the comparison of calculated values of α_1 and α_2 with theoretical (see Sect. 6.2, Tables 6.4, 6.5 and 6.6) values of the short-range order parameters of all superstructures that may be formed in nonstoichiometric interstitial compounds it is possible to hypothesize the type and symmetry of ordered phases of niobium, tantalum, hafnium and titanium carbides.

Table 11.9. Coefficients of the temperature dependencies of the magnetic susceptibility χ of niobium, tantalum and hafnium carbides in disordered and ordered states, temperature of nonequilibrium irreversible disorder→order transition, T_{trans}^{noneq}, and temperature of equilibrium reversible order↔disorder transition, T_{trans}

Carbide	Disordered state $\chi = a + bT^2$		T_{trans}^{noneq}	Ordered state T_{trans}	$\chi = a + bT^2$	
	$a \times 10^7$ ($cm^3\,g^{-1}$)	$b \times 10^{14}$ ($cm^3\,g^{-1}\,K^{-2}$)	(K)	(K)	$a \times 10^7$ ($cm^3\,g^{-1}$)	$b \times 10^{14}$ ($cm^3\,g^{-1}\,K^{-2}$)
$NbC_{0.81}$	−1.13	8.26	1170	–	−2.21	6.68
$NbC_{0.83}$	−1.20	8.63	1120	–	−1.79	4.47
$NbC_{0.88}$	−0.60	4.40	1230	–	−0.95	4.09
$NbC_{1.00}$	1.30	0	–	–	–	–
$TaC_{0.82}$	−1.01	3.80	980	1080	−1.09	3.39
$TaC_{0.83}$	−0.76	2.95	970	1110	−0.92	2.51
$TaC_{0.85}$	−0.65	1.87	1000	1130	−0.79	2.25
$TaC_{0.90}$	−0.24	0.49	–	–	–	–
$TaC_{0.98}$	0.68	−1.80	–	–	–	–
$HfC_{0.62}$	1.06	2.843	–	–	–	–
$HfC_{0.71}$	0.50	4.058	~840	870–890	0.41	3.673
$HfC_{0.78}$	−0.18	3.248	~840	970–1000	−0.23	3.010
$HfC_{0.80}$	−0.71	2.803	–	–	–	–
$HfC_{0.83}$	−0.76	3.937	~840	960-980	−0.78	2.654
$HfC_{0.87}$	−0.99	3.069	–	–	–	–
$HfC_{0.90}$	−1.20	3.313	–	–	–	–
$HfC_{0.96}$	−1.48	4.599	–	–	–	–

Table 11.10. Cluster probabilities $\lambda_i P_i$, short-range order parameters α_1 and α_2 and long-range order parameters η for ordered niobium, tantalum and hafnium carbides NbC_y, TaC_y and HfC_y

Carbide	Cluster probabilities $\lambda_i P_i$				Short-range order parameters		Type of ordered phase	Space group	Long-range order parameter
	$\lambda_0 P_0$	$\lambda_1 P_1$	$\lambda_2 P_2$	$\lambda_3 P_3$	α_1 ±0.005	α_2 ±0.005			η
$NbC_{0.81}$	0.06	0.75	0.05	0.15	−0.153	−0.137	Nb_6C_5	$C2/m$	0.87–0.92
$NbC_{0.83}$	0.14	0.70	0.04	0.12	−0.132	−0.126	Nb_6C_5	$C2/m$	0.80–0.82
$NbC_{0.88}$	0.40	0.48	0.03	0.09	−0.064	−0.040	Nb_6C_5	$C2/m$	0.39–0.49
$TaC_{0.82}$	0.23	0.47	0.06	0.24	−0.080	−0.086	Ta_6C_5	$C2/m$?	0.65–0.68
$TaC_{0.83}$	0.26	0.46	0.05	0.23	−0.063	−0.080	Ta_6C_5	$C2/m$?	0.57–0.64
$TaC_{0.85}$	0.33	0.44	0.05	0.18	−0.059	−0.043	Ta_6C_5	$C2/m$?	0.44–0.52
$HfC_{0.71}$	0.01	0.24	0.25	0.50	−0.207	−0.001	M_3C_2	$P2$, $Immm$	0.876
$HfC_{0.78}$	0.15	0.38	0.15	0.33	−0.123	−0.001	M_3C_2	$P2$, $Immm$	0.616
$HfC_{0.83}$	0.27	0.44	0.08	0.21	−0.079	−0.023	M_6C_5	$C2/m$, $P3_1$	0.34–0.63

Table 11.11. Coefficients of the temperature dependencies of the magnetic susceptibility χ of titanium carbide TiC_y in disordered and ordered states, temperature of nonequilibrium irreversible disorder→order transition, T_{trans}^{noneq}, and temperature of equilibrium reversible order↔disorder transition, T_{trans}

Carbide	Disordered state $\chi = a + bT^2$		Ordered state			
			T_{trans}^{noneq}	T_{trans}	$\chi = a + bT^2$	
	$a \times 10^7$ ($cm^3\,g^{-1}$)	$b \times 10^{14}$ ($cm^3\,g^{-1}\,K^{-2}$)	(K)	(K)	$a \times 10^7$ ($cm^3\,g^{-1}$)	$b \times 10^{14}$ ($cm^3\,g^{-1}\,K^{-2}$)
$TiC_{0.50}$	22.46	11.08	882–903	973–1073	21.30	16.19
$TiC_{0.55}$	19.66	13.03	877–968	1026–1148	18.84	12.69
$TiC_{0.59}$	17.91	13.19	849–885	988–1029	16.90	14.34
$TiC_{0.63}$	17.80	14.08	873–912	946–1043	16.80	7.07
$TiC_{0.68}$	16.65	17.39	821–872	943–1115	14.98	18.71
$TiC_{0.69}$	15.95	15.57	773–843	973–1073	13.63	14.48
$TiC_{0.81}$	10.19	8.15	753–803	837–944	9.91	6.06
$TiC_{0.83}$	9.16	6.62	–	–	–	–
$TiC_{0.90}$	5.68	2.58	–	–	–	–
$TiC_{0.925}$	4.56	1.96	–	–	–	–
$TiC_{0.94}$	3.92	1.93	–	–	–	–
$TiC_{1.00}$	1.62	4.56	–	–	–	–

Table 11.12. Cluster probabilities $\lambda_i P_i$, short-range order parameters α_1 and α_2 and long-range order parameters η for ordered titanium carbide TiC_y

Carbide	Cluster probabilities $\lambda_i P_i$							
	$\lambda_0 P_0$	$\lambda_1 P_1$	$\lambda_2 P_2$	$\lambda_3 P_3$	$\lambda_4 P_4$	$\lambda_5 P_5$	$\lambda_6 P_6$	$\lambda_7 P_7$
$TiC_{0.50}$	0	0	0	0.159	0.688	0	0.153	0
$TiC_{0.55}$	0	0.118	0	0.064	0.730	0.088	0	0
$TiC_{0.59}$	0	0.129	0.020	0.262	0.589	0	0	0
$TiC_{0.63}$	0	0.177	0.127	0.299	0.397	0	0	0
$TiC_{0.68}$	0	0.299	0.050	0.431	0.220	0	0	0
$TiC_{0.69}$	0	0.410	0.005	0.315	0.270	0	0	0
$TiC_{0.81}$	0.186	0.488	0.064	0.262	0	0	0	0

Carbide	Short-range order parameters		Type of ordered phase	Space group	Long-range order parameter η
	α_1 ±0.005	α_2 ±0.005			
$TiC_{0.50}$	0	−0.793	Ti_2C	$R\bar{3}m$, $Fd\bar{3}m$	0.891
$TiC_{0.55}$	0	−0.700	Ti_2C	$R\bar{3}m$, $Fd\bar{3}m$	0.832
$TiC_{0.59}$	0.004	−0.668	Ti_2C	$R\bar{3}m$, $Fd\bar{3}m$	0.804
$TiC_{0.63}$	−0.055	−0.406	Ti_3C_2	$C222_1$	0.68–0.92
$TiC_{0.68}$	−0.053	−0.394	Ti_3C_2	$C222_1$	0.64–0.88
$TiC_{0.69}$	−0.011	−0.441	Ti_3C_2	$C222_1$	0.29–0.92
$TiC_{0.81}$	−0.093	−0.096	Ti_6C_5	$C2/m$, $P3_1$, $C2$	0.72–0.73

In niobium and tantalum carbides, and also in carbides $HfC_{0.83}$ and $TiC_{0.81}$ both short-range order parameters α_1 and α_2 are negative, are close in absolute value, and are larger than −0.2. This combination of parameters α_1 and α_2 corresponds to a superstructure of type M_6C_5 with monoclinic or trigonal symmetry. In carbides $HfC_{0.71}$ and $HfC_{0.78}$ parameters $\alpha_1 < 0$ and $\alpha_2 \cong 0$. This combination of parameters α_1 and α_2 is possible only for an ordered phase of M_3C_2 type with the orthorhombic (space group *Immm*) or monoclinic (space group *P*2) structure.

In carbides $TiC_{0.50}$, $TiC_{0.55}$ and $TiC_{0.59}$ parameters $\alpha_1 \cong 0$ and $\alpha_2 < -0.65$. This combination of α_1 and α_2 applies only to the ordered phase Ti_2C with trigonal (space group $R\bar{3}m$) or cubic (space group $Fd\bar{3}m$) symmetry. Indeed, in these superstructures Ti_2C of a stoichiometric composition ($y_{st} = 1/2$) the limiting values of short-range order parameters $\alpha_1(y_{st}, \eta_{max})$ and $\alpha_2(y_{st}, \eta_{max})$ are 0 and −1 (see Table 6.5). In carbides $TiC_{0.63}$, $TiC_{0.68}$ and $TiC_{0.69}$ both short-range order parameters are negative ($-0.06 \le \alpha_1 \le -0.01$, $\alpha_2 \cong -0.4$). This is possible only if a rhombic (space group $C222_1$) superstructure of M_3C_2 type is formed (see Table 6.6). The relation $\alpha_2 < \alpha_1 < 0$ is always fulfilled in this superstructure M_3C_2 and the limiting values of parameters $\alpha_1(y_{st}, \eta_{max})$ and $\alpha_2(y_{st}, \eta_{max})$ for a stoichiometric composition ($y_{st} = 2/3$) are equal to −1/8 and −1/2.

Parameters of superstructural short-range order α_j are related to the long-range order parameter η in the nonmetallic sublattice of MC_y by (6.21):

$$\alpha_j(y,\eta) = -\eta^2 \frac{(1-y_{st})(m_{21}^{(j)} - y_{st})}{y(1-y)},$$

where $m_{21}^{(j)}$ is the coefficient equal to the relative number of occupied sites in the j-th coordination sphere with a vacant site at the center. Values of the coefficient $m_{21}^{(j)}$ in the first and second coordination spheres of the superstructures types M_2C, M_3C_2 and M_6C_5 at hand are given in Table 6.7.

The parameters α_1 and α_2 determined from measurements of the magnetic susceptibility characterize the general (superstructural and correlational) short-range order. If the contribution of the correlation short-range order is assumed to be negligibly small, from (6.21) it is possible to estimate the long-range order parameter η in samples of niobium, tantalum, hafnium and titanium carbides after susceptibility measurements. Values of the short-range order parameter α_1 in the first coordination sphere were taken to calculate η for carbides $HfC_{0.71}$ and $HfC_{0.78}$, because in the ordered phase M_3C_2 (space groups *Immm* and *P*2) the parameter $\alpha_2 \equiv 0$ and is independent of η. Short-range order parameters in the first and second coordination spheres were used to calculate η for the other carbides. Calculated values of the parameter η (Tables 11.10 and 11.12) are smaller than the maximum value of the long-range order parameter η_{max} in fully ordered carbides (η_{max} is 0.97, 0.996, 0.72, 0.984, 0.996, and 0.90 for carbides $NbC_{0.81}$, $NbC_{0.83}$, $NbC_{0.88}$, $TaC_{0.82}$, $TaC_{0.83}$, and $TaC_{0.85}$; 0.87, 0.66 and 0.996 for hafnium carbides $HfC_{0.71}$, $HfC_{0.78}$ and $HfC_{0.83}$; 1.00, 0.90, 0.82, 0.945, 0.96, 0.93, and 0.972 for titanium carbides TiC_{050}, $TiC_{0.55}$, $TiC_{0.59}$, $TiC_{0.63}$, $TiC_{0.68}$, $TiC_{0.69}$, and $TiC_{0.81}$ respectively). Thus, a relatively large, but not a maximum degree of ordering is achieved in nonstoichiometric carbides during susceptibility measurements.

Temperature dependencies of the susceptibility $\chi(T)$ obtained for non-stoichiometric carbides NbC_y, TaC_y, HfC_y and TiC_y, and conclusions drawn with respect to the type of superstructures formed in these carbides [15, 56–58, 151–160] agree well with phase diagrams of Nb–C, Ta–C, Hf–C, and Ti–C systems (see Sect. 10.3).

A method proposed [56] to analyze the magnetic susceptibility and estimate parameters of short-range order is based on the assumption that short-range order is absent in disordered nonstoichiometric compounds. However, short-range order is preserved even at a temperature higher than the order–disorder transition point T_{trans} when long-range order vanishes. This is confirmed, for example, by studies of short-range order in disordered carbides of titanium, vanadium and niobium by the method of diffuse scattering of electrons or neutrons [34, 177–181]. Therefore, to obtain a more correct estimate of short-range order parameters from the magnetic susceptibility, one has to take into account short-range order in a disordered phase. This short-range order can be determined in additional experiments on elastic scattering of neutrons.

11.6 Superconductivity

Almost all nonstoichiometric compounds with the $B1$ structure are superconductors having a superconducting transition temperature T_c up to 17 K. Superconductivity properties of those compounds are sensitive to the composition: compounds of the stoichiometric composition $MX_{1.0}$, which are free of vacancies, possess the highest T_c in the disordered state. The temperature T_c decreases quickly with growing concentration of structural vacancies. The effect of nonstoichiometry on superconductivity properties of the compounds at hand has been studied in sufficient detail and generalized more than once (see, e.g., [182, 183]). However, the interrelation between superconductivity and ordering of nonstoichiometric compounds has received scanty study. It is only in the last ten to fifteen years that several papers dealing with T_c of ordered nonstoichiometric carbides appeared. In what follows we shall take a closer look at the results of those works.

Literature sources report contradictory data on the superconductivity of nonstoichiometric titanium carbide TiC_y. Near-stoichiometric titanium carbide has a superconducting transition temperature $T_c < 0.05$ K [182, 183]. The superconducting transition temperature T_c of nonstoichiometric carbides and nitrides decreases rapidly with diminishing concentration of nonmetal interstitial atoms (carbon or nitrogen). However, researchers [184], who examined the electrical resistivity ρ of nonstoichiometric carbides TiC_y (y = 0.91, 0.83, 0.69, 0.515 and 0.46) in the temperature interval from 1.5 to 300 K, found an effect pointing to the superconducting transition.

Samples of TiC_y were prepared [184] by sintering of metallic titanium powders and acetylene carbon black. The synthesis temperature and the cooling rate of the synthesized samples were not specified. At a temperature of 4.2 to 300 K the dependence $\rho(T)$ of all the TiC_y samples exhibited a linear behavior and $\partial\rho/\partial T \leq$ (1–10) × 10^{-2} μΩ cm K^{-1}. A sharp drop of ρ was observed for $TiC_{0.515}$ and $TiC_{0.46}$ samples, while the critical transition temperature was 3.42 and 3.32 K respectively. Additional annealing of carbide $TiC_{0.46}$ at 1970 K for 4 h led to a still more abrupt drop of electrical resistivity.

Temperature dependencies of electrical resistivity of nonstoichiometric carbides TiC_y (y = 0.45, 0.48, 0.50, 0.55, and 0.63) at a temperature from 1.4 to 300 K were also studied in [185]. The samples were prepared from titanium powders and carbon black by the method of self-propagating high-temperature synthesis at 2000 K. The electrical resistivity ρ decreased by 3–7% as the temperature was reduced from 300 to 4.2 K. The ρ value dropped sharply (by 40% for $TiC_{0.45}$, $TiC_{0.48}$ and $TiC_{0.50}$; 60% for $TiC_{0.55}$; and 10% for $TiC_{0.63}$) in the interval from 1.4 to 3.0 K. After the samples were annealed at 1400 K and then quenched to 300 K, the dependencies $\rho(T)$ of carbides $TiC_{0.45}$, $TiC_{0.48}$, $TiC_{0.50}$ and $TiC_{0.55}$ exhibited a smeared superconducting transition. When the same samples were again heated to 1300 K and cooled slowly (at a rate of 5 K min^{-1}) to 300 K, the superconductivity was lost. In accordance with [185], quenching of carbide

$TiC_{0.45-0.55}$ from 1500–2000 K gives rise to superconductivity with $T_c \approx 3$ K. Researchers [185] hold to the opinion that the superconducting phase is a high-temperature disordered phase of nonstoichiometric titanium carbide. This inference fully contradicts all experimental data on cubic carbides, according to which T_c decreases rapidly as the composition of disordered carbide deviates from stoichiometry. Carbides corresponding to the lower boundary of the homogeneity interval do not possess superconductivity properties at all [183].

The results reported in [184, 185] were not confirmed later. Probably, superconductivity was due to partial oxidation of the titanium carbide samples during heat treatment, which was performed under insufficiently high vacuum. True, cubic oxide TiO has $T_c \approx 2.5$ K [182]. It should be remembered also that T_c can be determined reliably from inductive and calorimetric measurements. Measurements of electrical resistivity yield a relatively high error. Studies [184, 185] had the considerable drawback that the structure of the test samples was poorly certified.

To verify the results obtained in [184, 185], researchers [135] measured the electrical resistivity of disordered and ordered carbides $TiC_{0.625}$ in the temperature interval between 1.6 and 300 K. A synthesized sample of carbide $TiC_{0.625}$ was rendered disordered by annealing at 1270 K and subsequent quenching to 300 K. Ordered carbide was produced by annealing at 1070 K for 4 h, cooling to 1020 K for 10 days, and annealing at 1020 K for 4 days. The crystal structure of disordered and ordered carbides was carefully analyzed. All heat treatment procedures and measurements were performed in a high vacuum of 1×10^{-4} Pa. Measurements of ρ did not reveal any superconducting effects down to 1.6 K. At $T < 10$ K the electrical resistivity ρ was virtually independent of temperature. Measurements of the heat capacity of disordered and ordered carbides $TiC_{0.625}$ in the interval between 0.5 and 5.0 K did not reveal any superconductivity down to a temperature of 0.6 K. Thus, most reliable data suggest the absence of the superconducting transition in nonstoichiometric carbide TiC_y above 0.5 K.

Interesting results concerning the superconductivity were obtained for niobium carbide NbC_y. Nonstoichiometric disordered carbide NbC_y with the $B1$ structure possesses a wide homogeneity interval from $NbC_{0.70}$ to $NbC_{1.00}$ (see Sect. 3.2). Near-stoichiometric carbide NbC_y ($0.98 < y < 1.00$) has a superconducting transition temperature about 11 K [169, 186–188] (Fig. 11.45). In the pioneering work dealing with superconductivity properties of niobium carbide it was shown that T_c decreases rapidly with growing concentration of structural vacancies [169, 186, 188]. Let it be noted that those measurements were made disregarding the arrangement of interstitial atoms in the nonmetallic sublattice. Later studies [15–18, 177, 189] showed that carbon atoms can be ordered and disordered in nonstoichiometric niobium carbide depending on particular thermal treatment conditions. Changes in the structure of NbC_y during ordering of carbon atoms and vacancies led to the rearrangement of electron and phonon spectra. In this connection, it is important to ascertain how atomic ordering in niobium carbide affects the temperature of transition to the superconducting state.

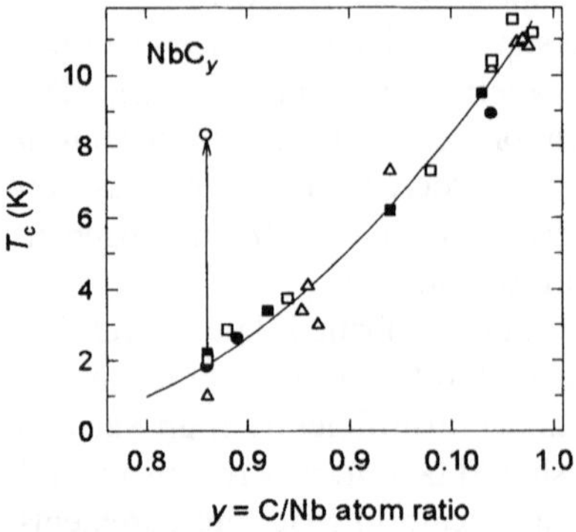

Fig. 11.45. Dependence of the superconductivity temperature T_c on the composition of disordered niobium carbide NbC_y: (●) data from [52,174], (□) data from [169], (Δ) data from [186], (■) data from [188]; (○) superconductivity temperature $T_c = 8.3$ K of disordered carbide $NbC_{0.83}$ (Nb_6C_5) [52, 173, 174]. The arrow shows how T_c increases from 2.1 to 8.3 K, i.e. nearly by a factor of four, as a result of ordering of carbide $NbC_{0.83}$

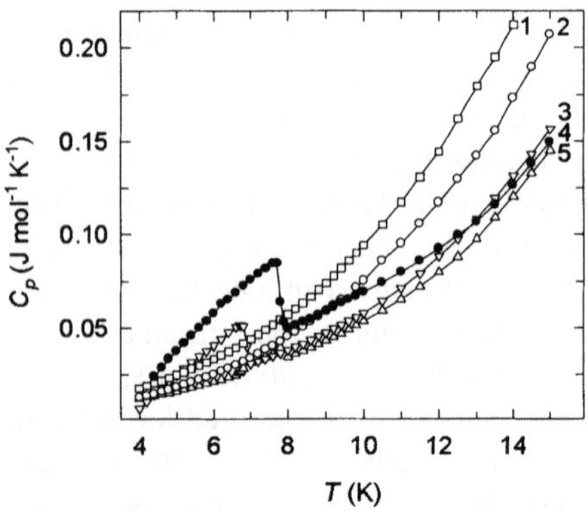

Fig. 11.46. Variation of the heat capacity of nonstoichiometric niobium carbide NbC_y at a temperature below 20 K [173, 174]: (1) partially ordered carbide $NbC_{0.845}$; (2) disordered carbide $NbC_{0.845}$; (3) ordered carbide $NbC_{0.83}$; (4) near-stoichiometric carbide $NbC_{0.97}$; (5) disordered carbide $NbC_{0.83}$. The abrupt changes of the heat capacity correspond to reversible transitions of the carbides from the normal state to the superconducting state

An effort to determine this effect was made by Dy and Williams [127] who measured the electrical resistivity (see Fig. 11.32) and the superconducting transition temperature of disordered carbide $NbC_{0.83}$ and partially ordered carbide Nb_6C_5 ($NbC_{0.83}$). In accordance with [127], disordered carbide $NbC_{0.83}$ has $T_c \approx$ 1.7 K, whereas partially ordered carbide Nb_6C_5 ($NbC_{0.83}$) of a similar composition has $T_c \approx 1.1$ K. Considering that researchers [127] could not prepare a fully ordered crystal of Nb_6C_5, their conclusion about smaller T_c of ordered niobium carbide is not well substantiated.

In contrast to [127], researchers [173] found that T_c of ordered carbide Nb_6C_5 ($NbC_{0.83}$) is higher than T_c of disordered niobium carbide of the same composition. This fact was attributed [173] to the translational symmetry of the nonmetallic sublattice in ordered carbide. No such symmetry is present in disordered carbide with smaller T_c. So, the problem concerning the effect of atomic ordering on superconductivity of nonstoichiometric niobium carbide was not solved conclusively in [127, 173]. It was assumed [52] that the high T_c of ordered carbide Nb_6C_5 may be due to the small dimensionality of some elements of the crystal structure. Ginzburg [190] analyzed the possibility of this relationship in theoretical terms. He thought that an additional condition for the appearance of superconductivity in three-dimensional structures may be ordering or formation of layered structures. Therefore, the study of the superconductivity in ordered nonstoichiometric compounds with the $B1$ structure may be viewed as modeling of low-dimensional superconductivity. Indeed, it was shown earlier [15–18] that ordering in nonstoichiometric niobium carbide leads to the formation of peculiar layered (or quasi-two-dimensional) superstructures (see Sect. 5.4.4). Results of a detailed study into the effect of the ordering on T_c of niobium carbide and discussion of possible causes of T_c variation can be found in [52, 174].

Temperature dependencies of the low-temperature heat capacity of near-stoichiometric carbide $NbC_{0.97}$ and ordered carbide $NbC_{0.83}$ exhibit abrupt changes in the heat capacity [52, 173, 174] (Fig. 11.46). These abrupt changes are associated with the transition of carbides from the normal to superconducting state. Values of the transition temperature T_c, which were measured using calorimetric and inductive methods, are given in Table 11.13.

Evaluating the energy gap width $\Delta(0)$ by the relation $\Delta(0) = 1.76k_BT_c$ yields 1.2 and 1.1 meV for carbide $NbC_{0.97}$ and ordered carbide $NbC_{0.83}$ respectively. These values are in good agreement with results of a tunneling experiment, according to which the gap width $\Delta(0)$ for $NbC_{0.96}$ is 1.6±0.1 meV [191].

The other samples of niobium carbide studied in [52, 174] (disordered carbides $NbC_{0.83}$ and $NbC_{0.845}$, and also partially ordered carbide $NbC_{0.845}$) did not change to the superconducting state when the temperature decreased to 4.0 K.

It should be specially emphasized that T_c of ordered carbide $NbC_{0.83}$ proved to be almost four times as high as T_c of disordered carbide having a similar composition (8.3 and 2.1 K respectively). The largest change ΔT_c caused by ordering of niobium carbide is equal to 6.2 K and is comparable in value with the change of T_c in the whole homogeneity interval of disordered carbide NbC_y (from $T_c < 1$ K for $NbC_{0.70}$ to $T_c = 11.1$ K for $NbC_{1.00}$). Moreover, the ordering-induced change $\Delta T_c = 6.2$ K exceeds the change of T_c of disordered carbide in the interval $NbC_{0.81}$–$NbC_{0.88}$ when the ordered phase Nb_6C_5 is formed: the temperature T_c for $NbC_{0.81}$ and $NbC_{0.88}$ is 1.5 and 4.4 K respectively and, consequently, $\Delta T_c \approx 2.9$ K.

Table 11.13. Parameters of electron and phonon spectra for nonstoichiometric niobium carbides NbC_y in disordered and ordered states [52]

Carbide	$NbC_{0.83}$	$NbC_{0.83}$	$NbC_{0.845}$	$NbC_{0.845}$	$NbC_{0.97}$
State	ordered	disordered	partially ordered	disordered	disordered
γ (mJ g-atom^{-1} K^{-2})	1.54±0.01	1.39±0.01	1.88±0.02	1.27±0.01	2.55±0.02
θ_D (K)	391±12	399±12	311±10	325±10	465±14
[1] T_c (K)	8.3	<4.2	<4.2	<4.2	8.9
[2] T_c (K)	7.3	<4.0	<4.0	<4.0	8.0
[3] T_c (K)	–	2.1	–	2.4	9.8
λ_c	0.69±0.04	0.48±0.03	–	0.52±0.03	0.67±0.04
$N(E_F)$ (eV^{-1} atom^{-1})	0.39±0.03	0.40±0.03	–	0.36±0.03	0.65±0.05

[1] T_c is measured by inductive method.
[2] T_c is measured by calorimetry method.
[3] Data [169, 188].

The tight-binding approximation, which allows for the interaction of electrons and phonons, is usually used to analyze superconductivity properties of carbides with the $B1$ structure. Actually, the electron–phonon interaction is taken into account by renormalizing the density of electron states at the Fermi level $N(E_F)$ using the electron–phonon coupling constant λ_c.

At a temperature lower than the characteristic Debye θ_D and Einstein θ_E temperatures, which correspond to maximum frequencies of acoustic and optical vibrations of atoms, the heat capacity of nonstoichiometric niobium carbide (per NbC_y molecule) in the normal (nonsuperconducting) state can be written as the sum of electron and lattice contributions

$$C_V(y,T) = (1+y)\left\{\frac{1}{3}\pi^2 k_B^2 N_A N(E_F)(1+\lambda_c)T + \frac{1}{2}\left[\frac{12\pi^4 k_B N_A}{5}\left(\frac{T}{\theta_D}\right)^3 + 3k_B N_A\left(\frac{\theta_E}{T}\right)^2 \exp\left(-\frac{\theta_E}{T}\right)\right]\right\}. \quad (11.28)$$

At a temperature below 20 K the contribution of optical vibrations to the heat capacity is negligibly small. Consequently, the Einstein temperature θ_E cannot be determined when the heat capacity of niobium carbide NbC_y at $T_c < T < 20$ K is described by (11.28). Expression (11.28) for the low-temperature heat capacity of conducting crystals MX_y is usually written as

$$C_V(y,T) = (1+y)(\gamma T + \beta T^3), \quad (11.29)$$

where γ is the electronic heat capacity coefficient equal to $\gamma = (\pi^2 k_B^2 N_A / 3)(1+\lambda_c)N(E_F)$ and $\beta = 6\pi^4 k_B N_A /(5\theta_D^3)$. Approximating the experimental values of the heat capacity by (11.29), it was possible to determine the electronic heat capacity coefficient γ and the Debye temperature θ_D for all NbC_y samples studied [52] (see Table 11.13).

The considerable difference of the magnetic susceptibility [15, 56–58] and the superconducting transition temperature in disordered and ordered niobium carbides [52, 173, 174] is an indication that electron and phonon spectra of these carbides change as a result of ordering. From experimental data it is possible to determine some parameters of the electronic structure of niobium carbide and find out how they change during ordering.

Possible mechanisms of the superconductivity (e.g. phonon or exciton mechanisms) can be analyzed using the general equation [190, 192] for T_c:

$$T_c = \theta \exp\left[-\frac{1+\lambda_c}{\lambda_c - \mu^*(1+\lambda_c)}\right], \tag{11.30}$$

where the parameters θ, λ_c and μ^* are expressed through some characteristics of the conductor under study using formulas analogous to those proposed by McMillan [193].

Discussing the phonon mechanism of superconductivity in niobium carbide, it is convenient to use McMillan's semi-empirical formula [193] as a particular case of formula (11.30):

$$T_c = \frac{\theta}{1.45}\exp\left[-\frac{1.04(1+\lambda_c)}{\lambda_c - \mu^*(1+0.62\lambda_c)}\right]. \tag{11.31}$$

McMillan found numerical values of the coefficients in (11.31) by analyzing the shape of the phonon spectrum of metallic niobium using Eliashberg's equations [194] (see also [195]). The pseudopotential μ^*, which characterizes the Coulomb repulsion of electrons, can be determined from the isotopic effect in superconductors by measuring T_c and the isotopic shift δT_c. Transition metals have a small pseudopotential μ^* equal to 0.10–0.15. Using (11.31), which holds for tight-binding superconductors, and T_c and θ_D values (see Table 11.13), it was possible to calculate the electron–phonon coupling constant λ_c. The Coulomb pseudopotential of carbide NbC_y was taken to be $\mu^* = 0.13$ in calculations. (This μ^* value is normally used in analyzing superconducting properties of carbides [127, 169].)

The density of electron states at the Fermi level $N(E_F)$ was calculated by the formula

$$N(E_F) = 3\gamma/[\pi^2 k_B^2 N_A(1+\lambda)] \tag{11.32}$$

using calculated values of the electronic heat capacity coefficient γ and the electron–phonon coupling constant λ_c. The calculation results for λ_c and $N(E_F)$ are given in Table 11.13.

From Table 11.13 it is seen that the density of electron states $N(E_F)$ decreases in disordered niobium carbide when the carbon concentration diminishes (the concentration of structural vacancies grows). Analogous results were obtained in

[169, 189]. This behavior of $N(E_F)$ with the NbC_y composition agrees well with the data on the magnetic susceptibility (see Sect. 11.5). Those data show that the density of electron states at the Fermi level decreases as the composition of disordered carbide changes from $NbC_{1.00}$ to $NbC_{0.83}$. In turn, measurements of the Hall coefficient showed that the concentration of conduction electrons increased with decreasing concentration of carbon in niobium carbide [196].

In carbides of group V transition metals the conduction band is filled less than half filled. The density of electron states at the Fermi level $N(E_F)$ can decrease and, simultaneously, the number of conduction electrons can rise during vacancy formation if a local peak of the density of states emerges in the conduction band below the Fermi energy. Indeed, the cluster calculation of the electronic structure of nonstoichiometric niobium carbide by the SCF-X_α method showed [50] that a narrow peak appeared in the energy spectrum after a carbon vacancy was formed. The said peak was located ~2 eV below the energy E_F. In accordance with [197], when the concentration of structural vacancies is sufficiently high, the energy spectrum of carbide NbC_y contains a vacancy band located below E_F and filled mostly with Nb4*d* electrons. So, if the carbon concentration of niobium carbide decreases, the density of electron states $N(E_F)$ diminishes and the fine structure of the energy spectrum below the Fermi energy is altered.

The variation of the electron–phonon coupling constant has a decisive effect on T_c during ordering [52, 174] (see Table 11.13). The sharp change of the electron–phonon coupling constant λ_c during ordering is probably due to the absence of translational symmetry in disordered niobium carbide and the presence of this symmetry in ordered carbide. In this case, the small T_c of partially ordered carbide $NbC_{0.845}$ can be explained by the fact that the composition of this carbide differs from the stoichiometric composition of the ordered phase Nb_6C_5 and, consequently, by the absence of complete order in the arrangement of atoms and vacancies.

The temperature T_c of near-stoichiometric $NbC_{0.97}$ is higher than that of ordered carbide $NbC_{0.83}$. This is probably due to a larger density of electron states $N(E_F)$ in carbide $NbC_{0.97}$. The electron–phonon interactions in ordered carbide $NbC_{0.83}$ and carbide $NbC_{0.97}$ are nearly equal (see Table 11.13). However, it is worth noting the following. The density of electron states $N(E_F)$ and the electron–phonon coupling constant λ_c were calculated as the tight-binding approximation. It is known [198] that in tight-binding superconductors the jump of the electronic contribution to the heat capacity, $(C_{se} - C_{ne})/C_{ne}$ (C_{se} and C_{ne} being the electronic heat capacity in the superconducting and normal states), exceeds 1.43 at $T = T_c$. The jump of the heat capacity of near-stoichiometric carbide $NbC_{0.97}$ and ordered carbide $NbC_{0.83}$ is much smaller at $T = T_c$ and equals 0.90 and 1.04 respectively. It stands to reason that a jump having an ideal shape cannot be obtained during measurements. However, even if corrections for the non-ideal shape are introduced, $(C_{se} - C_{ne})/C_{ne} < 1.43$ for the niobium carbides studied.

In principle, a small jump of the heat capacity may be due to heterogeneity of the samples. Indeed, if only part of a sample passes to the superconducting state, the ratio $(C_{se} - C_{ne})/C_{ne}$ calculated without taking account of the phase composition proves to be underestimated. However, the behavior of the heat capacity of carbide $NbC_{0.97}$ and ordered carbide $NbC_{0.83}$ at $T < T_c$ permits rejecting this assumption: If the temperature decreases below T_c, the heat capacity tends quickly to zero and the non-superconducting phase does not make a linear contribution to the heat capacity. Moreover, homogeneity of the carbide $NbC_{0.97}$ and ordered carbide $NbC_{0.83}$ is confirmed by X-ray phase analysis. It should be noted that a small jump of the heat capacity at $T = T_c$ is observed for carbides of other transition metals too. What may be the cause? In this respect, only qualitative considerations can be applied to the possible mechanism of the superconductivity in niobium carbide.

In superconductors with $B1$ and $A15$ structures the interelectron attraction and formation of the Cooper pairs are due to the interaction with the lattice. In other words, electrons interact through the exchange of phonons. Assuming the phonon mechanism of superconductivity and using the tight-binding approximation, researchers [52, 174] determined λ_c and $N(E_F)$. It should be remembered however that these quantities were calculated taking volume-averaged properties of the crystal. This calculation method is quite valid for a disordered nonstoichiometric carbide with an isotropic crystal structure. The $N(E_F)$ value calculated by the said method for ordered nonstoichiometric carbide characterizes the density of electron states averaged over the whole volume of an anisotropic crystal. However, the actual distribution of the density of states $N(E_F)$, which depends on the specific distribution of atoms in the lattice of an ordered compound, may be inhomogeneous in the crystal volume. The spatial distribution may have minima and maxima.

Ordered niobium carbide Nb_6C_5 ($NbC_{0.83}$) has a peculiar layered crystal structure [15–18]. In this connection, a quasi-two-dimensional model of superconductivity was proposed [52, 174] to account for superconductivity properties of ordered and disordered niobium carbides. The essence of this model is as follows.

Let us consider in more detail the crystal structure of stoichiometric niobium carbide $NbC_{1.00}$. In the direction $[111]_{B1}$ this structure can be pictured as successively alternating close-packed planes $(111)_{B1}$ that contain either metal or nonmetal atoms (Fig. 11.47*a*).

When the composition deviates from stoichiometry, part of the nonmetallic sublattice sites in disordered niobium carbide become vacant. The crystal structure is distorted locally and complete nonmetallic planes formed of carbon atoms only are replaced by defective nonmetallic planes that contain both carbon atoms and vacancies (Fig. 11.47*b*). Vacancies are arranged statistically in disordered carbide NbC_y and, consequently, the directions of local atomic displacements are random. Consequently, almost all atomic planes are distorted.

The disorder–order phase transition in niobium carbide causes a fundamentally different mutual arrangement of carbon atoms and vacancies. As was mentioned in Sect. 5.4.4, the formation of the ordered phase Nb_6C_5 is associated with three wave-vector stars $\{\boldsymbol{k}_9\}$, $\{\boldsymbol{k}_4\}$ and $\{\boldsymbol{k}_3\}$, and is described by three long-range order parameters η_9, η_4, and η_3 [199, 200]. If one considers the nonmetallic sublattice only, the structure of fully ordered niobium carbide has the specific feature that complete and ordered defective nonmetallic planes alternate successively in the direction $[111]_{B1}$ (see Fig. 11.47*c*). The parameter η_9 is responsible for the alternation sequence of those planes and the redistribution of carbon atoms and vacancies between the planes. The parameters η_4 and η_3 are responsible for the mutual arrangement of carbon atoms and vacancies in the nonmetallic planes. The last parameters have no effect on the relative concentration of carbon atoms and vacancies in those planes [200].

Assume that the superconductivity in niobium carbide is due to the presence of type *I* layers (see Fig. 11.47), which comprise two adjacent metallic planes and a complete nonmetallic plane located between them. Assume also that vacancies formed in the nonmetallic planes impede the superconductivity properties of the quasi-two-dimensional layers. Let us discuss the way the superconductivity properties of niobium carbide should change in this quasi-two-dimensional model.

Stoichiometric carbide $NbC_{1.00}$, which contains quasi-two-dimensional layers of type *I* only, has the largest density of electron states $N(E_F)$ in the disordered state (see Fig. 11.47*a*). The increase in the vacancy concentration of disordered carbide is followed by a monotonic drop of $N(E_F)$ (see Table 11.13) and the appearance of defective quasi-two-dimensional layers of type *II* (Fig. 11.47*b*).

Obviously, the density of electron states in complete quasi-two-dimensional layers of type *I* is much higher than $N(E_F)$ in defective quasi-two-dimensional layers of type *II*. Therefore, when the composition of niobium carbide deviates from stoichiometry, T_c should decrease rapidly. This inference is in good agreement with the results reported elsewhere [52, 169, 173, 174, 186, 188].

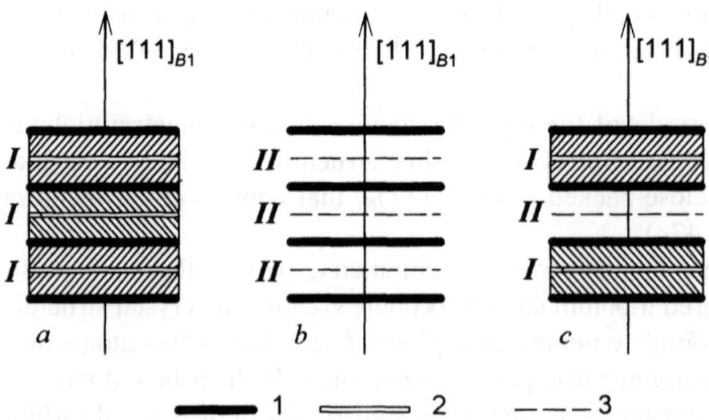

Fig. 11.47. Mutual arrangement of atomic planes $(111)_{B1}$ in stoichiometric carbide $NbC_{1.00}$ (*a*), nonstoichiometric disordered carbide NbC_y (*b*), and fully ordered carbide Nb_6C_5 ($NbC_{0.83}$) (*c*) [52]: (1) plane formed by niobium atoms; (2) complete (without vacancies) nonmetallic plane; (3) defective (with vacant sites) nonmetallic plane; (*I*) quasi-two-dimensional complete layer; (*II*) quasi-two-dimensional defective layer

From Fig. 11.47*c* it is seen that complete layers of type *I* with good superconductivity are formed in ordered niobium carbide. They are analogous to those formed in defect-free stoichiometric niobium carbide. Let it be noted that a maximum degree of long-range order η_9 is required for complete nonmetallic planes to be formed in niobium carbide NbC_y with carbon concentration $5/6 \le y \le 1$. Considering the high concentration of structural vacancies, the formation of complete nonmetallic planes in carbide NbC_y with $y < 5/6$ is hardly probable even at a maximum degree of long-range order. When superconducting layers of type *I* with a maximum density of electron states $N(E_F)$ appear in nonstoichiometric niobium carbide, T_c should approach the T_c of stoichiometric carbide, as was actually observed in experiments [52, 174].

Thus, researchers [52, 174] hold to the opinion that a quasi-two-dimensional layered structure is formed in nonstoichiometric niobium carbide during ordering. Ordered quasi-two-dimensional carbide Nb_6C_5 ($NbC_{0.83}$) has $T_c = 8.3$ K, which is by a factor of nearly four higher than $T_c = 2.1$ K of disordered carbide $NbC_{0.83}$. One may think that the increase in T_c, which is caused by formation of a quasi-two-dimensional structure in niobium carbide, is connected with periodic local peaks of the density of electron states $N(E_F)$ corresponding to complete layers of type *I* with good superconductivity [52].

The model of the two-dimensional distribution of conduction electrons suggests that niobium carbide NbC_y with a small concentration of structural vacancies and ordered carbide $NbC_{0.83}$ should have nearly equal temperatures of transition to the superconducting state. However, experiments show that T_c of ordered niobium carbide is a little lower. This fact may be attributed to two factors. First, the degree of long-range order in ordered carbide $NbC_{0.83}$ is by far not a maximum. In this case some complete nonmetallic planes become defective and the number of complete superconducting layers decreases. Second, even if the degree of long-range order in $NbC_{0.83}$ is a maximum, all superconducting layers of type *I* are adjacent to defective nonmetallic planes (see Fig. 11.47*c*). Analogous lines of reasoning were adopted in studies [201, 202] dealing with experimental and theoretical examination of the twinning plane in niobium bicrystals. The twinning plane represents a unique example of a physical two-dimensional system: only one monolayer of atoms is in the crystallographically favorable position. It was shown [201, 202] that the superconductivity properties of the two-dimensional layer were largely impaired if the twinning plane was surrounded by metal in the normal (nonsuperconducting) state.

The study of a single crystal of $NbC_{0.87}$ [203] confirmed that T_c of nonstoichiometric niobium carbide increases after annealing. Carbide $NbC_{0.87}$, which was quenched from 1600 K, had $T_c \approx 4$ K. The superconducting transition temperature was as high as 8.6 K after the same carbide was annealed at 1200–1400 K. The data [203] are fully explained in terms of the model of quasi-two-dimensional superconductivity proposed in [52, 174].

The results obtained in [52] and their explanation bring about some thoughts about the possibility of increasing T_c of carbides of other transition metals. Earlier it was shown [1, 18, 133, 200] that layered superstructures like M_6C_5 analogous to Nb_6C_5 are formed in vanadium and tantalum carbides too. It may be assumed that these compounds will have, similar to niobium carbide, a higher T_c in the ordered state than in the disordered one.

The effect of ordering on T_c of tantalum carbide was studied in [9, 204]. The superconducting transition temperature of stoichiometric carbide $TaC_{1.00}$, disordered carbides $TaC_{0.96}$, $TaC_{0.94}$, $TaC_{0.92}$, $TaC_{0.90}$, $TaC_{0.87}$, and $TaC_{0.83}$, and also ordered carbides $TaC_{0.87}$ and $TaC_{0.83}$ was measured at $T < 20$ K using the inductive method. The degree of long-range order in ordered carbides $TaC_{0.87}$ and $TaC_{0.83}$ was not a maximum.

Measurements showed that the T_c of disordered carbide TaC_y decreased rapidly from 10.2 K for $TaC_{1.00}$ to 1.5 K for $TaC_{0.83}$. The results [9, 204] agree well within the measurement error with the data of [169, 186, 188, 205] (Fig. 11.48). The T_c value of ordered carbides $TaC_{0.87}$ and $TaC_{0.83}$ proved to be slightly higher (by 0.2 and 0.6 K respectively) than T_c of disordered carbides having a similar composition. The observed effect of ordering was inappreciable and was larger than the measurement error equal to 0.4 K only in the case of $TaC_{0.83}$. Considering this, only a slight tendency to an increase in T_c of nonstoichiometric tantalum carbide may be assumed as a result of ordering. The different effect of ordering on T_c of niobium and tantalum carbides is probably due to different superstructures formed in these carbides.

Indeed, a commensurate superstructure of M_6X_5 type is formed in niobium carbide during ordering. Quasi-two-dimensional layers of types *I* and *II* alternate successively in this structure in the direction $[111]_{B1}$ (see Fig. 11.47*c*). The presence of quasi-two-dimensional layers of type *I* ensures that T_c of ordered carbide is higher than T_c of disordered niobium carbide, which does not contain such layers.

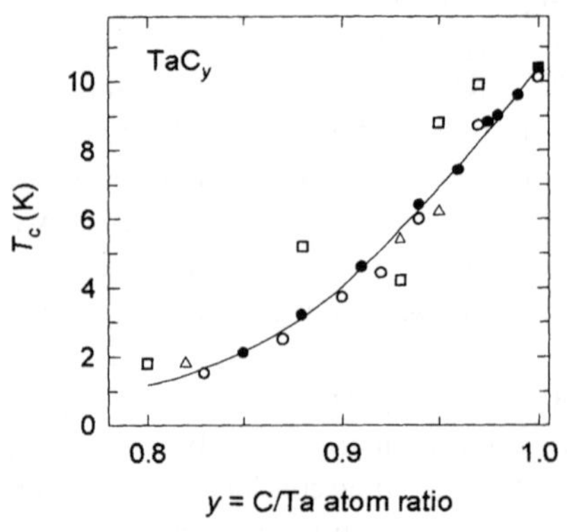

Fig. 11.48. Superconductivity temperature T_c for disordered tantalum carbide TaC_y: (○) [9, 204], (□) [169], (•) [186], (Δ) [188], (■) [205]

An incommensurate superstructure is formed in carbide TaC_y [9, 204, 206]. Even if the degree of long-range order is a maximum, quasi-two-dimensional layers of type *I* having good superconductivity occur in this superstructure 8–9 times less than in the commensurate superstructure M_6X_5. However, the analysis of the basic lattice constant [8] (see Sect. 11.1) and the heat capacity [207–209] showed that the maximum degree of order was not achieved in the tantalum carbide studied. This means that the relative number of complete quasi-two-dimensional layers ensuring superconductivity is much smaller in incommensurate ordered tantalum carbide than in the ideal commensurate superstructure M_6X_5. Therefore, the T_c of ordered carbide $TaC_{0.83}$ having an incommensurate superstructure and the T_c of disordered carbide with a similar composition differ little.

If a commensurate superstructure type M_6X_5 with a nearly maximum degree of order is produced in the homogeneity interval of the ordered phase of tantalum carbide $TaC_{0.79}$–$TaC_{0.89}$, the increase in T_c can be expected to be comparable with the increase observed during ordering of niobium carbide.

The solid solution of niobium carbide and nitride $NbC_{0.25}N_{0.75}$ has the highest T_c = 17.8 K among superconductors with the *B*1 structure [210]. Up to now, possible ordering of carbon and nitrogen atoms in the nonmetallic sublattice of niobium carbonitride has not been considered. Three nitrogen planes and one carbon plane or nonmetallic planes with an ordered distribution of carbon and nitrogen atoms may alternate successively in a hypothetical ordered layered quasi-two-dimensional structure of $NbC_{0.25}N_{0.75}$ in the direction $[111]_{B1}$. Such an ordering may probably lead to the increase in the superconducting transition temperature of this compound.

Thus, the quasi-two-dimensional model [52, 174] offered an explanation of why the T_c of ordered niobium carbide is higher than the T_c of disordered niobium carbide. However, some experimental data contradict this model.

11.7 Heat Capacity

Thermodynamic properties of nonstoichiometric compounds have long been analyzed disregarding ordering of interstitial atoms. Only the composition dependence of interstitial atoms was considered only. The study of thermodynamic properties as a function of the distribution of interstitial atoms and vacancies in the crystal lattice of nonstoichiometric compounds presents a much more complicated experimental task. It has been found by now that ordering affects the heat capacity of nonstoichiometric carbides of titanium, vanadium, niobium and tantalum. The effect of ordering on the heat capacity shows up differently in different temperature intervals. For this reason, before we proceed to a discussion of experimental results, let us consider a model describing the heat capacity of nonstoichiometric compounds.

11.7.1 Superposition Model of the Heat Capacity

Strongly nonstoichiometric carbides, nitrides, oxides and their solid solutions contain a high concentration of collectivized electrons and represent conductors of metallic type.

The molar heat capacity at constant volume, C_V, of a conducting crystal at a temperature higher than the superconducting transition temperature can be presented as the sum of the phonon (lattice) C_{ph} and electron C_{el} contributions

$$C_V = n(C_{ph} + C_{el}) , \quad (11.33)$$

where n is the number of atoms in a molecule ($n = 1 + y$ for nonstoichiometric compounds MX_y).

The electronic heat capacity (in terms of one g-atom) is proportional to the density of electron states at the Fermi level $N(E_F)$. In the case of crystals with a strong electron–phonon interaction, among which the nonstoichiometric compounds at hand are classified, one should consider additionally the renormalization of the density of electron states using the electron–phonon coupling constant λ_c [193]: $N^*(E_F) = (1 + \lambda_c)N(E_F)$, where $N^*(E_F)$ is the measured density of states and $N(E_F)$ is the density of electron states calculated disregarding the electron–phonon interaction (band density of states). The renormalization of the density of states leads to the corresponding renormalization of the electronic heat capacity, which varies linearly with temperature. Consequently, the electronic heat capacity (in terms of one g-atom) acquires the form

$$C_{el} = \frac{\pi^2 k_B^2 N_A}{3}(1+\lambda_c)N(E_F)T . \quad (11.34)$$

It stands to reason that the renormalization of the electronic heat capacity caused by the electron–phonon interaction is possible only at a temperature lower than the Debye temperature θ_D, i.e. $T \leq \theta_D$. If $T >> \theta_D$, most electron states near the Fermi level, which contribute to C_{el}, are insensitive to the electron–phonon interaction. If one compares the coefficients of T in the electronic heat capacity C_{el} for low and high temperatures, it is possible, in principle, to find the λ_c value. Unfortunately, such investigations of nonstoichiometric compounds have not been performed.

It was shown [1–3, 209, 211–213] that the lattice component of the heat capacity of nonstoichiometric compounds can be described by a superposition model [214].

The natural frequencies of a macroscopic crystal are distributed continuously. Consequently, the internal energy can be expressed through the spectral distribution $g(\omega)d\omega$ as

$$U = \int \varepsilon(\omega,T)g(\omega)d\omega . \quad (11.35)$$

Hence, the lattice thermal properties of a crystal are fully determined by the spectral distribution of natural vibrations.

In accordance with [214], the spectrum, which provides correct values of the lattice component of the heat capacity at low and high temperatures and ensures a fundamentally correct description of the heat capacity at intermediate temperatures, can be constructed as a superposition of Debye and Einstein spectra.

In the Einstein model, the spectrum contains only one frequency ω_E. Each atom of a crystal behaves itself as an isotropic spatial oscillator and, therefore, is equivalent to three linear oscillators having the same frequency ω_E. In this case the spectral density of a crystal comprising N atoms has the form

$$g_E(\omega) = 3N\delta(\omega - \omega_E), \tag{11.36}$$

where

$$\delta(\omega - \omega_E) = \begin{cases} 0 & \text{when } \omega \neq \omega_E \\ 1 & \text{when } \omega = \omega_E \end{cases}$$

is the Dirac delta function. Thus, the Einstein model describes the monochromatic spectrum of natural oscillations.

From the theory of elasticity it is known that three branches of acoustic vibrations are possible in a continuous medium. Debye replaced the crystal lattice by a continuous solid and assumed that optical branches of natural vibrations were absent, while acoustic branches coincided and were characterized by a linear law of dispersion: $\omega = ck$ ($\boldsymbol{k}$ being a wave vector and c the propagation rate of elastic vibrations). One more assumption due to Debye consists in the limitation of the elastic spectrum by some maximum frequency ω_D. Natural vibrations of a crystal comprising N atoms total $3N$. Consequently, considering the last assumption, the equality

$$\int_0^{\omega_D} g_D(\omega)d\omega = 3N \tag{11.37}$$

should be fulfilled for a discrete crystal. The elastic spectrum is proportional to ω^2, i.e. $g_D(\omega) = A\omega^2$. Considering this and integrating (11.37), it is possible to find the coefficient $A = 9N/\omega_D^3$ and the Debye function of the frequency distribution:

$$g_D(\omega) = 9N\frac{\omega^2}{\omega_D^3}. \tag{11.38}$$

If a crystal is formed by atoms of two or more species, its vibration spectrum differs considerably from the vibration spectrum of a continuous elastic medium. In particular, the vibration frequency in a one-dimensional chain of alternating atoms of two species having mass M and m is described by the function

$$\omega^2 = \frac{f}{mM}\left[m + M \pm \sqrt{m^2 + M^2 + 2mM\cos\varphi}\right], \tag{11.39}$$

where f is the elasticity coefficient; $\varphi = ka = 2\pi a/\lambda$; k is the wave number; a is the spacing (the distance between two consecutive like particles); and λ is the wavelength. From (11.39) it is seen that two frequencies correspond to each value of the wave number. If unlike atoms vibrate in phase ($\varphi = \pi$) and the minus sign precedes the root in (11.39), then

$$\omega_{(-)}^2 = \frac{2f}{M}, \tag{11.40}$$

i.e. vibrations corresponding to the frequency branch $\omega_{(-)}$ are analogous to vibrations of the chain of heavy atoms only. This branch of vibrations is referred to as the acoustic branch. When λ varies between ∞ and 0 (or φ between 0 and π), the frequency $\omega_{(-)}$ changes from 0 to a maximum value $(2f/M)^{1/2}$, i.e. analogously to the Debye frequency distribution. Thus, the Debye model describes acoustic vibration branches.

The other branch of vibrations is called the optical or Born branch. If the plus sign precedes the root in (11.39), the frequency is

$$\omega_{(+)}^2 = \frac{2f(m+M)}{mM} \tag{11.41}$$

for very long waves $\lambda \to \infty$, $k \to 0$ and $\varphi \to 0$.

If $M >> m$, then

$$\omega_{(+)}^2 = \frac{2f}{m}. \tag{11.42}$$

When $k = \pi/a$ and $\varphi = \pi$, (11.38) yields the same value of the frequency $\omega_{(+)}^2 = 2f/m$. This means that at $M >> m$ the frequency $\omega_{(+)}$ depends little on the wave number. In this case, vibrations of the optical branch are monochromatic. Considering the value of the wave number k, the acoustic and optical vibration frequencies are equal to:

$$\left.\begin{aligned} \omega_{(-)} &= \left(\frac{2f}{M}\right)^{1/2} \left|\sin\left(\frac{ka}{2}\right)\right| \\ \omega_{(+)} &= \left\{\frac{2f}{m}\left[1+\frac{m}{M}\cos^2\left(\frac{ka}{2}\right)\right]\right\}^{1/2} \end{aligned}\right\}. \tag{11.43}$$

Analogous results were obtained for the three-dimensional crystal lattice.

The comparison of (11.40) and (11.42) shows that when $M >> m$, the optical and acoustic branches are largely separated. Optical frequencies are grouped in one place and provide a sharp peak in the spectral distribution. Obviously, the optical branch of the spectrum can be roughly described in terms of the Einstein model, while acoustic branches are roughly described by the Debye model. Thus, the natural vibration spectrum of a crystal, which is formed by atoms of two species having different masses $M >> m$, can be approximated as a superposition of the Debye and Einstein spectra. This approximation is quite applicable to carbides, nitrides and oxides of transition metals, because in these compounds the atomic mass of the metal is 3–20 times larger than the atomic mass of the nonmetal.

If a minimum-volume translational cell of a crystal contains σ atoms, the total number of natural vibrations corresponding to a certain wavelength will be 3σ. Three frequencies belong to acoustic branches, while the other $(3\sigma-3)$ frequencies correspond to optical branches. Considering this, contributions of the Debye and Einstein terms to the frequency distribution function will be proportional to the relative number of the corresponding branches, i.e.

$$g(\omega)=\frac{3}{3\sigma}g_{\text{D}}(\omega)+\frac{3\sigma-3}{3\sigma}g_{\text{E}}(\omega) = 3N\left[\frac{1}{\sigma}\frac{3\omega^2}{\omega_{\text{D}}^3}+\frac{\sigma-1}{\sigma}\delta(\omega-\omega_{\text{E}})\right]. \quad (11.44)$$

The lattice heat capacity can be presented as the mean energy of a linear oscillator $\varepsilon(\omega,T) = (\hbar\omega/2)\coth(\hbar\omega/2k_{\text{B}}T)$ and the function $g(\omega)$ in the form

$$C_{\text{ph}} = \int\frac{\partial\varepsilon(\omega,T)}{\partial T}g(\omega)d\omega \quad (11.45)$$

or, considering (11.44), the heat capacity of one g-atom is equal to

$$C_{\text{ph}} = \frac{1}{\sigma}C_{\text{D}}\left(\frac{\theta_{\text{D}}}{T}\right)+\frac{\sigma-1}{\sigma}C_{\text{E}}\left(\frac{\theta_{\text{E}}}{T}\right). \quad (11.46)$$

In nonstoichiometric compounds MX_y with the $B1$ structure the number of sites in a minimum translational cell is $\sigma = 2$, while the number of atoms in a molecule is $n = 1 + y$. Therefore

$$C_{\text{ph}} = \frac{1+y}{2}\left[C_{\text{D}}\left(\frac{\theta_{\text{D}}}{T}\right)+C_{\text{E}}\left(\frac{\theta_{\text{E}}}{T}\right)\right], \quad (11.47)$$

where

$$C_{\text{E}} = 3Nk_{\text{B}}\left(\frac{\theta_{\text{E}}}{T}\right)^2\frac{\exp(\theta_{\text{E}}/T)}{[\exp(\theta_{\text{E}}/T)-1]^2} \quad (11.48)$$

and

$$C_{\text{D}} = 9Nk_{\text{B}}\left(\frac{T}{\theta_{\text{D}}}\right)^3\int_0^{\theta_{\text{D}}/T}\frac{x^4\exp x\,dx}{(\exp x-1)^2} \quad (11.49)$$

or

$$C_{\text{D}} = 3Nk_{\text{B}}\left[12\left(\frac{T}{\theta_{\text{D}}}\right)^3\int_0^{\theta_{\text{D}}/T}\frac{x^3dx}{\exp x-1}-\frac{3(\theta_{\text{D}}/T)}{\exp(\theta_{\text{D}}/T)-1}\right]. \quad (11.50)$$

In (11.48) $\theta_E = \hbar\omega_E/k_B$, while in (11.49) and (11.50) $x = \hbar\omega_D/k_BT$ and $\theta_D = \hbar\omega_D/k_B$.

Thus, considering (11.33), (11.34) and (11.47), the heat capacity of one mole of nonstoichiometric compound MX_y ($n = 1 + y$ and $\sigma = 2$) can be written in the general form as

$$C_V(y,T) = (1+y)\left\{\frac{\pi^2 k_B^2 N_A}{3}(1+\lambda_c)N(E_F)T + \frac{1}{2}\left[C_D\left(\frac{\theta_D}{T}\right) + C_E\left(\frac{\theta_E}{T}\right)\right]\right\}. \tag{11.51}$$

Fundamentally, (11.51) holds for any temperature, but the electron, Debye and Einstein contributions to the heat capacity are not equal in various temperature intervals. Formula (11.51) can be rearranged taking into account values of these contributions in each temperature interval. In particular, when $\theta_D/T >> 1$ or $\theta_D/T \to \infty$, i.e. at low temperature

$$\int_0^\infty \frac{x^4 \exp x dx}{(\exp x - 1)^2} = \frac{4\pi^4}{15} \quad \text{and} \quad \int_0^\infty \frac{x^3 dx}{\exp x - 1} = \frac{\pi^4}{15}$$

in (11.49) and (11.50). Therefore the Debye contribution to the heat capacity has the known form

$$C_D = \frac{12\pi^4 k_B N_A}{5}\left(\frac{T}{\theta_D}\right)^3 \quad \text{when } \theta_D/T >> 1. \tag{11.52}$$

In the low-temperature interval at $T_c < T < 20$ K ($\theta_D/T >> 1$ and $\theta_E/T >> 1$), when the contribution of optical vibrations is negligibly small, (11.51) transforms to (11.29).

In the temperature interval 20 K $\le T \le$ 60 K when $\theta_E/T >> 1$, the heat capacity can be found from the formula

$$C_V(y,T) = (1+y)\left\{\frac{\pi^2 k_B^2 N_A}{3}(1+\lambda_c)N(E_F)T + \frac{1}{2}\left[C_D(\theta_D/T) + 3k_B N_A\left(\frac{\theta_E}{T^2}\right)^2 \exp\left(-\frac{\theta_E}{T}\right)\right]\right\}. \tag{11.53}$$

Let it be noted that in this temperature interval the contribution of the electronic heat capacity (the first term in (11.53)) to the total heat capacity of a crystal is much smaller than the lattice contribution.

At higher temperatures (60 K $\le T \le$ 300 K) the electron contribution to the heat capacity becomes negligibly small, whereas the contribution of optical vibrations increases rapidly to be comparable with the contribution of acoustic vibrations. In this case (11.51) transforms to

$$C_V(y,T) = \frac{1+y}{2}\left[C_{\mathrm{D}}\left(\frac{\theta_{\mathrm{D}}}{T}\right) + C_{\mathrm{E}}\left(\frac{\theta_{\mathrm{E}}}{T}\right)\right]. \tag{11.54}$$

Expression (11.54) holds for higher temperatures from 300 K to $0.6T_{\mathrm{m}}$ (T_{m} being the melting temperature).

At a very high temperature $T > 0.8T_{\mathrm{m}}$ the electron contribution to the heat capacity again becomes appreciable and (11.51) is used to describe the heat capacity in this temperature interval.

A measured quantity in adiabatic calorimetry of solids is C_p. Therefore the correction ($C_p - C_V$) should be introduced when the measured heat capacity is approximated by the dependencies (11.51)–(11.54). The relationship between C_p and C_V in solids can be established as follows.

Let us consider the entropy S as a function of temperature T and pressure p. The full entropy differential has the form $dS = (\partial S/\partial T)_p dT + (\partial S/\partial p)_T dp$, whence

$$\left(\frac{\partial S}{\partial T}\right)_V = \left(\frac{\partial S}{\partial T}\right)_p + \left(\frac{\partial S}{\partial p}\right)_T \left(\frac{\partial p}{\partial T}\right)_V. \tag{11.55}$$

Considering $C_V = T(\partial S/\partial T)_V$ and $C_p = T(\partial S/\partial T)_p$, relation (11.55) can be written as

$$C_V = C_p + T\left(\frac{\partial S}{\partial p}\right)_T \left(\frac{\partial p}{\partial T}\right)_V \tag{11.56}$$

or, in accordance with the identity $(\partial T/\partial p)_V (\partial p/\partial V)_T (\partial V/\partial T)_p \equiv -1$ and the Maxwell equation $(\partial S/\partial p)_T = -(\partial V/\partial T)_p$, as

$$C_V = C_p + T\left(\frac{\partial V}{\partial T}\right)_p^2 \left(\frac{\partial V}{\partial p}\right)_T. \tag{11.57}$$

The relation between C_V and C_p can be written in terms of the temperature coefficient of volume expansion $\beta_T = (\partial V/V\partial T)_p$ ($\beta \approx 3\alpha$, where α is the temperature coefficient of linear expansion), the isothermal compressibility $\kappa_T = -(\partial V/V\partial p)_T$ ($\kappa_T = 1/K$, where K is the uniform compression modulus), and the molar volume $V = M/d$ (M being the molar mass and d the density):

$$C_p - C_V = \frac{9\alpha^2 VT}{\kappa_T} \equiv 9\alpha^2 VKT \equiv \frac{9\alpha^2 MKT}{d}. \tag{11.58}$$

Since the coefficient $\alpha \to 0$ as $T \to 0$, then $C_p \to C_V$ at $T \to 0$. Conversely, the correction $(C_p - C_V)$ increases with elevating temperature.

11.7.2 Heat Capacity of Nonstoichiometric Carbides at a Temperature Below 300 K

Let us discuss the data concerning the effect of ordering on the heat capacity of nonstoichiometric compounds at a temperature below 30 K.

The effect of ordering on the heat capacity of titanium carbide $TiC_{0.625}$ at $T <$ 5 K was studied in [135]. Measurements showed that at $T <$ 1.5 K the heat capacity of ordered trigonal (space group $R\bar{3}m$) carbide $TiC_{0.625}$ is nearly 3% smaller than the heat capacity of disordered carbide $TiC_{0.625}$. In the interval of 1.5–5.0 K heat capacities of the ordered and disordered carbides are equal. The heat capacity difference $\Delta C(T) = C(y,\eta,T) - C(y,0,T)$ was described by the relation $\Delta C(T) = \Delta\gamma T + \Delta\beta T^3$, where $\Delta\gamma = (0.18 \pm 0.03) \times 10^{-3}$ J mol^{-1} K^{-2} and $\Delta\beta = (0.6 \pm 1.7) \times 10^{-6}$ J mol^{-1} K^{-4}. From [135] it follows that ordered carbide $TiC_{0.625}$ has a smaller coefficient of the electronic heat capacity γ and a lower density of electron states at the Fermi level $N(E_F)$ (by 0.08 ± 0.02 eV^{-1} atom^{-1}) than disordered carbide with a similar composition.

The heat capacity of disordered and ordered niobium carbides NbC_y at a temperature below 30 K was studied in [52, 174]. The heat capacity was measured with a step of 0.1 K. The mean measurement error at a given temperature in the interval from 4 to 10 K was 2%. Temperature dependencies of the heat capacity of carbides NbC_y in different structural states at $T <$ 20 K are shown in Fig. 11.46. It is seen that within the given temperature interval the heat capacity of ordered carbides $NbC_{0.83}$ and $NbC_{0.845}$ is higher than the heat capacity of disordered carbides having similar compositions.

Researchers [15] were the first to study the effect of the structural state on the heat capacity of a nonstoichiometric compound (niobium carbide NbC_y) within the temperature interval 78–300 K. The effect of ordering on the heat capacity of niobium carbide is described in detail in [211, 212]. Later an analogous study was performed with respect to tantalum carbide [207–209].

The heat capacity of disordered and ordered nonstoichiometric carbides of niobium NbC_y and tantalum TaC_y was measured using the adiabatic method at a temperature from 78 to 300 K with a step from 3 to 5 K. The mean measurement error at a given temperature did not exceed 0.5%. The heat capacity polytherms $C_p(T)$ of the disordered and ordered carbides at hand looked like smooth curves. Smoothed values of the heat capacity and calculated values of the enthalpy and the entropy of disordered and ordered carbides NbC_y and TaC_y are given in Tables 11.14–11.24.

Considering the data of [52, 174], the heat capacity and thermodynamic functions of carbides $NbC_{0.83}$ (Table 11.16), $NbC_{0.97}$ and $NbC_{0.98}$ (Table 11.19) are given for the temperature interval from 4 to 300 K, while the heat capacity of carbide $NbC_{0.845}$ (Table 11.17) is shown only for the low-temperature interval from 4 to 30 K. Within the temperature interval studied, the measured dependencies $C_p(T)$ of disordered carbides of niobium NbC_y (y = 0.75, 0.81, 0.83,

0.845, 0.88, 0.93, 0.97, 0.98) and tantalum $TaC_{1.00}$ are in good agreement with the data of [215, 216].

It is seen from the Tables 11.15–11.18 and 11.21–11.23 that the heat capacity of ordered carbides is higher than the heat capacity of disordered carbides having similar compositions in the whole temperature interval studied.

Table 11.14. Heat capacity, entropy and enthalpy of disordered niobium carbides $NbC_{0.75}$ and $NbC_{0.93}$ in the temperature interval from 80 to 300 K [209, 211, 212]

T (K)	$NbC_{0.75}$			$NbC_{0.93}$		
	C_p	$S_T^0 - S_{80}^0$	$H_T^0 - H_{80}^0$	C_p	$S_T^0 - S_{80}^0$	$H_T^0 - H_{80}^0$
	(J mol^{-1} K^{-1})		(J mol^{-1})	(J mol^{-1} K^{-1})		(J mol^{-1})
80	9.99	0	0	9.89	0	0
90	11.63	1.273	108.2	11.74	1.249	108.2
100	13.12	2.577	232.1	13.49	2.577	234.4
110	14.51	3.893	370.3	15.13	3.940	377.5
120	15.81	5.212	522.0	16.70	5.325	536.8
130	17.07	6.258	686.5	18.22	6.722	711.5
140	18.28	7.838	863.3	19.68	8.126	901.0
150	19.45	9.139	1051.9	21.10	9.533	1105.0
160	20.58	10.43	1252.1	22.46	10.94	1322.9
170	21.68	11.71	1463.5	23.78	12.34	1554.2
180	22.74	12.98	1685.7	25.03	13.74	1798.3
190	23.76	14.24	1918.2	26.23	15.12	2054.7
200	24.74	15.48	2160.8	27.38	16.50	2322.9
210	25.68	16.71	2413.0	28.46	17.86	2602.1
220	26.56	17.93	2674.2	29.48	19.21	2891.9
230	27.41	19.13	2944.2	30.45	20.54	3191.7
240	28.21	20.32	3222.3	31.36	21.86	3500.8
250	28.97	21.48	3508.3	32.22	23.16	3818.8
260	29.68	22.64	3801.6	33.02	24.44	4145.0
270	30.36	23.76	4101.8	33.78	25.70	4479.1
273.16	30.56	24.11	4198.1	34.00	26.09	4586.3
280	30.99	24.88	4408.6	34.49	26.94	4820.5
290	31.59	26.01	4721.6	35.16	28.16	5168.9
298.16	32.05	26.86	4981.0	35.67	29.14	5457.6
300	32.15	27.06	5040.3	35.78	29.36	5523.6

Table 11.15. Heat capacity, entropy and enthalpy of niobium carbide $NbC_{0.81}$ in disordered and ordered states in the temperature interval from 80 to 300 K [209, 211, 212]

T (K)	Disordered state			Ordered state		
	C_p	$S^0_T - S^0_{80}$	$H^0_T - H^0_{80}$	C_p	$S^0_T - S^0_{80}$	$H^0_T - H^0_{80}$
	(J mol^{-1} K^{-1})		(J mol^{-1})	(J mol^{-1} K^{-1})		(J mol^{-1})
80	9.26	0	0	9.66	0	0
90	10.99	1.170	101.4	11.41	1.240	105.5
100	12.61	2.413	219.5	13.05	2.529	227.9
110	14.13	3.687	353.3	14.60	3.846	366.2
120	15.59	4.980	502.0	16.07	5.180	519.6
130	16.98	6.283	664.9	17.50	6.523	687.5
140	18.34	7.592	841.6	18.87	7.881	870.8
150	19.64	8.902	1031.5	20.20	9.218	1064.8
160	20.90	10.21	1234.3	21.48	10.56	1273.2
170	22.12	11.52	1449.5	22.70	11.90	1494.2
180	23.28	12.81	1676.5	23.88	13.23	1727.2
190	24.40	14.10	1915.0	25.00	14.56	1971.6
200	25.46	15.38	2164.4	26.06	15.86	2227.0
210	26.47	16.65	2424.2	27.07	17.16	2492.7
220	27.43	17.90	2693.7	28.02	18.44	2768.3
230	28.33	19.14	2972.6	28.92	19.71	3053.1
240	29.19	20.37	3260.3	29.76	20.96	3346.5
250	29.99	21.58	3556.3	30.56	22.19	3648.2
260	30.75	22.77	3860.1	31.30	23.41	3957.6
270	31.46	23.94	4171.2	32.00	24.60	4274.2
273.16	31.67	24.31	4271.1	32.21	24.98	4375.7
280	32.13	25.10	4489.3	32.66	25.78	4597.5
290	32.76	26.23	4813.8	33.27	26.93	4927.2
298.16	33.24	27.15	5083.0	33.70	27.89	5200.5
300	33.35	27.36	5144.5	33.85	28.10	5262.9

Table 11.16. Heat capacity, entropy and enthalpy of niobium carbide $NbC_{0.83}$ in disordered and ordered states in the temperature interval from 4 to 300 K [52, 209, 211, 212]

T (K)	Disordered state			Ordered state		
	C_p	$S^0_T - S^0_0$	$H^0_T - H^0_0$	C_p	$S^0_T - S^0_0$	$H^0_T - H^0_0$
	(J mol^{-1} K^{-1})		(J mol^{-1})	(J mol^{-1} K^{-1})		(J mol^{-1})
1	2	3	4	5	6	7
4.0	0.012	0.009	0.020	0.007	0.004	0.009
5.0	0.016	0.012	0.035	0.025	0.007	0.025
6.0	0.021	0.016	0.053	0.041	0.013	0.057
6.2	0.023	0.017	0.058	0.044	0.015	0.066
6.4	0.024	0.017	0.062	0.048	0.016	0.075

(*Continued*)

Table 11.16 (continued)

1	2	3	4	5	6	7
6.6	0.025	0.018	0.067	0.050	0.017	0.085
6.7	0.027	0.019	0.070	0.051	0.018	0.090
6.74	–	–	–	0.051	0.019	0.092
6.8	0.029	0.019	0.072	0.051	0.019	0.095
6.9	0.030	0.020	0.075	0.040	0.020	0.100
7.0	0.031	0.020	0.078	0.031	0.020	0.103
7.2	0.033	0.021	0.085	0.032	0.021	0.109
7.4	0.035	0.022	0.092	0.034	0.022	0.116
7.6	0.036	0.023	0.099	0.035	0.023	0.123
7.8	0.040	0.024	0.106	0.036	0.024	0.130
8.0	0.035	0.024	0.114	0.038	0.025	0.137
8.2	0.036	0.025	0.121	0.039	0.026	0.145
8.4	0.038	0.026	0.128	0.041	0.027	0.153
8.6	0.040	0.027	0.136	0.043	0.028	0.162
8.8	0.041	0.028	0.144	0.045	0.029	0.171
9.0	0.043	0.029	0.152	0.047	0.030	0.180
10	0.054	0.034	0.200	0.058	0.035	0.233
11	0.065	0.040	0.260	0.071	0.041	0.297
12	0.080	0.046	0.332	0.088	0.048	0.377
13	0.098	0.053	0.420	0.108	0.056	0.474
14	0.120	0.061	0.529	0.132	0.065	0.594
15	0.145	0.070	0.662	0.156	0.075	0.737
16	0.175	0.081	0.822	0.186	0.086	0.908
17	0.206	0.092	1.011	0.219	0.098	1.111
18	0.241	0.105	1.235	0.254	0.112	1.348
19	0.282	0.119	1.495	0.295	0.126	1.622
20	0.326	0.134	1.799	0.345	0.143	1.942
30	1.097	0.391	8.360	1.160	0.412	6.853
40	2.371	0.872	25.42	2.495	0.925	27.01
50	3.897	1.562	56.62	4.068	1.645	59.58
60	5.780	2.433	104.7	5.824	2.537	108.8
70	8.047	3.493	173.8	7.968	3.583	177.6
80	10.02	4.703	264.3	10.02	4.786	267.8
90	11.76	5.985	373.3	11.80	6.070	377.0
100	13.37	7.309	499.0	13.47	7.401	503.5
110	14.88	8.655	640.4	15.04	8.759	646.1
120	16.31	10.01	796.4	16.55	10.13	804.1
130	17.69	11.37	966.5	17.99	11.52	976.8
140	19.02	12.75	1152.1	19.39	12.91	1165.8

(*Continued*)

Table 11.16 (continued)

1	2	3	4	5	6	7
150	20.31	14.09	1346.8	20.74	14.29	1364.6
160	21.55	15.44	1556.2	22.04	15.67	1578.6
170	22.75	16.78	1777.7	23.29	17.04	1805.3
180	23.90	18.12	2011.1	24.48	18.41	2044.2
190	25.01	19.44	2255.7	25.61	19.76	2294.7
200	26.06	20.75	2511.1	26.68	21.10	2556.2
210	27.06	22.04	2776.7	27.70	22.43	2828.2
220	28.01	23.33	3052.1	28.66	23.74	3110.1
230	28.91	24.59	3336.8	29.56	25.03	3401.2
240	29.75	25.84	3630.1	30.41	26.31	3701.1
250	30.55	27.07	3931.7	31.20	27.57	4009.2
260	31.31	28.29	4241.1	31.95	28.81	4325.0
270	32.01	29.49	4557.8	32.64	30.03	4648.1
273.16	32.22	29.96	4659.4	32.85	30.41	4751.6
280	32.68	30.65	4881.3	33.30	31.22	4977.8
290	33.31	31.81	5211.3	33.91	32.40	5314.0
298.18	33.78	32.77	5484.9	34.38	33.38	5592.4
300	33.89	32.99	5547.4	34.49	33.60	5656.0

Table 11.17. Heat capacity, entropy and enthalpy of niobium carbide $NbC_{0.845}$ in disordered and ordered states in the temperature interval from 4 to 30 K [52]

T (K)	Disordered state			Ordered state		
	C_p	$S_T^0 - S_0^0$	$H_T^0 - H_0^0$	C_p	$S_T^0 - S_0^0$	$H_T^0 - H_0^0$
	(J mol^{-1} K^{-1})		(J mol^{-1})	(J mol^{-1} K^{-1})		(J mol^{-1})
4	0.013	0.010	0.021	0.018	0.015	0.032
5	0.018	0.013	0.036	0.025	0.020	0.053
6	0.025	0.017	0.058	0.033	0.025	0.082
7	0.034	0.022	0.087	0.044	0.031	0.121
8	0.045	0.027	0.127	0.057	0.038	0.171
9	0.059	0.033	0.178	0.074	0.046	0.236
10	0.076	0.040	0.245	0.094	0.054	0.320
12	0.117	0.057	0.439	0.144	0.076	0.556
14	0.173	0.079	0.726	0.212	0.103	0.913
16	0.249	0.107	1.144	0.290	0.136	1.413
18	0.335	0.141	1.727	0.390	0.176	2.088
20	0.440	0.182	2.501	0.508	0.223	2.981
22	0.571	0.230	3.509	0.641	0.277	4.124
24	0.720	0.286	4.797	0.792	0.339	5.559
26	0.879	0.350	6.393	0.960	0.409	7.309
28	1.061	0.422	8.332	1.144	0.487	9.409
30	1.247	0.501	10.646	1.340	0.572	11.885

Table 11.18. Heat capacity, entropy and enthalpy of niobium carbide $NbC_{0.88}$ in disordered and ordered states in the temperature interval from 80 to 300 K [209, 211, 212]

T (K)	Disordered state			Ordered state		
	C_p	$S_T^0 - S_{80}^0$	$H_T^0 - H_{80}^0$	C_p	$S_T^0 - S_{80}^0$	$H_T^0 - H_{80}^0$
	(J mol^{-1} K^{-1})		(J mol^{-1})	(J mol^{-1} K^{-1})		(J mol^{-1})
1	2	3	4	5	6	7
80	10.39	0	0	10.54	0	0
90	12.17	1.328	112.9	12.32	1.346	114.5
100	13.82	2.698	243.0	13.97	2.731	246.0
110	15.37	4.088	389.1	15.50	4.135	393.4
120	16.83	5.489	550.1	16.95	5.546	555.8
130	18.23	6.892	725.5	18.35	6.959	732.3
140	19.60	8.293	914.7	19.70	8.369	922.6
150	20.91	9.690	1117.3	21.00	9.772	1126.2
160	22.18	11.08	1332.7	22.26	11.17	1342.5
170	23.40	12.46	1560.7	23.48	12.56	1571.3
180	24.58	13.83	1800.6	24.65	13.93	1812.0
190	25.71	15.19	2052.1	25.77	15.29	2064.2
200	26.78	16.54	2314.7	26.85	16.64	2327.4
210	27.81	17.87	2587.7	27.86	17.98	2601.0
220	28.78	19.19	2870.7	28.83	19.30	2884.6
230	29.70	20.47	3163.2	29.75	20.60	3177.6
240	30.57	21.77	3464.2	30.61	21.89	3479.4
250	31.39	23.04	3774.4	31.43	23.16	3789.7
260	32.16	24.29	4092.3	32.20	24.41	4107.9
270	32.89	25.52	4417.6	32.92	25.64	4433.6
273.16	33.10	25.90	4521.9	33.13	26.02	4538.1
280	33.57	26.72	4749.9	33.60	26.84	4766.3
290	34.21	27.91	5088.9	34.24	28.06	5105.6
298.16	34.70	28.90	5369.9	34.73	28.99	5386.8
300	34.81	29.12	5434.1	34.84	29.20	5451.1

The heat capacity polytherms at 60 K $\leq T \leq$ 300 K were described by the superposition model discussed in Sect. 11.7.1. An application software package including random search by the Monte Carlo method [217], the simplex method due to Nelder and Mead [218], and the Davidson gradient method [219] was used for the correct description of the temperature dependence of the heat capacity of niobium and tantalum carbides by the function (11.54) allowing for the correction (11.58).

Table 11.19. Heat capacity, entropy and enthalpy of disordered niobium carbides $NbC_{0.97}$ and $NbC_{0.98}$ in the temperature interval from 4 to 300 K [52, 211, 212]

T (K)	$NbC_{0.97}$			$NbC_{0.98}$		
	C_p	$S_T^0 - S_0^0$ (J mol^{-1} K^{-1})	$H_T^0 - H_0^0$ (J mol^{-1})	C_p	$S_T^0 - S_0^0$ (J mol^{-1} K^{-1})	$H_T^0 - H_0^0$ (J mol^{-1})
1	2	3	4	5	6	7
4.0	0.016	0.009	0.022	0.017	0.013	0.028
5.0	0.037	0.016	0.048	0.032	0.018	0.053
6.0	0.058	0.024	0.096	0.058	0.026	0.097
7.0	0.076	0.035	0.163	0.085	0.037	0.169
7.2	0.079	0.037	0.179	0.090	0.039	0.187
7.4	0.082	0.039	0.195	0.095	0.041	0.206
7.6	0.084	0.041	0.212	0.100	0.043	0.226
7.8	0.064	0.043	0.227	0.105	0.046	0.246
8.0	0.050	0.045	0.238	0.110	0.050	0.266
8.2	0.051	0.046	0.248	0.115	0.052	0.288
8.4	0.053	0.047	0.259	0.119	0.055	0.311
8.6	0.055	0.048	0.270	0.123	0.058	0.335
8.8	0.057	0.050	0.281	0.127	0.061	0.360
9.0	0.059	0.051	0.292	0.131	0.064	0.387
10.0	0.069	0.058	0.357	0.150	0.079	0.527
11.0	0.080	0.065	0.431	0.167	0.094	0.686
11.2	0.082	0.066	0.447	0.169	0.097	0.719
11.4	0.084	0.068	0.463	0.152	0.100	0.752
11.6	0.087	0.069	0.481	0.135	0.102	0.780
11.8	0.090	0.071	0.498	0.116	0.104	0.805
12	0.093	0.072	0.517	0.097	0.106	0.827
13	0.107	0.080	0.617	0.112	0.115	0.931
14	0.127	0.089	0.733	0.127	0.124	1.051
15	0.150	0.098	0.872	0.145	0.133	1.187
16	0.177	0.109	1.035	0.165	0.143	1.342
17	0.211	0.121	1.228	0.186	0.154	1.517
18	0.252	0.134	1.460	0.210	0.165	1.715
19	0.297	0.149	1.734	0.236	0.177	1.938
20	0.351	0.165	2.057	0.350	0.192	2.231
30	1.359	0.459	9.753	1.204	0.511	10.41
40	2.676	1.021	29.62	2.080	0.951	25.93
50	4.200	1.778	63.84	4.063	1.617	56.16
60	6.103	2.705	115.0	6.238	2.554	107.9
70	8.452	3.824	187.7	8.307	3.672	180.7
80	10.44	5.092	282.9	10.17	4.904	273.2
90	12.32	6.432	396.9	12.08	6.213	384.5
100	14.08	7.822	529.0	13.87	7.579	514.3

(Continued)

Table 11.19 (continued)

1	2	3	4	5	6	7
110	15.73	9.243	678.1	15.55	8.980	661.5
120	17.30	10.68	843.4	17.15	10.41	825.0
130	18.81	12.13	1024.0	18.70	11.84	1004.4
140	20.27	13.57	1219.5	20.19	13.28	1198.9
150	21.69	15.02	1429.4	21.64	14.73	1408.1
160	23.05	16.46	1653.1	23.03	16.17	1631.5
170	24.36	17.90	1890.3	24.37	17.60	1868.5
180	25.62	19.33	2140.2	25.66	19.03	2118.8
190	26.83	20.75	2402.5	26.90	20.45	2381.6
200	27.97	22.15	2676.6	28.07	21.86	2656.5
210	29.06	23.54	2961.9	29.19	23.26	2942.9
220	30.10	24.92	3257.8	30.25	24.64	3240.1
230	31.07	26.28	3563.7	31.25	26.01	3547.7
240	31.99	27.62	3879.1	32.19	27.36	3865.0
250	32.86	28.95	4203.5	33.09	28.70	4191.5
260	33.68	30.26	4536.3	33.92	30.01	4526.6
270	34.45	31.54	4877.0	34.71	31.31	4869.8
273.16	34.68	31.94	4986.3	34.94	31.71	4980.0
280	35.17	32.81	5225.1	35.45	32.59	5220.8
290	35.84	34.05	5580.2	36.15	33.84	5578.8
298.18	36.36	35.07	5874.6	36.69	34.77	5872.8
300	36.48	35.31	5941.9	36.81	35.07	5943.7

Table 11.20. Heat capacity, entropy and enthalpy of tantalum carbide $TaC_{1.00}$ in the temperature interval from 80 to 300 K [209]

T (K)	C_p	$S_T^0 - S_{80}^0$ (J mol^{-1} K^{-1})	$H_T^0 - H_{80}^0$ (J mol^{-1})	T (K)	C_p	$S_T^0 - S_{80}^0$ (J mol^{-1} K^{-1})	$H_T^0 - H_{80}^0$ (J mol^{-1})
80	15.41	0	0	210	30.03	21.38	3028.2
90	17.01	1.91	162.3	220	30.94	22.80	3333.0
100	18.38	3.77	334.4	230	31.81	24.20	3646.8
110	19.61	5.58	529.4	240	32.64	25.57	3969.0
120	20.76	7.34	731.4	250	33.44	26.92	4299.5
130	21.86	9.04	944.5	260	34.19	28.25	4637.6
140	22.94	10.70	1168.5	270	34.91	29.54	4983.2
150	24.00	12.32	1403.3	273.16	35.13	29.95	5093.8
160	25.06	13.90	1648.6	280	35.59	30.83	5335.7
170	26.09	15.46	1904.3	290	36.24	32.08	5694.9
180	27.12	16.98	2170.4	298.16	36.74	33.10	5992.7
190	28.11	18.47	2446.6	300	36.85	33.33	6060.4
200	29.09	19.93	2732.6	–	–	–	–

Table 11.21. Heat capacity, entropy and enthalpy of tantalum carbide $TaC_{0.81}$ in disordered and ordered states in the temperature interval from 80 to 300 K [209]

T (K)	Disordered state			Ordered state		
	C_p	$S_T^0 - S_{80}^0$	$H_T^0 - H_{80}^0$	C_p	$S_T^0 - S_{80}^0$	$H_T^0 - H_{80}^0$
	(J mol^{-1} K^{-1})		(J mol^{-1})	(J mol^{-1} K^{-1})		(J mol^{-1})
80	14.29	0	0	14.45	0	0
90	15.72	1.765	150.2	15.86	1.774	151.7
100	16.96	3.493	313.8	17.10	3.511	316.8
110	18.09	5.159	489.1	18.22	5.195	493.4
120	19.14	6.778	675.3	19.28	6.824	681.0
130	20.15	8.353	871.8	20.31	8.407	878.8
140	21.16	9.883	1078.4	21.31	9.955	1087.2
150	22.15	11.38	1295.0	22.31	11.46	1305.3
160	23.13	12.83	1521.4	23.29	12.92	1533.3
170	24.09	14.26	1757.6	24.25	14.37	1771.1
180	25.04	14.76	2003.3	25.20	15.78	2018.5
190	25.96	17.05	2258.3	26.13	17.17	2274.8
200	26.85	18.41	2522.3	27.02	18.54	2541.1
210	27.71	19.73	2795.2	27.88	19.87	2815.1
220	28.53	21.04	3076.5	28.70	21.20	3098.6
230	29.33	22.34	3365.7	29.49	22.48	3389.4
240	30.07	23.59	3662.8	30.25	23.77	3688.2
250	30.79	24.83	3967.2	30.96	25.01	3994.3
260	31.48	26.05	4278.5	31.63	26.24	4307.5
270	32.12	27.25	4596.4	32.27	27.45	4626.9
273.16	32.31	27.63	4698.2	32.46	27.82	4729.5
280	32.72	28.44	4920.7	32.88	28.63	4952.6
290	33.29	29.14	5250.8	33.45	29.79	5284.3
298.16	33.75	30.52	5524.3	33.89	30.72	5559.7
300	33.84	30.73	5586.4	33.98	30.93	5622.0

Table 11.22. Heat capacity, entropy and enthalpy of tantalum carbide $TaC_{0.84}$ in disordered and ordered states in the temperature interval from 80 to 300 K [209]

T (K)	Disordered state			Ordered state		
	C_p	$S_T^0 - S_{80}^0$	$H_T^0 - H_{80}^0$	C_p	$S_T^0 - S_{80}^0$	$H_T^0 - H_{80}^0$
	(J mol^{-1} K^{-1})		(J mol^{-1})	(J mol^{-1} K^{-1})		(J mol^{-1})
80	14.44	0	0	15.20	0	0
90	15.93	1.794	152.1	16.64	1.877	159.4
100	17.24	3.533	318.1	17.93	3.698	332.4
110	18.44	5.235	496.6	19.12	5.465	517.6
120	19.58	6.891	686.8	20.26	7.185	714.6
130	20.69	8.510	888.2	21.37	8.841	922.7

(Continued)

Table 11.22 (continued)

1	2	3	4	5	6	7
140	21.77	10.08	1100.4	22.47	10.47	1141.8
150	22.83	11.61	1323.5	23.55	12.05	1371.9
160	23.88	13.13	1557.1	24.61	13.61	1612.8
170	24.91	14.60	1801.0	25.64	15.13	1864.0
180	25.90	16.05	2055.1	26.65	16.62	2125.5
190	26.87	17.48	2319.0	27.62	18.10	2396.8
200	27.80	18.88	2592.4	28.54	19.54	2677.7
210	28.69	20.26	2874.9	29.43	20.95	2967.6
220	29.54	21.62	3166.1	30.27	22.34	3266.1
230	30.35	22.94	3465.5	31.07	23.70	3572.9
240	31.12	24.25	3772.9	31.83	25.04	3887.4
250	31.85	25.55	4087.8	32.54	26.35	4209.3
260	32.53	26.80	4409.7	33.21	27.64	4538.0
270	33.18	28.04	4738.4	33.84	28.92	4873.3
273.16	33.38	28.42	4843.5	34.03	29.30	4980.5
280	33.79	29.26	5073.2	34.43	30.15	5214.7
290	34.37	30.45	5414.0	34.99	31.37	5561.2
298.16	34.81	31.41	5696.4	35.42	32.35	5849.1
300	34.91	31.55	5760.5	35.51	32.56	5914.4

Table 11.23. Heat capacity, entropy and enthalpy of tantalum carbide $TaC_{0.88}$ in disordered and ordered states in the temperature interval from 80 to 300 K [209]

T (K)	Disordered state			Ordered state		
	C_p	$S_T^0 - S_{80}^0$	$H_T^0 - H_{80}^0$	C_p	$S_T^0 - S_{80}^0$	$H_T^0 - H_{80}^0$
	(J mol^{-1} K^{-1})		(J mol^{-1})	(J mol^{-1} K^{-1})		(J mol^{-1})
80	14.62	0	0	14.71	0	0
90	16.17	1.814	154.2	16.26	1.814	155.0
100	17.54	3.591	322.8	17.63	3.600	324.6
110	18.79	5.320	504.5	18.88	5.349	507.2
120	19.98	7.012	698.4	20.08	7.041	702.1
130	21.24	8.657	904.1	21.24	8.695	908.7
140	22.27	10.27	1121.1	22.38	10.31	1126.9
150	23.38	11.84	1349.5	23.50	11.89	1356.3
160	24.47	13.39	1588.8	24.60	13.45	1596.8
170	25.54	14.89	1838.8	25.67	14.97	1848.1
180	26.57	16.39	2099.4	26.70	16.47	2110.0
190	27.57	17.85	2370.1	27.70	17.94	2382.1
200	28.53	19.29	2650.7	28.67	19.38	2664.0
210	29.45	20.43	2940.6	29.58	20.80	2955.3
220	30.32	22.10	3239.5	30.46	22.20	3255.5

(*Continued*)

Table 11.23 (continued)

1	2	3	4	5	6	7
230	31.15	23.46	3546.9	31.29	23.56	3564.3
240	31.94	24.81	3862.4	32.07	24.92	3881.2
250	32.68	26.13	4185.5	32.81	26.24	4205.6
260	33.38	27.42	4515.9	33.51	27.54	4537.3
270	34.05	28.69	4853.0	34.17	28.83	4875.7
273.16	34.25	29.09	4960.9	34.37	29.22	4984.0
280	34.67	29.95	5196.7	34.79	30.08	5220.6
290	35.25	31.17	5546.3	35.37	31.31	5571.5
298.16	35.71	32.16	5835.8	35.82	32.31	5861.9
300	35.81	32.37	5901.6	35.92	32.52	5927.9

Table 11.24. Heat capacity, entropy and enthalpy of disordered tantalum carbides $TaC_{0.93}$ and $TaC_{0.96}$ in the temperature interval from 80 to 300 K [209]

T (K)	$TaC_{0.93}$			$TaC_{0.96}$		
	C_p	$S_T^0 - S_{80}^0$	$H_T^0 - H_{80}^0$	C_p	$S_T^0 - S_{80}^0$	$H_T^0 - H_{80}^0$
	(J mol^{-1} K^{-1})		(J mol^{-1})	(J mol^{-1} K^{-1})		(J mol^{-1})
80	15.26	0	0	15.64	0	0
90	16.81	1.891	160.5	17.17	1.921	164.2
100	18.16	3.735	335.4	18.50	3.802	342.7
110	19.40	5.520	523.3	19.70	5.625	533.8
120	20.57	7.266	723.2	20.83	7.389	736.5
130	21.71	8.946	934.6	21.92	9.104	950.3
140	22.82	10.61	1157.3	23.00	10.76	1174.8
150	23.92	12.22	1391.0	24.06	12.39	1410.2
160	25.01	13.79	1635.8	25.11	13.97	1656.1
170	26.07	15.34	1891.1	26.15	15.52	1912.4
180	27.11	16.86	2157.1	27.17	17.05	2179.0
190	28.11	18.35	2433.2	28.16	18.55	2455.7
200	29.08	19.82	2719.2	29.12	20.02	2742.0
210	30.01	21.26	3014.6	30.05	21.46	3037.9
220	30.90	22.68	3318.8	30.94	22.87	3342.8
230	31.74	24.07	3632.4	31.79	24.27	3656.6
240	32.55	25.44	3963.9	32.60	25.65	3978.5
250	33.31	26.78	4283.2	33.37	26.99	4308.4
260	34.03	28.10	4619.8	34.11	28.31	4645.9
270	34.71	29.39	4963.6	34.80	29.62	4990.4
273.16	34.91	29.80	5073.5	35.01	30.64	5100.7
280	35.33	30.67	5313.9	35.46	30.90	5341.7
290	35.95	31.91	5670.4	36.08	32.15	5699.4
298.16	36.42	32.92	5965.7	36.56	33.15	5995.8
300	36.52	33.16	6032.9	36.66	33.38	6063.2

The approximation was highly accurate: the root-mean-square deviation σ_C of the measured values of the heat capacity from the calculated values was 0.05–0.07 J mol^{-1} K^{-1} for NbC_y and TaC_y samples having different compositions. Numerical values of the Debye θ_D and Einstein θ_E temperatures were also found from calculations. By way of example, Fig. 11.49 presents experimental data on the heat capacity of tantalum carbide $TaC_{0.93}$ and their approximation by the function (11.54) allowing for the correction (11.58). As seen from Fig. 11.49, the experimental heat capacity is very accurately described by the superposition of the Debye C_D and Einstein C_E contributions.

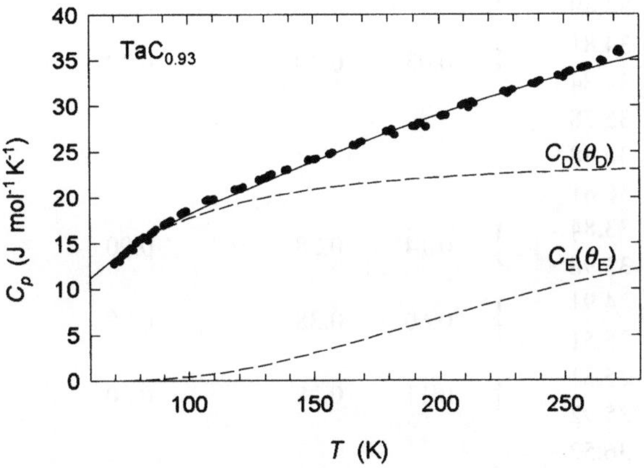

Fig. 11.49. Experimental dependence of the heat capacity $C_p(T)$ of tantalum carbide $TaC_{0.93}$ approximated by the function (11.54) taking into account the correction (11.58) [209]: (•) experimental values of the heat capacity. The solid line shows the calculated dependence $C_p(T)$. The dotted line shows the Debye C_D and Einstein C_E contributions to the heat capacity

The Debye temperature θ_D is determined most frequently from low-temperature data ($T < 30$ K) using the T^3 law, which is a limiting case of the dependence (11.54) at low temperature. However, in this temperature interval the Einstein contribution to the heat capacity (i.e. the contribution of optical vibrations) is negligibly small and the temperature θ_E cannot be determined. Using data on the heat capacity at a temperature from 60 to 300 K, it was possible to determine θ_D and θ_E without referring to simplified dependencies like T^3.

The analysis of the data shows that the temperature θ_E is much higher than θ_D for all samples of carbides NbC_y and TaC_y (Table 11.25). This is in good agreement with the mutual position of optical and acoustic vibration branches, which are separated by a wide forbidden gap resulting from largely different masses of Nb and C atoms (8 times) or Ta and C atoms (15 times). The temperatures θ_E and θ_D decrease slightly because of ordering (Table 11.25).

Table 11.25. Debye and Einstein temperatures θ_D and θ_E and heat capacity C_p of niobium and tantalum carbides NbC_y and TaC_y in disordered (1) and ordered (2) states at a temperature of 300 K

Carbide	State	θ_D (K)	θ_E (K)	$C_p(300)$ ($J\ mol^{-1}\ K^{-1}$)	$\Delta C_p(300)$ ($J\ mol^{-1}\ K^{-1}$) experimental	calculation $\eta(300) = \eta_{trans}$	calculation $\eta(300) = \eta_{max}$
$NbC_{0.75}$	1	347	844	32.15			
$NbC_{0.81}$	1	375	825	33.35	0.50	0.55	1.31
$NbC_{0.81}$	2	365	801	33.85			
$NbC_{0.83}$	1	358	823	33.89	0.60	0.56	1.39
$NbC_{0.83}$	2	359	789	34.49			
$NbC_{0.88}$	1	355	825	34.81	0.03	0.40	0.72
$NbC_{0.88}$	2	352	825	34.84			
$NbC_{0.93}$	1	375	813	35.78			
$NbC_{0.97}$	1	366	820	36.48			
$NbC_{0.98}$	1	377	827	36.81			
$TaC_{0.81}$	1	256	848	33.84	0.14	0.38	0.90
$TaC_{0.81}$	2	252	840	33.98			
$TaC_{0.84}$	1	258	818	34.91	0.60	0.38	0.88
$TaC_{0.84}$	2	243	788	35.51			
$TaC_{0.88}$	1	261	809	35.81	0.11	0.28	0.50
$TaC_{0.88}$	2	259	803	35.92			
$TaC_{0.93}$	1	256	824	36.52			
$TaC_{0.96}$	1	252	848	36.66			
$TaC_{1.00}$	1	263	875	36.85			

The effect of ordering on the temperature dependence of the heat capacity of strongly nonstoichiometric compounds MX_y with the $B1$ structure was explained in [208, 209]. Some postulates of the order-parameter functional method [1, 2, 43], which is described comprehensively in Chap. 9, were used. Ordering was described using a basis cluster figure in the form of an octahedron comprising six sites of the fcc nonmetallic sublattice, which represented the nearest environment of a metal atom located at the center of the octahedron.

In accordance with [1], the temperature–concentration dependence of the enthalpy of nonstoichiometric compound MX_y is generally a function of the composition y, the temperature T, and the long-range order parameter η. If the enthalpy $(H_T^0 - H_{298}^0)$ of disordered ($\eta = 0$) compound MX_y is presented as a power series

$$(H_T^0 - H_{298}^0)^{\text{disord}}_{MX_y} = H_0^*(T) + yH_1^*(T) + \sum_{n=2}^{R^{(s)}} y^n H_n^*(T)\,, \tag{11.59}$$

then, in accordance with [2, 220, 221], the enthalpy of a nonstoichiometric compound with any degree of long-range order η has the form

$$(H_T^0 - H_{298}^0)_{MX_y}^{ord} = H_0^*(T) + yH_1^*(T) + \sum_{n=2}^{R(s)} P_0^{(n)}(y,\eta)H_n^*(T), \quad (11.60)$$

where $H_n^*(T)$ stands for the coefficients of expansion of the enthalpy in the power series of y; $P_0^{(n)}(y,\eta)$ is the probability of a complete n-subfigure of the basis cluster. As noted in Sect. 9.5, the expansion of the enthalpy $(H_T^0 - H_{298}^0)$ of a disordered nonstoichiometric compound is a quadratic function of y:

$$(H_T^0 - H_{298}^0)_{MX_y}^{disord} = H_0^*(T) + yH_1^*(T) + y^2H_2^*(T). \quad (11.61)$$

In this case, formula (11.60) becomes

$$(H_T^0 - H_{298}^0)_{MX_y}^{ord} = H_0^*(T) + yH_1^*(T) + P_0^{(2)}(y,\eta)H_2^*(T). \quad (11.62)$$

If an ordered phase type $M_{2t}X_{2t-1}$ appears in nonstoichiometric compound MX_y, the probability $P_0^{(2)}$ is calculated by (9.94).

Differentiating (11.61) and (11.62) with respect to the temperature T, it is possible to find the heat capacity of a compound MX_y in disordered and ordered states:

$$C_p(y,0,T) = \frac{\partial H_0^*(T)}{\partial T} + y\frac{\partial H_1^*(T)}{\partial T} + y^2\frac{\partial H_2^*(T)}{\partial T}, \quad (11.63)$$

$$C_p(y,\eta,T) = \frac{\partial H_0^*(T)}{\partial T} + y\frac{\partial H_1^*(T)}{\partial T} + P_0^{(2)}\frac{\partial H_2^*(T)}{\partial T} + H_2^*(T)\frac{\partial P_0^{(2)}}{\partial T}. \quad (11.64)$$

It is worth noting that the disordered state of nonstoichiometric compounds is thermodynamically metastable at a temperature below the disorder–order transition point T_{trans}, but can be easily achieved thanks to quenching (see Sect. 3.2).

An ordered phase of M_6C_5 type is formed in niobium NbC_y and tantalum TaC_y carbides [207–209, 211, 212]. The probability $P_0^{(2)} = y^2 - \eta^2/36$ for the M_6C_5 phase. Considering this and (11.63) and (11.64), the difference of the heat capacities $\Delta C_p(y,\eta,T) = C_p(y,\eta,T) - C_p(y,0,T)$ of ordered and disordered compounds is

$$\Delta C_p(y,\eta,T) = -\frac{\eta^2}{36}\frac{\partial H_2^*(T)}{\partial T} - H_2^*(T)\frac{\eta}{18}\frac{\partial \eta}{\partial T}. \quad (11.65)$$

Let us take a closer look at (11.65). In Sect. 9.5 (see also [1, 2, 43, 44, 210, 211]) it was shown that the free energy of disordered strongly nonstoichiometric compounds is described by (9.71). The parameters of this formula are the energy coefficients $F_0(T)$, $F_1(T)$ and $F_2(T)$. A necessary condition for ordering in nonstoichiometric compounds is a positive value of the coefficient $F_2(T)$, i.e. $F_2(T) > 0$. In particular, ordering is possible if the coefficient F_2 is independent of temperature and is positive, i.e. $F_2 = const > 0$. This case is realized if the coefficient $H_2^*(T) \equiv 0$ in (11.61) and, consequently, $\partial H_2^*(T)/\partial T \equiv 0$. This corresponds to a linear dependence of the heat capacity on the composition of a nonstoichiometric compound. Thus, from (11.65) it follows that the heat capacities of compound

MX_y in ordered and quenched disordered states may be equal ($\Delta C_p \equiv 0$ if $F_2 = const > 0$).

In the most general case, the coefficient $H_2^*(T) \neq 0$ (except at $T = 298$ K when all the coefficients H_0^*, H_1^* and H_2^* are zero as follows from (11.61) and (11.62)) and $\partial H_2^*(T)/\partial T \neq 0$. At a temperature below the disorder–order transition temperature T_{trans} the derivative $\partial\eta/\partial T$ is negative, because the degree of order rises with decreasing temperature. Considering this, from (11.65) it follows that the sign of ΔC_p is determined by values of $H_2^*(T)$ and $\partial H_2^*(T)/\partial T$. The derivative $\partial H_2^*(T)/\partial T$ is the coefficient of the quadratic term in the expansion (11.63) of the heat capacity of disordered compound MX_y.

If $C_p(y,0,T)$ is a convex function of y, then $\partial H_2^*(T)/\partial T < 0$ and $H_2^*(T) < 0$. In this case the first term is positive and the second term is negative in (11.65). As the temperature rises, the first term diminishes and the second term increases in absolute value. Consequently, the difference ΔC_p depends on temperature. At 300 K the difference $\Delta C_p > 0$ and the heat capacity of an ordered nonstoichiometric compound is higher than that of a disordered compound. As the temperature increases, ΔC_p decreases and may acquire a negative value.

If $C_p(y,0,T)$ is a concave function of y, then $\partial H_2^*(T)/\partial T > 0$ and $H_2^*(T) > 0$. Using analogous lines of reasoning, it is seen that at $T = 300$ K the difference $\Delta C_p < 0$. As the temperature rises, ΔC_p increases and may acquire a positive value near T_{trans}.

Taking into account the above analysis, it is possible to discuss in more detail the results of investigations into the heat capacity of nonstoichiometric carbides NbC_y (Fig. 11.50) and TaC_y (Fig. 11.51). An examination of these figures shows that the dependence $C_p(y)$ of disordered carbides is a convex function of y. Therefore at $T = 300$ K the heat capacity of ordered carbides is higher than the heat capacity of quenched disordered carbides having a similar composition, i.e. $\Delta C_p(300) > 0$. Experimental results confirm this conclusion (see Table 11.25).

Comparing experimental and calculated values of $\Delta C_p(300)$, it is possible to estimate the degree of long-range order in the ordered niobium and tantalum carbides studied. The $\Delta C_p(300)$ value was calculated using two versions.

By the first version, it was assumed that a maximum degree of long-range order is achieved in the carbide samples at 300 K, i.e. $\eta(300) = \eta_{max}$. In this case the derivative $\partial\eta/\partial T = 0$, because the long-range order parameter $\eta(300)$ is a constant. Within the homogeneity interval of the M_6C_5 phase, which is formed in carbides NbC_y and TaC_y, the $\eta_{max}(y)$ value is determined from formula (11.5) where $t = 3$.

By the second version, it was assumed that the long-range order parameter at 300 K remained unchanged at the disorder–order transition temperature thanks to the low diffusion mobility of atoms, i.e. $\eta(300) = \eta_{trans}$ (the η_{trans} value is calculated by (9.104)). Clearly, $\partial\eta/\partial T = 0$ in this approximation too.

The comparison of the experimental values and the $\Delta C_p(300)$ values calculated by formula (11.65) (see Table 11.25) shows that the difference $\Delta C_p(300)$ obtained in the approximation $\eta(300) = \eta_{trans}$ approaches most closely the experimental data.

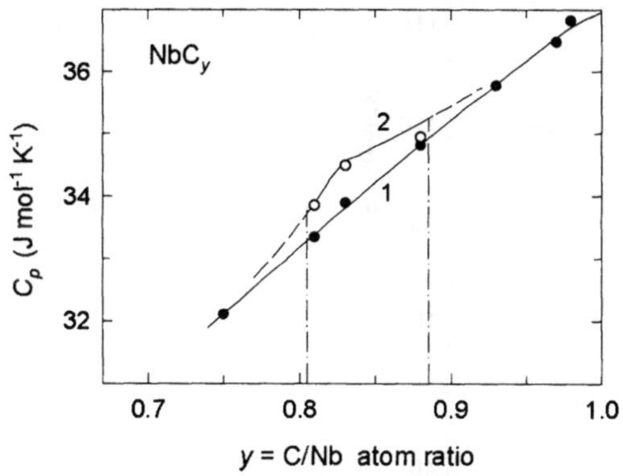

Fig. 11.50. Heat capacity of niobium carbide NbC_y at 300 K in disordered (1) and ordered (2) states [209, 211, 212]: (•) and (○) denote experimental C_p values for disordered and ordered carbide NbC_y. The solid line (2) corresponds to the heat capacity of ordered carbide $C_p(y,\eta,T) = C_p(y,0,T) + \Delta C_p(y,\eta,T)$ calculated in the approximation $\eta(300) = \eta_{trans}$. The vertical dotted lines show the boundaries of the homogeneity interval of ordered phase Nb_6C_5

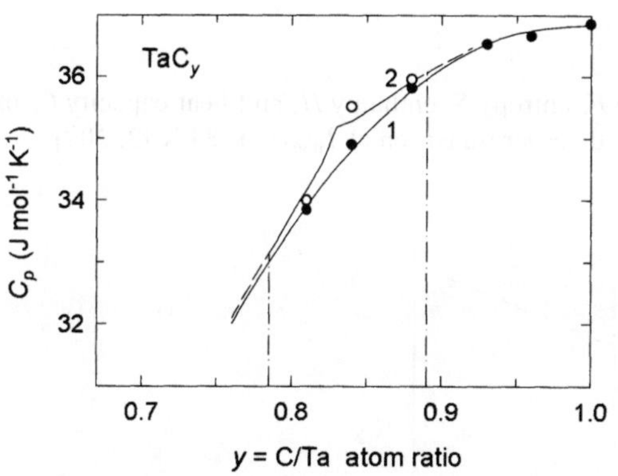

Fig. 11.51. Heat capacity of tantalum carbide TaC_y at 300 K in disordered (1) and ordered (2) states [209]: (•) and (○) denote experimental C_p values for disordered and ordered carbide TaC_y. The solid line (2) corresponds to the heat capacity of ordered carbide $C_p(y,\eta,T) = C_p(y,0,T) + \Delta C_p(y,\eta,T)$ calculated in the approximation $\eta(300) = \eta_{trans}$. The vertical dotted lines show the boundaries of the homogeneity interval of ordered phase Ta_6C_5

Using literature data [2, 207, 211, 212, 215, 222–225] on the formation enthalpy, entropy, and temperature dependencies of the enthalpy and the heat capacity of disordered niobium and tantalum carbides, one can calculate the main thermodynamic characteristics of ordered phases in this compounds and determine changes in thermodynamic parameters during the disorder–order transition.

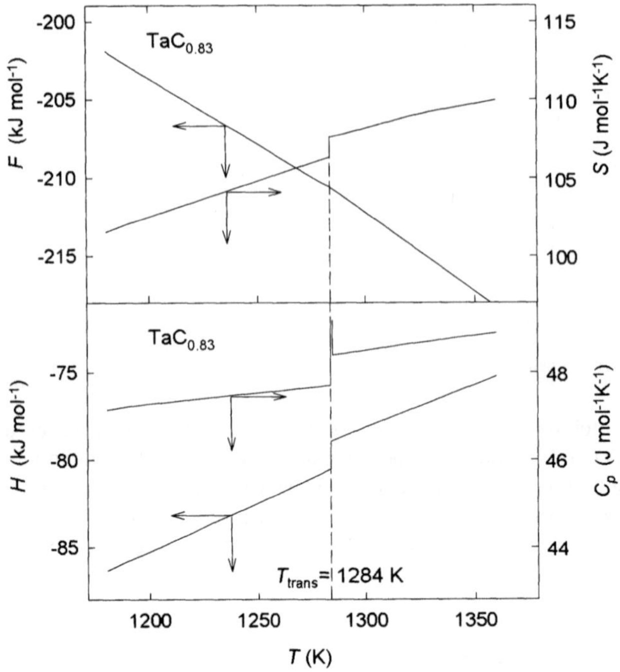

Fig. 11.52. Variation of the free energy *F*, entropy *S*, enthalpy *H*, and heat capacity C_p of tantalum carbide $TaC_{0.83}$ upon the order–disorder transition at T_{trans} = 1284 K [2, 209]

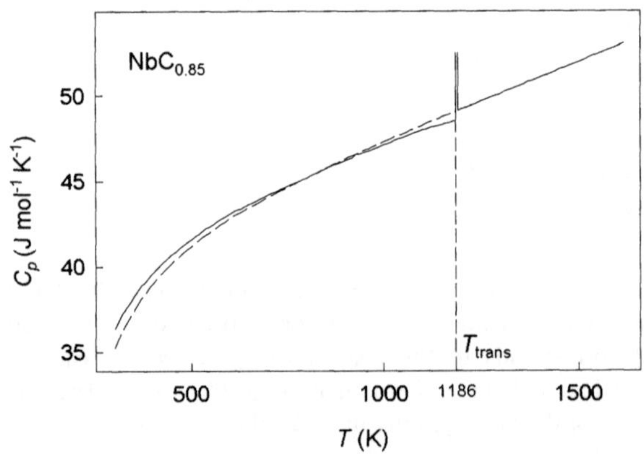

Fig. 11.53. Heat capacity of niobium carbide $NbC_{0.85}$ in equilibrium ordered and quenched nonequilibrium disordered states at $T < T_{trans}$ = 1186 K, and in an equilibrium disordered state at $T > T_{trans}$ [2, 209]. The solid line shows the heat capacity for equilibrium states. The dotted line shows the heat capacity for the quenched nonequilibrium state

Figure 11.52 presents the free energy F, the entropy S, the heat capacity C_p, and the enthalpy $H = \Delta H^0_{f298} + (H^0_T - H^0_{298})$ (ΔH^0_{f298} being the formation enthalpy) of tantalum carbide $TaC_{0.83}$ in thermodynamically equilibrium ordered (at $T < T_{trans} = 1284$ K) and equilibrium disordered (at $T > T_{trans}$) states. The free energy F exhibits a break and the temperature dependencies of the entropy and the enthalpy contain jumps at the transition temperature. This is characteristic of phase transitions of the first kind. The heat capacity has a discontinuity at the transition point.

The disordered state of strongly nonstoichiometric compounds is easily preserved as metastable at a temperature below T_{trans}. The analysis of the heat capacity difference $\Delta C_p(y,\eta,T)$ for nonstoichiometric compounds in equilibrium ordered and quenched nonequilibrium disordered states showed that the difference $\Delta C_p(y,\eta,T)$ can reverse sign with varying temperature. Figure 11.53 presents calculated dependencies $C_p(T)$ for niobium carbide $NbC_{0.85}$. The heat capacity of carbide $NbC_{0.85}$ in equilibrium ordered and nonequilibrium disordered states is shown at a temperature $T < T_{trans} = 1186$ K. The heat capacity of the same carbide in the equilibrium disordered state is shown at $T > T_{trans}$. It is seen that at 300 K the difference $\Delta C_p > 0$ (the heat capacity of ordered carbide is higher than the heat capacity of disordered carbide). As soon as the transition temperature T_{trans} is reached, the heat capacity exhibits a break. This is indicative of the order–disorder phase transition of the first kind.

Thus, calculations suggest that at 300 K the heat capacity of ordered niobium and tantalum carbides should be larger than the heat capacity of the same disordered carbides. The calculation results are in good agreement with the observed increasing C_p of carbides NbC_y and TaC_y during ordering.

11.7.3 Heat Capacity of Nonstoichiometric Carbides at a Temperature Above 300 K and Near the Point of Order–Disorder Transition

Pioneering direct measurements of the heat capacity of strongly nonstoichiometric carbides near the disorder–order transition temperature were made as late as in 1997. By now, variation of the heat capacity during disorder–order and order–disorder transitions has been detected experimentally for nonstoichiometric carbides of titanium [132, 226] and vanadium [10, 20, 21, 139].

Researchers [132, 226] measured the heat capacity of titanium carbides $TiC_{0.54}$, $TiC_{0.58}$ and $TiC_{0.62}$. The samples were synthesized by hot pressing of powder mixtures of $TiC_{0.98}$ and metallic titanium in an atmosphere of high-purity argon at a temperature of 1800 to 2000 K and a pressure of 23 to 25 MPa. All samples were homogeneous and contained only a disordered cubic phase TiC_y. To produce the ordered state in the synthesized samples, they were annealed for 340 h while the temperature was reduced smoothly from 1070 to 770 K. The samples were cooled from 770 to 300 K at a rate of 1 K min^{-1}. Superstructure reflections appeared in X-ray patterns of carbides $TiC_{0.54}$, $TiC_{0.58}$ and $TiC_{0.62}$ after annealing. The diffraction analysis [132] showed that annealed carbide $TiC_{0.54}$ contained the ordered cubic (space group $Fd\bar{3}m$) phase Ti_2C and annealed carbide $TiC_{0.58}$ included cubic and trigonal (space group $R\bar{3}m$) ordered phases of type Ti_2C and some quantity of ordered rhombic (space group $C222_1$) phase Ti_3C_2. The main phase of annealed carbide $TiC_{0.62}$ is the ordered rhombic phase Ti_3C_2.

The heat capacity was measured at a temperature from 300 to 1300 K in an atmosphere of extra-pure argon using a differential scanning calorimeter type Netzsch DSC404. Measurements were made at the heating and cooling rate of 20 K min^{-1} with a step of 5 K. Calibration measurements were performed beforehand under similar conditions to ensure precise determination of the transition heat for TiC_y. High-purity aluminum Al and gold Au were used for calibration. The reference sample in calorimetric experiments was sapphire.

When the heat capacity was measured during heating, the dependencies $C_p(T)$ of annealed carbides $TiC_{0.54}$, $TiC_{0.58}$ and $TiC_{0.62}$ contained effects associated with an equilibrium order–disorder transition. Effects related to an equilibrium transition from the high-temperature disordered state to the low-temperature ordered state were observed during cooling.

When annealed carbide $TiC_{0.54}$ was heated from 300 to 1000 K, its heat capacity C_p increased smoothly thanks to the excitation of the phonon subsystem. A sharp jump of the heat capacity was observed at the transition temperature $T_{trans} \approx 1000$ K (Fig. 11.54). This jump is related to the equilibrium Ti_2C (space group $Fd\bar{3}m$) $\rightarrow$ $TiC_{0.54}$ order–disorder transformation. The peak of the heat capacity near T_{trans} is symmetric rather than λ-shaped (Fig. 11.55). The symmetric shape of the observed peaks is more characteristic of phase transitions of the first rather than second kind. Moreover, the weak hysteresis of the C_p is observed during cooling. This fact also confirms that this transition is rather of the first kind. A temperature hysteresis in the region of the order–disorder transition was observed earlier in the electrical resistivity of carbide TiC_y [121, 130] (see Figs. 11.24 and 11.25). The lattice constant a_{B1} of the basic phase with the $B1$ structure increases a little during ordering in TiC_y (see Fig. 11.5). These facts taken together allow treating the reversible Ti_2C (space group $Fd\bar{3}m$) $\leftrightarrow$ $TiC_{0.54}$ order–disorder transformation as a weak phase transition of the first kind with a small latent heat of transformation.

The dependencies $C_p(T)$ of annealed carbides $TiC_{0.58}$ and $TiC_{0.62}$ (see Fig. 11.54) are shaped similarly to the dependence $C_p(T)$ of carbide $TiC_{0.54}$. The peak of the heat capacity of carbide $TiC_{0.58}$ can be presented as a superposition of two contributions (Fig. 11.55), which correspond to the $Ti_2C \rightarrow TiC_y$ and to the $Ti_3C_2 \rightarrow TiC_y$ transformations. The peaks corresponding to these contributions are wider than the peaks of the heat capacity for carbides $TiC_{0.54}$ and $TiC_{0.62}$. Probably, annealed carbide $TiC_{0.58}$ contains ordered phases Ti_2C and Ti_3C_2, in which the degree of long-range order is smaller than in single-phase ordered carbides.

The heat capacity peak with the largest symmetry is observed in the region of the order–disorder transition of carbide $TiC_{0.62}$ (Fig. 11.55) forming the Ti_3C_2 phase. The symmetry analysis (see Sect. 5.5) suggests that the $Ti_3C_2 \leftrightarrow TiC_y$ transformation may be a phase transition of the first kind only.

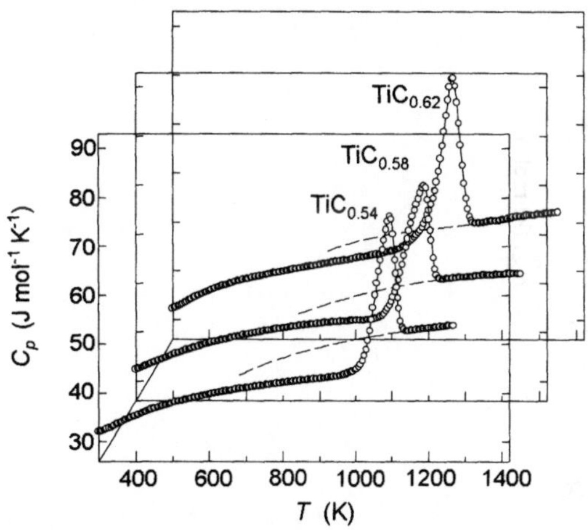

Fig. 11.54. Variation of the heat capacity $C_p(T)$ of annealed (ordered) nonstoichiometric titanium carbides $TiC_{0.54}$, $TiC_{0.58}$ and $TiC_{0.62}$ upon heating [132]. The dotted line shows the heat capacity of carbides TiC_y in a quenched nonequilibrium disordered state at $T < T_{trans}$

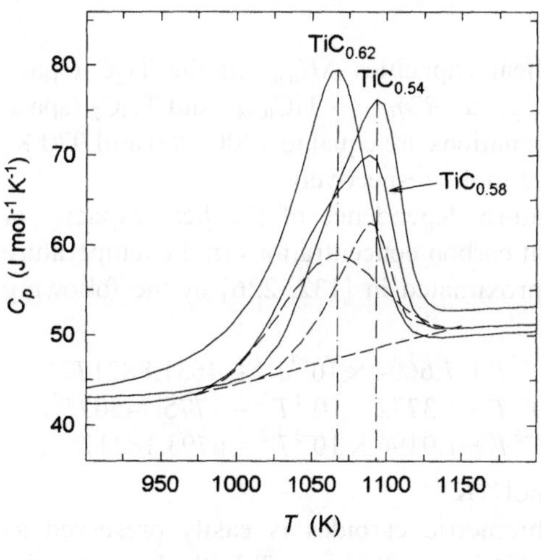

Fig. 11.55. Heat capacity of carbides $TiC_{0.54}$, $TiC_{0.58}$ and $TiC_{0.62}$ near the peaks corresponding to the transition from the disordered to ordered state [132]. Dotted curves show superposition contributions to the heat capacity of carbide $TiC_{0.58}$ corresponding to $Ti_2C \rightarrow TiC_y$ and $Ti_3C_2 \rightarrow TiC_y$ transitions

Figure 11.56 presents the variation of the enthalpy $(H_T^\circ - H_{298}^\circ)$ and the heat capacity $C_p(T)$ of carbide $TiC_{0.62}$ in the region of the phase transition. A jump is observed in the enthalpy $(H_T^\circ - H_{298}^\circ)$ at the transition temperature T_{trans}.

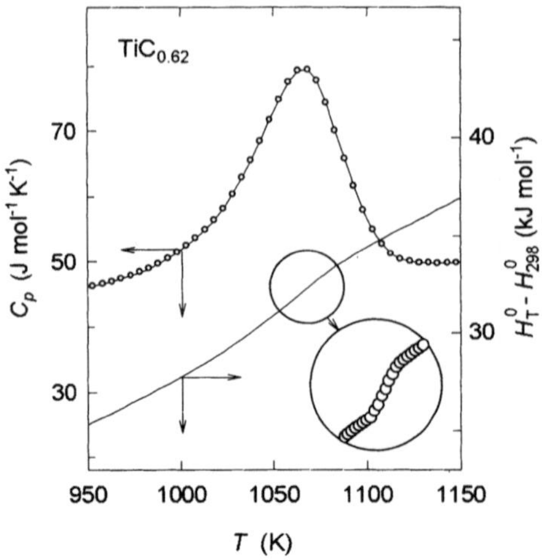

Fig. 11.56. Variation of the heat capacity $C_p(T)$ and enthalpy $(H^\circ_T - H^\circ_{298})$ of carbide $TiC_{0.62}$ in the region of the reversible Ti_3C_2 (space group $C222_1$) $\leftrightarrow$ $TiC_{0.62}$ order–disorder phase transition [132, 226]: $T_{trans} = 970 \pm 10$ K and $\Delta H_{trans} = 1.8 \pm 0.2$ kJ mol^{-1}

Measured temperatures T_{trans} and heat capacities ΔH_{trans} of the Ti_2C (space group $Fd\bar{3}m$) $\leftrightarrow$ $TiC_{0.54}$, Ti_2C (space group $R\bar{3}m$) $\leftrightarrow$ $TiC_{0.58}$ and Ti_3C_2 (space group $C222_1$) $\leftrightarrow$ $TiC_{0.62}$ phase transformations are equal to 1000, 980 and 970 K, and 1.5 ± 0.2, 1.5 ± 0.2 and 1.8 ± 0.2 kJ mol^{-1} respectively.

Experimental data on the temperature dependence of the heat capacity of ordered titanium carbides with different carbon concentrations in the temperature interval from 300 to 1000 K were approximated in [132, 226] by the following equations:

$TiC_{0.54}$: $C_p(T) = 46.5538 + 1.1566 \times 10^{-3}T + 7.6604 \times 10^{-7}T^2 - 4631.8871T^{-1}$.
$TiC_{0.58}$: $C_p(T) = 31.1767 + 2.7000 \times 10^{-3}T - 1.3772 \times 10^{-5}T^2 - 1775.7430T^{-1}$.
$TiC_{0.62}$: $C_p(T) = 60.8874 - 2.7600 \times 10^{-3}T + 1.9196 \times 10^{-5}T^2 - 6793.1821T^{-1}$.

Measured heat capacity is given in J mol^{-1} K^{-1}.

The disordered state of nonstoichiometric carbides is easily preserved as metastable at $T < T_{trans}$. From Fig. 11.54 it is seen that near T_{trans} the heat capacity of quenched disordered carbides TiC_y is larger than the heat capacity of the same carbides in the equilibrium ordered state. This behavior of C_p of strongly nonstoichiometric carbides MC_y in the region of the order–disorder transition (a higher heat capacity of disordered carbide and a jump of the heat capacity at the transition temperature) is in full agreement with the theoretical inferences in Sect. 11.7.3 concerning the change of $\Delta C_p(y,\eta,T)$ with temperature and was predicted earlier [2, 208, 209].

In general, results [132] of the study into the crystal structure and the heat capacity of titanium carbide TiC_y ($0.54 \le y \le 0.62$) showed that ordering of TiC_y, which is followed by formation of cubic (space group $Fd\bar{3}m$) and trigonal (space group $R\bar{3}m$) superstructures of the Ti_2C type and the rhombic (space group $C222_1$) superstructure Ti_3C_2, takes place at a temperature from 970 to 1000 K as a transformation close to phase transitions of the first kind. Results [132] agree with the theoretical phase diagram of the Ti–C system [44, 165] (see Fig. 10.4) and the experimental phase diagram of the Ti–C system (see Fig. 10.6), which was proposed in [130] and takes into account ordering of nonstoichiometric titanium carbide TiC_y.

The heat capacity of nonstoichiometric vanadium carbides $VC_{0.79}$, $VC_{0.83}$ and $VC_{0.87}$, which were prepared by the method of hot pressing, was measured within the temperature interval of 300 to 1600 K [10, 20, 21, 139]. The heat capacity was measured by the same method as the one used to measure C_p of nonstoichiometric titanium carbide. To produce vanadium carbides in the ordered state, the synthesized samples were annealed while the temperature was reduced smoothly from 1200 to 300 K. The ordered phase V_6C_5 was detected in annealed carbide $VC_{0.79}$. Annealed carbide $VC_{0.83}$ contained the ordered phase V_6C_5 and 15–20% ordered phase V_8C_7. Annealed carbide $VC_{0.87}$ had the ordered cubic (space group $P4_332$) phase V_8C_7 only.

The heat capacity of annealed VC_y samples was measured while heating the samples from 300 to 1500–1600 K and then while cooling the samples from the maximum heating temperature to 300 K.

Measurements of the heat capacity C_p of annealed carbide $VC_{0.87}$ showed that C_p increased smoothly as the temperature was elevated from 300 to 1350 K. A sharp jump (break) of the heat capacity (Fig. 11.57) was observed at a temperature of ~1380 K. This jump is related to the equilibrium $V_8C_7 \rightarrow VC_y$ order–disorder transformation and is characteristic of phase transitions of the first kind.

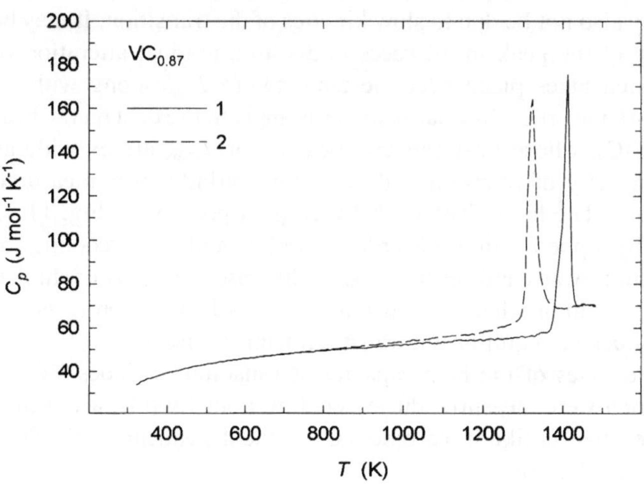

Fig. 11.57. Variation of the heat capacity C_p of annealed vanadium carbide $VC_{0.87}$ during heating (1) and cooling (2) [10, 139]

An analogous picture is observed during cooling from 1500 K. The only distinction is that the heat capacity peak, which is connected with the equilibrium disorder–order phase transition, is shifted to the region of lower temperatures (Fig. 11.57). The temperature interval between the heat capacity peaks of the forward and reverse runs is the region of metastability. The presence of this region also points to the first kind of reversible disorder–order phase transition, which is connected with the appearance of the ordered phase V_8C_7.

It is worth noting that near the transition temperature T_{trans} the heat capacity of disordered carbide is considerably higher than the heat capacity of ordered carbide. A similar effect is observed for the heat capacity of nonstoichiometric titanium carbide (see Fig. 11.54). Let it be emphasized that the temperature dependencies of the heat capacity of carbide $VC_{0.87}$ (during heating and cooling) exhibit only one peak. This is an indication that only one ordered phase V_8C_7 is formed in carbide $VC_{0.87}$ in this temperature interval. This observation is in agreement with the results of X-ray and microscopic studies, according to which only the cubic ordered phase V_8C_7 is present in annealed carbide $VC_{0.87}$.

The observed peak of the heat capacity has a relatively symmetric shape typical of phase transitions of the first kind. The first kind of the $VC_{0.87} \leftrightarrow V_8C_7$ phase transition is also confirmed by direct measurements of the phase transition heat ΔH_{trans} equal to 2.06 kJ mol^{-1} [227, 228] or 1.54 ± 0.62 kJ mol^{-1} [138]. Both peaks are very wide and have near the same width of about 20 K during heating and cooling (see Fig. 11.57). The same width of peaks means that the heating rate of 20 K min^{-1} has not affected both order–disorder and disorder–order transformations. Kinetic has made no limitation on these processes during measurements and observed specific features of the shape of the heat capacity peak should reflect equilibrium temperature dependence of long-range and short-range order parameters. The large width of the peak measured in [10, 20, 21, 139] at heating rate of 20 K min^{-1} may also not be due to slow kinetics of the transition. It may be thought that smearing of the peak in all cases is due to a gradual alteration of short-range order, which takes place near the temperature T_{trans} along with an abrupt change in long-range order. Similar comments apply to the data on the heat capacity of carbides TiC_y, whose heat capacity peaks near T_{trans} are as wide as 30–40 K. The heat capacity of titanium and vanadium carbides was measured under similar conditions. The large distance between peak position in Fig. 11.57 could be explained by upperheating of ordered phase and undercooling of disordered phase during forward and revers runs. In this case the purity of studied specimens should be very high what prevent nucleation and formation of either ordered or disordered nuclei at equilibrium transition temperature.

Temperature dependencies of the heat capacity of vanadium carbides $VC_{0.83}$, $VC_{0.79}$ and $VC_{0.87}$, which were measured during cooling from 1500 K, are shown in Fig. 11.58. It is seen that unlike in carbide $VC_{0.87}$, the dependencies $C_p(T)$ of carbides $VC_{0.79}$ and $VC_{0.83}$ have two peaks.

Let us discuss first the temperature dependence of the heat capacity of carbide $VC_{0.83}$.

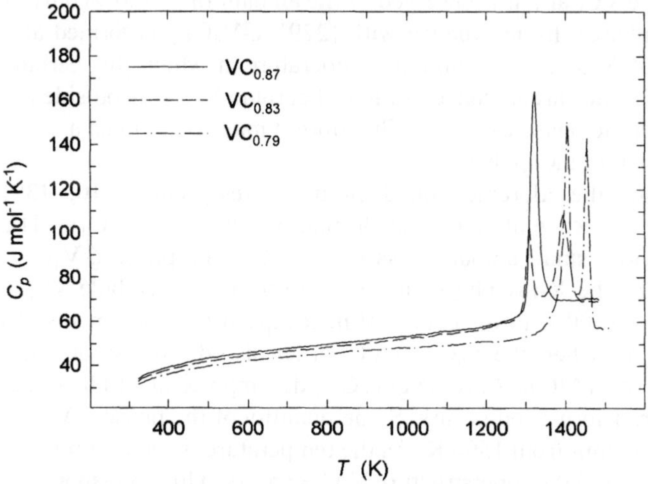

Fig. 11.58. Comparison of the effects of phase transitions observed in temperature dependencies of the heat capacity $C_p(T)$ of carbides $VC_{0.87}$, $VC_{0.83}$ and $VC_{0.79}$ during cooling from 1500 K [10, 139]

When disordered carbide $VC_{0.83}$ is cooled from 1500 K, the first jump of the heat capacity is observed in the interval from 1440 to 1400 K. As the temperature is reduced further, the second peak appears in the dependence $C_p(T)$ in the temperature interval from 1330 to 1300 K (Fig. 11.58). The carbon concentration of carbide $VC_{0.83}$ corresponds to the stoichiometric composition of the ordered phase V_6C_5. Therefore it may be assumed that the first jump of the heat capacity at 1400–1440 K is associated with the $VC_{0.83} \rightarrow V_6C_5$ transition. The next transformation, which occurs at 1330 K, may be the $V_6C_5 \rightarrow V_8C_7$ order–order transition. This conclusion is confirmed by the fact that the X-ray pattern of annealed carbide $VC_{0.83}$ contains superstructure lines due to both the phase V_6C_5 and the cubic ordered phase V_8C_7.

When annealed carbide $VC_{0.79}$ is heated, the temperature dependence of the heat capacity also exhibits two peaks at ~1440 K and ~1475 K. Both peaks are retained in the dependence $C_p(T)$ during cooling from 1500 K, but the intensity of the low-temperature peak is enhanced insignificantly (Fig. 11.58). The location of the low-temperature peak in the dependence $C_p(T)$ of carbide $VC_{0.79}$ nearly coincides with the location of the high-temperature peak in the dependence $C_p(T)$ of carbide $VC_{0.83}$. Therefore it may be assumed that the former peak is related to the reversible $V_6C_5 \leftrightarrow VC_y$ order–disorder transition. This assumption correlates with structural data: the X-ray pattern of annealed carbide $VC_{0.79}$ contains superstructure reflections of the ordered phase V_6C_5 only.

The high-temperature peak in the curve $C_p(T)$ of carbide $VC_{0.79}$ is probably due to the rhombic phase ζ-V_4C_{3-x} of vanadium carbide. Carbide ζ-V_4C_{3-x} is not an ordered phase of carbide VC_y and it is observed in the amount of 15–20 % only in the presence of other phases. In accordance with [229], ζ-V_4C_{3-x} is formed at a temperature below 2150 K and the minimum temperature at which this carbide was detected is 1540 K. The carbon concentration of carbide $VC_{0.79}$ is outside the homogeneity interval of the phase ζ-V_4C_{3-x}. Therefore it may contain just a small (less than 3–5 %) amount of the ζ-phase.

Taking into account the aforementioned findings, researchers [10, 139] proposed the following sequence of phase transformations in carbide $VC_{0.79}$. The ordered phase V_6C_5 and a small amount (less than 5%) of the phase ζ-V_4C_{3-x}, which is not detected by the X-ray phase analysis, are formed in carbide $VC_{0.79}$ during slow heating to ~1400 K in the course of heat capacity measurements. As carbide $VC_{0.79}$ is heated further, the $V_6C_5 \rightarrow VC_y$ order-disorder transition takes place at a temperature of ~1440 K. Carbide ζ-V_4C_{3-x} decomposes at ~1480 K, i.e. the ζ-$V_4C_{3-x} \rightarrow VC_y$ transformation occurs. Some quantity of the phase ζ-V_4C_{3-x} is formed first during cooling from 1500 K. As the temperature is reduced further, carbide $VC_{0.79}$ is ordered and the superstructure V_6C_5 appears. This corresponds to the low-temperature peak of the heat capacity (Fig. 11.58).

In general, calorimetric investigations [10, 20, 21, 139] showed that ordered phases V_6C_5 and V_8C_7 are formed in nonstoichiometric vanadium carbide as a phase transition of the first kind.

Discontinuities in the temperature dependencies of the heat capacity $C_p(T)$ of nonstoichiometric carbide $TiC_{0.82}$, nitride $TiN_{0.82}$ and titanium carbonitrides $Ti(C_xN_{1-x})_{0.82}$ ($x = 0.4$ and 0.6) at a temperature from 940 to 1070 K were observed in [230]. The heat capacity was measured using the method of differential scanning calorimetry within the temperature interval of 300–1300 K both during heating and cooling. The cooling and heating rates were 20 K min^{-1}. The temperature dependencies of the heat capacity exhibited an abrupt increase in C_p with elevating temperature or an abrupt decrease in C_p with reducing temperature near the discontinuity point. No peak of the heat capacity was observed. The absence of the C_p peak is most probably connected with the fact that the heat capacity was measured every 50 K, i.e. with a very large temperature step.

The discontinuity temperature of the heat capacity of titanium carbide $TiC_{0.82}$ is 1065 K during heating and 1034 K during cooling. These temperatures are nearly 70 K higher than the transition temperature T_{trans} determined for carbides $TiC_{0.54}$, $TiC_{0.58}$ and $TiC_{0.62}$ [132, 226]. The discontinuity temperature of C_p of titanium nitride $TiN_{0.82}$ is 973 K during heating and 939 K during cooling. The heat capacity of carbonitride $Ti(C_{0.4}N_{0.6})_{0.82}$ exhibits discontinuities at 1051 K during heating and 1000 K during cooling. A discontinuity in the heat capacity of carbonitride $Ti(C_{0.6}N_{0.4})_{0.82}$ occurs at 1075 K during heating. The temperature of the heat capacity discontinuity for the last carbonitride is not specified in [230].

The standard X-ray examination of $Ti(C_xN_{1-x})_{0.82}$ samples, which were quenched or cooled from a temperature above or below the discontinuity point of C_p, did not reveal any differences in diffraction patterns. If the observed discontinuities of the heat capacity are connected with ordering, then, as researchers [230] believe, the structure of the ordered phase differs from the structure of the ordered cubic carbonitride phase described in [231] and from the structure of the ordered phase Ti_2N (δ-$TiN_{0.50}$) [232–234]. Let it be noted that results [231] are extremely uncertain and doubtful.

In accordance with [44, 165] and the theoretical calculations described in Chap. 10, ordered phases Ti_6C_5 and Ti_6N_5 are formed in titanium carbide TiC_y and nitride TiN_y. The relative concentration of interstitial atoms in ideal ordered phases of the M_6X_5 type is $y = 0.833$. Considering this, it may be assumed that the effects found [230] for the heat capacity of carbide $TiC_{0.82}$, nitride $TiN_{0.82}$ and carbonitrides $Ti(C_xN_{1-x})_{0.82}$ are due to ordering of interstitial nonmetal atoms and vacancies in these compounds by the M_6X_5 pattern. Given this ordering pattern, carbon and nitrogen atoms are located at sites of the common lattice and vacancies arrange themselves at sites of the vacancy sublattice in titanium carbonitrides $Ti(C_xN_{1-x})_{0.82}$. It is worth noting that when ordering of titanium carbonitrides $Ti(C_xN_{1-x})_{0.63}$ [235, 236] is accompanied by formation of a cubic (space group $Fd\bar{3}m$) or trigonal (space group $R\bar{3}m$) structure, carbon and nitrogen atoms are also assumed to locate at sites of the common sublattice of interstitial atoms.

Thus, the effects detected [230] in the dependencies $C_p(T)$ of carbide $TiC_{0.82}$, nitride $TiN_{0.82}$ and carbonitrides $Ti(C_xN_{1-x})_{0.82}$ at a temperature of about 1000 K attest indirectly to the formation of ordered phases Ti_6C_5 , Ti_6N_5 and $Ti_6(C_xN_{1-x})_5$.

11.8 Microhardness

Nonstoichiometric interstitial compounds, especially carbides and nitrides, are inferior in hardness to only diamond. Thanks to their high hardness, they have been widely used in production of cutting tools and wear-resistant coatings. However, not all factors that influence hardness of these compounds have been analyzed in sufficient detail so far.

Measurements of hardness represent the most popular and available method for testing mechanical properties.

Hardness characterizes resistance of materials to plastic deformation during indentation and is determined mainly by the yield stress σ_y. On the one hand, hardness reflects the binding energy and specific features of the crystal symmetry. On the other hand, it depends directly on the microstructure (grain size, presence of fine precipitates, concentration of dislocations, etc.). The microstructure depends, in turn, on particular conditions of thermal treatment or on the deformation effect. Thus, it is a difficult task to pinpoint factors responsible for hardness alteration. Nonstoichiometry and ordering, which are inherent in the strongly nonstoichiometric compounds under discussion, render this task still more complex.

The dependence of microhardness on the composition of carbides MC_y has received the most study so far.

The variation of microhardness H_V in the homogeneity intervals of carbides TiC_y, ZrC_y, HfC_y, VC_y, NbC_y and TaC_y with the *B*1 structure is shown in Fig. 11.59.

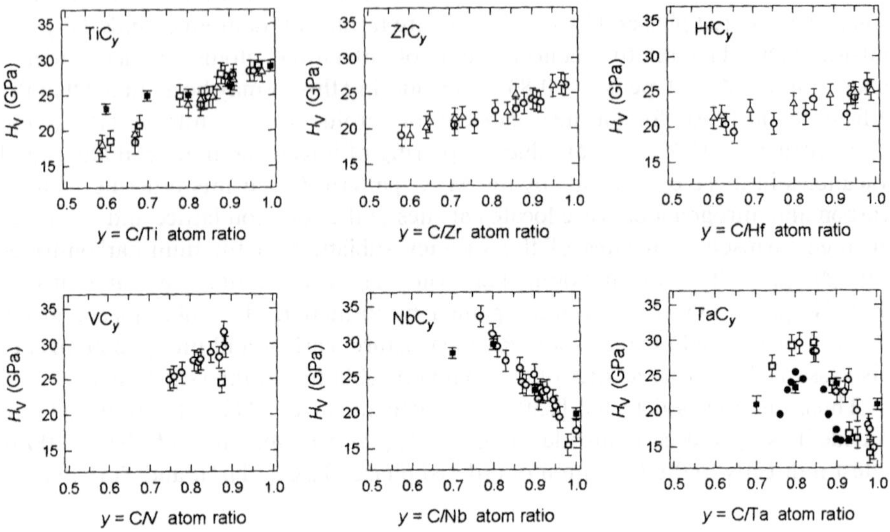

Fig. 11.59. Variation of the microhardness H_V in homogeneity intervals of disordered cubic carbides TiC_y [237, 238], ZrC_y and HfC_y [237], VC_y [122], NbC_y [122, 238], TaC_y [22, 122, 238]: (○), (Δ) and (□) are the samples produced [122, 237] by gaseous carburization of metallic fibers in methane CH_4, arc melting, and hot pressing respectively; (■) samples produced [238] by hot pressing; (•) TaC_y samples produced by gaseous carburization of tantalum fibers [22]

Figure 11.59 presents the most reliable literature data [122, 237] on H_V of these carbides, results [22] for carbide TaC_y, and data [238] for carbides TiC_y, NbC_y and TaC_y. Ramqvist [122, 237] made all measurements using monophase samples with minimum porosity. The samples were prepared by three methods: gas carburization of metallic fibers of ~0.25 mm diameter in methane CH_4; hot pressing of powders of presynthesized carbides; and arc melting of pellets compacted from powder mixtures of a metal and carbon.

The microhardness of carbides of group IV transition metals is the highest for carbides $M^{(IV)}C_{1.0}$ having a stoichiometric composition. The microhardness diminishes with decreasing concentration of carbon *y*. As the carbon concentration decreases, the microhardness of carbides NbC_y and TaC_y first grows, attains a maximum value of ~33 GPa for $NbC_{0.77\text{-}0.79}$ and ~29 GPa for $TaC_{0.81\text{-}0.84}$, and finally decreases to 26–28 GPa at the lower boundary of the homogeneity interval of the cubic phase (Fig. 11.59). Vanadium carbide VC_y has a maximum microhardness equal to 29–31 GPa at the upper boundary of the homogeneity interval corresponding to carbide $VC_{0.875}$. Microhardness diminishes with decreasing concentration of carbon in VC_y.

The microhardness of titanium carbide TiC_y has been examined most comprehensively among all carbides. This carbide has one of the widest homogeneity intervals from $TiC_{0.48}$ to $TiC_{1.00}$ of all known compounds. It may be in the disordered or ordered state depending on the composition, synthesis conditions, and heat treatment. Mechanical properties of titanium carbide TiC_y change significantly in the homogeneity interval depending on the carbon concentration. However, experimental data reported in the literature are often ambiguous and contradictory. This statement especially applies to microhardness, which can vary by a factor of 1.5–2 as measured by different researchers for carbides with one and the same carbon concentration. There is a wealth of literature data concerning the dependence of microhardness H_V on the composition of titanium carbide.

Researchers [239] measured the microhardness H_V of TiC_y samples prepared by solid-phase sintering of titanium hydride and carbon with the addition of 3 mass.% metallic nickel. In accordance with [239], TiC_y samples with $y > 0.32$ contained one phase, although a variety of experimental data suggest that two phases are present in the region of the phase diagram of the Ti–C system at $y = C/Ti < 0.48$. The homogeneity interval of TiC_y widens with elevating temperature. However, even at T = 1000 K carbide $TiC_{0.36}$ corresponds to the lower boundary of the homogeneity interval of the cubic phase under equilibrium conditions [44]. Considering this, one may think that microhardness, which was measured [239] on TiC_y samples with $y < 0.50$, refers to the two-phase region. The size of grains which were used for microhardness measurements was not specified in [239]. Judging by the synthesis method (sintering with the liquid phase of Ni), the grain size d_g was not less than 60–80 μm.

In accordance with [239] and later measurements [25, 237, 238, 240–246], the microhardness of titanium carbide TiC_y increases with growing concentration of carbon and is as high as ~28.5 ± 1.0 GPa for stoichiometric carbide $TiC_{1.00}$ (Fig. 11.60). Results of a detailed study into the microhardness of TiC_y [237] are remarkable. Three batches of thoroughly certified single-phase samples of TiC_y with $0.575 \le y \le 0.980$ were prepared in that study by the methods of gas carburization of metallic titanium fibers in methane CH_4, hot pressing of carbide powders, and arc melting of powder mixtures of titanium and carbon. The microhardness H_V was measured at a load of 0.05 kg. Concentration dependencies of microhardness $H_V(y)$ of TiC_y samples prepared by different methods are nearly coincident (Fig. 11.60). It was shown [237] that microhardness was enhanced by 3 and 5.5 GPa when the concentration of oxygen and nitrogen impurities in TiC_y increased from 0.02 to 0.3 and 1.5 mass.% respectively. The growth of H_V with increasing concentration of the oxygen impurity in titanium carbide was noted earlier in [240].

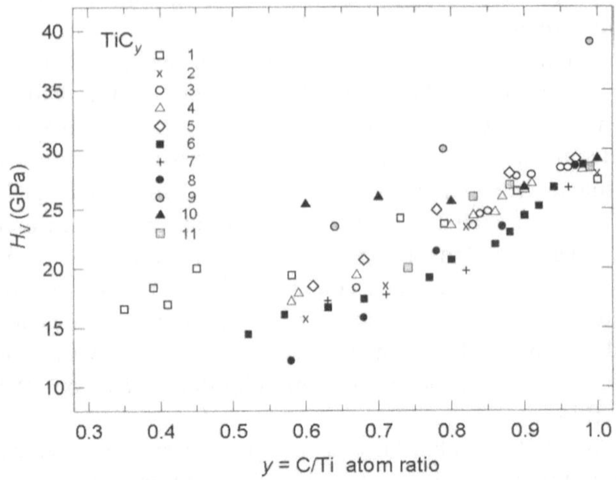

Fig. 11.60. Dependence of the microhardness H_V on the composition of titanium carbide TiC_y according to [25, 237, 239–245]: (1) [239]; (2) [240]; (3), (4) and (5) microhardness measured [237] on TiC_y samples produced by gas carburization of titanium in methane CH_4, arc melting, and hot pressing respectively; (6) [241]; (7) [242]; (8) [243]; (9) [244]; (10) [25]; (11) [245]

Microhardness of TiC_y ($0.48 \le y \le 0.98$) was studied [241, 246] on samples prepared by solid-phase sintering of titanium and carbon, arc and zone melting, and deposition from the gaseous phase during the interaction between $TiCl_4$ and $C_6H_5CH_3$ in hydrogen. The load was 0.06 to 0.35 kg. The synthesized samples were not annealed. In accordance with [241, 246], microhardness of carbide TiC_y exhibits a very weak dependence on the load and is almost independent of the production method. The last statement agrees with the conclusions drawn in [237].

Microhardness of TiC_y was the largest for single crystals of $TiC_{0.64}$, $TiC_{0.79}$ and $TiC_{0.99}$ [244]. Microhardness was measured [244] in the $(100)_{B1}$ plane of single crystals of titanium carbide at different angles α between the indenter diagonal and the direction <100> in the measurement plane. The load was 0.5 kg. The polar anisotropy of hardness was observed for all three samples: minima of H_V corresponded to the angles $\alpha = 0$, $\pi/4$ and $\pi/2$, while maxima of H_V were observed at the angles α equal to $\sim\pi/8$ and $\sim 3\pi/8$. The anisotropy decreased with increasing nonstoichiometry: the difference between the maximum and minimum values of H_V was 6.5 GPa for $TiC_{0.99}$, 3.3 GPa for $TiC_{0.79}$, and just 1.2 GPa for $TiC_{0.64}$.

Measurements of microhardness on the $(100)_{B1}$ face of monocrystalline carbide $TiC_{0.94}$ (the load of 0.1 and 0.2 kg) showed that H_V increased from 24.2 ± 1.6 to 25.8 ± 1.0 GPa after this carbide was annealed at a temperature of 2273 K for 2 h [247]. The microhardness of $TiC_{0.6}$, $TiC_{0.7}$ and $TiC_{0.8}$ samples was found [25] to increase by 1.2–1.7 GPa after annealing at 1000 and 900 K for 15 h at each temperature. The samples were synthesized by solid-phase sintering. The post-annealing increase in microhardness was explained [25] by possible ordering of nonstoichiometric titanium carbides and refinement of grains. Researchers [248] examined samples of carbide TiC_y produced by arc melting and found, in contradiction to [25, 247], that annealing of the TiC_y samples was followed by a decrease in H_V by 1.5–5.5 GPa. For the purpose of corroborating their results, researchers [248] refer to the data on microhardness obtained in [246] assuming erroneously that H_V was measured [246] for annealed TiC_y samples.

The size of grains of the carbide phase, which were used for H_V measurements, is not specified in [25, 237–241, 246, 248]. However, microhardness values are comparable only if the size of the indentation is smaller than the grain size. Otherwise grain boundaries affect the microhardness value and H_V depends on the grain size. Grain sizes and measured values of microhardness of carbide TiC_y are specified in [242, 243, 245]. Researchers [242] prepared samples by hot pressing of a mixture of carbide and titanium hydride at 3000–3200 K and subsequent homogenizing annealing at 3100–3200 K for 2 h. The high synthesis temperature was conducive to the growth of grains, which were as large as 750 μm in $TiC_{0.96}$. The grain size decreased with increasing concentration of carbon in TiC_y and was 110 μm in $TiC_{0.96}$. Samples of TiC_y were produced [243] by sintering of ready powders of titanium carbide at a temperature from 1600 K for $TiC_{0.58}$ to 2200 K for $TiC_{0.97}$. The mean size of grains was 14–20 μm. In [245] all samples of titanium carbide were prepared by solid-phase vacuum sintering under similar conditions. The grain size decreased quickly from ~65 to 5 μm in the series from $TiC_{0.74}$ to $TiC_{0.99}$. From the data [245] it follows that H_V of $TiC_{0.88}$ and $TiC_{0.99}$ samples could not be measured on one grain, because at a load of 0.1 kg the indentation diagonal was larger than the grain size.

The above analysis of the literature data shows that the different behavior of microhardness of titanium carbide is due to the following factors: the presence of impurities; the different structural states depending on a particular synthesis method and heat treatment conditions; the different microstructures depending on the grain size and grain boundaries; and the different loads used during measurements. Let it be noted that most experimental studies overlooked the fact that a carbide could have different (ordered or disordered) structural states.

The effect of nonstoichiometry on microhardness of disordered titanium carbide TiC_y was studied taking into account the size of the carbide phase grains in [249]. Samples of titanium carbide TiC_y ($0.50 \leq y \leq 1.00$) were synthesized using two methods: (1) solid-phase sintering of $TiC_{0.94}$ powders, gas carbon black and metallic titanium at 2000 K for 6 h under a vacuum of 0.001 Pa with intermediate grinding of the products after 3-hour sintering; (2) hot pressing of powder mixtures of $TiC_{0.98}$ and metallic titanium in an atmosphere of high-purity argon (the synthesis conditions are summarized in Table 11.26). The synthesized samples were placed in evacuated quartz ampoules and were annealed additionally at 1200 K for 4 h. Then the ampoules with the samples were quenched in water to produce the ordered state. The quenching cooling rate was ~250 K min^{-1}.

Table 11.26. Composition, synthesis conditions (temperature T, duration t and pressure of hot pressing p), porosity P, grain size d_g, and basic lattice constant a_{B1} of the specimens of disordered titanium carbide TiC_y [249]

Carbide	Synthesis conditions			Composition (mas. %)				P	d_g	a_{B1}
	T (K)	t (h)	p (MPa)	Ti	C_{bond}	O	N	(%)	(μm)	(nm)
$TiC_{0.50}$	[2] 2000	6.0	–	88.42	11.18	0.24	0.15	< 1	34	0.43017
$TiC_{0.52}$	[3] 1773	0.5	20	88.29	11.51	0.05	0.06	< 1	29	0.43057
$TiC_{0.54}$	[3] 1773	0.5	25	87.87	11.96	0.08	0.06	< 1	26	0.43068
$TiC_{0.55}$	[2] 2000	6.0	–	87.32	12.08	0.20	0.07	4.9	24	0.43072
$TiC_{0.58}$	[3] 1773	0.5	25	87.13	12.71	0.11	0.07	< 1	17	0.43105
$TiC_{0.59}$	[2] 2000	6.0	–	86.62	12.90	0.21	0.08	10.9	16	0.43114
$TiC_{0.62}$	[3] 1923	0.5	23	86.31	13.43	0.08	0.07	< 1	25	0.43152
$TiC_{0.63}$	[2] 2000	6.0	–	86.05	13.58	0.30	0.06	14.4	15	0.43160
$TiC_{0.68}$	[2] 2000	6.0	–	84.45	14.37	0.76	0.07	16.6	15	0.43174
$TiC_{0.68}$	[3] 2173	0.5	30	85.26	14.63	0.10	0.07	< 1	31	0.43198
$TiC_{0.69}$	[2] 2000	6.0	–	84.38	14.70	0.30	0.08	23.9	10	0.43183
$TiC_{0.81}$	[2] 2000	6.0	–	82.14	16.75	0.18	0.08	21.1	5	0.43248
$TiC_{0.83}$	[2] 2000	6.0	–	82.02	17.03	0.38	0.07	24.0	4	0.43260
$TiC_{0.83}$	[3] 2173	0.5	30	82.45	17.24	0.14	0.07	< 1	28	0.43254
$TiC_{0.85}$	[3] 2173	0.5	30	82.18	17.51	0.12	0.07	< 1	20	0.43260
$TiC_{0.90}$	[2] 2000	6.0	–	80.79	18.30	0.45	0.08	31.4	2	0.43256
[1] $TiC_{0.94}$	[2] 2000	6.0	–	79.68	18.84	0.31	0.07	32.3	< 2	0.43259
[1] $TiC_{0.97}$	[2] 2000	6.0	–	79.47	19.35	0.22	0.08	31.6	< 1	0.43265
$TiC_{0.98}$	[3] 2473	0.5	35	80.02	19.69	0.08	0.07	< 1	42	0.43258

[1] Concentration of free carbon in these specimens did not exceed 0.01 mass. %. Other specimens did not contain free carbon.
[2] Vacuum solid-phase sintering.
[3] Hot pressing.

The composition and the content of gas impurities in the samples are given in Table 11.26. The concentration of metal impurities did not exceed 0.02 mass.%. All the quenched samples were homogeneous and contained only a disordered phase TiC_y with the $B1$ structure. The load used during the microhardness measurements was 0.05, 0.1, 0.3 and 1.0 kg. To determine grain boundaries, the surface of polished samples was pickled in a ($2HNO_3 + 6HF + 3H_2O$) mixture. Metallographic examination confirmed that all samples contained one phase. The grain size was determined by the secant method.

Results [249] show that given similar conditions of thermal treatment and synthesis, the grain size d_g in carbide TiC_y increases as the composition deviates from stoichiometry (see Table 11.26). Grains grow (collective recrystallization) through chemical diffusion, i.e. a diffusion process, when the concentration gradient is absent but the chemical potential gradient is not zero. The difference of chemical potentials at grains with different dimensions is due to the contribution of the surface energy to the total energy of a grain. The relative value of this contribution is the larger, the smaller the grain size. Consequently, all other conditions being equal, fine grains possess an excess energy with respect to the energy of coarse grains. From this it follows that the thermodynamic cause of the spontaneous growth of grains in a homogeneous polycrystalline material is a decrease in the total energy of the system during the growth process.

The self-diffusion parameters in carbide TiC_y are independent of the carbon concentration [250], whereas the self-diffusion coefficient of carbon increases as the composition of titanium carbide deviates from stoichiometry [251]. Diffusion-controlled processes in carbides are determined by the effective diffusion coefficient rather than by partial self-diffusion coefficients of the components. Considering the data [250, 251] obtained for carbide TiC_y, the effective diffusion coefficient should decrease with increasing carbon concentration. Therefore the growth of grains by diffusion should be smallest in stoichiometric carbide and largest in carbide TiC_y near the lower boundary of the homogeneity interval. One should remember however that recrystallization is a more complex process than diffusion and is not reduced to the latter. The activation energy Q characterizes collective recrystallization. In accordance with [252], the change of the grain size d_g during collective recrystallization of sintered porous carbides is described by the relation

$$d_g^3 - d_{g0}^3 = kt\exp(-Q/RT), \tag{11.66}$$

where d_{g0} is the initial size of a grain. From (11.66) it follows that if all other conditions are equal, the grain size is the smaller, the higher the activation energy. In the homogeneity interval of carbide TiC_y the activation energy Q changes between 98 and 145–188 kJ mol^{-1} in the series from $TiC_{0.58-0.62}$ to $TiC_{0.96-0.98}$ [253, 254], i.e. Q grows as the composition approaches stoichiometry. This means that recrystallization is more intensive in nonstoichiometric carbides. Therefore, the grain size in titanium carbides sintered under similar conditions is the larger, the lower the concentration of carbon. This conclusion is in good agreement with experimental findings (see Table 11.26). For example, the grain size in samples produced by solid-phase sintering decreases from 34 μm for $TiC_{0.50}$ to $d_g < 2$ μm for $TiC_{0.97}$. Measured values of microhardness of disordered carbide TiC_y ($0.5 \leq y \leq 1.0$) in the homogeneity interval of a cubic phase having the $B1$ structure are depicted in Fig. 11.61. As a general trend, microhardness H_V increases with growing carbon concentration of carbide TiC_y. This is caused by changes in the electronic structure of TiC_y. As the composition of carbides of group IV transition metals MC_y deviates from stoichiometry, the density of electron states at the Fermi level $N(E_F)$ rises, while the Fermi level shifts towards larger energies to the band of M–M interactions [170, 255, 256]. This means a decrease in the covalent component and an increase in the metallic component of a combined chemical bond in TiC_y. This process shows up as a decrease in hardness, i.e. an increase in plasticity with decreasing y. However, the general tendency of H_V variation with y is accompanied by some specific features.

The microhardness H_V of sintered and hot-pressed samples rises sharply near the lower boundary of the homogeneity interval of TiC_y in the interval $TiC_{0.50}$–$TiC_{0.55}$. The further increase in y is followed by a monotonic growth of H_V, which is interrupted by some valley or plateau in the interval $TiC_{0.7}$–$TiC_{0.9}$. This feature of H_V is more pronounced for sintered samples than for hot-pressed samples. An analogous feature of H_V in the same interval of TiC_y compositions was observed earlier [237, 239, 245] (see Fig. 11.60). This feature was less pronounced or had the form of a plateau when the composition of TiC_y samples changed in large steps $\Delta y = 0.1$ [239, 245]. A clear-cut minimum of H_V was observed by Ramqvist [237] who measured microhardness of ten TiC_y samples with steps $\Delta y \approx 0.01$ in the interval $TiC_{0.77}$–$TiC_{0.88}$ (see Fig. 11.59).

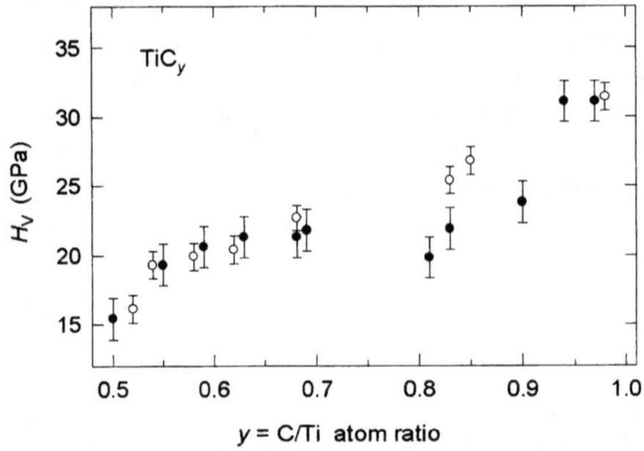

Fig. 11.61. Dependence of the microhardness H_V on the composition of disordered titanium carbide TiC_y samples produced by solid-state vacuum sintering (•) and hot pressing (○) [249]. The measurement load and time were 0.1 kg and 10 s respectively

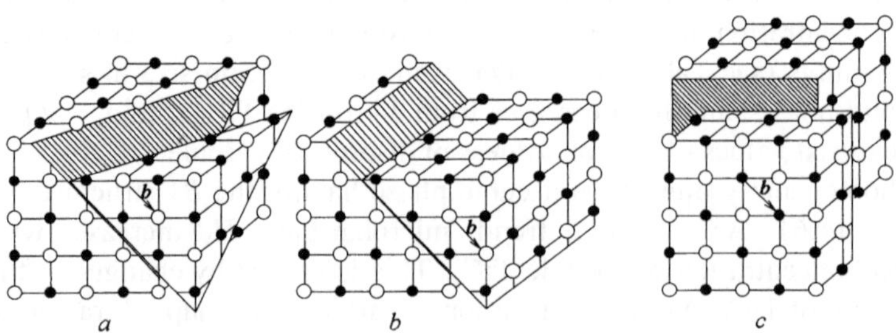

Fig. 11.62. Slip systems in disordered titanium carbide TiC_y with a basic $B1$ type structure: (a) slip system $\{111\}\langle110\rangle$ contributing most to plastic deformation at high ($T > 900$ K) temperatures; (b) slip system $\{110\}\langle110\rangle$ whose contribution to plastic deformation is almost independent of the TiC_y carbide composition; (c) slip system $\{100\}\langle110\rangle$ whose initiation requires the highest stress, $\boldsymbol{b} = \langle110\rangle$ is the vector showing the slip direction in planes with the largest reticular atomic density

Causes of the observed anomalous behavior of $H_V(y)$ in nonstoichiometric carbide TiC_y are discussed in [249].

Hardness characterizes the resistance of materials to plastic deformation during indentation and is connected with the resistance of materials to movement of dislocations. Microplasticity is localized near the mark left by the indenter during H_V measurements and shows up as slip lines. The analysis of slip lines and dislocation etch patterns of titanium carbide showed [257, 258] that at a temperature above ~900 K the main slip system is {111}<110>, i.e. plastic deformation in planes {111} in the direction <110> (Fig. 11.62*a*). The slip system {110}<110> was also detected [259–261] in TiC_y at a temperature below 870 K (Fig. 11.62*b*). It was noted [260] that the slip system {111}<110> is less pronounced in TiC_y at ~300 K, although this system dominates at $T > 900$ K. The study of microplasticity of $TiC_{0.64}$, $TiC_{0.}$and $TiC_{0.99}$ single crystals by the microhardness and electron microscopic methods showed [244] that at room temperature dislocations move in slip systems {110}<110> (Fig. 11.62*b*) and {100}<110> (Fig. 11.62*c*). At a high temperature deformation occurs mainly by the slip system {111}<110>.

The stresses necessary for slip by systems {111}<110> and {110}<110> are nearly equal in value. The least favorable is the slip system {100}<110>, which requires the highest stresses. Earlier studies assumed that the deviation from stoichiometry has no effect on the slip system in carbides. However, an experimental study [244] showed this assumption to be untrue. When the carbon concentration y of carbide TiC_y increases, plastic deformation is enhanced, and the slip system {111}<110> gains, while the slip system {100}<110> loses in significance [244]. The relative significance of the slip system {110}<110> for plastic deformation of TiC_y remains unchanged at different carbon content. Three TiC_y samples ($y = 0.64$, 0.79 and 0.99) were analyzed in [244] and, therefore, it was impossible to detect the anomalous plastic deformation in the interval $0.8 < y < 0.9$.

The anomalous decrease in microhardness, which was observed earlier and noted in [249], and the corresponding increase in plastic deformation of TiC_y at $0.8 < y < 0.9$ is probably due to the dominance of the slip system {111}<110> and a constant contribution of the slip system {110}<110> in this interval of y. Such dominance is possible in the presence of additional M–M interactions. Additional M–M interactions result from the appearance of a local peak of the density of electron states located below the Fermi energy in the electron energy spectrum of TiC_y at $0.8 \leq y < 0.9$. A local vacancy peak of the density of states was found experimentally in electron spectra of such nonstoichiometric compounds as TiN_y, ZrN_y, and NbC_y [51, 262–264]. The occurrence of this peak was confirmed by theoretical calculations [50, 197, 265, 266].

Self-consistent calculation of the band structure of nonstoichiometric titanium carbide $TiC_{0.75}$ (Ti_4C_3) and stoichiometric carbide $TiC_{1.0}$ [171, 172], which was performed by the method of augmented plane waves (APW), showed that the presence of structural vacancies leads to the formation of two, rather than one, local vacancy peaks of the density of states: 0.4 eV below and 0.8 eV above the

Fermi energy E_F (Fig. 11.63). Since the APW method has certain limitations with respect to the description of defective crystals, calculation [171] was made for a hypothetical ordered structure of $TiC_{0.75}$ (Ti_4C_3). However, this circumstance has no impact on the conclusion that deviation of the TiC_y composition from stoichiometry is followed by the appearance of local vacancy peaks in the electron spectrum and additional M–M interactions [171, 172, 267].

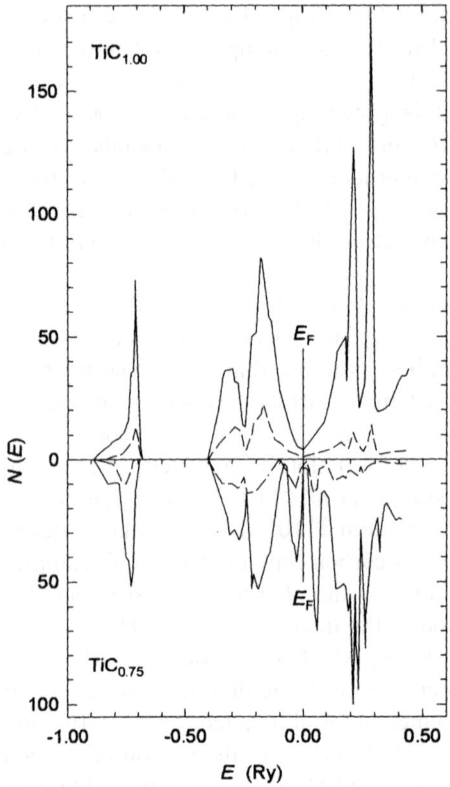

Fig. 11.63. Density of electron states $N(E)$ in stoichiometric titanium carbide $TiC_{1.0}$ and nonstoichiometric carbide $TiC_{0.75}$ (the density of states $N(E)$ is given as the number of states per energy value of one Rydberg and per cubic unit cell) [171]. The density curves are matched by the C2*s* band bottom. The Fermi level of carbide $TiC_{0.75}$ was taken as the zero energy. The solid line denotes the full density of electron states. The dotted line shows the density of states due to electrons of carbon atoms

Thus, results [249] in combination with literature data allow us to state the following. The observed minimum of microhardness of disordered titanium carbide TiC_y with $0.8 \le y < 0.9$ results directly from the fact that in the interval $TiC_{0.8}$–$TiC_{0.9}$ plastic deformation is realized predominantly through the slip system {111}<110>. The dominance of the slip system {111}<110> over other slip systems in this interval is explained by the relative increase in the contribution of M–M interactions, which is caused by the emergence of local vacancy peaks in the electron spectrum of TiC_y, as compared to covalent M–C interactions.

For a long time, data concerning the effect of ordering on microhardness of nonstoichiometric carbide TiC_y were unavailable. The different microhardness of quenched and annealed samples of TiC_y containing 33–43 at.% carbon was first detected in [248]. It was found that a monoclinic ordered phase Ti_2C is formed and microhardness decreased by 1.5–5.5 GPa when TiC_y samples ($0.50 \le y < 0.75$) are annealed by reducing the temperature from 1873 to 973 K for 100 h. The decrease in microhardness and cracking resistance of TiC_y upon transition from the disordered to ordered state is noted in [268]. In contradiction to [248, 268], Lipatnikov *et al.* [25] found that ordering of annealed carbides TiC_y ($y = 0.6$, 0.7 and 0.8) is accompanied by a small increase in microhardness. The change of the grain size during annealing of carbide TiC_y was not determined [25, 248, 268].

The effect of a phase transition from the disordered to ordered state on the microstructure and microhardness of titanium carbide TiC_y ($0.5 \le y \le 0.97$) is studied in [269]. TiC_y samples with different carbon content were synthesized by solid-phase sintering of $TiC_{0.94}$ powders, gas carbon black and titanium at 2000 K for 6 h under a vacuum of 0.001 Pa. The synthesized samples were annealed at 1200 K for 4 h and then were quenched to produce the disordered state.

The quenched samples contained only a disordered phase TiC_y with the *B*1 structure. Microhardness of TiC_y samples was measured under a load of 0.1 kg for 10 s. After measurements, disordered samples underwent long-term annealing to produce the ordered state. The samples in evacuated quartz ampoules were annealed for 278 h while the temperature was reduced from 1073 to 923 K and then for 167 h at a temperature of 873 K. The X-ray analysis of annealed $TiC_{0.50}$, $TiC_{0.55}$ and $TiC_{0.59}$ samples revealed superstructure reflections corresponding to cubic (space group $Fd\bar{3}m$) and trigonal (space group $R\bar{3}m$) ordered structures of Ti_2C type. Titanium reflections are present in X-ray patterns of carbide $TiC_{0.50}$. X-ray patterns of carbides $TiC_{0.63}$ and $TiC_{0.68}$ exhibited, in addition to lines of the ordered phase Ti_2C, reflections corresponding to the ordered phase Ti_3C_2 with rhombic (space group $C222_1$) symmetry. The microscopic examination of carbide $TiC_{0.50}$ confirmed formation of the ordered cubic phase Ti_2C and showed that domains of this phase appeared at carbide grain boundaries. The grain size in annealed carbide $TiC_{0.50}$ increase almost twice from ~34 to ~78 μm (Table 11.27).

The metallographic analysis of quenched and annealed titanium carbide $TiC_{0.55}$ also showed that the grain size increased after annealing. Some grains of the annealed sample have lamellar precipitates of another phase whose symmetry differs from the symmetry of the initial disordered phase. The domains of the ordered phase have a lamella- or prism-like shape, which is characteristic of crystals with trigonal syngony. As opposed to the disordered cubic phase, these domains have interference color, i.e. are anisotropic. In accordance with [269], the ordered phase in carbide $TiC_{0.55}$ is trigonal.

Table 11.27. Change of grain size d_g of nonstoichiometric titanium carbide TiC_y as a result of annealing [269]

Carbide	Grain size d_g (μm)	
	before annealing	after annealing
$TiC_{0.50}$	34	78
$TiC_{0.55}$	24	48
$TiC_{0.59}$	16	35
$TiC_{0.63}$	15	29
$TiC_{0.68}$	15	28
$TiC_{0.69}$	10	18
$TiC_{0.81}$	5	14
$TiC_{0.83}$	4	10
$TiC_{0.90}$	2	7
$TiC_{0.94}$	2	6
$TiC_{0.97}$	1	4

Generally, study of the microstructure showed that disorder–order phase transformations in nonstoichiometric titanium carbide TiC_y ($y < 0.7$) begin at boundaries and on the surface of grains of disordered carbide and extend to the bulk of crystallites with time.

The authors [269] hold to the opinion that the post-annealing changes in the microstructure of titanium carbide samples are due to the appearance of ordered phases: cubic (space group $Fd\bar{3}m$) phase Ti_2C in carbide $TiC_{0.90}$, trigonal (space group $R\bar{3}m$) phase Ti_2C in carbide $TiC_{0.55}$, and rhombic (space group $C222_1$) phase Ti_3C_2 in carbide $TiC_{0.63\text{-}0.68}$.

Measured values of microhardness H_V of quenched disordered and annealed ordered samples of TiC_y [269] are shown in Fig. 11.64. As the carbon concentration increases from $y = 0.5$ to $y \approx 0.7$, the microhardness of quenched and annealed samples of TiC_y rises. A local minimum of H_V is observed in the interval $0.80 < y < 0.85$ and then the microhardness grows again. In the interval $0.5 \le y \le 0.7$ concentration dependencies of microhardness $H_V(y)$ of disordered and ordered samples of TiC_y coincide within the measurement error. When $y > 0.8$, the dependence $H_V(y)$ of annealed TiC_y samples is located below the analogous dependence for quenched samples.

This behavior of microhardness can be explained as follows. The most significant increase in the grain size by 2.5–3.5 times after long-term annealing was observed in carbides TiC_y with a relatively high ($y > 0.8$) carbon concentration (see Table 11.27). The analysis of the crystal structure did not reveal the formation of any ordered phase in this interval of TiC_y compositions. Consequently, the lower microhardness of annealed carbides TiC_y ($y > 0.8$) is directly connected with the increase in the crystallite size. Indeed, a decrease in microhardness with increasing grain size is observed in a variety of materials: from metals to oxides, carbides and other compounds [270, 271].

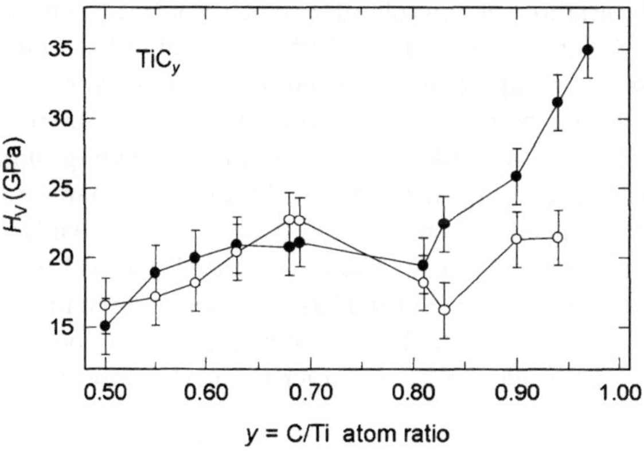

Fig. 11.64. Dependence of the microhardness H_V on the composition of disordered (•) and ordered (○) titanium carbide TiC_y [269]

In the interval $TiC_{0.50}$–$TiC_{0.70}$ the grain size also increased after annealing, but no marked impairment of microhardness of annealed $TiC_{0.50}$–$TiC_{0.70}$ samples was observed as compared to microhardness of quenched samples (Fig. 11.64). Diffraction and microscopic examinations showed that ordering took place in this interval of titanium carbide compositions. One may think that ordering and growth of grains mutually compensate their effect on microhardness. This means that the transition from the disordered to ordered state causes an increase in microhardness. This conclusion correlates with results [25, 272], according to which annealing and annealing-induced ordering lead to an insignificant increase in microhardness of nonstoichiometric titanium and niobium carbides.

Thus, low-temperature annealing is followed by the growth of grains in the whole homogeneity interval of titanium carbide and is accompanied by the formation of ordered phases in carbide TiC_y with $y < 0.7$. Microhardness decreases as a result of grain growth and increases thanks to ordering. These effects are mutually compensated in annealed $TiC_{0.50}$–$TiC_{0.70}$ carbides. Therefore microhardness values of quenched disordered and annealed ordered samples of titanium carbide TiC_y ($y < 0.7$) coincide within the measurement error. In the interval of TiC_y compositions, in which ordering does not take place ($y > 0.8$), annealing leads only to a decrease in microhardness because of the grain growth. Generally, the microhardness of carbide TiC_y decreases with growing grain size after annealing. Microhardness is enhanced due to the increase in the carbon concentration and the transition from the disordered to ordered state. Unfortunately, the size of domains of ordered phases and the change of this size depending on the annealing time were not specified in [269]. These data are very useful for the correct interpretation of experimental results concerning the change of microhardness of nonstoichiometric compounds after annealing and ordering.

The authors [27] reported variation of microhardness of disordered carbide $TiC_{0.59}$ after annealing at 970 K during 140 h (Fig. 11.65). Annealing led to the formation of the trigonal (space group $R\bar{3}m$) ordered phase Ti_2C and was accompanied by the growth of the phase domains from 1–6 to 50–100 nm. Microhardness of carbide $TiC_{0.59}$ was 20.4 ± 1.0 GPa before annealing and decreased to 16.8 ± 0.6 GPa after annealing. It is noteworthy that microhardness of carbide $TiC_{0.59}$ decreased after annealing from 20.4 ± 2.0 to 18.2 ± 2.0 GPa (see Fig. 11.64) and the grain size of the basic phase increased twice [269]. Researchers [27] did not specify the grain size of the basic phase, but domains of the ordered phase increased by a factor of 10–20 after annealing. Therefore the change in microhardness ΔH_V of carbide $TiC_{0.59}$ is larger than ΔH_V observed in [269].

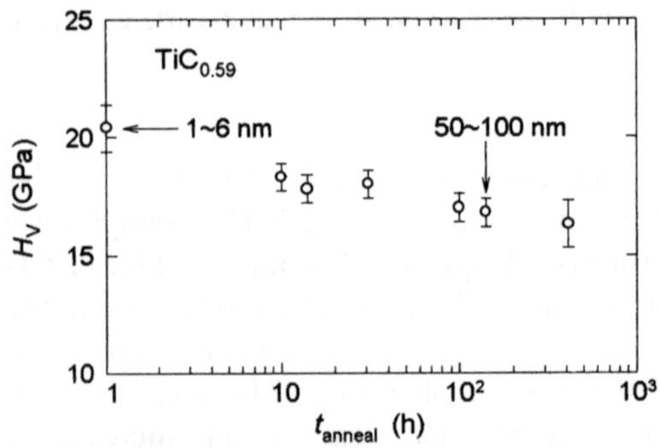

Fig. 11.65. Dependence of the microhardness H_V of titanium carbide $TiC_{0.59}$ on the annealing time t at a temperature of 970 K [27]. Arrows indicate the size of domains corresponding to the annealing time

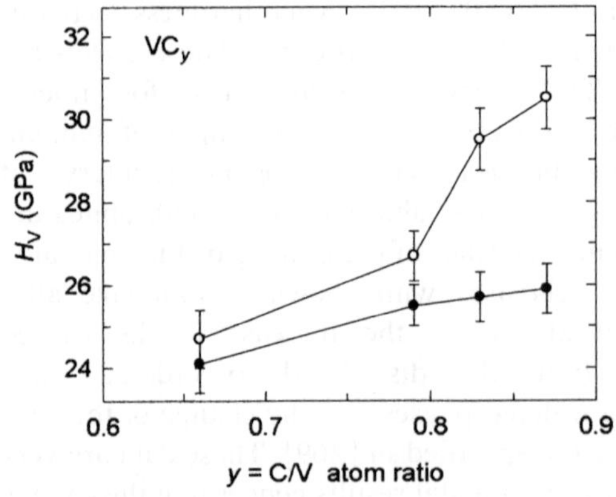

Fig. 11.66. Microhardness H_V of nonstoichiometric vanadium carbide VC_y in disordered (•) and ordered (○) states [21]

The effect of the domain size of the ordered phase V_6C_5 on the hardness of ordered vanadium carbide $VC_{0.84}$ was analyzed in [273]. Monocrystalline samples were prepared by the method of zone melting. The temperature of the VC_y–V_6C_5 disorder–order transition was 1520–1570 K. Ordered samples with different sizes of the ordered phase domains were prepared by quenching of single crystals from a temperature $T \geq T_{trans}$ to 870 K at a cooling rate of 225 K s^{-1} or from $T > T_{trans}$ to 1370 K at a cooling rate of 0.8, 9.5 and 62.5 K s^{-1}. Then the samples were annealed slowly while decreasing the temperature to 300 K at a rate of 4 K s^{-1}. Electron-microscopic examination showed that domains of the ordered phase V_6C_5 grow in the disordered parent lattice of the carbide. The size of domains in samples quenched at a rate of 225, 62.5, 9.5 and 0.8 K s^{-1} is 0.3, 2.0, 4.5 and 7.5 μm respectively. The microhardness of these samples is 25.0 ± 0.3, 22.9 ± 0.4, 22.1 ± 0.5 and 21.9 ± 0.5 GPa. In accordance with [273], the increase in the size d_g of domains of the ordered phase is accompanied by a decrease in microhardness H_V by the Hall–Petch law $H_V \approx H_0 + kd^{-1/2}$. Microhardness of a sample of fully ordered carbide with grains about 17 μm in size is 23.6 GPa, which is larger than the H_V value of disordered carbide. It was shown [273] that the microhardness of fully ordered carbide $VC_{0.84}$ is a maximum when domains of the ordered phase V_6C_5 are nearly 6 nm in size. In accordance with [273], both ordering and the decrease in the size of the ordered phase domains lead to an increase in microhardness. However, the domain size has a greater effect on microhardness.

The ordering-induced increase in microhardness of nonstoichiometric vanadium carbide VC_y was detected in [20, 21]. Samples of carbides $VC_{0.66}$, $VC_{0.79}$, $VC_{0.83}$ and $VC_{0.87}$ were synthesized by hot pressing. The synthesized samples were quenched to produce the disordered state. To obtain ordered carbide samples, the synthesized samples underwent long-term annealing at a temperature from 900 to 700 K. The diffraction analysis confirmed the formation of ordered phases after annealing.

Measurements [20, 21] showed that the microhardness of annealed ordered samples of VC_y was higher than the microhardness of quenched disordered samples. The largest increase in microhardness from 25.9 to 30.5 GPa was observed for carbide $VC_{0.87}$, in which the phase V_8C_7 (Fig. 11.66) was formed during ordering.

In 1999 researchers [45, 96] managed to produce a nanostructured powder of vanadium carbide $VC_{0.875}$. To this end, they used the abrupt increase (by 0.0004 nm) in the basic lattice constant of nonstoichiometric vanadium carbide $VC_{0.875}$ during ordering (see Sect. 11.1). The powder nanoparticles represent strongly bent plate-disks 400–600 nm in diameter and 15–20 nm thick. The bulk of the nanoparticles is comprised of ordered carbide V_8C_7 with a high degree of long-range order. A nanocrystalline powder of vanadium carbide was compacted cold under a pressure of 10 MPa and was then sintered in a vacuum of 0.001 Pa at a temperature of 400–2000 K with a step of 100 K. The holding time was 2 h at each temperature. The sintered samples were ground and Vickers microhardness was measured at a load of 200 and 500 g and a loading time of 10 s.

Microhardness did not exhibit any dependence on the load within the measurement error. The microhardness H_V was 60–80 GPa, which is 2–3 times larger than the microhardness of coarse-grained disordered vanadium carbide (29 GPa under a load of 0.1 and 0.2 kg) and approaches the microhardness of diamond. It is known that at 300 K the microhardness of nanomaterials is 2–7 times higher than H_V of common polycrystalline materials [270]. Researchers [45] think that the higher microhardness of the carbide $VC_{0.875}$ sample prepared from a nanopowder may be explained by the Hall–Petch law, i.e. may be due to the decrease in the grain size. Considering results [273], it may be assumed that ordering also contributes to the increase in microhardness of nanocrystalline carbide $VC_{0.875}$. Generally, the formation of an ordered nanostructure in vanadium carbide $VC_{0.875}$ causes the microhardness to increase by a factor of 2–3.

From the results discussed here it follows that the effect of the atomic ordering on the microhardness of nonstoichiometric compounds is very complicated. This complicated behavior comes from the change of the grain sizes of the basic phase and the domain sizes of the ordered phase which takes place simultaneously during high-temperature annealing.

References

1 A. I. Gusev, A. A. Rempel: *Structural Phase Transitions in Nonstoichiometric Compounds* (Nauka, Moscow 1988) 308 pp. (in Russian)

2 A. I. Gusev: *Physical Chemistry of Nonstoichiometric Refractory Compounds* (Nauka, Moscow 1991) 286 pp. (in Russian)

3 A. A. Rempel: *Effects of Ordering in Nonstoichiometric Interstitial Compounds* (Nauka, Ekaterinburg 1992) 232 pp. (in Russian)

4 A. A. Rempel: Atomic and vacancy ordering in nonstoichiometric carbides, Uspekhi Fiz. Nauk **166**, 33–62 (1996) (in Russian). (Engl. transl.: Physics – Uspekhi **39**, 31–56 (1996))

5 A. I. Gusev, A. A. Rempel, G. P. Shveikin: Structural stability and the boundaries of a homogeneity region of nonstoichiometric carbides, Doklady Akademii Nauk SSSR **298**, 890–894 (1988) (in Russian). (Engl. transl.: Doklady Phys. Chem. **298**, 121–125 (1988))

6 A. I. Gusev: Structural stability boundaries for nonstoichiometric compounds, Phys. Stat. Sol. (a) **111**, 443–450 (1989)

7 A. I. Gusev, A. A. Rempel: The boundaries of the structural stability of nonstoichiometric carbides, Zh. Neorgan. Khimii **34**, 556–561 (1989) (in Russian). (Engl. transl.: Russ. J. Inorg. Chem. **34**, 309–312 (1989))

8 V. N. Lipatnikov, A. A. Rempel, A. I. Gusev: Effect of nonstoichiometry and ordering on the basal structural parameters of the monocarbides of niobium and tantalum, Izv. AN SSSR. Neorgan. Materialy **26**, 2522–2526 (1990) (in Russian). (Engl. transl.: Inorganic Materials **26**, 2171–2175 (1990))

9 A. I. Gusev, A. A. Rempel, V. N. Lipatnikov: Incommensurate ordered phase in nonstoichiometric tantalum carbide, J. Phys.: Condens. Matter **8**, 8277–8293 (1996)

10 V. N. Lipatnikov, A. I. Gusev, P. Ettmayer, W. Lengauer: Phase transformation in nonstoichiometric vanadium carbide, J. Phys.: Condens. Matter **11**, 163–184 (1999)
11 L. V. Zueva, A. I. Gusev: Effect of nonstoichiometry and ordering on the period of the basis structure of cubic titanium carbide, Fiz. Tverd. Tela **41**, 1134–1141 (1999) (in Russian). (Engl. transl.: Physics of the Solid State **41**, 1032–1038 (1999)
12 E. K. Storms, R. I. McNeal: The vanadium–vanadium carbide system, J. Phys. Chem. **66**, 1402–1408 (1962)
13 R. Lorenzelli, I. de Dieuleveult: Existence d'une surstructure dans le carbure de thorium sous-stoechiometrique: $ThC_{0.76}$, J. Nucl. Mater. **29**, 349–353 (1969)
14 T. Satow: Electrical resistivity, magnetic susceptibility, and thermal expansion of thorium monocarbide, J. Nucl. Mater. **21**, 434–344 (1967)
15 A. A. Rempel, A. I. Gusev: *Ordering in Nonstoichiometric Niobium Monocarbide* (Urals Scientific Centre, Sverdlovsk 1983) 68 pp. (in Russian)
16 A. I. Gusev, A. A. Rempel: Ordering in the carbon sublattice of nonstoichiometric niobium carbide, Fiz. Tverd. Tela **26**, 3622–3627 (1984) (in Russian). (Engl. transl.: Sov. Physics – Solid State **26**, 2178–2181 (1984))
17 A. A. Rempel, A. I. Gusev: Order–disorder phase transition in nonstoichiometric niobium carbide, Kristallografiya **30**, 1112–1115 (1985) (in Russian). (Engl. transl.: Sov. Physics – Crystallography **30**, 648–650 (1985))
18 A. I. Gusev, A. A. Rempel: Order–disorder phase transition channel in niobium carbide, Phys. Stat. Sol. (a) **93**, 71–80 (1986)
19 T. Athanassiadis, N. Lorenzelli, C. H. de Novion: Diffraction studies of the order-disorder transformation in V_8C_7, Ann. Chim. France **12**, 129–142 (1987)
20 V. N. Lipatnikov, P. Ettmayer: Effect of vacancy ordering on structure and some properties of vanadium carbide, in Proc. of the 14th International Plansee Seminar **2**, G. Kneringer, P. Rodhammer and P. Wilhartitz (Eds.), (Plansee AG, Reutte 1997) pp. 485–497
21 V. N. Lipatnikov, W. Lengauer, P. Ettmayer, E. Keil, G. Groboth, E. Kny: Effects of vacancy ordering on structure and properties of vanadium carbide, J. Alloys Comp. **261**, 192–197 (1997)
22 G. Santoro: Variation of some properties of tantalum carbide with carbon content, Trans. Met. Soc. AIME **227**, 1361–1368 (1963)
23 V. Moisy-Maurice: *Structure atomique des carbures non-stoechiometriques de metaux de transition*, Rapport CEA-R-5127 (Commissariat à l'Energie Atomique, Gif-sur-Yvette (France) 1981). 184 pp.
24 V. Moisy-Maurice, N. Lorenzelli, C. H. de Novion, P. Convert: High temperature neutron diffraction study of the order–disorder transition in TiC_{1-x}, Acta Metall. **30**, 1769–1779 (1982)
25 V. N. Lipatnikov, A. A. Rempel, A. I. Gusev: Atomic ordering and hardness of nonstoichiometric titanium carbide, Int. J. Refract. Met. Hard Mater. **15**, 61–64 (1997)
26 A. N. Emel'yanov: Temperature conductivity of nonstoichiometric titanium carbide in the region of order–disorder phase transition, Teplofizika Vysok. Temp. **28**, 269–278 (1990) (in Russian)
27 S. Tsurekawa, H. Yoshihaga: Identification of long-range ordered structure in $TiC_{0.59}$ by transmission electron-microscopy, J. Japan. Inst. Metals **56**, 133–141 (1992)
28 A. V. Karpov, V.P. Kobyakov, E.A. Chernomorskaya: Dilatation of nonstoichiometric titanium carbide in the region of order–disorder phase transition, Neorgan. Materialy **31**, 655–659 (1995) (in Russian)

29 A. V. Karpov, V. P. Kobyakov: Order–disorder phase transition in $TiC_{0.55}$, Fiz. Tverd. Tela **38**, 1638–1640 (1996) (in Russian)
30 A. V. Karpov, V. P. Kobyakov: Order–disorder phase transition in $TiC_{0.55}$, Teplofizika Vysok. Temp. **34**, 965–968 (1996) (in Russian). (Engl. transl.: High Temp. **34**, 950–953 (1996))
31 F. C. Nix, D. MacNair: A dilatometric study of the order–disorder transformation in copper-gold alloys, Phys. Rev. **60**, 320–329 (1941)
32 P. Wright, K. F. Goddard: Lattice parameter and resistivity study of order in the alloy $CuAu_3$, Acta Metall. 7, 757–761 (1959)
33 A. E. Vol, I. K. Kagan: *Structure and Properties of Binary Metal Systems* **3** (Nauka, Moscow 1976) pp. 87–89 (in Russian)
34 V. Moisy-Maurice, C. H. de Novion, A. N. Christensen, W. Just: Elastic diffuse neutron scattering study of the defect structure of $TiC_{0.76}$ and $NbC_{0.73}$, Solid State Commun. **39**, 661–665 (1981)
35 T. H. Metzger, J. Peisl, R. Kaufmann: X-ray determination of local atomic displacements around carbon vacancies in NbC_{1-c} single crystals, J. Phys. F: Metal Phys. **13**, 1103–1113 (1983)
36 R. Kaufmann, O. Meyer: Ion channeling in NbC_{1-c}: determination of local atomic displacements around carbon vacancies, Phys. Rev. B **28**, 6216–6226 (1983)
37 T. Priem, B. Beuneu, C. H. de Novion, J. Chevrier, F. Livet, A. Finel, S. Lefevbre: Short-range order and static displacements in non-stoichiometric transition metal carbides and nitrides, Physica B **156–157**, 47–49 (1989)
38 T. Priem: *Etude de l'ordre a courte distance dans les carbures et nitrures non-stoechiometriques de metaux de transition par diffusion diffuse de neutrons*, Rapport CEA-R-5499 (Commissariat à l'Energie Atomique, Centre d'Etudes Nucléaires de Saclay, Gif-sur-Yvette (France) 1989) 162 pp.
39 T. E. Harris: Lower bound for the critical probability in a certain percolation process, Proc. Cambridge Philosoph. Soc. **56**, 13–20 (1960)
40 M. E. Fisher: Critical probabilities for cluster size and percolation problems, J. Math. Phys. **2**, 620-627 (1961)
41 A. L. Efros: *Physics and Geometry of Disorder* (Nauka, Moscow 1982) 175 pp. (in Russian)
42 H. Kesten: *Percolation Theory for Mathematicians* (Birkhäuser, Boston-Basel-Stuttgart 1982) 423 pp.
43 A. I. Gusev: Atomic ordering and the order parameter functional method, Philosoph. Mag. B **60**, 307–324 (1989)
44 A. I. Gusev, A. A. Rempel: Phase diagrams of metal–carbon and metal–nitrogen systems and ordering in strongly nonstoichiometric carbides and nitrides, Phys. Stat. Sol. (a) **163**, 273–304 (1997)
45 A. A. Rempel', A. I. Gusev: Nanostructure and atomic ordering in vanadium carbide, Pis'ma v ZhETF **69**, 436–442 (1999) (in Russian). (Engl. transl.: JETP Letters **69**, 472–478 (1999))
46 V. M. Cherkashenko, A. A. Rempel, A. I. Gusev: Influence of vacancy ordering on X-ray emission spectra of nonstoichiometric niobium carbides, Fiz. Tverd. Tela **27**, 1387–1390 (1985) (in Russian). (Engl. transl.: Sov. Physics – Solid State **27**, 838–840 (1985))
47 V. M. Cherkashenko, V. V. Shumilov, E. Z. Kurmaev, A. A. Rempel: The X-ray spectra of niobium and the electronic structure of ordered niobium carbides in the

nonstoichiometry range, Zh. Neorgan. Khimii **36**, 474–479 (1991) (in Russian). (Engl. transl.: Russ. J. Inorg. Chem. **36**, 265–268 (1991))
48 E. Z. Kurmaev, M. P. Butsman, V. A. Trofimova, V. A. Gubanov, A. L. Ivanovskii, Yu. G. Zainulin, G. P. Shveikin: Study of valent state of titanium atoms in nonstoichiometric carbides on X-ray emission spectra and on the data of MO LCAO calculation, Zh. Struct. Khimii **22**, 172–174 (1981) (in Russian)
49 K. Schwarz, N. Rösch: Effects of carbon vacancies in NbC on superconductivity, J. Phys. C: Sol. State Phys. **9**, L443–L447 (1976)
50 G. Reis, H. Winter: Electronic structure of vacancies in refractory compounds and its influence on T_c, J. Phys. F: Metal Phys. **10**, 1–7 (1980)
51 H. Höchst, P. Steiner, S. Hüfner, C. Politis: XPS valence band spectra of NbC_x, Z. Phys. **37**, 27–31 (1980)
52 A. I. Gusev, A. A. Rempel: Superconductivity in disordered and ordered niobium carbide, Phys. Stat. Sol. (b) **151**, 211–224 (1989)
53 A. Z. Men'shikov, E. Z. Kurmaev: Influence of nearest environment of atoms on X-ray emission spectra, Fiz. Metall. Metalloved. **41**, 748–756 (1976) (in Russian)
54 A. A. Rempel, A. I. Gusev: NMR spectra of ^{93}Nb in ordered niobium carbide, Fiz. Tverd. Tela **25**, 3169–3171 (1983) (in Russian). (Engl. transl.: Sov. Physics – Solid State **25**, 1827–1828 (1983))
55 A. A. Rempel, A. I. Gusev, M. Yu. Belyaev: ^{93}Nb NMR study of an ordered and a disordered non-stoichiometric niobium carbide, J. Phys. C: Sol. State Phys. **20**, 5655–5666 (1987)
56 A. I. Gusev, A. A. Rempel: A study of the atomic ordering in the niobium carbide using the magnetic susceptibility method, Phys. Stat. Sol. (a) **84**, 527–534 (1984)
57 A. A. Rempel, L. B. Dubrovskaya, A. I. Gusev, G. P. Shveikin: Effect of ordering on the magnetic susceptibility of niobium carbide, Izv. AN SSSR. Neorgan. Materialy **21**, 596–599 (1985) (in Russian). (Engl. transl.: Inorganic Materials **21**, 514–518 (1985))
58 A. I. Gusev, A. A. Rempel: Anomalous changes in the magnetic susceptibility due to ordering of niobium carbide, Fiz. Tverd. Tela **27**, 1528–1530 (1985) (in Russian). (Engl. transl.: Sov. Physics – Solid State **27**, 920–921 (1985))
59 S. R. Barman, D. D. Sarma: Electronic structure of TiO_x $(0.8 < x < 1.3)$ with disordered and ordered vacancies, Phys. Rev. B **49**, 16141–16148 (1994)
60 S. Bartkowski, M. Neumann, E. Z. Kurmaev, V. V. Fedorenko, S. N. Shamin, V. M. Cherkashenko, S. N. Nemnonov, A. Winiarski, D. C. Rubie: Electronic structure of titanium monoxide, Phys. Rev. B **56**, 10656–10667 (1997)
61 G. Hörmandinger, J. Redinger, P. Weinberger, G. Hobiger, P. Herzig: The influence of vacancies on the electronic structure of TiO, Sol. State Commun. **68**, 467–470 (1988)
62 M. A. Blokhin, I. G. Shveitser: *Handbook on X-ray Spectroscopy* (Nauka, Moscow 1982) 376 pp. (in Russian)
63 G. H. Schalder, A. M. Boring, P. Weinberger, A. Gonis: Electronic structure of stoichiometric and off-stoichiometric tantalum carbide, Phys. Rev. B **38**, 9538–9544 (1988)
64 L. Ramqvist, K. Hamrin, G. Johensson, U. Gelias, C. Nordling: VC, NbC, and TaC with varying carbon content studied by ESCA, J. Phys. Chem. Sol. **31**, 2669–2672 (1970)
65 H. Ihara, M. Hirabayashi, H. Nakagawa: Electronic band structures and X-ray photoelectron spectra of ZrC, HfC and TaC, Phys. Rev. B **14**, 1707–1714 (1976)

66 G. R. Gruzalski, D. M. Zenner: Defect states in substoichiometric tantalum carbide, Phys. Rev. B **34**, 3841–3848 (1986)
67 A. A. Rempel, A. K. Sinelnichenko: X-ray photoelectron spectra of nonstoichiometric tantalum carbide, Metallofizika **13**, 61–70 (1991) (in Russian). (Engl. transl.: Phys. Metals **11**, 352–364 (1992))
68 V. P. Zhukov, A. L. Ivanovskii, V. A. Gubanov, G. P. Shveikin, G. Veber: X_α-cluster calculations of electronic structure of hafnium and tantalum carbides and nitrides, Zh. Neorgan. Khimii **25**, 318–326 (1980) (in Russian)
69 J. P. Landesman, G. Treglia, P. Turchi, F. Ducastelle: Electronic structure and pairwise interactions in substoichiometric transition metal carbides and nitrides, J. de Physique **46**, 1001–1015 (1985)
70 P. Blaha, K. Schwarz, F. Kubel, K. Yvon: Chemical bonding in refractory metal compounds with 8, 9 and 10 valence electrons, J. Sol. State Chem. **20**, 199–206 (1987)
71 H. Ihara: *Electronic Structures of the Transition Metal Carbides and Borides Studies by X-ray Photoelectron Spectroscopy and Band Calculation*, Res. of Electrotechn. Lab. No. 775 (Electrotechnical Laboratory, Tokyo 1977) 127 pp.
72 V. I. Nefedov: *X-ray Electron Spectroscopy of Chemical Compounds* (Khimiya, Moscow 1984) 256 pp. (in Russian)
73 E. Z. Kurmaev, V. M. Cherkashenko, A. A. Rempel, A. I. Gusev, A. Moewes, D. L. Ederer: Synchrotron X-ray emission spcctroscopy of N_3-spectra in tantalum and nonstoichiometric tantalum carbide, in *Abstracts of 3rd Russian-German Seminar on Electron and X-ray Spectroscopy*, Yekaterinburg, September 15-19, 1999 (Institute of Metal Physics, Yekaterinburg 1999) pp. 50–51
74 E. Browne, J. M. Dairiki, R. E. Doebler: *Table of Isotopes*, 7th edn. C. M. Lederer, V. S. Shirley (Eds.) (John Wiley & Sons, New York 1978) 1523 pp.
75 R. M. Nieminen, M. J. Manninen: Positrons in imperfect solids: theory, in *Positron in Solids* P.Hautojärvi (Ed.) (Springer, Berlin 1979) pp. 145–195
76 A. A. Rempel, H.-E. Schaefer, M. Forster, A. I. Girka: Atomic defects in transition metal carbides and SiC studied by positron annihilation, in *Covalent Ceramics II: Non-oxides*, A. R. Barron, G. S. Fischman, M. A. Fury, A. F. Hepp (Eds.), USA Mater. Res. Soc. Symp. Proc. **327**, Boston, Massachusetts, November 29 – December 2, 1993 (Materials Research Society, Pittsburgh (Pennsylvania) 1994) pp. 301–304
77 M. J. Puska, M. Šob, G. Brauer, T. Kornonen: First-principles calculation of positron lifetimes and affinities in perfect and imperfect transition metal carbides and nitrides, Phys. Rev. B **49**, 10947–10958 (1994)
78 J. Störmer J, A. Goodyear, W. Anwand, G. Brauer, P. G. Coleman, W. Triftshäuser: Silicon carbide: a new positron moderator, J.Phys.: Condens. Matter **8**, L89–L94 (1996)
79 G. Brauer, W. Anwand, E.-M. Nicht, J. Kuriplach, M. Šob, N. Wagner, P. G. Coleman, M. J. Puska, T. Korhonen: Evaluation of some basic positron-related characteristics of SiC, Phys. Rev. B **54**, 2512–2517 (1996)
80 R. Würschum, A. Seeger: Diffusion–reaction model for the trapping of positrons in grain boundaries, Philosoph. Mag. **73**, 1481–1501 (1996)
81 H.-E. Schaefer: Investigation of thermal equilibrium vacancies in metals by positron annihilation, Phys. Stat. Sol. (a) **102**, 47–65 (1987)
82 R. Würschum, W. Bauer, K. Maier, A. Seeger, H.-E. Schaefer: Positron lifetime in semiconductors, J. Phys.: Condens. Matter **1**, A33–A48 (1989)

83 R. Würschum, H.-E. Schaefer: Interfacial free volumes and atomic diffusion in nanostructured solids, in *Nanomaterials: Synthesis, Properties, and Applications* A. S. Edelstein, R. C. Cammarata (Eds.) (Institute of Physics, Bristol 1996) pp. 277–301

84 R. Würschum, K. Reimann, S. Gruß, A. Kübler, P. Scharwaechter, W. Frank, O.Kruse, H. D. Carstanjen, H.-E. Schaefer: Structure and diffusional properties of nanocrystalline Pd, Philosoph. Mag. B **17**, 407–417 (1997)

85 H.-E. Schaefer, R. Würschum, R. Schwarz, D. Slobodin, S. Wagner: Amorphous hydrogenated Si studied by positron lifetime spectroscopy, Appl. Phys. A **40**, 145–149 (1986)

86 R. Würschum, P. Farber, R. Dittmar, P. Scharwaechter, W. Frank, H.-E. Schaefer: Thermal vacancy formation and self-diffusion in intermetallic Fe_3Si nanocrystallites of nanocomposite alloys, Phys. Rev. Lett. **24**, 4918–4921 (1997)

87 H.-E. Schaefer, M. Forster: As-grown metal oxides and electron-irradiated Al_2O_3 studied by positron lifetime measurements, Mater. Sci. Engineer. A **109**, 161–167 (1989)

88 A. A. Rempel, M. Forster, H.-E. Schaefer: Positron lifetime in carbides with *B*1 structure, Doklady Akad. Nauk SSSR **326**, 91–97 (1992) (in Russian). (Engl. transl.: Sov. Physics Doklady **37**, 484–487 (1992))

89 A. A. Rempel, M. Forster, H.-E. Schaefer: Positron lifetime in non-stoichiometric carbides with a *B*1 (NaCl) structure, J. Phys.: Condens. Matter **5**, 261–266 (1993)

90 A. A. Rempel, L. V. Zueva, V. N. Lipatnikov, H.-E. Schaefer: Positron lifetime in the atomic vacancies of nonstoichiometric titanium and vanadium carbides, Phys. Stat. Sol. (a) **169**, R9–R10 (1998)

91 A. A. Rempel, A. P. Druzhkov, V. N. Lipatnikov, A. I. Gusev, S. M. Klotsman, G. P. Shveikin: Angular correlation of annihilation radiation in nonstoichiometric tantalum carbide, Doklady Akad. Nauk SSSR **300**, 92–95 (1988) (in Russian). (Engl. transl.: Sov. Physics Doklady **33**, 357–359 (1988))

92 A. A. Rempel, A. P. Druzhkov, A. I. Gusev: Positron annihilation in tantalum and its carbide, Fiz. Metall. Metalloved. **68**, 371–279 (1989) (in Russian). (Engl. transl.: Phys. Met. Metallogr. **68**, 59–68 (1989))

93 A. A. Rempel, A. P. Druzhkov, A. I. Gusev: Angular correlation of annihilation radiation in transition metals and their carbides, Fiz. Tverd. Tela **32**, 1333–1338 (1990) (in Russian). (Engl. transl.: Sov. Physics – Solid State **32**, 779–782 (1990))

94 P. Kirkegaard, M. Eldrup, O. E. Mogensen, N. J. Pedersen: Program system for analysing positron lifetime spectra and angular correlation curves, Comp. Phys. Commun. **23**, 307–335 (1981)

95 M. J. Puska, M. Šob, G. Brauer, T. Korhonen: Positron annihilation characteristics in perfect and imperfect transition metal carbides and nitrides, J.Physique IV **5**, C1-135–C1-142 (1995)

96 A. A. Rempel, A. I. Gusev, O. V. Makarova, S. Z. Nazarova: Physical and chemical properties of nanostructured vanadium carbide, Perspectiv. Materialy No. 6, 9–15 (1999) (in Russian)

97 G. Brauer, W. Anwand, E. M. Nicht, P. G. Coleman, A. P. Knights, H. Schut, G. Kögel, N. Wagner: Positron Studies of polycrystalline TiC, J. Phys.: Condens. Matter **7**, 9091–9099 (1995)

98 A. Seeger, F. Banhart: On the systematics of positron lifetimes in metals, Phys. Stat. Sol. (a) **102**, 171–179 (1987)

99 W. Bauer, J. Briggmann, H.-D. Carstanjen, S. Connel, W. Decker, J. Diehl, K. Maier, J. Major, H.-E. Schaefer, A. Seeger, H. Stoll, E. Widmann: The stuttgart positron beam, its performance and recent experiments, Nuclear Instr. Methods in Phys. Res. B **50**, 300–306 (1990)
100 A. A. Rempel, H.-E. Schaefer: Irradiation-induced atomic defects in SiC studied by positron annihilation, Apl. Phys. A **61**, 51–53 (1995)
101 A. A. Rempel, R. Würschum, H.-E. Schaefer: Atomic defects in hexagonal tungsten carbide studied by positron annihilation, Phys. Rev. B **61**, 5945–5948 (2000)
102 F. Seitz, J. S. Koehler: Displacement of atoms during irradiation, in *Solid State Physics. Advances in Research and Applications* F. Seitz, D. Turnbull (Eds.) (Academic Press, New York 1956) pp. 305–449
103 J. Morillo, C. H. de Novion, J. Dural: Neutron and electron radiation defects in titanium and tantalum monocarbides: an electrical resistivity study, Radiation Effects **55**, 67–78 (1981)
104 D. Gosset, J. Morillo, C. Allison, C. H. de Novion: Electron irradiation damage in stoichiometric and substoichiometric tantalum carbides TaC_x. Part I. Threshold displacement energies, Radiation Effects and Defects in Solids **118**, 207–224 (1991)
105 M. E. Straumanis, H. W. Li: Gitterkonstanten, Ausdehnungskoeffizienten, Dichten, Fehlordnung und Aufbau der Phase Titan–Oxid, Z. Anorg. Allgem. Chem. **305**, 143–147 (1960)
106 S. Andersson, B. Collen, U. Kuylenstierna, A. Magneli: Phase analysis studies on the titanium–oxygen system, Acta Chem. Scand. **11**, 1641–1647 (1957)
107 R. L. Zafra, W. Joyner: Temperature effect on positron annihilation in condensed matter, Phys. Rev. **112**, 19–29 (1958)
108 K. Maier, H. Metz, D. Herlach, H.-E. Schaefer, A. Seeger: Formation and migration energies of monovacancies in tantalum, Phys. Rev. Lett. **39**, 484–487 (1977)
109 N. F. Miron, S. N. Maev: Positron spectroscopy of defects in vanadium, niobium and molibdenum after neutron irradiation, Fiz. Metall. Metalloved. **66**, 1159–1162 (1988) (in Russian)
110 V. I. Goldanskii: *Chemistry of the Positron and Positronium* (Nauka, Moscow 1968) 174 pp. (in Russian)
111 G. Dlubek, W. Rechner, O. Brummer: Contribution to the parametrization of the angular correlation curves of the 2γ-positron annihilation radiation, Exper. Tekn. Phys. **25**, 289–297 (1977)
112 D. J. Hudson: *Lectures on Elementary Statistics and Probability* (CERN, Geneva 1963) 101 pp.; D. J. Hudson: *Statistics lectures II: Maximum Likelihood and Least Squares Theory* (CERN, Geneva 1964) pp. 103–211
113 U. A. Arifov (Ed.): *Positron Diagnostics* (Fan, Tashkent 1978) 190 pp. (in Russian)
114 P. Marksteiner, P. Weinberger, A. Neckel, R. Zeller, P. H. Dederichs: Electronic structure of substoichiometric carbides and nitrides of titanium and vanadium, Phys. Rev. B **33**, 812–822 (1986)
115 P. Marksteiner, P. Weinberger, A. Neckel, R. Zeller, P. H. Dederichs: Electronic structure of substoichiometric carbides and nitrides of zirconium and niobium, Phys. Rev. B **33**, 6709–6717 (1986)
116 V. I. Trefilov, E. A. Zhurakovskii, V. V. Nemoshkalenko, Ya. V. Zaulichnyii, A. P. Shpak, B. P. Mamko, A. A. Lisenko: State of carbon atom in different crystal modification of free carbon and in refractory carbides, Metallofizika **10**, 18–25 (1988) (in Russian)

117 V. S. Mikhalenkov: Diagnostics of vacancy clusters in metals by electron-positron annihilation method, Metallofizika **5**, 44–60 (1983) (in Russian)
118 A. A. Rempel: Structural vacancy in nonstoichiometric carbides of IV and V groups transition metals, in *Solid State Chemistry*, G. P. Shveikin (Ed.) (Ural Division of the Russian Academy of Sciences, Ekaterinburg 1997) pp. 57–69 (in Russian)
119 W. S. Williams: Scattering of electrons by vacancies in nonstoichiometric crystals of titanium carbide, Phys. Rev. A **135**, 505–510 (1964)
120 Y. Ishizawa, S. Otani, H. Nozaki, T. Tanaka: Carbon–vacancy concentration dependencies of electrical properties of NbC_x single crystal, J. Phys.: Condens. Matter **4**, 8593–8598 (1992)
121 V. N. Lipatnikov, A. I. Gusev: Dependence of the resistivity of nonstoichiometric titanium carbide TiC_y on the density and distribution of carbon vacancies, Pis'ma v ZhETF **70**, 284–289 (1999) (in Russian). (Engl. transl.: JETP Letters **70**, 294–300 (1999))
122 L. Ramqvist: Variation of hardness, resistivity and lattice parameter with carbon content of group 5b metal carbides, Jernkontorets Annaler. **152**, 467–475 (1968)
123 J. R. Cooper, R. L. Hansler: Variation of electrical resistivity of cubic tantalum carbide with composition, J. Chem. Phys. **39**, 248–249 (1963)
124 R. Steinitz, R. Resnick: Electrical and magnetic properties of compositions in the tantalum-carbon system, J. Appl. Phys. **37**, 3463–3471 (1966)
125 G. Santoro, R. T. Dolloff: Hall coefficient of tantalum carbide as a function of carbon content and temperature, J. Appl. Phys. **39**, 2293–2296 (1968)
126 L. B. Dubrovskaya, I. I. Matveenko, P. V. Geld: Influence of temperature and composition on electrical conductivity of β- and γ-phases of tantalum–carbon system, Fiz. Metall. Metalloved. **20**, 243–250 (1965) (in Russian)
127 L. C. Dy, W. S. Williams: Resistivity, superconductivity and order–disorder transformations in transition metal carbides and hydrogen-doped carbides, J. Appl. Phys. **53**, 8915–8927 (1982)
128 S. Otani, T. Tanaka, Y. Ishizawa: Electrical resistivities in single crystals of TiC_x and VC_x, J. Mater. Sci. **21**, 1011–1014 (1986)
129 V. Christoph, J. Richter, W. Schiller: Influence of partial long-range and short-range orders on the electrical resistivity of substitutional alloys, Phys. Stat. Sol. (b) **100**, 595–602 (1980)
130 V. N. Lipatnikov, A. Kottar, L. V. Zueva, A. I. Gusev: Disorder–order phase transformations and electrical resistivity of a nonstoichiometric titanium carbide, Fiz. Tverd. Tela **40**, 1332–1340 (1998) (in Russian). (Engl. transl.: Physics of the Solid State **40**, 1211–1218 (1998))
131 V. A. Vlasov, I. A. Karimov, L. V. Kustova: Order–disorder phase transition in nonstoichiometric titanium carbide, Izv. AN SSSR. Neorgan. Materialy **22**, 231–233 (1986) (in Russian)
132 V. N. Lipatnikov, A. I. Gusev: Effect of ordering on the structure and specific heat of nonstoichiometric titanium carbide, Pis'ma v ZhETF **69**, 631–637 (1999) (in Russian). (Engl. transl.: JETP Letters **69**, 669–675 (1999))
133 A. I. Gusev, A. A. Rempel: Superstructures of non-stoichiometric interstitial compounds and the distribution functions of interstitial atoms, Phys. Stat. Sol. (a) **135**, 15–58 (1993)
134 A. N. Emel'yanov: Features of order–disorder phase transition in nonstoichiometric carbides of transition metals, Fiz. Tverd. Tela **38**, 3678–3682 (1996) (in Russian)

135 N. Lorenzelli, R. Caudron, J. P. Landesman, C. H. de Novion: Influence of the ordering of carbon vacancies on the electronic properties of $TiC_{0.625}$, Solid State Commun. **59**, 765–769 (1986)
136 N. Obata, N. Nakasawa: Superlattice formation in zirconium–carbon system, J. Nucl. Mater. **60**, 39–42 (1976)
137 L. W. Shacklette, W. S. Williams: Influence of order-disorder transformations on the electrical resistivity of vanadium carbide, Phys. Rev. B **7**, 5041–5053 (1973)
138 G. N. Emmons, W. S. Williams: Thermodynamics of order-disorder transformations in vanadium carbide, J. Mater. Sci. **18**, 2489–2602 (1983)
139 V. N. Lipatnikov, A. I. Gusev, P. Ettmaier, W. Lengauer: Order–disorder phase transformations and specific heat of nonstoichiometric vanadium carbide, Fiz. Tverd. Tela **41**, 529–536 (1999) (in Russian). (Engl. transl.: Physics of the Solid State **41**, 474–480 (1999))
140 V. S. Neshpor, S. S. Ordan'yan, A. I. Avgustinnik, M. B. Khusidman: Influence of chemical composition on electrical and thermophysical properties of ziconium and niobium carbides in their homogeneity intervals, Zh. Prikl. Khimii **37**, 2375–2379 (1964) (in Russian)
141 I. I. Matveenko, L. B. Dubrovskaya, G. D. Bogomolov, V. G. Zubkov, P. V. Geld: Electrical conductivity of titanium oxicarbides, Izv. AN SSSR. Neorgan. Materialy **6**, 1190–1191 (1970) (in Russian)
142 V. G. Zubkov, I. I. Matveenko, L. B. Dubrovskaya, G. D. Bogomolov, P. V. Geld: Neutron diffraction study of structure of titanium oxicarbides, Doklady Akad. Nauk SSSR **191**, 323–325 (1970) (in Russian)
143 H. Bittner, H. Goretzki: Magnetische Untersuchungen der Carbide TiC, ZrC, HfC, VC, NbC und TaC, Monatsh. Chem. **93**, 1000–1004 (1962)
144 M. I. Lesnaya, V. F. Nemchenko, I. M. Vinitskii, V. Ya. Naumenko: Temperature dependence of magnetic susceptibility of titanium monocarbide in homogeneity region, Izv. AN SSSR. Neorgan. Materialy **13**, 840–843 (1977) (in Russian)
145 A. S. Borukhovich, L. B. Dubrovskaya, I. I. Matveenko, P. V. Geld: Magnetic susceptibility and energy band structure of zirconium monocarbide, Phys. Stat. Sol. **36**, 97–102 (1969)
146 M. Ishikawa, L. E. Toth: Spezifische Warme and magnetische Suszeptibilitat von Vanadincarbiden VC_x, Monatsh. Chem. **103**, 492–502 (1972)
147 I. I.Matveenko, L.B.Dubrovskaya, P.V.Geld, M.G.Tretnikova: Magnetic susceptibility of cubic niobium carbide, Izv. AN SSSR. Neorgan. Materialy **1**, 1062–1064 (1965) (in Russian)
148 A. S. Borukhovich, L. B. Dubrovskaya, I. I. Matveenko, P. V. Geld: Magnetic susceptibility of niobium and tantalum monocarbides at low temperatures, Fiz. Tverd. Tela **11**, 830–832 (1969) (in Russian)
149 Ya. E. Genkin, I. A. Milovanova, A. V. Lyakutkin: Magnetic susceptibility of some compounds of niobium with boron, carbon and nitrogen, Izv. AN Kazakh. SSR. Seriya fiz.-mat. No. 2, 12–16 (1976) (in Russian)
150 L. B. Dubrovskaya, I. I. Matveenko: Magnetic properties of cubic tantalum carbide, Fiz. Metall. Metalloved. **19**, 199–204 (1965) (in Russian)
151 L. V. Zueva, S. Z. Nazarova, V. N. Lipatnikov, A. I. Gusev: Effect of atom–vacancy ordering on magnetuc susceptibility of nonstoichiometric titanium carbide, Doklady Akad. Nauk **365**, 82–86 (1999) (in Russian). (Engl. transl.: Doklady Phys. Chem. **365**, 65–69 (1999))

152 L. V. Zueva, V. N. Lipatnikov, A. I. Gusev: Magnetic susceptibility and ordering of titanium carbide, Zh. Fiz. Khimii **73**, 2163–2169 (1999) (in Russian). (Engl. transl.: Russ. J. Phys. Chem. **73**, 1952–1959 (1999))
153 A. N. Zyryanova, S. Z. Nazarova, A. I. Gusev: Magnetic susceptibility anomaly as evidence for ordering of nonstoichiometric hafnium carbide HfC_y, Doklady Akad. Nauk **359**, 348-353 (1998) (in Russian). (Engl. transl.: Doklady Phys. Chem. **359**, 91–96 (1998))
154 A. N. Zyryanova, A. I. Gusev: Magnetic susceptibility and ordering in nonstoichiometric hafnium carbide, Zh. Fiz. Khimii **72**, 2234–2242 (1998) (in Russian). (Engl. transl.: Russ. J. Phys. Chem. **72**, 3034–3041 (1998))
155 A. I. Gusev, A. N. Zyryanova: Atomic-vacancy ordering and magnetic susceptibility of nonstoichiometric hafnium carbide, Pis'ma v ZhETF **69**, 2096–301 (1999) (in Russian). (Engl. transl.: JETP Letters **69**, 324–329 (1999))
156 A. I. Gusev, A. N. Zyryanova: Ordering and magnetic susceptibility of non-stoichiometric hafnium carbide, Phys. Stat. Sol. (a) **177**, 419–437 (2000)
157 V. N. Lipatnikov, A. I. Gusev, A. A. Rempel, G. P. Shveikin: Effect of structural transition on magnetic susceptibility of tantalum carbide, Doklady Akad. Nauk SSSR **297**, 849–853 (1987) (in Russian). (Engl. transl.: Sov. Physics Doklady **32**, 988–990 (1987))
158 V. N. Lipatnikov, A. A. Rempel, A. I. Gusev: The influence of ordering of the vacancies on the magnetic susceptibility of tantalum carbide, Zh. Neorgan. Khimii **33**, 1860–1863 (1988) (in Russian). (Engl. transl.: Russ. J. Inorg. Chem. **33**, 1058–1059 (1988))
159 V. N. Lipatnikov, A. I. Gusev: Magnetic susceptibility of TaC_y in the disordered and ordered states, in *Solid State Chemistry*, P. V. Geld (Ed.) (Ural Polytechnical Institute, Sverdlovsk 1988) pp. 144–149 (in Russian)
160 A. I. Gusev, A. A. Rempel, V. N. Lipatnikov: Magnetic susceptibility and atomic ordering in tantalum carbide, Phys. Stat. Sol. (a) **106**, 459–466 (1988)
161 J. P. Landesman, A. N. Christensen, C. H. de Novion, N. Lorenzelli, P. Convert P. Order-disorder transition and structure of the ordered vacancy compound Nb_6C_5: powder neutron diffraction studies, J. Phys. C: Solid State Phys. **18**, 809–823 (1985)
162 A. I. Gusev: Phase diagrams of ordered nonstoichiometric hafnium carbide and titanium nitride, Doklady Akad. Nauk **322**, 918–923 (1992) (in Russian). (Engl. transl.: Doklady Phys. Chem. **322**, 84–88 (1992))
163 A. I. Gusev, A. A. Rempel: Calculation of phase diagrams of interstitial compounds, J. Phys. Chem. Solids **55**, 299–304 (1994)
164 A. I. Gusev, A. A. Rempel: Atomic ordering and phase equilibria in strongly nonstoichiometric carbides and nitrides, in *Materials Science of Carbides, Nitrides and Borides*, Y. G. Gogotsi, R. A. Andrievski (Eds.) (Kluwer Academic Publishers, Dordrecht 1999). pp. 47–64
165 A. I. Gusev: Order–disorder transformations and phase equilibrium in strongly nonstoichiometric compounds, Uspekhi Fiz. Nauk **170**, 3–40 (2000) (in Russian). (Engl. transl.: Physics – Uspekhi **43**, 1–37 (2000))
166 L. B. Dubrovskaya, S. Z. Nazarova, E. T. Kachkovskaya: Magnetic susceptibility of ordered titanium carbide, Izv. AN SSSR. Neorgan. Materialy **20**, 783–785 (1984) (in Russian)
167 P. W. Selwood: *Magnetochemistry*, 2nd edn. (Interscience Publ., New York 1956) 435 pp.

168 S. V. Vonsovskii: *Magnetism* (Nauka, Moscow 1971) 1032 pp. (in Russian)
169 L. B. Dubrovskaya, A. G. Rabin'kin, P. V. Geld: Nature of superconducting niobium and tantalum monocarbides, ZhETF **62**, 300–306 (1972) (in Russian)
170 P. V. Geld, V. A. Tskhai, A. S. Borukhovich, L. B. Dubrovskaya, I. I. Matveenko: Conduction band of IVa and Va subgroup transition metal carbides (I), Phys. Stat. Sol. (b) **42**, 85–93 (1970)
171 J. Redinger, R. Eibler, P. Herzig, A. Neckel, R. Podloucky, E. Wimmer: Vacancy induced changes in the electronic structure of titanium carbide. 1. Band structure and density of states, J. Phys. Chem. Solids **46**, 383–398 (1985)
172 J. Redinger, R. Eibler, P. Herzig, A. Neckel, R. Podloucky, E. Wimmer: Vacancy induced changes in the electronic structure of titanium carbide. 2. Electron densities and chemical bonding, J. Phys. Chem. Solids **47**, 387–393 (1986)
173 A. A. Rempel, A. I. Gusev, E. M. Gololobov, N. A. Prytkova, Zh. M. Tomilo: Influence of ordering on the superconducting transition temperature of nonstoichiometric niobium carbide, Fiz. Tverd. Tela **28**, 279–281 (1986) (in Russian). (Engl. transl.: Sov. Physics – Solid State **28**, 153–154 (1986))
174 A. A. Rempel, A. I. Gusev, E. M. Gololobov, N. A. Prytkova, Zh. M. Tomilo: Specific heat, superconductivity, and order-disorder transition in nonstoichiometric niobium carbide, Zh. Fiz. Khimii **61**, 1761–1766 (1987) (in Russian). (Engl. transl.: Russ. J. Phys. Chem. **61**, 919–922 (1987))
175 A. A. Rempel, A. I. Gusev: Short-range order in ordered alloys and interstitial phases, Fiz. Tverd. Tela **32**, 16–24 (1990) (in Russian). (Engl. transl.: Sov. Physics – Solid State **32**, 8–13 (1990))
176 A. A. Rempel, A. I. Gusev: Short-range order in superstructures, Phys. Stat. Sol. (b) **160**, 389–402 (1990)
177 J. Billingham, P. S. Bell, M. H. Lewis: Vacancy short-range order in substoichiometric transition metal carbides and nitrides with the NaCl structure. Electron diffraction studies of short-range ordered compounds, Acta Crystallogr. A **28**, 602-606 (1972)
178 M. Sauvage, E. Parthe: Vacancy short-range order in substoichiometric transition metal carbides and nitrides with the NaCl structure. Numerical calculation of vacancy arrangement, Acta Crystallogr. A **28**, 607–616 (1972)
179 M. Sauvage, E. Parthe, W. B. Yelon: Neutron diffraction measurement of diffuse scattering in $VC_{0.75}$, Acta Crystallogr. A **30**, 597–599 (1974)
180 C. H. de Novion, V. Maurice: Order and disorder in carbides and nitrides, J. Physique Colloq. (France) **38**, C7-211–C7-220 (1977)
181 K. Hiraga, M. Hirabayashi: Long-range and short-range order in interstitial compounds M_2X with special reference to V_2C and Nb_2C, J. Physique Colloq. (France) **38**, C7-224–C7-226 (1977)
182 L. Toth: *Transition Metal Carbides and Nitrides* (Academic Press, New York and London 1971) 280 pp.
183 S. V. Vonsovskii, Yu. A. Izyumov, E. Z. Kurmaev: *Superconductivity of Transition Metals, Their Alloys and Compounds* (Nauka, Moscow 1977) 384 pp. (in Russian)
184 V. S. Neshpor, V. P. Nikitin, V. I. Novikov: Superconductivity and residial resistivity of titanium carbide, Izv. AN SSSR. Neorgan. Materialy **7**, 1743–1747 (1971) (in Russian)
185 T. G. Utkina, Yu. S. Karimov, A. S. Rogachev: About possibility of existence of high-temperature superconducting phase of titanium carbide near the lower boundary of the homogeneity region, Fiz. Tverd. Tela **26**, 286–288 (1984) (in Russian)

186 A. L. Giorgi, E. G. Szklarz, E. K. Storms, A. L. Bowman, B. T. Matthias: Effect of composition on the superconducting transition temperature of tantalum carbide and niobium carbide, Phys. Rev. **125**, 837–838 (1962)
187 R. H. Willens, E. Buehler, B. T. Matthias: Superconductivity of transition metal carbides, Phys. Rev. **159**, 327–330 (1967)
188 L. E. Toth, M. Ishikawa, Y. A. Chang: Low-temperature heat capacity of superconducting niobium and tantalum carbides, Acta Met. **16**, 1183–1187 (1963)
189 A. N. Christensen: Vacancy order in Nb_6C_5, Acta Chem. Scand. A **39**, 803–804 (1985)
190 V. L. Ginzburg: Problem of high-temperature superconductivity, Uspekhi Fiz. Nauk **95**, 91–110 (1968); **101**, 185–215 (1970) (in Russian)
191 L. W. Shaklette, L. G. Radosevich, W. S. Williams: Gap energy of superconducting niobium carbide, Phys. Rev. B **4**, 84–87 (1971)
192 L. N. Bulaevskii, V. L. Ginzburg, G. F. Zharkov, D. A. Kirzhnits, Yu. V. Kopaev, E. G. Maksimov, D. I. Khomskii: *Problem of High-Temperature Superconductivity*, V. L. Ginzburg (Ed.) (Nauka, Moscow 1977) 400 pp. (in Russian)
193 W. L. McMillan: Transition temperature of strong coupled superconductors, Phys. Rev. **167**, 331–334 (1968)
194 G. M. Eliashberg: Interaction of electrons with oscillation of lattice in superconductor, ZhETF **38**, 976–996 (1960) (in Russian)
195 P. Morel, P. W. Anderson: Calculation of the superconducting state parameters with retaded electron-phonon interaction, Phys. Rev. **125**, 1263–1271 (1962)
196 A. S. Borukhovich, P. V. Geld, V. A. Tskhai, L. B. Dubrovskaya, I. I. Matveenko: Conduction band of IVa and Va subgroup transition metal carbides – II, Phys. Stat. Sol. (b) **45**, 179–187 (1971)
197 M. Morinaga, K. Sato, A. Aoki, H. Adachi, J. Harada: Existence of a vacancy band in electronic structures of niobium metal compounds, Philosoph. Mag. **47**, 107–111 (1983).
198 A. V. Svidzinskii: *Space-inhomogeneous Problems of Theory of Superconductivity* (Nauka, Moscow 1982) 312 pp. (in Russian)
199 A. A. Rempel, V. N. Lipatnikov, A. I. Gusev: Distribution function of carbon atoms and order parameters in ordered carbides of the type M_6C_5 , Kristallografiya **32**, 1032–1033 (1987) (in Russian). (Engl. transl.: Sov. Physics – Crystallography **32**, 607–608 (1987))
200 A. I. Gusev, A. A. Rempel: Vacancy distribution in ordered Me_6C_5-type carbides, J. Phys. C: Solid State Phys. **20**, 5011–5025 (1987)
201 I. N. Khlustikov, S. I. Moskvin: Superconductivity of twinned plane and topological phase transition in two-dimension superconducting system, ZhETF **89**, 1846–1856 (1985) (in Russian)
202 A. I. Buzdin, N. A. Khvorikov: Analysis of superconductivity of twinned plane in tin and niobium, ZhETF **89**, 1857–1869 (1985) (in Russian)
203 Yu. S. Karimov, T. G. Utkina: Superconductivity of nonstoichiometric niobium carbide, Pis'ma v ZhETF **51**, 468–470 (1990) (in Russian)
204 A. I. Gusev, A. A. Rempel, V. N. Lipatnikov: Incommensurate superlattice and superconductivity in tantalum carbide, Fiz. Tverd. Tela **33**, 2298–2305 (1991) (in Russian). (Engl. transl.: Sov. Physics – Solid State **33**, 1295–1299 (1991))
205 N. Pessal, R. E. Gold, H. A. Johansen: A study of superconductivity in interstitial compounds, J. Phys. Chem. Solids **29**, 19–38 (1968)

206 A. A. Rempel, V. N. Lipatnikov, A. I. Gusev: The superstructure in nonstoichiometric tantalum carbide, Doklady Akad. Nauk SSSR **310**, 878–882 (1990) (in Russian). (Engl. transl.: Sov. Physics Doklady **35**, 103–106 (1990))

207 V. N. Lipatnikov, A. A. Rempel, A. I. Gusev: Specific heat of tantalum carbide in states with different degrees of order, Fiz. Tverd. Tela **31**, 285–287 (1989) (in Russian). (Engl. transl.: Sov. Physics – Solid State **31**, 1818–1819 (1989))

208 A. I. Gusev, A. A. Rempel, V. N. Lipatnikov: Thermodynamic model of atomic ordering. 5. Specific heat of non-stoichiometric compounds in different structural states, Zh. Fiz. Khimii **64**, 2343–2348 (1990) (in Russian). (Engl. transl.: Russ. J. Phys. Chem. **64**, 1265–1267 (1990))

209 A. I. Gusev, A. A. Rempel, V. N. Lipatnikov: Heat capacity of niobium and tantalum carbides NbC_y and TaC_y in disordered and ordered states below 300 K, Phys. Stat. Sol. (b) **194**, 467–482 (1996)

210 B. T. Matthias: Transition temperatures of superconductors, Phys. Rev. **94**, 874-876 (1953)

211 A. A. Rempel, A. I. Gusev: Specific heat of niobium carbide in different structure states, Zh. Fiz. Khimii **62**, 2897–2901 (1988) (in Russian). (Engl. transl.: Russ. J. Phys. Chem. **62**, 1509–1511 (1988))

212 A. A. Rempel, A. I. Gusev: Heat capacity of niobium carbide in different structural states, Phys. Stat. Sol. (a) **113**, 353–358 (1989)

213 S. Z. Nazarova, A. I. Gusev: Heat capacity of $Nb_xTa_{1-x}C$ solid solutions, Izv. AN SSSR. Neorgan. Materialy **24**, 747–751 (1988) (in Russian). (Engl. transl.: Inorganic Materials **24**, 626–629 (1988))

214 G. Leibfried: *Gittertheorie der Mechanischen und Thermischen Eigenschaften der Kristalle* (Springer-Verlag, Berlin 1955)

215 T. A. Sandenaw, E. K. Storms: Heat capacities of $NbC_{0.702}$, $NbC_{0.825}$, $NbC_{0.980}$ and Nb_2C below 320 K, J. Phys. Chem. Solids **27**, 217–218 (1966)

216 K. K. Kelley: Specific heat at low temperatures of tantalum oxide and tantalum carbide, J. Amer. Chem. Soc. **62**, 818–819 (1940)

217 M. Ciguani: Metodi Monte-Carlo e sequenze pseudo casuali, Rendiconti del Seminario Matematico e Fisico di Milano **34**, 99–129 (1964)

218 J. A. Nelder, R. Mead: Simplex method for function minimization, Comput. J. **7**, 308–313 (1965)

219 W. C. Davidon: Variance algorithm for minimization, Comput. J. **10**, 406–410 (1968)

220 A. I. Gusev, A. A. Rempel: Calculating the energy parameters for CV and OPF methods, Phys. Stat. Sol. (b) **140**, 335–346 (1987)

221 A. I. Gusev, A. A. Rempel, V. N. Lipatnikov: Thermodynamic model of atomic ordering. 3. Calculation of cluster energies, Zh. Fiz. Khimii **61**, 916–921 (1987) (in Russian). (Engl. transl.: Russ. J. Phys. Chem. **61**, 476–479 (1987))

222 E. J. Huber, E. L. Head, C. E. Holley, E. K. Storms, N. H. Krikorian: The heat of combustion of niobium carbides, J. Phys. Chem. **65**, 1846–1849 (1961)

223 P. V. Geld, F. G. Kusenko: Enthalpy and heat capacity of niobium oxides and carbides at high temperature, Metallurgiya i Toplivo No. 2, 79–86 (1960) (in Russian)

224 E. J. Huber, E. L. Head, C. E. Holley, A. L. Bowman: The heat of formation of tantalum carbides, J. Phys. Chem. **67**, 794–796 (1963)

225 A. S. Bolgar, E. A. Guseva, V. A. Gorbatuk, V. V. Fesenko: Enthalpy and heat capacity of tantalum carbide in homogeneity interval, Poroshkovaya Metallurgiya No. 4, 60–62 (1968) (in Russian)

226 V. N. Lipatnikov, A. I. Gusev: High-temperature heat capacity and order–disorder phase transformations in nonstoichiometric titanium carbide, Phys. Stat. Sol. (b) **212**, R11–R12 (1999)
227 N. M. Volkova, P. V. Geld, S. I. Alyamovskii: Phase transformation of highest vanadium carbide, Zh. Neorgan. Khimii **10**, 1758–1759 (1965) (in Russian). (Engl. transl.: Russ. J. Inorg. Chem. **10**, 960–961 (1965))
228 N. M. Volkova, P. V. Geld: Temperature dependencies of enthalpy of vanadium carbide, Trudy Inst. Khimii Ural Fil. Akad. Nauk SSSR **14**, 41–46 (1967) (in Russian)
229 H. Wiesenberger, W. Lengauer, P. Ettmayer: Reactive diffusion and phase equilibria in the V–C, Nb–C, Ta–C and Ta–N systems, Acta Mater. **46**, 651–666 (1998)
230 W. Lengauer, S. Binder, K. Aigner, P. Ettmayer, A. Guillou, J. Debuigne: Solid state properties of group IVb carbonitrides, J. Alloys Comp. **217**, 137–147 (1995)
231 M. P. Arbuzov, S. Ya. Golub, B. V. Khaenko: Study of the phases of the Ti–TiC–TiN system, Izv. AN SSSR. Neorgan. Materialy **14**, 1442–1445 (1978) (in Russian)
232 C. Lobier, J. P. Marcon: Etude et structure d'une nouvelle phase du souns-nitrure de titane Ti_2N, Compt. Rend. Acad. Sci. Paris. Ser. C. **268**, 1132–1135 (1969)
233 E. Etchessahar, Sohn Yong-Un, M. Harmelin, J. Debuigne: The Ti–N system: kinetic, calorimetric, structural and metallurgical investigations of the δ'-$TiN_{0.51}$ phase, J. Less-Common Met. **167**, 261–281 (1991)
234 W. Lengauer: A study of δ'-TiN_{1-x} formation in temperature gradient diffusion couples, J. Alloys Comp. **179**, 289–297 (1992)
235 V. T. Em, I. Karimov, I. S. Latergaus: Influence of nitrogen on characteristics of order-disorder phase transition in TiC_x , Metallofizika **9**, 113–114 (1987) (in Russian)
236 V. T. Em, M. Yu. Tashmetov: The structure of the ordered phase in rocksalt type titanium carbide, carbidenitride, and carbidehydride, Phys. Stat. Sol. (b) **198**, 571–575 (1996)
237 L. Ramqvist: Variation of lattice parameter and hardness with carbon content of group 4b metal carbides, Jernkontorets Annaler. **152**, 517–523 (1968)
238 A. A. Rempel, A. I. Gusev, L.-M. Berger, V. Richter: Nonstoichiometric transition metal carbides – synthesis, properties and potential areas of application, in *Covalent Ceramics II: Non-oxides*, A. R. Barron, G. S. Fischman, M. A. Fury, A. F. Hepp (Eds.), USA Mater. Res. Soc. Symp. Proc. **327**, Boston, Massachusetts, November 29 – December 2, 1993 (Materials Research Society, Pittsburgh (Pennsylvania) 1994) pp. 183-188
239 A. E. Koval'skii, T. G. Makarenko: Dependence of titanium carbide microhardness on carbon content, Zh. Tekhn. Fiz. **23**, 265–266 (1953) (in Russian)
240 I. Cadoff, J. P. Nielsen, E. Miller: Properties of arc-melted versus powder metallurgy titanium carbide, in *Papers Presented on 2nd Plansee Seminar "De Re Metallica"*, Plansee Proc. 1955 (Metallwerk Plansee GmbH, Reutte 1956) pp. 50–55
241 J.-L. Chermant: Du carbure de titane: réactions de formation et propriétés mécaniques, Rev. Int. Hautes Temp. et Refract. **6**, 299-312 (1969)
242 I. I. Spivak, R. A. Andrievski, V. N. Rystsov, V. V. Klimenko: Creep of titanium monocarbide in the homogeneity interval, Poroshkovaya Metallurgiya No. 7, 69–74 (1974) (in Russian)
243 Yu. G. Tkachenko, S.S.Ordan'yan, V. K. Yulyugin, D. Z. Yurchenko, G. S. Tabatadze, I. B. Panteleev: Characteristics of friction and features of deformation of TiC in contant zone, Poroshkovaya Metallurgiya No. 6, 45–51 (1979) (in Russian)

244 E. Breval: Microplasticity at room temperature of single-crystal titanium carbide with different stoichiometry, J. Mater. Sci. **16**, 2781–2788 (1981)
245 V. G. Bukatov: Study of physical and mechanical properties of refractory metal carbides and some their alloys, Ph. D. Thesis, Institute of Steel and Alloys, Moscow (1979) (in Russian)
246 J.-L. Chermant, P. Delavignette, A. Deschanvres: Etude des bandes de precipitation dans le carbure de titane sous stoechiométrique, J. Less-Common Met. **21**, 89–101 (1970)
247 V. S. Sinelnikova, T. I. Shtukaturova, L. V. Strashinskaya, T. I. Shaposhnikova, G. S. Burkhanov, V. A. Kuzmishev: Composition and structure of single crystal of titanium carbide, Poroshkovaya Metallurgiya No. 8, 53–57 (1981) (in Russian)
248 M. P. Arbuzov, B. V. Khaenko, E. T. Kachkovskaya: Study of ageing and ordering of titanium carbide, Fiz. Metall. Metalloved. **44**, 1240–1244 (1977) (in Russian)
249 V. N. Lipatnikov, L. V. Zueva, A. I. Gusev: Microhardness and grain size of disordered nonstoichiometric titanium carbide, Neorgan. Materialy **35**, 1330–1336 (1999) (in Russian). (Engl. transl.: Inorganic Materials **35**, 1137–1143 (1999))
250 S. Sarian: Diffusion of ^{44}Ti in TiC_x, J. Appl. Phys. **40**, 3515–3520 (1969)
251 S. Sarian: Diffusion of carbon in TiC, J. Appl. Phys. **39**, 3305–3310 (1968)
252 R. A. Andrievski, A. G. Lanin, G. A. Rymashevskii: *Strength of High-Melting Compounds* (Metallurgiya, Moscow 1974) 232 pp. (in Russian)
253 G. V. Samsonov, S. A. Bozhko: Study of recrystallization at sintering of niobium and titanium carbide powders, Poroshkovaya Metallurgiya No. 7, 31–35 (1969) (in Russian)
254 S. A. Bozhko, G. V. Samsonov: Collective recrystallization in titanium carbide, Izv. AN SSSR. Neorgan. Materialy **7**, 59–63 (1971) (in Russian)
255 J. Klima: Density of states of substoichiometric TiC_x, J. Phys. C: Solid State Phys. **12**, 3691–3698 (1979)
256 L. M. Huisman, A. E. Carlsson, C. D. Gelatt, H. Ehrenreich: Mechanisms for energetic-vacancy stabilization: TiO and TiC, Phys. Rev. B **22**, 991–1006 (1980)
257 W. S. Williams, R. D. Shaal: Elastic deformation, plastic flow, and dislocations in single crystals of titanium carbide, J. Appl. Phys. **33**, 955–962 (1962)
258 G. E. Hollox, R. E. Smallman: Plastic behavior of titanium carbide, J. Appl. Phys. **37**, 818–823 (1966)
259 E. Breval: Fracture and plastic deformation of titanium carbide at room temperature, Scand. J. Metall. **10**, 51–54 (1981)
260 D. J. Rowcliffe, G. E. Hollox: Hardness anisotropy, deformation mechanisms, and brittle-to-ductile transition in carbides, J. Mater. Sci. **6**, 1270–1276 (1971)
261 D. K. Chatterjee, M. G. Mendiratta, H. A. Lipsitt: Deformation behavior of single crystals of titanium carbide, J. Mater. Sci. **14**, 2151–2156 (1979)
262 H. Höchst, R. O. Bringans, P. Steiner, Th. Wolf: Photoemission study of the electronic structure of stoichiometric and substoichiometric TiN and ZrN, Phys. Rev. B **25**, 7183–7191 (1982)
263 L. Porte, L. Roux, J. Hanus: Vacancy effects in the X-ray photoelectron spectra of TiN_x , Phys. Rev. B. **28**, 3214–3224 (1983)
264 L. Porte: Electronic structure of non-stoichiometric zirconium nitrides ZrN_x, Solid State Commun. **50**, 303–306 (1984)
265 P. Pecheur, G. Toussaint, E. Kauffer: Electronic structure of carbon vacancy in NbC, Phys. Rev. B **29**, 6606–6613 (1984)

266 P. Herzig, J. Redinger, R. Eibler, A. Neckel: Vacancy induced changes in the electronic structure of titanium nitride, J. Sol. State Chem. **70**, 281–294 (1987)
267 K. Schwarz: Band structure and chemical bonding in transition metal carbides and nitrides, Critical Rev. Solid State Mater. Sci. **13**, 211–257 (1987)
268 O. V. Bakun, O. N. Grigor'ev, B. V. Khaenko: Microstructure and some strength properties of ordered phase of titanium carbide, Doklady Akad. Nauk Ukrain. SSR. Seriya A No. 12, 62–65 (1983) (in Russian)
269 L. V. Zueva, V. N. Lipatnikov, A. I. Gusev: Ordering effects on the microstructure and microhardness of nonstoichiometric titanium carbide TiC_y, Neorgan. Materialy **36**, 836–840 (2000) (in Russian). (Engl. transl.: Inorganic Materials **36**, 695–698 (2000))
270 A. I. Gusev: Effects of the nanocrystalline state in solids, Uspekhi Fiz. Nauk **168**, 55–83 (1998) (in Russian). (Engl. transl.: Physics – Uspekhi **41**, 49–76 (1998))
271 A. I. Gusev: *Nanocrystalline Materials: Production and Properties* (Ural Division of the Russian Academy of Sciences, Ekaterinburg 1998) 200 pp. (in Russian)
272 O. P. Sivak: Atomic ordering in niobium carbides, Ph. D. Thesis, Institute of Problems of Material Science, Kiev (1988) 194 pp. (in Russian)
273 R. H. J. Hannik, M. J. Murray: The effect of domain size on the hardness of ordered $VC_{0.84}$, Acta Metal. **20**, 123–131 (1972)

Subject Index

Printing (Computer to Film): Saladruck Berlin
Binding: Stürtz AG, Würzburg

Springer Series in

MATERIALS SCIENCE

Editors: R. Hull   R. M. Osgood, Jr.   H. Sakaki   A. Zunger

1 **Chemical Processing with Lasers***
By D. Bäuerle

2 **Laser-Beam Interactions with Materials**
Physical Principles and Applications
By M. von Allmen and A. Blatter
2nd Edition

3 **Laser Processing of Thin Films and Microstructures**
Oxidation, Deposition and Etching of Insulators
By. I. W. Boyd

4 **Microclusters**
Editors: S. Sugano, Y. Nishina, and S. Ohnishi

5 **Graphite Fibers and Filaments**
By M. S. Dresselhaus, G. Dresselhaus, K. Sugihara, I. L. Spain, and H. A. Goldberg

6 **Elemental and Molecular Clusters**
Editors: G. Benedek, T. P. Martin, and G. Pacchioni

7 **Molecular Beam Epitaxy**
Fundamentals and Current Status
By M. A. Herman and H. Sitter 2nd Edition

8 **Physical Chemistry of, in and on Silicon**
By G. F. Cerofolini and L. Meda

9 **Tritium and Helium-3 in Metals**
By R. Lässer

10 **Computer Simulation of Ion-Solid Interactions**
By W. Eckstein

11 **Mechanisms of High Temperature Superconductivity**
Editors: H. Kamimura and A. Oshiyama

12 **Dislocation Dynamics and Plasticity**
By T. Suzuki, S. Takeuchi, and H. Yoshinaga

13 **Semiconductor Silicon**
Materials Science and Technology
Editors: G. Harbeke and M. J. Schulz

14 **Graphite Intercalation Compounds I**
Structure and Dynamics
Editors: H. Zabel and S. A. Solin

15 **Crystal Chemistry of High-T_c Superconducting Copper Oxides**
By B. Raveau, C. Michel, M. Hervieu, and D. Groult

16 **Hydrogen in Semiconductors**
By S. J. Pearton, M. Stavola, and J. W. Corbett

17 **Ordering at Surfaces and Interfaces**
Editors: A. Yoshimori, T. Shinjo, and H. Watanabe

18 **Graphite Intercalation Compounds II**
Editors: S. A. Solin and H. Zabel

19 **Laser-Assisted Microtechnology**
By S. M. Metev and V. P. Veiko
2nd Edition

20 **Microcluster Physics**
By S. Sugano and H. Koizumi
2nd Edition

21 **The Metal-Hydrogen System**
By Y. Fukai

22 **Ion Implantation in Diamond, Graphite and Related Materials**
By M. S. Dresselhaus and R. Kalish

23 **The Real Structure of High-T_c Superconductors**
Editor: V. Sh. Shekhtman

24 **Metal Impurities in Silicon-Device Fabrication**
By K. Graff 2nd Edition

25 **Optical Properties of Metal Clusters**
By U. Kreibig and M. Vollmer

26 **Gas Source Molecular Beam Epitaxy**
Growth and Properties of Phosphorus Containing III–V Heterostructures
By M. B. Panish and H. Temkin

27 **Physics of New Materials**
Editor: F. E. Fujita 2nd Edition

28 **Laser Ablation**
Principles and Applications
Editor: J. C. Miller

* The 2nd edition is available as a textbook with the title: *Laser Processing and Chemistry*

MIX
Papier aus verantwortungsvollen Quellen
Paper from responsible sources
FSC® C105338

If you have any concerns about our products,
you can contact us on
ProductSafety@springernature.com

In case Publisher is established outside the EU,
the EU authorized representative is:
Springer Nature Customer Service Center GmbH
Europaplatz 3, 69115 Heidelberg, Germany

Printed by Libri Plureos GmbH
in Hamburg, Germany